ENCYCLOPEDIA
OF
GENETICS

ENCYCLOPEDIA OF GENETICS

EDITORS-IN-CHIEF
Sydney Brenner
Jeffrey H. Miller

ASSOCIATE EDITORS
William J. Broughton
Malcolm Ferguson-Smith
Walter Fitch
Nigel D. F. Grindley
Daniel L. Hartl
Jonathan Hodgkin
Charles Kurland
Elizabeth Kutter
Terry H. Rabbitts
Ira Schildkraut
Lee Silver
Gerald R. Smith
Ronald L. Somerville

ACADEMIC PRESS

A Harcourt Science and Technology Company

San Diego   San Francisco   New York   Boston   London   Sydney   Tokyo

Academic Press
A division of Harcourt, Inc.
Harcourt Place, 32 Jamestown Road, London NW1 7BY, UK
http://www.academicpress.com

Academic Press
A division of Harcourt, Inc.
525 B Street, Suite 1900, San Diego, California 92101-4495, USA
http://www.academicpress.com

ISBN 0-12-227080-0

Library of Congress Catalog Number: 2001089059
A catalogue record for this book is available from the British Library

Access for a limited period to the online version of the *Encyclopedia of Genetics* is included in the
purchase price of the print edition.

This online version has been uniquely and persistently identified by the Digital Object Identifier
(DOI)

10.1006/rwgn.2001

By following the link
http://dx.doi.org/10.1006/rwgn.2001

from any Web Browser, buyers of the *Encyclopedia of Genetics*
will find instructions on how to register for access.

If you have any problems with accessing the online version, e-mail
idealreferenceworks@harcourt.com

Typeset by Kolam Information Services Pvt Limited, Pondicherry, India
Printed and bound in Spain by Grafos SA Arte Sobre Papel, Barcelona
01 02 03 04 05 06 GF 9 8 7 6 5 4 3 2 1

Jonathan Hodgkin
University of Oxford
Genetics Unit
Department of Biochemistry
South Parks Road
Oxford
OXI 3QU
UK

Charles Kurland
Uppsala University
Munkarpsv. 21
SE 243 32, Hoor
Sweden

Elizabeth Kutter
The Evergreen State College
Lab I
Olympia
WA 98505
USA

Terry H. Rabbitts
MCR Laboratory of Molecular Biology
Division of Protein and Nucleic Acid Chemistry
Hills Road
Cambridge
CB2 2QH
UK

Ira Schildkraut
New England Biolabs Inc.
32 Tozer Road
Beverly
MA 01915
USA

Lee Silver
Princeton University
Department of Molecular Biology
Princeton
NJ 08544-1014
USA

Gerald R. Smith
Fred Hutchinson Cancer Research Center
1100 Fairview Avenue North, A1-162
Seattle
WA 98104-1024
USA

Ronald L. Somerville
Purdue University
Department of Biochemistry
West Lafayette
IN 47907-1153
USA

Preface

In the spring of 1953, shortly after Watson and Crick's discovery of the double-helical structure of DNA, I found myself at dinner next to the famous geneticist R.A. Fisher. When I asked him what he thought would be that structure's implication for genetics, he replied firmly "None!" He may have taken this myopic view, because to him genetics was an abstract mathematical subject with laws that were independent of the physical nature of genes. At the time, it was an esoteric subject taught only to a few biologists and regarded as largely irrelevant to medicine. Archibald Garrod's seminal book *Inborn Errors of Metabolism* had made little impact, perhaps because no-one knew what genes were made of or how mutations acted.

The first recognition that mutations act on proteins came in 1949 when Linus Pauling and his collaborators published their paper on "Sickle cell anemia, a molecular disease." They found that patients suffering from this recessively inherited disease had an abnormal hemoglobin that differed from the normal form by the elimination of two negative charges. Eight years later, Vernon Ingram showed that this was due to the replacement of a single glutamic acid residue in each of the identical half-molecules of hemoglobin by a valine. Ingram's discovery was the first specific evidence of the chemical effect of a mutation. It marked the birth of molecular genetics and molecular medicine and it started the transformation of genetics to its central position in molecular biology, biochemistry, and medicine today. Even so, people who have studied genetics as part of their curriculum may not have become familiar with all the hundreds of specialized terms that geneticists have coined. To them and many others at the periphery of genetics this *Encyclopedia* will prove most useful.

Like the *Encyclopedia Britannica* it is a cross between a dictionary and a text book. Hybrids are defined in three lines, while Jonathan Hodgkin's brilliant exposition of past and present research on *Drosophila's* rival, the minute nematode worm *Caenorhabditis elegans*, occupies nearly six pages. This entry is intelligible to non-specialists, but that is not true of some others that were "Greek" to me.

The *Encyclopedia* includes biographies of many of the pioneers, from Gregor Mendel to Ernst Mayr, the great evolutionist who actually contributed several entries. Many of the other contributors are also pioneers, even though they are not yet old enough to figure among the biographies. Sydney Brenner has written the entry on the genetic code which he himself helped to discover 40 years ago; David Weatherall has written on thalassemia to whose exploration he has devoted a lifetime; Malcolm Ferguson-Smith has written on human chromosomal anomalies on which he is the world authority; he and some others made such varied contributions to their fields that they figure among many of the 1650-odd entries. Sydney Brenner, one of the two editors-in-chief, may be the only contributor who does not need the *Encyclopedia*.

What of the future of genetics? Its applications are likely to multiply, and many of the applicants coming from other fields will welcome the *Encyclopedia*. An increasing number of applications will be in medicine. I have heard predictions that in future every newborn child will have its genes screened and the results imprinted on a computer chip that he or she will carry for life. Recorded on it would be all genetic anomalies, susceptibilities to diseases, and intolerance of drugs. In case of illness or accident, that chip would activate an algorithm that automatically prescribes the correct treatment. Unfortunately for such utopias, medicine is more complex. Weatherall has shown elsewhere that the single recessive disease thalassemia is a multiplicity of different diseases and that the same genotype may give rise to widely different phenotypes, depending on environmental and other factors. This complexity arises even when single point mutations do not necessarily lead to disease, but merely to susceptibilities to disease, as in α_1-antitrypsin deficiency or with certain abnormal hemoglobins. The complexity is much greater still in multifactorial diseases like schizophrenia or diabetes. For many reasons good medicine will continue to require wide knowledge, mature judgement, empathy, and wisdom.

There have been glib predictions that the mapping of the human genome will allow most genetic diseases to be cured by either germline or somatic gene therapy, but the former is too risky and the latter is proving extremely difficult and costly. The risks of germline therapy are illustrated by the attempt to create a genetically modified monkey. Scientists injected the gene for the green fluorescent protein from a jellyfish into 222 monkey eggs. After fertilizing them with monkey sperm, they incubated them and implanted a pair each into the wombs of 20 surrogate monkey mothers. Only five of the implants resulted in pregnancies. Only three monkey babies were born and only one of them carried the jellyfish gene, but the monkey does not fluoresce, because the gene, though incorporated into its chromosomes, fails to be expressed. It will be argued that technical improvements may lead to a 100% success rate, but this kind of gene transfer has now been practised for some years in mice and other animals, and it has remained a haphazard affair that would be criminal to apply to humans. Human cloning carries similar risks.

After many failures, somatic gene therapy of a potentially fatal human genetic disease recently succeeded for the first time. A French team cured two baby boys of severe combined immunodeficiency (SCID)-XI disease. They infected cells extracted from the boys' bone marrow with cDNA containing the required gene coupled to a retrovirus-derived vector, and then re-injected them back into the boys' bone marrow. The therapy restored normal immune function that was still intact 10 months later. Another interesting development was the restoration of normal function to the muscle of a dystrophic mouse after injection of fragments of the giant muscular dystrophy gene. On the other hand, it has so far been impossible to cure some of the most common genetic diseases: thalassemia, sickle cell anemia, or cystic fibrosis, because it has proved extremely difficult to express the genes in the correct place in the patients' chromosomes and get them to express the required protein in sufficient quantities in the right tissues. If gene therapy has a bright future, it does not look as though it is just round the corner.

Before the completion of the Human Genome Project, identification of the genes for some human inherited diseases required truly heroic efforts. The search for the Huntington's disease gene occupied up to a hundred people for about 10 years. The same work could now be accomplished by few people in a fraction of the time. This is one of the Human Genome Project's important medical benefits. Others may be the rapid identification of promising new drug targets against diseases ranging from high blood pressure to a variety of cancers, the epidemiology of alleles linked to susceptibility to various diseases and improved basic understanding of human physiology and pathology.

Agriculture offers the greatest scope for applied genetics, but distrust of genetically modified foods has blinded the public to its potential benefits and its vital importance for the avoidance of widespread famines later in this century. Since the early 1960s, food available per head in the developing world has increased by 20% despite a doubling of the population. This outstanding success has been achieved by the introduction of crops improved by crossing and by intensive application of fertilizers, pesticides, and weed-killers. Even so, there are 800 million hungry people and 185 million seriously malnourished pre-school children in the developing world. It is unlikely that the methods that have raised cereal yields hitherto will allow them to be raised again sufficiently to reduce these distressing numbers. Since most fertile land is already intensively cultivated, scientists are trying to introduce genes into crops that would allow them to be grown on poorer soils and in harsher climates, and to make existing crops more nutritious. In the tropics, fungi, bacteria, and viruses still cause huge harvest losses. Scientists are trying to introduce genes that will confer resistance to some of these pests, enabling farmers to use fewer pesticides. Genetically modified plants offer our best hope of feeding a world population that is expected to double in the next 50 years. It will be tragic if the present outcry over genetically modified foods will discourage further research and development in this field. If this *Encyclopedia* helps to promote better public understanding of genetics, this might be the best remedy against irrational fears.

Max F. Perutz
MRC Laboratory of Molecular Biology
Hills Road, Cambridge
CB2 2QH
UK

Introduction

Genetics, the study of inheritance, is fundamental to all of biology. Living organisms are unique among all natural complex systems in that they contain within their genes an internal description encoded in the chemical text of DNA. It is this description and not the organism itself which is handed down from generation to generation and understanding how the genes work to specify the organism constitutes the science of genetics. Furthermore, this constancy is embedded in a vast range of diversity, from bacteria to ourselves, all having arisen by changes in the genes. Understanding evolution is also part of genetics, and an area which will benefit from our increasing ability to determine the complete DNA sequences of genomes. In some sense, these sequences contain a record of genome history and we have now learnt that many genes in our genomes can be found in other organisms, quite unlike us. Indeed some are much the same as those found in bacteria, and can be viewed as molecular fossils, preserved in our genomes.

Although "like begets like" must be one of the oldest observations of mankind, it was only in the 19th century that major scientific advances began. Charles Darwin put forward his theory of the origin of the species by natural selection but he lacked a credible theory of the mechanism of inheritance. He believed in blending inheritance which meant that variation would be continually removed and he was therefore compelled to introduce variation in each generation as an inherited acquired character. Gregor Mendel, working at the same time, discovered the laws of inheritance and showed how the characteristics of the organism could be accounted for by factors which specified them. Mendels' work was rediscovered in 1900 independently by Correns, de Vries and Tchermak and soon after this, Bateson coined the term "gene" for the Mendelian factors and called the science, "genetics."

During the first 50 years of the twentieth century, there was a stream of important discoveries in genetics. We came to understand the relation between genes and chromosomes, and the connection between recombination maps and the physical structure of chromosomes. However, what the genes were made of and what they did remained a mystery until 1953 when Watson and Crick proposed the double helical structure of DNA, which at one blow unified genetics and biochemistry and ushered in the modern era of molecular biology.

Genetics and especially the molecular approach to it is now a pervasive field covering all of biology. In the *Encyclopedia of Genetics*, we have tried to draw together the many strands of what is still a rapidly expanding field, to present a view of all of genetics. This has been a five year effort by over 700 expert authors from all around the world. We have tried to ensure that the breadth of the work has not compromised the depth of the articles, and we hope that readers will be able to find accurate and up-to-date information on all major topics of genetics. We have also included articles on the history of the field as well as the impact of the applications of genetics to medicine and agriculture.

When we began this work, the sequencing of complete genomes was still in its infancy and the sequencing of the human genome was thought to be far in the future. Technological advances and the concentration of resources have brought this to fruition this year, and genetics is a subject very much in the public eye. We hope that at least some of the articles will also be of value to those who are not professional biological or medical scientists, but want to discover more about this field. Many of the articles contain lists for further reading, and the online version of the *Encyclopedia* also includes hypertext links to original articles, abstracts, source items, databases and useful websites, so that readers can seamlessly search other appropriate literature.

We would like to acknowledge the efforts of the Associate Editors, who worked hard in commissioning individual contributors to prepare cutting-edge articles, and who reviewed and edited the manuscripts in a timely manner. Our thanks also go to the Publishers, Academic Press, and the outstanding staff for their commitment, resourcefulness and creative input; and in particular Tessa Picknett, Kate Handyside, Peronel Craddock and the production team, who helped to make this a reality.

Sydney Brenner, Jeffrey H. Miller
Editors-in-Chief

Guide to use

Structure of the *Encyclopedia*

The material in the *Encyclopedia* is arranged as a series of entries in alphabetical order. These range greatly in size, from glossary items, definitions, and short articles, through to major (five pages or more) full articles, as appropriate.

To help you realize the full potential of the material in the *Encyclopedia* we have provided several features to help you find the topic of your choice.

The Contents List

Your first point of reference will probably be the contents list. The complete contents list appearing in all four volumes will provide you with both the volume number and page number of the entry.

Scanning Through the Text

Alternatively you may choose to browse through a volume using the alphabetical order of the entries as your guide. To assist you in identifying your location within the *Encyclopedia*, running headlines indicate the first entry on left-hand pages and the last entry on right-hand pages.

Dummy Entries

You will find 'dummy entries' where obvious synonyms exist, occasional species names/common names are used, or for entries where a topic may be covered within another entry. For example: If you wished to locate the entry which discusses **DNA Code**, the dummy heading would lead you to the entry on **Genetic Code**. The dummy entry for **Fruit Fly** takes you to ***Drosophila melanogaster.***

Cross-References

All of the articles in the *Encyclopedia* have been extensively cross-referenced. To direct the reader to other entries on related topics, the section immediately before the References entitled *See also* in most articles lists related entry titles. For example, the entry **Genetic Code** includes the following cross-references: *See also*: **Adaptor Hypothesis, Codons, Variable Codons, Universal Genetic Code.**

Index

The complete index for the four volumes is provided at the back of the last volume. This will help you to find entries for specific topics that do not appear as such in the list of contents. Index entries differentiate between material that is a whole article, is part of an article, is a figure or contains data presented as a table. Further guidance on the use of the index is supplied on the opening page of the index.

Color Plates

The color figures for each volume have been grouped together in a plate section. The location of this section is cited at the end of the contents list.

Contributors

A full list of contributors and their addresses is given at the beginning of each volume.

Abbreviations

A	adenine	CDC	cell division cycle
aa	amino acid	CDK	cyclin-dependent kinase
aa-tRNA	amino acyl-tRNA	cDNA	complementary DNA
ABC	ATP binding cassette	CDS	coding sequence
AD	Alzheimer's disease	CEN	centromere
ADA	adenosine deaminase	CF	cystic fibrosis
ADC	adenocarcinoma	CFTR	cystic fibrosis transmembrane
ADH	alcohol dehydrogenase		conductance regulator
ADP	adenosine 5'-diphosphate	CGH	comparative genome hybridisation
AFA	acromegaloid facial appearance (syndrome)	CHO	Chinese hamster ovary (cells)
		CJD	Creutzfeldt–Jacob disease
AFB	aflatoxin B	CL	cutis laxa
AFP	alpha-fetoprotein	cM	centimorgan
AHA	acute hemolytic anemia	CMD	congenital muscular dystrophy
AIDS	acquired immunodeficiency syndrome	CML	chronic myeloid leukemia
		CMS	cytoplasmic male sterility
AIL	advanced intercross line	CoA	coenzyme A
Ala	alanine	Col	colicin
ALL	acute lymphoblastic leukemia	COP	coat protein
AMCA	aminomethyl coumarin acetic acid	CPE	cytoplasmic polyadenylation element
AMH	anti-Müllerian hormone	cR	centiray
AML	acute myeloid leukemia	CR	conserved region
Amp	ampicillin	CRC	colorectal adenocarcinoma
AMP	adenosine 5'-monophosphate	Cre	cyclization recombination
A_n	polyadenylation	CREB	cAMP response element binding factor
AP	apyrimidinic (site)	CRP	cAMP receptor protein
APC	adenomatous polyposis coli	CT	cholera toxin
APO-	apolipoprotein-	ctDNA	chloroplast DNA
APOBEC	apo-B mRNA editing cytidine deaminase	ctf	cotransformation frequency
		CTP	cytidine triphosphate
APP	amyloid precursor protein	CV	coefficient of variation
ARF	ADP-ribosylation factor	CVS	chorionic villus sampling
Arg	arginine	Cys	cysteine
ARS	autonomous replication sequence	Da	dalton
Asn	asparagine	DI	dentinogenesis imperfecta
ASO	allele specific oligonucleotide	DIC	differential inference
Asp	aspartic acid	DMD	differentially methylated domain
AT	ataxia telangiectasia	DMD	Duchenne muscular dystrophy
ATP	adenosine 5'-triphosphate	DMEM	Dulbecco's Modified Eagle's Medium
BAC	bacterial artificial chromosome	DMI	deviation from Mendelian inheritance
BER	base excision repair	DMSO	dimethylsulfoxide
BIC	Breast Cancer Information Corp	dN	deoxynucleotide
BIME	bacterial interspersed mosaic element	DNA	deoxyribonucleic acid
BMD	Becker muscular dystrophy	DNP	deoxyribonucleoprotein
bp	base pair	dNTP	deoxyribonucleotide triphosphate
BR	Balbiani rings	ds	double stranded
BS	Bloom's syndrome	DSBR	double strand break repair
BSE	bovine spongiform encephalopathy	EBN	endosperm balanced number
BWS	Beckwith–Wiedemann syndrome	EBV	Epstein–Barr virus
C	cytosine	ECM	extracellular matrix
C-	carboxyl-	EDS	Ehlers–Danlos syndrome
CAF	chromatin assembly factor	EF	elongation factor
cAMP	cyclic AMP	EGF	epidermal growth factor
CAP	catabolite activator protein	ELISA	enzyme linked immunosorbant assay
CD	campomelic dysplasia	EMBL	European Molecular Biology Laboratories
CD	circular dichroism		

EN	early nodule	HGMP	Human Genome Mapping Project
EN	endonuclease	His	histidine
ENU	N-ethyl-N-nitrosourea	HIV	human immunodeficiency virus
EP	early promotor	HLA	human leukocyte antigen
ER	endoplasmic reticulum	HLH	helix–loop–helix
eRF	eukaryal release factor	HMC	5′-hydroxymethyl-cytosine
ES	embryonic stem (cells)	HMG	high mobility group
ESI	electrospray ionization	HMW	high molecular weight
ESS	evolutionarily stable strategy	HNPCC	hereditary nonpolyposis colorectal cancer
EST	expressed sequence tagged		
F-factor	fertility-factor	hnRNP	heterogeneous nuclear RNP
FA	fluctuating asymmetry	HPLC	high-performance liquid chromatography
FACS	fluorescence-activated cell sorter		
FAD	flavin-adenine dinucleotide	HPV	human papillomavirus
FADH$_2$	reduced FAD	hsp	heatshock protein
FAK	focal adhesion kinase	HTH	helix–turn–helix
FAP	familial adenoma polyposis	HTLV	human T-cell leukemia virus
FDS	first-division segregation	HV-I, HV-II	hypervariable regions I, II
FFI	familial fatal insomnia	HW equilibrium	Hardy–Weinberg equilibrium
FGF	fibroblast growth factor	IBD	identical by descent
FH	familial hypercholesterolemia	IBS	identical by state
FIGE	field inversion gel electrophoresis	ICM	inner cell mass
FISH	fluorescent *in situ* hybridization	ICSI	intracytoplasmic sperm injection
FITC	fluorescein isothoicyanate	IES	internal eliminated sequences
FRDA	Friedreich's ataxia	IF	initiation factor
FRET	fluorescence energy resonance transfer	Ig	immunoglobulin
FSH	follicle stimulating hormone	IGF	insulin-like growth factor
FSHMD	facioscapulohumeral muscular dystrophy	IHF	integration host factor
		IL	interleukin
G	guanine	ILAR	Institute for Laboratory Animal Research
G-6-P	glucose-6-phosphate		
G-banding	Giemsa-banding	Ile	isoleucine
G-proteins	GTP-binding proteins	IMAC	immobilized metal ion affinity chromatography
GAP	GTPase activating protein		
GDB	genome database	IN	integrase (protein)
GDP	guanosine diphosphate	INR	initiator region
GEF	guanine nucleotide exchange factor	IPTG	isopropylthiogalactoside
GF	growth factor	IS	insertion sequence
GFP	green fluorescent protein	ISH	*in situ* hybridization
GIST	gastrointestinal stromal tumors	ISR	induced systemic resistance
Glu	glutamic acid	IVF	*in vitro* fertilization
Gly	glycine	IVS	intervening sequence
gp	glycoprotein	kb	kilobase
GPCR	G-protein-coupled receptor	KL	kit ligand
gRNA	guide RNA	KO	knockout
GSD	Gerstmann–Sträussler disease	KSS	Kearns–Sayre syndrome
GSS	Gerstmann–Sträussler–Scheinker syndrome	Lac	lactose
		LBC	lampbrush chromosome
GTP	guanosine triphosphate	LCR	locus control region
HA	hemagglutinin	LD	linkage disequilibrium
HAT	histone acetyl transferase	LDL	low-density lipoprotein
Hb	hemoglobin	Leu	leucine
hCG	human chorionic gonadotrophin	LH	luteinizing hormone
HCL	hairy cell leukemia	LHSI/II	light harvesting system I/II
HDAC	histone deacetylase	LINE	long interspersed nuclear element
HDGS	homology-dependent gene silencing	LMC	local mate competition
HDL	high-density lipoprotein	LMW	low molecular weight
Hfr	high frequency recombination	LOD	logarithm of the odds (score)
HGMD	Human Gene Mutation Database	LOH	loss of heterozygosity

Lox	locus of X-over	NCS	nonchromosomal striped
LPS	lipopolysaccharide	NER	nucleotide excision repair
LRC	local resource competition	NHL	non-Hodgkin lymphoma
LRE	local resource enhancement	NK	natural killer (cells)
LRR	leucine-rich repeat	NMD	nonsense-mediated mRNA decay
LTR	long terminal repeat	NMR	nuclear magnetic resonance
Lys	lysine	NOR	nucleolus organizing region
m-BCR	minor breakpoint cluster region	NOS	nitric oxide synthetase
M-BCR	major breakpoint cluster region	NPC	nasopharyngeal carcinoma
M-phase	meiosis/mitosis phase	NPC	nuclear pore complex
MAP	microtubule associated protein	NR	nuclear reorganization
MAP	mitogen-activated protein	nt	nucleotide
MAPK	mitogen-activated protein kinase	OI	osteogenesis imperfecta
MAR	matrix-attached region	OMIM	Online Mendelian Inheritance in Man
Mb	megabase	OPMD	oculopharyngeal muscular dystrophy
MBP	myelin-based protein	ORC	origin of replication complex
MC'F	micro-complement fixation	ORF	open reading frame
MCR	mutation cluster region	*Ori*	origin (of replication)
MCS	multiple cloning site	*Ori*T	origin of transfer
MDS	myelodysplastic syndrome	OTU	operational taxonomic unit
Mel	maternal-effect embryonic lethal	PAA	propionic acidemia
MELAS	mitochondrial encephalomyopathy, lactic acidosis, and stroke-like symptoms	PAC	P1 artificial chromosome (vector)
		PAC	prostate adenocarcinoma
		PAGE	polyacrylamide gel electrophoresis
MEN	multiple endocrine neoplasia	PAI	pathogenicity islands
MERRF	myoclonus epilepsy with ragged red fibers	PAPP	pregnancy-associated plasma protein
		PAR	pseudoautosomal region
Met	methionine	*Pax*	paired box-containing genes
MFH	malignant fibrous histiocytoma	PBP	penicillin binding protein
MFS	Marfan syndrome	Pc	polycomb
MGD	Mouse Genome Database	PCO	polycystic ovarian (disease)
MGF	mast cell growth factor	PCR	polymerase chain reaction
MGI	Mouse Genome Informatics	PCT	plasmacytoma
MHC	major histocompatibility complex	PDGF	platelet derived growth factor
MIC	minimum inhibitory concentration	PE	phosphatidylethanolamine
Mis	Müllerian-inhibiting substance	PEP	phosphoenolpyruvate
MLP	major late promotor	PEV	position effect variegation
MLS	myxoid liposarcoma	PFGE	pulsed-field gel electrophoresis
mM	millimolar	PGPR	plant growth-promoting Rhizobacteria
MMC	maternally inherited myopathy and cardiomyopathy		
		PH	plekstrin homology
MMR	mismatch repair	PH	polyhedron
MOI	multiplicity of infection	Phe	phenylalanine
MPS	mucopolysaccharidosis	Pi	inorganic phosphate
MRCA	most recent common ancestor	PI	phosphatidylinositol
MRD	minimal residual disease	PIP_2	phosphatidylinositol-4,5-bisphosphate
mRNA	messenger RNA		
MS	mass spectroscopy	PKA	protein kinase A
MSI	microsatellite instability	PKU	phenylketonuria
mtDNA	mitochondrial DNA	PMF	proton motive force
Mu element	mutator element	PMS	postmeiotic segregation
N-	amino-	Pol	polymerase
NAD	nicotinamide adenine dinucleotide	PR	protease
NADH	reduced NAD	Pro	proline
NADP	nicotinamide adenine dinucleotide phosphate	PS	phosphatidylserine
		PS I/II	photosystem I/II
NAP	nucleosome assembly protein	PTC	premature termination codon
NBU	nonreplication Bacteroides units	PTGS	posttranscriptional gene silencing
ncRNA	noncoding RNA	Q-(banding)	quinacrine

QTL	quantitative trait loci	SNP	single nucleotide polymorphism
R	resistance (eg Amp^R; ampicillin resistance)	snRNA	small nuclear RNA
		snRNP	small nuclear ribonucleoprotein
R-plasmids	resistance plasmids	SOD	superoxide dismutase
RA	retinoic acid	SRE	sterol response element
Ram	ribosomal ambiguity	SRP	signal recognition particle
RB	retinoblastoma	SRY	sex-determining region Y
RCC	renal cancer cell	ss	single stranded
RCL	round cell liposarcoma	SSB	single strand binding (protein)
RCS	recombinant congenic strains	SSLP	simple sequence length polymorphism
rDNA	recombinant DNA	SSR	simple tandem sequence repeats
RDR	recombination-dependent replication	STR	short tandem repeats
RED	repeat expansion detection	STS	sequence tagged sites
RER	rough endoplasmic reticulum	su	subunit
REV	reticuloendotheliosis virus	SU	surface (viral)
RF	release factor	SV40	simian virus 40
RF	replicative form	T	thymine
RFLP	restriction fragment length polymorphism	*Taq*	*Thermus aquatus*
		TBP	TATA-box binding protein
RI	recombinant inbred	TCA	tricarboxylic acid
Rif	rifampicin	TCR	T-cell receptor
RIM	reproductive isolating mechanism	TDF	testis-determining factor
RIP	repeat induced point (mutation)	TE	transposable elements
RM	restriction modification	TEL	telomere
RN	recombination nodule	TEM	transmission electron microscopy
RNA	ribonucleic acid	Ter	terminator
RNAi	RNA interference	Tet	tetracyclin
RNP	ribonucleoprotein	TF	transcription factor
ROS	reactive oxygen species	TGF	transforming growth factor
RRF	ribosome recycling factor	TGS	transcriptional gene silencing
rRNA	ribosomal RNA	Thr	threonine
RSS	recombination signal sequence	TIM	translocase of the inner membrane
RSV	Rous sarcoma virus	TIMPS	tissue inhibitor of metalloproteinases
RT	reverse transcriptase	TIR	terminal inverted repeats
RTK	receptor tyrosine kinase	TK	thymidine kinase
S-phase	synthesis phase	Tm	melting temperature
SAR	scaffold-attached region	TM	transmembrane
SAR	systemic acquired resistance	TMV	tobacco mosaic virus
SBT	shifting balance theory	TN	transposon
SC	synaptonemal complex	TNF	tumor necrosis factor
SCE	sister chromatid exchange	TOM	translocase of the outer membrane
SCF	stem cell factor	topo	topoisomerase
scRNA	small cytoplasmic RNA	T_{opt}	optimum temperature
SDP	strain distribution pattern	TRD	transmission ratio distortion
SDS	second-division segregation	TRiC	T-complex polypeptide ring complex
SDS-PAGE	sodium dodecyl-sulfate polyacrylamide gel electrophoresis	TRITC	tetramethyl rhodamine isothiocyanate
		tRNA	transfer RNA
SEM	scanning electron microscopy	Trp	tryptophan
sen DNA	senescent DNA	TSD	Tay–Sachs disease
Ser	serine	TSE	transmissible spongiform encephalopathy
SER	smooth endoplasmic reticulum		
SF	steroidogenic factor	TSG	tumor supressor gene
SH (domains)	Src homology (domains)	TSS	transcription start site
SI	self-incompatibility	Tyr	tyrosine
SINE	short interspersed nuclear element	U	uracil
SIV	simian immunodeficiency virus	Ub	ubiquitin
SL	spliced leader	UPD	uniparental disomy
SLF	steel factor	URF	unidentified reading frame
snoRNA	small nucleolar RNA	USS	uptake signal sequence

UTI	upper respiratory tract infection	WHO	World Health Organization
UTP	uridine triphosphate	WS	Werner syndrome
UTR	untranslated region	WT	wild-type
UV	ultraviolet	WT1	Wilm's tumor 1
V gene	variable gene	XIC	X-inactivation center
Val	valine	Xist	X-inactive specific transcript
VEGF	vascular endothelial growth factor	XP	xeroderma pigmentosum
VHL	Von Hippel–Lindau disease	YAC	yeast artificial chromosome
VNTR	variable number of tandem repeats	ZP	zona pellucida
VWF	Von Willebrand factor		

Contributors

S A Aaronson
The Derald H Ruttenberg Cancer Center, Mount Sinai Medical Center, 1425 Madison Avenue, New York, NY 10029, USA

J M Adams
Division of Molecular Genetics of Cancer, The Walter and Eliza Hall Institute of Medical Research, Post Office Royal Melbourne Hospital, Melbourne, VIC 3050, Australia

T H Adams
Monsanto Company, 62 Maritime Drive, Mystic, CT 06355, USA

S Adhya
Laboratory of Molecular Biology, National Cancer Institute, National Institutes of Health, Bethesda, MD 20892, USA

N A Affara
Department of Pathology, University of Cambridge, Cambridge, CB2 1QP, UK

J D Aitchison
Institute for Systems Biology, 4225 Roosevelt Way, Seattle, WA 98105, USA

S-I Aizawa
Department of Biosciences, Teikyo University, 1-1 Toyosatodai, Utsunomiya 320–8551, Japan

Z Alavidze
G Eliava Institute of Bacteriophages, Microbiology and Virology, Tbilisi, 380060, Georgia, USA

D G Albertson
Cancer Research Institute, University of California–San Francisco, San Francisco, CA 94143, USA

K H Albrecht
The Jackson Laboratory, Bar Harbor, ME 04609, USA

R W Alexander
Department of Chemical Biology, Wake Forest University, Winston-Salem, NC 27109, USA

P Alifano
Dipartimento di Biologia e Patologia Cellulare et Moleculare, Università de Napoli Federico, I-80131 Naples, Italy

P Aman
Department of Pathology, Lundberg Laboratory for Cancer Research, Goteborg University, S-41345 Göteborg, Sweden

M Ambrose
Department of Cancer Cell Biology, Harvard School of Public Health, Boston, MA 02215, USA

G F Ames
Division of Biochemistry and Molecular Biology, University of California–Berkeley, Berkeley, CA 94720, USA

P Anderson
Department of Genetics, University of Wisconsin–Madison, Madison, WI 53706, USA

H Antoun
RSVS Pavillian Charles–Eugene Marchand, Université Laval, Québec, GIK 7P4, Canada

C F Aquadro
Department of Molecular Biology and Genetics, Biotechnology Building Cornell University, Ithaca, NY 14853, USA

K Ardlie
Genomics Collaborative, 99 Evie Street, Cambridge, MA 02139, USA

M Arkin
Sunesis Pharmaceuticals, 3696 Haven Avenue, Redwood City, CA 94063, USA

J Arnold
Department of Genetics, University of Georgia, Athens, GA 30602, USA

K Artzt
Section of Molecular Genetics and Microbiology, Institute of Cell and Molecular Biology, University of Texas at Austin, Austin, TX 78712, USA

A Ashworth
Breakthrough Breast Cancer Research Centre, Chester Beatty Laboratories, Institute of Cancer Research, London, SW3 6JB, UK

M A Asmussen
Department of Genetics, University of Georgia, Athens, GA 30602, USA

K J Aufderheide
Department of Biology, Texas A & M University, Austin, TX 77843, USA

J Austin
Department of Molecular Genetics and Cell Biology, University of Chicago, Chicago, IL 60637, USA

J C Avise
Department of Genetics, University of Georgia, Athens, GA 30602, USA

R Baer
Department of Pathology, Institute of Cancer Genetics, College of Physicians and Surgeons of Columbia University, New York, NY 10032, USA

J-L Baert
Institut de Biologie de Lille, F-59021 Lille, France

A Bafico
Mount Sinai Medical Center, The Derald H Ruttenberg Cancer Center, 1425 Madison Avenue, New York, NY 10029, USA

B Bain
Department of Haematology, Imperial College School of Medicine, St Mary's Hospital, London W2 1NY, UK

T T Baird Jr
Department of Pharmaceutical Chemistry, University of California–San Francisco, San Francisco, CA 94143, USA

A Balmain
Cancer Research Institute, University of California–San Francisco, San Francisco, CA 94143, USA

W M Barnes
Department of Biochemistry and Molecular Biophysics, Washington University Medical School, St Louis, MO 63110, USA

M A Barrand
Department of Pharmacology, University of Cambridge, Cambridge, CB2 1QJ, UK

G S Barsh
Department of Pediatrics and Genetics and the Howard Hughes Medical Institute, Stanford University School of Medicine, Beckman Center, Stanford, CA 94305, USA

D P Bartel
The Whitehead Institute for Biochemical Research and Department of Biology, Massachusetts Institute of Technology, Cambridge, MA 02142, USA

R Baumeister
Genzentrum, D-81377 München, Germany

C Beamish
T4 Lab, The Evergreen State College, Olympia, WA 98505, USA

H-A Becker
Max-Planck Institut für Züchtungsforschung, D-50829 Köln, Germany

K M Beckingham
Department of Biochemistry and Cell Biology, Rice University, Houston, TX 77251, USA

C V Beechey
Mammalian Genetics Unit, Medical Research Council, Harwell, Didcot, OX11 0RD, UK

J D Beggs
The Wellcome Centre for Cell Biology, Institute of Cell and Molecular Biology, University of Edinburgh, Edinburgh, EH9 3JR, UK

M Belfort
Wadsworth Center, New York State Department of Health, Albany, NY 12201, USA

P N Benfey
Department of Biology, New York University, New York, NY 10003, USA

J L Bennetzen
Department of Biological Sciences, Purdue University, West Lafayette, IN 47907, USA

D E Bergstrom
The Jackson Laboratory, Bar Harbor, ME 04609, USA

A J Berk
Molecular Biology Institute, University of California–Los Angeles, Los Angeles, CA 90095, USA

M K B Berlyn
E. coli Genetic Stock Center, Department of MCD Biology, Yale University, New Haven, CT 06520, USA

A Bernardi
Laboratoire d'Enzymologie, CNRS, PGE, 1 Avenue de la Terrasse, F-91198 Gif sur Yvette, France

A Berns
The Netherlands Cancer Institute, NL-1066 CX, Amsterdam, The Netherlands

A Bernstein
Program in Molecular Biology and Cancer, Samuel Lunenfeld Research Institute, Mount Sinai Hospital, Toronto, ON, M5G 1X5, Canada

S S Bhattacharya
Department of Molecular Genetics, Institute of Ophthalmology, University College London, London, EC1V 9EL, UK

D L Black
Department of Microbiology and Molecular Genetics, Howard Hughes Medical Institute, University of California–Los Angeles, Los Angeles, CA 90024, USA

R D Blank
University of Wisconsin Medical School, Madison, WI 53792, USA

T Blumenthal
Department of Biochemistry and Molecular Genetics, University of Colorado School of Medicine, Denver, CO 80262, USA

F Bonhomme
Génome, Populations, Interactions UMR 5000, Université de Montpellier, F-34095 Montpellier, France

M I Borges-Walmsley
Division of Infection and Immunity, Institute of Biomedical and Life Sciences, University of Glasgow, Glasgow, G11 6NN, UK

P Boursot
Génome, Populations, Interactions UMR 5000, Université de Montpellier, F-34095 Montpellier, France

Y Boyd
Institute for Animal Health, Compton, RG20 7NN, UK

B Braaten
Department of Molecular, Cellular and Developmental Biology, University of California–Santa Barbara, Santa Barbara, CA 93106, USA

W J Brammar
Department of Biochemistry, University of Leicester, Leicester, LE1 7RH, UK

B Brembs
Department of Neurobiology and Anatomy, The University of Texas–Houston Medical School, Houston, TX 77030, USA

P J Brennan
Department of Pathology and Laboratory of Medicine, University of Pennsylvania School of Medicine, Philadelphia, PA 19104, USA

C Brenner
Faculté de Médecine, Université Libre de Bruxelles, Bruxelles, Belgium

S Brenner
Molecular Sciences Institute, University of California–Berkeley, Berkeley, CA 94704, USA

N J Brewin
Genetics Department, John Innes Centre, Colney, Norwich, NR4 7UH, UK

B A Bridges
MRC Cell Mutation Unit, University of Sussex, Brighton, BN1 9RH, UK

M H Brilliant
Department of Pediatrics, University of Arizona School of Medicine, Tucson, AZ 85724, USA

A Brinker
Max Planck Institut für Biochemie, D-82152 Martinsried, Germany

J Brookfield
Institute of Genetics, University of Nottingham, Greens Medical Centre, NG7 2UH, UK

W J Broughton
Department of Plant Molecular Biology, University of Geneva, CH-1292 Chambésy, Geneva, Switzerland

K E Browman
Bristo-Myers-Squibb Co., Neuroscience and Genitourinary Drug Discovery, Wallingford, CT 06492, USA

M S Brown
Department of Molecular Genetics, University of Texas Southwestern Medical Center, Dallas, TX 75235, USA

T A Brown
Department of Biomolecular Sciences, University of Manchester Institute of Science and Technology, Manchester, M60 1QD, UK

V Brown
Department of Genetics, Emory University School of Medicine, Atlanta, GA 30322, USA

M E Bruce
Neuropathogenesis Unit, Institute for Animal Health, Edinburgh, EH9 3JF, UK

M L Budarf
Division of Human Genetics, Department of Pediatrics, University of Pennsylvania School of Medicine, Philadelphia, PA 19104, USA

L Bülow
Department of Pure and Applied Chemistry, Center for Chemistry and Chemical Engineering, S-22100 Lund, Sweden

A Burchell
Tayside Insititute of Child Health, Ninewells Hospital and Medical School, University of Dundee, Dundee, DD1 9SY, UK

R R Burgess
McArdle Laboratory for Cancer Research, University of Wisconsin–Madison, Madison, WI 53706, USA

T R Bürglin
Department of Cell Biology, University of Basel, CH-4056 Basel, Switzerland

A Bürkle
Department of Gerontology, University of Newcastle upon Tyne, Newcastle upon Tyne, NE4 6BE, UK

J Burn
Department of Human Genetics, University of Newcastle upon Tyne, Newcastle upon Tyne, NE2 4AA, UK

D W Burt
Roslin Institute, Roslin, Midlothian, EH25 9PS, UK

C Caldas
CRC Department of Oncology and The Wellcome Trust
Centre for Molecular Mechanisms in Disease, Cambridge
Institute for Medical Research, University of Cambridge,
Cambridge, CB2 2XY, UK

A Campbell
Department of Biological Sciences, Stanford University,
Stanford, CA 94305, USA

S A Camper
Department of Human Genetics, University of Michigan
Medical School, Ann Arbor, MI 48109, USA

E Canaani
Department of Molecular Cell Biology, Weizmann
Institute of Science, 76100 Rehovot, Israel

E P M Candido
Department of Biochemistry and Molecular Biology,
University of British Columbia, Vancouver, British
Columbia, V6T1Z3, Canada

J C Carey
Department of Pediatrics, Obstetrics, Gynecology and
Nursing, Health Science Center, University of Utah,
Salt Lake City, UT 84132, USA

F Carneiro
Faculty of Medicine and IPATIMUP, University of Porto,
4200 Porto, Portugal

A T C Carpenter
Department of Genetics, University of Cambridge,
Cambridge, CB2 3EH, UK

D Carr
The Wellcome Trust, 210 Euston Road, London,
NW1 2BE, UK

D Carroll
Department of Biochemistry, University of Utah
School of Medicine, Salt Lake City, UT 84132, USA

G Caruana
Program in Molecular Biology and Cancer, Samuel
Lunenfeld Research Institute, Mount Sinai Hospital,
Toronto, Ontario, M5G 1X5, Canada

S M Case
Department of Biology, Salem State College, Salem,
MA 01970, USA

M Cashel
Laboratory of Molecular Genetics, NICHD,
National Institutes of Health, Bethesda, MD 20892,
USA

L A Casselton
Department of Plant Sciences, University of Oxford,
Oxford, OX1 3RB, UK

A Castells
Department of Gastroenterology, Barcelona University,
Barcelona 8036, Spain

D Catovsky
Academic Department of Haematology and
Cytogenetics, Royal Marsden Hospital, London,
SW3 6JJ, UK

B M Cattanach
MRC Mammalian Genetics Unit, Harwell, Didcot,
OX11 0RD, UK

R Chaganti
Memorial Sloan-Kettering Cancer Center, 1275 York
Avenue, New York, NY 10021, USA

R Chakraborty
Human Genetics Center, University of Texas School of
Public Health, Houston, TX 77225, USA

M Chalfie
Department of Biological Sciences, Columbia University,
New York, NY 10027, USA

M Chandler
Laboratoire de Microbiologie et de Génétique
Moléculaire, CNRS, F-31620 Toulouse, France

A C Chandley
20 Comely Bank, Edinburgh, EH4 1AL, UK

F Chedin
Department of Pathology and of Biochemistry, Norris
Comprehensive Cancer Center, University of Southern
California School of Medicine, Los Angeles 90089, CA,
USA

R Chetelat
Department of Vegetable Crops, University of California–
Davis, Davis, CA 95616, USA

A D Chisholm
Department of Biology, University of California–Santa
Cruz, Santa Cruz, CA 95064, USA

S Chong
New England Biolabs Inc., 32 Tozer Road, Beverly,
MA 01915, USA

J Chory
Plant Biology Laboratory, Howard Hughes Medical
Institute and The Salk Institute for Biological Studies,
La Jolla, CA 92037, USA

G A Churchill
The Jackson Laboratory, Bar Harbor, ME 04609, USA

G Churchward
Department of Microbiology and Immunology,
Emory University, Atlanta, GA 30322, USA

A G Clark
Department of Biology, Pennsylvania State University, University Park, PA 16802, USA

S G Clark
Skirball Institute of Biomolecular Medicine, New York University School of Medicine, New York, NY 10016, USA

M A Cleary
Department of Molecular Biology, Princeton University, Princeton, NJ 08544, USA

E H Coe Jr
Plant Genetics Research Unit, United States Department of Agriculture-ARS, University of Missouri, Columbia, MO 65211, USA

N Coleman
Department of Molecular Histopathology, Addenbrooke's Hospital, University of Cambridge, Cambridge, CB2 2QQ, UK

V P Collins
Addenbrooke's Hospital, University of Cambridge, Cambridge, CB2 2QQ, UK

N C Comfort
Center for History of Recent Science, George Washington University, Washington, DC 20052, USA

J M Connor
Department of Medical Genetics, University of Glasgow, Glasgow, G12 8QQ, UK

F Constantini
Department of Genetics and Development, College of Physicians and Surgeons, Columbia University, New York, NY 10032, USA

A Cooke
Department of Pathology, University of Cambridge, Cambridge, CB2 1QP, UK

C S Cooper
Institute of Cancer Research, Haddow Laboratories, Belmont, Sutton, SM2 5NG, UK

V Cormier-Daire
Department of Genetics, Hopitâl Necker Enfants Malades, Paris, France

B C Coughlin
Proceedings of the National Academy of Sciences, Washington, DC 20007, USA

A Coulson
The Sanger Centre, Hinxton, Cambridge, CB10 ISA, UK

D W Cox
Department of Medical Genetics, University of Alberta, Edmonton, Alberta, T6G 2H7, Canada

M M Cox
Department of Biochemistry, University of Wisconsin–Madison, Madison, WI 53706, USA

T M Cox
Department of Medicine, Addenbrooke's Hospital, University of Cambridge, Cambridge, CB2 2QQ, UK

J Coyne
Department of Ecology and Evolution, University of Chicago, Chicago, IL 60637, USA

J C Crabbe
Portland Alcohol Research Center, Department of Veterans Affairs Medical Center, Oregon Health Sciences University, Portland, Oregon 97201, USA

C S Craik
Department of Pharmaceutical Chemistry, University of California–San Francisco, San Francisco, CA 94143, USA

K A Crandall
Department of Zoology, Brigham Young University, Provo, UT 84602, USA

J F Crow
Genetics Laboratory, University of Wisconsin–Madison, Madison, WI 53706, USA

T J Crow
Department of Psychiatry, Warneford Hospital, University of Oxford, Oxford OX3 7JX, UK

J Cruz-Reyes
Department of Biological Chemistry, The Johns Hopkins University School of Medicine, Baltimore, MD 22105, USA

M B Cruzan
Department of Ecology and Evolutionary Biology, University of Tennessee, Knoxville, TN 37996, USA

A K Csink
Department of Biological Sciences, Carnegie Mellon University, Pittsburgh, PA 15213, USA

J Z Dalgaard
Marie Curie Research Institute, The Chase, Oxted, RH8 0TL, UK

A Danchin
Pasteur Research Centre, Hong Kong University, Pokfulam, Hong Kong

G Daniels
Bristol Institute for Transfusion Sciences, Southmead Road, Bristol, BS10 5ND, UK

A Darvasi
Department of Ecology, Systematics and Evolution, Silberman Institute of Life Sciences, The Hebrew University of Jerusalem, Jerusalem 91904, Israel

M T Davisson
The Jackson Laboratory, Bar Harbor, ME 04609, USA

L De Gregorio
Department of Pediatrics and Institute for Molecular Genetics, University of California–San Diego, La Jolla, CA 92093, USA

Y de Launoit
Faculté de Médecine, Université Libre de Bruxelles, Bruxelles, Belgium

D C DeLuca
Department of Biochemistry and Molecular Biology, University of Arkansas for Medical Sciences, Little Rock, AR 72205, USA

P Demant
Division of Molecular Biology, The Netherlands Cancer Institute, NL-1066 CX Amsterdam, The Netherlands

J Dénarié
Laboratorie de Biologie Moléculaire des Relations Plantes-Microorganismes, INRA/CNRS, F-31326 Castanet, France

K M Derbyshire
Division of Infectious Disease, Wadsworth Center, New York State Department of Health, and Department of Biomedical Sciences, School of Public Health, Albany, NY 12201, USA

S A des Etages
Pfizer Inc., Genetic Technologies, Groton, CT 06340, USA

R J Desnick
Department of Human Genetics, Mount Sinai School of Medicine, New York, NY 10029, USA

K M Devos
John Innes Centre, Colney, Norwich, NR4 7UH, UK

J E Donelson
Department of Biochemistry, University of Iowa, Iowa City, IA 52242, USA

D Donnai
University Department of Medical Genetics and Regional Genetics Service, St Mary's Hospital, Manchester, MI3 0JH, UK

R L Dorit
Department of Biology, Yale University, New Haven, CT 06511, USA

K Douglas
Department of Biology, University of Michigan, Ann Arbor, MI 48109, USA

W F Dove
McArdle Laboratory for Cancer Research, University of Wisconsin–Madison, WI 53706, USA

D M Downs
Department of Bacteriology, University of Wisconsin–Madison, Madison, WI 53706, USA

J J Doyle
Department of Plant Biology, L H Bailey Hortorium, Cornell University, Ithaca, NY 14853, USA

J W Drake
National Institute of Environmental Health Sciences, Irvine, NC 27709, USA

M E Dresser
Program in Molecular and Cell Biology, Oklahoma Medical Research Foundation, Oklahoma City, OK 73104, USA

M Driscoll
Department of Molecular Biology and Biochemistry, Nelson Laboratories, The State University of New Jersey, Rutgers, Piscataway, NJ 08854, USA

K Drlica
Public Health Research Institute, New York, NY 10016, USA

K R Dronamraju
Foundation for Genetic Research, Houston, TX 77227, USA

T Dunckley
Department of Molecular and Cellular Biology, Howard Hughes Medical Institute, University of Arizona, Tucson, AZ 85721, USA

J Dvořák
Department of Agronomy and Range Sciences, University of California–Davis, Davis, CA 95616, USA

B D Dyer
Department of Biology, Wheaton College, Norton, MA 02766, USA

M J S Dyer
Department of Haematology, Leicester Royal Infirmary, University of Leicester, Leicester, LE2 7LX, UK

S M Eacker
Departments of Genetics, University of Washington, Seattle, WA 98195, USA

W F Eanes
Department of Ecology and Evolution, State University of New York, Stony Brook, NY 11794, USA

J Eberwine
Department of Pharmacology, University of Pennsylvania Medical Center, Philadelphia, PA 19104, USA

W Eckhart
The Salk Institute for Biological Studies, La Jolla, CA 92037, USA

A W F Edwards
Gonville and Caius College, University of Cambridge,
Cambridge, CB2 1TA, UK

R Edwards
Duck End Farm, Park Lane, Dry Drayton, Cambridge
CB3 8DB, UK

J C J Eeken
Department of Radiation Genetics and Chemical
Mutagenesis, Leiden University Medical Center,
2300 RC Leiden, The Netherlands

R Egel
Department of Genetics, University of Copenhagen,
DK-1353 Copenhagen K, Denmark

A L Eggler
Department of Biochemistry, University of
Wisconsin–Madison, Madison, WI 53706, USA

L Ehrman
Division of Natural Sciences, State University of New
York, Purchase, NY 10577, USA

E M Eicher
The Jackson Laboratory, Bar Harbor, ME 04609, USA

T H Eickbush
Department of Biology, University of Rochester,
Rochester, NY 14627, USA

R C Eisensmith
Institute for Gene Therapy and Molecular Medicine,
Mount Sinai School of Medicine, New York,
NY 10029, USA

W S El-Deiry
Howard Hughes Medical Institute, University
of Pennsylvania School of Medicine, Philadelphia,
PA 19104, USA

N A Elgerdy
Department of Pathology, South Manchester
University Hospital Trust, Manchester,
M23 9LT, UK

R P Elinson
Department of Biological Sciences, Bayer School of
Natural and Environmental Sciences, Duquesne
University, Pittsburgh, PA 15282, USA

A E H Emery
Department of Neurology, Royal Devon and Exeter
Hospital, Exeter, EX2 5DW, UK

J A Endicott
Laboratory of Molecular Biophysics, Department
of Biochemistry and Oxford Centre for
Molecular Sciences, University of Oxford, Oxford, OX1
3QU, UK

W R Engels
Laboratory of Genetics, University of Wisconsin–
Madison, Madison, WI 53706, USA

M Estelle
Section of Molecular and Development Biology, Institute
of Cellular and Molecular Biology, University of
Texas–Austin, Austin TX 78712, USA

E P Evans
Mammalian Genetics Unit, Medical Research Council,
Harwell, Didcot, OX11 ORD, UK

T C Evans Jr
New England Biolabs Inc., 32 Tozer Road, Beverly,
MA 01915, USA

W J Ewens
Department of Biology, University of Pennsylvania,
Philadelphia, PA 19104, USA

J R Fabian
7 Partridgeberry Lane, Keene, NH 03431, USA

B A Fane
Department of Veterinary Science and Microbiology,
University of Arizona, Tucson, AZ 85721, USA

C Fankhauser
Department of Molecular Biology, University of Geneva,
CH-1211 Genève 4, Switzerland

N Fedoroff
Biotechnology Institute, Pennsylvania State University,
University Park, PA 16802, USA

M Feiss
Department of Microbiology, University of Iowa College
of Medicine, Iowa City, IA 52242, USA

T V Feldblyum
The Institute for Genomic Research, 9712 Medical
Center Drive, Rockville, MD 20850, USA

A C Ferguson-Smith
Department of Anatomy, University of Cambridge,
Cambridge, CB2 3DY, UK

M A Ferguson–Smith
Department of Clinical Veterinary Medicine, University of
Cambridge, Cambridge, CB3 OES, UK

A R Fersht
Department of Chemistry, University of Cambridge,
Cambridge, CB2 1EW, UK

W Filipowicz
Friedrich Miescher Institute, CH-4002 Basel, Switzerland

T M Finan
Department of Biology, McMaster University, Hamilton,
Ontario, L8S 4K1, Canada

J R S Fincham
Division of Biology, University of Edinburgh, Edinburgh, EH10 5RY, UK

H Firth
Department of Medical Genetics, Addenbrooke's Hospital, University of Cambridge, Cambridge, CB2 2QQ, UK

W Fitch
Department of Ecology and Evolutionary Biology, University of California–Irvine, Irvine, CA 92697, USA

J Folkman
Department of Surgery, The Children's Hospital and Harvard Medical School, Boston, MA 02115, USA

J Ford
Division of Oncology, Stanford University School of Medicine, Stanford, CA 94305, USA

J Forejt
Institute of Molecular Genetics, Academy of Sciences of the Czech Republic, 14220 Praha 4, Czech Republic

S L Forsburg
Molecular Biology and Virology Laboratory, The Salk Institute for Biological Studies, La Jolla, CA 92037, USA

J A Fossella
Laboratory of Neurobiology and Behavior, Rockefeller University, New York, NY 10021, USA

P L Foster
Department of Biology, Indiana University, Bloomington, IN 47405, USA

M Frame
Beatson Institute for Cancer Research, CRC Beatson Laboratories, Garscube Estate, Switchback Road, Bearsden, Glasgow, G62 1BD, UK

J Frampton
Weatherall Institute of Molecular Medicine, John Radcliffe Hospital, University of Oxford, Oxford, OX3 9DS, UK

W N Frankel
The Jackson Laboratory, Bar Harbor, ME 04609, USA

R Frankham
Key Centre for Biodiversity and Bioresources, Department of Biological Sciences, Macquarie University, Sydney, NSW 2109, Australia

C M Fraser
Institute for Genomic Research, 9712 Medical Center Drive, Rockville, MD 20850, USA

G R Fraser
Department of Clinical Genetics, John Radcliffe Hospital, Oxford, OX3 7LJ, UK

K Fredga
Department of Pediatrics and Institute for Molecular Genetics, University of California, San Diego, La Jolla, CA 92093, USA

S Fredriksson
Department of Pure and Applied Chemistry, Center for Chemistry and Chemical Engineering, Lund University, S-22100 Lund, Sweden

E C Friedberg
Laboratory of Molecular Pathology, Department of Pathology, University of Texas Southwestern Medical Center, Dallas, TX 75235, USA

T Friedmann
Center for Molecular Genetics, University of California–San Diego School of Medicine, La Jolla, CA 92093, USA

D J Futuyma
Department of Ecology and Evolution, State University of New York, Stony Brook, NY 11794, USA

G L Gabor Miklos
GenetixXpress Proprietary Ltd, Palm Beach, Sydney, NSW 2108, Australia

J Gallant
Uppsala University, SE-24332 Hoor, Sweden

J I Garrels
Proteome Division, Incyte Genomics, Beverly, MA 01915, USA

S M Gartler
Departments of Medicine and Genetics, University of Washington, Seattle, WA 98195, USA

A Gavalas
Department of Development Neurobiology, National Institute for Medical Research, Mill Hill, London NW7 1AA, UK

V Gavrias
Monsanto Co., Mystic, CT 06355, USA

A F Gazdar
Hamon Center for Therapeutic Oncology Research, University of Texas Southwestern Medical Center, Dallas, TX 75390, USA

M L Gennaro
Public Health Research Institute, New York, NY 10016, USA

P Gepts
Department of Agronomy and Range Science, University of California–Davis, Davis, CA 95616, USA

J German
Department of Pediatrics, Weill College of Medicine, Cornell University, Ithaca, NY 10021, USA

E Gherardi
Department of Oncology, MRC Centre, Addenbrooke's Hospital, Cambridge, CB2 2QH, UK

F Giannelli
Division of Medical and Molecular Genetics, Guy's, King's, and St Thomas's School of Medicine, King's College London, London, SE1 9RT, UK

G N Gill
University of California–San Diego School of Medicine, La Jolla, CA 92093, USA

M A Gilson
Wadsworth Center, New York State Department of Health, Albany, NY 12201, USA

A C Glasgow
Department of Microbiology, University of Georgia, Athens, GA 30602, USA

M Goldman
Department of Biology, San Francisco State University, San Francisco, CA 94132, USA

M Goldschmidt-Clermont
Department of Molecular Biology, University of Geneva, CH-1211 Genève 4, Switzerland

M Goldsmith
Department of Biological Sciences, University of Rhode Island, Kingston, RI 02881, USA

J L Goldstein
Department of Molecular Genetics, University of Texas Southwestern Medical Center, Dallas, TX 75235, USA

E S Golub
Lykeion Corp., Solana Beach, CA 92075, USA

M F Goodman
Department of Biological Sciences and Chemistry, University of Southern California, Los Angeles, CA 90089, USA

J A Goodrich
Department of Chemistry and Biochemistry, University of Colorado–Boulder, Boulder, CO 80309, USA

J Goodship
Department of Human Genetics, University of Newcastle upon Tyne, Newcastle upon Tyne, NE2 4AA, UK

D M Gordon
Division of Biology and Zoology, Australian National University, Canberra, ACT, Australia

I I Gottesman
245 Terrell Road West, Charlottesville, VA 22901, USA

C S Grant
Department of Surgery, The Mayo Clinic, Rochester, MN, USA

J-P Gratia
Microbial Genetics and Ecology Unit, Brussels University, School of Medicine, B-1070 Brussels, Belgium

J A M Graves
Comparative Genomics Research Group, Research School of Biological Sciences, The Australian National University, Canberra, ACT 2601, Australia

M W Gray
Department of Biochemistry and Molecular Biology, Dalhousie University, Halifax, Nova Scotia, B3H 4H7, Canada

M I Greene
Department of Pathology and Laboratory of Medicine, University of Pennsylvania School of Medicine, Philadelphia, PA 19104, USA

N Gregersen
Research Unit for Molecular Medicine, Aarhus University Hospital and Faculty of Health Sciences, Skejby Hospital, DK-8200 Aarhus N, Denmark

P Gresshoff
Department of Botany, University of Queensland, Brisbane, QLD 4072, Australia

M A Griep
Department of Chemistry, University of Nebraska, Lincoln, NE 68588, USA

T Grigliatti
Department of Biological Sciences, University of British Columbia, Vancouver, British Columbia, V6T 1Z4, Canada

N D F Grindley
Department of Molecular Biophysics and Biochemistry, Yale University, New Haven, CT 06520, USA

A P Grollman
Department of Pharmacological Sciences, State University of New York, Stony Brook, NY 11794, USA

R Grosschedl
Gene Center and Institute of Biochemistry, University of Munich, D-81377 Munich, Germany

P Gruss
Max Planck Institute for Biophysical Chemistry, Department of Molecular Cell Biology, D-37077 Göttingen, Germany

J B Gurdon
Wellcome CRC Institute, Tennis Court Road, Cambridge, CB2 1QR, UK

B S Guttman
Lab I, The Evergreen State College, Olympia,
WA 98505, USA

N Haites
Department of Medical Genetics, University of Aberdeen,
Aberdeen, UK

S E Halford
Department of Biochemistry, University of Bristol,
Bristol, B58 1TD, UK

R M Hall
Division of Biomolecular Engineering, CSIRO Molecular
Science, Sydney Laboratory, North Ryde, NSW 1670,
Australia

J L Hamerton
Department of Human Genetics, University of Manitoba,
Winnipeg, Manitoba, R3E 0W3, Canada

P Hanawalt
Department of Biological Sciences, Stanford University,
Stanford, CA 94305, USA

D B Haniford
Department of Biochemistry, University of Western
Ontario, London, Ontario, N6A 5C1, Canada

H Harada
Division of Gastroenterology, University of Pennsylvania,
Philadelphia, PA 19104, USA

S C Hardies
Department of Biochemistry, University of Texas
Health Science Center, San Antonio, TX 78229, USA

J C Harper
Department of Obstetrics and Gynaecology, University
College London, London WCIE 6HX, UK

R Harris
Department of Medical Genetics, University of
Manchester, Manchester, M13 9PT, UK

C J Harrison
Department of Haematology, Royal Free Hospital,
London, NW3 2QG, UK

D J Harrison
Department of Pathology, University of Edinburgh,
Edinburgh, UK

D L Hartl
Department of Organismic and Evolutionary
Biology, Harvard University, Cambridge, MA 02138,
USA

F U Hartl
Max Planck Institut für Biochemie, D-82152 Martinsried,
Germany

M Hasegawa
Institute of Statistical Mathematics, 4-6-7 Minami Avenue,
Minato Ku, Tokyo 106-8569, Japan

R Haselkorn
Department of Molecular Genetics and Cell Biology,
School of Medical Sciences, University of Chicago,
Chicago, IL 60637, USA

P S Hasleton
Department of Pathology, South Manchester University
Hospital Trust, Manchester, M23 9LT, UK

N Hastie
MRC Human Genetics Unit, Western General Hospital,
Edinburgh, EH4 2XU, UK

P J Hastings
Department of Molecular and Human Genetics, Baylor
College of Medicine, Houston, TX 77030, USA

P M Hawkey
Public Health Bacteriology, Public Health Laboratory,
Birmingham Heartlands Hospital, Bordesley Green East,
Birmingham, B9 5SS, UK

R S Hawley
Section of Molecular and Cellular Biology, University of
California–Davis, Davis, CA 95616, USA

T Hazelrigg
Department of Biological Sciences, Columbia University,
New York, NY 10027, USA

J K Heath
Department of Biochemistry, University of Birmingham,
Birmingham, B15 2TT, UK

J F Heidelberg
Institute for Genomic Research, Rockville, MD 20850,
USA

D R Helinski
University of California–San Diego, La Jolla, CA 92093,
USA

J G Hengstler
Institut für Toxikologie, Obere Zahlbacher Strasse 67,
D-55131 Mainz, Germany

T M Henkin
Department of Microbiology, Ohio State University,
Columbus, OH 43210, USA

R K Herman
Department of Genetics and Cell Biology, University of
Minnesota, St Paul, MN 55108, USA

I Herskowitz
Department of Biochemistry and Biophysics,
University of California–San Francisco, San Francisco,
CA 94143, USA

R Hesketh
Department of Biochemistry, University of Cambridge, Cambridge, CB2 1QW, UK

J S Heslop-Harrison
Department of Biology, University of Leicester, Leicester, LE1 7RH, UK

C Heyting
Laboratorium voor Erfelijkheidsleer, NL-6700 HB Wageningen, The Netherlands

P Hieter
Centre for Molecular Medicine and Therapeutics, Department of Medical Genetics, University of British Columbia, Vancouver, British Columbia, V5Z 4H4, Canada

D Higgins
Department of Biochemistry, University College, Cork, Ireland

K L Hill
Department of Microbiology, Immunology and Molecular Genetics, University of California–Los Angeles, Los Angeles, CA 90095, USA

W G Hill
Institute of Cell, Animal and Population Biology, University of Edinburgh, Edinburgh, EH9 3JT, UK

M M Hingorani
Laboratory of DNA Replication, Rockefeller University, New York, NY 10021, USA

M J Hobart
Department of Biological Sciences, De Montfort University, Leicester, LE1 5BH, UK

J Hodgkin
Genetics Unit, Department of Biochemistry, University of Oxford, Oxford, OX1 3QU, England

B Hohn
Friedrich Miescher Institute, CH-4058 Basel, Switzerland

K E Holsinger
Department of Ecology and Evolutionary Biology, University of Connecticut, Storrs, CT 06269, USA

T R Hoover
Department of Microbiology, University of Georgia, Athens, GA 30602, USA

J C Hu
Department of Biochemistry and Biophysics, Texas A & M University, Austin, TX 77843, USA

C Huang
Department of Biological Chemistry, The Johns Hopkins University School of Medicine, Baltimore, MD 21205, USA

R E Huber
Department of Biological Sciences, University of Calgary, Calgary, Alberta, T2N 1N4, Canada

N Hughes-Jones
Formerly member of the Medical Research Council's Mechanisms in Immunopathogy Unit, MRC Centre, Cambridge, CB2 2XY, UK

M Hülskamp
Botanik III, Universität Köln, D-50931 Köln, Germany

M A Hultén
Department of Biological Sciences, University of Warwick, Coventry, CV4 7AL, UK

D M Hunt
Division of Molecular Genetics, Institute of Ophthalmology, University College London, London, EC1V 9EL, UK

S M Huson
Department of Clinical Genetics, John Radcliffe Hospital, Oxford, OX3 7LJ, UK

P G Isaacson
Department of Histopathology, Royal Free and University College School of Medicine, University College London, London, WCIE 6BT, UK

W E Jack
New England Biolabs Inc, 32 Tozer Road, Beverly, MA 01915, USA

I Jackson
MRC Human Genetics Unit, Western General Hospital, Edinburgh, EH4 2XU, UK

P A Jacobs
Wessex Regional Genetics Laboratory, Salisbury District Hospital, Salisbury, SP2 8BJ, UK

E Jansen
Laboratory for Molecular Oncology, Center for Human Genetics, University of Leuven and Flanders Interuniversity for Biotechnology, B-3000, Leuven, Belgium

N G J Jaspers
Department of Cell Biology and Genetics, Erasmus University, NL-3000 DR Rotterdam, The Netherlands

M Jayaram
Section of Molecular Genetics, University of Texas–Austin, Austin, TX 78712, USA

R C Johnson
Department of Biological Chemistry, University of California–Los Angeles, Los Angeles, CA 90095, USA

R Johnson
Department of Biochemistry and Molecular Biology, M D Anderson Cancer Center, University of Texas–Houston, Houston, TX 77030, USA

E M Jorgensen
Department of Biology, University of Utah, Salt Lake City, UT 84112, USA

T H Jukes
Formerly of Department of Integrative Biology, University of California–Berkeley, Berkeley, CA 94720, USA

L W Jurata
Gene Expression Laboratory, The Salk Institute for Biological Studies, La Jolla, CA 92037, USA

M M Kaback
Department of Pediatrics, Children's Hospital and Health Center, San Diego, CA 92123, USA

C I Kado
Department of Plant Pathology, University of California–Davis, Davis, CA 95616, USA

A Kallioniemi
National Human Genome Research Institute, National Institutes of Health, Bethesda, MD 20892, USA

D K Kalousek
Cytogenetics and Embryopathology Laboratory, British Columbia Children's Hospital, Vancouver, British Columbia, V6H 3VH, Canada

C Kane
Department of Molecular and Cell Biology, University of California–Berkeley, Berkeley, CA 94720, USA

D K R Karaolis
Department of Epidemiology and Preventive Medicine, University of Maryland School of Medicine, Baltimore, MD 21201, USA

J Karn
MRC Laboratory of Molecular Biology, Addenbrooke's Hospital, Cambridge, CB2 2QH, UK

R M T Katso
Ludwig Institue for Cancer Research, 91 Riding House Street, London, W1W 8BT, UK

E A Kellogg
Department of Biology, University of Missouri–St Louis, St Louis, MO 63121, USA

T Kelly
Department of Molecular Biology and Genetics, The Johns Hopkins University School of Medicine, Baltimore, MD 21205, USA

K J Kemphues
Department of Molecular Biology and Genetics, Cornell University, Ithaca, NY 14853, USA

M G Kidwell
Department of Ecology and Evolutionary Biology, University of Arizona, Tucson, AZ 85721, USA

M Kimmel
Statistics Department, Rice University, Houston, TX 77251, USA

H Kishino
University of Tokyo, 113-8657 Tokyo, Japan

A J S Klar
Gene Regulation and Chromosome Biology Laboratory, National Cancer Institute at Frederick, Frederick, MD 21702, USA

G Klein
Microbiology and Tumor Biology Center, Karolinska Institute, Stockholm, S-17177, Sweden

L A Klobutcher
Department of Biochemistry, University of Connecticut Health Center, Farmington, CT 06030, USA

J W Kloepper
Department of Entomology and Plant Pathology, Auburn University, AL 36849, USA

M A Koch
Institute of Botany, University of Agricultural Sciences, A-1180 Vienna, Austria

Y Kohara
Center of Genetic Resources Information, National Institute of Genetics, Mishima, 411-8540, Shizuoka-ken, Japan

G S Kopf
Center for Research on Reproduction and Women's Health, University of Pennsylvania School of Medicine, Philadelphia, PA 19104, USA

H C Korswagen
Hubrecht Laboratory, Netherlands Institute for Development Biology, NL-3584 CT Utrecht, The Netherlands

D C Krakauer
Institute for Advanced Study, Princeton University, Princeton, NJ 08540, USA

M Kreitman
Department of Ecology and Evolutionary Biology, University of Chicago, Chicago, IL 60637, USA

K N Kreuzer
Department of Microbiology and Biochemistry, Duke University Medical Center, Durham, NC 27710, USA

C B Krimbas
Department of Philosophy and History of Science, University of Athens, Athens 16771, Greece

H B Krishnan
Department of Plant Pathology, United States Department of Agriculture-ARS, University of Missouri, Columbia, MO 65211, USA

R Krumlauf
Stowers Institute for Medical Research, Kansas City, MO 64110, USA

A Kumar
Department of Molecular Cellular and Developmental Biology, Yale University, New Haven, CT 06520, USA

C Kurland
Uppsala University, S-24332 Hoor, Sweden

M Kutateladze
G Eliava Institute of Bacteriophages, Microbiology and Virology, Tbilisi, 380060, Georgia

E Kutter
Lab1, The Evergreen State College, Olympia, WA 98505, USA

C P Kyriacou
Department of Genetics, University of Leicester, Leicester, LEI 7RH, UK

B N La Du
Department of Pharmacology, University of Michigan Medical School, Ann Arbor, MI 48109, USA

M Labouesse
Institut de Génétique et de Biologie Moléculaire et Cellulaire, CNRS/INSERM/ULP, BP163, F-67404 Illkirch, France

S A Lacks
Department of Biology, Brookhaven National Laboratory, Upton, NY 11973, USA

B C Lamb
Department of Biology, Imperial College, London, SW7 2AZ, UK

A Landy
Department of Biology and Medicine, Brown University, Providence, RI 02912, USA

R A LaRossa
E.I. du Pont de Nemours & Co., Experimental Station, Wilmington, DE 19880, USA

D S Latchman
Windeyer Institute of Medical Sciences, University College London Medical School, London, WIP 6DB, UK

F Latif
Department of Medical Genetics, University of Birmingham, Birmingham, B15 2TT, UK

J Y-K Lau
3098 Holly Hall, Houston, TX 77054, USA

J Laurén
Molecular/Cancer Biology Laboratory, University of Helsinki, SF00014, Helsinki, Finland

J Laval
Institut Gustav Roussy, CNRS, F-94805 Villejuif, France

P Laybourn
Department of Biochemistry and Molecular Biology, Colorado State University, Fort Collins, CO 80523, USA

P Leder
Department of Genetics, Harvard Medical School, Boston, MA 02115, USA

J Y Lee
Whitehead Institute for Biomedical Research and Department of Biology, Massachusetts Institute of Technology, Cambridge, MA 02142, USA

G Lefranc
Laboratoire d'ImmunoGénétique Moléculaire, Institut de Génétique Humaine, Université Montpellier II, F-34396 Montpellier, France

M-P Lefranc
Laboratoire d'ImmunoGénétique Moléculaire, Institut de Génétique Humaine, Université Montpellier II, F-34396 Montpellier Cedex 5, France

J W Lengeler
Fachbereich Biologie/Chermie, AG Genetik, University of Osnabrück, D-49069 Osnabrück, Germany

R E Lenski
Center for Microbial Ecology, Michigan State University, East Lansing, MI 48824, USA

P A Lessard
Department of Biology, Massachusetts Institute of Technology, Cambridge, MA 02138, USA

G Levan
CMB Genetics, University of Göteborg, S-40530 Göteborg, Sweden

J Levinton
Department of Ecology and Evolution, State University of New York, Stony Brook, NY 11794, USA

S B Levy
Center for Adaptation Genetics and Drug Resistance, Tufts University School of Medicine, Boston, MA 02111, USA

D Lew
Duke University Medical Center, Durham, NC 27710, USA

R Lewis
7 Harvest Drive, Scotia, New York, NY 12302, USA

R C Lewontin
Museum of Comparative Zoology, Harvard University, Cambridge, MA 02138, USA

S W L'Hernault
Department of Biology, Emory University, Atlanta, GA 30322, USA

W-H Li
Department of Ecology and Evolution, University of Chicago, Chicago, IL 60637, USA

M R Lieber
Department of Pathology and of Biochemistry, Norris Comprehensive Cancer Center, University of Southern California School of Medicine, Los Angeles, CA 90089, USA

A Liljas
Molecular Biophysics, Center for Chemistry and Chemical Engineering, Lund University, S-22100, Lund, Sweden

D M J Lilley
CRC Nucleic Structure Research Group, Biochemistry Department, University of Dundee, Dundee, DD1 4HN, UK

J Limon
Department of Biology and Genetics, Medical University of Gdansk, 80-210 Gdansk, Poland

G J Lithgow
School of Biological Sciences, University of Manchester, Manchester, M13 9PT, UK

A C Lloyd
MRC Laboratory for Cell Biology, University College London, London, WC1E 6BT, UK

A Long
Department of Ecology and Evolutionary Biology, University of California–Irvine, Irvine, CA 92697, USA

W-E Lönning
Max Planck Institut für Züchtungsforschung, D-50829 Köln, Germany

E J Louis
Department of Biochemistry and Molecular Biology, University of Leicester, Leicester, LE1 7RH, UK

P S Lovett
Department of Biological Sciences, University of Maryland, Catonsville, MD 21228, USA

S T Lovett
Department of Genetics and Molecular Biology, Brandeis University, Waltham, MA 02254, USA

D Low
Department of Molecular, Cellular and Development Biology, University of California–Santa Barbara, Santa Barbara, CA 93106, USA

K B Low
Radiobiology Laboratories, Yale University, New Haven, CT 06520, USA

P Lu
Department of Chemistry, University of Pennsylvania, Philadelphia, PA 19104, USA

S Lusetti
Department of Biochemistry, University of Wisconsin–Madison, Madison, WI 53706, USA

L Luzzatto
Instituto Nazionale per la Ricerca sul Cancro, 16132 Genova, Italy

J Lyndal York
University of Arkansas for Medical Sciences, Little Rock, AR 72205, USA

M F Lyon
MRC Mammalian Genetics Unit, Harwell, Didcot, OX11 0RD, UK

H C Macgregor
Department of Biology, University of Leicester, Leicester, LE1 7RH, UK

J Mager
Department of Genetics, University of North Carolina, Chapel Hill, NC 27599, USA

M E Magnello
Wellcome Trust Centre for the History of Medicine at University College London, London, NW1 1AD, UK

T Magnuson
Department of Genetics, School of Medicine, University of North Carolina, Chapel Hill, NC 27599, USA

E Maher
Department of Medical Genetics, University of Birmingham, Birmingham, B15 2TT, UK

E M Maine
Department of Biology, Syracuse University, Syracuse, NY 13244, USA

A Maitra
Department of Pathology, University of Texas Southwestern Medical Center, Dallas, TX 75390, USA

N Maizels
Department of Immunology, University of Washington
Medical School, Seattle, WA 98195, USA

K D Makova
Department of Ecology and Evolution, University of
Chicago, Chicago, IL 60637, USA

S Malcolm
Molecular and Clinical Genetics Institute,
Institute for Child Health, London,
WC1N 1EH, UK

S Maloy
Department of Microbiology, University of Illinois,
Urbana, IL 61801, USA

T Maniatis
Department of Molecular and Cellular Biology, Harvard
University, Cambridge, MA 02138, USA

A Mansouri
Department of Molecular Cell Biology, Max Planck
Institute for Biophysical Chemistry, D-37077 Göttingen,
Germany

R Marcus
Department of Haematological Medicine, MRC Centre,
Cambridge, CB2 2QH, UK

A Martinez-Arias
Department of Genetics, University of Cambridge,
Cambridge, CB2 3EA, UK

P H Masson
Laboratory of Genetics, University of Wisconsin–
Madison, Madison, WI 53706, USA

C Mathew
Division of Medical and Molecular Genetics, Guy's, King's
and St Thomas's School of Medicine, King's College
London, London, SE1 9RT, UK

J Mathur
Botanik III, Universität Köln, D-50931 Köln, Germany

L Maxson
College of Liberal Arts, University of Iowa, Iowa City,
IA 52242, USA

E Mayr
Museum of Comparative Zoology, Cambridge,
MA 02138, USA

S McKee
Clinical Genetics Unit, Birmingham Women's Hospital,
Birmingham, B15 2TG, UK

D W Meinke
Department of Botany, Oklahoma State University,
Stillwater, OK 74078, USA

F Meins Jr
Friedrich Miescher Institute, Maulbeerstrasse 66, Basel
CH-4058, Switzerland

M Melnick
University of California, Los Angeles, CA 90089,
USA

S K Merickel
Department of Biological Chemistry, University of
California–Los Angeles, Los Angeles, CA 90095,
USA

J Merriam
Department of Biology, University of California–Los
Angeles, Los Angeles, CA 90095, USA

J P Métraux
Département de Biologie, Institut de Biologie Végétale,
Université de Fribourg, CH-1700 Fribourg,
Switzerland

J Michiels
Centre of Microbial and Plant Genetics, Kasteelpark
Arenberg 20, B-3001 Heverlee, Belgium

C J Migeon
Department of Pediatrics, The Johns Hopkins Hospital,
Baltimore, MD 21287, USA

H I Miller
Hoover Institution, Stanford University, Stanford,
CA 94305, USA

J H Miller
Department of Microbiology and Molecular Genetics,
University of California–Los Angeles, Los Angeles,
CA 90024, USA

O J Miller
Center for Molecular Medicine and Genetics, Wayne
State University School of Medicine, Detroit, MI 48201,
USA

L Mindich
Public Health Research Institute, New York University,
New York, NY 10016, USA

J B Mitton
Department of Environmental, Population and
Organismic Biology, University of Colorado–Boulder,
Boulder, CO 80309, USA

P B Moens
Department of Biology, York University, Downsview,
Ontario, M3J 1P3, Canada

G Morata
Centro de Biología Molecular, Consejo Superior de
Investigaciones Científicas, Universidad Autónoma de
Madrid, 28049 Madrid, Spain

I Mori
Laboratory of Molecular Neurobiology, Division of Biological Science, Graduate School of Science, Nagoya University, 464-8602 Nagoya, Japan

H C Morse
National Institute of Allergy and Infectious Diseases, National Institutes of Health, Bethesda, MD 20892, USA

R K Mortimer
Department of Molecular and Cell Biology, University of California–Berkeley, Berkeley, CA 94720, USA

N E Morton
Genetic Epidemiology Research Group, Human Genetics, Princess Anne Hospital, University of Southampton, Southampton, S016 5YA, UK

R C Moschel
Chemistry of Carcinogenesis Laboratory, National Cancer Institute at Frederick, Frederick, MD 21702, USA

R Mottus
Department of Biological Sciences, University of British Columbia, Vancouver, British Columbia, V6T 1Z4, Canada

L D Mueller
Department of Ecology and Evolutionary Biology, University of California–Irvine, Irvine, CA 92697, USA

B Müller-Hill
Institut für Genetik, Universität Köln, D-50931 Köln, Germany

L M Mulligan
Departments of Pediatrics and Pathology, Queen's University, Kingston, Ontario, K7L 3N6, Canada

E J Murgola
Department of Molecular Genetics, The University of Texas M.D. Anderson Cancer Center, Houston, TX 77030, USA

K Nagai
MRC Laboratory of Molecular Biology, Cambridge, CB2 2QH, UK

M Nei
Department of Biology, Pennsylvania State University, University Park, PA 16802, USA

F C Neidhardt
Department of Microbiology and Immunology, University of Michigan, Ann Arbor, MI 48109, USA

H C M Nelson
Department of Biochemistry and Biophysics, University of Pennsylvania School of Medicine, Philadelphia, PA 19104, USA

K E Nelson
Institute for Genomic Research, 9712 Medical Center Drive, Rockville, MD 20850, USA

C Neuhauser
Department of Ecology, Evolution and Behavior, University of Minnesota, St Paul, MN 55108, USA

G Newton
The Wellcome Trust, 210 Euston Road, London, NW1 2BE, UK

W C Nierman
The Institute for Genomic Research, 9712 Medical Center Drive, Rockville, MD 20850, USA

M E M Noble
Laboratory of Molecular Biophysics, Department of Biochemistry and Oxford Centre for Molecular Sciences, University of Oxford, Oxford, OX1 3QU, UK

K M Noll
Department of Molecular and Cell Biology, University of Connecticut, Storrs, CT 06269, USA

C J Norbury
Imperial Cancer Research Fund, Institute of Molecular Medicine, University of Oxford, Oxford, OX3 7LJ, UK

M A Nowak
Institute for Advanced Study, Princeton University, Princeton, NJ 08540, USA

W L Nyhan
Department of Pediatrics and Institute for Molecular Genetics, University of California–San Diego, La Jolla, CA 92093, USA

S J O'Brien
Laboratory of Genomic Diversity, National Cancer Institute, Frederick Cancer Research and Development Center, Frederick, MD 21701, USA

K O'Connell
Laboratory of Molecular Biology, University of Wisconsin–Madison, Madison, WI 53706, USA

C J O'Kane
Department of Genetics, University of Cambridge, Cambridge, CB2 3EH, UK

S L O'Kane Jr
Department of Biology, University of Northern Iowa, Cedar Falls, IA 50614, USA

O O'Neill
Newnham College, Cambridge, CB3 9DF, UK

J L H O'Riordan
14 Northampton Park, London, N1 2PJ, UK

R Oberbauer
Department of Internal Medicine III, University of Vienna, A-1090 Vienna, Austria

F Oesch
Institut für Toxikologie, Obere Zahlbacher Strasse 67, D-55131 Mainz, Germany

R F Ogle
Department of Obstetrics and Gynecology, University College London, London, WC1E 6HX, UK

T Ohta
National Institute of Genetics, Mishima, 411-8540 Shizuoka-ken, Japan

R Olby
Department of the History and Philosophy of Science, University of Pittsburgh, Pittsburgh, PA 15260, USA

G J Olsen
Department of Chemistry and Life Sciences, University of Illinois, Urbana, IL 61801, USA

T L Orr-Weaver
Whitehead Institute for Biomedical Research and Department of Biology, Massachusetts Institute of Technology, Cambridge, MA 02142, USA

A Orth
Génome, Populations, Interactions UMR 5000, Université de Montpellier, F-34095 Montpellier, France

E A Ostrander
Fred Hutchinson Cancer Research Center, 1100 Fairview Ave N, Seattle, WA 98109, USA

R A Padua
Hematology Department, University of Wales College of Medicine, Cardiff, CF14 4XN, UK

R Palacios
Nitrogen Fixation Research Center, National University of México, Cuernavaca, Mor. CP 62210, Mexico

P Pamilo
Department of Biology, University of Oulu, SF-90014 Oulu, Finland

V E Papaioannou
Department of Genetics and Development, College of Physicians and Surgeons of Columbia University, New York, NY 10032, USA

J Parker
Department of Microbiology, Southern Illinois University Carbondale, Carbondale, IL 62901, USA

R Parker
Department of Molecular and Cellular Biology, Howard Hughes Medical Institute, University of Arizona, Tucson, AZ 85721, USA

H C Passmore
Department of Biological Sciences, Rutgers University, Piscataway, NJ 08855, USA

J Paszkowski
Friedrich Miescher Institute, CH-4002 Basel, Switzerland

M L Pato
Department of Microbiology, University of Colorado Health Sciences Center, Denver, CO 80262, USA

I T Paulsen
Institute for Genomic Research, 9712 Medical Center Drive, Rockville, MD 20850, USA

W J Pavan
Laboratory of Genetic Disease Research, National Human Genome Research Institute, National Institutes of Health, Bothesda, MD 20892, USA

P L Pearson
Division of Medical Genetics, Wilhelmina Children's Hospital, Utrecht University Medical Center, NL-3508 AB, Utrecht, The Netherlands

L Peltonen
Department of Human Genetics, University of California–Los Angeles, School of Medicine, Los Angeles, CA 90095, USA

D Penny
Institute of Molecular Sciences, Massey University, Palmerston North, New Zealand

R R Pera
Departments of Urology, Obstetrics and Gynecology and Reproductive Sciences and Physiology, University of California–San Francisco, San Francisco, CA 94143, USA

J J Perona
Department of Chemistry and Biochemistry, University of California–Santa Barbara, Santa Barbara, CA 93106, USA

X Perret
Laboratoire de Biologie Moléculaire des Plantes Supérieures, Université de Genève, CH-1292 Chambésy, Genève, Switzerland

M M R Petit
Laboratory for Molecular Oncology, Center for Human Genetics, University of Leuven and Flanders Interuniversity Institute for Biotechnology, B-3000 Leuven, Belgium

J F Petrosino
Department of Molecular and Human Genetics, Baylor College of Medicine, Houston, TX 77030, USA

J Phelan
Department of Organismic Biology, University of California, Los Angeles, CA 90095, USA

E Pianka
Section of Integrative Biology, University of Texas–Austin, Austin, TX 78712, USA

S H Pilder
Department of Anatomy and Cell Biology, Temple University School of Medicine, Philadelphia, PA 19104, USA

J M Pipas
Department of Biological Sciences, University of Pittsburgh, Pittsburgh, PA 15260, USA

R H A Plasterk
Hubrecht Laboratory, Netherlands Institute for Development Biology, NL-3584 CT Utrecht, The Netherlands

E Pollak
Department of Statistics, Iowa State University, Ames, IA 50011, USA

J W Pollard
Departments of Developmental and Molecular Biology and Obstetrics and Gynecology and Women's Health, Albert Einstein College of Medicine, NY 10461, USA

D Pomeranz Krummel
Structural Studies Division, Medical Research Council, Laboratory of Molecular Biology, Cambridge, CB2 2QH, UK

F M Pope
Institute of Medical Genetics, University Hospital of Wales, Cardiff, CF4 4XN, UK

M Potter
Laboratory of Genetics, National Cancer Institute, National Institutes of Health, Bethesda, MD 20892, USA

J Poulton
Department of Paediatrics, John Radcliffe Hospital, University of Oxford, Oxford, OX3 9DU, UK

P M Pour
UNMC, Eppley Cancer Center and Department of Pathology and Microbiology, University of Nebraska Medical Center, Omaha, NE 68198, USA

W Powell
Scottish Crop Research Institute, Invergowrie, Dundee, DD2 5DA, UK

D Prangishvili
Department of Microbiology, Universität Regensburg, D-93053 Regensburg, Germany

T Prout
Department of Evolution and Ecology, University of California–Davis, Davis, CA 95616, USA

R E Pyeritz
Department of Human Genetics, University of Pennsylvania School of Medicine, Philadelphia, PA 19104, USA

P Rabbitts
Department of Oncology, Addenbrooke's Hospital, Cambridge, CB2 2QH, UK

T H Rabbitts
Division of Protein and Nucleic Acid Chemistry, MRC Laboratory of Molecular Biology, Cambridge, CB2 2QH, UK

T N K Raju
Department of Pediatrics, University of Illinois–Chicago, Chicago, IL 60612, USA

E A Raleigh
New England Biolabs, 32 Tozer Road, Beverly, MA 01915, USA

J T Reardon
Department of Biochemistry and Biophysics, University of North Carolina School of Medicine, Chapel Hill, NC 27599, USA

G D Recchia
Department of Biochemistry, University of Oxford, Oxford, OX1 3QU, UK

R J Redfield
Department of Zoology, University of British Columbia, Vancouver, British Columbia, V6T 1Z4, Canada

R H Reeves
Department of Physiology, Johns Hopkins University School of Medicine, Baltimore, MD 21205, USA

L J Reha-Krantz
Department of Biological Sciences, University of Alberta, Edmonton, Alberta, T6G 2E9, Canada

S Reichheld
Department of Biology, McMaster University, Hamilton, Ontario, L8S 4K1, Canada

W Reik
Programme in Developmental Genetics, Babraham Institute, Cambridge, CB2 4AT, UK

B J Reinhart
Department of Molecular Biology, Massachusetts General Hospital, Boston, MA 02114, USA

R I Richards
Centre for Medical Genetics, Department of Cytogenetics and Molecular Genetics, Women's and Children's Hospital, North Adelaide, SA 5006, Australia

J P Richardson
Department of Chemistry, Indiana University, Bloomington, IN 47405, USA

N Richardson
Institute of Cancer Research, Chester Beatty Laboratories, London SW3 6JB, UK

R W Ridge
Division of Natural Sciences, International Christian University, Osawa, Mitaka, Tokyo 181-8585, Japan

P Riggs
New England Biolabs, 32 Tozer Road, Beverly, MA 01915, USA

D L Rimoin
Medical Genetics Birth Defects Center, Steven Spielberg Pediatrics Research Center, Cedars–Sinai Medical Center and University of California–Los Angeles School of Medicine, Los Angeles, CA 90048, USA

E M Rinchik
Oak Ridge National Laboratory, Oak Ridge, TN 37831, USA

J L Rinkenberger
Department of Anatomy, University of California–San Francisco, San Francisco, CA 94143, USA

L S Ripley
Department of Microbiology and Molecular Genetics, University of Medicine and Dentistry of New Jersey, New Jersey Medical School, Newark, NJ 07103, USA

F T Robb
Center of Marine Biotechnology, University of Maryland Biotechnology Institute, Baltimore, MD 21202, USA

J-D Rochaix
Department de Biologie Moléculaire, Université de Genève, CH-1211 Genève, Switzerland

C H Rodeck
Department of Obstetrics and Gynecology, University College London, London, WC1E 6HX, UK

T H Roderick
The Jackson Laboratory, Bar Harbor, ME 04609, USA

B A Roe
Department of Chemistry and Biochemistry, University of Oklahoma, Norman, OK 73019, USA

F J Rohlf
Department of Ecology and Evolution, State University of New York, Stony Brook, NY 11794, USA

C Ronson
Department of Microbiology, University of Otago, Dunedin, New Zealand

W A Rosche
Department of Biological Science, University of Tulsa, Tulsa, OK 74104, USA

S M Rosenberg
Department of Molecular and Human Genetics, Biochemistry, and Molecular Virology and Microbiology, Baylor College of Medicine, Houston, TX 77030, USA

J R Roth
Department of Biology, University of Utah, Salt Lake City, UT 84112, USA

N Rougier
Department of Anatomy, University of California–San Francisco, San Francisco, CA 94143, USA

A E Rougvie
Department of Genetics, Cell Biology and Development, University of Minnesota, St Paul, MN 55108, USA

M P Rout
The Rockefeller University, New York, NY 10021, USA

J D Rowley
Section of Hematology and Oncology, University of Chicago Medical School, Chicago, IL 60637, USA

C A Royer
Centre de Biochimie Structurale, Université de Montpellier, F-34090 Montpellier, France

D C Rubinsztein
Department of Medical Genetics, Wellcome Trust Centre for Molecular Mechanisms in Disease, Cambridge Institute for Medical Research, University of Cambridge, Cambridge, CB2 2XY, UK

F Ruddle
Department of Molecular Genetics, Yale University, New Haven, CT 06520, USA

A K Rustgi
Division of Gastroenterology, University of Pennsylvania, Philadelphia, PA 19104, USA

I Ruvinsky
Department of Molecular Biology, Massachusetts General Hospital, Boston, MA 02114, USA

G Ruvkun
Department of Molecular Biology, Massachusetts General Hospital, Boston, MA 02114, USA

H Saedler
Max Planck Institut für Züchtungsforschung, D-50829 Köln, Germany

N Saitou
Laboratory of Evolutionary Genetics, National Institute of Genetics, Mishima, 411-8540 Shizuoka-ken, Japan

M Salas
Centro de Biología Molecular 'Severo Ochoa' (CSIC-UAM), Universidad Autónoma, Canto Blanco, E-28049 Madrid, Spain

L D Samson
Department of Cancer Cell Biology, Harvard School of Public Health, Boston, MA 02215, USA

A Sancar
Department of Biochemistry and Biophysics, University of North Carolina School of Medicine, Chapel Hill, NC 27599, USA

M R Sanderson
Randall Centre for Molecular Mechanisms of Cell Function, Guy's, King's and St Thomas's Hospitals School of Biomedical Sciences, London SE1 1UL, UK

T Sasaki
Rice Genome Research Program, National Institute of Agrobiological Resources, Tsukuba, Ibaraki 305-8602, Japan

B Sauer
Oklahoma Medical Research Foundation, Oklahoma City, OK 73104, USA

R Savarirayan
Victorian Clinical Genetics Service, Royal Children's Hospital, Melbourne, VIC 3052, Australia

T Schedl
Department of Genetics, Washington University, School of Medicine, St Louis, MO 63110, USA

K Schesser
Department of Cell and Molecular Biology, Section of Immunology, Lund University, Lund, S-22184, Sweden

I Schildkraut
New England Biolabs, 32 Tozer Road, Beverly, MA 01915, USA

P Schimmel
Department of Molecular Biology, Skaggs Institute for Chemical Biology, The Scripps Research Institute, La Jolla, CA 92037, USA

T Scholl
Myriad Genetic Laboratories, Salt Lake City, UT 84108, USA

B E Schoner
Lilly Research Laboratories, Lilly Corporate Center, Indianapolis, IN 46285, USA

K Schüler
Institut für Pflanzengenetik und Kulturpflanzenforschung Gatersleben, Genbank Aussenstelle, D-18190 Gross Lüsewitz, Germany

R M Schultz
Division of Biochemistry, Department of Cell Biology, Neurobiology and Anatomy, Stritch School of Medicine, Loyola University Chicago, Maywood, IL 60153, USA

J Scott
Hammersmith Hospital, Imperial College School of Medicine, London, W12 ONN, UK

A G Searle
MRC Mammalian Genetics Unit, Harwell, Didcot, OX11 0RD, UK

D Seemungal
The Wellcome Trust, 210 Euston Road, London NW1 2BE, UK

S Segal
Department of Pediatrics and Medicine, The Children's Hospital of Philadelphia, Philadelphia, PA 19104, USA

A M Segall
Department of Biology and Molecular Biology Institute, San Diego State University, San Diego, CA 92182, USA

M F Seldin
Department of Biological Chemistry and Medicine, University of California–Davis, Davis, CA 95616, USA

M J Seller
Division of Medical and Molecular Genetics, Guy's, King's and St Thoma's Hospitals School of Medicine, London, SE1 9RT, UK

V Sgaramella
Dipartimento di Biologia Cellulare, Università della Calabria, 87030 Arcavacata di Rende CS, Italy

H B Shaffer
Department of Ecology and Evolutionary Biology, University of California, Davis, Davis, CA 95616, USA

P Sham
Institute of Psychiatry, King's College London, London, SE5 8AF, UK

P M Sharp
Institute of Genetics, University of Nottingham, Nottingham, NG7 2UH, UK

D J Sherratt
Department of Biochemistry, University of Oxford, Oxford, OX1 3QU, UK

J Shulman
Department of Molecular and Medical Genetics, University of Toronto, Toronto, Ontario, M5G 1X5, Canada

D O Sillence
Department of Clinical Genetics, The New Children's Hospital, Paramatta, NSW 2124, Australia

L Silver
Department of Molecular Biology, Princeton University, Princeton, NJ 08544, USA

E H Simon
Department of Biological Sciences, Purdue University, West Lafayette, IN 47907, USA

R W Simons
Department of Microbiology, Immunology and Molecular Genetics, University of California–Los Angeles, Los Angeles, CA 90095, USA

A Sinclair
Department of Paediatrics, University of Melbourne and Murdoch Children's Research Institute, Royal Children's Hospital, Melbourne, VIC 3052, Australia

R S Singh
Department of Biology, McMaster University, Hamilton, Ontario, L8S 4K1, Canada

A J Sinskey
Department of Biology, Massachusetts Institute of Technology, Cambridge, MA 02138, USA

T R Skopek
Genetic and Cellular Toxicology Department, Merck Research Laboratories, West Point, PA 19486, USA

B J Smith
Department of Molecular Biology, Princeton University, Princeton, NJ 08544, USA

D W Smith
Department of Biology and Center for Molecular Genetics, University of California–San Diego, San Diego, CA 92093, USA

G R Smith
Fred Hutchinson Cancer Research Center, Seattle, WA 98104, USA

J W Snape
John Innes Centre, Colney, Norwich, NR4 7UH, UK

P H A Sneath
Department of Microbiology and Immunology, University of Leicester, Leicester, LE1 9HN, UK

M Snyder
Department of Molecular Cellular and Developmental Biology, Yale University, New Haven, CT 06520, USA

B Sollner-Webb
Department of Biological Chemistry, The Johns Hopkins University School of Medicine, Baltimore, MD 21205, USA

D Solter
Department of Developmental Biology, Max Planck Institute of Immunology, D-79108 Freiburg, Germany

R L Somerville
Department of Biochemistry, Purdue University, West Lafayette, IN 47907, USA

L C Sowers
Department of Pediatrics and Molecular Medicine, City of Hope National Medical Center, Duart, CA 91010, USA

B T Spear
Department of Microbiology and Immunology, University of Kentucky, Chandler Medical Center, Lexington, KY 40536, USA

T P Speed
Division of Genetics and Bioinformatics, The Walter and Eliza Hall Institute of Medical Research, Melbourne, VIC 3050, Australia

R A Spritz
Human Medical Genetics Program, University of Colorado Health Sciences Center, Denver, CO 80262, USA

N K Spurr
SmithKline Beecham Pharmaceuticals, Harlow, CM19 5AW, UK

G Stacey
Center for Legume Research, University of Tennessee, Knoxville, TN 37996, USA

D Stadler
Department of Genetics, University of Washington, Seattle, WA 98195, USA

C Staehelin
Laboratoire de Biologie Moléculaire des Plantes Supérieures, Université de Genève, CH-1292 Chambésy/Genève, Switzerland

F W Stahl
Institute of Molecular Biology, University of Oregon, Eugene, OR 97403, USA

N Standart
Department of Biochemistry, University of Cambridge, Cambridge, CB2 1QW, UK

G S Stent
Department of Molecular and Cell Biology, University of California–Berkeley, Berkeley, CA 94720, USA

C L Stewart
Laboratory of Cancer and Developmental Biology, National Cancer Institute at Frederick, Frederick, MD 21702, USA

A Stoltzfus
Center for Advanced Research in Biotechnology, 9600 Gudelsky Drive, Rockville, MD 20874, USA

J Stougaard
Laboratory of Gene Expression, Department of
Molecular and Structural Biology, University of Aarhus,
DK-8000 Aarhus C, Denmark

L D Strausbaugh
Department of Molecular and Cell Biology, University of
Connecticut, Storrs, CT 06269, USA

J C Strefford
The Orchid Cancer Appeal, St Bartholomew's Medical
College, London, EC1M 6BQ, UK

L Stubbs
Genome Center, Lawrence Livermore National
Laboratory, Livermore, CA 94550, USA

Y Sugimoto
Department of Pharmacology, University of Pennsylvania
Medical Center, Philadelphia, PA 19104, USA

J Sullivan
Department of Microbiology, University of Otago,
Dunedin, New Zealand

K F Sullivan
Department of Cell Biology, The Scripps Research
Institute, La Jolla, CA 92037, USA

W C Summers
Department of Therapeutic Radiobiology, Yale University,
New Haven, CT 06520, USA

A T Sumner
Saileyknowes Court, North Berwick, EH39 4RG, UK

M Susman
Laboratory of Genetics, University of
Wisconsin–Madison, Madison, WI 53706, USA

G R Sutherland
Department of Cytogenetics and Molecular Genetics,
Women's and Children's Hospital, Centre for Medical
Genetics, North Adelaide, SA 5006, Australia

E Szathmáry
Department of Plant Taxonomy and Ecology, Eötvös
University and Collegium Budapest (Institute for
Advanced Study), H-1014 Budapest, Hungary

M D Szczelkun
Department of Biochemistry, University of Bristol,
Bristol, B58 1TD, UK

T Szczepański
Department of Immunology, Erasmus University,
Rotterdam, University Hospital Rotterdam, NL-3000 DR
Rotterdam, The Netherlands

D Szymanski
Department of Agronomy, Purdue University, West
Lafayette, IN 49707, USA

F Tata
Department of Genetics, University of Leicester,
Leicester, LE1 7RH, UK

S Tavaré
Department of Mathematics, University of Southern
California, Los Angeles, CA 90089, USA

N Tavernarakis
Department of Molecular Biology and Biochemistry, The
State University of New Jersey, Rutgers, Piscataway, NJ
08854, USA

A M R Taylor
CRC Institute for Cancer Studies, University of
Birmingham, Birmingham, B15 2TT, UK

C Tease
Department of Biological Sciences, University of
Warwick, Coventry, CV4 7AL, UK

G Theissen
Department Molekulare Pflanzengenetik, Max Planck
Institut für Züchtungsforschung, D-50829 Köln, Germany

E Thomas
Cold Spring Harbor Laboratory, Cold Spring Harbor,
NY 11724, USA

G Thomson
Department of Integrative Biology, University of
California–Berkeley, Berkeley, CA 94720, USA

W E Timberlake
Ōcaté Technical Consulting, PO Box 127, Ocate,
NM 87734, USA

T D Tlsty
Department of Pathology and University of California–
San Francisco Comprehensive Cancer Center, University
of California–San Francisco, San Francisco, CA 94143,
USA

J L Tolmie
Duncan Guthrie Institute of Medical Genetics, Yorkhill,
Glasgow, G3 8SJ, UK

M Tracey
Department of Biological Sciences, Florida International
University, Miami, FL 33199, USA

A Travers
MRC Laboratory of Molecular Biology, Cambridge,
CB2 2QH, UK

L-C Tsui
Department of Genetics, The Hospital for Sick Children,
Toronto, Ontario, M5G 1X8, Canada

P J Turek
Department of Urology, University of California–San
Francisco, San Francisco, CA 94143, USA

C Turleau
Service de Cytogénétique, Hôpital Necker Enfants
Malades, F-75743 Paris, France

A B Ulrich
Eppley Cancer Center and Department of Pathology and
Microbiology, University of Omaha Medical Center,
Omaha, NE 68198, USA

F D Urnov
Sangamo Biosciences, Point Richmond Technical Center,
Richmond, CA 94804, USA

D W Ussery
Institute of Biotechnology, Technical University of
Denmark, DK-2800 Lyngby, Denmark

M K Uyenoyama
Department of Zoology, Duke University, Durham,
NC 27708, USA

W J M Van de Ven
Laboratory for Molecular Oncology,
Center for Human Genetics, University of Leuven and
Flanders Interuniversity Institute for Biotechnology,
B-3000 Leuven, Belgium

E van den Berg
Department of Medical Genetics, University of
Groningen, NL-9713 AW Groningen,
The Netherlands

V H J van der Velden
Department of Immunology, Erasmus University
Rotterdam and University Hospital Rotterdam, NL-3000
DR, Rotterdam, The Netherlands

J Van Dommelen
Centre of Microbial and Plant Genetics, Kasteelpark
Arenberg 20, B-7001 Heverlee, Belgium

J J M van Dongen
Department of Immunology, Erasmus University
Rotterdam, NL-3000 DR Rotterdam,
The Netherlands

R A Van Etten
Center for Blood Research, Harvard Medical School,
Boston, MA 02115, USA

M Van Montagu
Department of Molecular Genetics, Ghent University,
B-9000 Ghent, Belgium

J Vanderleyden
Centre of Microbial and Plant Genetics, Kasteelpark
Arenberg 20, B-3001 Heverlee, Belgium

A Varshavsky
Division of Biology, California Institute of Technology,
Pasadena, CA 91125, USA

A R Venkitaraman
CRC Department of Oncology and The Wellcome Trust
Centre for Molecular Mechanisms in Disease,
Cambridge Institute for Medical Research, Cambridge,
CB2 2XY, UK

I M Verma
Laboratory of Genetics, The Salk Institute for Biological
Studies, La Jolla, CA 92037, USA

M Vidal
Dana-Farber Cancer Institute, Department of Genetics,
Harvard Medical School, Boston, MA 02115

T F Vogt
Department of Pharmacology, Merck and Co., Inc.,
Merck Research Laboratories, West Point,
PA 19486, USA

M J Wade
Department of Biology, Indiana University, Bloomington,
IN 47405, USA

G P Wagner
Department of Ecology and Evolutionary Biology, Yale
University, New Haven, CT 06520, USA

B T Wakimoto
Departments of Genetics, University of Washington,
Seattle, WA 98195, USA

A R Walmsley
Division of Infection and Immunity, Institute of Biomedical
and Life Sciences, University of Glasgow, Glasgow,
G12 8QQ, UK

S T Warren
Department of Biochemistry, Genetics and Pediatrics,
University School of Medicine, Atlanta, GA 30322, USA

P M Wassarman
Department of Biochemistry and Molecular Biology,
Mount Sinai School of Medicine, New York, NY 10029,
USA

M D Waterfield
Ludwig Institute for Cancer Research, 91 Riding House
Street, London, W1P 8BT, UK

R K Wayne
Department of Organismic Biology, Ecology and
Evolution, University of California–Los Angeles,
Los Angeles, CA 90095, USA

D J Weatherall
Institute of Molecular Medicine, John Radcliffe Hospital,
University of Oxford, Oxford, OX3 9DU, UK

N F Weeden
Department of Plant Sciences and Plant Pathology,
Montana State University, Bozeman, MT 59717, USA

G M Weinstock
Department of Microbiology and Molecular Genetics, University of Texas–Houston Medical School, Houston, TX 77225, USA

A E Weis
Department of Ecology and Evolutionary Biology, University of California–Irvine, Irvine, CA 92697, USA

R D Wells
Institute of Biosciences and Technology, Texas A & M University System Health Science Center, Houston, TX 77030, USA

Z Werb
Department of Anatomy, University of California–San Francisco, San Francisco, CA 94143, USA

T Werner
Institute of Mammalian Genetics, GSF-National Research Center for Environment and Health, D-85764 Neuherberg, Germany

S A West
Institute of Cell, Animal and Population Biology, University of Edinburgh, Edinburgh, EH9 3JT, UK

J A White
MRC Human Biochemical Genetics Unit, The Galton Laboratory, University College London, London, NW1 2HE, UK

D J Wieczorek
Department of Microbiology, University of Iowa College of Medicine, Iowa City, IA 52242, USA

D E Wilcox
Duncan Guthrie Institute of Medical Genetics, University of Glasgow, Glasgow, G3 8SJ, UK

E O Wiley
Department of Ecology and Evolutionary Biology, University of Kansas, Lawrence, KS 66045, USA

A O M Wilkie
Institute of Molecular Medicine, University of Oxford, John Radcliffe Hospital, Oxford, OX3 9DS, UK

L B Willis
Department of Biology, Massachusetts Institute of Technology, Cambridge, MA 02138, USA

H Winking
Institut für Biologie, Medical University of Lübeck, D-2400 Lübeck, Germany

M E Winkler
Infectious Diseases Research, Lilly Research Laboratories, Lilly Corporate Center, Indianapolis, IN 46285, USA

†A P Wolffe
Formerly of Sangamo Biosciences, Point Richmond Technical Center, Richmond, CA 94804, USA

J Wolstenholme
School of Biochemistry and Genetics, University of Newcastle upon Tyne, Newcastle upon Tyne, NE2 4AA, UK

H Y Wong
Museum of Natural History, University of Oxford, Oxford, OX1 3PW, UK

S L C Woo
Institute for Gene Therapy and Molecular Medicine, Mount Sinai School of Medicine, New York, NY 10029, USA

A H Wyllie
Department of Pathology, University of Cambridge, Cambridge, CB2 1QP, UK

M-Q Xu
New England Biolabs, 32 Tozer Road, Beverly, MA 01915, USA

Y Yamamoto
Department of Genetics, Hyogo College of Medicine, 1-1 Mukogawa-cho, Nishinomiya, 663-8501 Hyogo, Japan

C Yanofsky
Department of Biological Sciences, Stanford University, Stanford, CA 94305, USA

K J Yook
Department of Biology, University of Utah, Salt Lake City, UT 84112, USA

Z-B Zeng
Department of Statistics, North Carolina State University, Raleigh, NC 27695, USA

D O Zharkov
Department of Pharmacological Sciences, State University of New York, Stony Brook, NY 11794, USA

A Zhelonkina
Department of Biological Chemistry, The Johns Hopkins University School of Medicine, Baltimore, MD 21205, USA

W Zillig
Max Planck Institute for Biochemistry, 82152 Martinsried,
Germany

F K Zimmermann
Goethestrasse 12, D-64372 Ober-Ramstadt,
Germany

G Zubay
Department of Biological Sciences, Columbia University,
New York, NY 10027, USA

J W Zyskind
Department of Biology, San Diego State University, San
Diego, CA 92182, USA

Contents

N

E.coli

See: *Escherichia coli*

Early Genes (in Phage Genomes)

E Kutter

doi: 10.1006/rwgn.2001.0392

Genes that are expressed immediately after phage infection are termed early genes. Their transcription generally requires only the machinery of the host, possibly augmented with proteins carried inside the phage particle. The products of these genes are primarily involved in restructuring the host cell to become an efficient factory for making new phages – blocking host nucleases, adapting the transcription or translation machinery, changing membrane properties, degrading host DNA or blocking its synthesis. Depending on the life cycle of the particular phage, preparation of the cellular machinery to make the phage DNA may involve either early genes or middle-mode genes, which do require new phage-encoded proteins for their expression.

See also: Bacteriophages

EBNA

G Klein

doi: 10.1006/rwgn.2001.1563

The 'Epstein–Barr virus determined nuclear antigen' or EBNA was discovered by Reedman and Klein (1973) by anticomplement immunofluorescence, following the staining of acetone-methanol fixed smears of EBV-carrying lymphoblastoid cell lines with the sera of EBV-positive human donors. All EBV-DNA-carrying but not EBV-negative cells show brilliant fine granular nuclear fluorescence. Unlike the T-antigens of the papova- and adenoviruses, EBNA remains associated with the chromosomes during mitosis. Later studies revealed that EBNA is a family of six proteins. For their nomenclature and function, see Epstein–Barr Virus (EBV).

Reference

Reedman BM and Klein G (1973) Cellular localization of an Epstein–Barr virus (EBV)-associated complement-fixing antigen in producer and non-producer lymphoblastoid cell lines. *International Journal of Cancer* 11: 499–520.

See also: Epstein–Barr Virus (EBV); Tumor Antigens Encoded by Simian Virus 40

Ectoderm

See: **Developmental Genetics**

Ectodermal Dysplasias

F M Pope

doi: 10.1006/rwgn.2001.0394

The ectodermal dysplasias form two main subgroups: those in which sweating is deficient and the absence of sweat glands coincides with varying degrees of hypodontia, and hair and nail deformities, and those in which sweating and teeth are normal, but in which brittle hair, nails, and palmoplantar hyperkeratosis occur. Whilst the former is X-linked recessive or autosomal recessive, the latter is usually autosomal dominant. Numerous syndromic variations occur and recently there has been substantial progress in the pathogenesis, with several candidate genes having been identified.

Pathogenesis

Unlike many other genodermatoses, such as the ichthyoses, Ehlers–Danlos syndrome (EDS), pseudoxanthoma elasticum, epidermolysis bullosa, and cutis laxa in which particular structural components are defective, the ectodermal dysplasias show defective organogenesis. In contrast to these simpler disorders, in which particular anatomical structures, such as blood vessels, bones, or ligaments have intrinsic weaknesses, in ectodermal dysplasias, the orchestration of organogenesis is disturbed. Thus instead of errors in simple structural proteins such as collagen or keratin in EDS or the ichthyoses, respectively, mutations that cause ectodermal dysplasia are generally caused by faults in orchestration proteins.

Anhidrotic Ectodermal Dysplasia

Given its X-linked inheritance, males are affected, while the affected females show partial forms. Hypotrichosis and abnormal teeth and sweat glands are consistent features, while affected females show reduced or distorted teeth and minor sweat gland or breast deficiency (Clarke, 1987). Not surprisingly, given lyonization heterozygotes have sweating deficiencies, which coincide with Blashko's lines. The chromosomal location at Xq12 was first deduced from a translocation and eventually this was shown to be syntenic with the mouse tabby locus. Eventually a gene responsible for epithelial-mesechymal signaling was cloned. This was homologous in mouse and human and, as well as a transmembrane domain, contained a 19 GlyXY collagenous repeat (Monreal et al., 1998). Although its function is unknown, it seems very possible that interaction with other matrix proteins is probably functionally important for epithelial-mesenchymal interactions. Short and long isoforms, with or without the collagenous domain, have been identified. Isoform II is functionally important in tooth, hair, and sweat gland morphogenesis. There is also some evidence that the protein ectodysplasin is a member of the TNF ligand family, which orchestrates epithelial–mesenchymal interactions, which in turn regulate epidermal appendage formation. Other similar transmembrane proteins with collagenous domains include collagen XVII. Another transmembrane protein is plakophilin, and mutations of this desmosomal protein cause another form of ectodermal dysplasia (McGrath et al., 1999).

Hidrotic Ectodermal Dysplasia (Clouston Syndrome)

In contrast to the anhidrotic variant, sweat glands, sebacious glands, and teeth are completely normal. Instead there is severe diffuse alopecia, with dystrophic nails and patchy hyperpigmentation. Variations include palmar–plantar hyperkeratosis, eyebrow hypoplasia, and mental retardation. Linkage analysis has mapped Clouston-like families to 13q11-12 with possible mutations of the connexin 30 family (Lamartine, et al., 2000).

Autosomal Anhidrotic Ectodermal Dysplasia (ED3)

Autosomal dominant

Several large American families have been described in which mild hair thinning, mild dental hypodontia, and variable hypohidrosis segregate with autosomal dominant inheritance. The skin is smooth and dry, eyebrows are atrophic, eyelashes and scalp hair are deficient, and sweating is confined to the axilae, palms and soles. Mutations have been identified in the human homolog of the mouse gene 'downless' (Majumdar et al., 1998).

Autosomal recessive

This is phenotypically indistinguishable from the X-linked form, except crucially that the inheritance is autosomal recessive, rather than X-linked. Mutations in the EDA1 gene are also missing. Instead an homologous gene in the same relation as a second mouse homolog to the tabby gene has been identified, of which in mouse the autosomal homolog (to tabby) is crinkled and 'downless' (Majumdar et al., 1998).

Hypohidrotic Ectodermal Dysplasia with Immunodeficiency

This variant which affects hair, teeth, and sweat glands is caused by mutations at the C-terminus of the KK-gamma gene. This is allelic to incontentia pigmentii (Zonana, 2000).

Other Variants

Other are variants include hypohidrotic ectodermal dysplasia with hypothyroidism and corpus callosum agenesis, and anhidrotic ectodermal dysplasia with cleft lip and palate. The latter has recently had mutations of the cell/cell adhesion protein PVL1 identified (Suzuki et al., 2000).

References

Clarke A (1987) Hypohidrotic ectodermal dysplasia. *Journal of Medical Genetics* 24: 659–663.

Lamartine et al. (2000) A 1.5 Mb map of the hidrotic ectodermal dysplasia (Clouston syndrome) gene region on human chromosome 13q11. *Genomics* 67: 232–236.

Majumdar K, Shawlot W and Schuster G et al. (1998) Yac rescue of downless locus mutations in mice. *Mammalian Genome* 9: 863–868.

McGrath JA, Hoeger PH and Christiano AM et al. (1999) Skin fragility and hypohydrotic ectodermal dysplasia resulting from ablation of plakophilin 1. *British Journal of Dermatology* 140: 297–307.

Monreal AW, Zonana J and Ferguson B (1998) Identification of a new splice form of the EDA1 gene permits detection of nearly all X-linked ectodermal dysplasia mutations. *American Journal of Human Genetics* 63: 1253–1255.

Suzuki K, Hu D and Bustos T et al. (2000) Mutations of PVRL1 encoding a cell–cell adhesion molecule/herpes virus receptor in cleft lip-palate ectodermal dysplasia. *Nature Genetics* 25: 427–430.

Zonana et al. (2000) *American Journal of Human Genetics* 67: 1555–1662.

See also: **Metabolic Disorders, Mutants**

Editing and Proofreading in Translation

J Parker

doi: 10.1006/rwgn.2001.0395

Translation, like the other steps in the flow of information from DNA to protein, is very accurate; the overall frequency of all types of translational errors is clearly less than one mistake per 1000 codons read. Part of this accuracy is achieved through various energy-dependent editing or proofreading functions which operate at the different steps involved in translation. These functions prevent the formation of a defective protein by preventing the formation of an error-containing intermediate or rejecting such an intermediate at some step in the pathway.

The terms 'editing' and 'proofreading' are often used as synonyms to describe mechanisms for preventing errors in translation. Such mechanisms can operate during the formation of aminoacyl-tRNA or during the selection of an aminoacyl-tRNA by the ribosome. In this entry, the term editing refers to such events during aminoacylation and proofreading refers to those that occur on the ribosome. Although these mechanisms work to increase accuracy, it is important to note that translation in normal cells is not maximized for accuracy, since mutations exist that increase accuracy above that seen in the wild-type. However, such mutations lead to a decrease in the growth rate, indicating that translational accuracy has been optimized for accuracy and growth rate.

Editing in Aminoacylation

Aminoacylation, the attachment of an amino acid to a tRNA, is typically a two-step process catalyzed by the aminoacyl-tRNA synthetases. The first step, termed 'activation,' is the formation of an aminoacyl-AMP (aminoacyl-adenylate) on the enzyme through the hydrolysis of ATP. The second step is the transfer of the activated amino acid residue from the adenylate to a tRNA in a reaction referred to as 'charging.' Editing can occur in either of these two reactions.

Because many of the amino acids are similar in structure, misactivation or mischarging of an amino acid by a given synthetase often involves a subset of amino acids structurally related to the cognate amino acid. It is believed that in some cases the binding energy of closely related amino acids leads to only approximately a 100-fold preference for the correct amino acid and that editing functions can increase this selectivity 1000-fold.

The molecular pathways involved in editing can differ for different aminoacyl-tRNA synthetases. For instance, in the case of valyl-tRNA synthetase from *Escherichia coli*, misactivated threonine is first charged to a tRNA^{Val} (releasing AMP) and then hydrolyzed and released from the tRNA. A more widely used pathway seems to be the direct hydrolysis of the misactivated amino acid, which also releases AMP but does not involve mischarging. This reaction has been shown to be important *in vivo* in the prevention of misincorporation of homocysteine by methionyl-tRNA synthetase. It appears that this reaction occurs approximately once for every 100 methionine residues incorporated. Note that such editing schemes result in hydrolysis of ATP and, therefore, occur at a metabolic cost to the organism.

The aminoacyl-tRNA synthetases must also recognize the correct tRNA. The accuracy of tRNA selection is several orders of magnitude greater than the selection of amino acids. Indeed, these enzymes are involved in the process whereby mature, nondefective tRNAs are identified and exported from the eukaryotic nucleus. This process is also referred to as 'proofreading' but is unrelated to the editing processes described above.

Editing at Ribosome

During elongation, aminoacyl-tRNAs are brought to a site on the ribosome containing the next codon to be translated (the A-site) as a ternary complex containing the aminoacyl-tRNA, guanosine triphosphate (GTP), and an elongation factor. As in the case of aminoacylation, the initial selection of the aminoacyl-tRNA on the ribosome would give approximately a 100-fold preference for the correct versus a nearly correct codon–anticodon interaction. This is, of course, far lower than the observed accuracy of protein synthesis. Accuracy is apparently enhanced by proofreading, a rechecking of the codon–anticodon action. As in the case with editing by aminoacyl-tRNA synthetases, proofreading is energy driven, in this case by the hydrolysis of GTP. Every time EF-Tu delivers an aminoacyl-tRNA to the A-site, there will be GTP hydrolysis. Proofreading would involve the rejection of an aminoacyl-tRNA from the A-site with concomitant GTP hydrolysis. Therefore, proofreading involves the hydrolysis of GTP in excess of what would be required per peptide bond formed.

The antibiotic streptomycin, which increases many types of translational errors, seems to affect both initial selection of aminoacyl-tRNA and proofreading on the ribosome. Streptomycin-resistant mutants of *E. coli* have an altered ribosomal protein S12. Many such mutants decrease the level of translational errors in the cells, i.e., they lead to hyperaccurate ribosomes. Evidence seems to indicate that this decrease is related to the initial selectivity of aminoacyl-tRNA by such mutants, not with an increase in proofreading.

It has been postulated that there exists another type of editing or proofreading on ribosomes, involving loss of the peptidyl-tRNA from the P-site after the misincorporation of an amino acid residue in a growing peptide chain. Loss of a peptidyl-tRNA is often referred to as drop-off. If a proofreading mechanism exists that involves drop-off, it would be very expensive for the cell if it was operative at any time after the first few peptide bonds had formed, since all the energy required for each of the previous elongation cycles would be lost.

See also: **Aminoacyl-tRNA Synthetases; Elongation; Elongation Factors; Mistranslation; RNA Editing in Animals; RNA Editing in Plants; Translation**

Effective Population Number

J F Crow

Copyright © 2001 Academic Press
doi: 10.1006/rwgn.2001.1421

The concept of effective population number was introduced by Sewall Wright in his pioneering paper 'Evolution in Mendelian populations' (Wright, 1931). Wright was interested in the effect of random changes in gene frequencies (random genetic drift) that occur in finite populations, especially small ones. In the absence of systematic changes (e.g., selection, mutation, migration) the process of mating and reproduction can be thought of as drawing a sample of genes from an infinite pool to which each parent had contributed equally. In a diploid population of size N, the sample contains $2N$ genes. The variance of this binomial

sampling process is $p(1-p)/2N$, where p is the frequency of the allele of interest in the parent generation. This expression formed a part of the general mathematical expression for allele frequency change formulated by Wright. Real populations do not conform to this idealization. The number of males and females may differ, the population size may fluctuate over time, and parents are not equally viable and fertile. To accommodate these problems, rather than write more elaborate equations, Wright introduced the concept of effective population number, N_e. N_e is a number calculated from an actual population of size N, that when substituted into the Wright equations leads to the same amount of random genetic drift as is occurring in the actual population. The population size is most appropriately assessed at the beginning of the reproductive period.

When the number of females, N_f, differs from the number of males, N_m, the effective population number is given by $1/N_e = 1/4N_f + 1/4N_m$. Notice that when N_f and N_m are each $N/2$, $N_e = N$ as expected. If the numbers of the two sexes differ, then the sex with the smaller number dominates the value of N_e. To consider an extreme example, the effective size of a polygamous population with one male and 100 females is about 4, much closer to the number of the rarer sex. When the population varies from time to time, the effective population number is the harmonic mean of the values at different times. Thus, if N_i is the population number at generation i, the effective population size averaged over t generations is given by $1/N_e = (1/t)\Sigma_i(1/N_t)$. Again, since N_t appears in the denominator, the smaller values dominate. Therefore, size bottlenecks are important factors in assessing the importance of random genetic drift in the history of a population. A situation that is often more important and more difficult to calculate occurs when the viability and fertility differ among the members of the parent generation. In the simplest case where the sexes are equally frequent, mating is at random (including a random amount of self-fertilization), and the population is neither increasing nor decreasing. In this case the effective population number is $N_e = (4N-2)/(2+\sigma^2)$, where N is the population number and σ is the standard deviation of the number of progeny per parent, counted as adults. Formulae adjusting for counting at the wrong stage are available (Crow and Morton, 1955).

This aspect of effective population number has had a great deal of research in recent years and increasingly complicated formulae have been developed to take into account such factors as separate sexes, unequal numbers of the sexes, increasing or decreasing population number, inbreeding, and population structure. The small population effect is manifested both in a decrease of heterozygosity and fluctuations in allele frequency. In a stationary population, the effective population numbers are the same for both effects, but in a growing or diminishing population they are different. For a review, see Caballero, 1994.

Usually the effective size of a population is less than the actual size. It clearly is if the population is censused at an early stage at which the death rate is high (as in most fish), but it is also true if the population is censused at the adult stage. Measured values range widely, but typically for most animals the effective number is between 1/4 and 3/4 of the census number.

Although this is the case for a population without structure, it is not necessarily true if the population is subdivided. When the total population is divided into subgroups between which migration is limited, the effective population number can be, and often is, greater than the census number of the whole population. This is found, for example, in prairie dog colonies.

Random genetic drift plays an important role in Sewall Wright's 'shifting balance' theory of evolution. It is also a key quantity in the neutral theory of molecular evolution of Motoo Kimura.

Further Reading

Provine WB (1986) *Sewall Wright and Evolutionary Biology.* Chicago, IL: University of Chicago Press.

Wright S (1968, 1969, 1977, 1978) *Evolution and the Genetics of Populations,* 4 vols. Chicago, IL: University of Chicago Press.

References

Caballero A (1994) Developments in the prediction of effective population size. *Heredity* 73: 657–679.

Crow JF and Morton NE (1955) Developments in the prediction of effective population size. *Evolution* 9: 202–214.

Wright S (1931) Evolution in Mendelian populations. *Genetics* 16: 97–159.

See also: Fitness Landscape; Kimura, Motoo; Shifting Balance Theory of Evolution; Wright, Sewall

Ehlers–Danlos Syndrome

F M Pope

Copyright © 2001 Academic Press
doi: 10.1006/rwgn.2001.0397

Ehlers–Danlos syndrome (EDS) is essentially a disorder of collagen connective tissue, in which skin is over-fragile and hyperelastic. First described by Tschernogobov and then separately by both Ehlers and Danlos (Beighton, 1993) at the end of the nineteenth century there are at least seven separate types, most of which are nonallelic. Recently a consensus committee have proposed a modified classification, which merges the old types I and II, separates type VII into two distinct subsets, and confines types V and VIII to minority status (Beighton *et al.,* 1998).

There are some very strong phenotypical correlations in EDS. In EDS types I/II, (the old gravis or mitis forms) missassembled collagen fibers form diagnostic cauliflowers. Clinically skin is extremely extensile and tears easily (**Figure 1**). Epicanthic folds, a mesomorphic build, with broadened hands and feet are common. Defects occur in either the COL5A1 and 5A2 genes, causing faulty type V collagen protein, although linkage excludes both of these genes in some families. Known mutations include exon skips and glycine substitutions.

EDS type IV is specifically caused by collagen III mutations (Pope *et al.,* 1996). Typically there is acral thinning and premature aging (face, hands, and feet), thereby overlapping with metageria (**Figure 2a,b**). Thin skin with prominent capillaries is especially widespread over the shoulders and upper chest and is generalized in the severest cases. Early talipes is also common. Light microscopy of skin shows dermal collagen depletion and elastic proliferation, whilst

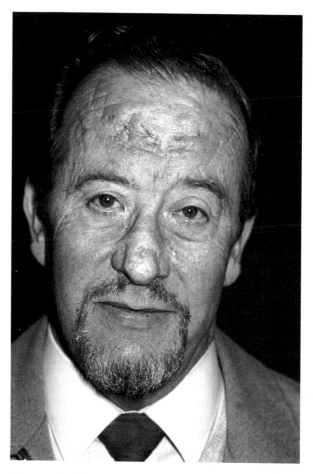

Figure 1 (See Plate 14) *Typical late facial scarring of EDS I/II.*

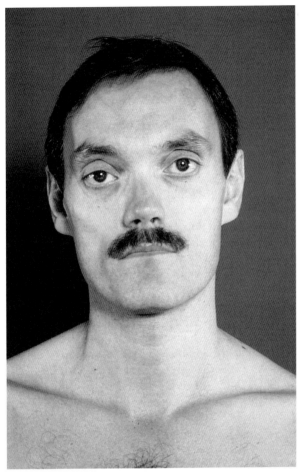

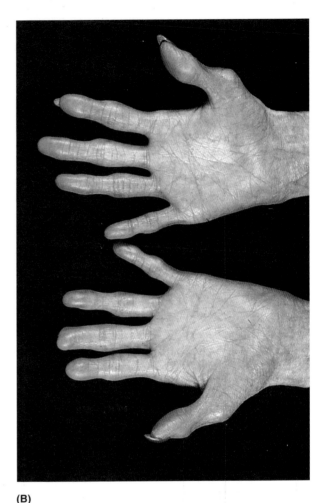

(A)   (B)

Figure 2 (See Plate 13) (A) *Face and (B) hands of acrogeric EDS IV patients. The large eyes and lobeless ears are typical, whilst the hands show pulp atrophy from terminal phalyngeal erosions.*

electron microscopy of skin at 60 000 magnification shows patchy irregularity of fibril size. Collagen III protein analysis nearly always shows depletion of procollagen, and collagen in the medium and over-modified mutant protein intracellularly. Most mutations are private and unclustered, although exon 24 is a hot spot for skipping errors, whilst first position glycine substitutions are also very common (Pope *et al.*, 1996). Helical 3′ mutations have the most abnormal external appearance and also the highest frequency of arterial rupture. The latter is commonest in the fourth decade of life, but patients are also at risk during adolescence and pregnancy. Blood pressure and aortic ultrasound monitoring are of unproven value.

EDS type VI is typified by extreme general laxity and hypotonia, and motor delay owing to ligamentous laxity is very common. The typical hypotonic child, has good power and normal muscle biopsy. Early scoliosis indicates surgical correction in adolescence. Blue sclerae and osteopenia overlap with osteogenesis imperfect (OI), giving severe joint laxity with EDS

VII (see below). EDS can be distinguished by abnormal type I collagen chemistry (type VII and OI) or electron microscopy of skin (normal in EDS VI). Type VIb lysyl hydroxylase deficiency can be monitored by gel electrophoresis of collagen type I proteins which migrate faster in affected patients, or by measuring urinary cross-links. The gene has 19 exons, and most mutations are double heterozygotes.

EDS type VII is caused by the persistent *N*-propeptides, either from faulty *N*-proteinase or individual structural mutations of exon 6 of COL1A1 or COL1A2 genes. Clinically congenital hip dislocation combines with generalized joint laxity. Fragile skin and the external general phenotype overlaps with EDS 1 and II, but is distinguished by abnormal collagen chemistry (types a and b), in which the retained uncleaved propeptide is retained either as an extra α-1 or α-2 chain. In type VIIc, both α-1 and 2 extensions remain (Pope and Burrows, 1997). Types a, b and c retain 1, 2 or 3 chains, with a gradation of clinical severity, greatest in type VIIc in which there is spectacular cutis laxa. Types a–c are also distinguishable by electron microscopy of the skin, which also reflects fibril mispacking caused by retained propeptides, ranging from angulated to swept-wing to hieroglyphic fibrils in type VIIc (Pope and Burrows, 1997). The EDS III/benign hypermobile subtype is the most common occurring in up to 10% of caucasian populations, and higher still in other races. It overlaps with all EDS subtypes, and also many other inherited connective tissue disorders, such as PXE, Marfan syndrome, Sticker syndrome, and certain chondrodysplasias etc.

Although EDS VIII was relegated to minority status (Beighton *et al.*, 1998), it also overlaps with EDS I/II and IV, in all of which periodontal recession and pretibial scarring with hemosiderosis occur. Essential criteria include autosomal transmission of premature gum recession and bone loss, with granulomatous hemosiderin-rich pretibial plaques. Normal type III collagen levels exclude EDS type IV, and an absence of cauliflowers eliminates type I/II.

Other minority EDS subtypes include type V. Only two families have been described with an external phenotype resembling EDS III/BHS, but with X-linked inheritance. Type IX EDS with fibronectin deficiency has been described only in one family. The new classification (Beighton *et al.*, 1998) combines types I and II, retains types III and IV, dividing types VI and VII into two separate types which split the enzymic from the two structural mutants. Its superiority to the old numerical classification is doubtful.

With the exception of EDS IV in which uterine rupture or lethal arterial fragility during delivery can be life-threatening, pregnancy is usually safe. In EDS IV, early hospital admission and bedrest with an elective cesarian section is advisable, preferably in a major medical centre with vascular surgical cover. Except for EDS I/II in which premature rupture of the membranes is common, all other EDS subtypes have perineal fragility, which can usually be minimized by controlled delivery, to avoid severe third-degree tears and later pelvic prolapse.

References

Beighton P (ed.) (1993) *McKusick's Inherited Defects of Connective Tissue*, 5th edn pp. 189–192. St Louis, MO: Mosby.

Beighton P, De Paepe A, Steinnmann B *et al.* (1998) Ehlers Danlos syndrome: revised nosology, Villefranche 1997. *American Journal of Medical Genetics* 77: 31–37.

Pope FM and Burrows NP (1997) Ehlers Danlos syndrome has varied molecular mechanisms (Syndrome of the Month). *Journal of Medical Genetics* 34: 400–410.

Pope FM, Narcisi P, Nicholls AC, Germain D, Pals G and Richards AJ (1996) COL3A1 mutations cause variable clinical phenotypes including acrogeria and vascular rupture. *British Journal of Dermatology* 135: 163–181.

See also: **Pseudoxanthoma Elasticum (PXE)**

Electron Microscopy

M E Dresser

Copyright © 2001 Academic Press
doi: 10.1006/rwgn.2001.0398

Electron microscopy is the image formation based on the interaction of high-energy electrons with a specimen, typically used for high magnifications to reveal details that cannot be resolved by light microscopy.

The first electron microscope was built based on the principles of light microscope optics by Ruska and colleagues in 1931, following the recognition that a magnetic lens can focus a beam of electrons much as a glass lens can focus a beam of light. Electrons have much shorter wavelengths than visible light and, because resolution in microscopy is inversely related to the wavelength of the illuminating beam, electron microscopy (EM) is in practice capable of resolving details of biological specimens that are ~100× smaller than those resolved by light microscopy – roughly 2 versus 200 nm. Resolution of even finer detail is possible in principle but is limited primarily by lens aberrations and by specimen damage from the electron beam.

The major contributions of EM to genetics have followed from signal developments in specimen preparation to overcome inherent problems. First, specimens must be durable against beam damage and able to withstand high vacuum. This is because electrons interact so strongly with matter that electron microscopes must operate under very high vacuum. Second, specimens generally require some treatment to increase their contrast. Biological molecules scatter electrons poorly, being composed of elements with low atomic numbers. Thus, heavy metal atoms generally are introduced as stains, as thin coatings, or as discrete spheres (see below) to provide the necessary contrast. Third, specimens must remain extremely thin, since electrons must pass through the specimen to create the image for the most common high-resolution form of electron microscopy, transmission electron microscopy (TEM). An important extension of this approach is to use tomographic techniques to extract two- and three-dimensional information from relatively thick sections examined in high voltage microscopes.

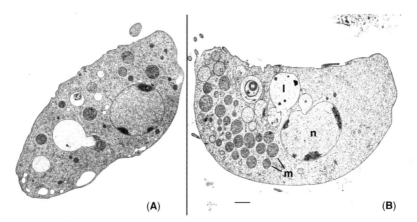

Figure 1 (above) Thin sections of ultrarapidly frozen, freeze-substituted *Dictyostelium* amoebae. (A) Wild-type cell. (B) CluA-cell, a mutant where mitochondria associate in a single large cluster. The scale bar represents 1 μm; nucleus (n), lysosome (l), and mitochondria (m). (Images were kindly provided by S. Fields and M. Clarke, Oklahoma Medical Research Foundation.)

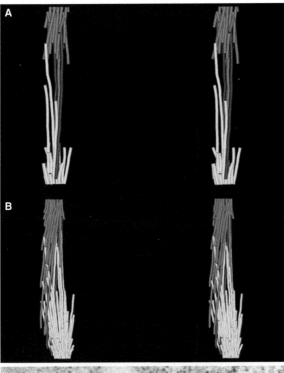
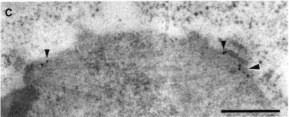
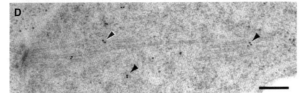

Determination of cytological phenotypes at high resolution generally depends on some variation of the following common sequence of methods:

1. Single cells, tissues, or whole organisms are 'fixed' to stabilize internal structures by cross-linking chemically using, for example, glutaraldehyde.
2. The samples are embedded by infiltration with hardening resins then sliced into 30–100 nm 'thin sections' using diamond knives mounted in instruments termed microtomes.
3. Contrast of the specimens is enhanced by staining with solutions of uranium, lead, or osmium salts (this step can occur earlier in the process).

Figure 2 (See Plate 16) (left) Mitotic spindles in the yeast *Saccharomyces cerevisiae*, analyzed using thin sections. Stereo three-dimensional reconstructions of mitotic spindles from wild-type (A) and a *cdc20* mutant (B). Light gray and dark gray lines represent microtubules; red lines represent microtubules that are continuous between the two poles. The *cdc20* cell division cycle mutant was grown at the nonpermissive temperature (36 °C) for 4 h, where these cells arrest in mitosis with an average spindle length of ~2.5 μm and contain many more microtubules than wild-type spindles of comparable lengths (see Winey *et al.*, 1995 and O'Toole *et al.*, 1997). Immuno-electron microscopy localization of Kar3-GFP (C) and Slk19-GFP (D) fusion proteins (arrowheads). Spindle microtubules appear as straight structures emanating into the nucleus from dense spindle pole bodies which are embedded in the nuclear envelope. Kar3-GFP is a motor enzyme of the kinesin family that localizes close to the spindle poles, whereas Slk19-GFP localizes to kinetochores and the spindle midzone (see Zeng *et al.*, 1999). The scale bars represent 250 nm.

4. The sections are mounted for viewing in the microscope on thin, stable support films, typically composed of carbon or plastic, laid over fine-mesh metal grids.
5. Images are captured photographically or electronically.

In some instances, cells are frozen then exposed to chemical fixatives at low temperatures in order to stabilize morphology; ultrarapid freezing prevents formation of ice crystals which damage ultrastructure (**Figure 1**). In single sections, specific molecules are localized by tagging them with enzymes that are used to develop an electron-dense precipitate or with discrete markers, such as colloidal gold spheres with diameters in the nanometer range, usually employing an antibody intermediate (hence the term 'immunolocalization'; **Figure 2C,D**). Three-dimensional information can be generated from thin sections by collecting and imaging serial sections then arranging the image series in a 'stack' to reconstruct the original sample architecture (**Figure 2A,B**).

Intracellular topology is visualized using freeze-fracture and deep-etch methods. Following freezing, cells are broken open and held under vacuum to allow sublimation of the ice to proceed until intracellular structures are exposed. The surface is then coated with heavy metal to produce a film imprinted with the surface topology. This film is examined using TEM after being separated from the underlying substrate.

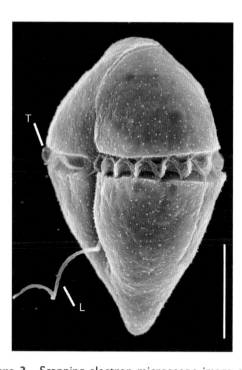

Figure 3 Scanning electron microscope image of the freshwater dinoflagellate *Gymnodinium acidotum*, showing the typical morphology and positioning of the two flagellae. The sinusoidal transverse flagellum (T) sits in an equatorial groove that encircles the cell; the longitudinal flagellum (L) projects from a longitudinal groove. The scale bar represents 10 μm. (Image was kindly provided by S. Fields, University of Oklahoma, S. R. Noble Electron Microscopy Laboratory.)

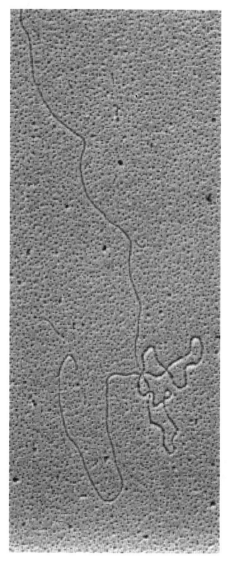

Figure 4 A single molecule of duplex DNA partially unwound by *Escherichia coli* RecBCD enzyme, visualized on a thin nitrocellulose support by rotary shadowing with platinum. RecBCD unwinds DNA and produces two single-stranded DNA loops which are relatively thick due to binding by single-strand DNA-binding protein present in the unwinding reaction. (Image was kindly provided by A. F. Taylor and G. R. Smith, Fred Hutchinson Cancer Research Center, Seattle, USA.)

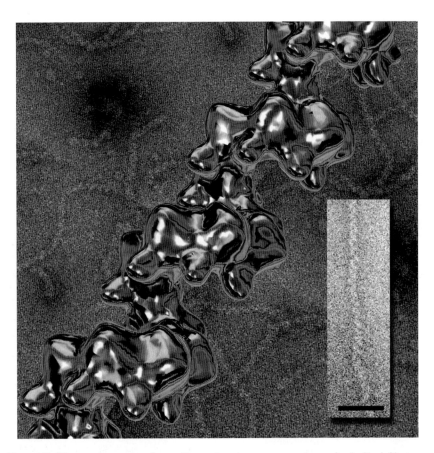

Figure 5 (See Plate 15) The surface of a three-dimensional reconstruction of a helical filament of human Rad51 protein on DNA is shown in gold in the foreground. In the background is an electron micrograph of the actual filaments that Rad51 protein forms on single-stranded DNA in the presence of ATP. (The Rad51 protein is from the laboratory of Dr Steve West, ICRF, UK.) The inset (right), a portion of such a Rad51-DNA filament (scale bar represents 400 Å) shows the very poor signal-to-noise ratio present in such images. To surmount this problem, the reconstruction has been generated using an algorithm for processing such images (Egelman (2000) *Ultramicroscopy* 85: 225–234) and involved averaging images of 7620 segments. The reconstruction shows that the filaments contain ~6.4 subunits per turn of a 99 Å pitch helix.

Surface topology of even larger structures, including whole organisms, is the province of scanning electron microscopy (SEM), which uses the back-scattered electrons from a scanning beam to produce an image. Inherently lower in resolution than TEM, SEM nevertheless is valuable for providing great depth of focus and uniformly clear images of large specimens (**Figure 3**).

Single molecules and molecular assemblies are visualized on thin, uniform support films by 'shadowing' where the biomolecules cause perturbations in an otherwise uniform heavy metal coating (**Figure 4**) or by 'negative staining' where the biomolecules are evident as the less dense areas in a puddle of dried heavy metal salt (**Figure 5**).

Future advances in the utility of electron microscopy are likely to derive from refinements in specimen preparation, preservation, and contrast enhancement as well as from developments in digital image acquisition, processing, analysis, and display.

References

O'Toole et al. (1997) *Molecular Biology of the Cell* 8: 1–11.
Winey et al. (1995) *Journal of Cell Biology* 129: 1601–1615.
Zeng et al. (1999) *Journal of Cell Biology* 146: 415–425.

See also: X-Ray Crystallography

Electrophoresis

See: Gel Electrophoresis, Pulsed Field Gel Electrophoresis (PFGE)

Electroporation

I Schildkraut

Copyright © 2001 Academic Press
doi: 10.1006/rwgn.2001.0400

Electroporation is the introduction of DNA molecules into cells by use of an electric current to temporarily make the cells permeable.

Elongation

A Liljas

Copyright © 2001 Academic Press
doi: 10.1006/rwgn.2001.0401

There are two processes in gene expression in which elongation is of primary interest. One is in transcription, the synthesis of RNA from a DNA template, and translation; the other is synthesis of a polypeptide from a messenger RNA on the ribosome. Both processes go through the phases of initiation, elongation, and termination.

Elongation in Transcription

Genomic DNA cannot be translated but has to be copied or transcribed into RNA by different RNA polymerases. Here the classic mechanism discovered by Watson and Crick applies. One strand of the double-stranded DNA (the negative one) is copied with Watson–Crick base-pairing into a positive strand of RNA. This occurs in the 5′ to 3′ direction. The double-stranded DNA is opened up in a 'bubble' that travels along the duplex during transcription. Here, a DNA–RNA hybrid is formed transiently. The process of transcription is in all cases strongly regulated. Some genes are transcribed frequently, whereas others are transcribed only rarely. Again some genes are transcribed in some brief period in the life of the cell, whereas others are copied more or less continuously.

Elongation in Translation on Ribosomes

The process of translation occurs on the ribosome. The ribosome is a complex of a few large rRNA molecules and between 50 and 90 different proteins. The ribosome is made up of two subunits (large and small) with different functions that dissociate from each other at the end of the process. Translation is traditionally divided into three steps: initiation, elongation, and termination. Soluble protein factors catalyze the process by binding to the ribosome transiently.

In each cycle of elongation, one amino acid is incorporated into the nascent peptide. There are three elongation factors in eubacteria, which catalyze two of the basic steps in translation: the binding of an aminoacyl-tRNA to the A-site, and the translocation of the peptidyl-tRNA from the A-site to the P-site. During this step, the mRNA is moved to expose the next codon in the ribosomal A-site. However, during the central event in elongation, peptidyl transfer, no protein factor is needed.

The recognition of the codon by the anticodon of the tRNA is a process that is done in several steps. In the initial selection, the anticodon of the aminoacyl-tRNA in complex with elongation factor Tu (EF-Tu) and GTP is matched against the codon in the A-site of the ribosome. When there is a good match, the ribosome induces EF-Tu to hydrolyze its bound GTP to GDP and phosphate. The EF-Tu/GDP complex has a conformation that has low affinity for the aminoacyl-tRNA and the ribosome; accordingly it dissociates. The aminoacyl moiety of the tRNA is, when bound to EF-Tu located far from the peptidyl transfer center but can reorient itself into the A-site of the ribosome, while retaining the interaction with its codon. This process coincides with the proofreading of the anticodon of the tRNA by the codon of the mRNA. An incorrect (noncognate) match of the anticodon to the codon increases the likelihood that the aminoacyl-tRNA will dissociate before its amino acid has reached the peptidyl transfer site of the ribosome.

Peptidyl transfer is catalyzed by the rRNA of the large subunit without direct assistance of ribosomal proteins or elongation factors. Once the aminoacyl moiety reaches the A-site of the peptidyl transfer site, the peptide on the peptidyl-tRNA in the P-site can be transferred to it. This leads to a peptidyl-tRNA in the A-site and a deacylated tRNA in the P-site.

The final step of elongation is the translocation of the peptidyl-tRNA from the A-site to the P-site and the movement of the mRNA by three nucleotides so that the next codon is exposed in the A-site. EF-G, which catalyzes this process, binds to the ribosome in complex with GTP. After translocation, it dissociates in complex with GDP. A surprising finding is that the ternary complex of EF-Tu with GTP and aminoacyl-tRNA has the same shape as EF-G. It remains possible that EF-G, when it dissociates from the ribosome, leaves an imprint into which the ternary complex fits exactly.

See also: **Messenger RNA (mRNA); Ribosomes; RNA Polymerase; Transcription; Translation**

Elongation Factors; Translation

J Parker

Copyright © 2001 Academic Press
doi: 10.1006/rwgn.2001.0402

Translational elongation factors are proteins that play two important roles during the elongation cycle of protein biosynthesis on the ribosome. First, elongation factors are involved in bringing aminoacyl-tRNA (aa-tRNA) to the ribosome during protein synthesis. Second, an elongation factor is involved in translocation, the step in elongation at which the peptidyl-tRNA is moved from one ribosomal site to another as the mRNA moves through the ribosome. Both steps result in the hydrolysis of guanosine triphosphate (GTP), and the conformation of the elongation factors changes depending on whether they are bound to GTP or to guanosine diphosphate (GDP). The elongation factors of archaea and bacteria (both are types of prokaryotes) and eukaryotes are similar in structure and function, as are the steps in protein biosynthesis in which they participate. The first part of this entry will deal with the elongation factors in bacteria, and later the factors found in other types of organisms will be discussed.

Factors Related to Aminoacyl-tRNA Binding in Bacteria

Elongation factor Tu (EF-Tu), when bound to GTP, brings aa-tRNA to the ribosome during the elongation phase of translation. When EF-Tu is bound to GTP, it has a high affinity for aa-tRNA and forms the ternary complex, EF–Tu–GTP-aa-tRNA. EF-Tu must recognize common features of all tRNAs and also recognize that the tRNA is aminoacylated. EF-Tu is one of the most abundant proteins in bacterial cells, often present as 5% of the total cell protein. In *Escherichia coli* there are more than five molecules of EF-Tu per ribosome, and most of the aa-tRNA in the cell is bound to EF-Tu.

The ternary complex has a high affinity for the ribosomal A-site, the site at which incoming aa-tRNA must be bound during the elongation step on the ribosome. If there is not a match between the aa-tRNA and the open codon at the A-site, the ternary complex leaves the ribosome. If there is a match, the aa-tRNA is delivered to the site, GTP is hydrolyzed, and EF-Tu-GDP is released from the ribosome. Another elongation factor, EF-Ts, is involved in a nucleotide exchange, whereby the GDP on the EF-Tu is replaced by GTP.

Interestingly, there is still disagreement on the number of GTPs hydrolyzed during binding of each aa-tRNA. Models of translation show the involvement of the classic ternary complex; however, some studies indicate that there are two molecules of EF-Tu bound per each aa-tRNA and two GTPs are consumed during the cycle.

The gene encoding EF-Tu is called *tuf*, and many bacteria have duplicate genes for this protein (*tufA* and *tufB*). In addition to antibiotic resistance mutants, some mutants of EF-Tu alter the error frequency of translation (such mutants are also known for the homologous protein in yeast). EF-Ts is encoded by the *tsf* gene.

Factors Related to Translocation in Bacteria

Translocation involves a conformational change of the ribosome during elongation, whereby the newly formed peptidyl-tRNA is moved from the ribosomal A-site to the P-site (the tRNA formerly occupying the P site is displaced to the E-site) and the next codon on the mRNA is moved into the A-site. Translocation, then, completes a cycle of elongation and positions the ribosome to accept the next incoming aa-tRNA. Translocation is catalyzed by the elongation factor EF-G, which is encoded by the *fus* gene. EF-G is bound to the ribosome as EF-G–GTP. The binding site for EF-G overlaps with that of EF-Tu and, fascinatingly, the structure of the EF-G mimics that of the ternary complex. The hydrolysis of the GTP seems to provide the energy for translocation, after which EF-G–GDP dissociates from the ribosome. Several mutants of EF-G are also known, and some of these also display altered accuracy of translation.

Other Factors in Bacteria

Almost certainly other protein factors, not yet completely characterized, are involved in translation. A protein called elongation factor P seems to function at an early step in protein synthesis, possibly in formation of the first peptide bond. The gene encoding this protein, *efp*, has been found throughout the bacteria. The homologous protein in eukaryotes is the initiation factor, eIF5A. There is also a separate EF-Tu-like elongation factor specifically for bringing selenocysteinyl-tRNA to the ribosome in response to a UGA codon in the appropriate context. Such a protein is also found in the archaea and eukaryotes.

Elongation Factors in Archaea and Eukaryotes

The eukaryotes have elongation factors that perform the same functions as EF-Tu, EF-Ts, and EF-G. The eukaryotic equivalent of EF-Tu is EF-1α, and there is high sequence conservation between EF-Tu and EF-1α. EF-1α is also one of the most abundant cytoplasmic proteins in eukaryotes. Genes for this protein are often present in more than one copy and may have cell-type or stage-specific regulation.

The eukaryotes have a complex of proteins, EF-1β, EF-1γ, and EF-1δ, which function in a nucleotide exchange reaction like that involving EF-Ts. The factors EF-1β and EF-1δ are closely related to each other, but none of these proteins is closely related to EF-Ts.

The eukaryotic equivalent of EF-G is called EF-2. Like EF-G, it is responsible for the GTP-dependent translocation step of the ribosome. It also contains a diphthamide residue, a unique posttranslational modification of a histidine residue, which is the cellular target for ADP ribosylation by diphtheria toxin.

Interestingly, the elongation factors of the archaea are more closely related to those of the eukaryotes than they are to those of bacteria, and, therefore, factors from the archaea are given the same nomenclature as those from eukaryotes. The only elongation factor in the archaea that is more closely related to a bacterial factor than to the one from the eukaryotes is the elongation factor that brings selenocysteinyl-tRNA to the ribosome.

As for the prokaryotes, there are almost certainly other eukaryotic protein factors involved in elongation. For instance, the fungi have a factor called EF-3 which has both ATPase and GTPase activities.

Effects of Antibiotics on Function of Elongation Factors

The elongation factors, or the steps in protein synthesis catalyzed by the elongation factors, are the targets of several different antibiotics, and some of these are well studied. The antibiotic kirromycin inhibits EF-Tu, blocking its exit from the ribosome. Kirromycin-resistant alleles of the *tuf* genes have also been isolated. Fusidic acid inhibits EF-G (and EF-2) by preventing it from leaving the ribosome. Mutants of EF-G are known which are resistant to fusidic acid, and they are responsible for the gene encoding this factor being termed *fus*. The aminoglycoside antibiotic kanamycin also inhibits translocation, and this antibiotic can be used to select mutants of EF-G (although these do not result in high-level resistance to the drug). Thiostrepton is a modified peptide antibiotic that binds to a site on 23S rRNA and inhibits elongation factor-dependent reactions in both archaea and bacteria.

Tetracycline inhibits protein synthesis by interfering with the binding of aa-tRNA to ribosomes. Bacterial resistance to the tetracyclines is mediated by two major mechanisms. One mechanism involves protection of ribosomes from the action of the antibiotic by one of a group of proteins whose N-terminal amino acid sequences are similar to those of elongation factors Tu and G.

The large-subunit ribosomal RNA contains a very highly conserved sequence which is cleaved by the antibiotic α-sarcin and modified by the antibiotic ricin, both of which abolish protein synthesis on eukaryotic ribosomes (and are somewhat less effective against prokaryotic ribosomes). These antibiotics block the functions of ribosomes dependent on elongation factors, apparently by blocking their binding to the ribosome. Sordarins are a new family of highly specific antifungal antibiotics which inhibit the action of fungal EF-2.

See *also*: **Translation**

Embryo Transfer

F Constantini

Copyright © 2001 Academic Press
doi: 10.1006/rwgn.2001.0406

'Embryo transfer' refers to the transplantation of a mammalian preimplantation embryo into the reproductive tract of a recipient female so that it may implant and continue to develop to birth. Mammalian embryos of many species can develop *in vitro* from fertilization to the blastocyst stage (approximately 100 cells), but at this point they must implant in the uterus in order for embryogenesis to proceed normally. For this reason, the ability to produce live young, or even mid-term fetuses, from isolated preimplantation embryos depended historically on the development of embryo transfer techniques. The first successful embryo transfer was performed in 1890 in the rabbit. However, the techniques of embryo transfer were not perfected and applied to a large number of mammalian species until the 1950s and 1960s, when methods for the efficient *in vitro* culture of preimplantation embryos were also developed. In 1978, this work culminated in the first birth of a human from a transferred embryo, which had been conceived by *in vitro* fertilization.

In laboratory mice, embryo transfer is usually performed by surgical methods. Mouse embryos at the one-cell stage or cleavage stages are transferred to the oviduct, while blastocysts, which are ready to implant, are transferred to the uterus. The recipient female mouse, or 'foster mother,' is first made 'pseudopregnant' by mating with a male that has been sterilized by vasectomy, so that her own eggs will not be fertilized and cannot compete with the transferred embryos. The gestational age of the donor embryos and the recipient must be synchronized, with optimal results occurring when the embryos are one day more advanced than the gestational age of the recipient. The recipient female is anesthetized and the oviduct or uterus is exposed through a small incision. Under a low-power stereomicroscope, the embryos to be transferred are loaded in a small volume of liquid into a fine glass pipette, which is inserted through a small hole in the bursa (the membrane covering the ovary and oviduct) into the infundibulum (the open end of the oviduct). In the case of a uterine transfer, a small hole is made in the side of the uterus with a sharp needle, and the transfer pipette is inserted through the hole into the uterine lumen. The embryos are then expelled and the incision is closed. Under optimal conditions, the rate of successful implantation and development of the embryos to term can exceed 90%.

In large animals and in humans, embryo transfer is usually performed using a transvaginal approach, in which the embryos are inserted through the cervix and into the uterus with a catheter.

The applications of embryo transfer are numerous and of great importance for basic research in genetics and experimental embryology, for animal husbandry and genetic manipulation of livestock, and for reproductive medicine in humans. Experiments in which the preimplantation embryo is physically manipulated (for example, by microsurgery, injection of cells, or cell lineage tracers) depend on embryo transfer to determine the effects of the treatment on the resulting fetus or animal. All of the powerful and widely used techniques for genetic manipulation (transgenesis) of the mouse and other animals involve the introduction of foreign genetic material into the preimplantation embryo, either in the form of purified DNA injected at the one-cell stage, embryonic stem cells introduced at the cleavage or blastocyst stages, or nuclei transplanted at the one-cell stage. Therefore, embryo transfer is required for these genetically manipulated embryos to develop into live animals. Rare or valuable strains of mice and other mammals are often preserved by embryo freezing (cryopreservation), and embryo transfer is used to revive these strains. In agriculturally important mammals, in addition to the aforementioned applications,

embryo transfer has been used for artificial twinning (by separating the blastomeres of two-cell embryos) and cloning (by nuclear transplantation), and for increasing the reproductive yield of valuable donors by inducing superovulation and transferring the embryos to multiple recipients. In humans, embryo transfer has made possible the recent advances in treatments for infertility, such as *in vitro* fertilization, intracytoplasmic sperm injection, and egg donation. Another application likely to increase in importance in the future is the diagnosis of inherited diseases at the preimplantation stage (using DNA isolated from one or a few cells), after which embryos selected for the absence of disease will be reimplanted.

See also: **Embryonic Stem Cells; Infertility**

Embryonic Development of the Nematode *Caenorhabditis elegans*

M Labouesse

Copyright © 2001 Academic Press
doi: 10.1006/rwgn.2001.0404

An overview of how the nematode *Caenorhabditis elegans* embryo develops is given below. Some background anatomical information is provided and the mechanisms that are involved in specifying the axes of the embryo and determining the fate of the first 28 cells to be born are addressed. A brief discussion on how the embryo generates its main tissues and organs then follows; and finally how the embryo acquires its shape and assembles its muscles is described. An effort has been made throughout to explain how our understanding *C. elegans* development aids us in our understanding of other animals.

Despite the huge evolutionary distance that separates *C. elegans* from most species, and its lack of the anatomical features (limbs, eyes, hair) that identify more familiar animals, *C. elegans* has undeniably illuminated some general principles that govern animal development. The reader who that doubts that *C. elegans* can relate to humans must realize that this relationship is at first deceptive if one considers our anatomy. *C. elegans*, which has muscles, nerves, skin, and gut, has brought a wealth of information at another level, the cellular level. This article aims to convey the notion that *C. elegans* has proven a remarkable model organism for studying the intracellular machinery at play that makes development

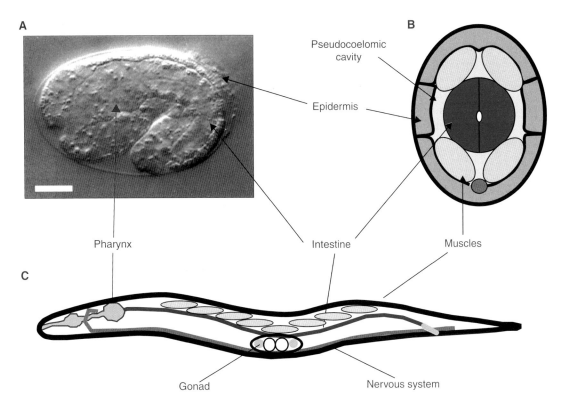

Figure 1 Anatomy of the nematode embryo. (A) Nomarski picture of the embryo at mid-embryogenesis; muscles, neurons, and the gonad cannot be distinguished in this picture. (B) Schematic drawing of a section through the embryo or the larva showing the positions of the main tissues. (C) Schematic drawing of the young hatchling. White scale bar in (A) is 10 μm. In the embryo (A) and the larva (C) anterior is to the left and dorsal is up. The color code for tissues and organs is the same in (B) and (C); note that only a subset of muscles are drawn and that only the main nerve is shown (but the cell bodies of neurons are not represented).

possible, for example, can tell the anterior from the posterior or generates functional muscles.

Anatomy of the *Caenorhabditis elegans* Embryo and Timing of its Development

The key features that make *C. elegans* so easy to study are its transparency, its extremely rapid rate of development, its simplicity, and the invariance of its division pattern. Each individual cell of the embryo can be visualized at all times in live specimens, so it is actually possible to watch a cell divide or migrate. Embryogenesis only lasts for 14 h at 25 °C, during which time a fertilized egg becomes a young larva with 558 cells[1] (for comparison the fruit fly *Drosophila* embryo has about 10^5 cells at the end of embryogenesis). These cells arise

in a reproducible and fixed pattern of cell divisions, migrations, and fusions. This feature together with a great deal of patient work allowed a group of scientists led by John Sulston to reconstitute the entire pattern of cell divisions from the zygote to the adult, which is now referred to as the *C. elegans* cell lineage.

The anatomy of the late embryo is very simple: it comprises two concentric tubes, an inner tube which corresponds to the digestive tract (pharynx, intestine, and rectum), and an outer tube which corresponds to the epidermis (skin). The precursors to the reproductive organ, the neurons, and the muscles lie between these tubes (**Figure 1**). There are no appendages or external organs, implying that a mutant must be identified based on its gross shape, on the aspect of its internal organs, on muscle activity, or on the presence and position of a specific cell.

C. elegans embryogenesis can be conveniently divided into three main stages (**Figure 2**). During the first 100 min, five divisions give rise to 28 cells (**Figure 3**). Gastrulation, which corresponds to the set of cell rearrangements that ultimately gives rise to the separation between the three germ layers, starts

[1] 1090 cells are generated during *C. elegans* embryonic and larval development, of which 131 die from apoptosis and 959 survive. Among those that survive some fuse, such that there are in fact 959 somatic nuclei in adult animals and 558 nuclei in young hatchlings but slightly fewer cells.

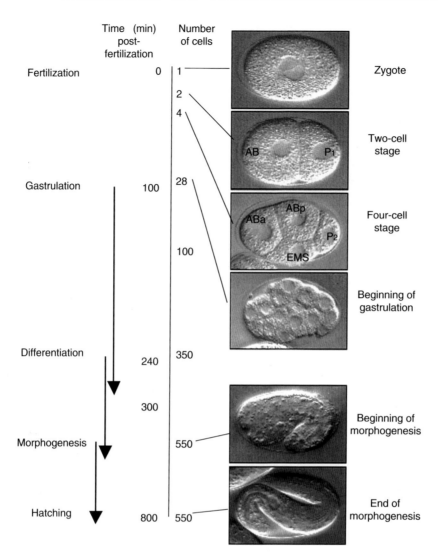

Figure 2 The main stages of *Caenorhabditis elegans* embryogenesis. The Nomarski pictures on the right show from top to bottom: a zygote after positioning of the two pronuclei at its centre; a two-cell embryo (note that the anterior AB blastomere is larger than the P_1 blastomere); a four-cell embryo (the blastomere names are indicated); a 28-cell embryo which is initiating gastrulation; an embryo at the beginning of elongation; an embryo at the end of elongation. Major embryonic events on the left and pictures on the right should be related to the time-scale and cell-number scale shown in the center. The scales are not linear.

at the 28-cell stage. The second stage corresponds to the time of gastrulation, organ formation, and initial differentiation; it takes 4 h and is accompanied by six further cycles of cell division. During the final stage, terminal differentiation and morphogenesis of the embryo from a ball of cells to a worm-shaped embryo occur with only very few additional cell divisions.

Methods Used to Examine *Caenorhabditis elegans* Embryos

Microscopy

Light microscopy using differential interference contrast optics (Nomarski optics) and the use of an autofluorescent protein (green fluorescent protein) fused

to the protein of interest play an essential role in analyzing mutant phenotypes. The development of time-lapse recording methods, in particular the increasing power of modern computers, is greatly facilitating the observation of embryos.

Embryological Methods

To assess the function of a particular cell in a normal embryo, it is possible to eliminate that cell using a laser microbeam focused onto the cell to be killed via the Nomarski microscope. It is possible to determine in which cell a given gene acts by removing the eggshell in order to separate the first blastomeres and by reassociating them in different combinations.

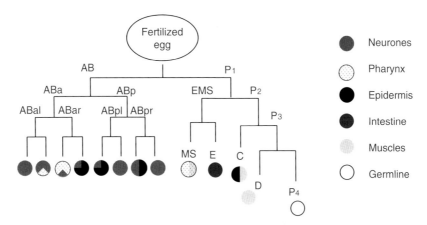

Figure 3 Early embryonic lineage and founder cells. This figure shows the beginning of the embryonic lineage. The vertical axis shows time, the horizontal axis shows divisions. The organ/tissue that is contributed to by each major branch of the lineage is symbolized by sectored circles which are roughly proportional to the number of cells produced. The color code is given on the right. The letters that follow blastomere names normally refer to the axis of divisions. For instance, ABal is the left daughter of the anterior daughter of AB (note however that AB divides along the dorso-ventral axis and not the anterior/posterior axis but that physical constraints push it to adopt an anterior position, hence its name).

Molecular Methods

The usual range of modern molecular tools, such as transgenes and reporter genes, is available to examine *C. elegans* embryos. A powerful tool to assess the embryonic function of genes predicted from the genome sequence (*C. elegans* was the first multicellular eukaryote whose genome was fully sequenced) is called RNA interference. In this method, double-stranded RNA specific for a target gene is introduced into embryos where it will efficiently and specifically inhibit the expression of the endogenous target gene, thereby creating a transient knockout of that gene.

Caenorhabditis elegans Embryos Define their Anterior/Posterior Axis at the One-Cell Stage

Animal embryos use different strategies to define their anterior/posterior (A/P) axis. In many species the oocyte is already polarized along the future A/P axis (e.g., in *Drosophila melanogaster*), or along a so-called animal/vegetal axis (e.g., in amphibians) which in some respects resembles the A/P axis. In such species, the cytoplasmic composition of the oocyte is not homogeneous and differs at both poles. In *D. melanogaster*, we know that specialized cells, called the nurse cells, deposit mRNAs encoding morphogens,[2] at the future anterior pole which will be transported to the posterior pole or will stay anteriorly. In contrast, the mature *C. elegans* oocyte is ovoid but has no apparent polarity. The sperm entry point provides the initial cue for A/P polarity. Analysis of the mechanism that transforms this asymmetric cue into A/P polarity has revealed an apparently well-conserved machinery used not only in embryos but also in many different polarized cells.

As in many species, fertilization triggers intense cytoplasmic movements. Internal cytoplasm flows toward the sperm entry point, while its direction is reversed at the cortex. Meanwhile, the female pronucleus migrates toward the male pronucleus until they meet; the two juxtaposed pronuclei move back to their final position slightly off the center of the long embryonic axis, which will be the A/P axis. The first division is asymmetric, giving rise to a large anterior daughter named AB and to a smaller posterior daughter named P_1.

The cytoplasmic flow, which can be visualized using Nomarski optics, is the manifestation of a complete reorganization of the cytoplasmic content of the zygote. Antibodies against constituents of the early embryo have been used to identify cytoplasmic granules, termed the P granules, which may carry germ cell determinants. They are initially uniformly distributed throughout the cytoplasm of the zygote and gradually accumulate at the posterior pole,[3] such that they are

---

[2]A morphogen is a factor (generally a protein) that can induce the formation of different structures depending on its concentration.

[3]Drugs that inhibit polymerization of the protein called actin prevent the accumulation of P granules posteriorly, demonstrating that movement of P granules occurs along the actin cytoskeleton.

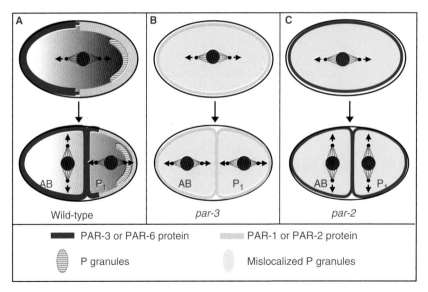

Figure 4 The A/P axis is defined by a polarizing mechanism. (A–C) Schematic representation of the major differences observed between wild-type (A), *par-3* (B) or *par-2* (C) embryos, at the one-cell (top drawings) and two-cell (bottom drawings) stages. In a wild-type embryo, (1) the zygote and then the posterior blastomere are polarized (asymmetric black shading), (2) PAR-3 and PAR-6 proteins are at the anterior cortex (red line) while PAR-1 and PAR-2 proteins are at the posterior cortex (brown line), (3) P granules (red and white dots) are located posteriorly and segregate to the P_1 blastomere, (4) the AB blastomere is larger than the P_1 blastomere and divides along the D/V axis. In a *par-3* mutant embryo (it would be the same in a *par-6* mutant embryo), (1) polarity is abolished, (2) PAR-1 and PAR-2 proteins are found all around the cortex, (3) P granules are uniformly distributed, (4) the first cleavage is symmetric and generates two blastomeres that divide along the A/P axis. In a *par-2* mutant embryo, the situation is similar, except that now (1) PAR-3 and PAR-6 proteins are found all around the cortex and (2) both AB and P_1 divide along the D/V axis. Anterior is to the left.

absent from the AB blastomere[4] after the first division. Thus, four criteria can be used to identify the initial polarity of the zygote: the asymmetric localization of P granules, the unequal size of AB and P_1, the fact that AB divides along the future dorso/ventral axis (D/V axis) while P_1 divides along the A/P axis, and the different fates[5] of AB and P_1 progenies (**Figure 4A**).

A breakthrough in the understanding of how the zygote acquires its A/P polarity came with the isolation of mutations that affect the distribution of P granules. These mutations define six maternal[6] genes, called *par-1* through *par-6* (*par* stands for partitioning

defective). Although *par* genes act in a common pathway, they can be subdivided into at least two groups. One group includes proteins localized at the anterior cortex (PAR-3 and PAR-6; see **Figure 4B**), the other includes proteins localized at the posterior cortex (PAR-1 and PAR-2; see **Figure 4C**). Genetic analysis has shown that PAR-1 is the final effector of this pathway. The nature of PAR proteins suggests that they act in a signaling process. In particular, PAR-1 and PAR-4 have protein kinase[7] domains, whereas PAR-3, PAR-6, and PAR-2 have protein–protein interaction modules suggesting that they could position or tether other proteins in specific places.

It is thought that PAR proteins act to interpret the polarity cue provided by the sperm, which brings two centrosomes. One model that is supported by the involvement of actin states that PAR proteins act by mediating local changes in the cytoskeleton. How they do so and what their immediate targets are is unknown. Whatever the mechanism, their activity is required to localize several cell fate determinants

[4]Cells in early embryos are called blastomeres.
[5]The fate of a blastomere refers to its pattern of division and the type of cells (for instance muscle versus neurons) it generates.
[6]A mutation is classified as maternal when the mother has to be a homozygous mutant to affect the embryo; it affects genes that are expressed in the oocyte prior to fertilization (the gene product is stored in the oocyte as an mRNA or a protein). A mutation is classified as zygotic when the embryo itself has to be a homozygous mutant to be affected; it corresponds to genes that are expressed in the embryo after the onset of embryonic transcription.

[7]Protein kinases add a phosphate group onto certain serine/threonine or tyrosine residues of other proteins, and by doing so modify their activity or their subcellular localization.

(including P granules) to the appropriate blastomeres. The demonstration that *Drosophila par-1* and *par-3* homologs play an active role in determining the polarity of the oocyte and of epithelial cells (see below for a description of epithelial cells) suggests that *par* genes correspond to an ancient mechanism used to polarize cells.

Cell–Cell Interactions Define the Dorso–Ventral and the Left–Right Axes

Remarkable as it is, fertilization is only the beginning of development. Subsequently, all animal embryos generate by rapid cleavage smaller cells with ever more restricted potentials, generally with no increase in embryonic volume. In all species, the initial egg cytoplasm is unequally partitioned and/or modified through signals sent by one group of cells to their neighbors, as occurs in amphibian and fish embryos, or else through the action of localized transcription factors, as occurs in the *Drosophila* embryo. In *C. elegans*, the 28 first blastomeres present at the onset of gastrulation acquire distinct fates both through cell–cell interactions and localized transcription factors. The dorso–ventral and left–right axes are specified during this early stage of embryogenesis through strategies that differ from those used in insects and vertebrates.

As described above, the first division along the A/P axis generates two cells with different potentials that divide perpendicular to each other. The axis of AB division defines the D/V axis. Due to physical constraints imposed by the eggshell the ventral daughter of AB becomes positioned anteriorly to the dorsal daughter (hence their names ABa and ABp). The ABa and ABp blastomeres subsequently divide along an axis that is neither the A/P nor the D/V axis and defines the L/R axis. Unlike the initial division, the AB division and then the ABa/ABp divisions are symmetric resulting in daughters that initially have equal potentials. For this reason and because the axes are defined when the embryo contains very few cells, establishing the axis becomes a matter of generating a difference between cells that are equivalent.

Dorso-Ventral Axis and Left–Right Axis

The eggshell causes the ABp blastomere, but not the ABa blastomere, to come in direct contact with the P_2 blastomere. A signaling cascade between P_2 and ABp sets the D/V axis by instructing ABp to become different from its sister ABa. This is achieved by a ligand (encoded by the gene *apx-1*) expressed at the surface of P_2 that interacts with a receptor (encoded by the gene *glp-1*) present in ABa and ABp (**Figure 5A**). As the APX-1 ligand is not diffusible, only ABp can receive it. In *apx-1* or *glp-1* mutant embryos, ABp is not

instructed and generates cells and tissues normally generated by ABa (**Figure 5B**).

The L/R axis is also specified through cell–cell interactions which occur when the embryo reaches the 12-cell stage. At that time, the MS blastomere is in contact only with a subset of ABa descendants, namely ABalp and ABara, but not in contact with their left/right relatives ABarp and ABala. In this case, a signaling cascade involving the gene *glp-1* again but a ligand of unknown nature sets the L/R axis by instructing ABara to become different from ABala and ABalp different from ABarp.

Generation of Cell Diversity until Gastrulation by Polarization

Starting with the division of EMS, ABal, ABar, ABpl and ABpr blastomeres (**Figure 3**), all cells divide along the A/P axis until gastrulation starts. During this time a common mechanism is repeatedly used to make the anterior daughter different from its posterior sister. This process involves the phosphorylation of a transcription factor that accumulates in an active form only in the nucleus of the anterior daughter. The transcription factor is encoded by the maternal gene *pop-1*, while the kinase involved in phosphorylating the POP-1 protein is the product of the maternal gene *lit-1* (a so-called MAP kinase). In *pop-1* mutant embryos, the anterior daughters adopt the fates of their posterior sisters. Conversely, in *lit-1* mutant embryos the posterior daughters adopt the fates of their anterior sisters, implying that *lit-1* is a negative regulator of *pop-1* in the posterior daughter. It is actually not known whether POP-1 phosphorylation prevents its accumulation, its stability, or its activity in the posterior daughter.

The process used to initiate the polarization of the anterior/posterior division is well understood in the EMS lineage (see **Figure 3**). Blastomere reassociation experiments together with the isolation of specific mutations have demonstrated that the EMS blastomere can generate intestinal cells only if it contacts the P_2 blastomere during a specific time-window during its cell cycle. In the absence of such a contact, the posterior daughter of EMS (E blastomere) fails to generate intestinal cells and adopts the MS fate. During this contact, a signaling pathway[8] polarizes EMS to ultimately induce the asymmetric localization of the POP-1 protein in the nucleus of MS, which is the

---

[8]This pathway involves all the classical components of a Wnt pathway, including a Wnt signal (encoded by the gene *mom-2*), a Frizzled-type receptor (encoded by the gene *mom-5*), a GSK-3 kinase (encoded by the gene *sgg-1*), a β-catenin homolog (encoded by the gene *wrm-1*), and finally a TCF/LEF transcription factor (encoded by the gene *pop-1*).

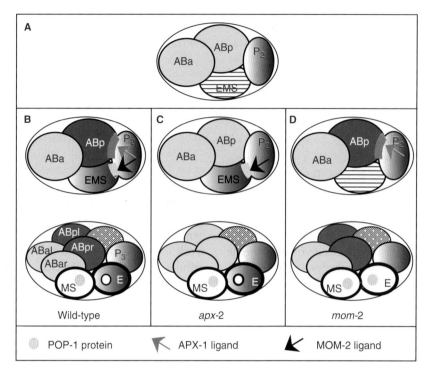

Figure 5 Defining the D/V axis and generating cell diversity in the early embryo. (A) At the four-cell stage, due to the activity of *par* genes, the EMS and P_2 blastomeres are different from each other and both are different from the ABa and ABp blastomeres; however, ABa and ABp are initially equivalent. (B, upper) In wild-type embryos, the APX-1 ligand in P_2 instructs ABp to become different from ABa, while the MOM-2 ligand in P_2 polarizes EMS (asymmetric red shading). (B, lower) In turn, polarization of EMS is interpreted when it divides in such a way that the POP-1 protein becomes nuclearly localized mainly in the MS blastomere; this allows the E blastomere to generate the intestine. (C) In *apx-1* mutant embryos, ABa and ABp remain identical and express the 'ABa fate.' (D) In *mom-2* mutant embryos, EMS is not polarized at the four-cell stage, hence MS and E both inherit nuclear POP-1, which causes them to be identical and to express the 'MS fate' (the intestine is not made).

anterior daughter of EMS (**Figure 5A**). The ligand is expressed in the P_2 blastomere, while its receptor is expressed in the EMS blastomere. In embryos lacking the ligand or its receptor, the posterior daughter E adopts the fate of its anterior sister MS (**Figure 5C**).

How POP-1 activity in the anterior daughter can ultimately contribute to generate cell fate diversity is beginning to be understood for the EMS lineage. Genetic and molecular analysis has shown that the maternal gene *skn-1* encodes a transcription factor that is present and essential in EMS, MS, and E blastomeres. Hence, at least two transcription factors will be active in MS, POP-1, and SKN-1, which will contribute to specify the 'MS fate,' while only one will be active in E, SKN-1 which will activate the endoderm specification program. Presumably POP-1 together with other transcription factors can similarly specify unique fates among the first 28 cells of the pregastrulation embryo. Genetic analysis in *C. elegans* has thus uncovered an entirely new mechanism to generate cell diversity, which has now also been shown to exist in vertebrates.

Totipotency of the Germline is Preserved by Repressing Gene Expression

The germline lineage needs to prevent premature differentiation to preserve its totipotency. Historically, nematodes were important in recognizing the special nature of the germline.[9] A major contribution of *C. elegans* to biology has been to show that the germline is set aside by repressing gene expression. The most compelling evidence has been provided by the isolation of maternal-effect mutations that lead to

---

[9]Theodor Boveri, a German embryologist working more than 100 years ago, was the first to observe chromosomes. He discovered that in the nematode *Parascaris aequorum* chromosomes become fragmented during embryogenesis in somatic tissue but not in germ cells. During this phenomenon, which has been called chromatin diminution, different somatic cells inherit different pieces of chromosomes. For some time chromatin diminution provided a plausible model for terminal differentiation. Now we know that in most species, including *C. elegans*, all cells inherit the same set of chromosomes.

the absence of the germline, either because the germline lineage becomes transformed into a somatic lineage or because germ cells die. Genes defined by these mutations are associated with P granules or the germline. They act in germline precursors to lock the chromatin in a repressed state (*mes-2* and *mes-6*[10]), to repress gene expression at the transcriptional level (*pie-1*) or at the translational level (*pos-1*, *mex-1*, and *mex-3*). By doing so, they probably prevent premature differentiation of germline precursors. Further evidence in support of the idea that germline precursors are transcriptionally inactive comes from the use of specific monoclonal antibodies that distinguish the active from the inactive pool of RNA polymerase II, the major enzyme involved in gene transcription. In *C. elegans* embryos, as well as in *Drosophila* embryos, it has been shown that germline precursors contain only an inactive form of RNA polymerase II.

Mechanisms Used to Generate Tissues and Organs

Zygotic genes that control the formation of several tissues and organs once gastrulation has been initiated have been characterized. These genes fall into two categories: those that specify organ/tissue identity and those that control organ/tissue differentiation. Genes that specify the identity of the intestine, the pharynx, and the epidermis[11] share the following characteristics: (1) when they are inactivated, the precursor cells from which the organ or tissue is derived adopt another fate, leading to the absence of the organ/tissue primordium; (2) they are expressed very soon after the onset of gastrulation (the intestine identity gene is even expressed prior to gastrulation); and (3) they can reprogram other cells to develop as if they were part of the organ/tissue. In other words, these 'identity genes' confer the potential to form the intestine, pharynx, or epidermis in a group of cells when gastrulation starts. Interestingly, homologous genes have been described in flies and vertebrates, which play similar roles. Therefore, it appears that, despite very different strategies for the early steps of embryogenesis, which reflect the necessity to adapt to very different environments, the genetic control of organ/

tissue formation has probably been conserved during evolution and may be very ancient.

Genes that are important for the differentiation of organs and tissues differ from organ/tissue 'identity genes' in two respects. First, their inactivation does not lead to the absence of the organ/tissue primordium but only to the abnormal differentiation of cells within the organ/tissue. Second, they are expressed slightly later than 'identity genes' and their expression depends on the latter. These genes are more numerous than 'identity genes,' and probably act together with them to activate all or a subset of terminal differentiation genes in the organ/tissue (e.g., genes that control specific muscle proteins). Further work in *C. elegans* should help in understanding the cellular and genetic steps that are essential to build organs and tissues.

Caenorhabditis elegans Embryo as a Model System in Cell Biology

C. elegans is particularly well suited to help analyze some cellular processes that are not necessarily specific to embryogenesis. Three of these will be briefly mentioned: the mechanics of cell division, the biology of epithelial cells, and the assembly of muscles.

The First Cell Cycle

As described before, fertilization of the oocyte induces completion of the female pronucleus meiosis, its migration toward the male pronucleus, movement of both pronuclei to the center of the embryo, spindle assembly, chromosome separation, and finally cytokinesis. These events can be easily monitored in live embryos because the zygote is a very large cell. Genetic analysis has shown that it is possible to individually affect each of these steps. There is no doubt that the *C. elegans* embryo will provide an invaluable system with which to analyze processes taking place more specifically during the first embryonic cell cycle (e.g., pronuclear migration) as well as those common to all cell divisions (e.g., spindle assembly).

Biology of epithelial cells

Epithelial cells are polarized and characterized by two main membrane domains, the apical surface facing the external environment and the basolateral surface facing the inside of the animal. Among other roles, they are essential to shape organs and tissues. A typical example is wound healing, which relies primarily on epidermal cells (which are epithelial) changing their shapes to extend over the wounded area. Genetic analysis in *C. elegans* has identified several genes that are important in controlling cell shape changes. During the second half of *C. elegans* embryogenesis, epidermal

---

[10]The genes *mes-2* and *mes-6* encode Polycomb-like proteins. In *Drosophila*, Polycomb is known to negatively regulate gene transcription by binding to chromatin and is involved in maintaining silent certain genes of the *Hox* complex in appropriate segments of the embryo.

[11]The names of these genes are *end-1* for the intestine, *pha-4* for the pharynx, and *elt-1* for the epidermis. They encode transcription factors.

cells, which initially have a square shape, stretch out along the A/P axis and narrow along the D/V axis resulting in a constriction of the internal contents of the embryo and its elongation along the A/P axis (see **Figure 2**). Contraction of the actin cytoskeleton within epidermal cells provides the driving force to undergo this dramatic cell shape change. As in vertebrate epithelial cells, actin is anchored to specialized junctions that separate the apical surface from the basolateral surface via a complex of proteins known as α-catenin, β-catenin, and cadherin (encoded by the genes *hmp-1*, *hmp-2*, and *hmr-1*, respectively). Mutations affecting the catenin/cadherin complex disrupt actin anchoring and prevent elongation of *C. elegans* embryos. Genes that regulate actin contraction during the process of elongation are also known about (for instance mutations in the gene *let-502* reduce the extent of elongation).

Assembly of Muscle Fibers

Knowledge of muscle function and assembly is particularly detailed in *C. elegans*. *C. elegans* muscle sarcomeres, which assemble during the second half of embryogenesis, are in most respects very similar to those observed in other species, except that muscle cells do not fuse. They include alternating thick filaments, which contain myosin, and thin filaments, which contain actin, tropomyosin, and troponin.

Many genes encoding proteins required for sarcomere assembly and most if not all those encoding structural sarcomeric components have been identified, generally by genetic analysis. In vertebrates, muscles are anchored to our bones; in *C. elegans*, they are anchored to the cuticle, which is secreted by epidermal cells at their apical surface and acts as an external skeleton (or exoskeleton). Mutations in genes encoding muscle-anchoring components lead to embryonic lethality, prevent full embryonic elongation (see above), and often lead to muscle integrity defects. Detailed analysis of these mutations suggests that sarcomeres are first assembled around a structure called the dense body in muscles (**Figure 6**), which itself attaches to a network of proteins in the space separating muscles from the underlying epidermis (the extracellular matrix). This network in turn is anchored to the cuticle through other proteins, some of which remain to be identified.

Conclusion

For a long time, nematodes were thought to be unique among animal species owing to their invariant lineage. Indeed, early blastomeres have a fixed fate in *C. elegans*, implying that if a blastomere is ablated the tissues that should normally be generated by this blastomere will be missing. In contrast, in many other species early

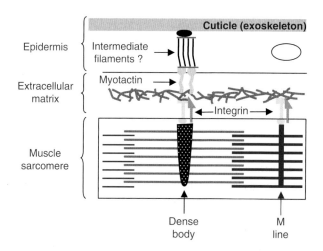

Figure 6 How the muscle is attached. Sarcomeres are formed by alternating thin (gray lines) and thick filaments (red lines). Within muscle cells the anchoring structure is called the dense body and consists of a complex formed by several proteins (vinculin, talin, and α-actinin), which interact with actin within thin filaments, and an integrin dimer (dark pink/pale pink pair; genes *pat-2* and *pat-3*) at the muscle membrane (some additional attachment is provided through so-called 'M lines' at the center of thick filaments). The integrin itself recognizes a protein from the extracellular matrix called perlecan (intertwined gray lines; gene *unc-52*). Within epidermal cells the anchoring structures are called fibrous organelles: they are made in part by a long transmembrane protein called myotactin (gray; gene *let-805*) that extends toward muscles and contacts in turn one or more proteins that probably run across the epidermal cytoplasm from the basal membrane to the cuticle. It is not yet known what provides attachment to the cuticle.

blastomeres do not have a fixed fate, such that if one is ablated cell–cell regulatory mechanisms compensate for this loss. Embryonic development in *C. elegans* is said to be 'mosaic,' whereas in other species it is said to be 'regulative.' Mosaic development was long thought to be strictly under the control of lineage-dependent transcription-based mechanisms. It is now clear that the invariance of *C. elegans* lineage does not preclude the existence of cell–cell interactions mediated by a ligand and its receptor (see above); in parallel it appears that classical models with a regulative mode of development also use transcriptional control. Furthermore, more primitive nematodes, which are generally marine nematodes, do not have a fixed lineage. Therefore, nematodes develop like all other animal species and their further study will be relevant to the understanding of human biology, particularly, as repeatedly emphasized throughout this article, to analyze cellular processes.

Further Reading

Bowerman B and Shelton CA (1999) Cell polarity in the early *Caenorhabditis elegans* embryo. *Current Opinion in Genetics and Development* 9: 390–395.

Brown NH (2000) Cell–cell adhesion via the ECM: integrin genetics in fly and worm. *Matrix Biology* 19: 191–201.

Costa M, Raich W, Agbunag C et al. (1998) A putative catenin–cadherin system mediates morphogenesis of the *Caenorhabditis elegans* embryo. *Journal of Cell Biology* 141: 297–308.

Gonczy P, Schnabel H, Kaletta T et al. (1999) Dissection of cell division processes in the one cell stage *Caenorhabditis elegans* embryo by mutational analysis. *Journal of Cell Biology* 144: 927–946.

Kemphues KJ (2000) PARsing embryonic polarity. *Cell* 101: 345–348.

Labouesse M and Mango SE (1999) Patterning the *C. elegans* embryo: moving beyond the cell lineage. *Trends in Genetics* 15: 307–313.

Rose LS and Kemphues KJ (1998) Early patterning of the *C. elegans* embryo. *Annual Review of Genetics* 32: 521–545.

Seydoux G and Strome S (1999) Launching the germline in *Caenorhabditis elegans*: regulation of gene expression in early germ cells. *Development* 126: 3275–3283.

Thorpe CJ, Schlesinger A and Bowerman B (2000) Wnt signalling in *Caenorhabditis elegans*: regulating repressors and polarizing the cytoskeleton. *Trends in Cell Biology* 10: 10–17.

Wissmann A, Ingles J and Mains PE (1999) The *Caenorhabditis elegans* mel-11 myosin phosphatase regulatory subunit affects tissue contraction in the somatic gonad and the embryonic epidermis and genetically interacts with the Rac signaling pathway. *Developmental Biology* 209: 111–127.

See also: Caenorhabditis elegans; Cell Lineage; Developmental Genetics; Developmental Genetics of Caenorhabditis elegans

Embryonic Development, Mouse

L Silver

Copyright © 2001 Academic Press
doi: 10.1006/rwgn.2001.0403

Early Embryonic Development Is Highly Plastic

For the purposes of scientific analysis, mammalian development is divided into two distinct stages of unequal length that are separated by the moment of implantation into the uterus. During the preimplantation phase, which lasts 4.5 days, the embryo is a free-floating object within the mother's body. Because it is naturally free-floating, the preimplantation embryo can be removed easily from its mother's body and cultured in a petri dish, where it can undergo genetic manipulation before it is placed back into a female where it can continue along the developmental path to a newborn animal. Once the embryo has undergone implantation, it can no longer be removed from its mother's body and remain viable. The accessibility of the preimplantation embryo provides the basis for a number of specialized genetic tools that are used to study mammalian development, including the production of transgenic animals and targeted mutagenesis.

The preimplantation phase starts with the zygote (the one-cell fertilized egg or embryo) at the time of conception. Development begins slowly with the first 22 h devoted to the expansion of the highly compacted sperm head into a paternal pronucleus that matches the size of the original egg (maternal) pronucleus. Once this process is completed, the embryo undergoes the first of four equal divisions, or cleavages, that increase the number of cells, over a period of 60 h, from one to 16 (see **Figure 1**).

Throughout this period, known as the cleavage stage, all of the cells in the developing embryo are equivalent and totipotent. The word totipotent is used to describe a cell that has not yet undergone differentiation, and still retains the ability, or potency, to produce every cell type present in the developing embryo and adult animal. The cleavage stage mammalian embryo is also called a morula.

As a consequence of totipotency, cleavage stage embryos can be broken into smaller groups of cells that each have the potential to develop into individual animals. The outcome of this process can be observed in humans with the birth of identical twins or, much more rarely, identical triplets. In the laboratory, scientists have obtained completely normal mice from

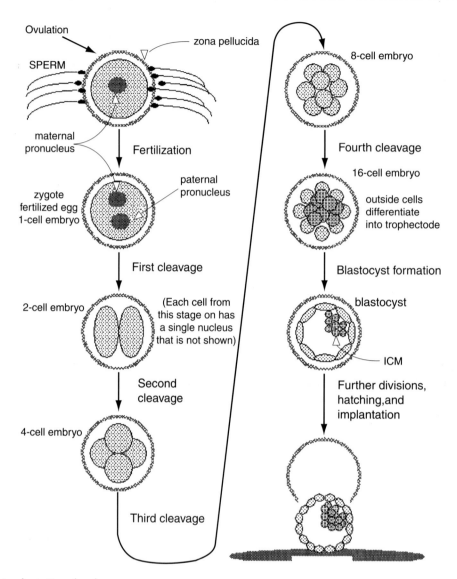

Figure 1   Preimplantation development.

individual cells that were dissected out of the four-cell-stage mouse embryo and placed back individually into the female reproductive tract. This experimental feat demonstrates the theoretical possibility of obtaining four identical clones from a single embryo of any mammalian species.

It is important to contrast the early developmental program of all placental mammals with that of other animals including the two model organisms *Caenorhabditis elegans* and *Drosophila melanogaster*. Identical twins can never be obtained from a single nematode or fly embryo. During nematode development, individual embryonic cells from the two-cell stage onward are highly restricted in their developmental potential or 'fate.' The fly egg is polarized even before it is fertilized and different cytoplasmic regions are devoted to supporting different developmental programs within the nuclei that end up in these

locations. Thus, half a nematode embryo or a half a fly embryo could never give rise to a whole animal.

Embryonic Differentiation and Postimplantation Development

During the 16-cell stage of mammalian embryogenesis, the first differentiative event occurs, and the developmental potency of individual cells finally becomes restricted. The cells on the outside of the embryo turn into a trophectoderm layer that will eventually take part in the formation of the placenta. Meanwhile, the cells on the inside compact into a small clump that remains attached to one spot along the inside of the trophectoderm sphere. This clump of cells is called, appropriately enough, the inner cell mass (ICM). The fetus will develop entirely from the ICM. At this stage of development, the embryo is called a blastocyst.

Two more rounds of cell division occur during the blastocyst stage before the embryo implants.

Throughout the process of normal preimplantation development, the embryo remains protected within the inert zona pellucida. Thus, there is no difference in size between the one-cell zygote and the 64-cell blastocyst. To accomplish implantation, the blastocyst must first 'hatch' from the zona pellucida, so that it can make direct membrane-to-membrane contact with the cells in the uterine wall. Implantation initiates the development of the placenta, which is a mixture of embryonic and initiates the development of the placenta, which is a mixture of embryonic and maternal tissue that mediates the flow of nutrients, in one direction, and waste products, in the other direction, between the mother and embryo. The placenta maintains this intimate connection between mother and fetus until the time of birth. The process of internal uterine development is a unique characteristic of all mammals other than the primitive egg-laying platypus.

With the development of the placenta, a period of rapid embryonic growth begins. Cells from the ICM differentiate into all three germ layers (endoderm, ectoderm, and mesoderm) during a stage known as gastrulation. The foundation of the spinal cord is put into place, and the development of the various tissues and organs of the adult animal is initiated. With the apperance of organs, the embryo is now called a fetus. The fetus continues to grow rapidly in size and in the mouse birth occurs at ~ 21 days after conception. Newborn mice remain dependent upon their mothers during a suckling period which can last another 18 to 25 days. By 5 to 6 weeks after birth, mice have reached adulthood and are ready to begin the reproductive cycle all over again.

See also: Developmental Genetics; Embryonic Stem Cells

Embryonic Stem Cells

L Silver

Present naturally in the very early mammalian embryo, embryonic stem (ES) cells are members of a special class of cells that have the potential to differentiate into every cell type present in the adult animal. In recent years, scientists have gained the ability to culture and grow ES cells (derived from embryos) *in vitro*, and also to convert somatic cells into embryonic stem cells. By definition, embryonic

stem cells are totipotent, which means they have the potential to differentiate into every cell type of an animal. Thus, the embryonic stem cell is operationally defined, rather than phenotypically defined. The only way in which an embryonic stem cell can be identified is in the generation of a complete animal from a single cell through the normal process of development.

See also: Cell Lineage

End Labeling

End labeling is a technique for adding a radioactively labeled group to one end (5′ or 3′) of a DNA strand.

See also: Autoradiography

Endoderm

See: Developmental Genetics

Endonucleases

The endonucleases are a group of enzymes that cleave nucleic acids at positions within the chain. Some act on both RNA and DNA (e.g., S1 nuclease, specific for single-stranded molecules). Ribonucleases (e.g., pancreatic, T1, etc.) are specific for RNA, and deoxyribonucleases for DNA. Bacterial restriction endonucleases are important in recombinant DNA technology for their ability to cleave double-stranded DNA at highly specific sites.

See also: Nuclease; Restriction Endonuclease

End-Product Inhibition

End-product inhibition is the process whereby a product of a metabolic pathway inhibits the activity of an enzyme that catalyzes an early step in the pathway.

See also: Enzymes

Enhancers

R Grosschedl

Copyright © 2001 Academic Press
doi: 10.1006/rwgn.2001.0410

Enhancers are operationally defined as *cis*-acting elements that augment the activity of a promoter in an orientation- and position-independent manner. Initially identified in a sea urchin histone gene and in the simian virus 40 (SV40) viral genome as regulatory elements that increase transcription from a promoter located at a distant position on the same DNA molecule transcriptional enhancers are found in most eukaryotic genes transcribed by RNA polymerase II. In lower eukaryotes such as yeast, upstream activator sequences (UAS) can also function at variable distances from the promoter and in either orientation. UAS are thus analogous to enhancers of higher eukaryotes, although they differ from enhancers in their inability to activate a promoter from downstream positions. In bacteria, a simple enhancer has been identified upstream of a promoter that is recognized by a specific form of RNA polymerase containing sigma factor 54.

Transcriptional enhancers of higher eukaryotes are typically composed of multiple modules that cooperate to augment gene expression. The modules consist either of binding sites for individual transcription factors or of composite binding sites for different transcription factors. The multiplicity and modularity of transcription factor binding sites in enhancers allow for combinatorial control and functional diversity. In addition, interactions between multiple enhancer-binding proteins can help to increase the accuracy of DNA sequence recognition in a large and complex genome.

The modularity of enhancers has been demonstrated by experiments in which individual modules of an enhancer have been multimerized to generate synthetic enhancers. Synthetic enhancers typically augment gene expression. However, they do not reproduce all regulatory properties of natural enhancers such as cell type specificity or inducibility. Insight into the modularity and functionality of natural enhancers is provided by experiments examining interactions between transcription factors bound at different modules. At some natural enhancers, multiple transcription factors have been shown to interact with each other, resulting in the assembly of a higher-order nucleoprotein complex, termed the 'enhanceosome.' The assembly of such complexes can be facilited by architectural proteins that have no activation potential by themselves but augment interactions between other enhancer-binding proteins and/or bend the DNA helix.

The mechanisms by which enhancers regulate transcription appear to be diverse. Transcriptional run-on experiments have shown that the SV40 enhancer increases the rate of transcription initiation from a linked promoter, suggesting that enhancer can regulate the recruitment and/or activity of RNA polymerase. This effect of enhancers appears to involve contacts between enhancer- and promoter-bound proteins in which the intervening DNA is looped out. Interactions between enhancer- and promoter-bound proteins by DNA looping have been visualized by electron microscopy. In support of a looping model, enhancers can activate promoters that are located on physically linked, but topologically uncoupled DNA molecules. In the simplest form of activation, the enhancer-binding protein of bacteria, Ntr-C, contacts the sigma-factor-54-containing RNA polymerase and induces the formation of an open complex in an ATP-dependent manner. In an analogous manner, activators bound at a eukaryotic enhancer may activate RNA polymerase via interactions with the polymerase-associated mediator complex or they may augment the recruitment of RNA polymerase by interactions with general transcription factors bound at the core promoter.

A second mechanism of enhancer function involves alterations of chromatin structure. Enhancers have been found to increase the accessibility of sequences in the context of chromatin. These chromatin alterations can be detected by an increased sensitivity of chromatin toward digestion with deoxyribonuclease I (DNase I) and by an increased accessibility of adjacent binding sites for transcription factors or restriction enzymes.

Although enhancers of higher eukaryotes are typically defined by their potential to activate promoters in tissue-culture transfection assays, enhancers alone are often inefficient in activating promoters in transgenic mice. Enhancers are also found as components of large and complex regulatory regions, known as locus control regions, which confer activation upon linked promoters independent of the chromosomal position in transgenic mice. In locus control regions, enhancers act in combination with other less-defined regulatory elements and regulate the activity of multiple genes that are located within a domain of a chromosome. These chromosomal domains represent structural entities that display increased DNase I sensitivity.

Enhancers can typically act on heterologous promoters. However, the interactions between enhancer and promoter can display specificity. In addition, the interactions between enhancers and promoters can be regulated by insulators, which act as boundaries of chromosomal domains. An insulator that is placed between an enhancer and a promoter blocks the interactions between these elements. Thereby, insulators help to impart promoter specificity in complex gene loci in which multiple promoters are located in the vicinity of an enhancer.

See *also*: Chromatin; *Cis*-Acting Proteins; Promoters

Enzymes

D C DeLuca and J Lyndal York

Copyright © 2001 Academic Press
doi: 10.1006/rwgn.2001.0414

Enzymes are catalysts that accelerate the rate of chemical reactions without permanent alteration to themselves. Virtually all enzymes are proteins or conjugated proteins, although some catalytically active RNAs have been identified. These catalytic/enzymatic activities are essential to the information and energy management requirements of a cell. Specific enzymatic activities are found within all cellular organelles. In comparison with classical catalysts of chemical reactions enzymes are characterized by: (1) higher reaction rates; (2) effectiveness under milder reaction conditions in terms of temperature, pressure, and pH; (3) greater reaction specificity in terms of the reactants, products, and the absence of undesirable side reactions; and (4) ability to be regulated either by reaction rate control, by catalyst concentration, or by specific small molecules.

A system of classification and nomenclature for enzymes has been established by the International Union of Biochemistry. This system places all enzymes into one of six major classes based on the type of reaction catalyzed. Each enzyme is uniquely identified by a four-digit classification number. This system is often usurped by trivial nomenclature that attempts to give some information concerning the reactants (substrates) involved and the type of reaction catalyzed. Such names usually end with the suffix 'ase.' For example, histidine decarboxylase removes CO_2 from histidine to form histamine.

The active site of an enzyme (sometimes referred to as the catalytic center) is that portion of the molecule that interacts with substrate and converts it into product. The initial step is the formation of an enzyme–substrate complex (ES). Two distinct models of how an enzyme binds its substrate have been proposed: the lock-and-key (complementary) model of Fischer and the induced fit (conformational change) model of Koshland. These models represent extreme cases. Different enzymes show features of both models. Amino acid side chain residues at the enzyme's active site interact chemically or physically with substrate to lower the energy required for the reaction to occur at physiological temperatures. Substrate specificity is determined by the chemical properties and spatial arrangement of the amino acid residues forming the active site of an enzyme. The restriction endonucleases illustrate enzyme substrate specificity. These enzymes are responsible for very specific cutting of DNA into unique fragments. Separation of these fragments provides a 'fingerprint' of an individual organism's DNA for unambiguous identification. Restriction enzymes play a critical role in the development of the field of biotechnology.

Some enzymes require the presence of small nonprotein units (cofactors), either inorganic ions, organic molecules, or both. The precursors for some organic molecules (coenzymes) are the vitamins. Coenzymes covalently attached to the enzyme are called prosthetic groups and cosubstrates if they undergo chemical modification during the reaction. An enzyme with its cofactor is called the holoenzyme; without the cofactor, the species is called an apoenzyme.

Isoenzymes (isozymes) are distinct forms of an enzyme that catalyze the same reaction but differ in physical or kinetic properties. Different isoenzymes are usually encoded by different genes and may occur in different tissues of an organism. For example, human creatine kinase exists as three isozymes that predominate in skeletal muscle, heart muscle, and brain tissue, respectively.

Molecules that act directly on an enzyme to reduce its catalytic activity are known as inhibitors. Many therapeutically useful drugs, pesticides, and herbicides are inhibitors of specific enzymes. Inhibitors are classified as either reversible or irreversible. The effect of reversible inhibitors may be overcome in various ways, whereas irreversible inhibitors lead to a state of permanent inactivity as they often form stable covalent bonds with reactive amino acid residues of the protein. Reversible inhibition is further subdivided into competitive and noncompetitive types. A competitive inhibitor is one whose effect is overcome by the addition of substrate. Noncompetitive inhibitors engage a site other than the catalytic site causing a conformational change altering catalytic activity. This state cannot be overcome by substrate addition.

Enzyme-catalyzed reactions are subject to a variety of exquisite control mechanisms. These are feedback inhibition of allosteric enzymes, covalent modification, proteolytic activation, and regulation of protein synthesis and breakdown. Feedback inhibition occurs in a metabolic pathway when an early enzyme in the pathway is inhibited by pathway end-product. Inhibition of the first step of a pathway conserves metabolic energy and prevents the unnecessary accumulation of metabolites. Since pathway end products may have little structural resemblance to the initial substrate of a pathway, the active site of the initial enzyme of a pathway may not bind the metabolic end product. Substances that bind at sites other than the substrate-binding site and cause a conformational change in the enzyme such that the activity is decreased are referred to as allosteric (other site) inhibitors. Enzymes that exhibit this behavior are called allosteric enzymes and, in some cases, can be activated by positive allosteric modifiers. Allosteric enzymes are often, but not always, multisubunit proteins.

Covalent modification of enzymatic activity can be either reversible or irreversible. Irreversible modification is illustrated by the partial proteolysis of the zymogen, chymotrypsinogen, to form the active digestive enzyme chymotrypsin. Reversible covalent modifications include phosphorylation, adenylylation, and disulfide reduction. Reaction sequences of this type serve as a rapid, reversible switch to turn a metabolic pathway on or off as required by the cell. This is illustrated by the interrelationship between kinases and phosphatases. Kinases phosphorylate enzymes and have a key role in regulation of metabolic pathways, cell cycle control, cellular proliferation, and in programmed cell death (apoptosis). Phosphatases counter the effects of protein kinases by removing phosphate, thereby serving as regulators of signaling by kinases. Both the kinases and phosphatases are known to be subject to hormonal regulatory control.

The ultimate control of enzyme activity is at the gene level. Since enzymes are proteins, the amount of an enzyme in a cell is regulated by factors that control gene expression. Such factors include hormones and some metabolic pathway end-products. For example, in the synthesis of the heme portion of hemoglobin excess heme represses at the gene level the synthesis of the first enzyme in the heme biosynthetic pathway. Enzymes subject to this type of control are usually very unstable and have a short lifetime in the cell.

The emergence of pharmacogenomics, identification of population subgroups that would benefit from a particular drug treatment, and toxicogenomics, identification of population subgroups that would exhibit adverse responses, illustrates the importance of understanding the interrelationship between genes and their product enzymes. For example, the genes that are differentially expressed in people sensitive to penicillin have been identified, cloned, and sequenced. About 150 genes have been identified as predictors of penicillin hypersensitivity. The categories of genes

induced include those associated with ribosomal, apoptosis-related, energy generation, and cell cycle regulatory enzymes but, surprisingly, not those enzymes associated with drug metabolism or detoxification. In addition to the practical value in terms of human health, multidisciplinary investigations of this type should provide a better understanding of the biological interrelationships within and between cells.

For substantial overviews of enzyme structure and function see Devlin (1997). More detailed information on enzyme structure and mechanism with an introduction to the current concepts of protein engineering can be found in Fersht (1999). For discussion of the chemical basis of enzymatic activity Jack Kyte's book (Kyte, 1995) is recommended.

References
Devlin TM (1997) *Biochemistry with Clinical Correlations*. New York: Wiley–Liss.
Fersht A (1999) *Enzyme Structure and Mechanism*, 3rd edn. New York: WH Freeman.
Kyte J (1995) *Mechanism in Protein Chemistry*. New York: Garland Publishing.

See also: **Cell Cycle; Gene Expression; Proteins and Protein Structure**

Epidermal Growth Factor (EGF)
R M T Katso and M D Waterfield

Copyright © 2001 Academic Press
doi: 10.1006/rwgn.2001.1564

Cellular activities are modulated in response to diverse extracellular stimuli from their surrounding environment. In multicellular organisms, growth factors represent a subset of external cues that program the cellular machinery to proliferate, differentiate, or die. Soluble growth factor peptide ligands bind to their cognate receptors and initiate a cascade of intracellular signals that culminate in an appropriate developmental response. Epidermal growth factor (EGF), the prototypic member of the EGF family of peptide growth factors, represents one such form of extracellular signals. The EGF family of peptide growth factors consists of 12 ligands or growth factors which can be broadly classified into five groups:

1. Growth factors that primarily interact with the EGF receptor erbB-1: EGF, transforming growth factor α (TGF-α); amphiregulin (AR); vaccinia

growth factor (VGF); shope fibroma growth factor (SFGF); myxoma virus growth factor (MGF).
2. The neuregulin or heregulin ligand families which primarily interact with erbB-3 and erbB-4: neregulin 1-α, β, 2-α (NRG-1α, NRG-1β, NRG2-α, NRG2-β).
3. Ligands that interact equally with erbB-1 and erbB-4: betacellulin (BTC); heparin-binding growth factor (HB-EGF).
4. Ligands that bind exclusively to erbB-4: neuregulin 3 and 4 (NRG3, NRG4).
5. Pan or broad specificity ligands that bind to erbB-1, erbB-3, or erbB-4: epiregulin (EPR).

EGF is synthesized as an inactive transmembrane precursor that is processed and released by proteolysis into the active soluble form that functions as a signal transducer. Six cysteine residues which define a three-loop secondary structure that is both required and sufficient for receptor binding and activation characterize the prototypic EGF ligand. Three poxvirus-encoded EGF-like factors (VGF, SFGF, and MGF) have been isolated. Vaccinia growth factor is synthesized as a transmembrane precursor glycoprotein after infection with the vaccinia virus. The tumorigenic viruses, myxoma virus and shope fibroma virus, encode MGF and SFGF as secreted peptides, respectively. Although, the viral encoded ligands have lower binding affinities than their mammalian counterparts, they exhibit equivalent mitogenicity.

The EGF family demonstrates distinct expression patterns; while EGF is found in most body fluids the other related family members are secreted as autocrine or paracrine factors and so generally act over short distances. EGF peptides exhibit distinct expression patterns that are either developmentally regulated or tissue specific. This is amply demonstrated by the highly regulated expression of HB-EGF in the uterine luminal epithelium 6–7 h prior to implantation of the egg into the uterus. In the adult organism, EGF peptides play essential roles in the proliferation and differentiation of the mammary gland (mammopoiesis) at puberty and mammary gland milk production (lactogenesis) during pregnancy. Targeted inactivation of the EGF ligands indicates that they have specific as well as overlapping roles in mammary gland development. For example absence of AR is associated with impaired mammary ductal morphogenesis, whilst inactivation of EGF and TGFα suggest that both factors are required for lactogenesis. The viral encoded EGF-like factors are not required for viral replication. Genetic inactivation studies suggest that the viral encoded EGF ligands are required for the enhancement of virulence and stimulation of cell proliferation at the primary site of infection; therefore they may

have a role in inflammatory responses. In general, the controlled expression of the EGF family of ligands appears to be one way of determining their signaling specificity. The significance of the regulated expression of the EGF family of ligands is underscored by the fact that aberrant expression of the EGF-related peptides underlies the pathogenesis of conditions such as cancer and inflammatory disease. Co-overexpression of the EGF-related peptides and their cognate receptors frequently occurs in human breast, pancreatic, endometrial, and ovarian carcinomas as well as in inflammatory conditions such as chronic pancreatitis. The deregulated expression of the growth factors results in an autocrine pathway that drives uncontrolled cell growth and maintains the neoplastic transformation.

The biological effects of the EGF ligand family are mediated by its cognate receptors, the erbB receptor tyrosine kinase family, which consists of four members: erbB-1 (commonly referred to as the EGF receptor); erbB-2 (also known as the neu or Her-2 receptor); erbB-3; and erbB-4. The multiple EGF ligands differentially induce certain receptor combinations probably because each ligand is bivalent, carrying not only a high-affinity site, but also a low- or broad-specificity site that determines the dimerization partner. The monomeric form of receptor tyrosine kinases is inactive, but upon growth factor binding, oligomerization primarily through homodimerization results in receptor auto- and transphosphorylation. The bivalent nature of the EGF peptides enables the simultaneous binding of two identical (homodimerization) or different (heterodimerization) erbB receptors. The dimerization or juxtapositioning of two erbB receptors results in the activation of the intrinsic tyrosine kinase activity and receptor auto- and transphosphorylation of specific tyrosine residues. The transphosphorylation event creates docking sites on the activated receptor, which initiate a diverse range of intercellular signaling events through the recruitment of signaling effectors. The recruitment is highly specific and is governed by tyrosine-phosphorylated modules in the juxta-membrane and carboxyl tail of the RTK containing primarily either Src-homology 2 (SH2) or phosphotyrosine-binding (PTB) motifs. As a result, several linear signaling cascades that culminate in regulation of gene expression are initiated. The EGF ligand family exhibits differential mitogenic potency and signaling potential. Both factors are inextricably linked to the composition of the homo- or heterodimeric receptor complex, which determines ligand dissociation rates, receptor recycling/degradation as well as the temporal duration of the signal. In addition, coupling of a given receptor to specific intracellular signaling proteins is modulated by the

EGF ligand dimerization partner and may indeed originate from differential receptor transphosphorylation. As a result the different cellular responses to the EGF family of peptide growth factors is due to the array of erbB receptors activated and the repertoire of signaling pathways that are engaged at the effector level.

See also: erbA and erbB in Human Cancer; Neu Oncogene; SH2 Domain; Signal Transduction

Epigenetics

F D Urnov and A P Wolffe[†]

Copyright © 2001 Academic Press
doi: 10.1006/rwgn.2001.0415

The textbook definition of an epigenetic phenomenon is "a mitotically and/or meiotically heritable change in gene function that cannot be explained by changes in DNA sequence" (reviews: Russo *et al.*, 1996; Chadwick and Cardew, 1998). To expand this succinct formula, epigenetics studies genetic censorship, i.e., instances of genome control where a particular locus is inactivated (or activated) in a very stable manner (through multiple mitotic divisions, sometimes for the entire life of the organism, and frequently through multiple generations, i.e., it cannot be changed even by meiosis!). Remarkably, the program to maintain the active or inactive state of a given locus is not contained in the primary DNA sequence of that locus, hence the etymology of the name 'epigenetics,' i.e., heritable phenomena that appear to occur on top of, or above, the sequence of the DNA.

Above and Beyond DNA

It is useful to distinguish epigenetic regulation of genome function from other instances of stable gene expression programs. For example, in metazoa, certain genes are permanently and exclusively activated in unique cell types (in eutherian mammals, globins are only expressed in erythroblasts, insulin, in cells of Langerhans islets of the pancreas, serum albumin, in hepatocytes, etc.) and are silenced in all other cell types. While stable, this regulation is not commonly referred to as epigenetic, because it is known to be due to action by stretches of regulatory DNA (promoters and enhancers) contained within those loci (in concert, of course, with a host of attending DNA-binding proteins). By contrast:

- A complete chromosome is transcriptionally inactivated (with the important exception of a single gene) in each cell of mammalian females, but this inactivation is not encoded for in the primary sequence of the unfortunate censored piece of DNA, but rather is determined by how many such chromosomes there are in the nucleus.
- In the genomes of eutherian mammals, many genes are 'imprinted,' i.e., only expressed from one copy (we are functionally hemizygous for a number of loci in our genome): which allele is censored into silence is entirely determined by whether it was inherited by its current genome of residence from the mother or the father; thus, it does not matter what the allele 'says,' but only where it came from.
- In certain fungi, the mating type ('gender') of a particular cell is not initially decided on by the cell's genotype, but rather is determined by events in its grandmother cell.
- Both in plants and in animals, genomes protect themselves from parasites such as transposable elements by maintaining them in stably silenced form. Quite contrary to expectation, however, this silencing is not determined by some unique primary sequence feature of the transposon or endogenous retrovirus, but rather the copy number of that articular sequence in the genome; thus, stretches of DNA sequence reiterated in a given nucleus more than a certain 'allowed' number of times are censored into silence not because they say something offensive (which they do, in an evolutionary sense, although the cell does not have a mechanism for sensing that), but because they occur more than a certain number of times (which the cell somehow does sense).

These, and many other, examples of epigenetic regulation of gene expression have been an understandable source of bewilderment and wonder for many years: it was very clear that conventional models of gene regulation were inadequate to explain, for instance, how fission yeast switch mating type, or how repeated DNA is silenced, or how gametes imprint particular loci, but virtually nothing was known about the underlying molecular mechanisms. The past 5 years have reversed this predicament quite emphatically, and epigenetics is no longer the proverbial 'black box' carrying the familiar "...and then a miracle occurs" logo as a euphemism for "we have not the slightest idea how this might work."

There follows a brief survey of the history of scholarship in epigenetics and then we consider, from a general molecular standpoint, what challenges a cell faces in creating a stable domain of gene expression. Detail of recent molecular evidence regarding the origin and functional impact of DNA methylation in the regulation of vertebrate genomes is described and aspects of

---

[†] deceased

chromosome structure that collude with DNA methylation in effecting epigenetic control are discussed. Possible mechanisms for the tagging of loci in taxa are also reviewed, e.g., arthropods, that do not appear to have DNA methylation in their genomes. Then several representative examples of epigenetic regulation, are presented, focusing in each case on the presumed evolutionary benefit reaped by the cell and the organism from effecting such a mode of gene control, and on recent molecular data that offer mechanistic explanations (reviews: Russo *et al.,* 1996; Chadwick and Cardew, 1998). In conclusion, there is a short perspective on the general applicability of epigenetic principles to genome control in eukarya.

Brief History of Research in Epigenetics

As is the case for virtually every branch of genetics, most major epigenetic phenomena were initially characterized in plants and insects. In the early 1950s, Barbara McClintock (Cold Spring Harbor Laboratory) discovered that the suppressor-mutator (*Spm*) transposable element in maize can be inactivated and kept silent for generations, until this silence is suddenly reversed, again in heritable fashion (McClintock, 1958). Soon afterwards, R. Alexander Brink (University of Wisconsin, Madison) reported that the penetrance of particular gene alleles controlling kernel color in maize is sometimes dependent on the genotype of the parent plant from which they were inherited, and not the genetic constitution of the plant currently carrying them (this remarkable phenomenon was dubbed 'paramutation'; Brink, 1958). In the 1930s, Charles Metz (Columbia University) discovered that female flies of the genus *Sciara* have a most unusual mechanism of sex inheritance: a given female has only daughters or only sons. A genetic explanation for this peculiarity was provided in 1960 by Helen Crouse (Columbia University) – once a student of Barbara McClintock – who realized that, in *Sciara*, chromosomes of paternal origin are somehow heritably 'marked' for elimination in future generation; she dubbed this 'chromosome imprinting' (Crouse, 1960).

The general characteristic that emerged from these seemingly unrelated observations was that, in utter contradiction to common-sense notions of a relationship between genotype and phenotype for a given organism, epigenetically regulated traits (be it seed color in corn or gender in flies) are sometimes defined by the genotypic environment that was experienced by particular alleles controlling that trait prior to being inherited by that organism. Since the primary DNA sequence of those alleles remained unchanged (McClintock, Brink, Metz, and Crouse did not know

that at the time of their pioneering studies, but we now do), something other than the sequence must have "tagged along" with the DNA to regulate its expression. The nature and mechanism of action of that something is the focus of this article.

The study of epigenetics on a single-cell level was launched by Mary Lyon's (Medical Research Council, UK) studies (in 1961) on coat color in the mouse: She insightfully combined earlier cytological observations by Susumu Ohno on the compaction of one X chromosome into a dense 'Barr body' with her own genetic analysis to propose that a stable and random inactivation of one of the X chromosomes must occur in females (Lyon, 1961). Thus, it became clear that, even in the lifespan of a given organism, significant portions of the genome can be entirely eliminated from expression programs, and that such elimination is not based on primary DNA sequence (if it were, then only a specific X chromosome would be inactivated, whereas the inactivation is random).

In 1975, after a sufficient number of phenomena of this sort had been reported in the literature to warrant attempts at a mechanistic explanation, A. Riggs (City of Hope NMC, USA) and, independently, Robin Holliday and J. Pugh (NIME, UK) proposed a role for DNA methylation in controlling vertebrate genomes (Holliday and Pugh, 1975; Riggs, 1975). By then, it had been established that the chemical modification by methyl groups of bases in double-stranded DNA of prokaryotic genomes plays an important role in the familiar restriction-modification pathways for host genome stability. While based on little to no experimental data, these investigators' proposals have withstood empirical testing remarkably well. As discussed in some length in the section "The little methyl that can", DNA methylation is particularly well suited to carrying the censor's epigenetic mark. Its prominence and ubiquity in mammalian genomes was revealed by Adrian Bird and Edward Southern, who in 1978 developed an ingenious method for its detection.

A short time later, Azim Surani (Wellcome Institute) and Davor Solter (Max Planck Institute) made an experimental observation with far-reaching consequences: by pronuclear transplantation, they showed that the artificial union of two haploid male genomes or of two haploid female genomes cannot sustain normal embryonic development. Thus, they reasoned, the contributions made by the two haploid chromosome sets to the final karyotype must be unequal, at least for some loci that are required during embryogenesis. A dedicated effort from a large number of researchers has now led to the firm realization that specific loci in mammalian genomes are stably silenced in a gender-specific manner (i.e., female gametes always repress subset X and male gametes, subset Y of the genome).

Because this repression is irreversible until the next passage through the germline, a gynogenetic or androgenetic embryo will be functionally null for subset X or Y, respectively; this results in lethality, because at least one active copy of each gene in both subsets is required for development. Thus, it became clear that the phenomenon of chromosome imprinting discovered by Helen Crouse in *Sciara* has a close evolutionary analog in mammals, except the parent-of-origin imprint applies not to entire chromosomes, but to smaller chromosomal domains, or even individual genes.

For many scholars of epigenetics and molecular biology, it was very intellectually gratifying when data from many laboratories obtained over the past 10 years provided a firm functional link between the methylation status of particular alleles and imprinting. Some of the most convincing of these observations came from the work done in Timothy Bestor's laboratory at Columbia University: a mouse genetically engineered to lack the major enzyme responsible for maintaining the genome in a methylated state (DNA methyltransferase-1) died owing to an extraordinary misregulation of epigenetic pathways, most notably X chromosome inactivation and the maintenance of transcriptional silencing at imprinted loci. Many of the functional pathways involved in the latter two phenomena were meticulously dissected in work by Rudolf Jaenisch's research group at MIT, and Shirley Tilghman's laboratory at Princeton University. An additional piece in the puzzle was filled in by the discovery in Adrian Bird's laboratory, at the University of Edinburgh, that mammalian genomes contain several proteins that appear to very selectively bind to methylated (epigenetically regulated) DNA loci (Bird and Wolffe, 1999). Subsequent work from the Bird laboratory, and from Alan Wolffe's research group at the NIH, has shown that some of these proteins are potent repressors of transcription (Bird and Wolffe, 1999). In an exciting development, this repression was revealed to be mechanistically based on the localized creation of an area of highly specialized, inaccessible chromosome structure; thus, a hypothetical mechanism whereby such a structure could propagate itself through multiple rounds of cell division became immediately apparent. Considering the progress made over the past few years, it is nevertheless remarkable how much in epigenetics remains obscure, unexplained, and occasionally beyond the pale of rational explanation; this promises many more decades of exciting data.

How to Keep Your State of Expression When All about You Are Losing Theirs

Gene and genome regulation is at its core a dynamic phenomenon: All of its key players are bound to each other, not through covalent or strong electrostatic bonds into permanent crystalline arrays, but rather through weaker-charge, hydrophobic, and van der Waals interactions. This is not a whim of nature, but a response to evolutionary pressure. The eukaryotic genome evolved to be rapidly responsive to a great variety of internal and external stimuli, so macromolecular complexes that control it do not associate with each other permanently, but rather engage in much more fluid interactions. While familiar textbook images of gene control in bacteria present static pictures – e.g., the *lac* repressor firmly bound to the operator in the *lac* operon – the reality is that many protein–DNA interactions that occur in the nucleus have relatively high off rates (i.e., complexes fall apart relatively easily, and then quickly reform, and then fall apart again).

Genomes in all taxa, however, have a firm need to impose on particular regions of themselves a relatively permanent state of activity; for example, so-called 'housekeeping genes' (i.e., genes whose products are indispensable for cell viability, such as enzymes involved in anabolic and catabolic pathways, proteins that are structural components of the cytoskeleton) have to be active at all times, as do tissue-specific genes (in the cognate tissue). Conversely, some regions of the genome – for example, invading genomic parasites such as transposable elements – must be kept in perpetual silence, because their spurious activation will lead to an intranuclear epidemic and the destruction of the genome.

A solution very commonly used by cells is to not rely on single proteins to regulate the expression of a particular gene, but many proteins; for example, tissue-specific genes are well known to be activated through the concerted binding and action of at least a dozen distinct factors bound to several 100 bp of DNA both in promoters (i.e., stretches of DNA next to the transcription start site) and enhancers (DNA more distant). Work from the laboratory of Tom Maniatis at Harvard has shown that, in the case of the human interferon-β gene, all these regulatory complexes coalesce into a "united we stand" type of structure called an 'enhanceosome.'

A serious problem arises, however, when the cell needs to divide and therefore replicates its DNA; DNA polymerase and its entourage move through the chromosome with all the subtlety of a military tank, erasing all nucleoprotein organization in their path. The carefully assembled regulatory complexes that sat over particular loci are therefore destroyed and must be recreated *de novo*, and in two copies, instead of just one that existed before replication. It is hardly surprising, therefore, that many tissue-specific genes are transiently deactivated when mature, differentiated cells are induced to divide (for example, proliferating hepatocytes stop expressing many liver-specific markers). In general, proliferating cells in multicellular organisms only very rarely express genes associated with the differentiated state, such luxury is only allowed to cells that are replicatively quiescent and can assemble regulatory complexes on DNA with no fear of being swept out of the way by a passing megadalton assembly of DNA polymerase. Thus, the first major challenge facing the cell in enabling stable domains of gene expression is a need to maintain them through repeated rounds of genomic replication.

There is an additional problem, however: even in cells that are in a state of proliferative arrest, a protein complex bound to DNA is no guarantee of stability. In large part this is because the eukaryotic nucleus contains significant quantities of "philandering" transcriptional regulators; not tethered to any particular DNA segment, they can spuriously affect a random gene through a "hit-and-run" mechanism. One possible kind of insurance against such accidents is the establishment of a particular kind of "fortified" regulatory structure at a given locus that would be impervious to such sporadic attacks. The second major challenge in stably controlling the genome is to minimize levels of regulatory noise due to spurious interactions.

As discussed in the following two sections, epigenetically regulated genes and loci offer a wonderful example of how parsimonious natural selection can be in molding regulatory pathways. The solution to both challenges is implemented via a remarkably elegant integration of simple biochemical mechanisms.

The Little Methyl that Can

Nature's answer to the first challenge – maintenance of structure in the face of replication – turns out to be a remarkably ancient one; the biochemical system used is found in all taxa studied and is commonly used to maintain genomic stability in the face of invaders or DNA replication.

As mentioned in the section "A brief history of research in epigenetics," the mid-1970s saw the concurrent emergence of two independent proposals that DNA methylation may be an important regulatory mechanism. Before elaborating on the remarkable series of discoveries about its functional role in the genome, it would be helpful to show a quick snapshot of the main agent in question: 5-methylcytosine (m^5C; **Figure 1**). Of the many different kinds of DNA methylation that occur in nature, the one most conspicuous in the genome of higher vertebrates is that on carbon atom 5 in cytosine (**Figure 1**) – we will focus on this, bearing in mind that other bases in the DNA are modified by methylation as well.

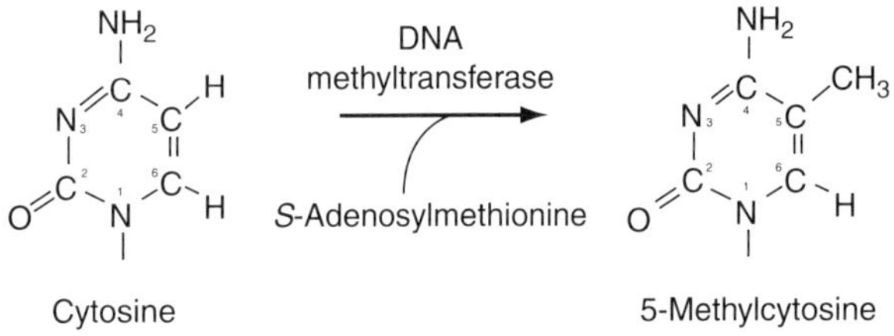

Figure 1 Cytosine and 5-methylcytosine.

Teleological reasoning is dangerous in describing biological phenomena, but it is nevertheless remarkable how well suited m^5C is to the task of being at the center of epigenetic regulation. This stems from several circumstances: (1) the C–C bond is not chemically labile, and thus provides desired stability to the regulatory pathway that exploits it; (2) carbon-5 of cytosine does not engage in Watson–Crick hydrogen bonding with the guanine on the other stand of DNA, thus this modification does not impede the formation of conventional DNA structure; and (3) that being said, the hydrogen atom normally attached to carbon-5 does project into the major groove of DNA, and, as such, is part of a recognition surface for the multitude of DNA-binding proteins that read the DNA sequence by scanning the electron orbital profiles of the bases in the major groove. It is hardly surprising, therefore, that replacing a hydrogen atom with a methyl group yields a highly distinctive statement in the molecular Braille of DNA, one that can be, and is, recognized as being markedly different from unmethylated cytosine.

To appreciate the final bit of m^5C biology relevant to epigenetics, we must mention a peculiarity to methylated cytosine in our genomes that has profound regulatory consequences: Its overwhelming majority occurs not on random cytosines, but only in the context of the dinucleotide 5'-CpG-3'. By itself, this would not be particularly significant, unless one appreciates the fact that CpG is quite the genomic oddity in mammals, for two reasons: (1) because there are $4^2 = 16$ possible dinucleotides, one would expect each one to represent $\sim 6.25\%$ of the human genome, and this simple math holds for all dinucleotides, except CpG, which is remarkably *rare* in mammalian genomes compared with all the others; and (2) within the sequence of the genome, most dinucleotides occur relatively randomly, except for CpG, which occurs in clusters (commonly called 'CpG islands'). Thus, as one browses through the narrative of mammalian DNA, CpG will be the rarest three-letter word, and, when it does occur, it will do so many times within the course of a single short passage (e.g., 100 times over a given 1000 bp), only to disappear again (i.e., occur once every 200 bp).

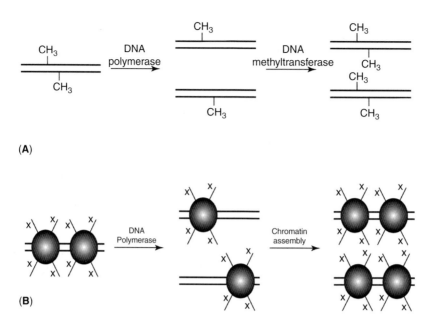

(A)

(B)

Figure 2 (A) The passage of the replication fork and reestablishment of methylation; (B) the replication of chromatin.

This is very significant for controlling epigenetic phenomena, because, in the double helix of DNA, the stretch around 5′-CpG-3′ is symmetric (i.e., the other strand reads 3′-GpC-5′); in fact, when human DNA is examined, all of the methylation is also symmetric:

$$5′ - \ldots m^5C - P - G \ldots - 3′$$
$$3′ - \ldots \quad G - P - m^5C \ldots - 5′$$

An immediate consequence of such symmetry is that methylation can be endlessly propagated on the DNA during replication; all that is needed for such maintenance is an enzymatic activity that will recognize the product of replicating methylated DNA – 'hemimethylated DNA' (i.e., DNA where one strand is methylated and the other is not) – and methylate the currently unmodified strand (**Figure 2**). As the reader will know, such systems exist in bacteria to facilitate replication-coupled DNA repair (when a postreplicative mismatch is encountered, the strand that is methylated is assumed to contain the correct sequence by default). Cells of higher vertebrates contain an enzyme called DNA methyltransferase-1 (DNMT1); its enzymatic specialty is to restore methylation to the other strand (i.e., perform the reaction shown in **Figure 2**). In wonderful testament to the evolutionary unity of life, bacterial, plant, and mammalian DNA methyltransferases are closely related to each other in primary sequence. One of the really interesting features of this system in our cells is how fast it is: DNMT1 is known to be targeted to 'replication foci' (i.e., portions of the nucleus where the DNA replication machinery is located and acts), and

remethylation occurs within 1 min of replication. Thus, the censors act quickly to suppress any unwanted information from being revealed! It is useful to recall at this point that DNMT1 is required for normal mouse development; thus, the organism takes the censors' job seriously.

The molecular weight of the modifying methyl groups is a combined miniscule 26 Da. Thus, in the language of DNA, methylation is little more than a diacritical mark such as the umlaut (e.g., ä or ü) in German, a tiny modification of the text; and yet its impact on molecular complexes that outsize it by 4–5 orders of magnitude (e.g., the RNA polymerase II holoenzyme) is quite powerful. This, perhaps, is not surprising, since in human languages these tiny accents, when placed over particular words, can change the meaning of entire passages (for example, in German, *schon*, meaning 'already' can become *schön*, meaning 'pretty'). In an analogous way, the behavior of the same DNA stretch in methylated and unmethylated form is dramatically different, because it means quite distinct things. How such alteration of meaning is thought to be effected is described in the next section.

Chromatin and Methylation: Large Effects from Small Causes

A general rule that helps understand the role of methylation in epigenetic control is as follows: methylated loci in our genomes are repressed (how epigenetic control works in organisms whose genomes lack

DNA methylation, e.g., arthropods such as the fruit fly *Drosophila melanogaster* and nematodes such as the round worm *Caenorhabditis elegans*, is a fascinating issue). Thus, if a particular gene has its promoter methylated, it becomes transcriptionally inert ('invisible' to RNA polymerase). DNA methylation is one of the most potent mechanisms for transcriptional repression known in biology today. Most significantly, we currently do not know of any way to reactivate a chromosomal locus that has been silenced by methylation except to remove the methyl residues.

This is not the academic issue that it might seem; advanced forms of human cancer are well known to have aberrant DNA methylation; for example, genes required for cell-cycle arrest, such as cyclin-dependent kinase inhibitors, are erroneously silenced in tumor cell lines because their promoters are methylated (which they never are in normal, noncancerous cells). In addition, as elaborated in the section "Repetition, mother of genetic silencing" up to one-third of our genomes consists of parasites (self-propagating DNA elements such as transposons). They are kept in check (i.e., silenced) by hypermethylation, and woe to the genome that unleashes the parasites within it.

How can DNA methylation – by the tiny 26 Da of methyl groups involved – be so powerful in antagonizing the transcriptional machinery? The answer to this question also takes care of the other challenge to enabling stable domains of gene expression: how to prevent noise. It turns out that methylation of a particular DNA stretch is read by the cell as a command to envelop it in a protective cocoon, a specialized protein structure that makes the DNA physically inaccessible to regulators. This shielding occurs in two steps: first, when a DNA stretch is methylated, it is immediately bound by specialized proteins, discovered by Adrian Bird's research group, called methylated DNA binding domains (MBDs) – these have the interesting and useful property of being highly selective for methylated DNA (for example, the best-studied MBD, a protein called MeCP2, will bind m^5CpG, but not CpG). Once bound, these MBDs attract other proteins, all of which can remodel chromatin, i.e., alter the structure of the chromosome around their binding site.

To appreciate how chromatin remodeling can lead to transcriptional repression, we must recall that there is no naked DNA in our cells, all of it is complexed with highly positively charged proteins called histones. Every 146 bp of DNA in the genome is wound around eight molecules of histones to form a nucleosome – the elementary building block of our chromosomes; the familiar "bead-on-a-string" array of nucleosomes winds around itself to form chromosomes. As one might expect, the cell uses chromatin to regulate its

genome: in a general sense, tighter assembly of DNA into mature chromatin leads to transcriptional repression. This "tightness" is, to some extent, regulated by changing the charge of the histones: dedicated enzymatic complexes called histone deacetylases can promote DNA binding to the histones by increasing the amount of positive charge found in the histone tails (stretches of the histone proteins that stick like tentacles outside of the spool of the nucleosome). The positively charged tails then envelop the phosphodiester backbone of DNA in a web of protein and make it inaccessible to other molecules. As discovered in the laboratories of Adrian Bird and Alan Wolffe, proteins that bind methylated DNA exist in various large complexes that include, among other things, histone deacetylases, and ATP-dependent molecular machines that make chromatin more compact.

Thus, a stretch of methylated DNA is bound by dedicated proteins (MBDs) that, in turn, target specialized enzymatic complexes (histone deacetylases), which build a wall of repressive chromatin between the DNA and the rest of the nucleus. Therefore, much like the pea from H.C. Andersen's famous fairytale *The Princess and the Pea*, the tiny methyl makes its presence known through layer upon layer of proteins that assemble over the DNA under its command.

Most significantly for our purpose, however: (1) methylation does not change the primary sequence of the DNA, and (2) it can be propagated endlessly through many rounds of DNA replication. How that manifests itself in the various epigenetic phenomena known to occur in nature is described in the following section, but at this point, it is useful to consider that, while DNA methylation itself is 'replication-resistant,' a particular chromatin structure associated with a hypermethylated locus will also probably segregate to the two nascent DNA strands in the aftermath of DNA replication fork passage (**Figure 2**), and thus also enforce a particular state of expression onto the daughter chromosomes.

Epigenetic Regulation in Action: The Rest Is Silence

From a few short reports in the 1950s and 1960s, epigenetics has blossomed into a very large field of study, populous enough to warrant *c.* 40 separate review chapters in a recent compendium (Russo *et al.,* 1996)! We will briefly survey the wondrous breadth of epigenetic phenomena.

Epigenetic Silencing of Entire Chromosomes
In the most extreme case of epigenetic regulation, an entire chromosome is eliminated from expression. The most common instance for this is in 'dosage

compensation': a ubiquitous mechanism whereby the organism ensures that males and females have an equal number of active alleles for all loci on their sex chromosomes. By way of a simple example, men express one set of X chromosome genes and are genotypically hemizygous for them; women also express a single set, but are genotypically diploid. This very interesting predicament comes about by the well-studied process of X chromosome inactivation: very early in development, female embryos randomly and permanently inactivate one of their X chromosomes. The silenced X becomes condensed (and forms the famous Barr body), replicates much later in S-phase of the cell cycle than its active homolog (a well-known feature of transcriptionally inert DNA), has its CpG islands hypermethylated, and its histones deacetylated (in comforting support of a role for both processes in epigenetic regulation). The key point to stress here is that the primary DNA sequence of the inactive X can be indistinguishable from that of the active X; thus, regulation in this system is enabled on a level "above the DNA."

The inactive X is transcriptionally silent, with a few very important exceptions, the most significant ones being genes found in the a small portion of the X chromosome, the *XIC* (the X-inactivation center). Of the genes in the *XIC*, the most interesting one is *Xist* (X-inactivated-specific transcript, pronounced "exist"). Its product is a 17-kb RNA that does not contain open reading frames and is presumed to function by physically coating the chromosome from which it is transcribed, and thereby inactivating all of it, except the gene for itself, which remains active. A great number of tantalizing hypotheses and datasets have been offered to explain the many questions surrounding this remarkable phenomenon: How is the chromosome to be inactivated chosen from the two that are active early in development? Why does only one X chromosome express *Xist*? How does a coat of RNA lead to hypermethylation and chromatin condensation? How does the *Xist* gene on the inactive X escape from being inactivated by its own product?

The organism whose study originated the term 'imprinting,' the fly *Sciara*, goes even further than mammals and inactivates an entire chromosomal set (Gerbi, 1986): In this organism, the female determines the sex of her progeny, and this is enabled by a truly remarkable process of chromosome elimination in the male: during spermatogenesis, in meiosis I, the entire paternal set of chromosomes condenses and is physically eliminated, leaving the spermatocyte with only its mother's chromosomes! Thus, the paternal chromosome set, when inherited by the male, carries an epigenetic imprint for the entirety of that male's lifespan, until such a mark instructs the gonads to eliminate

them. How this occurs is largely unknown (insects are not known to have CpG methylation in their genomes), but it is interesting that in other insects such as *Drosophila* dosage-compensated, epigenetically regulated sex chromosomes have defined alterations in chromatin structure and histone tail acetylation.

Fetal Growth as a *Casus Belli*: Imprinting in Mammals

As mentioned in "A brief history of research in epigenetics," the maternal and paternal genome make unequal contributions to the genomic output of their joint product, the progeny: for a certain number of genes, one copy in our genome is inactivated for the duration of our lifespan. Appropriately borrowing the term from Helen Crouse's study of *Sciara*, loci regulated in this way are referred to as 'imprinted.' It would be helpful to explain the terminology used in this field: a gene is called 'maternally expressed' if organisms always use (i.e., transcribe) the allele they inherited from their mother; conversely, a 'paternally expressed' gene is one in which the allele inherited from the father is the one that is active. Imprinting, while of immense academic and general intellectual interest, has medical relevance: several human disorders, including Prader–Willi and Angelman syndromes, are caused by a misregulation of imprinted loci.

Of the many questions that spring to mind regarding imprinting, we will briefly address three: (1) On a mechanistic level, how is the difference in expression between the two alleles effected? (2) How do male and female gonads ensure correct imprinting in future generations and distinguish sets of paternally and maternally expressed genes when producing gametic precursors? (3) What selection pressure could have lead to the evolution of such a peculiar mode of gene regulation?

1. The difference in expression is maintained by keeping the imprinted loci in a state of differential methylation, such that the expressed allele is demethylated, and the repressed allele is hypermethylated. We emphasize that the primary DNA sequence of the two alleles can be identical, and yet profound differences in expression levels are observed. A genetic ablation of pathways leading to DNA methylation abrogates correct regulation on many imprinted loci in the mouse genome. One very interesting recent development has been the discovery, by the research groups of Shirley Tilghman at Princeton and Gary Felsenfeld at the NIH, that for the H19/Igf2 imprinted locus, the effect of such differential methylation is to control the ability of a protein called CTCF to associate with a regulatory element

found in this area (CTCF binding is prevented by methylation). Interestingly, the role of CTCF binding is to speed the spread of regulatory information along the chromosome, i.e., CTCF enables boundary, or insulator function in this locus. In other cases, the effect of methylation is to drive the creation of a repressive chromatin structure over the genes being regulated.

2. Few things in epigenetics are the cause of greater wonder and mystery than the establishment of methylation patterns relevant to imprinting during gametogenesis. By way of example, consider a paternally expressed gene in a human female: of the two alleles she has in her genome, the allele she inherited from her father (let us designate it as ♂) is demethylated and active; the allele she inherited from her mother (♀) is methylated and inactive. During oogenesis, the following happens ('m' stands for 'methylated'):

Paternally expressed gene in an ovary: ♀^{m}♂ → ♀^{m}♂^{m}

Thus, in her germ cells, the maternal allele is kept methylated, and the paternal allele is methylated *de novo* (by definition, her children must inherit this allele from her in inactive, methylated form). Remarkably, a maternally expressed locus in that same woman undergoes the exact opposite process:

Maternally expressed gene in an ovary: ♀♂^{m} → ♀♂

(i.e., the cell takes the allele this woman inherited from her father, and demethylates it; again, by defination, this woman's children have to receive this gene from her in active, demethylated form).

3. During spermatogenesis in the male, all maternally expressed genes are methylated and all paternally expressed genes are demethylated. This process is called 'resetting of gametic marks,' and we have little beyond very weak conjecture on how it is enabled. How can the cell possibly scan the vast narrative of its entire genome for all loci that are differentially methylated on two homologous nonsister chromatids? Once these loci have all been found, whatever quasi-miraculous machine lies behind this search, what can be the mechanism whereby the cell determines whether this particular allele pair is maternally or paternally expressed (this cannot be based on the difference in methylation between the two alleles, of course), and decides whether to demethylate or hypermethylate both alleles? As if this was not mysterious enough, how can such programs be perfectly reversed in a gender-specific way (i.e., the same locus is treated in a diametrically opposite way in males versus females)?

However this works, it clearly does, and quite well; but how does organism benefit? A very attractive hypothesis proposed by David Haig and colleagues is that the actual embryo in which imprinting is manifested does not benefit from it at all, and that, instead, genomic imprinting is the manifestation of a genomic arms race, a tug-of-war between its two parents. This 'conflict theory' is based on the known unequal contribution parents make to their child in mammals: A father's investment is frequently minimal, but he does have an interest in having a healthy baby, a mother carries the fetus for the duration of gestation, but cannot afford to devote all her resources to the development of this particular infant, because she needs to reproduce again. Thus, fathers have an agenda: the embryo must grow as large as possible and obtain as many resources from the mother as it can, all for the cause of propagating the father's genes. The mother's agenda is more balanced: it cannot allow a given fetus to squander away too many of her resources, because she has other pregnancies in the future to consider. How can these agendas be implemented in molecular terms? *Ergo* imprinting: according to the theory, paternally expressed genes promote embryo growth, while maternally expressed genes stymie it. In the embryo, a genome-wide tug-of-war thus occurs: paternally expressed genes attempt to make the embryo grow larger, while maternally expressed genes try to do the exact opposite.

This elegant theory received a great deal of attention and experimental testing, and has largely withstood these empirical trials. One of its most interesting predictions is that imprinting should have disappeared in monogamous mammals (i.e., animals that mate for life), because parents have an equal interest in their progeny (the father cares about the mother's welfare, because she is the sole carrier of his children). Shirley Tilghman's laboratory tracked down a species of monogamous mouse (which was hard to find, because true monogamy is very, very rare among mammals), *Peromyscus polionotus*, and discovered that imprinting has been preserved. There are sufficient data from other studies, however, to keep the conflict theory as the best explanation we currently have for the utility of imprinting.

Repetition, Mother of Epigenetic Silencing

The examples of epigenetic regulation discussed up to now all involved pathways for the control of the organism's own genes. In addition, a wide variety of examples from fungi, plants, and animals all point to a major role for epigenetic silencing in preserving the stability of the genome and protecting it from being swamped by genomic parasites.

For humans and other mammals, this is not an academic issue at all (our genomes are only 5% exons of active genes, and a lofty 35% intact or mutated genomic invaders such as transposons). A simple illustration of the gravity of this matter was provided in a study of genomic stability in interspecific hybrids by J. Marshall-Graves and colleagues: the offspring of two different species of wallaby had its genome practically destroyed by an explosion of endogenous retroelements (these were succesfully kept in check in the parent species, but the hybrid failed to recognize transposons of heterologous origin as a threat).

The overwhelming bulk of endogenous genomic parasites are silenced by DNA methylation; one very interesting side effect of such repression is the known tendency of m^5CpG to spontaneously deaminate and yield TpG. Such point mutations, while relieving methylation-driven repression, irreversibly inactivate open reading frames within the retroelement required for its propagation! A major question in the study of this 'genome defense' pathway is the mechanism whereby the genome recognizes repetitive DNA within itself and targets the DNA methylation machinery to it. It has been suggested that such recognition occurs during gametogenesis, when repetitive DNA will tend to associate during homologous chromatid pairing in meiosis.

It is important to appreciate that the silencing of repetitive DNA is a phenomenon that occurs in all eukaryotic taxa; for example, it is well known that, of the many rDNA repeats found in the genome of the budding yeast *Saccharomyces cerevisiae*, only a few are transcriptionally active. Similar processes occur in such filamentous fungi as *Ascobolus* and *Neurospora* (Wolffe and Matzke, 1999), where repetitive DNA is actively sought out and epigenetically inactivated in processes termed 'repeat-induced point mutation' (RIP) and 'methylation induced premeiotically' (MIP). In some cases, the probable utility to the cell of such silencing is not in the abrogation of transcription *per se* – after all, rRNA, for example, is essential for viability – but in the suppression of DNA recombination capacity (repetitive DNA is a dangerous site for interchromatid recombination, because it leads to genomic instability). In other cases, the genome's capacity to seek out and inactivate repetitive DNA is clearly a defense mechanism (although somewhat of an inefficient one, since mammals, armed to the genetic teeth with methylation, have 10 times as many genomic parasites as invertebrates, which do not methylate their genomes). From a clinical standpoint, the irreversible inactivation by hypermethylation of transgenes introduced into organisms during gene therapy is, however, a poignant illustration of the power of such defense.

Epigenetic Inheritance as a Violation of Mendelian Principles

Whatever the mechanism whereby epigenetic activation or repression is enabled, one of its most salient features is that traits controlled epigenetically frequently exhibit non-Mendelian inheritance patterns.

A classic example is the paramutation phenomenon that affects maize kernel color: the R locus controls pigment formation, with the R^r allele producing dark kernels and the R^{st} allele, stippled kernels. This would be yet another case of the exceptional utility of color inheritance in providing textbook illustrations of Mendelian inheritance, if it were not for the following: a testcross of an R^r/R^r plant yields all dark kernels and of an R^{st}/R^{st}, all stippled, in full agreement with expectation. In overwhelming contradiction to common sense, a testcross of an R^r/R^{st} plant yields all stippled kernels, even though 50% of them are genotypically R^r. We now know that the r allele is somehow epigenetically modified (weakened) by the act of its passage through a heterozygotic environment that contains an st allele, but the mechanistic details are not well understood (Russo *et al.*, 1996).

As mentioned earlier, studies of epigenetic phenomena were initiated by Barbara McClintocks's experiments on the *Spm* transposon in maize (which, incidentally, also affects pigment formation in the kernel), in which she showed that activity of this transposon can fluctuate through generations, and that it can become epigenetically inactivated and reactivated (not surprisingly, the trait affected by the transposon insertion fails to comply with Mendelian segregation rules). Subsequent work from Nina Fedoroff's laboratory at Penn State University has shown this regulation to be due to alterations in the methylation status of a stretch of the *Spm* promoter that contains a very high percentage of G/C residues – a hypermethylated transposon is heritably inactivated until passage through a nucleus containing an active, demethylated copy of the *Spm* transposon leads to demethylation and activation (Russo *et al.*, 1996).

A final, wonderful example of the extraordinary power of epigenetic regulation in effecting non-Mendelian inheritance comes from fission yeast *Schizosaccharomyces pombe*. Work from Amar Klar and his colleagues has unraveled the very elegant mechanism whereby this organism switches mating type: after meiosis, haploid spores reacquire the capacity to mate by assuming one of two mating types ('plus' and 'minus'; **Figure 3A**). The mating type is defined though a DNA recombination event in the spore; only spores of opposite type can mate. Remarkably, when a single spore of the plus mating type divides twice, of its four granddaughters, three remain plus and one switches to minus. We now know this

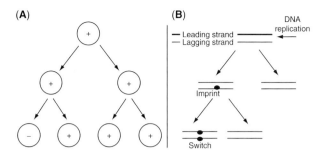

Figure 3 The epigenetic regulation of mating-type switching in fission yeast. (A) After meiosis, haploid spores assume a 'plus' or 'minus' mating type; (B) after invasion of the mating-type locus by a DNA replication fork, strand-specific epigenetic modification occurs in the grandmother cell.

to be enabled by a strand-specific epigenetic modification that occurs in the grandmother. As shown in **Figure 3B**, the invasion from a specific direction by a DNA replication fork of the mating-type locus creates an inherent asymmetry whereby the 'top strand' is replicated by the leading strand mechanism and the 'bottom' strand, by the lagging strand mechanism (i.e., via Okazaki fragments). As a consequence, a strand-specific epigenetic modification is introduced into the bottom strand; this modification is passed on to one of the two daughter cells and, when that daughter replicates its own DNA, one of its two progeny will inherits an epigenetically modified DNA double helix, which leads to the initiation of recombination and mating-type switching.

Epigenetic Regulation: Old Curiosity Shop?

Is epigenetic regulation merely an intellectually amusing curiosity, or does it illuminate principles in gene control of general relevance? While several very specialized systems certainly use epigenetic regulation, its broad applicability is also clear. We present three brief examples.

Work from Kim Nasmyth's laboratory at the Institute of Molecular Pathology in Vienna examined the regulation of the budding yeast *HO* endonuclease gene. Using high-resolution analysis, these investigators made the remarkable observation that the effects of certain transcriptional regulators on the activity of this gene persist long after the regulators themselves have left the DNA. The likeliest explanation for this epigenetic memory is that the regulator effects a stable modification of chromatin structure over the gene promoter, and that the structure itself is stable enough to confer regulation of the gene.

Studies from the laboratories of Renato Paro, Vincenzo Pirrotta, and others have investigated the

regulation of homeotic genes in *Drosophila*. These are required for proper body-plan development during embryogenesis, and are expressed in stable fashion in specific segments of the embryo. Biochemical and genetic analysis showed that a class of proteins termed 'Polycomb' form large-scale repressive, self-propagating complexes that epigenetically silence homeotic genes, and that proteins of the trithorax group act in similar fashion, but with opposite functional effects, i.e., genes become stably activated.

Finally, it is useful to recall that the impact of epigenetic regulation on the function of the human genome clearly extends beyond imprinted loci and the inactivated X chromosome. For example, recent genetic data have shown that humans with mutations of the methylated DNA-binding protein MeCP2 develop a progressive and debilitating developmental and neurological disorder called Rett syndrome. Thus, epigenetic regulatory pathways control many more aspects of our genome's behavior than we currently appreciate.

References
Bird AP and Wolffe AP (1999) Methylation-induced repression: belts, braces, and chromatin. *Cell* 99: 451–454.
Brink RA (1958) Basic of a genetic change which invariably occurs in certain maize heterozygotes. *Science* 127: 1182–1183.
Chadwick DJ and Cardew G (eds) (1998) *Epigenetics*. Chichester, UK: John Wiley.
Crouse HV (1960) The controlling element in sex chromosome behaviour in *Sciara*. *Genetics* 45: 1429–1443.
Gerbi SA (1986) Unusual chromosome movements in sciarid flies. In: Hennig W (ed.) *Results and Problems in Cell Differentiation*. Berlin: Springer-Verlag.
Holliday R and Pugh JE (1975) DNA modification mechanisms and gene activity during development. *Science* 187: 226–322.
Lyon MF (1961) Gene action in the X-chromosome of the mouse. *Nature* 190: 372–373.
McClintock B (1958) The suppressor–mutator system of control of gene action in maize. *Carnegie Institution of Washington Year Book* 60: 469–476.
Riggs AD (1975) X inactivation, differentiation, and DNA methylation. *Cytogenetics and Cell Genetics* 14: 9–25.
Russo VEA, Martienssen RA, Riggs AD (eds) (1996) *Epigenetic Mechanisms of Gene Regulation*. Plainview, NY: Cold Spring Harbor Laboratory Press.
Wolffe AP and Matzke MA (1999) Epigenetics: regulation through repression. *Science* 286: 481–486.

See also: Chromatin; CpG Islands; Dosage Compensation; Gene Expression; Gene Silencing; Housekeeping Gene; Imprinting, Genomic; X-Chromosome Inactivation; XIST

Episome

K B Low

doi: 10.1006/rwgn.2001.0416

The term episome was introduced (Jacob and Wollman, 1958) to describe an accessory genetic element (e.g., in *Escherichia coli* bacteria) similar to a plasmid, but which had the additional ability to become integrated into the chromosome semistably and, furthermore, could at some point dissociate from the chromosome and again replicate independently, or could even be totally eliminated (cured) from the cell. The definition was also intended to include lysogenic bacteriophages that could either integrate into the chromosome and persist as prophages or replicate extrachromosomally and produce new bursts of phage particles on cell lysis. A plasmid in contrast, was defined (Lederberg, 1952) as an accessory, extra-chromosomal, independently replicated element. Most plasmids or bacteriophages do not provide any essential function for the survival of the cell except in special circumstances, such as a plasmid that carries a gene conferring antibiotic resistance when the cell is growing in the presence of the antibiotic. The episome concept was initially very useful in pointing out that two types of element, namely the temperate bacterio-phage lambda and the F sex factor for conjugal fertility, could both spend part of their life history integrated into the chromosome, even though they were first discovered as chromosome-independent entities, and chromosomal integration was not previously known to occur for plasmids or bacteriophages. The term episome was sometimes also used for analogous mammalian systems such as the simian virus SV40.

In recent years, the term episome has evoked less meaning and has been used less. This is because an increased spectrum of recombination events between plasmids and chromosomes has been observed, spanning a vast range of frequency and host cell dependence, and depending more or less (or not at all) on special recombination functions to facilitate the integration and/or excision events. Thus, it is difficult, if not impossible, to distinguish an episome from a plasmid based on some arbitrarily defined frequency of integration/excision, etc. Depending on the particular element, either the term plasmid or lysogenic bacteriophage generally suffices to cover the range of naturally occurring elements of this type.

References

Jacob F and Wollman EL (1958) Les épisomes, éléments géné-tiques ajoutés. *Comptes Rendues de l'Académie des Sciences* 247: 154–156.

Lederberg J (1952) Cell genetics and hereditary symbosis. *Physiological Review* 32: 403–430.

See also: **Bacteriophages; Plasmids**

Epistasis

G A Churchill

doi: 10.1006/rwgn.2001.0417

In its original usage (Bateson, 1909), epistasis referred to the masking or unmasking of the effects of allelic substitution at one locus by the allelic state at a second locus. In modern usage, epistasis refers to any relationship of nonadditive interaction between two or more genes in their combined effects on a phenotype. Epistasis is only defined in the context of genetic variation at multiple loci. This variation may be natural or experimental.

Epistasis is an important concept in biochemical genetics, population genetics, and quantitative genetics. Although its definition varies somewhat across these fields, the underlying concept is that the effects of allelic substitution at one gene can be dependent on the allelic state of another gene or genes. In biochemical genetics, analysis of epistatic relationships can be used to assign genes to pathways and to define the order of gene action within a pathway. In population genetics, epistasis plays a role in theories of fitness and adaptation. In quantitative genetics, epistasis has taken on a broader meaning that encompasses any nonadditive interaction among genes and it is often identified with the interaction term in analysis of variance. Defining the scale of measurement is important when considering epistasis as some statistical interactions can be removed by a change of scale, e.g., multiplicative effects can be converted to additive effects by taking logarithms. Epistatic interactions can be synergistic (greater than additive) or antagonistic (less than additive). When two genes interact statistically it is implied that they must also interact physically, either through direct (protein–protein) interaction or indirectly through a network of inter-acting gene products. Thus statistical epistasis can provide insights into the genetic architecture under-lying complex phenotypes.

Biochemical Epistasis

Gene products often act together in pathways and networks. Examples include biosynthetic pathways, signal transduction pathways, and transcriptional regulation networks. Epistasis analysis of mutant alleles provides a means to assign genes to pathways and to determine their order of action. Suppressor and enhancer screens are often used to identify epistatic mutations. A suppressor is a mutation at a second site that causes a reversion to the wild-type phenotype and thus masks the mutation at the first site. An enhancer is a second mutation that has a novel phenotype, thus unmasking an effect that cannot be observed in either of the single mutants. Synthetic lethality is a special case in which both of the single mutants are viable but the double mutant is not. This can occur when the two loci belong to parallel pathways driving an essential function as illustrated in scheme (1):

$$
\begin{array}{c}
\mathbf{A} \\
\mathbf{S} \underset{\searrow}{\overset{\nearrow}{}} \underset{\nearrow}{\overset{\searrow}{}} \mathbf{P} \\
\mathbf{B}
\end{array}
\qquad (1)
$$

Here **A** and **B** represent two enzymes, either of which can convert substrate **S** into product **P**. If a loss-of-function mutation occurs at either **A** or **B**, the other gene can still provide the function. In a positive regulatory signaling pathway (Scheme 2), loss of function at any step will result in the inability of the system to respond (**R**) to a signal **S**. The single mutants and the double mutant all have the same phenotype:

$$\mathbf{S} \rightarrow \mathbf{A} \rightarrow \mathbf{B} \rightarrow \mathbf{R} \qquad (2)$$

In a negative regulatory pathway (scheme 3), a loss of function at **B** will lead to a noninducible response **R** and a loss of function at **A** will lead to a constitutive response. The double mutant will behave like the single mutation in **B**:

$$\mathbf{S} \rightarrow \mathbf{A} \dashv \mathbf{B} \rightarrow \mathbf{R} \qquad (3)$$

Although epistasis analysis explicitly involves only two loci, it can be applied repeatedly to combinations of loci to elucidate the structure of larger pathways and networks.

Statistical Epistasis

Epistasis in Biometrical Genetics

Cockerham, 1954 introduced the idea of partitioning the genetic variance from inbred line crosses into additive, dominance, and epistatic components. The contrasts defining this partitioning are shown in **Table 1**. Variance components have proven to be useful in predicting the response of a population to selective pressure and have been successfully applied in breeding programs. The epistatic variance components reflect an average effect over many genes on the phenotype distribution in a population. Physiological epistasis can contribute to both additive and dominance components of variance, but physiological epistasis must be present in order to generate statistical epistasis.

Epistasis in Quantitative Trait Analysis

The availability of polymorphic DNA marker loci distributed throughout the genomes of many organisms enables us to track the inheritance of specific loci in line crosses and in pedigrees. Methods for analyzing quantitative trait inheritance using marker data are often based on single gene models that do not allow for epistasis. In the presence of large epistatic effects, these methods may fail to detect important loci or may produce misleading results by a statistical phenomenon known as Simpson's paradox. Epistasis can be detected using the F test for interaction in a two-way analysis of variance. However, this test can require large sample sizes to achieve reasonable power. In addition, corrections required to avoid potential false results when searching through all locus pairs further restrict the power. These difficulties may explain the paucity of reports of epistasis in the quantitative traits literature.

This measured genotype approach can be used to make specific predictions about the phenotype of individuals based on their genotype, whereas biometrical analysis can only make statements about population averages.

Table 1 Analysis of variance contrasts for variance components in diallele cross

Component	AABB	AABb	AAbb	AaBB	AaBb	Aabb	aaBB	aaBb	aabb
Additive	2	2	2	0	0	0	−2	−2	−2
Additive	2	0	−2	2	0	−2	2	0	−2
Dominance	1	1	1	−2	−2	−2	1	1	1
Dominance	1	−2	1	1	−2	1	1	−2	1
A×A	1	0	−1	0	0	0	−1	0	1
A×D	1	−2	1	0	0	0	−1	2	−1
D×A	1	0	−1	−2	0	2	1	0	−1
D×D	1	−2	1	−2	4	−2	1	−2	1

Table 2 Epistasis in mouse model of cleft palate

clf1	clf2 BB	Bb	bb
AA	–	–	–
Aa	–	–	–
aa	–	+	++

Epistasis in Complex Disease Traits

Many common diseases show familial aggregation but do not follow simple Mendelian patterns of inheritance. Evidence for epistasis has been reported for traits of medical importance including cancer, hypertension, kidney disease, epilepsy, and alcoholism. In addition, there are many genetic modifiers of disease phenotypes that alter the severity of a trait depending on the genetic background in which they occur. Background effects are an example of epistasis.

Epidemiological studies of a common birth defect, cleft lip and palate, in human populations have suggested that a single major gene with incomplete penetrance may be responsible for this condition. In a mouse model (Juriloff, 1995), the condition appears to be determined by an epistatic interaction between two loci, *clf1* and *clf2*, as shown in **Table 2**. When the *clf1* genotype is aa, the *clf2* heterozygote shows a mild form of the condition and the *clf2* bb homozygotes show a more severe form. It is conjectured that these two genes have partially overlapping functions and that the recessive alleles are loss-of-function. In this example, a model of epistatic interaction provides a testable prediction about the molecular mechanism. Epistasis can be particularly difficult to unravel in outbred human populations, but is often more amenable to analysis in the context of a model organism. Construction of special inbred lines (congenic or nearly isogenic lines) is of further use in the analysis of epistasis by reducing the complexity of the genetic background.

Epistasis in Population Genetics

The quantitative theory of population genetics, as introduced by Fisher, 1918, is based on models of additive genetics in which epistatic effects are represented as a 'noise' term. However, epistasis is known to play a key role in a number of evolutionary processes. Epistasis in traits related to fitness of an individual can lead to the existence of multiple fitness peaks and multiple stable equilibria for gene frequencies in a population. This idea is central to Wright's (1930, 1980) shifting balance theory. Wright proposed that population subdivision can lead to the evolution

of coadapted gene complexes. Incompatibilities among sets of genes can lead to genetic isolation and speciation. It is interesting to note that epistasis in quantitative traits has been observed more often in crosses between widely diverged strains than in crosses between closely related strains.

Epistasis can be beneficial. If epistasis is present in a population that has been reduced to a very small size, inbreeding leads to an increase in additive variance. Thus hidden variation is exposed to selection and rapid adaptation can occur following a bottleneck. In asexually reproducing populations, the gradual accumulation of deleterious mutations (an effect known as Muller's ratchet) can be slowed significantly by the presence of epistasis. Finally, the theoretical advantage of sexual reproduction requires that deleterious mutations should occur frequently and that their effects should be synergistic.

Future of Epistasis

What sorts of phenotypes will tend to show epistatic effects? Transcriptional regulation of gene expression is complex, involving both positive and negative regulation of multiple factors with varying degrees of specificity. Gibson, 1996 demonstrated that inherent properties of such systems lead to epistatic and pleiotropic effects. Traits that are closely related to direct regulation of one or a few genes are more likely to reveal epistasis than are morphological traits that depend on the cumulative effects of many genes for their expression. The availability of molecular markers and technology for monitoring gene expression opens up new possibilities for unraveling the network of biochemical mechanisms underlying the relationship between phenotype and genotype. As our ability to study the effects of genes at the biochemical level improves, so will our understanding of the mechanisms underlying epistasis. Modern molecular techniques are helpful in reuniting the biochemical and statistical descriptions of this ubiquitous phenomenon.

References

Bateson W (1909) The progress of genetics since the rediscovery of Mendel's papers. *Progressus rei botanicae* 1: 368–418.
Cockerham CC (1954) An extension of the concept of partitioning hereditary variance for analysis of covariances among relatives when epistasis is present. *Genetics* 39: 859–882.
Fisher RA (1918) The correlation between relatives on the supposition of Mendelian inheritance. *Transactions of the Royal Society of Edinburgh* 2: 399–433.
Gibson G (1996) Epistasis and pleiotropy as natural properties of transcriptional regulation. *Theoretical Population Biology* 49: 58–89.

Juriloff DM (1995) Genetic analysis of the construction of the AEJ. A congenic strain indicates that nonsyndromic CL (P) in the mouse is caused by two loci with epistatic interaction. *Journal of Craniofacial Genetic Development Biology* 15: 1–12.

Wright S (1931) Evolution in Mendelian populations. *Genetics* 10: 97–159.

Wright S (1980) Genic and organismic selection. *Evolution* 34: 825–843.

See also: **Adaptive Landscapes; Genetic Load; QTL (Quantitative Trait Locus)**

Epstein–Barr Virus (EBV)

G Klein

Copyright © 2001 Academic Press
doi: 10.1006/rwgn.2001.1565

Epstein–Barr virus (EBV) is a human lymphotropic herpesvirus that is carried in a latent, essentially nonpathogenic state by 80–90% of all humans. It belongs to the gamma herpesvirus subfamily and is regarded as the prototype lymphocryptovirus. Such viruses have only been found in Old World primates. Humans are the exclusive natural host for EBV. Each of the other Old World primate species is infected with closely related lymphocryptoviruses and is resistant to infection with human EBV. In immunologically naive New World primate species, EBV can cause fatal lymphoproliferative disease.

In humans, the virus is mainly transmitted by the saliva. In low socioeconomic groups early childhood infection is the rule, followed by seroconversion but no identified disease. Under good hygienic conditions, where the primary infection is often postponed to the teens or to adulthood, the first encounter with the virus leads to mononucleosis, a self-limiting lymphoproliferative disease, in about half of the cases. The other half undergoes silent seroconversion. In immunodeficient patients mononucleosis may follow a progressive course. EBV-carrying immunocytomas occur in iatrogenically (e.g., transplant recipients), congenitally (e.g., X-linked lymphoproliferative syndrome), or infection (e.g., HIV) based immunosuppressive states, with fatal outcome. They can be cured by adoptive immunotherapy with appropriate reactive and histocompatible T-cells.

Similarly to other herpesviruses, EBV has a toroid-shaped protein core, wrapped with DNA, a nucleocapsid with 162 capsomeres, a protein tegument between the nucleocapsid and the envelope, and an outer envelope with external glycoprotein spikes. The major EBV capsid proteins are 160, 47, and 28 kDa in size, packaged with a number of minor virion proteins. The most abundant EBV envelope and tegument proteins are 350/220 and 152 kDa in size, respectively. The EBV genome is carried by the virion as a linear, double-stranded 172-kbp DNA.

The interactions of EBV with the human host are seemingly paradoxical. It is the most highly transforming known virus. It turns resting B lymphocytes regularly into immunoblasts that can give rise to immortal lymphoblastoid cell lines (LCLs). The EBV-transformed immunoblasts closely mimic IL-4 and anti-CD40 activated blasts morphologically and with regard to their repertoire of activation markers.

In spite of its high transforming ability, EBV induces or contributes to malignant disease only exceptionally. These exceptions can be seen as biological accidents at the level of the host (like immunosuppression as already mentioned) or of the cell (oncogene activation). The second, related paradox concerns the relationship between the virally infected lymphocyte and the host organism. EBV-transformed immunoblasts are highly immunogenic. Mononucleosis can be seen as a somewhat chaotic but nevertheless efficient rejection reaction. *In vitro* exposure of T cells to autologous EBV-transformed immunoblasts generates CD8+ killer T cells that lyse their specific targets with an equally high efficiency as allogeneic T cells can kill MHC class I incompatible targets. In spite of the highly efficient elimination of the proliferating immunoblasts, the virus regularly succeeds in establishing its permanent latency in the B cell compartment itself, without causing either proliferation or rejection of its carrier cell. Both paradoxes have been resolved by the analysis of the viral strategy.

Viral Expression Phenotypes

Like other herpesviruses, EBV can enter latent (non-lytic) or lytic interactions with its host cell. The lytic cycle is only different in detail, but not in principle, from other herpesviruses. The nonlytic, growth transforming interactions are specific for EBV.

The course of the primary infection has been mainly studied in normal B lymphocytes. The virus uses a B-cell-specific membrane component, CD21, also known as CR2, or as the B-cell-specific complement (C3d) receptor, as its receptor. Following its attachment to CD 21, the viral envelope fuses with the host cell membrane and its DNA is internalized. The linear viral genome circularizes 12–16 h after entry and amplifies to 40–50 episomal copies. The infected B cell is activated like after mitogen exposure and turns into an immunoblast. Viral transcription starts at the Wp promoter, at the time of circularization. A giant message is generated out of which monocistronic messages for six nuclear proteins, EBNA 1–6 (alternative names: EBNA1, EBNA2, EBNA3A, B, C, and EBNA LP), are spliced. EBNA2 and EBNA5 (alternative name: EBNA-LP) are expressed first, reaching their peak level in 24–32 h. EBNA2 transactivates a gamut of cellular genes, including immunoblast-associated activation markers and the virally encoded membrane proteins LMP1, LMP2A, and LMP2B. Meanwhile, the transcriptional start of the EBNAs switches to the Cp promoter, as a rule. All six EBNAs remain expressed in the immortalized lymphoblastoid cell lines (LCLs) that emerge. All nine growth transformation associated genes (6 EBNAs and 3 LMPs) are expressed by 32 hours.

Function of the Growth Transformation Associated Proteins

Six of nine proteins expressed in lymphoblastoid cell lines, EBNA1, EBNA2, EBNA3 (alternative name: EBNA3A), EBNA5 (alternative name: EBNA-LP), EBNA6 (alternative name: EBNA3C), and LMP1, are essential for immortalization. Their function is only incompletely known.

EBNA1 is a sequence-specific DNA-binding protein that interacts with the latent replication origin (oriP) of the virus. This binding is essential for the maintenance of the EBV genomes as circular episomes and for their replication in synchrony with cellular DNA synthesis. EBNA2 is a transcription factor. It is essential for the initiation of immunoblastic transformation and for the maintenance of the immortalized state. It activates the Cp promoter that generates the polycistronic message for the six EBNAs. It also activates the viral LMP1/LMP2 promoter and numerous cellular genes. It is noteworthy that the EBNA2 responsive LMP1/LMP2 promoter element works, like several EBNA2-induced cellular genes, only in B lymphocytes. EBNA2 interacts with the transcriptional regulator, RBPJk, also called CBF1, a DNA-binding cellular protein that activates, in turn, CD23, other immunoblast markers, and B cell survival factors.

The EBNA 3 family (EBNA3A, B and C, alternative names: EBNA 3, 4 and 6) encode similar motifs, including binding sites for RBPJk, a leucine zipper, acidic domains, proline- and glutamine-rich repeats, and several arginine or lysine residues, responsible for nuclear translocation. The full significance of these interactions remains to be elucidated, but it is noteworthy that RBPJk belongs to a conserved group of proteins linked to the Notch signaling pathway. Ligand-elicited signaling by Notch can influence differentiation and proliferative responses. All three members of the EBNA 3 group are related. Only EBNA3 (EBNA 3A) and EBNA6 (EBNA3C) but not EBNA4 (EBNA3B) are essential for transformation.

EBNA6 (EBNA3C) is a transcriptional activator. It upregulates cellular genes like CD21 and viral genes like LMP1. Insertion of an amber stop codon after aa 365 results in recombinants incapable of B-cell immortalization.

All three members of the EBNA 3 family are preferred targets for cytotoxic T cell responses. It may be therefore inferred that all three, including EBNA4, would have been eliminated, were they not essential for the viral strategy.

EBNA5 (EBNA-LP) is one of the earliest viral proteins expressed after primary B cell infection. It is required for the induction of cyclin D2, in cooperation with EBNA2. The length of its repetitive part (W repeat) varies between different EBV isolates. This can be exploited for tracing the origin of viral substrains. EBNA5 colocalizes with the hsp 70, PML, and retinoblastoma (Rb) proteins in virally transformed immunoblast nuclei.

The major cell-membrane-associated protein, LMP1, can transform immortal rodent fibroblasts in vitro and is therefore regarded as a viral oncogene. It forms patches and caps on the villous surfaces of lymphoblastoid cells. It has a short, cytoplasmic, N-terminal hydrophilic part and six transmembrane loops, followed by the C-terminal cytoplasmic part of the protein. The number of transmembrane loops is not critical. Important functions are associated with the C-terminal part which has to be anchored to the membrane by the hydrophobic segment. The structure of LMP1 is similar to some ion-channel proteins. LMP1 induces many of the changes associated with EBV transformation of B lymphocytes, such as cell clumping, and the parallel increase of villous projections, vimentin expression, cell surface expression of CD23 and other activation markers, MHC class II proteins, IL-10, and the cell adhesion molecules LFA1, ICAM1, and LFA3. It also upregulates several adhesion molecules on B cells, a calcium-dependent protein kinase, bcl-2, and NFkB.

The cytoplasmic domain of LMP1 interacts with cellular proteins that mediate cytoplasmic signaling from the TNFR family. LMP1 aggregates interact with TNFR (tumor necrosis factor receptor)–TRAF (TNFR associated factor) aggregates to form large complexes. Through this mechanism, LMP1 can cause constitutive cell growth, inhibit apoptosis, and activate NFkB.

The transmembrane domains and the carboxy terminus are essential for primary B lymphocyte growth transformation. The first 44 amino acids of the transmembrane domain interact with a protein, LAP1, homologous to the TNFR-associated factors (TRAFs).

LMP1 associates with LAP1 in B lymphoblastoid lines and with an EBV-induced cell protein, EB16, which is the human homolog of the murine TRAF1, implicated in cell growth and NFkB activation.

LMP2A and B (TP1 and 2)

The first exons of these two membrane proteins are unique while all other exons are shared. They encode 12 hydrophobic integral membrane sequences and a 27aa hydrophilic domain. Both proteins colocalize in the plasma membrane with LMP1. LMP2A associates with tyrosine kinases of the src family and can modulate transmembrane signal transduction. LMP2A and B are not required for immortalization. Importantly, LMP2A blocks the switch from latent to lytic infection in B lymphocytes and is therefore believed to contribute to the maintenance of latency.

The EBERs, two EBV-encoded small RNAs, are expressed in virtually all EBV-carrying cells. They are the most abundant EBV products in latent infection and are therefore the preferred targets for the immunohistochemical detection of EBV-carrying cells by *in situ* hybridization. They are localized in the cell nucleus where they form a complex with the cellular La protein. The EBERs are not essential for lymphocyte transformation and their function is unknown.

Program Switches and Viral Strategy

Three major forms of latency have been identified in EBV-carrying growth transformed and/or neoplastic cells. Phenotypically representative type I Burkitt lymphoma (BL) cells express a monocistronic EBNA1 message, initiated from the Qp promoter. In addition to EBNA1, they express the EBERs, but none of the other growth transformation associated viral products (except occasionally LMP2). This expression pattern is referred to as latency I and is also found in latently infected normal B cells of healthy seropositive persons. Latency II is similar to latency I, in that EBNA1 is expressed from the Qp promoter and EBNA 2–6 are not expressed. LMP1 and 2 are constitutively expressed, however. It is found in nasopharyngeal carcinoma (NPC) and in most other EBV-carrying non-B cells. In latency III all six EBNAs and all three LMPs are expressed. This program is only used in immunoblasts, such as freshly transformed B cells, established LCLs, BL lines that have drifted to a more immunoblastic (type III) phenotype, proliferating B cells in mononucleosis, and in the immunoblastomas that arise in immunodefective persons.

The choice between these three main programs (and minor variants that will not be discussed here) thus depends on the host cell phenotype. The Wp/Cp-initiated giant message from which all six EBNA mRNAs are spliced is thus only used in cells with an immunoblastic phenotype. The LMP1 promoter is repressed in B cells, but this repression can be overridden by EBNA2. In latency I, Wp/Cp are inactive, Qp is active. EBNA2 is not made and LMP1 is, therefore, repressed. Non-B cells permit constitutive LMP1 expression in the absence of EBNA2, as a rule.

The scenario of the primary B cell infection starts with the massive induction of immunoblast proliferation. The majority of the virus-carrying blasts are rejected after a couple of weeks (see below). A small fraction of the EBV-carrying immunoblasts are believed to switch to long-lived memory cells with a resting B cell phenotype. Concurrently, they switch their EBV expression pattern to the more restricted type I program. The virus thus hides from immune rejection in memory B cells. There is no evidence for other sites of latent viral persistence. Ablation of the bone marrow eradicates the resident virus in bone marrow transplant recipients. This is consistent with the exclusive hemopoetic localization of the resident virus.

The lytic cycle can be induced in many but not all EBV-carrying cell lines by phorbol esters, butyrate, hydroxyurea and, in some B cell lines, by anti-Ig antibodies. The lytic cycle is initiated by the activation of the BZLF1 (also called Z, or Zebra) gene, a viral transactivator of multiple early genes. *In vivo*, infectious virus matures in the keratinizing cells of the pharyngeal epithelium. Oral hairy leukoplakia, frequently observed in AIDS and other immunosuppressed patients, is a macroscopically visible focus of productive EBV infection. It is curable by antiherpes drugs, e.g., acyclovir.

Several EBV genes expressed during the lytic cycle are closely homologous to cellular genes. The immediate early lytic switch gene, BZLF1, is closely related to the jun/fos family of transcriptional activators. The early gene BHRF1 resembles anti apoptotic bcl-2 gene structurally and functionally. The late gene BCRF1 is nearly identical to human IL-10.

Immune Responses

EBV-transformed immunoblasts are highly immunogenic for autologous T cells. Several immune effectors and CD8+ T cells react to them with an equally intense proliferation and cytotoxic response as to allogeneic MHC class I incompatible cells. In the autologous T anti EBV-B mixed lymphocyte culture, one or two of the growth transformation associated EBV-encoded proteins (with the exception of EBNA1) are chosen as the main targets, depending on the MHC class I allotypes of the responder that serve as the preferential restriction specificities. Other effectors, such as NK cells, LAK-type cells, and macrophages are also mobilized and a variety of lymphokines are released in the course of mononucleosis, but the efficient rejection is probably largely due to the CD8+ CTL. EBNA3, 4, and 6 and LMP2 are the most frequent rejection targets.

The exemption of EBNA1 from being targeted by the CD8+ T cells is due to the long glycine–alanine repeat that inhibits the proteasome–ubiquitin-dependent processing of EBNA1, as long as it is in the normal *cis* position. This exceptional handling of EBNA1 can be seen in relation to the fact that it is the only EBV-encoded protein that can be expressed irrespectively of the cellular phenotype. This is also one of the main reasons why the memory B cells that carry latent virus escape the "attention" of the immune system.

Disease Association

EBV is the causative agent of infectious mononucleosis and of the immunoblastomas that arise in immunosuppressed patients, such as transplant recipients, congenital immunodeficiencies, particularly the X-linked lymphoproliferative syndrome (XLP, an inherited immunodeficiency syndrome that preferentially effects the EBV-specific immune surveillance mechanism), and in HIV-infected persons. The virus is associated with 98% of endemic Burkitt lymphomas, but is only present in about 20% of the sporadic cases. All Burkitt lymphomas carry the chromosomal Ig/myc translocation, however, that is believed to provide the proliferative drive of the tumor. Multiple viral genomes are present in 100% of low differentiated or anaplastic nasopharyngeal carcinomas (see Nasopharyngeal Carcinoma (NPC)). They are also present in 50% of Hodgkin's lymphomas, a variable but usually low percentage of T-cell lymphomas (except midline granulomas where the association is 100%), NK-cell leukemias, a small fraction of gastric adenocarcinomas, and leimyosarcomas that arise in immunosuppressed (e.g., HIV-infected) patients. The role of EBV in these malignant diseases is not clear.

Further Reading

Farrell PJ (1995) EBV immortalizing genes. *Trends in Microbiology* 3: 105–109.

Hayward SD (1990) Immortalization by EBV. *Epstein–Barr Virus Report* 6: 151–157.

Kieff E (1986) Epstein–Barr Virus and its replication. In: Fields HN *et al.* (eds) *Field's Virology*, 3rd edn, pp. 2343–2396. Philadelphia, PA: Raven.

Klein G (ed.) (1987) *Advances in Viral Oncology*. New York: Raven.

Longnecker R (2000) EBV latency. *Advances in Cancer Research* 79: 175–200.

See also: **Burkitt's Lymphoma; EBNA; Hodgkin's Disease; Nasopharyngeal Carcinoma (NPC)**

Equilibrium

M A Asmussen

Copyright © 2001 Academic Press
doi: 10.1006/rwgn.2001.1237

A system is said to be at equilibrium if it is no longer changing. Equivalently, an equilibrium is a state at which the system will remain, baring any perturbations away from it. The stability of an equilibrium hinges on whether the system returns to the equilibrium state following a perturbation. A stable equilibrium can be either locally or globally stable, depending on whether it shows stability after only small or arbitrarily large perturbations, respectively.

Mathematical Criteria

To formalize the concepts of equilibrium and stable and unstable equilibrium for a system determined by a single variable, suppose the value of the variable x (say the gene frequency at a locus) changes through time such that after one generation its new value x' is some function, $f(x)$, of its value x in the previous generation. The variable x then changes from one generation to the next according to the recursion equation:

$$x' = f(x) \tag{1}$$

Equilibrium

At 'equilibrium,' the variable x is no longer changing, i.e., the change in x after one generation, $\Delta x = x' - x$, equals zero. An equilibrium of this system is thus a value $\hat{x}$ for x satisfying the mathematical condition, $f(x) = x$.

Locally Stable Equilibrium

An equilibrium state $\hat{x}$ is 'locally stable' if the system always returns to it following a slight perturbation. This will hold if whenever the value of the variable x is near $\hat{x}$, its next value x' is closer to $\hat{x}$ or, equivalently, after one generation the deviation from the equilibrium $\hat{x}$ is less than it was previously. Formally, local stability of $\hat{x}$ requires that:

$$|x' - \hat{x}| < |x - \hat{x}| \text{ whenever } x \cong \hat{x} \tag{2}$$

It is shown below that this condition holds if at $x = \hat{x}$ the derivative (rate of change) of the new value of x with respect to the old has magnitude less than 1, i.e.:

$$-1 < \frac{df(\hat{x})}{dx} < 1 \tag{3}$$

Under this condition, the equilibrium $\hat{x}$ is locally stable because the value of the variable x will always return to the equilibrium value $\hat{x}$ if it is perturbed slightly from that equilibrium.

Globally Stable Equilibrium

An equilibrium $\hat{x}$ is called 'globally stable' if the value of the variable x always converges through time to $\hat{x}$ from all possible (nonequilibrium) starting values, no matter how far away the system initially is from this equilibrium.

Unstable Equilibrium

An equilibrium $\hat{x}$ is called 'unstable' if the system moves away from it following a slight perturbation. This will hold if for values of the variable x near $\hat{x}$, its next value x' is farther from $\hat{x}$ or, equivalently, after one generation the deviation from the equilibrium $\hat{x}$ is more than it was previously. Formally, instability of $\hat{x}$ requires that the inequality in equation (2) fails for x values near $\hat{x}$. This will be the case whenever at $x = \hat{x}$ the derivative (rate of change) of the new value of x with respect to the old has magnitude greater than 1, i.e.:

$$\frac{df(\hat{x})}{dx} > 1 \text{ or } \frac{df(\hat{x})}{dx} < -1 \qquad (4)$$

Under this condition, the equilibrium $\hat{x}$ is unstable because the value of the variable x moves away from the equilibrium value $\hat{x}$ if it is perturbed slightly from that equilibrium.

Example

A simple selection model provides a useful example of the concepts of equilibrium and stability. Consider an autosomal locus with two alleles, A_1 and A_2, in a population of haploid organisms where the frequency of the A_1 allele in newborn individuals is x and the frequency of the alternate allele A_2 is $1 - x$. Suppose a fraction f_1 of newborns carrying the A_1 allele and a fraction f_2 of newborns carrying the A_2 allele survive to reproduce where $f_1 \neq f_2$, and that this is the only evolutionary force acting on this locus.

The frequency of the A_1 allele after one generation of selection is readily derived by working through a complete generation of this organism. To assist with this derivation, let us deal in terms of numbers, using N to denote the current number of newborn individuals. The number of new adults of each type after selection is then simply the number of newborns of that type that survive to reproduce, as shown in **Table 1**.

The frequency of the A_1 allele in the new adults is simply the fraction of adults carrying that allele:

$$\frac{f_1 x N}{f_1 x N + f_2 (1 - x) N}$$

Since the differential survival rates are assumed to be the only force acting on this genetic locus, this expression also gives the new frequency of the A_1 allele in the new generation of zygotes. After canceling the common factor N, the new allele frequency simplifies to:

$$x' = \frac{f_1 x}{f_1 x + f_2 (1 - x)} = f(x) \qquad (5)$$

To find the equilibria in this system, we first find after some straightforward algebra that the change in allele frequency after each generation of selection is:

$$\Delta x = x' - x = \frac{f_1 x}{f_1 x + f_2 (1 - x)} - x$$
$$= \frac{(f_1 - f_2) x (1 - x)}{f_1 x + f_2 (1 - x)} \qquad (6)$$

Since at equilibrium $\Delta x = x' - x = 0$, we conclude that there are two equilibrium allele frequencies, $\hat{x} = 0$ (fixation for A_2, with only A_2 alleles and no A_1 alleles) and $\hat{x} = 1$ (fixation for A_1, with only A_1 alleles and no A_2 alleles).

To determine when these two equilibria are locally stable, we differentiate the right-hand side of the recursion equation (5) which yields:

$$\frac{df(x)}{dx} = \frac{f_1 f_2}{[f_1 x + f_2 (1 - x)]^2}$$

This derivative is f_1/f_2 at $\hat{x} = 0$ and f_2/f_1 at $\hat{x} = 1$. Since the two survival rates, f_1 and f_2, are nonnegative fractions, we conclude that fixation for $A_2 (\hat{x} = 0)$ is locally stable and fixation for $A_1 (\hat{x} = 1)$ is unstable if $0 \leq f_1 < f_2$, while fixation for $A_1 (\hat{x} = 1)$ is locally stable and fixation for $A_2 (\hat{x} = 0)$ is unstable if $0 \leq f_2 < f_1$. In other words, fixation for a given allele is locally stable if individuals carrying that allele have a higher survival rate than the other type in the population. Fixation for the allele with the lower survival is unstable.

In this particular biological system, each equilibrium is actually globally stable whenever it is locally

Table 1 Generation cycle under selection

	Type of individual	
	A_1	A_2
Number of newborns	xN	$(1-x)N$
Survival rate	f_1	f_2
Number of new adults	$f_1 xN$	$f_2(1-x)N$

stable because analysis of the sign of allele frequency change in equation (6) shows that the frequency of the A_1 allele will always steadily increase to 1 if $0 \leq f_2 < f_1$ and will steadily decrease to 0 if $0 \leq f_1 < f_2$. Thus, the frequency of the allele conferring the higher survival rate will always increase to 1 (and that of the allele with the lower survival rate will always decline to zero) under the simple selection scheme for haploid populations considered here. Some sample trajectories showing how the frequency of the A_1 allele changes through successive generations under various parameter values are shown in **Figure 1**. In **Figure 1A**, fixation for the A_1 allele is unstable and its frequency always monotonically decreases to 0. In **Figure 1B**, fixation for the A_1 allele is locally (and globally) stable and its frequency always monotonically increases to 1.

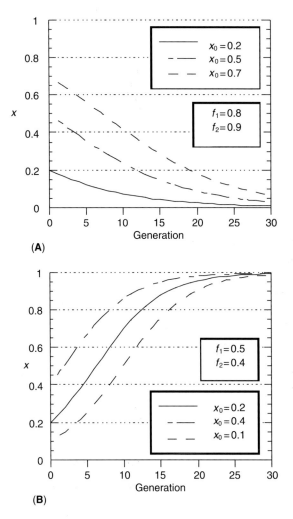

(A)

(B)

Figure 1 Trajectories through time in generations of the frequency of the A_1 allele for various initial frequencies (x_0) for A_1 and survival rates (f_1, f_2) of the two alleles.

Derivation of Local Stability Condition

The local stability criterion in equation (3) follows by noting that if the variable x starts near the equilibrium value $\hat{x}$, its new value x' after one generation in equation (1) can be approximated by the tangent line to the function $f(x)$ at the point $x = \hat{x}$. The latter is also the first-order Taylor polynomial approximation to the function $f(x)$ near $\hat{x}$. Under this linear approximation, we have:

$$x' \cong f(\hat{x}) + \frac{df(\hat{x})}{dx}(x - \hat{x}) \text{ for } x \cong \hat{x}$$

Remembering that $f(\hat{x}) = \hat{x}$ at any equilibrium for this system, we immediately find that:

$$x' - \hat{x} \cong \frac{df(\hat{x})}{dx}(x - \hat{x}) \text{ for } x \cong \hat{x}$$

and thus the condition in equation (2) for local stability reduces to the criterion given in equation (3).

See also: **Balanced Polymorphism; Hardy–Weinberg Law**

Equilibrium Population

K E Holsinger

Copyright © 2001 Academic Press
doi: 10.1006/rwgn.2001.0418

When a population is in equilibrium both genotype and allele frequencies remain constant from one generation to the next. If a population satisfies the conditions necessary to ensure that genotypes are in Hardy–Weinberg proportions, it follows that it is also in equilibrium. Even if a population does not satisfy the Hardy–Weinberg conditions, however, it may still be in equilibrium. The frequency of recessive alleles preventing individual monkeyflower plants from producing pollen, for example, is likely to represent a balance between the tendency of natural selection to eliminate the recessive allele and recurrent mutation that tends to increase its frequency. A population in which such forces are balanced might be said to be in dynamic equilibrium.

Populations in the Absence of Selection

It is easiest to understand the concept of an equilibrium population by considering what happens to allele frequencies in a very large population from one generation to the next. Suppose that the frequency of an allele, A_1, at a particular locus is p and the frequency of the alternative allele at this locus, A_2, is $q(q = 1 - p)$. If there are no differences among individuals in the probability that they survive or in the numbers of offspring that they produce and if there is no mutation, then clearly we will have the same number of A_1 and A_2 alleles in the next generation as we have in this generation. Putting it another way:

$$p_{t+1} = p_t$$

where p_t refers to the allele frequency in the current generation and p_{t+1} refers to the allele frequency in the next generation. A population is in equilibrium whenever $p_{t+1} = p_t$.

Suppose we now allow for the possibility that mutation can occur. Then we would normally expect the allele frequency in the next generation to be different from the allele frequency in the present generation. Specifically, imagine that A_1 mutates to A_2 with a frequency μ and that A_2 mutates to A_1 with a frequency ν. Then:

$$p_{t+1} = (1 - \mu)p_t + \nu(1 - p_t)$$

Clearly, p_{t+1} will not normally equal p_t, so the population is not at equilibrium. But what if $p_t = \nu/(\mu + \nu)$? It is not hard to verify that p_{t+1} will also equal $\nu/(\mu + \nu)$. Thus, $p_{t+1} = p_t$ and the population is at equilibrium. At this equilibrium the rate at which A_1 alleles give rise to A_2 alleles ($\mu\, p_t$) is equal to the rate at which A_2 alleles give rise to A_1 alleles ($\nu\,(1 - p_t)$) so there is no net change in the frequency of either allele.

Populations Undergoing Selection

In a population undergoing selection the situation is a bit more complicated, but the basic principle is the same. A population is in equilibrium if allele frequencies do not change from one generation to the next. Norway rats in Great Britain, for example, evolved partial resistance to the blood anticoagulant warfarin that has been used for rat control since World War II. The resistance results from a mutation in a gene that would normally be deleterious. When warfarin is present homozygotes for the susceptibility allele (SS) survive only 68% as often and homozygotes for the resistance alleles (RR) survive only 37% as often as heterozygotes (SR). Because heterozygotes are the most likely to survive, natural selection maintains both alleles in the population. Moreover, the population will evolve from any initial allele frequency of S to a frequency of 0.66. Once that allele frequency is attained, it will remain constant. When the frequency

of S equals 0.66, in other words, the population is in equilibrium.

Warfarin resistance is an example of a general pattern of selection known as heterozygote advantage or overdominance. The relative survival abilities of the genotypes are referred to as relative fitnesses. Whenever a population is large and heterozygotes are more likely to survive than homozygotes, natural selection maintains both alleles in the population in which the frequency of A_1 is

$$p = (1 - w_{22})/(2 - w_{11} - w_{22})$$

where w_{11} is the probability that the genotype homozygous for A_1 survives relative to the probability that the heterozygous genotype survives, and w_{22} is the probability that the genotype homozygous for A_2 survives relative to the probability that the heterozygous genotype survives.

Often mutations cause deleterious effects on the individuals that carry them. If it were not for the fact that mutation is introducing new copies of these deleterious alleles, natural selection would tend to eliminate them from populations. If the mutations recur repeatedly, however, the population will approach an equilibrium where the rate at which natural selection eliminates deleterious alleles is exactly balanced by the rate at which mutation reintroduces them, a phenomenon often called mutation–selection balance. If the relative survival probabilities of the favorable homozygote, the heterozygote, and the deleterious homozygote are denoted as 1, 1-hs, and 1-s, respectively, when the deleterious allele is completely recessive its frequency is:

$$q = (\mu/s)^{1/2}$$

When the deleterious allele is expressed in heterozygotes its frequency is:

$$q = (\mu/hs)$$

Stationarity in Finite Populations

The frequency of alleles may change from one generation to the next in small populations simply because of random chance, a process referred to as genetic drift. Over time genetic drift would lead to the loss of genetic variability within populations. In fact, a population would lose a fraction $1/2N_e$ of the genetic variability it contains every generation if genetic drift were the only process affecting the population, where N_e is the effective population size. Just as recurrent mutation to a deleterious allele can prevent its elimination by natural selection, however, mutation can prevent the loss of genetic variability from small populations.

About $2N_e\mu$ alleles are introduced into a diploid population of size N_e every generation by mutation, but there is no equilibrium between mutation and drift in the same sense as there is between mutation and selection. In a small population where allele frequencies are subject to drift they will tend to change in every generation. Nonetheless, the 'probability' that a population will have a particular allele frequency will eventually stop changing. When it does we say that the population has reached stationarity.

Stationarity in a small population is the analog of equilibrium in large ones. Although it may be very difficult to calculate the probability that a population has a particular allele frequency, populations will almost always approach stationarity if rates of mutation, selection, and migration remain constant and if the population persists for a long enough period of time: about $4N_e$ generations, on average.

Applicability of Equilibrium Concepts

In real populations of plants or animals it is rarely, if ever, the case that rates of mutation, migration, and selection remain constant for long periods. As a result, real populations are rarely, if ever, exactly at equilibrium (or at stationarity, if small). Nonetheless, the features of equilibrium populations play an important role in evolutionary theory, both because sometimes the variation in evolutionary forces is small enough that the assumption of equilibrium is not far wrong and because the investigation of equilibrium conditions allows us to infer the direction in which evolution is likely to proceed. Even if we knew only that rats heterozygous for warfarin resistance were more likely to survive than those homozygous for either allele, we could predict that natural selection would tend to maintain both alleles in populations exposed to warfarin. Neither will this part of the outcome be affected if the survival probabilities of genotypes differ from one generation to the next, provided that heterozygotes are alway most likely to survive.

In many genetic models of the evolutionary process two types of equilibria are encountered: stable equilibria and unstable equilibria. Although a population with an allele frequency that matches the frequency of either type of equilibrium will not change in later generations, populations tend to evolve away from unstable equilibria and tend to evolve toward stable equilibria. Small differences between a population's allele frequency and the allele frequency at an unstable equilibrium are magnified from one generation to the next, while allele frequency differences from a stable

equilibrium are decreased in every generation. Similarly, small populations will tend to change in ways that cause them to have allele frequencies that are associated with high probabilities at stationarity.

Further Reading

Crow JR and Kimura M (1970) *An Introduction to Population Genetics Theory.* Minneapolis, MN: Burgess.

Hartl DL and Clark AG (1997) *Principles of Population Genetics*, 3rd edn. Sunderland, MA: Sinauer Associates.

May RM (1985) Evolution of pesticide resistance. *Nature* 315: 12–13.

Willis JH (1999) The contribution of male-sterility mutations to inbreeding depression in *Mimulus guttatus. Heredity* 83: 337–346.

See also: **Effective Population Number; Equilibrium; Fitness; Genetic Drift; Hardy–Weinberg Law; Heterogenote**

erbA and erbB in Human Cancer

J Y-K Lau

Copyright © 2001 Academic Press
doi: 10.1006/rwgn.2001.1566

The tyrosine kinase pathway constitutes a very important cellular signal transduction pathway. Tyrosine kinases can be grouped into two classes: receptor tyrosine kinases and nonreceptor tyrosine kinases (without extracellular binding domains). When cellular tyrosine phosphorylation is enhanced, for instance, by a growth factor to the receptor tyrosine kinase, this triggers a cascade of downstream signals, thereby affecting many different cellular functions. Importantly, many of the cellular tyrosine kinases are frequently products of proto-oncogenes and their aberrant expression has been associated with many different human cancer types. One of the best-studied families of tyrosine kinases is the epidermal growth factor receptor (EGFR) family. The erbB family consists of four different types of receptor tyrosine kinase, including *erbB-1* (also known as EGFR), *erbB-2* (*HER-2/neu*), *erbB-3* (*HER-3*), and *erbB-4* (*HER-4*). The first two types have been well-studied and characterized in human cancer.

Amplification/overexpression of *erbB-1* and *erbB-2* has been associated with different types of human cancer, for example, breast cancer, lung cancer, and head and neck squamous cell carcinoma. erbB-1 is a transmembrane tyrosine kinase receptor. The erbB-1 protein is composed of two cysteine-rich extracellular domains and an intracellular tyrosine kinase domain. It shares extensive sequence homology with *erbB-2*. erbB-1 is expressed throughout development and in a variety of cell types. Several ligands, such as TGF-α and amphiregulin, can bind to the 170-kDa cell-surface erbB-1, resulting in activation of its intrinsic kinase activity. In the presence of its ligands, over-expression of erbB-1 can transform the mouse fibroblast cells indicating its potential role in oncogenesis. In some types of human cancer, the expression level of erbB-1 is significantly associated with the tumor stage and size. In addition, antibodies against erbB-1 have been shown to inhibit tumor growth in experimental studies. This indicates that erbB-1 may play a significant role in oncogenesis in some human cancer types.

The *erbB-2* gene encodes a transmembrane protein of 185 kDa. erbB-2 has intrinsic tyrosine kinase activity. Amplification/overexpression of the *erbB-2* oncogene was found in 20–30% of cases of human breast cancer. Its overexpression is also found in ovarian, lung, gastric, and oral cancer with high frequency, suggesting that *erbB-2* overexpression may play an important role in the development of human cancer. In an experimental model, transfection of the normal *erbB-2* gene into cells expressing *erbB-2* at low levels can enhance metastatic potential by promoting multiple steps associated with metastasis such as cell migration rate and *in vitro* invasive ability. Unlike erbB-1, no ligands directly binding to the erbB-2 protein have been clearly identified. A mutation at the transmembrane domain or its overexpression can result in constitutively activated erbB-2. This is likely due to the enhanced formation and stabilization of the receptors, allowing the protein to be in the activated state. When erbB-2 protein is activated, it can interact with many different cellular proteins such as mitogen-activated protein (MAP) kinase, Shc, PLC-γ, and GAP, PI3 kinase mediating the signal transduction pathway.

Members of the erbB family have been shown to be able to form heterodimer and transphosphorylate in response to NDF (also known as heregulin) or EGF. It has been found that the *erbB-3* gene product is a receptor for NDF and coexpression of *erbB-2* and *erbB-3* reconstitutes a high-affinity receptor for NDF. NDF can also stimulate cell proliferation in breast and ovarian cancer cell lines. A recent report has shown that NDF can stimulate mitogenesis in NIH 3T3 cells that express either *erbB-3* or *erbB-4*, but not transformation. However, when the cells expressing either erbB-1 or erbB-2 are coexpressed with either erbB-3 or erbB-4, NDF can induce cellular transformation. These data indicate that different members of the EGFR family may have different

signaling pathways. Also, the findings imply that erbB-1 and erbB-2 may play an important role in cellular transformation.

With the elucidation of the significant role of erbB-1 and erbB-2 in the pathogenesis of human cancer, different approaches have been used to target the signal transduction pathway of these two oncogenes and their expression level in cancer cells. For example, in erbB-2, a recombinant humanized monoclonal antibody, herceptin, has been used in different clinical trials with some encouraging results. In a series of *in vitro* and animal experiments, adenovirus E1A protein successfully repressed the expression level of erbB-2, and there was significant improvement in the survival of those mice with erbB-2-overexpressing tumors that were treated with E1A. In addition, tyrosine kinase inhibitors, such as tyrphostin and emodin, have been used to block the erbB-1 and erbB-2 tyrosine kinase activities. Thus, different strategies that target either at the expression level of these oncogenes or their signaling transduction pathways have been employed with considerable success, providing a novel and hopefully better therapeutic option for those suffering with cancer.

See also: **Cancer Susceptibility; Oncogenes**

Error Catastrophe

C Kurland and J Gallant

doi: 10.1006/rwgn.2001.0419

The error catastrophe is a conjecture in search of experimental verification. The initial form of the conjecture arose at an early stage in the analysis of the genetic code, when Leslie Orgel contrived the following syllogism. (1) Translation of the genetic code must be afflicted by some nonzero frequency of error. (2) The devices that translate the genetic code are themselves proteins (e.g., aminoacyl-tRNA synthetases, translation factors, ribosome proteins) and they will themselves contain errors. (3) Therefore, error rates of gene expression are intrinsically unstable because they are autocatalytic. This conclusion follows from the supposition that the error frequencies of translation are enhanced by the errors already incorporated into the proteins of the translational apparatus itself. Accordingly, the more errors that have been incorporated into the proteins of the translation system, the more errors the translation system will make. Clearly, the catastrophic implication is that at some point the autocatalysis of translation errors will get out of hand

and the translation system will be unable to generate canonical gene products. Thus, a destructive, positive feedback loop fueled by the errors of translation leads inexorably to the death of cells.

According to the original formulation of the error catastrophe, the question is not whether such an error catastrophe will occur. It is taken to be inevitable. Rather, the question is how long it takes before the catastrophe erupts. Of course, the underlying appeal of this scenario is that it provides a simple, molecular explanation of senescence and death at the cellular level.

Orgel and others soon recognized that the error catastrophe is not inevitable if the magnitude of the coupling between errors in proteins and errors of translation is sufficiently small. In other words, if the feedback between successive rounds of translation errors is contained within sufficient bounds, the error rate of translation will be stable, i.e., not inclined to catastrophes.

From the 1960s through the 1980s, a great variety of studies were aimed at testing the prediction that aging (either in whole organisms or in cultured cells, a favorite model system) is accompanied by increasing errors in protein synthesis. The overwhelming majority of technically adequate studies detected no such increase. A smaller number of studies sought to evaluate the formal characteristics of error feedback. Most of these studies, utilizing bacteria, demonstrated that the error feedback term was indeed small, and that normal as well as artificially enhanced translation error frequencies are stable.

The failure to detect signs of the error catastrophe either in aging test subjects or under conditions of experimentally enhanced translational error led to a frustrating situation for the experimentalist. A proponent could always argue that it had not been proven that the error catastrophe never occurs. The experimentalist, realizing that it is in principle impossible to find such a proof, was then obliged to return to theory in order to find out what made the error catastrophe so elusive.

The underlying assumption of the error catastrophe is that errors of protein construction inexorably increase the errors of protein function. Such a state of affairs is encountered if the accuracy of protein function has evolved to an absolute maximum. At such a maximum, changes in protein structure can not improve the accuracy of function. Rather, structural changes can only make the accuracy of function worse or at best leave it unchanged. In contrast, it has been known for many years that hyperaccurate mutations in ribosomes are easily obtained by selection with antibiotics such as streptomycin. These ribosomes contain alterations in their proteins. If ribosomal proteins can mutate so that translation is carried out at much higher accuracy levels than that supported by wild-type ribosomes, the canonical ribosomes are not operating at maximum accuracy.

Since some mutations can increase translation accuracy while others decrease it, the net effect of errors in the constructions of ribosomes could well be to cancel each other out. This would account for the apparent stability of the error rates of both wild-type and error-prone mutant ribosomes. It would also account for the measurements of substitution errors in ribosomal proteins which suggest that in a normal ribosome, containing roughly 7800 amino acids in its proteins, there are an average of three to ten erroneous amino acid substitutions. Thus, no two ribosomes are completely alike and the bacterial cell's entire ribosome population normally provides an experiment in error feedback. These ribosomes do not generate a catastrophic cascade of errors in translation, presumably because the influence of these errors is to neutralize one another.

See also: **Aging, Genetics of; Ribosomes**

Erythroblastosis Fetalis

R F Ogle and C H Rodeck

Copyright © 2001 Academic Press
doi: 10.1006/rwgn.2001.0420

Historical Background

The first recorded case of hemolytic disease of the newborn was described in 1609, but it was not until 1932 that hydrops fetalis, jaundice, and kernicterus were shown to be part of the same disease associated with hemolytic anemia, extramedullary erythropoiesis, hepatomegaly, and erythroblastosis. These features were collectively known as erythroblastosis fetalis. By 1939, Levine and Stetson had demonstrated the involvement of the Rh antigen, but it was not until 1954 that Chown proved that the fetal hemolysis was caused by the production of maternal anti-RhD alloimmune antibodies. Since 1970 prevention of the disease has been possible by routinely giving anti-RhD γ-globulin to nonsensitized RhD-negative mothers immediately following the birth of a RhD-positive child. The antibody removes fetal cells from the maternal circulation before they can cause sensitization.

Rh Blood-Group System

Depending on the presence or absence of D antigen on the red blood cell surface, individuals are classified as Rh-positive or Rh-negative. In addition to the D antigen there are two other major antigens, the C/c and E/e antigens, which have important clinical implications, the only difference being that there is no apparent d antigen, where 'd' refers to an absence of 'D'. In the case of Cc and Ee, both the upper and lower case letters indicate the presence of serologically definable antigen.

The genes encoding these three sets of antigens are inherited together rather than randomly. Earlier investigators therefore proposed a single gene where recombination would only rarely occur. When individuals totally lack these antigens, they usually have membrane instability, suggesting that the Rh antigens have major physiological importance. The ethnic incidence of the RhD-negative phenotype varies considerably, being about 15% in Caucasians, 35% in Basques, and virtually zero in Asiatic Chinese and Japanese.

The nomenclature of the Rh system is confusing, but the most common system used is the Fisher–Race system which was based on the theory that the Rh system locus consists of three genes with antithetical alleles C/c, D/d, and E/e. The haplotypes are described in triplets, Cde, cde, and cDE being the most frequent.

Molecular Basis of Rh Antigens

There were found to be approximately 60 000 Rh polypeptides per erythrocyte. When the isolated Rh polypeptides were digested and analyzed by electrophoresis, the variations in the degradation patterns indicated that RhD is distinct from the C/c and E/e polypeptides, with the former having a molecular weight of 31.9 kDa and the latter two each having a molecular weight of 33.1 kDa. These studies therefore showed that the Rhc, RhD, and RhE polypeptides were very similar though distinct proteins.

Cloning of the Rh polypeptides was complicated as the monoclonal antibodies that had been developed to identify the different Rh antigen sites on the red cell membrane were not suitable for identification of Rh polypeptides expressed from cDNA expression libraries. Oligonucleotide probes for isolating Rh cDNAs were designed from partial amino acid sequence data of isolated polypeptides. In 1990 two groups of workers (Avent *et al.*, 1990; Cherif-Zahar *et al.*, 1990) used the polymerase chain reaction (PCR) with oligonucleotide primers from segments of N-terminal amino acid sequence to amplify cDNA templates prepared from thalassemic spleen erythroblasts or peripheral reticulocytes. These PCR products were then hybridized to commercially available cDNA libraries. The open reading frame sequences were found to be identical for each group and *in situ* hybridization

confirmed localization to chromosome 1p34.3–1p36.1, which had previously been suggested by linkage data some years earlier. The first cDNA clone proved to encode both the C/c and E/e proteins. By 1992 Le Van Kim and colleagues (Le Van Kim, 1992) reported the isolation of the RhD polypeptide, and through restriction fragment length polymorphism analysis showed RhD-positive individuals had two polypeptide genes, and RhD-negative individuals had just one. The conclusion was that the RH gene locus consists of two highly homologous, closely linked genes, one of which encodes both the C/c and E/e proteins. The other gene encodes the RhD protein, which is absent in RhD-negative individuals.

Structure of the Rh cDNAs

The RhD and RhEe cDNAs consist of 10 exons, exons 1–9 being almost identical. Exon 10 of the RhD cDNA contains regions of divergence with an Alu repeat element. Subsequently it was demonstrated that the RhCcEe gene encoded both the E/e and C/c polypeptides by differential splicing of the primary mRNA transcript. The RhE polypeptide is synthesized from a full-length transcript of the RhCcEe gene. The 417-amino acid polypeptide is the same length and has a very similar sequence to the RhD polypeptide. The difference between the antithetical E and e epitopes depends on a single point mutation in exon 5 at position 226, substituting an alanine in the E polypeptide for a proline in the e polypeptide.

The Cc polypeptides are synthesized from at least two different truncated transcripts that have exons 4, 5, and 6, or exons 4, 5, and 8 spliced out. The transcripts are identical to one another and to the E/e polypeptide at the N-terminus. The C-terminus is either identical to the same region of E/e but has reverse orientation in the membrane or has a novel protein sequence as a consequence of the introduction of frameshift by the splicing of exon 8. The difference between C and c is due to a series of six point mutations in exons 1 and 2, two being silent and four that result in amino acid substitutions.

The molecular basis of a number of rare RhD positive/negative variants has now been identified. Many are due to substitution of parts of the RhD gene sequence into the RhCcEe gene or vice versa to form 'hybrid' genes.

Function of the Rh Polypeptides

Even though the predominant interest regarding Rh polypeptides is in their role as antigens, it is probable that they play a crucial role in the physiology of red cell membranes that is quite unrelated to their antigenicity. Their function is clearly defined but the multiple membrane-spanning domains of the Rh polypeptides suggest a transportor protein. Erythrocytes from all of the common Rh phenotypes are normal, but the membrane defects seen in those with the Rhnull phenotype provide some clues to their function. Rhnull individuals have a mild to moderate hemolytic anemia (never severe) suggesting that the Rh system may act as a fine-tuning mechanism in membrane stability; however, the exact mechanisms are not clear.

Prenatal Determination of the RhD Phenotype

Approximately 56% of Rh-positive subjects are heterozygous for the D antigen. When the mother is RhD-negative and the father is heterozygous RhD-positive there is a 50% chance that the fetus will be RhD-negative and so not at risk for erythroblastosis fetalis. Previously in clinical practice, prenatal determination of the RhD phenotype where the father was heterozygous for the trait involved fetal blood sampling with serological Rh typing, resulting in a 1–2% fetal loss rate and a 40% risk of fetomaternal hemorrhage, which may also have increased the risk of sensitization. An alternative method is serial amniocentesis for quantitation of bilirubin in amniotic fluid. This technique is unable to distinguish an RhD-positive fetus that is mildly affected from an RhD-negative one. It also potentially exposes the fetus to multiple invasive procedures.

The ideal strategy for prenatal determination of fetal RhD phenotype would be to generate a pair of PCR primers that only amplified a specific region of the RhD gene without cross hybridization to the RhCeEe gene or any other gene. A second amplification, ideally of the RhCcEe gene, should be performed in duplex. This strategy is difficult to design in practice as there is a high degree of homology between the RhCcEe and D genes. Exon 10 of the RHD gene demonstrates a region of divergence with a copy of the Alu repeat motif which is found in a large number of other genes and noncoding sequences. Bennett *et al.*, 1993 reported the first use of PCR for prenatal determination of fetal RhD using primers at the 5′ extreme end of exon 10 in RHD. The 5′ primer lay within a region of 100% homology between RhCcEe and RHD, but it was the 3′ primer, designed to an RhD-specific region, that gave the specificity to the reaction. A control primer from sequences in exon 7, which amplified a 134-bp product from both RhCcEe and RHD, acted as a control in the duplex reaction.

The original report was from 15 samples of amniotic fluid cells with the fetal RhD type being also confirmed on fetal blood sampling. Several groups of workers have used different primers but it would appear that the original primers of exons 7/10s are able to predict consistently the RhD serotype in all cases. Nevertheless, it is still important to use two different primer sets, designed from a different part of the RhCcEe gene and D genes, and this can be combined successfully in a single multiplex reaction. Trophoblastic tissue has been shown not to express the RhD antigen so is not useful as a form of prenatal diagnostic test for genotyping.

Noninvasive or Minimally Invasive Prenatal Determination of RhD Type

As the maternal RhD-negative DNA should not act as a template for PCR, PCR should only amplify a product if there is fetal RhD-positive DNA present. Several groups have attempted to extract fetal DNA from the maternal circulation with variable success, possibly in some circumstances due to the rapid clearance of RhD-positive cells from the circulation of sensitized RhD-negative women. There is evidence that extraction of fetal DNA by fluorogenic PCR analysis from the maternal circulation is reliable from the second trimester, being less reliable in the first trimester where samples give false negative results, presumably due to the low concentration of fetal DNA in the maternal plasma at the time. Other noninvasive tests, such as harvesting fetal cells from endocervical mucus, have not been shown to be sufficiently reliable to be used in clinical practice.

Preimplantation Determination of RhD Type

For sensitized women with heterozygous partners who have experienced recurrent miscarriages or serial transfusions and are unable to cope with future affected pregnancies and also for those who have had severely affected pregnancies prior to conventional therapy, preimplantation determination of embryonic RhD type after *in vitro* fertilization and before embryonic transfer is a possible option. To perform molecular diagnosis of RhD from DNA present in a single human diploid cell requires amplification by 'nested PCR.' A low number of cycles is used with an outer set of oligonucleotide primers. The second round of PCR is then performed on a small aliquot from the first reaction, using a higher number of amplification than that in the first reaction with primers internally nested. A common primer rather than two sets of primers for duplex PCR in the outer reaction of nested PCR reduces the incidence of locus-specific amplification when used in a single-cell diploid genome.

To differentiate the two genes, one inner primer was designed to anneal with RhD sequences and the second to RhCcEe sequences, but at different yet overlapping sites, so resulting in an amplification product of different size from that of the RhD gene. This method would eliminate the risk of an RhD-positive embryo being missed.

References

Avent ND, Ridgell K, Tanner MJA and Anstee VJ (1990) cDNA cloning of a 30 kDa erythrocyte membrane protein associated with Rh blood group antigen expression. *Biochemical Journal* 271: 821.

Bennett PR, Le Van Kim C, Collin Y *et al.* (1993) Prenatal determination of fetal RhD type by DNA amplification. *New England Journal of Medicine* 329: 607.

Cherif-Zahar B, Bloy C, Levankim C *et al.* (1990) Molecular cloning and protein structure of a human blood group Rh polypeptide. *Proceedings of the National Academy of Sciences, USA* 87: 6243.

Le Van Kim C, Mouro I, Cherif-Zahar B *et al.* (1992) Molecular cloning and primary structure of the human blood group RhD polypeptide. *Proceedings of the National Academy of Sciences, USA* 89: 10: 925.

See also: **Rh Blood Group Genes**

ES Cells

See: **Embryonic Stem Cells**

Escherichia coli

F C Neidhardt

Copyright © 2001 Academic Press
doi: 10.1006/rwgn.2001.0422

The science of genetics has benefited from concentrated studies on a relatively small number of living systems – so-called paradigm or model organisms. Examples include the laboratory (or house) mouse (*Mus musculus*), the fruit fly (*Drosophila melanogaster*), the nematode worm (*Caenorhabditis elegans*), the protozoan Paramecium (*Paramecium aurelia*), and the bread mold (*Neurospora crassa*). The enteric bacterium *Escherichia coli* has been among the model organisms of genetics ever since the middle of the twentieth century. Study of *E. coli* and its viruses has contributed much information to fundamental genetics, including the nature of the genetic material, the molecular definition of genes, and the mechanisms of their function and regulation. The biotechnology industry was founded on the basis of discoveries about the genetics of *E. coli*, and the organism itself continues to serve many important roles in biotechnology processes.

Escherichia coli and Its Life

E. coli is a rod-shaped bacterium measuring a few microns in length and 0.5 µm wide. Being a prokaryote, it lacks a nuclear membrane. Its 4290 genes reside on a single circular, double-stranded DNA molecule tightly packed within the cytosol of the cell (**Figure 1**). *E. coli* grows rapidly in simple media (generation time of the order of 1 h) and reproduces by binary fission. A double-membrane envelope gives the cell a gram-negative staining characteristic. Because it is a facultative anaerobe (able to grow anaerobically by fermentation and aerobically by oxidation), it is admirably suited for its main ecological niche – the intestine of humans and other animals, where it universally constitutes a part of the normal flora. The genus to which *E. coli* belongs was named after Theodor Escherich, an early bacteriologist. *E. coli* is said to be an enteric bacterium because its major habitat is the intestine (enteron) of humans and other animals. There are several other species of *Escherichia*, but none comes close to sharing the research spotlight with *E. coli*. The closest relatives to *E. coli* seem to be the several species of the genus *Shigella*, many of which are human pathogens, but it is also quite similar to the mouse and human pathogen *Salmonella*.

Although resident within the ileum (rather than the colon, as its name might imply), cells of *E. coli* must perforce survive conditions external to animals sufficiently to insure successful passage from one individual to another. Humans are colonized almost immediately after birth, generally by the *E. coli* strain inhabiting the mother; every few months another replaces the particular resident strain from the environment. A few strains are pathogenic; some cause genitourinary infections, and some are responsible for traveler's diarrhea. Great attention has been directed toward the exceptional strains that produce a potent toxin that can produce a fatal or near fatal septicemia when ingested from contaminated water or food.

How *Escherichia coli* Became a Paradigm for Genetic Studies

That *E. coli*, or any bacterium for that matter, should turn out to be a preeminent subject in explorations of genetics is exceedingly odd, for until the mid-twentieth century there were scientists who

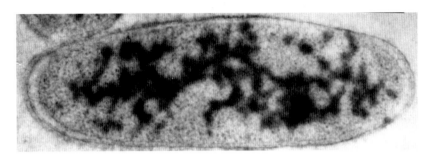

Figure 1 Thin section of *Escherichia coli*. The DNA was immunostained, revealing the nucleoid as the convoluted central area. (Reproduced with permission from Kellenberger E (1996) Structure and function at the subcellular level. In: Neidhardt FC *et al.* (eds) Escherichia coli *and* Salmonella: *Cellular and Molecular Biology*, 2nd edn, ch. 4. Washington, DC: ASM Press.)

questioned whether bacterial inheritance follows the same rules that govern plants and animals.

Do Bacteria Follow the Standard Rules of Genetics?

There were good reasons for suspecting that bacterial inheritance was too specialized to serve as a model for general cellular genetics. The cardinal rule of genetics, that like begets like, seemed often violated. A population (called 'a culture') of bacteria produced under one set of growth conditions might differ in subtle ways (enzyme content, antigenic characteristic, etc.) and not-so-subtle ways (cell size and gross chemical composition) from a culture of the same organism grown in a second environment. Many of the properties of the cells in the second medium would disappear when these cells were grown to produce offspring in the original medium. This extreme plasticity of bacterial cells raised the question of what role heredity played in these very small cells. This doubt was reinforced by the lack of convincing cytological evidence that bacteria had chromosomes and assorted them by mitosis.

The easy manner in which bacteria acquired and lost characteristics depending on their growth medium was largely explained when it became recognized that the genetic makeup of a bacterial cell (its genotype) determines a wide range of possible appearances (phenotypes). Gene expression is greatly influenced by the environment. The enzymatic constitution of these cells depends on the activation and repression of the individual genes of their genome in response to chemical signals from the environment.

But there was a second problem. Some properties acquired by the population in a particular environment were retained during subsequent growth in other environments; that is, they appeared to breed true, as do mutations in higher organisms. Exposure of a bacterial population to a deleterious agent, for example, led usually to the growth of cells resistant to

that agent. In these cases it seemed that environmental components might induce mutations that favor growth in that environment, quite contrary to the well-established principle of Mendelian genetics that specific mutations are not directed by the environment. This was not a preposterous notion, since bacteria grow by binary fission and thus there is no distinction between germ cells and somatic cells; each bacterium passes on to its daughter cells whatever changes may have occurred to its genetic material.

But in 1943 Luria and Delbrück concluded, from measurements of the distribution of cells resistant to bacteriophage T1, that spontaneous mutations occur at random in a growing *E. coli* population. In 1952 Joshua and Esther Lederberg, by means of replica-plating, isolated mutants resistant to streptomycin without ever exposing the cells to that agent. By these and other ingenious ways of demonstrating the existence of spontaneous mutants in a population, microbiologists became persuaded by the mid-twentieth century that bacterial mutations occur essentially randomly within individual cells, and that the environment plays a large role in selecting chance mutants that have a growth advantage. What had not been sufficiently appreciated earlier was that selective pressures could bring about changes in population composition very quickly in organisms growing exponentially, with generation times measured in minutes rather than months or years.

The uneasiness about bacterial inheritance had not prevented some fundamental discoveries in bacterial genetics even before the issue of mutations was settled. In 1944, Oswald Avery, Colin MacCloud, and Maclyn McCarty demonstrated that the 'transforming principle' discovered by Frederick Griffith (1928), which conferred new properties on *Streptococcus pneumoniae*, was DNA. Alfred Hershey and Martha Chase in 1952 verified the conclusion that DNA and not protein was the genetic material by showing that it was only the DNA of bacteriophage T2 that is injected into the host *E. coli* cells which proceed to produce a new crop of phage.

Escherichia coli as a Model Organism for Genetic Studies

The very characteristics that provided the early puzzles about phenotypic plasticity and mutability made *E. coli* enormously valuable once it was realized that its genetics would model that of plants and animals. The small size of these cells, their rapid growth rate, and the extensive phenotypic influence by the environment provided geneticists with powerful tools. Small size meant that many millions, even billions of individuals could be studied in a single experiment. Rapid growth meant that many generations could be produced within a single day. The ability to grow these cells in chemically diverse media and at different temperatures made it possible for biochemical genetics to flourish. The latter characteristic opened the door to the biochemistry of how genes function and how inheritance works, and also brought genetic analysis to bear on discovering the biochemical nature and workings of the cell. As the structure and function of DNA and the nature of the genetic code became known, the biochemical genetics of *E. coli* evolved into the field of molecular genetics. Soon thereafter recombinant DNA techniques were developed, aided greatly by studies of *E. coli* and its restriction enzymes.

Of particular importance was the early realization of the rich opportunities for genetic studies provided by the many kinds of bacteriophage (bacterial viruses, or 'phage') to which *E. coli* is susceptible.

Contributions of *Escherichia coli* to Genetics

Once the advantages of working with *E. coli* and its bacteriophage were appreciated, genetic studies advanced quickly. In fact, the great success of studies on *E. coli* relates to the possibility of bringing the power of genetic analysis to any problem studied in this organism. Accordingly, the contributions to biology made through genetic studies with *E. coli* are extremely impressive. A few examples chosen for variety will illustrate the riches harvested over the second half of the twentieth century:

1. Biochemical pathways. In the period from 1950 to 1965, the enzymatic steps in the synthesis of amino acids and nucleotides were established in *E. coli*, largely through the powerful tool of mutant analysis; this accomplishment provided the framework for understanding the biosynthetic pathways of all organisms.
2. Definition of the cistron. Intensive study of one genetic locus (rII) of the bacteriophage T4 enabled Benzer in the early 1960s to define the term 'gene' with great precision and to distinguish between genes as units of mutation, of recombination, and of function. Benzer's work moved the concept of the genetic material from an image of genes being like beads on the chromosome string to a depiction closely approximating our current molecular understanding of the gene as a segment of a linear DNA molecule. He introduced the term 'cistron' (defined operationally by the *cis/trans* test) as the unit of heredity that encodes a single polypeptide chain.
3. Regulation of gene function. The elements of gene regulation were first indicated by the monumental genetic and biochemical study of the *lac* genes of *E. coli* led by Jacques Monod and François Jacob. To their work, and that of their many international collaborators, we owe the discovery of: regulatory genes and their protein products, operator regions of DNA where the regulators work, and mRNA transcripts which carry information to the ribosomes for making proteins. Negative regulation by repressor proteins was quickly followed by recognition of positive regulation by activator proteins. Genetic analysis of the regulation of the *trp* operon uncovered still a third mode of regulation, attenuation, which functions by alterations in the secondary structure of a leader sequence of mRNA. Other studies with *E. coli* have uncovered additional means by which this bacterium controls its genes, leading to the conclusion that any step leading from a gene to its ultimate cellular function can serve as a control point in this organism.
4. Genetic code. Mutational studies in the early 1960s with the rII locus of *E. coli* bacteriophage T4 provided the first experimental evidence that the genetic code probably related triplets of nucleotide bases to individual amino acids. In the same era, work with the bacteriophage T4 coat protein and with the *trp*A gene of *E. coli* independently demonstrated the colinearity of gene and polypeptide.
5. Global gene control systems. Realization of the hierarchical nature of gene regulatory networks came about through discovery of regulons (groups of genes controlled by the same regulatory protein) and modulons (groups of operons and regulons that are subject to a common control system). So-called global control systems, which govern the activity of dozens or even hundreds of independent genes, have been characterized in *E. coli*; heat shock, catabolite repression, the stringent response, and emergency repair of DNA damage are well-studied examples.
6. DNA repair and recombination. How damage to DNA is repaired, and how recombination occurs,

has received much attention in *E. coli* genetics. Repair of damage by ultraviolet (UV) radiation has been intensively studied and has led to the discovery of an excision-based repair of minor damage and an inducible system that handles major DNA damage. The latter, called the SOS (or distress) system, involves a dozen or so genes that collectively halt cell division, depress metabolism, and catalyze an efficient, but error-prone, repair of the damaged DNA.

7. Gene cloning *in vivo* and *in vitro*. The use of *E. coli* to clone genes of interest, whatever their source, began in the 1960s with sophisticated use of viral (e.g., bacteriophages M13 and lambda) and plasmid (e.g., ColE1 and its derivative pBR322) vectors. Used alone or, more commonly, combined with *in vitro* recombinant DNA techniques, these cloning procedures have continued to make it possible to isolate individual genes, obtain them in multiple copies *in vivo*, and express their products for further study. Many specialized techniques have been developed in *E. coli* to aid physiological and genetic exploration of cell processes, of which the most common may be fusion of genes of interest to reporter genes (such as *lacZ*) with easily recognized or measured products.

Inheritance in *Escherichia coli*

The formal genetics of *E. coli* (i.e., inheritance, as distinguished from the biochemical nature, action, and regulation of genes) includes the origin of genetic variability, the mechanisms of intercellular genetic exchange, the intracellular mobility of genes, the nature of auxiliary genetic elements, the genetic structure of populations, and evolution.

Mutational Studies

The advantage of working with millions of cells of short generation time is attractive to scientists interested in the nature of mutations and mutagenic agents. Measurement of mutation rates is rather straightforward with bacterial cultures, so different agents and treatments can be assessed for their ability to increase the frequency of mutations. The specific biochemical changes in DNA induced by different chemical mutagens and by physical agents such as UV and X-ray radiation have been characterized, and the means by which the cell repairs the damage has been intensively studied. Interestingly, the issue of Darwinian versus Lamarckian acquisition of mutations has arisen anew in *E. coli*, but with a slightly different twist. The question today is not whether mutations occur at random and are then subjected to environmental

selection: that certainly is true; the question is whether, under stressful circumstances, the rate of mutation is ever increased in favor of mutations relevant to the environmental stress. Experimental results have shown that mutations that restore function in a mutant *lacZ* gene increase during starvation, particularly if lactose is present; other examples of so-called adaptive evolution have been uncovered. As suggested by Margaret Wright, a mechanism can be envisioned by which environmental stress can increase the mutation rate of genes related to relief of that stress; such genes are commonly induced or derepressed in this circumstance, leading to the formation of transcription bubbles, where locally single-stranded DNA could be more vulnerable to damage. This possibility is under active investigation.

The usefulness of mutants was greatly increased by introduction of the technique of conditionally expressed mutations. These mutations, which are expressed only under specified conditions such as high temperature permit the isolation and growth of mutants defective in growth-essential genes. The mutant cells can be grown under permissive conditions and the effect of the loss of the gene's function studied under the restrictive condition.

Suppressors, which partially reverse the effect of a mutation, provide another approach to the isolation of mutants in essential functions. Suppressors can be mutant transfer RNA molecules that at low frequency mistakenly insert an amino acid at a nonsense codon produced by mutation (nonsense suppression), or insert the correct amino acid at an incorrect codon (missense suppression). The antibiotic streptomycin, at low concentrations or in resistant mutants, causes misreading of the genetic code, and this property can be used to isolate and grow streptomycin-dependent mutants in essential genes. The sequencing of the *E. coli* chromosome (consult the web site http://www.genetics.wisc.edu/) has made it possible to clone and mutate each of the 4290 genes (or open reading frames) to study its function.

Genetic Exchange

At the heart of genetic analysis is the ability of the investigator to execute crosses, that is, to mate two individuals of differing phenotype and observe the phenotypic and genetic properties of the offspring. Bacteria are haploid and reproduce by binary fission, each cell dividing into two daughter cells when its mass has doubled. Bacterial geneticists therefore had to search for means to carry out crosses outside the normal reproductive cycle of these cells. For years bacterial geneticists searched in vain for some way to perform crosses with *E. coli*. Eventually three processes were discovered, all of which involve one-way transfer of DNA from a donor cell to a recipient cell.

Conjugation

In 1946, Joshua Lederberg and Edward Tatum demonstrated that two particular strains that had multiple nutritional requirements (auxotrophic mutants) different from each other would, when mixed together, give rise to cells able to grow without any nutritional supplement. These so-called prototrophic recombinants had a complete set of wild-type genes, so it was logical to think that a mating had occurred by cell fusion. But this turned out not to be the case. Cell contact between the two 'parental' strains was necessary, but, as shown by William Hayes in 1952, the two wild-type recombinants did not arise by fusion of a pair of the two different auxotrophs. Rather, one of the strains, the donor, transferred its DNA into the other, or recipient strain, by a process called conjugation.

How conjugation brings about the transfer of bacterial genes is an odd story having to do with plasmid biology. Plasmids are autonomously replicating, circular, double-stranded DNA molecules, much smaller than the chromosome, found in great variety within *E. coli* and probably all bacteria. They confer a great range of properties on the cell. In *E. coli*, 'maleness,' the ability to transfer DNA to a recipient, is related to the presence of the F (for fertility) plasmid within the donor cell. This plasmid carries genes that can bring about the plasmid's transfer from one cell to another. Among many related functions of these genes is the ability to produce a hair-like protein structure called a sex pilus, which helps the male cell (called donor or F$^+$) capture the female cell and maintain a conjugation bridge through which the F plasmid DNA passes. The transfer is initiated by a single-strand break that occurs at a site called *oriT* (origin of transfer). The linearized strand is driven into the female cell (called recipient or F$^-$) by a special mode of replication called transfer replication of the plasmid. The strand entering the recipient cell directs the synthesis of its complementary strand, the completed plasmid DNA circularizes, and its genes become functional. The formerly F$^-$ cell grows a sex pilus and is now functionally a donor, male cell. The donor cell remains F$^+$ because the strand not transferred directs synthesis of its complement. Mixing a population of F$^+$ cells with one of F$^-$ cells results in the massive conversion of the latter to F$^+$ (**Figure 2**).

Transfer of F does not transfer chromosomal genes from donor to recipient. A variation of this process is responsible. Every so often, in a population of F$^+$ cells, the plasmid DNA and the chromosome fuse and form a cointegrated DNA molecule. A cell with

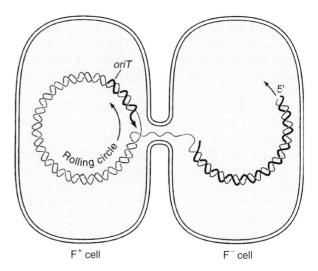

Figure 2 Transfer of F plasmid from an F$^+$ to an F$^-$ cell. Formation of a mating pair triggers transfer replication of F. One strand is nicked at *oriT*, then replication (at arrowhead) occurs by a rolling-circle mechanism. The newly synthesized DNA displaces a preexisting single strand of F, which enters the F$^-$ cell, where its complementary strand is synthesized. (Reproduced with permission from Neidhardt FC *et al.* (1990) *Physiology of the Bacterial Cell: A Molecular Approach*. Sunderland, MA: Sinauer Associates.)

this cointegrated DNA is called an Hfr cell (for <u>h</u>igh <u>f</u>requency of <u>r</u>ecombination). When it encounters an F$^-$ cell, conjugation is initiated as usual by the breaking of the cointegrate DNA at the normal *oriT* site, but in this case entry of a segment of the F genome into the F$^-$ cell brings along the integrated bacterial chromosome. Although a portion of the *E. coli* chromosome enters the F$^-$ cell, the conjugation bridge ruptures long before the entire chromosome and the remaining segment of the F genome can be transferred, so the recipient remains F$^-$.

Hfr cells, though a very small proportion of any F$^+$ population, are the ones responsible for bacterial gene transfer. A pure population of Hfr cells mixed with F cells gives rise to large numbers of recombinants.

Conjugation provided two independent measures of the relative locations of genes from which a genetic linkage map could be constructed for *E. coli*. The frequencies with which genes were separated by crossover events during conjugation provided one measure; the time of entry of genes into the F$^-$ cells, in experiments in which the conjugation process was interrupted by vigorous blending of the mating mixtures at intervals after mixing the Hfr and F$^-$ populations, provided independent information.

One of the early signs that the *E. coli* chromosome is circular came from conjugation experiments employing Hfr strains, with the F plasmid integrated at different sites of the chromosome and in different orientations. The different patterns of gene transmission could be interpreted only with the assumption of a circular linkage map resulting from a circular physical chromosome.

Later physical studies bore out this interpretation. When the F genome occasionally excises from its residence on the chromosome of Hfr cells, the excision is not always precise, and some adjacent bacterial DNA becomes part of the newly formed F plasmid. Such hybrid plasmids, called F′ to signal the presence of bacterial genes, are widely employed for genetic studies, including complementation and *cis/trans* tests, or whenever heterodiploids are useful.

Valuable as it has been for genetic studies on *E. coli*, conjugation must be regarded not as a bacterial process designed for genetic exchange, but as an accidental consequence of the peculiar properties of the F plasmid.

Transduction

While a plasmid is the mediating factor in genetic transfer by conjugation, viruses (bacteriophages) bring about the transfer of DNA from donor to recipient cells in the process called transduction. First discovered in *Salmonella* by Norton Zinder in 1948, transduction quickly became both a tool for genetic analysis and a phenomenon with which to investigate virus–cell interactions in *E. coli* as well.

E. coli bacteriophages are either virulent or temperate. Virions (viral particles) of the former kind infect cells by injection of their DNA (or RNA), take over the host's synthetic apparatus, and direct the production of a new crop of virions associated usually with the lysis of the infected cell. Temperate bacteriophages may initiate a lytic process, but can also produce a lysogen which is an infected cell that carries in quiescent form a copy of the bacteriophage genome (a prophage), either physically integrated into the cell's chromosome or maintained as a plasmid. The lysogen and its offspring can grow indefinitely with this viral passenger, but occasionally in a population of lysogens viral multiplication and virion production will be triggered. Transduction is mediated by temperate phages, and they do so in one of two broad ways. In generalized transduction, any given gene of the host cell has an equal probability of being packaged, by mistake, into the protein capsules of the new virions, forming a pseudovirion (viral particles containing bacterial instead of viral DNA). Infection of a bacterial population with pseudovirions results in the injection of the bacterial DNA and with subsequent recombination with the recipients' chromosome. Generalized transduction occurrs commonly with bacteriophage P1, but can occur with any bacteriophage that forms its mature virions by a process called headful packaging of DNA.

Specialized transduction occurs with those bacteriophages that have a chromosomally integrated prophage. When lysogens of this sort are induced to the lytic process, imprecise excision of the prophage DNA occasionally leads to the incorporation into a virion of a small segment of bacterial DNA along with the truncated phage DNA. Only genes that border the integrated prophage can be picked up in this way, and hence the name 'specialized transduction' or 'restricted transduction' is used for this process. Geneticists have learned, however, to engineer the prophage integration site in order to produce transducing virions of their genes of interest.

Transformation

Transformation is a bacterial process in which DNA released into the environment by the lysis of some cells is directly taken up by other cells and recombined with their DNA. Many bacterial species (notably *Streptococcus pneumoniae* and *Hemophilus influenzae*) have natural mechanisms for the uptake of DNA and are thus said to be competent. Despite its widespread use as a host cell in recombinant DNA technology, involving the necessary uptake of hybrid plasmids, *E. coli* has no functional mechanism for transformation, i.e., it is not naturally competent. Treatment with salt and temperature shocks, or electroporation, must be employed to bring about entry of DNA into *E. coli* and thereby achieve artificial transformation.

Gene Transpositions

How genes move within and between chromosomes of *E. coli* has been an area of great interest for geneticists interested in the mechanism of gene rearrangements, and for medical microbiologists exploring the development and spread of antibiotic-resistance among bacteria. The considerable intracellular mobility of genes within *E. coli* is the result in large measure of transposable elements. Transposable elements are genetic elements that have the ability to catalyze their own movement (transposition), with or without replication, from one DNA site to another, on either the same or a different DNA molecule. The simplest transposable elements, called insertion sequence elements (IS elements), are small (approximately 1000 bp) segments of DNA consisting of terminal inverted-repeat sequences bordering a few genes that encode enzymes for transposition. There are six different IS elements found in the *E. coli* chromosome, each in several copies. IS elements contain only genes for their own transposition, and thus are not readily detectable genetically unless they happen to transpose to a new site within a gene, thereby inactivating it. The second broad class of transposable element consists of transposons, which are segments of DNA-containing genes beyond those needed for transposition, frequently genes encoding enzymes for antibiotic resistance. Many transposons include IS elements at their ends. Transposons promote many types of DNA rearrangements.

Extrachromosomal Genomes

As noted in our discussion of the F plasmid, most if not all *E. coli* strains found in nature contain one or more different plasmids as auxiliary genomes. The variety of cellular properties associated with plasmids goes far beyond fertility and includes production of toxins (including bacteriocins that kill other bacteria), resistance to antimicrobial agents and other toxic chemicals, and especially properties associated with virulence. As a general rule, the plasmids of *E. coli* are involved in interactions of these cells with their environment rather than with metabolism and growth.

Population Structure and Evolution

Study of the genetic structure of populations of *E. coli* and of the origin and evolution of this organism is of relatively recent origin, dating only from the early 1970s, with the publication of studies of electrophoretic variability of proteins in a large number of strains from around the world. Because of their replication by binary fission the structure of bacterial populations is essentially clonal, i.e., populations consist of clones of immense numbers of organisms with an exclusive common ancestor. But recombination following intercellular transfer of genes (mediated by plasmids, viruses, or direct uptake of DNA) modifies this clonal inheritance. One task of population geneticists is to evaluate the contribution of recombination through genetic exchange to *E. coli* evolution. Current work benefits greatly from nucleotide sequence information, including whole genome analysis.

Current Genetic Studies in *Escherichia coli*

The continued use of *E. coli* in fundamental cell research as well as in applied processes in biotechnology derives in large measure from the ease of genetic manipulation of this organism. Genetic analysis provides the major tool for the current study of advanced cellular functions such as motility and chemotaxis, pathogenesis, and cell division in *E. coli*. The information being generated in the field of bioinformatics, with contributions from genomic and proteomic research, is encouraging attempts on the one hand to construct models of the living *E. coli* cell, and on the

other hand to understand the origins and evolution of this model organism. With the complete sequence of the genome of three important strains of *E. coli* known, the outlook for further discoveries is bright.

References

Birge EA (2000) *Bacteria and Bacteriophage Genetics*, 4th edn. New York: Springer-Verlag.

Brock TD (1990) *The Emergence of Bacterial Genetics*. Plainview, NY: Cold Spring, Harbor Laboratory Press.

Miller J (1992) *A Short Course in Bacterial Genetics*. Plainview, NY: Cold Spring Harbor Laboratory Press.

Neidhardt FC Curtiss R III, Ingraham JL *et al.* (eds) (1996) *Escherichia coli and Salmonella: Cellular and Molecular Biology*, 2nd edn. Washington, DC: ASM Press.

See also: Bacterial Genes; Bacterial Genetics; Conjugation, Bacterial; Rolling Circle Replication

ESS

See: **Evolutionarily Stable Strategies**

Established Cell Lines

See: **Cell Lines, Tissue Culture**

Ethics and Genetics

O O'Neill

Copyright © 2001 Academic Press
doi: 10.1006/rwgn.2001.0424

Genetics, Ethics, and Eugenics

All discussion of ethics and genetics takes place in the shadow of abusive use of supposed genetic knowledge in the early and middle years of the twentieth century, especially (but not only) in Nazi Germany. The so-called 'eugenic movement' sought to improve the genetic characteristics of populations, either by encouraging the supposedly "genetically superior" to have children, or by preventing the supposedly "genetically inferior" from doing so. At its worst, eugenic prejudices led to forced detentions and sterilizations and even to extermination, particularly of mentally ill persons and of racial minorities. These were only the most serious aspects of a more pervasive lack of respect for persons and their rights. The complicity of many doctors with Nazi eugenics constitutes a massive and much-studied example of dereliction of professional duties.

The explosive growth in understanding of human genetics at the end of the twentieth century has led to public fears that new, sounder genetic knowledge might be used for eugenic purposes, yet also to public demands that any medical benefits of this knowledge be made rapidly available. A realistic look at the ethical issues raised by genetic knowledge at the start of the twenty-first century reveals a wide spectrum of issues, of potential benefits and dangers, and of ethical difficulties, as well as numerous efforts to devise regulatory structures that will guarantee that the new genetic knowledge is used only for ethically acceptable purposes.

Is Genetics Ethically Distinctive?

Many, but not all, of the ethical problems raised by the new genetics resemble other problems in medical ethics and research ethics. However, genetics also raises some distinctive ethical problems. Some of these arise because genetic information is intrinsically familial, rather than attached solely to individual patients; others because it can sometimes be used to make very long-term predictions (predicting late-onset illness early in life); yet others because it can be used for nonmedical purposes (such as insurance). A more general range of concerns arises from the widespread sense that genetic knowledge and its use may alter our sense of self and family relationships in ways that are hard to foresee.

Genetics and Research Ethics

Genetic research on human subjects can raise a number of distinctive problems. One common problem is that researchers (and sometimes experimental subjects) may acquire genetic information that also pertains to relatives who have not consented to any investigation and need not be made aware of its results. All accounts of research ethics insist that prior consent must be obtained from individual research subjects, and that data obtained must remain confidential. This individualistic position is challenged when the results of investigation are relevant not only to an individual but to a family. Genetic research may also raise distinctive ethical problems if it 'medicalizes' characteristics previously accepted as natural variation.

Genetics and Medical Ethics

The ethical problems arising in clinical genetics are numerous, and mostly similar to those arising in other areas of medicine. When genetic tests are used for diagnosis, when genetic conditions are treated, even if by use of somatic gene therapy, the ethical problems arising will mainly be those that recur throughout medicine. Typical problems will be those of conveying difficult information to patients and their families with adequate care, ensuring that genuinely informed consent to tests and treatment is obtained, preserving confidentiality, and identifying best available treatments (particularly when some treatments are new, risky, or expensive).

However, when genetic tests yield either certain (or probabilistic) information about relatives, or about late-onset conditions, distinctive additional ethical problems can arise. Should consent from certain relatives be sought before genetic tests are undertaken? Should relatives have a right to receive genetic information obtained from others, but which pertains to them? Should unexpected information about undisclosed paternity or nonpaternity be divulged? When, if ever, should genetic tests for late-onset conditions be done on individuals (children, noncompetent adults) who cannot consent for themselves?

Genetics and Reproductive Ethics

By contrast, genetics raises numerous distinctive ethical problems in human reproduction. Prenuptial, preconception, and preimplantation genetic testing (for those using *in vitro* fertilization (IVF)) are all used in various communities or jurisdictions to enable those who otherwise risk having a child with genetic disease to eliminate this risk by avoiding conceiving such children. More controversially, prenatal testing followed by abortion of affected fetuses can be and is used for the same purpose. Genetic tests can also be used to settle paternity either prenatally or later.

Some people fear that these possibilities might revive the old eugenic agenda, others that these practices will lead to lack of respect for those who suffer genetic diseases. One fairly common view is that "negative" uses of genetic tests to avoid disease are permissible, but that their "positive" use to have "designer babies" with genes chosen for reasons other than avoiding disease is wrong. This position is problematic insofar as the boundary between disease and undesired characteristics is blurred.

Germline gene therapy, which would eliminate the genes for certain diseases not only for a patient but for descendants, remains more controversial. Eliminating a gene associated with harmful effects in certain cases might also eliminate beneficial effects it has for carriers or in combination with other factors.

Genetics and Social Issues

Genetic information may be of value not only to patients and their families, and to would-be parents, but more widely. Insurers have argued that they need genetic test information to calculate risks and set premiums more accurately. There has been public worry that those whose test results indicate particularly high risks of disease or early death could be priced out of health or life insurance, so creating a 'genetic underclass.'

In practice, there is so far limited evidence of the actuarial implications of most genetic variations. Risk levels are accurately established only for some serious single-gene disorders. Moreover, even for single-gene disorders, genetic tests for early-onset conditions add little of actuarial value, since information about these conditions is generally included in medical records. If insurers were permitted to request disclosure of all genetic test results (let alone to require that tests be taken) complex ethical problems could arise, particularly in the areas of privacy and data protection.

Genetic test results can also be of relevance in numerous other social contexts. For example, they can be used forensically to identify criminals and to eliminate innocent suspects. They may be of interest to employers who want to know whether employees face particular health risks. Evidently, in these and other contexts, protection of individual rights and control both of genetic testing and of the use of test results will be ethically and politically sensitive, and demand effective regulation.

Ethics and NonHuman Genetics

Genetic information about nonhuman animals has long been seen as valuable: witness the breeding of pedigree animals. As in human genetics, advances in nonhuman genetics have raised additional issues. Some of the most contentious new issues in this area have been about genetic modification or engineering of animals. When this is done to treat human disease without harm to animals ('pharming': e.g., producing sheep that express human insulin in their milk) there is considerable public acceptance. When it is done for research purposes or harms animals there is considerable public opposition and unease (the engineering of animal models for human disease: 'oncomouse'). The prospect of using the organs of genetically modified animals for transplant to humans (xenotransplantation) has aroused both public eagerness about possible benefits and public anxiety on safety and other ethical grounds.

A mixture of ethical concerns also surrounds the genetic modification of plants. Some point to the possibility of harm to the environment, to nonhuman animals, or to human consumers of genetically modified plants (e.g., by crops with built-in insecticide); others point to the possible benefits to the environment, to nonhuman animals, and to humans, for example by reducing the use of insecticides and herbicides and from nutritional improvements.

Further Reading

Adams M (ed.) (1990) *The Wellborn Science: Eugenics in Germany, France, Brazil and Russia.* New York: Oxford University Press.

Agar N (1995) Designing babies: morally permissible ways to modify the human genome. *Bioethics* 9(1): 1–15.

Barkan E (1992) *The Retreat of Scientific Racism: Changing Concepts of Race in Britain and the United States Between the World Wars.* Cambridge: Cambridge University Press.

Brownsword R, Cornish WR and Llewellyn M (1998) *Law and Human Genetics: Regulating a Revolution.* Oxford: Hart Publishing.

Friedman JM (1991) Eugenics and the "New Genetics". *Perspectives in Biology and Medicine* 35(1): 145–154.

Harris J (1993) Is gene therapy a form of eugenics? *Bioethics* 7(2/3): 178–187.

Human Genetics Commission http:/www.hgc.gov.uk/

Kevles D (1985) *In the Name of Eugenics: Genetics and the Uses of Human Heredity.* New York: Knopf.

Kitcher P (1996) *The Lives to Come: The Genetic Revolution and Human Possibilities.* New York: Simon & Schuster.

National Reference Centre for Bioethics Literature http:/www.georgetown.edu/research/nrcb1/ir/kwd2000.htm.

Nuffield Council on Bioethics http://www.nuffield.org.uk/bioethics/

Rollin B (1995) *The Frankenstein Syndrome: Ethical and Social Issues in the Genetic Engineering of Animals.* Cambridge: Cambridge University Press.

See also: Gene Therapy, Human; Genetic Counseling

Ets Family

C Brenner, J-L Baert, and Y de Launoit

Copyright © 2001 Academic Press
doi: 10.1006/rwgn.2001.1567

The Ets family of eukaryotic transcription factors characterized by a strongly conserved DNA-binding domain, called the domain ETS, is composed of more than 30 members and classified in 13 subfamilies depending on the sequence identity of this latter domain as well as on the conservation of other domains/motifs. The founding member of the Ets family, *ets*-1, was discovered in the early 1980s as part of the tripartite oncogene of the E26 avian erythroblastosis virus. In 1990, Ets proteins are found to activate transcription of genes by binding a sequence-specific site in the promoter/enhancer of these target genes. Most Ets proteins are transcriptional activators but some have been characterized as repressors.

Evolutionary Relatedness

Ets genes are conserved throughout the metazon species ranging from diploblastic organisms to *Drosophila* and vertebrates, but they are absent from the genome of plants and yeast. Phylogenetic analyses indicate that the *ets* genes in contemporary species are derived from an ancestral gene early in metazoan evolution. The amplification of such families of transcription factors is viewed as a critical step in the evolution of multicellular animals, including higher vertebrates.

The ETS Domain

The ETS domain identifies all Ets proteins as sequence-specific DNA-binding proteins. This motif, composed of 85 amino acids, forms a winged helix–turn–helix tertiary structure, which allows Ets proteins to interact with an approximately 10 bp long DNA element containing a GGAA/T central core. This recognition motif is present in a vast majority of promoters and enhancers.

Regulating Domains

The other transcriptional modulating domains of the Ets proteins display only very few sequence identities, but are characterized by domains enriched in certain amino acids, i.e., proline, glutamine, or acidic residues. The variability lies in the number and the composition of these domains.

Biological Importance

Biological Role

The Ets factors are expressed in almost all tissues of the organism and control a vast number of target genes. It is expected that different Ets proteins regulate the expression of distinct target genes, thus generating biological specificity. Due to the experimental difficulty of demonstrating this target gene selection, there are only tentative lists that link putative target genes to Ets regulators. Several Ets proteins play an important role in regulating mammalian hematopoiesis and a number of other developmental processes. For example, Ets-1 plays a critical role in the differentiation of hemopoietic stem cells and Tel is critical for fetal angiogenesis. Ets proteins are also implicated in the development and regulation of the immune system, and also in the regulation of genes controlling the cell cycle, neural differentiation, and apoptosis.

Implication of Ets in Cancer

DNA rearrangements in the loci encoding several Ets proteins are associated with tumorigenic processes. Chimeric proteins that contain domains of Ets proteins have been identified in certain types of leukemia, such as B-type childhood acute lymphoblastic leukemia (ALL) and in Ewing's tumors. This chromosomal translocation fuses a fragment of an *ets* gene to an unrelated gene that results in the expression of a chimeric oncoprotein.

Ets proteins are also implicated in the appearance and/or evolution of certain types of cancer and some of them have their own oncogenic potential. In many cancers, there is an overexpression of one or several Ets proteins. For example, Ets-1 is overexpressed in invasive cancer, while the PEA3 subfamily is overexpressed in breast carcinoma.

Regulation of Ets transcription factors

Regulation of DNA Binding

Many Ets transcription factors are subject to autoregulatory mechanisms, which inhibit their DNA-binding activity by domains outside the ETS domain. This may function to prevent promiscuous DNA binding by these transcription factors because of their relatively nonstringent DNA-binding specificity. Moreover, posttranslational modifications, such as phosphorylation, represent other potential mechanisms for regulating DNA binding.

Interactions with Co-Regulatory Partners

Ets proteins interact not only with the basal transcriptional complex, but also with other gene-specific transcription factors. For example, direct physical interaction between Ets factors and b-ZIP proteins represents a conserved mechanism for regulating gene expression in a variety of lymphoid and nonlymphoid cell types.

Ets proteins also functionally cooperate with various transcriptional coactivators, such as the histone acetylase CBP/p300 that modulates the chromatin structure.

Regulation by Signal Transduction Pathways

Differential phosphorylation of transcription factors by signal transduction pathways plays a major role in gene expression. Many Ets transcription factors have been demonstrated to be direct targets of the mitogen activated protein kinase (MAPK) pathway. Most phosphorylation sites are located in the ETS domain and so phosphorylation may play a dual role in regulating transcriptional activation and DNA binding.

Further Reading

Graves B and Petersen JM (1998) Specificity within the ets family of transcription factors. *Advances in Cancer Research* 75: 1–56.

Sharrocks AD, Brown AL, Ling Y and Yates PR (1997) The ets domain transcription factor family. *International Journal of Biochemistry and Cell Biology* 29: 1371–1387.

See also: **Signal Transduction; Transcription**

Euchromatin

A T Sumner

Euchromatin is the term for those parts of chromosomes, generally the greater proportion, which show a normal cycle of decondensation at the end of mitosis. Most genes are in euchromatin, which, however, also contains a very high proportion of nongenic DNA.

See also: **Chromatin; Heterochromatin**

Eugenics

See: **Ethics and Genetics**

Eukaryotes

A eukaryote is an organism whose cells have chromosomes with nucleosomal structure, separated from the cytoplasm by a nuclear envelope and exhibit functional compartmentalization in distinct cytoplasmic organelles.

See also: **Prokaryotes**

Eukaryotic Genes

T M Picknett and S Brenner

Eukaryotic genes may include additional sequences that exist within the coding region, interrupting the protein coding sequence. These introns are excised from the messenger RNA, which represents the exon (coding) regions.

In the interrupted genes of eukaryotes, most introns appear to serve no function, and are removed during gene expression. However, some exceptions exist, notably in the yeast mitochondrion, where an intron itself codes for the synthesis of a protein that functions independently from the protein encoded by the exons.

Not all eukaryotic genes are interrupted. Some correspond directly to the protein product as found in prokaryotes.

See also: **Introns and Exons**

Euploid

J R S Fincham

Euploid is the term that denotes the condition of having a complete normal set of chromosomes, or a multiple thereof. Thus the term includes haploid (n), diploid ($2n$), triploid ($3n$), etc. Organisms that are not euploid ($2n + 1$, $2n - 1$, etc.) are called aneuploid.

See also: **Aneuploid; Diploidy; Polyploidy; Triploidy**

Evolution

B Guttman

Evolution is the process through which organisms change into new types over time, as individuals gradually diverge from one another during the course of their reproduction. The fact that evolution has occurred (and continues to occur) is well documented in an enormous fossil record, and it is attested to by studies of comparative anatomy and comparative molecular structure (for instance, amino-acid sequences of homologous proteins in various species). As the geneticist Theodosius Dobzhansky observed, nothing in biology makes sense except in the light of evolution. The evolutionary history of life on earth may be summarized by a complex branching tree, a phylogeny, showing the relationships of all species to one another and the probable course of their evolution, although, of course, phylogenies are subject to continuing revision and modification like any other pieces of scientific information. At least in eukaryotes, similar phylogenetic trees are generated whether one bases them on the relationships between morphological structures or the sequences of ribosomal RNA or of widely distributed proteins like cytochrome c.

Theory of Evolution

The theory of evolution is based on the fundamental nature of organisms. Organisms are genetic systems (Guttman, 1999). That is, they are self-reproducing systems that operate on the basis of instructions encoded in their genomes, and so during the course of reproduction, parental genomes must be replicated to produce new copies for the offspring. But replication is inherently an error-prone process, and mutation continuously introduce genetic novelties. Furthermore, sexual reproduction entails the reshuffling of chromosomes into different combinations, so individuals acquire variant genomes, thus giving them different traits. Organisms inhabit ecosystems that afford various opportunities for obtaining the resources they need (energy, raw materials, living spaces, etc.). Each organism, with its particular combination of traits, has a particular ability to exploit those resources – in other words, to adapt to a particular ecological niche – and thus experiences a certain level of reproductive success, which is generally measured by an organism's *fitness*. Those with the highest reproductive success have, by definition, the highest fitness, and thus are most successful in passing on their particular genotypes. Thus, organisms are subject to a process of natural selection as those that are most fit for a particular way of life – those that are best adapted to a particular niche – are most successful in reproducing and their genotypes become most common. This, as Darwin recognized, is the most central and most critical process governing evolution, even though other factors also intervene. (The description of the process clearly has a certain tautological character.)

The Modern Synthesis

While some major features of evolution, and especially the centrality of natural selection, became clear from Darwin's great work, a modern consensus on the process only emerged in the early decades of the twentieth century. By about 1940, it was possible to outline a modern synthetic theory combining the essential discoveries of Mendelian genetics with a mathematical analysis of genes in populations as outlined by R.A. Fisher, S. Wright, and J.B.S. Haldane,

and with the observations of taxonomists and field naturalists, as outlined most clearly by Ernst Mayr for animals and by G.L. Stebbins for plants. Since that time, some features of the modern theory of have been challenged, and enormous detail has been added, but the theory as a whole remains successful and intact.

Evolution is fundamentally a population phenomenon. Individuals do not evolve. Populations do. Studies of morphological, genetic, and biochemical features have shown that natural populations harbor enormous variation, and it is generally believed that this variability is the basis for further evolution. It acts as a kind of genetic insurance, a buffer that allows the population to maintain itself by adapting to future environmental changes and perils. (A major concern for the survival of endangered species is the severe reduction in their genetic variability as a result of their reduced populations.) All natural populations are highly polymorphic, at least at a genetic and biochemical level. The classic observations by Dobzhansky and his associates of natural populations of *Drosophila* demonstrated that these populations carry many chromosome types, identified by the inversions they carry, and that the frequencies of chromosome types vary geographically, apparently reflecting subtle adaptations to different environments. Furthermore, the relative frequencies of different chromosomes may change regularly throughout the year, reflecting adaptations to conditions that change with the seasons. It is generally believed that the mere recombination of allelic differences already in a natural population (ignoring further variations created by mutation) is adequate to produce considerable novelty and thus considerable raw material for natural selection in the future.

The evolutionary process is commonly divided into three phases. Microevolution refers to the relatively small changes that occur within populations and individual species; speciation refers to the process in which a single species divides into two or more; and macroevolution refers to the larger changes observed over much longer times as organisms of quite different forms develop. This description revolves to a degree around the concept of a species, which is in itself a matter of considerable controversy at present. Contemporary thinking has been shaped largely by the biological species concept delineated most clearly by Mayr: a species is a series of populations that are actually or potentially capable of interbreeding with one another. This definition is only relevant to sexually reproducing organisms. The concept of a species may be meaningless for those that reproduce asexually, since such organisms are related only by an ever-expanding family tree of cell division after cell division, augmented by occasional lateral genetic transfer, often mediated by viruses. The features of organisms on the many branches of this tree may diverge from one another without limit or may be kept somewhat confined by continuing selection. The biological species concept has been applied most consistently and successfully to certain groups of animals; it may be applied with some difficulty in plants, which often are able to reproduce in more plastic ways and to hybridize with one another quite freely. This conception of a species has been challenged by the phylogenetic species concept, which is much more difficult to define but says, essentially, that a species shall be considered a distinct branch of a phylogenetic tree that can be distinguished morphologically or genetically. The issue may be more anthropological (that is, reflective of the human need to categorize objects neatly) than biological; it is clear in any case that 'species' by any conception have diverged from one another in the past and continue to do so.

Speciation

As described by Mayr and others, speciation probably occurs primarily through geographic isolation. Two populations are said to be sympatric if their ranges overlap and allopatric if they do not. Speciation in many well-documented instances has evidently occurred when one population of a species becomes isolated from the rest. During the time of its isolation, it acquires differences that result in reproductive isolation once the populations again become sympatric. Reproductive isolating mechanisms may entail ecological factors, such as occupying slightly different habitats so prospective mates do not come into contact; temporal factors, such as breeding at different times; and physical barriers to reproduction such as chromosomal rearrangements, incompatibility between sperm and eggs, or failure of hybrid embryos to develop. The records of intense speciation in the past are quite obvious in archipelagos; the ground finches (Geospizinae) of the Galapagos Islands or the honey-creepers (Drepanididae) of the Hawaiian Islands show how one original species has apparently diverged into a considerable variety of species, occupying different ecological niches, as populations probably became quite isolated from one another on different islands.

While allopatric speciation may be a common process in animals, many plants have evolved through genetic events that may occur sympatrically. Plants appear to be much more plastic genetically than animals, and plant development seems to be much more tolerant of major changes in the genome, such as the loss or addition of whole chromosomes and changes from a diploid to a triploid or tetraploid condition (or even higher ploidy). Related plant species often have genomes related by such large changes. A great deal of plant evolution has been explained by introgressive hybridization, in which related species hybridize and one or more chromosomes of one parent species becomes incorporated into the genome of the other, eventually resulting in a third species with features derived from both parents.

Detailed mathematical analysis of the behavior of genes in populations has shown how the frequencies of alleles may be changed by mutation or by various regimes of selection. The rate at which allele frequencies may change depends strongly on the size of the population, leading Sewall Wright to point out that in small populations gene frequencies may change rapidly in directions not determined by natural selection. This phenomenon, called genetic drift, may be very important in speciation. The individuals that become isolated in the first place may themselves have genotypes different from the average genotypes of the parent population – the founder effect; furthermore, genetic drift within the small isolated population may produce just those differences that make for eventual reproductive isolation.

Extinction

The fossil record reveals three major patterns in evolution: speciation, extinction, and phyletic evolution. Speciation has already been described. Extinction is clearly a major feature of evolution. Although a few species have apparently persisted for very long times (in relatively stable environments, such as the depths of the ocean), most species have appeared in the fossil record, have persisted for periods on the order of 100 000 years to a few million years, and then have become extinct. The paleontologist G.G. Simpson estimated that 99.9% of all species have become extinct. Phyletic evolution refers to a gradual change in morphology in a certain direction; for instance, hominid (human) evolution, while involving apparent instances of speciation, has also entailed a gradual increase in height and cranial capacity, and certain trends in anatomical details. However, some paleontologists have proposed that phyletic evolution is illusory and that evolution is more properly described as punctuated equilibrium – that is, a species generally endures with little or no change until it becomes extinct, but occasional instances of speciation occur rapidly, so it may appear that a single species has gradually changed. This may be a non-issue. Instances of both phyletic evolution and punctuated equilibrium can apparently be documented in the fossil record.

The synthetic theory pictured evolution as being driven largely by gradual selection of alleles with small effects, over relatively long times. This viewpoint has been challenged, by champions of punctuated equilibrium with rapid speciation and by proposals that more drastic genetic events might be responsible for quite dramatic changes in morphology. It is clear that very small effects, as demonstrated by selection experiments with animals such as *Drosophila*, can account for the large morphological changes observed in fossil series (Stebbins and Ayala, 1981). Furthermore, speciation that appears to be rapid on the geological time scale may actually require tens of thousands of years, a period perfectly consistent with small, slow genetic events. On the other hand, studies of developmental genetics have revealed genes, such as homeotic genes, that govern major morphological changes, and the growing marriage of developmental biology with evolution may reveal ways that rapid evolutionary change might result from changes in these regulatory genes.

Microbial Evolution

There are clearly instances of lateral transfer, where viruses, for example, carry genes from one species to another or insert themselves in the middle of a gene making substantial changes. That may further cloud the picture at times, and it clearly plays a far greater role in complexities of microbial evolution. The current sequencing of large numbers of microbial genomes is facilitating rapid growth in our understanding of that process, presenting very different pictures of the early stages of cellular evolution and the development of the three kingdoms than those based primarily on ribosomal RNA data. The possibility has been raised that viruses may even provide windows into some of the ancient organisms that disappeared in the bottleneck of the 'last common ancestor' of the three kingdoms; only a fraction of the genes of the large viruses look like anything seen to date in cellular organisms, and a significant number of similarities have been seen between genes of bacteriophages and eukaryotic viruses. There is much evidence, at least, that most families of viruses are very ancient in origin and have coevolved with their various hosts.

Further Reading

Dobzhansky TS (1941) *Genetics and the Origin of Species*. New York: Columbia University Press.

Huxley JS (1942) *Evolution: The Modern Synthesis*. New York: Harper & Sons.

Mayr E (1942) *Systematics and the Origin of Species*. New York: Columbia University Press.

Stebbins GL (1950) *Variation and Evolution in Plants*. New York: Columbia University Press.

References

Guttman BS (1999) *Biology*. Dubuque, IA: WCB/McGraw-Hill.

Stebbins GL and Ayala FJ (1981) Is a new evolutionary synthesis necessary? *Science* 213: 967–971.

See also: **Darwin, Charles; Dobzhansky, Theodosius; Evolutionary Rate; Speciation; Wallace, Alfred Russel**

Evolution of Gene Families
L Silver

Copyright © 2001 Academic Press
doi: 10.1006/rwgn.2001.0433

Genomic Complexity Increases by Gene Duplication and Selection for New Function

Mice, humans, the lowly intestinal bacterium *Escherichia coli*, and all other forms of life evolved from the same common ancestor that was alive on this planet a few billion years ago. We know this is the case from the universal use of the same molecule – DNA – for the storage of genetic information, and from the nearly universal genetic code. But *E. coli* has a genome size of 4.2 megabases (Mb), while the mammalian genome is nearly 1000-fold larger at ~3000 Mb. If one assumes that our common ancestor had a genome size that was no larger than that of the modern-day *E. coli*, the obvious question one can ask is where did all of our extra DNA come from?

The answer is that our genome grew in size and evolved through a repeated process of duplication and divergence. Duplication events can occur essentially at random throughout the genome and the size of the duplication unit can vary from as little as a few nucleotides to large subchromosomal sections that are tens, or even hundreds, of megabases in length. When the duplicated segment contains one or more genes, either the original or duplicated copy of each is set free to accumulate mutations without harm to the organism since the other good copy with an original function will still be present.

Duplicated regions, like all other genetic novelties, must originate in the genome of a single individual and their initial survival in at least some animals in each subsequent generation of a population is, most often, a simple matter of chance. This is because the addition of one extra copy of most genes – to the two already present in a diploid genome – is usually tolerated without significant harm to the individual animal. In the terminology of population genetics, most duplicated units are essentially neutral (in terms of genetic selection) and thus, they are subject to genetic drift, inherited by some offspring but not others derived from parents that carry the duplication unit. By chance, most neutral genetic elements will succumb to extinction within a matter of generations. But even when a duplicated region survives for a significant period of time, random mutations in what were once-functional genes will almost always lead to nonfunctionality. At this point, the gene becomes a pseudogene. Pseudogenes will be subject to continuous genetic drift with the accumulation of new mutations at a pace that is so predictable (~0.5% divergence per million years) as to be likened to a 'molecular clock.' Eventually, nearly all pseudogene sequences will tend to drift past a boundary where it is no longer possible to identify the functional genes from which they derived. Continued drift will act to turn a once-functional sequence into a sequence of essentially random DNA.

Miraculously, every so often, the accumulation of a set of random mutations in a spare copy of a gene can lead to the emergence of a new functional unit – or gene – that provides benefit and, as a consequence, selective advantage to the organism in which it resides. Usually, the new gene has a function that is related to the original gene function. However, it is often the case that the new gene will have a novel expression pattern – spatially, temporally, or both – which must result from alterations in *cis*-regulatory sequences that occur along with codon changes. A new function can emerge directly from a previously functional gene or even from a pseudogene. In the latter case, a gene can go through a period of nonfunctionality during which there may be multiple alterations before the gene comes back to life. Molecular events of this class can play a role in 'punctuated evolution' where, according to the fossil or phylogenetic record, an organism or evolutionary line appears to have taken a 'quantum leap' forward to a new phenotypic state.

Duplication by Transposition

With duplication acting as such an important force in evolution, it is critical to understand the mechanisms by which it occurs. These fall into two broad categories: (1) transposition is responsible for the dispersion of related sequences; (2) unequal crossing-over is responsible for the generation of gene clusters. Transposition refers to a process in which one region of the genome relocates to a new chromosomal location. Transposition can occur either through the direct movement of original sequences from one site to another or through an RNA intermediate that leaves the original site intact. When the genomic region itself (rather than its proxy) has moved, the 'duplication' of

genetic material actually occurs in a subsequent generation after the transposed region has segregated into the same genome as the originally positioned region from a nondeleted homolog. In theory, there is no upper limit to the size of a genomic region that can be duplicated in this way.

A much more common mode of transposition occurs by means of an intermediate RNA transcript that is reverse-transcribed into DNA and then inserted randomly into the genome. This process is referred to as retrotransposition. The size of the retrotransposition unit – called a retroposon – cannot be larger than the size of the intermediate RNA transcript. Retrotransposition has been exploited by various families of selfish genetic elements, some of which have been copied into 100 000 or more locations dispersed throughout the genome with a self-encoded reverse transcriptase. But, examples of functional, intronless retroposons – such as *Pgk2* and *Pdha2* – have also been identified. In such cases, functionality is absolutely dependent upon novel regulatory elements either present at the site of insertion or created by subsequent mutations in these sequences.

Duplication by Unequal Crossing-Over

The second broad class of duplication events result from unequal crossing-over. Normal crossing-over, or recombination, can occur between equivalent sequences on homologous chromatids present in a synaptonemal complex that forms during the pachytene stage of meiosis in both male and female mammals. Unequal crossing-over – also referred to as illegitimate recombination – refers to crossover events that occur between nonequivalent sequences. Unequal crossing-over can be initiated by the presence of related sequences – such as highly repeated retroposon-dispersed selfish elements – located nearby in the genome. Although the event is unequal, in this case, it is still mediated by the homology that exists at the two nonequivalent sites.

So-called nonhomologous unequal crossovers can also occur, although they are much rarer than homologous events. They are "so-called" because even these events may be dependent on at least a short stretch of sequence homology at the two sites at which the event is initiated. The initial duplication event that produces a two-gene cluster may be either homologous or nonhomologous, but once two units of related sequence are present in tandem, further rounds of homologous unequal crossing-over can be easily initiated between nonequivalent members of the pair as illustrated in the **Figure 1**. Thus, it is easy to see how clusters can expand to contain three, four, and many more copies of an original DNA sequence.

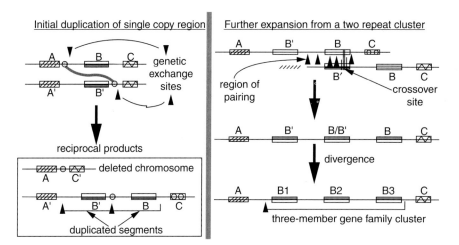

Figure 1 Unequal crossing-over generates gene families. The left side illustrates an unequal crossing-over event and the two products that are generated. One product is deleted and the other is duplicated for the same region. In this example, the duplicated region contains a second complete copy of a single gene (B). The right side illustrates a second round of unequal crossing-over that can occur in a genome that is homozygous for the original duplicated chromosome. In this case, the crossover event has occurred between the two copies of the original gene. Only the duplicated product generated by this event is shown. Over time, the three copies of the B gene can diverge into three distinct functional units of a gene family cluster.

In all cases, unequal crossing-over between homologs results in two reciprocal chromosomal products: one will have a duplication of the region located between the two sites and the other will have a deletion that covers the same exact region (**Figure 1**). It is important to remember that, unlike retrotransposition, unequal crossing-over operates on genomic regions without regard to functional boundaries. The size of the duplicated region can vary from a few base pairs to tens or even hundreds of kilobases and it can contain no genes, a portion of a gene, a few genes, or many.

Genetic Exchange between Related DNA Elements

There are many examples in the genome where genetic information appears to flow from one DNA element to other related – but nonallelic – elements located nearby or even on different chromosomes. In some special cases, the flow of information is so extreme as to allow all members of a gene family to coevolve with near-identity as in the case of ribosomal RNA genes. In at least one case – that of the class I genes of the major histocompatibility complex (*MHC* or *H2*) – information flow is unidirectionally selected, going from a series of 25 to 38 nonfunctional pseudogenes into two or three functional genes. In this case, intergenic information transfer serves to increase dramatically the level of polymorphism that is present at the small number of functional gene members of this family.

Information flow between related DNA sequences occurs as a result of an alternative outcome from the same exact process that is responsible for unequal crossing-over. This alternative outcome is known as intergenic gene conversion. Gene conversion was originally defined in yeast through the observation of altered ratios of segregation from individual loci that were followed in tetrad analyses. These observations were fully explained within the context of the Holliday model of DNA recombination which states that homologous DNA duplexes first exchange single strands that hybridize to their complements and migrate for hundreds or thousands of bases. Resolution of this 'Holliday intermediate' can lead with equal frequency to crossing-over between flanking markers or back to the *status quo* without crossing-over. In the latter case, a short single strand stretch from the invading molecule will be left behind within the DNA that was invaded. If an invading strand carries nucleotides that differ at any site from the strand that was replaced, these will lead to the production of heteroduplexes with base pair mismatches. Mismatches can be repaired (in either direction) by specialized 'repair enzymes' or they can remain as-is to produce non-identical daughter DNAs through the next round of replication.

By extrapolation, it is easy to see how the Holliday model can be applied to the case of an unequal crossover intermediate which can be resolved in one of two directions with equal probability. With one resolution, unequal crossing-over will result; with the alternative resolution, gene conversion can be initiated between nonallelic sequences. Remarkably, information transfer – presumably by means of gene conversion – can also occur across related DNA sequences that are even distributed to different chromosomes. There have been numerous modifications of the Holliday model – including those proposed by Meselson and Rading – that allow a better fit to the actual data, and there is still lack of consensus on the some of the details involved. However, the central feature of the Holliday model – single-strand invasion, branch migration, and duplex resolution – is still considered to provide the molecular basis for gene conversion.

See also: **Gene Conversion; Holliday's Model; Major Histocompatibility Complex (MHC); Molecular Clock; Unequal Crossing Over**

Evolutionarily Stable Strategies

E Pianka

Copyright © 2001 Academic Press
doi: 10.1006/rwgn.2001.0431

When natural selection acts on several different alternative behaviors, the most optimal should be favored. If costs and benefits of alternatives depend on choices made by other individuals, optimal solutions are not always as obvious as they are in simpler situations. An evolutionarily stable strategy, or ESS, is a mathematical definition for an optimal choice of strategy under such conditions.

Interactions between two individuals can be depicted as a mathematical game between two players. A branch of mathematics, called game theory, seeks to find the best strategy to play in any given carefully defined game. The central problem of game theory is to find the best strategy to take in a game that depends on what other players are expected to do.

Originally used in studies of economics and human conflicts of interest, game theoretical thinking was first used in biology by Hamilton (1967) to study evolution of sex ratios. Later, game theory was explicitly applied to behavioral biology by Maynard Smith (1972) and Maynard Smith and Price (1973). Maynard Smith coined the term ESS for a refinement of the Nash equilibrium used by economists to define a solution to a game.

The notion of a Nash equilibrium makes some tacit assumptions about rational foresight on the part of the player. An ESS must meet a stricter set of requirements than Nash equilibria; the mathematical difference boils down to whether a tie between strategies leads to a new strategy being considered better. An ESS attempts to define conditions under which blind evolution will return to the strategy in question, rather than requiring rational foresight to dissuade the exploration of alternatives.

An ESS is a strategy that cannot be beaten by any other strategy. An individual adopting it outperforms any individual adopting any alternative tactic. No other strategy can outperform an ESS. Individuals adopting an ESS tactic have a higher reproductive success than individuals adopting other tactics. Such an unbeatable tactic can go to fixation (100%) in a population and such a population cannot be invaded by any other tactic. Inevitably, an ESS ends up encountering itself more often than it confronts any other strategy, and it must therefore perform better against itself than any other strategy can perform against it.

Game theory involves conflicts of interest in which the value of a given action by a decision maker depends both on its own choices as well as on those of others. A 'payoff' matrix of values of outcomes is postulated based on the respective behaviors of two or more contestants under all possible situations. Payoffs are frequency dependent. Decision rules that represent an evolutionarily stable solution to such an evolutionary game constitute an ESS (Axelrod and Hamilton, 1981).

As an example, consider a well-known game theoretical model called the 'prisoner's dilemma.' In this hypothetical situation, two partners in crime have been arrested. The police interrogate each person alone. Each party could cooperate with the other and steadfastly refuse to squeal on their friend. If both cooperate and remain silent, the authorities cannot establish guilt and both get off scott free (loyalty pays off). Alternatively, each could betray their partner and confess. Now consider respective rewards and punishments received by each partner for making each decision. If only one party confesses while the other remains quiet, this betrayal is rewarded by giving the confessor a light sentence for providing 'state's evidence' and testifying as to the guilt of their loyal silent partner, who is then found guilty and receives a much longer prison term (he gets the 'sucker's payoff'). However, if both partners tell, the authorities put both on trial and both receive moderate, but not long, sentences of imprisonment. In a 'zero sum' game, all losses add up to equal all gains. Not so in this game, where each partner can gain considerably without as much loss to the other (indeed, by working together, both could escape conviction altogether). But they are not allowed to work together and neither knows what the other will do.

Here then, is the classic 'prisoner's dilemma': each prisoner must decide what to do without knowing what decision the other will make. What is the best strategy? Confess to the crime! Any attempt to cooperate could lead to the 'sucker's payoff,' but confession results either in a light sentence or a moderate one. Avoid the worst situation. In such a symmetric nonzero sum game, both partners betray the other's confidence and both do moderate 'time.' Although both partners would have been better off if they had cooperated, the best solution for each person individually in isolation is to defect rather than take the risk of being loyal but being betrayed and ending up with the inglorious 'sucker's payoff.'

The 'prisoner's dilemma' game involves just one decision. Suppose instead, that participants interact repeatedly and that each knows that the other will be encountered again and again. Now many decisions must be made in sequence. In such a situation, "the future can cast a long shadow backwards onto the present" (Axelrod, 1984). Cooperation can evolve under such a long-term situation. Consider the evolutionary game "tit-for-tat," the rules of which are cooperate on the first encounter but then copy the behavior of the other player on all subsequent encounters. Using this strategy, a player always cooperates on its first encounter. But, if player B defects, player A retaliates on its next move. In a population composed of a mixture of players with a variety of behavioral strategies, an individual employing the tit-for-tat strategy does well. When interacting with cooperative individuals, players always cooperate to the mutual advantage of both. If the other player does not cooperate, the two may then retaliate all the time, and the tit-for-tat player will receive none of the advantages of cooperation. The initial attempt at cooperation will incur only a minor cost. The tit-for-tat strategy is most profitable, quickly spreading to fixation. When the entire population employs the tit-for-tat strategy, it cannot be invaded by individuals employing most other tactics – tit-for-tat is normally an ESS (but see below for an exception).

Axelrod (1984) identified three behavioral tendencies that would favor the evolution of cooperation: (1) being 'nice' (never first to defect); (2) being 'provocable' (retaliate against defection); and (3) being 'forgiving.' The first two are the hallmarks of tit-for-tat. The third, allowing bygones to be bygones and resuming cooperation is the strategy known as 'generous tit-for-tat,' unusual in that it can invade tit for tat under certain conditions. Possession of these three behavioral traits make it more likely that both parties will reap the benefits of mutual cooperation. Many highly social animals do indeed display these three behaviors.

The above examples illustrate 'pure' strategies: always adopt a single, best rule of behavior. Such an outcome often arises in contests with just two contestants. However, when an individual must play against an entire population of other individuals, ESS solutions are often 'mixed,' with probabilistic rules determining the chosen strategy. In a particular situation, be a bully with probability p but be cowardly with probability q. At equilibrium, a fraction p of the population will be bullies and another fraction q will be cowards, with each tactic doing equally well overall. Overall benefit to all bullies equals overall benefit for all cowards. If the proportions in the population deviate toward too many bullies, cowards outperform bullies, whereas if there are too many cowards, bullies perform better. This is the classic hawk–dove game. Sex ratios are similar: if males are in short supply, on average an individual male will contribute more genes to the next generation than an individual female (and vice versa if females are scarce). These are also examples of frequency-dependent selection.

ESS rules can also be 'conditional,' taking a form like "if hungry, be a bully, but if satiated be a coward" (Enquist, 1985). In the real world, most behaviors are probably closely attuned to such immediate environmental situations. Often, combatants are not equal, leading to conditional rules, such as "fight if I'm bigger" but "flee if I'm smaller" (Hammerstein, 1981). Such rules lead to pecking orders with larger animals dominant over smaller ones. Because even the winner can be injured in a fight, fights are best avoided by both contestants if the outcome is already relatively certain. Often, ritualized appeasement behaviors and postures are adopted by the loser, effectively curtailing aggressive behaviors of winners. Indeed, fights only make evolutionary sense when two contestants are closely matched and each is equally likely to win (Enquist and Leimar, 1983). In such a situation, fights escalate and serious injuries can occur. Often the loser gives up abruptly and flees, but holds its stance almost as a bluff, right up until the end. Among many animals, residents typically win in encounters with vagrants – the first animal to arrive seems to acquire ownership and the motivation to defend its turf. Game theory easily accommodates such flexible behavior (Maynard Smith and Parker, 1976). The ESS approach has been particularly useful in analyzing the evolution of communication (Johnstone, 1997).

References

Axelrod R (1984) *The Evolution of Cooperation.* New York: Basic Books.

Axelrod R and Hamilton WD (1981) The evolution of cooperation. *Science* 211: 1390–1396.

Enquist M (1985) Communication during aggressive interactions with particular reference to variation in choice of behaviour. *Animal Behavior* 33: 1152–1161.

Hamilton WD (1967) Extraordinary sex ratios. *Science* 156: 477–488.

Hammerstein P (1981) The role of asymmetries in animal contests. *Animal Behaviour* 29: 193–205.

Johnstone RA (1997) The evolution of animal signals. In: *Behavioural Ecology: An Evolutionary Approach*, 4th edn, pp. 155–178. Oxford: Blackwell Scientific Publications.

Maynard Smith J (1972) *On Evolution*. Edinburgh: Edinburgh University Press.

Maynard Smith J and Price GR (1973) The logic of animal conflict. *Nature* 246: 15–18.

Maynard Smith J (1982) *Evolution and the Theory of Games*. Cambridge: Cambridge University Press.

Maynard Smith J and Parker GA (1976) The logic of asymmetric contests. *Animal Behaviour* 24: 159–175.

See also: **Evolution; Hamilton's Theory**

Evolutionary Rate

R E Lenski

Copyright © 2001 Academic Press
doi: 10.1006/rwgn.2001.0432

An evolutionary rate is used to describe the dynamics of change in a lineage across many generations. The changes of interest may be in the genome itself or in the phenotypic expression of underlying genetic events.

For example, one might be interested in the evolutionary rate during the domestication of corn (*Zea mays*) from its teosinte ancestor (*Z. parviglumis* or a related species). One would first need an estimate of the time since their divergence from a common ancestor, which in this case is approximately 7500 years ago based on archeological evidence from Mesoamerica (where corn was domesticated). The evolutionary rate of genetic change could be ascertained by comparing DNA sequences, ideally for several genes from a number of individuals of each species. To a first approximation, the rate of change can be expressed as the number of differences, per base pair sequenced, per year of divergence, where the time of divergence of two lineages is twice the time since their common ancestor. In practice, several issues may necessitate more complex analyses, such as the possibility that a single difference in DNA sequence may reflect multiple evolutionary changes; this particular effect is most pronounced when the sequences are highly divergent.

The evolutionary rate of phenotypic change could be obtained by comparing the values of one or more traits of interest, such as the number of seeds produced per ear or the concentration of oil in the seeds. These traits may depend on environmental influences, such as soil fertility, as well as on genetic changes; it is therefore important that the corn and teosinte plants be grown under the same conditions to isolate the effect of evolutionary changes in genotype from the direct effects of environment. The rate of change in a given phenotype could then be calculated as the difference in the average value of the trait in the two species, divided by twice the time of divergence. Note, however, that this calculation may give a misleading picture, as the common ancestor may not have had a trait value intermediate to the values in modern corn and teosinte. Indeed, it is likely in this case that the ancestor was much more like present-day teosinte, with most of the phenotypic change having occurred as a consequence of rapid evolution of corn under domestication.

Evolutionary rates differ quite substantially from one case to the next, and for a variety of reasons. In the broadest terms, evolutionary change at the genetic level depends on the interplay of several processes, including mutation, which produces new genetic variation, and natural selection, which influences the fate of any particular genetic variant. A few examples serve to illustrate two of the most important factors that influence rates of genetic evolution.

Replication and Repair

All cellular organisms have DNA as their hereditary material, but some viruses, including HIV (which causes AIDS) and influenza virus, use RNA instead. These RNA viruses undergo extremely rapid sequence evolution because RNA replication lacks the proofreading and repair processes that increase the fidelity of DNA replication. Even among the DNA-based bacteria, there exist mutants that are defective in DNA repair, and these 'mutators' should evolve much faster at the level of their DNA sequence.

Functional Constraints

A mutation may be deleterious, neutral, or beneficial in terms of its effect on an organism's reproductive success. Deleterious and neutral mutations are both very common, whereas beneficial mutations are much rarer and thus have less effect on variation in evolutionary rates at the genetic level. Because of the redundancy of the genetic code, some point mutations in protein-encoding genes (especially those at the third position in a codon) will not actually alter the amino acid sequence of the protein. Such synonymous mutations are therefore likely to be neutral. By contrast, nonsynonymous mutations cause a change from one amino acid to another, and such mutations often have deleterious consequences for the protein's function and, ultimately, the organism's performance. The extent to which nonsynonymous mutations are deleterious depends on their particular position within a gene as well as on the particular gene. Mutations that alter critical positions in a protein's structure are usually more harmful than those that affect a less crucial site. Evolutionary approaches can be used to identify conserved sequences, which in turn suggest potentially important features of protein structure and function. Among different genes, those that encode essential and highly constrained proteins can tolerate fewer mutations than those that encode less constrained proteins, which may accept a wider range of mutations without compromising the organism's performance. For example, the rate of amino acid substitution in fibrinopeptides (proteins involved with blood-clotting) is more than 100 times faster than the corresponding rate in histones (proteins used to package DNA in eukaryotic chromosomes).

Neutral mutations serve as a sort of benchmark for understanding evolutionary rates, and they lead to the notion of a 'molecular clock' to describe genetic evolution. Population genetic theory shows that the expected evolutionary rate of genetic change for neutral mutations depends only on the underlying rate at which these mutations occur, and not on population size or natural selection. This simple result can be understood as follows. Let μ be the rate of neutral mutation and N be the population size, so that each generation $2N\mu$ new neutral mutations arise in a diploid population. Because they are neutral, each of these mutations has no greater or lesser chance of eventually being substituted in the population than any other of the $2N$ alleles present at a locus; in other words, a neutral mutation has a probability of $1/2N$ of becoming substituted. Given these considerations, the overall rate of substitution of neutral mutations is $2N\mu \times 1/2N = \mu$. In other words, the rate of genetic evolution would, in the case of neutral mutations, behave like a stochastic molecular clock which ticks at the rate μ.

Rates of phenotypic evolution are even more complex and variable. Whereas the balance between neutral and deleterious mutations is especially important for understanding rates of genetic evolution, neither of these classes is thought to play much role in phenotypic evolution – neutral mutations because they have no outward manifestation, and deleterious mutations because they will be eliminated by natural selection. Instead, phenotypic evolution depends on beneficial mutations, which are rare but extremely important because they provide raw material for organisms to adapt evolutionarily to their environments. Species that live in environments that hardly change over long periods of time typically show very slow rates of phenotypic evolution. Such organisms have presumably run out of ways to become better adapted to their environment, accounting for their phenotypic stasis. The horseshoe crab (*Limulus polyphemus*) is one such 'living fossil'; its outward appearance is very similar, although not identical, to fossils from more than 200 million years ago. At the other extreme, organisms in new environments often experience different selective agents and constraints from their ancestors, thus promoting rapid phenotypic evolution as they adapt genetically to their new environment. The conspicuous differences between domesticated plants and animals and their wild progenitors provide many examples of very rapid change. Another interesting example is the morphological divergence of Darwin's finches (*Geospiza* spp.) in the Galapagos Islands, where these birds experienced an environment different from their mainland ancestor. A critical factor in their rapid evolution was their release from competition with other species, which presented the island populations with the opportunity to fill ecological roles that would otherwise not have been available.

See also: Genetic Drift; Molecular Clock; Mutators; Natural Selection; Retroviruses

Ewing's Tumor

N Coleman

Copyright © 2001 Academic Press
doi: 10.1006/rwgn.2001.1568

Ewing's tumor is a malignant neoplasm of bone and soft tissues, known by several alternative names, including peripheral primitive neuroectodermal tumor, neuroepithelioma, and Askin's tumor (when affecting the chest wall). The tumor usually develops in children and adolescents. The neoplastic cells are primitive, although there is varying evidence of neuroectodermal differentiation. The cells contain characteristic chromosomal translocations, producing fusions between the *EWS* gene at chromosome 22q12 and several members of the Ets family of transcription factors, most frequently FLI1 at chromosome 11q24. The Ews–Ets family fusion genes are likely to contribute to neoplastic progression by induction of a range of secondary transforming genes.

See also: Ets Family

Exchange

J R S Fincham

Copyright © 2001 Academic Press
doi: 10.1006/rwgn.2001.0434

In the context of genetics, this usually means exchange of segments between chromosomes. Exchanges between equivalent segments of paired homologous chromosomes occur regularly in meiosis, and occasionally in mitosis (see Crossing-Over). Exceptionally, crossing-over can occur between chromosomes which are paired out of register, to give products of unequal size (see Unequal Crossing Over).

Exchanges between nonhomologous segments to give structurally rearranged chromosomes occur as rare aberrations, the frequency of which is greatly increased by chromosome-breaking agents such as X-rays (see Segmental Interchange).

The term exchange may also refer to the point in a single recombinant chromosome or nucleic acid molecule where the sequence switches from one parental type to the other. Thus a bacteriophage particle emerging from a mixedly infected bacterial cell may be said to have undergone one or more exchanges in its genome without any implication as to the nature, reciprocal or nonreciprocal, of the recombination process.

See also: **Crossing-Over; Genetic Recombination; Segmental Interchange; Unequal Crossing Over**

Exchange Pairing

See: **Exchange, Segmental Interchange**

Excision Repair

J T Reardon and A Sancar

Copyright © 2001 Academic Press
doi: 10.1006/rwgn.2001.0437

Faithful maintenance of the genome is important for survival of both the species and the individual. While stability is the hallmark of genome maintenance, the DNA molecule itself is susceptible to alterations as it is the target for a variety of reactive molecules that damage and modify DNA. Typically we are not aware that such damage has occurred because cells have mechanisms for the error-free removal of DNA damage and restoration of the DNA molecule to its original unmodified state. If DNA damage is not removed, mutations (permanent changes in the genetic code) may result, and mutations in critical genes are important events in cancer initiation and progression.

DNA Damage

At some point in time virtually all cells are exposed to endogenous and environmental agents that damage the genome. Genetic damage is a rare event as cells possess multiple mechanisms for eliminating or neutralizing genotoxic substances before they damage DNA. But damage does occur and sources of modification include rare misincorporation events during DNA replication, normal cellular metabolism involving oxygen and water which generate DNA-damaging free radicals, and extracellular sources such as environmental chemicals and sunlight (UV radiation). Damage may include base modification or cleavage of the phosphodiester backbone such that RNA and DNA polymerases are blocked at the lesion (modified base or strand break) and unable to translocate along the helix, thus interrupting normal DNA replication and RNA transcription.

General Comments on DNA Repair

In both prokaryotes and eukaryotes, a major cellular mechanism for the removal of DNA damage is nucleotide excision repair (excision repair), an enzymatic pathway that recognizes and corrects a wide spectrum of structural anomalies (DNA lesions) ranging from bulky, helix-distorting adducts to nonhelix-distorting lesions. The modifications that transform normal bases into damaged bases corrected by nucleotide excision repair are so diverse that it is unlikely that a specific chemical structure is recognized. Rather, it appears that any abnormal DNA structure that destabilizes (denatures) the double helix is recognized as damage both in *Escherichia coli* and human cells.

The primary function of nucleotide excision repair is removal of bulky adducts generated by chemicals or UV radiation, while base excision repair is the major pathway for correction of non-helix-distorting lesions such as those introduced by ionizing radiation or cellular metabolic events. Additional pathways exist for direct reversal of certain types of damage (e.g., photolyase and methyltransferase), correction of mismatched bases, removal of interstrand crosslinks, and repair of DNA strand breaks. Excision repair involves removal of a damaged nucleotide by dual incisions bracketing the lesion; this is accomplished by a multisubunit enzyme referred to as the excision nuclease or excinuclease. The basic mechanism of excision repair involves: (1) damage recognition; (2) subunit assembly; (3) dual incisions that result in excision of the damage-containing oligomer; (4) resynthesis to fill in the gap; and (5) ligation to regenerate an intact molecule.

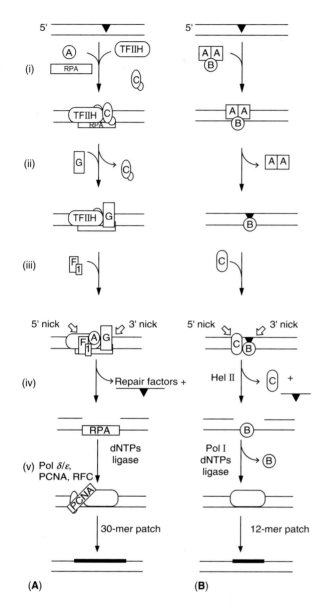

Figure 1 Models for nucleotide excision repair in humans (A) and in *Escherichia coli* (B). (i) DNA damaged by UV radiation or chemicals is a substrate for excision repair and repair is initiated by recognition of the damage and formation of a stable complex at the damage site. This is accomplished by RPA, XPA, XPC, and TFIIH in humans and by UvrA and UvrB in *E. coli*. (ii) Each system has a protein that functions as a molecular matchmaker. In humans XPC•HR23B recruits XPG to the preincision complex and in *E. coli* UvrA delivers UvrB to the damage site; both XPC and UvrA then leave, having matched the 3′ endonuclease to the damaged DNA. (iii) In both pathways the last member of the excinuclease to assemble is the enzyme responsible for the 5′ incision event. This is the XPF•ERCC1 heterodimer in humans and UvrC in *E. coli*. (iv) Dual incisions follow rapidly and the damage-containing oligomer is released as the excision nuclease dissociates. In humans this is accomplished without additional proteins, while *E. coli* requires an accessory repair helicase (Hel II, product of the UvrD gene) to release the damage-containing oligomer and UvrC. (v) In humans resynthesis of the gap is accomplished by polymerase δ and ε, their accessory factors RFC and PCNA, and a DNA ligase to generate a 30-nucleotide repair patch. In *E. coli* a 12-mer patch is resynthesized by polymerase I and ligated to the parental DNA.

Excision Repair in *Escherichia coli*

UvrA, UvrB, and UvrC constitute the *E. coli* excision nuclease, (A)BC excinuclease. UvrA binds specifically to both damaged DNA and UvrB and, by virtue of these interactions, delivers UvrB to the damage site (see **Figure 1**). UvrA, a molecular matchmaker, then dissociates from the UvrB–DNA complex. UvrC interacts with UvrB bound to DNA and the two acting together make the 3′ incision, and then UvrC makes the 5′ incision on the damaged strand. These concerted reactions begin with hydrolysis at the fourth or fifth bond 3′ to the damage, followed within a fraction of a second by incision at the eighth phosphodiester bond 5′ to the lesion. UvrD is a helicase that releases both UvrC and the 12–13 nucleotide-long damage-containing oligomer. Repair is completed by DNA polymerase I, which synthesizes the repair patch and

displaces UvrB, and by DNA ligase, which ligates the newly synthesized DNA to the parental DNA.

Excision Repair in Mammalian Cells

Biological Relevance of Excision Repair
The physiological importance of nucleotide excision repair is illustrated by a rare human hereditary disease, xeroderma pigmentosum (XP), caused by mutations in any of seven genes named *XPA* through *XPG*. XP patients are extremely sensitive to sunlight and have an increased incidence of skin and certain internal cancers. Cultured cells derived from XP patients are hypersensitive to both killing and mutation induction by UV light and chemicals and, biochemically, this hypersensitivity has been correlated with defects in nucleotide excision repair.

Mechanism of Nucleotide Excision Repair
In contrast to the three-subunit (A)BC excinuclease employed in *E. coli*, the human excision nuclease utilizes 15 polypeptides in six repair factors for the basal steps, which include damage recognition and

sequential assembly of subunits leading to the dual incision event. These six factors are XPA, RPA, TFIIH (XPB and XPD plus four additional polypeptides), XPC•HR23B, XPG, and XPF•ERCC1. Following damage recognition by XPA and RPA, XPC and TFIIH are recruited to the damage site (see **Figure 1**) to form the first stable preincision complex. The initial, localized helical denaturation resulting from DNA damage is extended both 5′ and 3′ by the helicase activities of two TFIH subunits, XPB and XPD. XPC helps to stabilize this open complex and, furthermore, XPC is a molecular matchmaker that dissociates after recruiting and positioning XPG 3′ to the DNA damage. The last factor to assemble is the XPF•ERCC1 heterodimer. Dual incisions follow rapidly with XPG nicking the DNA at the sixth ± 3 phosphodiester bond 3′ to the damage and XPF•ERCC1 hydrolyzing at the 20th ± 5 bond 5′ to the lesion. The 24–32 nucleotide-long oligomer containing the damaged base is released from the DNA (excision) and repair factors rapidly dissociate following the dual incision event leaving a gapped substrate. In subsequent steps, DNA polymerases δ and ε and their accessory factors, PCNA and RFC, assemble at the gapped molecule and the undamaged strand is used as a template for precise resynthesis of the DNA. The repair patch size matches the size of the excision gap and, when the gap is filled to the 3′ end, the repair patch is ligated to the parental DNA by a ligase.

Further Reading

Petit C and Sancar A (1999) Nucleotide excision repair from *E. coli* to man. *Biochimie* 81: 15–25.

Sancar A (1996) DNA excision repair. *Annual Review of Biochemistry* 65: 43–81.

Wood R D (1996) Excision repair in eukaryotes. *Annual Review of Biochemistry* 65: 135–167.

See also: DNA Repair; Xeroderma Pigmentosum

Exon

See: **Introns and Exons**

Exonucleases

Exonucleases are enzymes that digest the ends of a piece of DNA. The nature of the digestion is usually specific (e.g. 5′ or 3′ exonuclease). Exonuclease III (exo III), for example, is used to prepare deletions in cloned DNA, or for DNA footprinting.

***See also:* Endonucleases; Footprinting; Nuclease**

Expression Vector

An expression vector is a vector designed for the expression of inserted DNA sequences propagated in a suitable host cell. The inserted DNA is transcribed and translated by the host's cellular machinery.

***See also:* Vectors**

Expressivity

J A Fossella

Expressivity refers to the variation seen among individuals expressing a particular trait or mutant phenotype. 'Variable expressivity' is the term used to describe a trait or mutant phenotype that fluctuates in degree or severity from individual to individual in a population. For example, all individuals of a population expressing a trait or mutant phenotype such as 'spotted' may show an identical number of spots. This would be an example of low or nonvariable expressivity. Alternatively, some individuals may have many spots while others only a few and many with an intermediate number of spots. This would be an example of variable expressivity, since all the individuals express the trait or mutant phenotype of 'spotted' but vary in the degree of spotting.

Expressivity is similar in meaning to 'penetrance' and the two terms are often used together when describing mutations. For example, certain weak alleles of the *W* locus seen in mice result in white coat color spots. These mutant alleles are said to show reduced penetrance and variable expressivity. The distinction between penetrance and expressivity is that penetrance refers to the genotype while expressivity refers to the phenotype. In this example, only some of the mice that carry the *W* /+ genotype show any spots at all. This is an example of reduced penetrance. Of the animals that show the spotted 'phenotype' however, some tend to show much spotting while others show very little spotting. This is an example of variable expressivity.

The phenomena of variable expressivity and reduced penetrance have a similar root cause. The phenotypic effects of a specific gene are highly contingent on the environmental conditions that exist during the development of an organism and during maturity. The effects of a specific gene are also dependent on other modifier genes in the same developmental or physiological pathway. Hence, variation in the environment and in modifier loci among individuals in a population may alter the phenotypic effects of a specific gene or mutation resulting in reduced penetrance and variable expressivity.

See also: **Penetrance; *W* (White Spotting) Locus**

Extranuclear Genes
See: **Cytoplasmic Inheritance**

F Factor

S M Rosenberg and P J Hastings

Copyright © 2001 Academic Press
doi: 10.1006/rwgn.2001.0454

The F (for fertility) factor is a conjugative plasmid of *Escherichia coli*. It was the first plasmid discovered and has been significant in the development and practice of bacterial genetics. Like other conjugative plasmids, the F factor encodes the machinery for its own conjugative transfer and for the transfer of other DNA molecules that contain transfer origins – specific sequences that allow them to be mobilized (recruited and transferred) by the F-encoded transfer proteins, during bacterial conjugation.

Structure of the F Factor

The F factor is 100 kb of duplex DNA with two replication-origin regions (**Figure 1**). The *oriV* or vegetative replication region contains two replication origins, one of which is used for bidirectional maintenance replication of the plasmid when it is not being transferred to another cell. *oriT*, the transfer origin, promotes a special mode of unidirectional, single-(leading) strand replication used during conjugative transfer of the F factor to another cell. The copy number control of the F factor is similar to that of the chromosome such that there are one or two copies per bacterial chromosome. This feature has made the F factor useful to workers wishing to perform complementation and dominance tests with their gene in a single copy replicon in *E. coli*. This allows creation of a state of partial diploidy (also called 'merodiploidy'). Originally, this was done by isolation of F′ plasmids: F factors that have incorporated often large segments of DNA from the bacterial chromosome by homologous recombination with the chromosome. Formation of F′ plasmids is described (below) in "Importance of the F Factor in Bacterial Genetics" (see **Figure 2**). Since the advent of recombinant DNA technology, smaller derivatives of the F factor have been constructed, including roughly 9-kb mini-F plasmids, containing just the *oriV* region, and the 55-kb pOX

plasmids, containing DNA from *oriV* clockwise to the far end of the transfer region.

The F factor encodes genes for sexual pili, thin rod-like structures with which F-carrying (male or donor) bacteria attach to F⁻ (female or recipient) cells for conjugative transfer. The F factor carries an operon of about 30 genes, encoding Tra proteins promoting transfer (**Figure 1**). Importantly for bacterial genetics, the F factor also contains four transposable genetic elements: two copies of the insertion sequence IS*3*, one IS*2*, and one transposon Tn*1000* (also called γδ). These elements are important in two respects. First, because they are also present in the *E. coli* chromosome, the transposable elements provide regions of DNA at which homologous recombination occurs between the F factor and the chromosome. The F

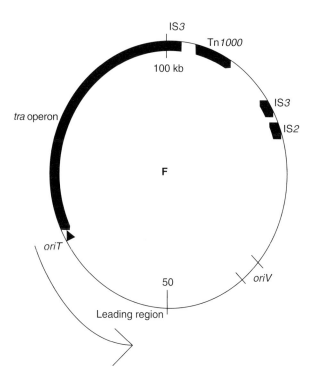

Figure 1 The F factor. The F factor is a 100-kb conjugative plasmid. The *tra* operon encodes functions required for conjugative transfer of the F factor. Transposable elements are indicated: IS*3*, IS*2*, and Tn*1000*, and the direction of transfer is indicated by the thin arrow. (Modified from Firth *et al.*, 1996.)

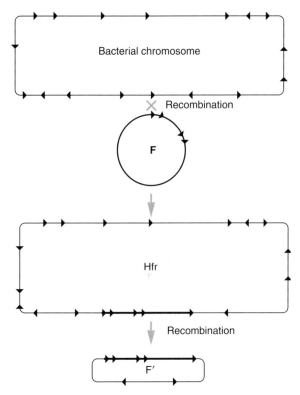

Figure 2 Formation of Hfr and F′ molecules by homologous recombination of the F plasmid with the bacterial chromosome. Transposable elements are represented as triangles, and single lines represent duplex DNA. The transposable elements present in the F plasmid provide regions of sequence identity with the *E. coli* chromosome and so allow the F plasmid to become incorporated into the chromosome via homologous recombination, to form an Hfr. Once incorporated, recombination may occur between transposable elements other than those that recombined upon integration of the F plasmid. This can produce an F′ plasmid.

factor can integrate into the chromosome, forming an Hfr strain by this route (**Figure 2**). The F factor is therefore an episome, that is, a replicon that can exist either outside, or integrated into, the bacterial chromosome. Second, the Tn*1000* insertion interrupts the *finO* (*fertility inhibition*) gene. In other, similar conjugative plasmids, the FinO protein represses expression of the *tra* or transfer operon genes such that they are inducible upon mating. In the F factor, their expression is constitutive.

Interesting F Factor Products that May Affect DNA Metabolism

Other genes carried by the F factor encode proteins that probably affect DNA metabolism in the recipient bacterium during conjugative transfer. The leading region of the F factor, that is, the region that is transferred first, encodes a single-stranded DNA binding protein, Ssb, a protein (PsiB) that inhibits the SOS response by modifying RecA protein, and Flm, the F leading maintenance protein (also called ParL and Stm). The F factor also encodes the Ccd plasmid addiction system. Plasmid addiction systems consist of a stable toxin protein and a labile antidote protein. If the plasmid is lost from a cell, degradation of the antidote leads to killing of the cell by the stable toxin. The Ccd toxin binds the topoisomerase DNA gyrase, resulting in it functioning like a double-strand endonuclease. Flm is part of a different plasmid addiction system, with a different postsegregational killing mechanism. The plasmid addiction systems, plus the infectivity of the F factor between cells, species, genera, and domains, give an impression of a selfish DNA element. Although the F was once considered a narrow-host-range conjugative plasmid, the discovery of its transfer to distantly related bacteria and even to yeast has changed this classification.

Conjugative Transfer

Cells carrying the F factor are called male or donor cells. They express long, rod-like pili on their surfaces and use these to attach to female cells for transfer of the F factor. Once attached, the pili retract, bringing the mating pair into close contact. The TraI endonuclease makes a single-strand nick at *oriT*, and, with its helicase activity, peels back the 5′ end to which it remains covalently bound. The 3′ end primes leading strand synthesis that displaces the 5′-ending strand. The displaced strand is transferred into the recipient cell. Whether the DNA is transferred through a pilus or via some other close contact is not yet clear. The synthesis and strand displacement end when the whole single-strand length of the circle has been displaced and the 3′ growing end again reaches *oriT*. TraI is hypothesized to nick again, releasing the end of the displaced strand, and to assist recircularization of the ends in the recipient cell. Meanwhile, the complement of the transferred single-strand is synthesized in the recipient cell, such that a duplex circle is reestablished. The recipient thus becomes an F-carrying male, and the donor remains male. The Tra proteins can act on other bacterial plasmids with similar origins of transfer, including the ColE1 plasmids (from which pBR322, pUC, and many other cloning vectors are derived). The process of recruiting and transferring other plasmids is called mobilization. The sites on those plasmids that allow mobilization are called *mob* and the nick site itself (*oriT*, which is necessary but not sufficient for transfer) has also been called *bom* and *nic*. pBR322 lacks *mob* but carries *bom*, and cannot be mobilized by the F factor unless a third

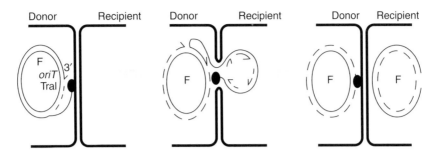

Figure 3 Conjugative transfer of the F plasmid. Each line represents a DNA strand; dashed lines represent newly synthesized DNA, and arrowheads 3′ ends. During transfer, the F-encoded TraI endonuclease cleaves one strand of DNA at the transfer origin, *oriT*, and remains covalently bound to the 5′ end. Leading-strand synthesis primed from the 3′ end displaces the cleaved strand, which is transferred into a recipient cell. Lagging-strand synthesis and recircularization occur in the recipient, regenerating an F plasmid there.

plasmid, apparently supplying *mob* function (ColK), is present. pUC plasmids contain neither *mob* nor *bom* sites and so cannot be mobilized.

Importance of the F Factor in Bacterial Genetics

The original isolate of *E. coli* K12, from Stanford, carried an F plasmid. When Edward Tatum turned to *E. coli* for generalization of his biochemical genetic studies with Beadle (which led to the "one gene, one enzyme" hypothesis) in the fungus *Neurospora*, he made auxotrophic mutants of *E. coli* K12. To bring about mutagenesis, he used large doses of radiation, which caused loss of the F factor in some of the derivative strains. Joshua Lederberg's interest in testing whether mating could occur between different *E. coli* auxotrophic mutant strains, to give proto-trophic recombinants (the selection of which he invented), led to his joining Tatum and using K12-derived strains (Lederberg and Tatum, 1946b) Because some of the strains had lost their F factor and others had not, Lederberg discovered mating and recombination in bacteria. In strains that retained the F factor, the F factor could integrate into the bacterial chromosome. The integrated F factor can transfer segments of chromosomal DNA contiguous with its integration site during conjugation, and these can be recombined into the recipient chromosome, resulting in the proto-trophic recombinant bacteria reported by Lederberg and Tatum in 1946 (Lederberg and Tatum, 1946b) (Hfr). The results encouraged the idea that bacteria, like other organisms, had genes, and led to much of our current understanding of DNA recombination.

Strains with the F factor integrated are called Hfr (<u>h</u>igh-<u>f</u>requency <u>r</u>ecombination) strains (Hfr). The integrated F factor can be excised from the chromosome using homologous recombination with the same insertion sequences used upon its integration to regenerate an F$^+$ plasmid (a wild-type F plasmid with no bacterial DNA incorporated into it). If different insertion sequences from the bacterial chromosomal DNA are used for direct repeat recombination excising the F factor, then the F factor brings with it chromosomal DNA, forming an F′ factor (**Figure 2**).

The discoveries by William Hayes, Elie Wollman, and François Jacob that the F factor is a plasmid, and the subsequent discoveries of other bacterial plasmids, made possible the development of plasmid vectors for molecular cloning (Hayes, 1952; Wollman *et al.*, 1956; Jacob and Wollman, 1958). Fs are important replicons used in single-copy gene-complementation experiments and in tests of dominance. As discussed in Hfr, the use of Hfrs for studies of bacterial recombination led to the characterization of mechanisms and proteins used in homologous genetic recombination in *E. coli*, and, because the DNA transferred in Hfr crosses is linear, the enzymes used in double-strand break-repair were illuminated in these studies. Descriptions of those proteins, bacterial recombination, and double-strand break-repair are given in *Rec* Genes, Recombination Pathways, RecA Protein and Homology, RecBCD Enzyme, Pathway, RuvAB Enzyme, RuvC Enzyme.

Further Reading

Brock TD (1990) *The Emergence of Bacterial Genetics*. Plainview, NY: Cold Spring Harbor Laboratory Press.

Firth N, Ippen-Ihler K and Skurray RH (1996) Structure and function of the F factor and mechanism of conjugation. In: Neidhardt FC, Curtiss III R, Ingraham JL *et al.* (eds) *Escherichia coli* and Salmonella: *Cellular and Molecular Biology*, 2nd edn, vol. 2, pp. 2377–2401. Washington, DC: ASM Press.

Holloway B and Low KB (1996) F-prime and R-prime factors. In: Neidhardt FC, Curtiss III R, Ingraham JL *et al.* (eds) *Escherichia coli* and Salmonella: *Cellular and Molecular Biology*, 2nd edn, vol. 2, pp. 2413–2420. Washington, DC: ASM Press.

Low KB (1996) Hfr strains of *Escherichia coli* K12. In: Neidhardt FC, Curtiss III R, Ingraham JL *et al.* (eds) Escherichia coli *and* Salmonella: *Cellular and Molecular Biology*, 2nd edn, vol. 2, 2402–2405. Washington, DC: ASM Press.

References

Hayes W (1952) Recombination in *E. coli* K12: unidirectional transfer of genetic material. *Nature* 169: 118–119.

Jacob F and Wollman EL (1958) Les épisomes, éléments génétiques ajoutés. *Comptes Rendus de l'Académie des Sciences* 242: 303–306.

Lederberg J and Tatum EL (1946a) Gene recombination in bacteria. *Nature* 158: 558.

Lederberg J and Tatum EL (1946b) Novel genotypes in mixed cultures of biochemical mutants of bacteria. *Cold Spring Harbor Symposia on Quantitative Biology* 11: 113–114.

Wollman EL Jacob F and Hayes W (1956) Conjugation and genetic recombination in *Escherichia coli* K12. *Cold Spring Harbor Symposia on Quantitative Biology* 21: 141–162.

See also: Conjugation, Bacterial; Conjugative Transposition; Hfr; Plasmids

F1 Generation

The F1 generation is the first generation resulting from a cross between two dissimilar parental lines.

See also: Mendelian Genetics; Mendelian Inheritance

F1 Hybrid

L Silver

The most obvious advantage of working with inbred strains is genetic uniformity over time and space. Researchers can be confident that the inbred animals of a particular strain used in experiments today are essentially the genetic equivalent of animals from the same strain used 10 years ago. Thus, the existence of inbred strains serves to eliminate the contribution of genetic variability to the interpretation of experimental results. However, there is a serious disadvantage to working with inbred animals in that a completely inbred genome is an abnormal condition with detrimental phenotypic consequences. The lack of genomic heterozygosity is responsible for a generalized decrease in a number of fitness characteristics including body weight, life span, fecundity, litter size, and resistance to disease and experimental manipulations.

It is possible to generate organisms that are genetically uniform without suffering the consequences of whole genome homozygosity. This is accomplished by simply crossing two inbred strains to each other. The resulting F1 hybrid organisms express hybrid vigor in all of the fitness characteristics just listed with an overall life span that will exceed that of both inbred parents. Furthermore, as long as both of the parental inbred strains are maintained, it will be possible to produce F1 hybrids between the two, and all F1 hybrids obtained from the same cross will be genetically identical to each other over time and space. Of course, uniformity will not be preserved in the offspring that result from an "intercross" between two F1 hybrids (see Intercross); instead random segregation and independent assortment will lead to F2 animals that are all genotypically distinct.

See also: Hybrid Vigor; Intercross

FAB Classification of Leukemia

B Bain

From 1976 onward, a French–American–British (FAB) cooperative group of hematologists formulated a series of classifications of acute myeloid leukemia (AML), acute lymphoblastic leukemia (ALL), the myelodysplastic syndromes, chronic lymphoid leukemias, and the leukemic phase of non-Hodgkin's lymphoma. These classifications were initially based only on cytology and cytochemistry, but immunophenotypic analysis was later incorporated. Subsequently it became apparent that several FAB categories of leukemia identified specific cytogenetic/molecular genetic entities, e.g., M3 AML (hypergranular promyelocytic leukemia) and L3 ALL (Burkitt's lymphoma-related acute leukemia). Other FAB categories included more than one specific cytogenetic/molecular genetic entity, e.g., M5 AML was found to include not only various acute monocytic/monoblastic leukemias associated with t(9;11)(p21-22;q23) and other translocations with an 11q23 breakpoint, but also the completely different entity, acute monoblastic leukemia associated with t(8;16)(p11;p13). The FAB classifications were important in advancing knowledge of hematological malignancies, since they provided a framework for cytogenetic and molecular genetic research and also, by providing widely accepted terminology and definitions, facilitated clinical trials and international collaboration.

See also: **Leukemia; WHO Classification of Leukemia**

Fabry Disease (α-Galactosidase A Deficiency)

R J Desnick

Copyright © 2001 Academic Press
doi: 10.1006/rwgn.2001.0443

Fabry disease, an X-linked lysosomal storage disease, is caused by the deficient activity of α-galactosidase A (α-Gal A; EC 3.2.1.22), a lysosomal exoglycohydrolase which catalyzes the hydrolysis of terminal α-galactosyl residues from glycosphingolipids, primarily globotriaosylceramide. The primary site of pathology is the vascular endothelium. Patients with the classic form of Fabry disease have no detectable α-Gal A activity and typically present in childhood with acroparesthesias, angiokeratoma, hypohidrosis, and characteristic corneal and lenticular opacities. With increasing age, the progressive glycosphingolipid deposition results in renal failure, cardiac disease, and strokes. Death usually results from vascular disease of the kidney, heart, or brain. Patients with the clinically milder 'cardiac variant' have residual α-Gal A activity and present in mid to late adulthood primarily with cardiac manifestations. The disorder is panethnic and its estimated incidence is about 1 in 40 000 males. Over 160 mutations in the α-Gal A gene that cause Fabry disease have been identified. Clinical trials of enzyme replacement therapy are underway and effective treatment may be available in the real future.

Further Reading

Desnick RJ, Ioannou YA and Eng CM (2001) In: Scriver CR, Beaudet AL, Sly WS and Valle D (eds) *The Metabolic and Molecular Bases of Inherited Disease*, pp 3733–3774. New York: McGraw-Hill.

Eng CM, Banikazemi M, Gordon R et al. (2001) A phase I/2 clinical trial of enzyme replacement in Fabry disease: pharmacokinetic, substrate clearance, and safety studies. *American Journal of Human Genetics* 68: 711–222.

See also: **Sex Linkage**

1-, 2-, 3-Factor Crosses

F W Stahl

Copyright © 2001 Academic Press
doi: 10.1006/rwgn.2001.0930

A genetic marker is a nucleotide sequence difference that has phenotypic consequences, so that its transmission to progeny of a genetic cross can be monitored. In a genetic cross, one, two, or three (or more) markers (factors) may distinguish the parents.

1-Factor Crosses

When only one marker distinguishes two, sexually reproducing eukaryotic parents, three Mendelian principles can be illustrated:

1. The F1, diploid offspring from a cross between two pure-breeding diploid parents (P, the parental generation), may resemble either one parent or the other, illustrating dominance/recessiveness.
2. The population of haploid cells (gametes) produced by meiosis in the F1 is composed equally of cells containing one or the other of the markers that distinguished the parents, illustrating Mendel's law of segregation.
3. The diploid generation resulting from the union of F1 gametes (F2) will have a phenotypic ratio of 3:1, favoring the type determined by the dominant gene. This ratio is expected if gametes unite with each other at random, without regard to their genotype.

2-Factor Crosses

When the P generation differs by two markers located in different genes, additional Mendelian principles are illustrated:

1. The frequency of recombinants among haploid cells produced by meiosis in the F1 illustrates Mendel's principle of independent assortment when, as is likely to be true, the two markers involved are on separate chromosomes. If the two markers are on the same chromosome, they may illustrate linkage, by the production of fewer than 50% recombinant haploid products of meiosis.
2. When the factors involved influence different phenotypes, the F2 usually manifests independent expression of those phenotypes. If the two genes are on separate chromosomes, this results in relative frequencies of the four phenotypes of 9:3:3:1, illustrating the mosaic nature of phenotype determination.

3-Factor Crosses

When the P generation differs by three markers, new principles emerge if the three markers are linked.

1. The recombination frequencies for the factors taken two at a time determine the order of the markers on the linkage map. In the absence of complications, this corresponds to the order of the markers on the chromosome and genetically defines the concept of locus.
2. The frequency of double crossovers may be different from that expected if simultaneous crossing-over in the two joint intervals were the result of statistically independent exchange events (interference).

Tetrad Analysis

Some fungi produce meiotic spore tetrads in which the order of the spores in the ascus reflects their origin via the two divisions of meiosis. The first two spores in the ascus are sister spores from the same second meiotic division, as are the last two spores. In a 1-factor cross, the frequency with which sister spores carry the same marker (first division segregation) is a measure of the distance of that marker from the centromere of the chromosome on which it is carried. If the frequency is close to 100%, the marker is close to its centromere. (For those species, like *Neurospora crassa*, that have eight ascospores by virtue of postmeiotic mitosis, substitute 'spore pairs' for 'spores' in the above.)

In organisms with unordered spore tetrads (e.g., *Saccharomyces*), linkage of a newly found marker to its centromere can be established when another marker tightly linked to a different centromere is available. In a 2-factor cross involving the new and the old markers, the frequency of tetratype tetrads is indicative of the degree of linkage under question. If the frequency of tetratype tetrads is close to zero, the new marker also is close to its centromere.

Variations on Mendel's Rules

Variations from the Mendelian expectations outlined above may be encountered.

In 1-factor crosses between pure-breeding parents, the phenotype of the F_1 may be intermediate between that of the parents (incomplete dominance).

In 2-factor crosses involving unlinked marker pairs each of which shows simple dominance, ratios other than 9:3:3:1 may be observed, implying that the phenotypes of the genes involved are not expressed independently of each other. For instance, a ratio of 9:3:4 implies that the genotype at one locus interferes with phenotypic expression at the other (epistasis).

In 1-factor crosses, examination of the four haploid products coming from individual acts of meiosis reveals occasional violations (gene conversions) of the expected 2:2 marker ratio.

In 2-factor crosses, conversion of one marker occurs independently of that of other markers unless the two markers are tightly linked, in which case they may undergo co-conversion.

In 3-factor crosses with linked markers, conversion at the central site is accompanied by a high rate of crossing-over of the flanking markers, implying that conversion and crossing-over are aspects of a common process.

1-, 2-, or 3-factor crosses may be conducted with bacteria or viruses with similar consequences.

See also: **Deletion Mapping; Epistasis; Gene Conversion; Gene Mapping; Incomplete Dominance; Interference, Genetic; Mapping Function; Marker; Tetrad Analysis**

Facultative Heterochromatin

See: **Heterochromatin**

Familial Fatal Insomnia (FFI)

See also: **GSD (Gerstmann–Straussler Disease)**

Familial Hypercholesterolemia

M S Brown and J L Goldstein

Copyright © 2001 Academic Press
doi: 10.1006/rwgn.2001.0444

Familial hypercholesterolemia (FH) is a prevalent autosomal codominant disorder that causes elevated blood cholesterol levels and premature heart attacks. The disease is caused by mutations in the gene encoding the low density lipoprotein (LDL) receptor, which removes LDL, the major cholesterol-carrying protein, from blood. FH heterozygotes (1 in 500 in most populations) have a 50% reduction in LDL receptors and a two- to threefold elevation in plasma LDL levels. They frequently experience heart attacks in the fifth decade. The rare FH homozygotes (1 in 1 million) manifest three- to eightfold elevations of plasma LDL, and they typically have heart attacks in childhood. The disorder has been observed in nearly every population of the world, placing FH among the most prevalent single-gene disorders in humans. More than 500 mutations in the LDL receptor gene have been defined by genomic analysis. The LDL receptor was the first cell-surface receptor that was recognized to carry a protein into cells by receptor-mediated endocytosis. Comparison of normal and mutant FH cells helped to establish the properties of this fundamental process, which is now known to be used for many purposes in all animal cells.

See also: **Genetic Diseases**

Fanconi's Anemia

C Mathew

Copyright © 2001 Academic Press
doi: 10.1006/rwgn.2001.0445

Fanconi's anemia is an autosomal recessive inherited disorder associated with progressive aplastic anemia, diverse congenital abnormalities, and a high incidence of acute myeloid leukemia. It is genetically heterogeneous, with seven complementation groups (A–G) having been described. The genes for six of these groups have been identified, but the sequences of the encoded proteins have not provided immediate insight into the functional pathway that is disrupted in this condition. Cells from patients are hypersensitive to DNA cross-linking agents such as mitomycin C, which suggests that the encoded proteins may be involved in the repair of DNA interstrand cross-links.

See also: **Leukemia, Acute**

Fate Map

Copyright © 2001 Academic Press
doi: 10.1006/rwgn.2001.1841

A fate map is a map of an embryo illustrating the adult tissues that will be derived from particular embryonic regions.

See also: **Cell Lineage; Embryonic Stem Cells**

Favism

L Luzzatto

Copyright © 2001 Academic Press
doi: 10.1006/rwgn.2001.0446

The term 'favism' is used to indicate a severe reaction occurring on ingestion of foodstuffs consisting of or containing the beans of the leguminous plant *Vicia faba* (fava bean, broad bean). The reaction manifests itself, within 6-24 h of the fava bean meal, with prostration, pallor, jaundice, and dark urine. These signs and symptoms result from (sometimes massive) destruction of red cells (acute hemolytic anemia), triggered by certain glucosides (divicine and convicine) present at high concentrations in the fava beans. These substances cause severe damage to red cells only if they are deficient in the enzyme glucose 6-phosphate dehydrogenase (or G6PD), therefore favism only occurs in people who have inherited G6PD deficiency (see Glucose 6-Phosphate Dehydrogenase (G6PD) Deficiency). Favism is more common and more life-threatening in children (usually boys) than in adults; however, once the attack is over a full recovery is usually made. In a person who is G6PD deficient favism can recur whenever fava beans are eaten, although whether this happens or not is greatly influenced by the amount of beans ingested and probably by many other factors. From the public health point of view, it has been proven that favism can be largely prevented by screening for G6PD deficiency and by education through the mass media.

See also: **Mutagens**

F-Duction

P J Hastings and S M Rosenberg

Copyright © 2001 Academic Press
doi: 10.1006/rwgn.2001.0447

F-duction is the same as sexduction, i.e., the high-frequency transfer of a segment of bacterial (for example, *Escherichia coli*) DNA incorporated into an F′ plasmid.

See also: **Sexduction**

Feline Genetics

S J O'Brien

Copyright © 2001 Academic Press
doi: 10.1006/rwgn.2001.0165

The Family Felidae includes 37 recognized species that range over five continents. Human fascination with these champions of predatory hunting has led to their deification in ancient Egypt and Asia, to domestication, to celebration in art and theology, and to voluminous literary and scientific descriptions dating from Tutankhamen's tomb and Marco Polo's chronicles. Biologists have studied the Felidae extensively, producing deep insight into an evolutionary process that favored these stunningly efficient carnivores, specialized or adapted in stealth, speed, and majesty, and unchallenged in the natural habitat until the rise of humankind. And nonspecialists treasure cats unashamedly; indeed "cat" is among the first words uttered or spelt by English-speaking children.

The Felidae is one of the eight families of the Carnivora order which began to evolve intrinsic specialities during the lower Eocene, some 40 million years ago. Today's feline species descend from an ocelot-sized ancestor named *Pseudailurus* from which a group of large saber-toothed cats and surviving wild cats emerged, largely in the last 12 million years. The saber-tooths disappeared rather recently in the Pleistocene (10–20 000 years ago) coincident with the latest ice ages which saw the extinction of quite a few other large mammals such as the mammoths, mastodons, dire wolves, and giant ground sloths.

Living cats, along with the hyena, mongoose, and civet families, comprise the aelurid (cat-like) side of the carnivore family tree, which was originally recognized by the presence of an ossified segment in the auditory bulla in the cranial inner ear. This is in contrast to arctoid (bear-like) carnivores that do not have it. Today a variety of additional morphological and DNA-based characters have affirmed the historical separation between the two Carnivora suborders. The cats share several adaptations inherited from their common ancestor, including blunt foreshortened face, large eyes with binocular color vision, retractible claws into a fur-covered sheath, and large sensitive ears. The tawny color range and patterning serves as adaptive camouflage for cats, three-quarters of which inhabit dense forests and live isolated solitary existences.

Pelage or coat display among the living cats varies in pattern (stripes in tigers, tawny solid in lions and pumas, marbled in clouded leopard, king cheetahs, and marbled cat), in pigmentation (albino in lions and tigers, black in leopards, jaguars, and jaguarundi), and in hair length (long hair in snow leopard, great mane in African male lions, short hair in jaguarundi). That the same pelage patterns seen among different felid species are also observed and selected within domestic cat breeds implies that the intrinsic pelage genetic diversity may have originated in the ancestors of all cats to be reinforced by natural selective pressures during species isolation.

Cat specialists recognize 36 wild cat species (**Figure 1**), and there is little disagreement on their identification (there is some; for example a few hold-outs consider the Iriomote cat a separate species, but most data classify this as a subspecies of leopard cat). Domestic cat is considered a separate species for ease of discussion even though domestic cat establishment from artificial selection of African wild cat in Egypt date to around 2000 BC.

Genus level relationships among the 36 species have been contentious with dozens of taxonomic opinions over the twentieth century ranging from a minimum of two genera (*Felis* and *Acinonyx* – cheetah) to a maximum of 19 genera. Molecular genetic data using a consensus of mitochondrial and nuclear gene comparisons have been a useful new approach that many believe will solve this difficult taxonomic puzzle. (The reason it is so difficult is because 36 wild species diverged in a relatively short evolutionary time period of 12 million years.) The gene comparisons cluster the species into three major lineages (subfamilies): (1) the ocelot lineage (7 species); (2) the domestic cat lineage (6 species); and (3) the pantherine lineage (23 species). Within these lineages, the cats assort into eight monophyletic groups; that is, each group displays evidence for a recent common ancestor subsequent to divergence from the older common ancestor for all modern cats, *Prionailuris*. One of these groups includes the five traditional *Panthera* species, lions, tiger, jaguar, leopard, and snow leopard, plus the clouded leopard, all descended from a 3-million-year-old ancestor. Another group joins cheetah, puma, and jaguarundi to an older 8–10-million-year-old split. The eight groups will likely represent a future genus proposal for the Felidae, one based on the imputed evolutionary history of species divergence, a type of pedigree of Felidae natural history.

Below the species level, geographical subdivision and isolation is probably the best way to identify subspecies, an important distinction that Charles Darwin considered as preludes to future species isolation. Molecular genetic tools are currently being applied to develop explicit DNA-based characters to recognize subspecies partitions and identification. Such studies usually reduce maximal members of subspecies to populations with explicit verifiable criteria for distinction. Thus leopard subspecies have been reduced from

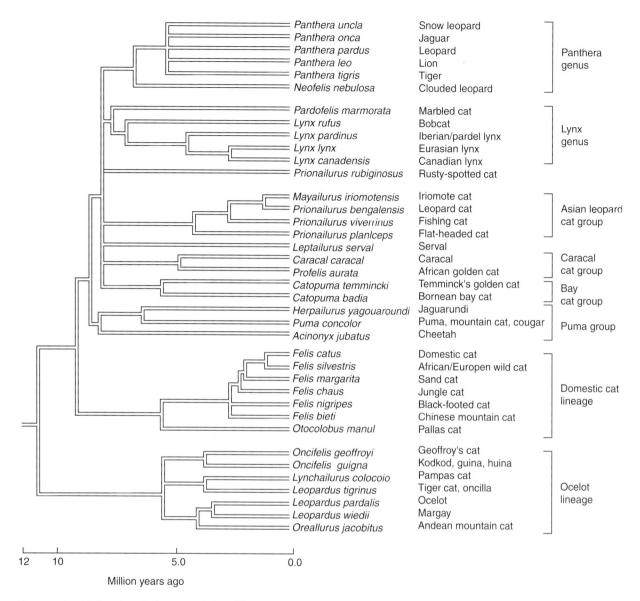

Panthera uncla	Snow leopard	
Panthera onca	Jaguar	
Panthera pardus	Leopard	Panthera
Panthera leo	Lion	genus
Panthera tigris	Tiger	
Neofelis nebulosa	Clouded leopard	
Pardofelis marmorata	Marbled cat	
Lynx rufus	Bobcat	
Lynx pardinus	Iberian/pardel lynx	Lynx
Lynx lynx	Eurasian lynx	genus
Lynx canadensis	Canadian lynx	
Prionailurus rubiginosus	Rusty-spotted cat	
Mayailurus iriomotensis	Iriomote cat	
Prionailurus bengalensis	Leopard cat	Asian leopard
Prionailurus viverrinus	Fishing cat	cat group
Prionailurus planlceps	Flat-headed cat	
Leptailurus serval	Serval	
Caracal caracal	Caracal	Caracal
Profelis aurata	African golden cat	cat group
Catopuma temmincki	Temminck's golden cat	Bay
Catopuma badia	Bornean bay cat	cat group
Herpailurus yagouaroundi	Jaguarundi	
Puma concolor	Puma, mountain cat, cougar	Puma group
Acinonyx jubatus	Cheetah	
Felis catus	Domestic cat	
Felis silvestris	African/Europen wild cat	
Felis margarita	Sand cat	
Felis chaus	Jungle cat	Domestic cat
Felis nigripes	Black-footed cat	lineage
Felis bieti	Chinese mountain cat	
Otocolobus manul	Pallas cat	
Oncifelis geoffroyi	Geoffroy's cat	
Oncifelis guigna	Kodkod, guina, huina	
Lynchailurus colocoio	Pampas cat	
Leopardus tigrinus	Tiger cat, oncilla	Ocelot
Leopardus pardalis	Ocelot	lineage
Leopardus wiedii	Margay	
Oreallurus jacobitus	Andean mountain cat	

12   10   5.0   0.0

Million years ago

Figure 1 Molecular phylogeny of the 37 species of cats based on nuclear and mitochondrial DNA divergence patterns.

27 to 8; tigers from 8 to 5, lions from 7 to 2, and pumas from 32 to 6. Since subspecies recognition forms the basis for conservation strategies and protective legislation, these distinctions gain added importance.

Sadly each of the 36 wild cat species is listed as endangered or threatened by IUCN and CITES, international bodies that monitor global conservation. The threat to Felidae survival is principally in three areas: (1) habitat loss owing to human development; (2) hunting and depredation owing to human protection; and (3) poaching for skins and internal organs of erroneously perceived medicinal/aphrodisiac benefit. The realization of the sorry state of Felidae species has spawned conservation initiatives across their range and continues to be a high priority worldwide to stop or reverse the extinction of these remarkable specimens.

See also: **Conservation Genetics**

Female Carriers

D E Wilcox

Copyright © 2001 Academic Press
doi: 10.1006/rwgn.2001.0448

The term 'female carrier' usually refers to females who are heterozygous for X-linked recessive disorders.

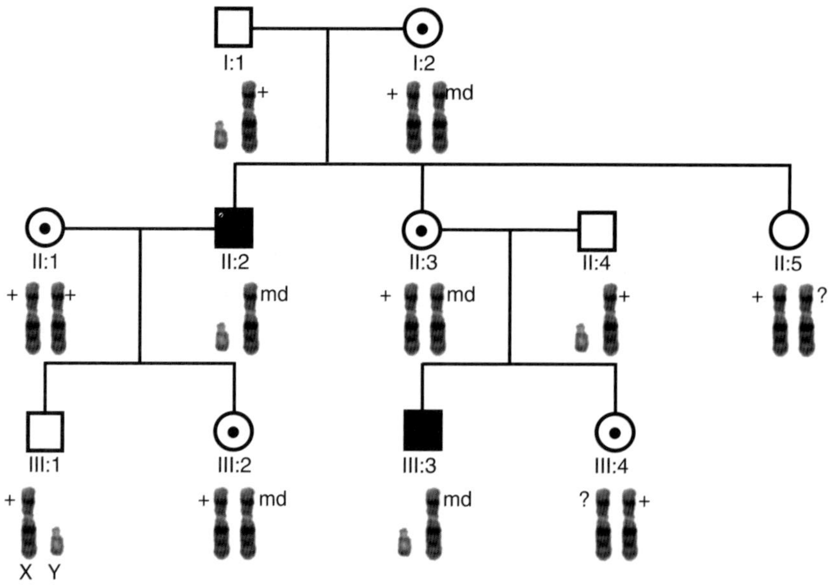

Figure 1 Example pedigree of Becker muscular dystrophy showing the arrangement of the alleles on each individual's sex chromosomes.

Most will have a healthy phenotype and so carrier detection is a major role of the genetic clinic. A female carrier of an X-linked recessive disorder has a 1 in 2 chance of passing the mutant allele to each child. Since the child has a 1 in 2 chance of being male, her chance of having an affected male is 1 in 4.

Although carrier females are heterozygous for any X-linked recessive trait they carry, only one allele is active in each cell. In early embryogenesis in females, one of each cell's X chromosomes is randomly and permanently inactivated. The mix in the tissues usually prevents the development of the full mutant phenotype but female carriers of an X-linked recessive trait are at risk of developing variable expression of the disorder. This contrasts with autosomal recessive traits where carrier heterozygotes have the normal phenotype.

Pedigree Analysis

Features of X-Linked Recessive Pedigrees

Figure 1 shows a family with X-linked recessive Becker muscular dystrophy (BMD), which is allelic to the commoner and more severe Duchenne form (DMD). The pedigree shows several features typical of X-linked recessive inheritance. Only males are affected. Affected males form a 'knight's move' pattern in the pedigree, i.e., they are related through healthy females. There is no apparent male to male transmission. An affected male such as II:2 cannot transmit the BMD to his sons as he gives them his Y chromosome and not the X chromosome with the

mutant allele. Sons inherit their X chromosome from their mother.

Identification of Obligate Carriers

After drawing the pedigree of an X-linked recessive disorder, it may be possible to infer which females, with a healthy phenotype, have heterozygote (carrier) genotype from their position in the family. Such females are obligate carriers and their identification is an important first step in assessing the carrier risks of the other females in the family. Obligate carriers are marked on the pedigree diagram by placing a dot in the middle of the pedigree symbol. In the pedigree in **Figure 1**, there are two affected males in generations II and III, II:2 and III:3. The older affected male's daughter, III:2, is an obligate carrier because her father must have given her his only X chromosome, which carries the mutation (md). Her aunt, II:3, is also an obligate carrier as she has an affected son and an affected brother. Since two separate new mutations in one family would be extremely rare, she must have inherited her brother's mutation from her mother, in order to pass it to her son. Her mother, I:2, must also be an obligate carrier as she has two offspring who have the mutation. The genetic situation of the first obligate carrier in a family is complex. She could have inherited the mutation, she could be a new mutation, or the mutation could have started in her ovaries (gonadal mosaicism). In the first situation, her sisters and aunts also have a carrier risk and genetic counselling should be offered. In the second and third situations, only her female descendants are at risk.

Identification of Females with a Carrier Risk
The healthy females II:5 and III:4 each have an obligate carrier mother. Their mothers have two X chromosomes and the chance of passing on the mutant allele (md) to their daughters is 1 in 2. If a female with a carrier risk has healthy son(s), the (conditional) risk for each son being born healthy can be used in a Bayes calculation to reduce her inherited (prior) carrier risk.

Pitfalls in Pedigree Analysis

Nonpaternity
Before assuming that the daughter of an affected male is an obligate carrier, the geneticist should make sure that the man is the biological father.

New mutation
Very occasionally, perhaps once in several hundred pedigrees, a separate new mutation can arise in a branch of a family with an existing mutation. Before assuming the intervening females are all obligate carriers, it is worth trying to confirm that the affected males in each branch of the family carry the same mutation.

Gonadal mosaicism
Mother of a sporadic affected male   If the mother of a sporadic affected male has no evidence of her son's mutation in her blood, it cannot be assumed that she is not a carrier. She could be a gonadal mosaic and carry the mutation in her ovaries. Although her mother, sisters, and aunts are not at risk, her daughters will have a carrier risk and she is at risk of passing the mutation to any future sons.

Parents of a sibship consisting of only carriers at the top of a pedigree   In a family where the first sibship known to carry the mutation consists of two or more carriers and no affected males, it is tempting to assume that the mutation has been inherited from the mother. If she has no mutation in her blood it cannot be assumed that she is a gonadal mosaic. The carrier sisters' father could be a gonadal mosaic. In this case, all sisters in the sibship, including any half sisters he has by another partner, have a very high carrier risk depending on the proportion of his gonads carrying the mutation. The mother of the carrier sisters is not a carrier and her aunts, sisters, and any children she has by another partner are not at risk. DNA linkage analysis can assist in determining the parental origin of the mutated X chromosome.

Daughters of normal transmitting males
In most X-linked recessive disorders, such as BMD or DMD, the mutation cannot be passed from a healthy male, such as III:1 in **Figure 1**, to his daughters. This is because the mutations tend to be fully penetrant as males are hemizygous for X-linked genes. However, some disorders do not follow classical Mendelian inheritance patterns. For example, in fragile X mental retardation syndrome, caused by an unstable amplified CGG trinucleotide repeat mutation in the FMR1 gene, some phenotypically normal males inherit a small amplified repeat, or premutation, from their mothers. The males then transmit this to all their daughters. These healthy obligate carriers are at risk of transmitting an expanded full mutation to their offspring because premutations are more unstable when transmitted through females than males. In fragile X syndrome all healthy males with a prior risk of inheriting the mutation should have their DNA screened before advising that their daughters are not carriers.

Carrier Tests

Direct Tests, which Confirm Carrier Status
Following pedigree analysis, a number of different types of test can help determine the carrier status of at risk females. The most accurate methods directly identify the mutation or the mutated gene product. It is preferable to confirm the mutation in the proband. Unfortunately, direct tests are currently not possible in all circumstances.

Conditional Tests, which Produce a Carrier Risk
Clinical geneticists can combine the results of several independent conditional carrier tests together with conditional pedigree information using a Bayes calculation to produce a final carrier risk. Where the mutation is not detectable, genetic linkage studies can help. These label the chromosome around the gene and track the segment of chromosome through individuals in several generations of the family. Linkage studies are not 100% accurate because of genetic recombination, which can cause the markers to switch chromosomes. Biochemical tests, not directly related to the trait's gene product, can also provide conditional risk information. Examples include serum creatine kinase in DMD and fibroblast very long chain fatty acids in adrenoleukodystrophy. Carrier/noncarrier risk ratios are available for various values of these substances.

Symptomatic Female Carriers: Underlying Genetic Mechanisms

Manifesting carriers are found in many X-linked recessive disorders such as DMD, in which about 1 in 40 carriers have some symptoms.

Nonrandom X-Inactivation

Studies have shown that most manifesting carriers of DMD have skewed X-inactivation with over 70% of the chromosomes, carrying the normal allele, inactivated. The muscles are a considerable proportion of total body mass and to manifest symptoms, a large shift in total body nonrandom X-inactivation needs to occur. Other traits such as color blindness have much smaller target tissues and a manifesting color blind carrier may have considerable skewing of X inactivation in the retina but still have overall random X inactivation.

Homozygosity

Homozygous mutant females will express the full phenotype of an X-linked recessive trait. Females may inherit a mutation from one parent who is affected or a carrier and the X chromosome from the normal parent undergoes a new mutation. Some disorders such as colour blindness and glucose-6-phosphate dehydrogenase deficiency are common, particularly in some communities. In this situation, homozygosity often occurs because the mutations are inherited from an affected father and a carrier mother.

Turner Syndrome (45,X0)

Females with Turner syndrome are hemizygous for all X-linked genes and manifest the full phenotype of any X-linked recessive disorder they carry. Females with partial deletions of the X chromosome will manifest symptoms of any recessive trait carried on the non-deleted 'normal' X, if the corresponding allele is not present on the deleted X.

X-Autosome Translocation

In a number of X-linked recessive disorders, some manifesting carriers have reciprocal X–autosome translocations. The breakpoint on the X chromosome disrupts the disorder's gene locus. In this situation, the normal X with the healthy allele selectively inactivates in all cells. This happens because inactivation of the translocated X would spread through the disrupted, nonfunctioning gene to the adjoining autosomal genes, which would be lethal.

Further Reading

Connor JM and Ferguson-Smith MA (1997) *Essential Medical Genetics*, 5th edn. Oxford: Blackwell Science.

Gelehrter TD, Collins FS and Ginsburg D (1998) *Principles of Medical Genetics*, 2nd edn. Bethesda, MD: Williams & Wilkins.

Online Mendelian Inheritance in Man: http://www.ncbi.nlm.nih.gov/omim/

University of Glasgow, Department of Medical Genetics, Encyclopaedia of Genetics pages contain a number of illustrations and animated diagrams to accompany this article: http://www.gla.ac.uk/medicalgenetics/encyclopedia.htm

See also: **Sex Linkage; Translocation; X-Chromosome Inactivation**

Feral

L Silver

Copyright © 2001 Academic Press
doi: 10.1006/rwgn.2001.0449

Although the success of many species throughout the world is dependent on their status as commensal species in some regions with appropriate environmental conditions, animals have reverted back to a noncommensal state, severing their dependence on humankind. Such animals are referred to as feral. The return to the wild can occur most readily with a mild climate, sufficient vegetation or other food source, and weak competition from other species. Feral mice, for example, have successfully colonized small islands off Great Britain and in the South Atlantic, and in Australia, *Mus musculus* has replaced some indigenous species. Although feral populations exist in North America and Europe as well, here they seem to be at a disadvantage relative to other small indigenous rodents such as *Apodemus* (field mice in Europe), *Peromyscus* (American deer mice), and *Microtus* (American voles). In some geographical areas, individual house mice will switch back and forth from a feral to a commensal state according to the season – in mid-latitude temperate zones, human shelters are much more essential in the winter than in the summertime.

See also: **Commensal; *Mus musculus***

Fertility, Mutations

See: **Infertility**

Fertilization

J C Harper

Copyright © 2001 Academic Press
doi: 10.1006/rwgn.2001.0451

Fertilization is a complex set of events involving the fusion of two gametes to produce a new individual. In pro duction of the gametes, the number of chromosomes are halved by meiosis. Oogenesis results in a large, complex oocyte that contains the proteins, enzymes, and other factors necessary for the first days of development. Spermatogenesis has to ensure that the sperm is able to travel through the female reproductive tract to meet and fertilize the oocyte.

Oogenesis is a complex process that starts during fetal life. At birth, females are born with primary oocytes arrested in meiosis I (diplotene stage) and no further oocytes will be produced. Each month only one oocyte will fully mature and just before ovulation, meiosis is resumed, the first polar body is extruded (containing one set of oocyte chromosomes), and the oocyte arrests in metaphase II. Meiosis is only complete upon fertilization. Maturation (M-phase) promoting factor (MPF) causes resumption of meiosis and is regulated by the *c-mos* gene. MPF is high during meiosis I and II.

Sperm are produced by the testis and become mature and motile as they travel through the epididymis. Sperm released on ejaculation have to travel through the cervical os, the uterus, and the fallopian tubes where hopefully an oocyte is waiting for fertilization. The mature sperm consists of a head piece, neck, and flagellum. The flagellum is responsible for initiation and maintenance of motility through the female reproductive tract. The head of the sperm contains the sperm DNA and is the area involved in recognition of the zona pellucida (the glycoprotein coat that surrounds the oocyte) and sperm–oocyte fusion. The head is covered by a membrane-bound vesicle called the acrosome. Before a sperm can fertilize an oocyte it needs to go through capacitation and the acrosome reaction. Capacitation occurs in the female reproductive tract and involves a range of poorly understood processes that do not alter the ultrastructure of the sperm, but enable the sperm to fertilize the oocyte.

At ovulation the oocyte is surrounded by a dense array of cumulus cells that play a vital role in oogenesis. The sperm swim through the cumulus cells until they reach the zona pellucida. It is thought that there are two stages to sperm binding to the human oocyte. The first involves the binding of the acrosome intact sperm to the primary binding site on the oocyte, termed ZP3. The acrosome matrix consists of a number of enzymes, the most important of which is acrosine, a serine protease that is packaged as the inactive pro-acrosin. The acrosome reaction exposes the second binding site on the sperm head that binds to the secondary binding site, ZP2, on the oocyte. The sperm is then able to penetrate the oocyte and two reactions are stimulated: the cortical reaction and oocyte activation.

The cortical reaction may be involved in blocking additional sperm penetrating the oocyte. *In vitro* analysis of human zygotes show many sperm bound to the zona pellucida but usually only one sperm fertilizes the oocyte. The cortical reaction involves vesicles of Golgi apparatus, which contain enzymes and mucopolysaccharides. The granules break open releasing their contents into the perivitelline space which causes the zona pellucida to harden. Oocyte activation is caused by calcium oscillations within the oocyte. Calcium oscillations occur in mammals and blocking this calcium increase with chelators blocks fertilization. Calcium causes a decrease of MPF. MPF needs to be low for the oocyte to exit meiosis I and II. There are two hypotheses to how sperm cause calcium oscillations. The first is the surface receptor mediated model, which compares the sperm to a giant ligand that binds to a receptor on the oocyte surface resulting in activation of the polyphosphoinositide pathway. The second is the soluble sperm factor hypothesis, which suggests that sperm contain a soluble factor that is released into the ooplasm causing calcium oscillations. This hypothesis would explain why intracytoplasmic sperm injection (ICSI) is able to work (see below).

Upon sperm entry, the oocyte chromosomes undergo the final stages of meiosis and the oocyte extrudes the second polar body. The fertilized oocyte, or zygote, contains two pronuclei; one from the oocyte and one from the sperm, and two polar bodies which are the waste product of oogenesis. The meiotic spindle of the oocyte breaks down and the sperm contributes factors that are involved in establishing the first mitotic spindle. The sperm produces astral microtubules that migrate through the ooplasm and pull the female pronucleus close to the male pronucleus. The zygote undergoes syngamy (the sun in the egg) where the nuclear membranes of the male and female pronuclei break down and the chromosomes condense separately and line up on the first mitotic spindle. The zygote cytoplasm will divide in half (cleavage) to give two identical daughter cells, each with a complete diploid set of chromosomes.

IVF

In vitro fertilization (IVF) is a technique developed for the treatment of some forms of infertility. The first successful birth was that of Louise Brown in 1978. IVF can be used for the treatment of several types of male and female infertility. In the female, one of the most common causes of infertility is tubal blockage, but IVF can also be used in the treatment of ovulatory disorders, including polycystic ovarian disease (PCO), immunological disorders such as antisperm antibodies, endometriosis, coital problems, and "unexplained" infertility where no etiological factor has been identified.

In an IVF treatment cycle the female's menstrual cycle is usually downregulated to block her normal cycle and injections of follicle stimulating hormone (FSH) are administered to stimulate multiple follicle development. Follicle development is tracked using ultrasound and when the follicles have developed sufficiently (i.e., three follicles over 18 mm), a single dose of hCG (human chorionic gonadotrophin) is administered to mimic the luteinizing hormone (LH) surge and ensure maturation of the oocytes. The oocytes are collected under light sedation with ultrasound guidance, by aspiration of the follicles. The oocytes can easily be identified under a dissecting microscope and are placed in a simple culture medium and stored at 37 °C. The male partner's sperm is collected and prepared to remove seminal plasma and enrich for motile sperm. This can be performed using the traditional swim-up technique in which medium is laid over the sperm and motile sperm swim up into the medium, or using density gradient centrifugation, which separates motile from immotile sperm. The prepared sperm are used to inseminate the oocytes (approximately 100 000 sperm ml^{-1} of culture medium). The following day after insemination, the cumulus cells are removed from the oocytes and the oocytes checked for fertilization. Normal fertilization can be seen by the presence of the male and female pronuclei. Normally fertilized oocytes (zygotes) are returned to 37 °C and examined over the next 1–2 days, during which stage they should undergo cleavage. Cleavage is the halving of the ooplasm to produce daughter cells or blastomeres. Human embryos are graded taking into account the size and shape of the blastomeres, the number of cell divisions, and the degree of fragmentation. Normally fertilized embryos can be transferred to the uterus of the female either on day 2 post insemination (2–4 cell stage) or day 3 (6–8 cell stage). Good-quality embryos can be cryopreserved for a future cycle. Recently, some groups have reported an improvement in the pregnancy rate and a decrease in the multiple pregnancy rate by using blastocyst transfer (day 5–6 of embryo development). This procedure has been hindered as very few human embryos develop to the blastocyst stage *in vitro*, but recent improvements in IVF culture medium have improved this. Most research in IVF is aimed at improving the IVF success rates. Other recent advances include assisted hatching and aneuploidy screening. Assisted hatching involves making a hole in the zona pellucida to ensure that the embryo can successfully hatch. Aneuploidy screening involves removing 1–2 blastomeres from the 6–10 cell embryo and testing for the chromosomes commonly involved in aneuploidy (13, 16, 18, X, and Y) so that embryos normal for these chromosomes are transferred.

For male infertility, if the sperm count is very low (the World Health Organization guideline for a normal sperm count is 20 million sperm ml^{-1}), intracytoplasmic sperm injection (ICSI) can be used. This technique involves injection of a single sperm into the cytoplasm of a metaphase II oocyte. The surrounding cumulus cells are removed and the oocyte is positioned with the first polar body at the 6 or 12 o'clock position to minimize damage to the meiotic spindle. A single sperm is taken up into a fine-bore pipette and the pipette inserted directly through the zona pellucida into the ooplasm. A small amount of ooplasm is gently aspirated into the pipette and the sperm is expelled. This procedure has proved to be very successful for most forms of male infertility.

In cases where there is no sperm in the ejaculate, it may be possible to aspirate sperm from the epididymis (MESA – microepididymal sperm aspiration or PESA – percutaneous epididymal sperm aspiration) in the cases of obstructive azoospermia, or the testis (TESA – testicular needle aspiration or TESE – open biopsy testicular extraction). More recently, several IVF centers worldwide have reported on the use of the injection of spermatids and some successful pregnancies have been obtained. However, spermatid injection is still controversial. In some cases of male infertility, a genetic reason for the infertility is known. Some men show deletions of regions of the long arm of the Y chromosome and will therefore pass infertility to all their sons. Other genetic abnormalities may be mutations in the androgen receptor gene, expansion or reduction in the triplet repeat in the androgen receptor gene, cystic fibrosis mutations which can lead to congenital absence of the vas deferens, and chromosomal translocations, which have been shown to be associated with an increase risk of infertility. If the genetic causes of the infertility are known, genetic counseling is required to ensure these families are aware of the risk of transmitting these abnormalities to their offspring.

See also: Fertilization, Mammalian; Genetic Counseling; Oogenesis, Mouse; Spermatogenesis, Mouse; Zygote

Fertilization, Mammalian

P M Wassarman

Copyright © 2001 Academic Press
doi. 10.1006/rwgn.2001.0452

Introduction to Mammalian Fertilization

Fertilization is the means by which sexual reproduction takes place in nearly all multicellular organisms and is fundamental to maintenance of life. It is defined as the process of union of two germ cells, egg and sperm, whereby the somatic chromosome number is restored and the development of a new individual exhibiting characteristics of the species is initiated. Both mammalian eggs and sperm are designed to ensure that fertilization takes place reliably. Accordingly, mechanisms are in place to support species-specific interactions between gametes and to prevent fusion of eggs with more than one sperm (polyspermy).

Egg Development

Oogenesis begins during fetal development when primordial germ cells are transformed first to oogonia (mitotic) and then to oocytes (meiotic). The pool of small, nongrowing oocytes, present at birth, is the sole source of unfertilized eggs in the sexually mature female mouse. These oocytes are arrested at the diplotene (dictyate) stage of the first meiotic prophase. Each oocyte ($\sim 15\,\mu m$ in diameter) is contained within a cellular follicle that grows concomitantly with the oocyte for about 2 weeks, from a single layer of a few epithelial-like cells to three layers of cuboidal granulosa cells by the time the oocyte has completed its growth ($\sim 80\,\mu m$ in diameter). Over several days, while the oocyte remains the same size, follicular cells undergo rapid division, increasing to more than 5×10^4 cells in the Graafian follicle. The follicle exhibits a fluid-filled cavity, or antrum, when it consists of $\sim 6 \times 10^3$ cells and, as the antrum expands, the oocyte takes up an acentric position surrounded by two or more layers of granulosa cells (cumulus cells).

Fully grown oocytes in Graafian follicle complete the first meiotic reductive division, called meiotic maturation, just prior to ovulation in response to a surge in the level of luteinizing hormone (LH). Oocytes progress to metaphase II of the second meiotic division, with separation of homologous chromosomes and emission of a first polar body, and become unfertilized eggs. Oocytes must complete meiotic maturation in order to be capable of being fertilized by sperm. The ovulated egg completes meiosis, with separation of chromatids and emission of a second polar body (i.e., becomes haploid, $1n$), only upon fertilization by sperm (sperm chromosomes restore a diploid, $2n$, state to the zygote).

Sperm Development

In mice, it takes ~ 35 days for each spermatogonial stem cell, already present in the fetus, to progress through meiosis as a spermatocyte, become four haploid spermatids ($1n$), and to be transformed into spermatozoa. Spermatogenesis takes place within the seminiferous epithelium lining the tubules of the testes and is supported by Sertoli cells, a secretory cell type and major site of testosterone action. Spermatozoa initially move passively from the seminiferous epithelium to the rete testis, to the epididymis, and to the vas deferens. During this period of transport, sperm become motile and fully functional. The final and essential maturation of sperm, called capacitation, occurs in the female genital tract following ejaculation. Capacitation involves removal of inhibitory factors from sperm, as well as biochemical changes in sperm proteins (e.g., tyrosine phosphorylation). Only capacitated sperm are capable of binding to eggs, undergoing exocytosis (acrosome reaction), and producing zygotes.

Eggs and Sperm in Oviduct

Eggs released from Graafian follicles enter the opening (ostium) of the oviduct (fallopian tube) and move to the lower ampulla region where fertilization takes place. It has been estimated that mouse eggs and sperm in the oviduct remain capable of being fertilized and giving rise to normal offspring for 8–12 h following ovulation. Typically, very few ovulated eggs are found in oviducts of mice (~ 10) and human beings (~ 1). Similarly, very few sperm are found at the site of fertilization (~ 100–150) as compared to the number of sperm deposited into the female reproductive tract ($\sim 10^7$); an extremely low percentage of ejaculated sperm make their way to the position of unfertilized eggs in the oviduct. It takes ~ 15 min for ejaculated mouse sperm and ~ 30 min for human sperm to traverse the female genital tract and reach the oviduct. Whether binding of mammalian sperm to eggs occurs due to a chance encounter in the oviduct or is promoted by a chemical gradient stimulus (chemotaxis) remains to be determined. Today, there is evidence for human sperm chemotaxis mediated by an egg follicular factor.

Pathway to Mammalian Fertilization

The pathway to fertilization in mice follows a compulsory order (**Figure 1**):

1. Capacitated, acrosome-intact sperm bind in a species-specific manner to the egg zona pellucida (ZP).
2. Bound sperm undergo the acrosome reaction (cellular exocytosis).
3. Acrosome-reacted sperm penetrate the ZP.
4. Sperm that penetrate the ZP bind to the egg plasma membrane.
5. Bound sperm fuse with the egg plasma membrane to form a zygote (fertilization is completed).
6. Following fusion, blocks to polyspermy are instituted.

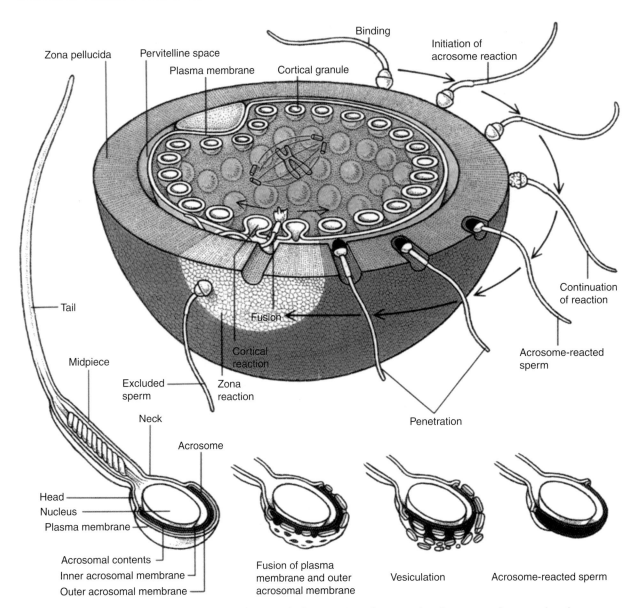

Zona pellucida
Pervitelline space
Plasma membrane
Cortical granule
Binding
Initiation of acrosome reaction

Tail

Continuation of reaction

Acrosome-reacted sperm

Midpiece

Fusion

Cortical reaction

Excluded sperm
Zona reaction

Neck

Penetration

Acrosome

Head
Nucleus
Plasma membrane

Acrosomal contents
Inner acrosomal membrane
Outer acrosomal membrane

Fusion of plasma membrane and outer acrosomal membrane

Vesiculation

Acrosome-reacted sperm

Figure 1 The mammalian fertilization pathway includes a series of steps taken in a compulsory order. Acrosome-intact sperm bind to sperm receptors in the zona pellucida (ZP) by using egg-binding proteins on the sperm head plasma membrane. Sperm then undergo the acrosome reaction (cellular exocytosis), penetrate through the ZP, and reach the perivitelline space between the ZP and plasma membrane. A single sperm then binds to and fuses with egg plasma membrane. Fusion of sperm and egg triggers the cortical reaction that, in turn, triggers the zona reaction. The zona reaction alters the properties of the ZP making it a barrier to other sperm and preventing polyspermic fertilization. (Adapted with permission from Wassarman, 1988.)

Some of the egg and sperm molecules that support each step in this pathway to fertilization have been identified and characterized. A description of these molecules and the manner in which they participate in mammalian fertilization follows below.

Binding of Sperm to Unfertilized Eggs

All mammalian eggs are surrounded by a thick extra-cellular coat, called the zona pellucida (ZP). Consequently, sperm must first bind to and then penetrate the ZP in order to reach and fuse with the egg plasma membrane (**Figure 2**). Removal of the ZP (e.g., by using acidic buffers or proteases) exposes the egg plasma membrane directly to sperm and, as a result, virtually eliminates any barriers to fertilization between species *in vitro*. This has made the 'hamster test' a routine method of assessing the fertilizing capacity of human sperm in *in vitro* fertilization (IVF) clinics.

The ZP consists of only a few glycoproteins, called ZP1–3, that are organized via noncovalent bonds into

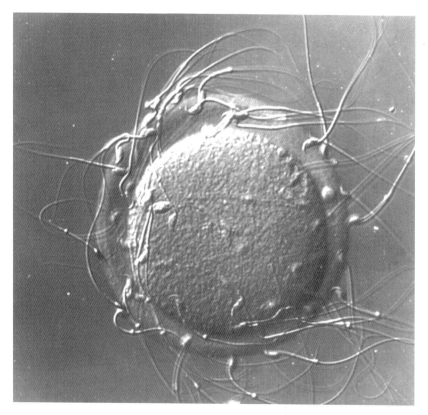

Figure 2 Light photomicrograph of mouse sperm bound to the ZP of an unfertilized mouse egg *in vitro*.

cross-linked filaments. Apparently, the ZP of eggs from all mammalian species, from mice to human beings, consists of ZP1–3. Even the vitelline layer surrounding eggs from many nonmammalian species, including fish, birds, and amphibia, contains glycoproteins structurally related to ZP1–3. Each glycoprotein possesses a unique polypeptide that is heterogeneously glycosylated with both asparagine-linked and serine/threonine-linked oligosaccharides. Genes encoding ZP1–3 polypeptides from a wide variety of mammalian species have been cloned and characterized. In addition, targeted mutagenesis of ZP genes has been carried out by using homologous recombination in embryonic stem (ES) cells and 'knockout' mice produced.

In mice, acrosome intact sperm bind exclusively to the glycoprotein ZP3, which is therefore called the sperm receptor. Sperm recognize and bind to specific oligosaccharides linked to serine residues in a region of ZP3 polypeptide near the C-terminus (encoded by exon-7). These oligosaccharides have been isolated and shown to possess sperm receptor activity. Thus, binding of mammalian sperm to eggs is another example of carbohydrate-mediated cellular adhesion. Whether species-specific binding of sperm to eggs can be attributed to changes in oligosaccharide structure (composition, sequence, linkage, and modification) is currently under investigation.

When acrosome-intact sperm bind to the ZP they do so by using one or more proteins associated with plasma membrane overlying the sperm head. These proteins recognize and bind to sperm receptors in the ZP and are called egg-binding proteins. Many such proteins have been described during the past 20 years. Some of these are integral membrane proteins, while others are peripheral proteins associated with integral membrane proteins. Examples of these include β-1,4-galactosyltransferase, sperm proteins −56 and −17, zonadhesin, spermadhesin, mannose- and galactose-binding proteins, and many others. Some of these candidate proteins can be considered to be lectins. It is unclear to what extent the diversity of these proteins is attributable to misleading experimental evidence. It is possible that a single class of sperm proteins may eventually emerge as the *bona fide* egg-binding protein in many, if not all, mammals.

Acrosome Reaction

The acrosome is a large secretory vesicle that appears in spermatids as a product of the Golgi apparatus and, in certain respects, is biochemically similar to a lysosome. It is located at the anterior portion of the sperm head, just under the plasma membrane and above the nucleus. Acrosomal membrane underlying the plasma membrane is called the 'outer' acrosomal membrane and that overlying the nucleus is called the 'inner' acrosomal membrane. During the acrosome reaction, multiple fusions occur between plasma membrane and outer acrosomal membrane at the anterior region of the sperm head. Extensive formation of hybrid membrane vesicles takes place. As a result, the egg ZP is exposed to the inner acrosomal membrane and acrosomal contents of bound sperm.

Among the many different acrosome reaction-inducers, which include progesterone, is the sperm receptor ZP3. It is now generally accepted that ZP3 is the natural agonist that initiates the acrosome reaction following binding of sperm to the ZP. It appears that multivalent interactions between egg-binding protein(s) and ZP3 may trigger this Ca^{2+}-dependent reaction. As in secretion by somatic cells, intracellular Ca^{2+} is necessary and sufficient to initiate the acrosome reaction. ZP3 stimulation of sperm activates voltage-sensitive T-type Ca^{2+} channels, resulting in depolarization of the sperm membrane from ~ -60 to ~ -30 mV, and increases intracellular Ca^{2+} concentration from ~ 150 to ~ 400 nM. It is likely that opening of sperm T-type channels leads to a sustained release of Ca^{2+} from an internal store, perhaps via inositol 3,4,5-triphosphate (IP3) and IP3 receptors. In addition to these changes, ZP3-stimulated sperm exhibit a transiently elevated pH (alkanization) that may activate Ca^{2+}/calmodulin-dependent adenyl cyclase, protein phosphatases, protein kinases, tyrosine kinases, and phospholipases.

It is clear that ZP3 stimulation of sperm also activates G proteins and activation of G_{i1} and G_{i2} accounts for the pertussis toxin sensitivity of the acrosome reaction. Participation of another G protein, $G_{q/11}$, has also been suggested. Receptors that activate G proteins have remained elusive, although aggregation of β-galactosyltransferase on the sperm head by ZP3 or antibodies has been reported to lead to activation of a pertussis toxin-sensitive G-protein complex and induction of the acrosome reaction.

Penetration of Zona Pellucida by Sperm

Only acrosome-reacted sperm can penetrate the ZP and fuse with egg plasma membrane. The course taken by sperm is indicated by a narrow slit left behind in the ZP of the fertilized egg. In mice, it takes ~ 15–20 min for acrosome-reacted sperm to penetrate the ZP and reach the egg plasma membrane. Until relatively recently, it was thought that the acrosomal serineprotease, acrosin, was essential for penetration of the egg ZP by bound, acrosome-reacted sperm. However, sperm from mice that are homozygous nulls for acrosin ($Acr^{-/-}$) penetrate the ZP and fertilize eggs, suggesting that acrosin may not be essential for these steps. On the other hand, the absence of acrosin does cause a delay in penetration of the ZP by sperm, which may

be due to a delay in dispersal of acrosomal proteins during the acrosome reaction. It is possible that other acrosomal proteases either replace acrosin or are themselves responsible for sperm penetration through the ZP. It should also be noted that sperm motility is an important contributing factor to ZP penetration and schemes have been suggested whereby sperm penetrate the ZP solely by mechanical shear force.

Fusion of Sperm and Egg

In mice, plasma membrane above the equatorial segment of acrosome-reacted sperm fuses with egg plasma membrane. Fusion between gametes nearly always involves egg microvillar membrane (i.e., all but the region where the second metaphase plate and first polar body are located), since it permits maximum apposition of sperm and egg. As proposed for other biological systems, localized dehydration at the site of membrane contact and establishment of hydrophobic interactions are critical steps for fusion. As mentioned earlier, there is little evidence for barriers to interspecies fertilization once sperm have penetrated the ZP and reached the plasma membrane. In most mammals, fusion of the sperm head with the egg is closely followed by entry of the sperm tail into the egg cytoplasm.

Several sperm proteins have been implicated in binding of sperm to and fusion of sperm with egg plasma membrane. One of these proteins, PH-30 or fertilin, has received the most attention. Fertilin is a heterodimer of α- and β-glycosylated subunits and is a member of the ADAM (contain a disintegrin and a metalloprotease domain) family of transmembrane proteins. Peptides based on sequences at the disintegrin domain of fertilin-β and, perhaps, fertilin-α, can prevent binding of sperm to eggs from which the ZP has been removed *in vitro*. It has been proposed that binding of acrosome-reacted sperm to egg plasma membrane is supported by interactions between fertilin's disintegrin domains and integrin (e.g., $\alpha_6\beta_1$) receptors on unfertilized eggs.

Fertilin-α possess a moderately hydrophobic sequence, ~ 17–25 amino acids long, in its cysteine-rich domain that may function as a fusion peptide following binding of acrosome-reacted sperm to egg plasma membrane. The peptide can be modeled as an α-helix having a strongly hydrophobic face (amphipathic helix), similar to several viral fusion peptides. Experimental evidence suggests that this peptide and related peptides can bind to membranes and induce fusion. Despite such evidence, sperm from mice that are homozygous null for fertilin-β and possess reduced levels of fertilin-α can fuse with egg plasma membrane in *in vitro* assays, albeit with reduced efficiency. This, as well as other observations with the null mice, suggests that either an additional fertilin-β-independent pathway to fusion exists or that fertilin-α and -β are not essential components of the gamete fusion pathway. In this context, it has been reported that, although the fertilin-α gene is expressed in humans, it does not produce a functional protein. Further experimentation should clear up these issues.

Prevention of Polyspermy Following Fertilization

Once an egg has fused with a single sperm to become a zygote it is imperative that no additional sperm fuse with the zygote's plasma membrane. In mammals this is achieved by immediate changes in the electrical properties of the plasma membrane ('fast block,' within seconds) and by slower changes in the properties of the ZP ('slow block,' within minutes). The latter is a result of the so-called 'zona reaction.'

The zona reaction occurs within minutes of fertilization. It is induced by the contents of cortical granules, small membrane-bound organelles that underly the egg plasma membrane, which are deposited into the ZP following the 'cortical reaction.' There are ~4000 cortical granules in each mouse egg. The cortical reaction involves fusion of cortical granule and plasma membranes with exocytosis of cortical granule contents into the ZP. Apparently, this occurs as a result of localized release of Ca^{2+} from egg cytoplasmic stores. Among the contents are a variety of enzymes and other proteins. These components cause a hardening of the ZP (i.e., a decrease in solubility), perhaps due to proteolytic modification of ZP2, and a loss of sperm binding, perhaps due to modification of ZP3 by glycosidases. Consequently, movement of bound sperm through the ZP and binding of additional sperm to the ZP are prevented.

Final Considerations

Reproduction of the species is a fundamental property of all living things. Fertilization activates the mammalian egg to initiate a complex program of development, transforming a single cell into a multicellular organism. Accordingly, development of eggs and sperm and the interactions between gametes that culminate in fertilization are highly regulated. Some of the egg and sperm molecules that participate in the fertilization pathway have been identified and their mechanisms investigated. This relatively new information has already contributed to our ability to control reproduction and will continue to have an impact on medical aspects of human reproduction for years to come.

Further Reading

Austin CR and Short RV (eds) (1982) *Reproduction in Mammals*, vol. 1, *Germ Cells and Fertilization*. Cambridge: Cambridge University Press.

Saling PM (1996) Fertilization: mammalian gamete interactions. In: Adashi EY, Rock JA and Rosenwaks Z (eds) *Reproductive Endocrinology, Surgery, and Technology*, vol. 1, 403–420. Philadelphia, PA: Lippincott-Raven.

Snell WJ and White JM (1996) The molecules of mammalian fertilization. *Cell* 85: 629–637.

Wassarman PM (1988) Fertilization in mammals. *Scientific American* 256 (December): 78–84.

Wassarman PM (ed.) (1991) *Elements of Mammalian Fertilization*, vols. 1 and 2. Boca Raton, FL: CRC Press.

Wassarman PM (1999) Mammalian fertilization: molecular aspects of gamete adhesion, exocytosis, and fusion. *Cell* 96: 175–183.

Yanagimachi R (1994) Mammalian fertilization. In: Knobil E and Neill JD (eds), *The Physiology of Reproduction*, vol. 1, 189–317. New York: Raven Press.

See also: Gametes; Oogenesis, Mouse; Spermatogenesis, Mouse

Filamentous Bacteriophages

E Kutter

Filamentous bacteriophages (phages) are long, flexible rods that are simply extruded through the surface of their host bacteria rather than killing them during productive infection. Each consists of a circular, single-stranded (ss) DNA molecule encased in a sheath. The best-studied are three related 'Ff' coliphages – f1, M13, and fd – and *Vibrio cholerae* phage, CTXφ, which encodes the cholera toxin (CT). Like most filamentous phages, they use the tip of a conjugative pilus as an initial receptor in recognizing their target bacteria; they thus are relatively specific for bacteria containing appropriate conjugative plasmids. Attachment occurs via the N-terminal of a specific protein, pIII, at one end of the phage. It leads to retraction of the pilus, probably by depolymerization into the membrane, bringing the phage tip into contact with the bacterial outer membrane proteins Tol Q, R, and A, which are required to translocate the DNA into the cytoplasm. As long as these three proteins are present, filamentous phages can infect at a very low efficiency even in the absence of a fertility plasmid.

The Ff phages are approximately 7 nm in diameter and 690 nm long, with a mass of 16.3 MDa, 13% of it DNA. The sheath contains 2700 molecules of a 50-amino acid protein, pVIII, which is highly α-helical. It has a basic C-terminal domain toward the DNA phosphate backbone, a hydrophobic central region, and an acidic N-terminal exposed to the outside; the molecules are arranged in single fashion. One tip contains five copies each of the pIII recognition protein (406 AA) and a 111-AA protein, pVI, forming knobs. The other end has about five copies each of pVII (33 AA) and pIX (32 AA). A 78-nucleotide hairpin loop of the DNA located at this latter end serves as the packaging signal. The capsid proteins reside in the inner membrane until assembly. pVIII is synthesized with a 23-AA signal sequence that is cleaved after it is inserted into the membrane, leaving the N-terminal domain in the periplasmic space. An 18-residue signal sequence helps pIII get through the membrane into the periplasmic space, remaining anchored to the inner membrane by a 23-AA C-terminal hydrophobic sequence. The other three capsid proteins simply contain membrane-spanning hydrophobic regions.

Six additional proteins are encoded in the Ff phage genomes, along with a regulatory intergenic region. Three of the proteins are required for replication: pII, 409 AA, a site-specific endonuclease required for both phage and replicative-form (RF) DNA synthesis; pX, 111 AA, synthesized from an internal start in gene II, needed for synthesis of the ss viral DNA; and pV (87 AA), a ss DNA-binding protein (SSB). A possible, specific hairpin site in the intergenic region of the entering viral DNA is recognized by the host SSB and RNA polymerase (RNAP), which synthesizes a primer used by the host DNA polymerase III to initiate synthesis of the double-stranded replicative form.

DNA pol I and ligase are needed to close the complementary strand. The DNA is then supercoiled by gyrase to become replicative form I (RFI), the template for transcription. Further replication is rather complex. The phage pII has to nick a specific site, the viral-strand origin, and the 3′-OH end thus formed acts as the primer for rolling-circle replication carried out by pol III and the host SSB and *rep* helicase. After one round of replication, pII cleaves and circularizes the single-stranded viral-strand 'tail,' which can initiate formation of a new RFI, while the RFI containing the new viral RNA strand is resealed and supercoiled to again act as a substrate for pII, as well as for transcription.

Once sufficient gpV accumulates in the cell, it binds cooperatively to some of the ssDNA molecules; pX somehow helps regulate the balance between progeny DNA and RF formation. Three additional, phage-encoded proteins aid in viral assembly. Outer-membrane protein pIV (405 AA), 'secretin,' is synthesized with a 21-AA signal sequence. Gene I encodes a 348-AA inner-membrane protein with its N-terminal 253 residues in the cytoplasm. An internal translational start produces pI* (108 AA), still containing the membrane and periplasmic domains of pI. As shown genetically, the cytoplasmic part of pI interacts with thioredoxin and the packaging signal during assembly, while the outer portion appears to interact with pIV in the outer membrane to form the extrusion passage. The pV helps form the DNA into a linear antiparallel structure facilitating assembly; about 1500 molecules of pV are required per phage for this process. Assembly of the virus particle can then be initiated by an interaction of the packaging signal with the cytoplasmic domain of pI, the membrane-associated pVII and pIX. During elongation and extrusion, pVIII from

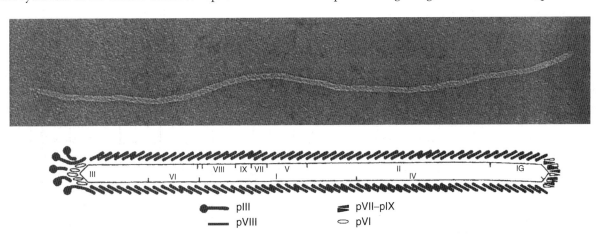

Figure 1 The f1 filamentous bacteriophage. Top: Electron micrograph of a negatively stained particle, with the pIII-VI end located on the left side. Bottom: Schematic representation of the positions of the structural proteins and DNA in the bacteriophage. (Adapted from Webster RE and Lopez J (1985) In: Casjens S (ed.) *Virus Structure and Assembly*, p. 235. Boston, MA: Jones & Bartlett.)

the membrane displaces the pV, with some sort of assistance from reduced host thioredoxin. When the end of the DNA is reached, pVI and pIII are added and the particle is released from the cell; the C-terminal region of pIII is particularly important for particle stability.

Two classes of filamentous phages have been described; within each class, the DNA is largely homologous, but phage Ike, prototypic of the second class, is only 55% homologous to the Ff phages. *Pseudomonas* phages Pf1 and Pf3 (in the second class?) have been shown to have a very different packaging of the DNA and protein than the Ff viral particles.

In 1996, Matthew Waldor and John Mekalanos showed that the structural genes for CT are actually encoded by a filamentous bacteriophage (designated CTXphi) related to coliphage M13 and f1. The CTX genome either replicates as a plasmid or integrates in the chromosome. CTX uses the toxin-coregulated pili (TCP) that are required for intestinal colonization as its receptor and infects *V. cholerae* cells within the gastrointestinal tracts of mice more efficiently than under laboratory conditions. Thus, the emergence of toxigenic *V. cholerae* involves horizontal gene transfer that may depend on *in vivo* gene expression. Although the genome of CTXphi closely resembles that of coliphage f1, CTXphi lacks a homolog of f1 gene IV; instead of encoding its own outer membrane 'secretin,' it uses *eps*D, the putative outer membrane pore for the host type II secretion system, which is also used for excreting the CT as well as protease and chitinase.

The fact that the length of the phage simply depends on the amount of DNA being packaged has made the filamentous phages popular as cloning vectors. Up to 6 kb of DNA can be inserted into appropriate intergenic regions without affecting packaging efficiency, and it is possible to put in significantly longer inserts. Either ss DNA or ds DNA can be readily obtained for various purposes. A variation on this theme has been the construction of 'phagemids,' vectors incorporating the intergenic packaging and replication signal of a filamentous phage in addition to a plasmid origin of replication. They thus replicate as plasmids until infected by a helper filamentous phage, which activates the phage origin of replication and provides the proteins to package the plasmid containing the clone into a transducing phage.

Filamentous phages have also been used extensively as vehicles for 'phage display' by cloning sequences encoding small peptides into the N-terminal region of pIII. Libraries made in this fashion can readily be screened for a large variety of binding activities. Larger proteins can be incorporated into specific places in either pIII or pVIII as long as they are being expressed from a plasmid in the infected cell and replace only a small fraction of the given protein molecules in the final phage.

Further Reading

Beltrán P, Delgado G, Navarro A, Trujillo F, Selander RK and Cravioto A (1999) Genetic diversity and population structure of *Vibrio cholerae*. *Journal of Clinical Microbiology* 37: 581–590.

Boyd EF, Heilpern AJ and Waldor MK (2000) Molecular analyses of a putative CTXphi precursor and evidence for independent acquisition of distinct CTXphi s by toxigenic *Vibrio cholerae*. *Journal of Bacteriology* 182: 5530–5538.

Boyd EF, Moyer KE, Shi L and Waldor MK (2000) Infectious CTXphi and the *Vibrio* pathogenicity island prophage in *Vibrio mimicus*: evidence for recent horizontal transfer between *V. mimicus* and *V. cholerae*. *Infection and Immunity* 68: 1507–1513.

Byun R, Elbourne LDH, Lan R and Reeves PR (1999) Evolutionary relationships of pathogenic clones of *Vibrio cholerae* by sequence analysis of four housekeeping genes. *Infection and Immunity* 67: 1116–1124.

Chakraborty S, Mukhopadhyay AK, Bhadra RK et al. (2000) Virulence genes in environmental strains of *Vibrio cholerae*. *Applied and Environmental Microbiology* 66: 4022–4028.

Davis BM and Waldor MK (2000) CTXphi contains a hybrid genome derived from tandemly integrated elements. *Proceedings of the National Academy of Sciences USA* 10.1073/pnas.140109997v1.

Davis BM, Kimsey HH, Chang W and Waldor MK (1999) The *Vibrio cholerae* O139 Calcutta bacteriophage CTXphi is infectious and encodes a novel repressor. *Journal of Bacteriology* 181: 6779–6787.

Faruque SM, Albert MJ and Mekalanos JJ (1998) Epidemiology, genetics, and ecology of toxigenic *Vibrio cholerae*. *Microbiology and Molecular Biology Reviews* 62: 1301–1314.

Faruque SM, Asadulghani, Abdul Alim ARM et al. (1998) Induction of the lysogenic phage encoding cholera toxin in naturally occurring strains of toxigenic *Vibrio cholerae* O1 and O139. *Infection and Immunity* 66: 3752–3757.

Faruque SM, Asadulghani, Rahman MM, Waldor MK and Sack DA (2000) Sunlight-induced propagation of the lysogenic phage encoding cholera toxin. *Infection and Immunity* 68: 4795–4801.

Fidelma Boyd E and Waldor MK (1999) Alternative mechanism of cholera toxin acquisition by *Vibrio cholerae*: generalized transduction of CTXphi by bacteriophage CP-T1. *Infection and Immunity* 67: 5898–5905.

Jiang SC, Louis V, Choopun N, Sharma A, Huq A and Colwell RR (2000) Genetic diversity of *Vibrio cholerae* in Chesapeake Bay determined by amplified fragment length polymorphism fingerprinting. *Applied and Environmental Microbiology* 66: 140–147.

Jiang SC, Matte M, Matte G, Huq A and Colwell RR (2000) Genetic diversity of clinical and environmental isolates of *Vibrio cholerae* determined by amplified fragment length polymorphism fingerprinting. *Applied and Environmental Microbiology* 66: 148–153.

Kimsey HH and Waldor MK (1998) *Vibrio cholerae* hemagglutinin/protease inactivates CTXphi. *Infection and Immunity* 66: 4025–4029.

Miao EA and Miller SI (1999) Bacteriophages in the evolution of pathogen–host interactions. *Proceedings of the National Academy of Sciences, USA* 96: 9452–9454.

Muniesa M and Jofre J (1998) Abundance in sewage of bacteriophages that infect *Escherichia coli* O157:H7 and that carry the shiga toxin 2 gene. *Applied and Environmental Microbiology* 64: 2443–2448.

Murley YM, Carroll PA, Skorupski K, Taylor RK and Calderwood SB (1999) Differential transcription of the tcpPH operon confers biotype-specific control of the *Vibrio cholerae* ToxR virulence regulon. *Infection and Immunity* 67: 5117–5123.

Nasu H, Iida T, Sugahara T et al. (2000) A filamentous phage associated with recent pandemic *Vibrio parahaemolyticus* O3:K6 strains. *Journal of Clinical Microbiology* 38: 2156–2161.

Schlör S, Riedl S, Blass J and Reidl J (2000) Genetic rearrangements of the regions adjacent to genes encoding heat-labile enterotoxins (eltAB) of enterotoxigenic *Escherichia coli* strains. *Applied and Environmental Microbiology* 66: 352–358.

Sharma C, Thungapathra M, Ghosh A et al. (1998) Molecular analysis of non-O1, non-O139, *Vibrio cholerae* associated with an unusual upsurge in the incidence of cholera-like disease in Calcutta, India. *Journal of Clinical Microbiology* 36: 756–763.

See also: **Bacteriophages; Capsid; Gene Rearrangements, Prokaryotic; Plasmids**

Filial Generations

L Silver

Filial generation is the term pertaining to a particular generation in a sequence of brother–sister matings that can be carried out to form an inbred strain. The first filial generation, symbolized as F_1, refers to the offspring of a cross between animals having nonidentical genomes. When F_1 siblings are crossed to each other, their offspring are considered to be members of the second filial generation or F_2, with subsequent generations of brother–sister matings numbered with integer increments.

See also: **F1 Hybrid; Inbred Strain**

Filter Hybridization

Filter hybridization is a technique for *in situ* solid-phase hybridization whereby denatured DNA is immobilized on a nitrocellulose filter and incubated with a solution of radioactively labeled RNA or DNA.

See also: *In situ* **Hybridization**

Fingerprinting
J H Miller

The chromatographic pattern of spots produced by proteolytic digestion of a protein followed by electrophoresis.

See also: **Proteins and Protein Structure**

First and Second Division Segregation
J R S Fincham

During the first division of meiosis in a diploid cell the chromosomes are each divided into chromatids, but sister chromatids remain attached together at the centromere. At first anaphase, the centromeres do not split, as in anaphase of mitosis; instead the centromeres of homologous chromosomes separate (segregate) from each other toward the two poles of the division spindle as wholes, each taking two chromatids with it. Centromeres always segregate at the first division of meiosis and do not split to allow their two halves to separate into different meiotic products until the second division.

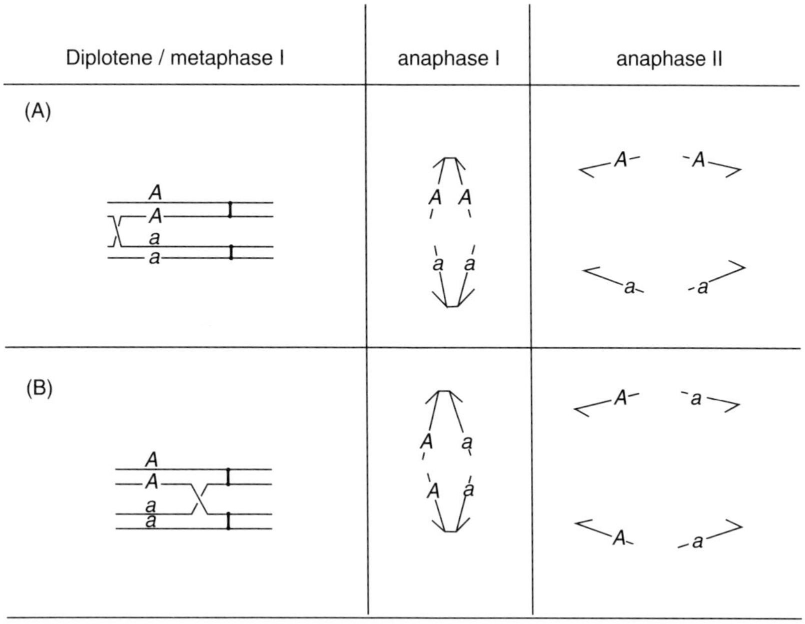

Figure 1 First division (A) or second division (B) segregation of alleles *A* and *a* depending on whether or not a crossover occurs between the *A/a* locus and the centromere (vertical bar in the left panel), the point at which sister chromatids remain connected until the second division of meiosis.

When two homologous chromosomes are distinguished by a genetic marker, say with allele *A* on one chromosome and allele *a* on the other, the *A–a* difference will segregate at the first division provided that the alleles remain attached to their original centromeres. However, when a single crossover, which always involves just one chromatid of each chromosome, occurs between the *A/a* locus and the centromere, different alleles become joined to the same centromere, and the anaphase separation at first anaphase will not be between *A-A* and *a-a* but rather between *A-a* and *A-a* (**Figure 1**). Then segregation of *A* from *a* will be delayed to the the second division.

The effect of two crossovers in the locus–centromere interval depends on whether the same chromatids are involved in the second crossover as in the first. If the same two cross over twice (a two-strand double), or if the second crossover involves the two chromatids not involved in the first crossover (a four-strand double), the effect in either case is to restore first division segregation. If one chromatid crosses over twice, two cross over once each, and the other not at all (three-strand double), the effect is second division segregation. So double crossovers give, on average, 50% second division segregation. An indefinitely large number of crossovers will give, on average, two-thirds second division segregation, a result most easily understood by imagining the four alleles as totally uncoupled from their centromeres and distributed two-and-two to the first division spindle poles at random. An *A* allele will then be twice as likely to be accompanied by an *a* allele as by the other *A* allele.

First and second division segregation can be distinguished by tetrad analysis (see Tetrad Analysis). When the marker is a gross chromosomal feature such as a large terminal deletion, first and second division segregation can also sometimes be seen under the microscope (**Figure 2**).

The second division frequency of a genetic marker is a measure of the frequency of its crossing-over with the centromere, and hence of the map length of the marker–centromere interval. To make second division segregation percentages equivalent to recombination percentages, on which map units (centimorgans, cM) are conventionally based, they need to be divided by two. This is because a single crossover in a marker centromere interval will always give second division segregation, whereas a single crossover in a marker–marker interval will recombine only two out of the four chromatids. In fact, second division segregation and recombination frequencies relate linearly to true map distance (total average number of crossovers per chromosome pair × 50) only when there is never more than one crossover in the interval concerned. Both measures approach a maximum value as the number of crossovers in the interval becomes large, and this maximum value is different in the two cases: 50% recombination and 67% second division segregation, which, without correction, would convert to 50 and 33.3 cM. Thus, as distance increases, both measures

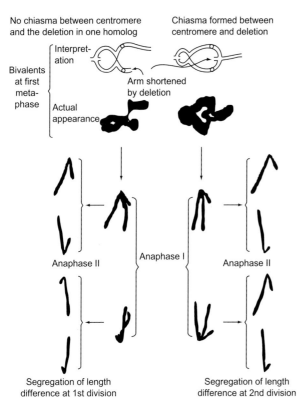

No chiasma between centromere and the deletion in one homolog

Chiasma formed between centromere and deletion

Bivalents at first meta-phase

Interpret-ation

Arm shortened by deletion

Actual appearance

Anaphase II   Anaphase I   Anaphase II

Segregation of length difference at 1st division

Segregation of length difference at 2nd division

Figure 2 First and second division segregation made visible in a lily heterozygous for a chromosome length difference. (Reproduced with permission from Fincham JRS (1983) *Genetics* after Brown and Zohary (1955) *Genetics* 40: 850.)

increasingly underestimate true map distance, but second division segregation does so to a greater extent.

See also: **Centimorgan (cM); Centromere; Crossing-Over; Map Distance, Unit; Meiosis; Tetrad Analysis**

FISH (Fluorescent *in situ* Hybridization)

J Read and S Brenner

Copyright © 2001 Academic Press
doi: 10.1006/rwgn.2001.2090

Fluorescent *in situ* hybridization (FISH) is a technique used to identify the chromosomal location of a particular DNA sequence. A DNA probe is fluorescently labeled and hybridized to denatured metaphase chromosomes spread out on glass slides.

See also: **Physical Mapping**

Fisher, R.A.

A W F Edwards

Copyright © 2001 Academic Press
doi: 10.1006/rwgn.2001.0459

Sir Ronald Fisher (1890–1962), the father of modern statistics, was for most of his life a professor of genetics, first in London and then at Cambridge. He made lasting contributions to mathematical and evolutionary genetics as well as to statistical theory applied to genetics, and experimented widely, studying especially linkage in the mouse and in polysomic plants and natural selection in the wild.

Fisher was born in London on 17 February 1890, the son of a fine-art auctioneer. His twin brother was stillborn. At Harrow School he distinguished himself in mathematics despite being handicapped by poor eyesight which prevented him working by artificial light. His teachers used to instruct him by ear, and Fisher developed a remarkable capacity for pursuing complex mathematical arguments in his head. This manifested itself later in life in an ability to reach a conclusion whilst forgetting the argument, to handle complex geometrical trains of thought, and to develop and report essentially mathematical arguments in English (only for students to have to reconstruct the mathematics later). Fisher's interest in natural history was reflected in the books chosen for special school prizes at Harrow, culminating in his last year in the choice of the complete works of Charles Darwin in 13 volumes.

Fisher entered Gonville and Caius College, Cambridge, as a scholar in 1909, graduating BA in mathematics in 1912. At college he instigated the formation of a Cambridge University Eugenics Society through which he met Major Leonard Darwin, Charles's fourth son and president of the Eugenics Education Society of London, who was to become his mentor and friend. Prevented from entering war service in 1914 by his poor eyesight, Fisher taught in schools for the duration of the war and in 1919 was appointed Statistician to Rothamsted Experimental Station, an agricultural station at Harpenden north of London. In 1933 he was elected to succeed Karl Pearson as Galton Professor of Eugenics (i.e., of Human Genetics, as it later became) at University College, London, and in 1943 he was elected Arthur Balfour Professor of Genetics at Cambridge and a Fellow of Gonville and Caius College. He retired in 1957 and spent his last few years in Adelaide, Australia, where he died of a postoperative embolism on 29 July 1962. His ashes lie under a plaque in the nave of Adelaide Cathedral.

Fisher married Ruth Eileen Guinness in 1917 and they had two sons and six daughters, and a baby girl who died young. He was elected a Fellow of the Royal Society in 1929 and was knighted in 1952 for services to science. He was the founding President of the Biometric Society, and served as President of the Royal Statistical Society, the International Statistical Institute, and the Genetical Society. He received many honorary degrees and memberships of academies, and the Royal, Darwin, and Copley Medals of the Royal Society.

Fisher made profound contributions to applied and theoretical statistics, to genetics, and to evolutionary theory. This account concentrates on genetics and evolution. Attracted to natural history at school, in his first term as an undergraduate at Cambridge Fisher bought Bateson's book *Mendel's Principles of Heredity*, with its translation of Mendel's paper. Before graduating he had already remarked on the surprisingly good fit of Mendel's data, and by 1916, encouraged by Leonard Darwin, he had completed the founding paper of biometrical genetics and the analysis of variance *The Correlation between Relatives on the Supposition of Mendelian Inheritance*, eventually published in 1918.

From his post of statistician at Rothamsted Fisher made advances which revolutionized statistics, but his advances in genetics and evolution were hardly less revolutionary. In a single publication in 1922 he proved that heterozygotic advantage in a diallelic system gives rise to a stable gene-frequency equilibrium, introduced the first stochastic model into genetics (a branching process), and initiated the study of gene-frequency distributions by means of the diffusion approximation, and in another paper he applied the method of maximum likelihood to the estimation of linkage for the first time. Other papers dealt with variability in nature, the evolution of dominance, and mimicry, and in 1926 he started his long association with E.B. Ford with whom he later measured the effect of natural selection in wild populations.

In 1930 Fisher's *The Genetical Theory of Natural Selection* was published, containing a wealth of new evolutionary arguments, from the fundamental theorem of natural selection to ideas about sexual selection, inclusive fitness, and parental expenditure. More than any other work *The Genetical Theory* established a firm basis for the modern view that evolution by natural selection is primarily a within-species phenomenon.

Taking up his appointment at University College in 1933, Fisher's pace did not slacken. Experimental organisms included mice, poultry, and the purple loosestrife, and even dogs, under the auspices of the Genetical Society. But it is in human genetics that he made the most lasting contribution. In 1935 he secured funds from the Rockefeller Foundation to establish a Blood-Group Serum Unit at his Galton Laboratory with the express purpose of initiating the construction of a linkage map for man, for he had already seen the connection between "Linkage studies and the prognosis of hereditary ailments" (to use the title of his lecture to the International Congress on Life Assurance Medicine in that year). Here is the intellectual origin of the Human Genome Project. At the same time Fisher, with J.B.S. Haldane and L.S. Penrose, was advancing the special statistical theory required in the estimation of human linkage.

In 1943 Fisher moved to Cambridge, where he was reunited with his colleagues from the Blood-Group Unit who had been evacuated there during the war. An immediate consequence was his brilliant solution of the Rhesus blood-group puzzle, involving three closely linked loci which between them explained the array of serological reactions which to everyone else had appeared chaotic: Fisher did for Rhesus what Mendel did for round and wrinkled.

After World War II ended in 1945 Fisher attempted to establish bacterial genetics in his Cambridge department and to retain for Cambridge the Blood-Group Unit, but without success. Work in his small department revolved around linkage in the mouse, and studies on purple loosestrife, wood sorrel, and primroses, always with a strong background of mathematical and statistical developments. His *Theory of Inbreeding* was published in 1949, and in 1950 he published the first paper applying a computer to a biological problem. Fisher retired from Cambridge in 1957.

Fisher was one of the great intellects of the twentieth century. In statistics he keeps company with Gauss and Laplace whilst in biology he has been compared with Charles Darwin as "the greatest of his successors." In the intersection of the two fields of statistics and biology he was the outstanding pioneer, and as the first person to recognize both the desirability and the practicability of constructing the human genome map he initiated one of the major scientific achievements of the century.

Further Reading

Box JF (1978) *R.A. Fisher: The Life of a Scientist*. New York: Wiley.

Edwards AWF (1990) R.A. Fisher: Twice professor of genetics: London and Cambridge or "A fairly well-known geneticist". *Biometrics* 46: 897–904.

Fisher RA (1971–74) *Collected Papers of R.A. Fisher*, vols. 1–5, ed. J.H. Bennett. Adelaide, SA: University of Adelaide Press.

See also: Fundamental Theorem of Natural Selection; Genetics; Natural Selection

Fitness

C B Krimbas

Copyright © 2001 Academic Press
doi: 10.1006/rwgn.2001.0460

Definition

Fitness is a concept that is often considered to be central to population genetics, demography, and the synthetic theory of evolution. In population genetics, it is technically a relative or absolute measure of reproductive efficiency or reproductive success. The absolute or Darwinian fitness of a certain genetic constitution living in a defined homogeneous environment would be equated with the mean number (or the expected number) of zygotes sufficiently similar to that produced during its entire lifetime, whereas relative fitness would be the measure of the reproductive efficiency of a certain genotype, as defined above, compared with that of another from the same population.

The term 'sufficiently similar' needs explanation: it does not mean that the offspring of a certain genotype would necessarily share the same genotype with their parent (actually, in many instances, Mendelian segregation would prohibit this); rather, it indicates that genetic effects seriously affecting fitness with one generation delay should be taken into consideration and should affect the value of fitness of the parent genotype exclusively responsible for these delayed effects. Thus, the *grandchildless* mutants in *Drosophila subobscura* and *D. melanogaster* have such an effect: female homozygotes for the mutant allele produce sterile offspring (regardless of the genotype of the male parent or that of the offspring). The reason for this is that in their fertilized eggs, the posterior polar cells are not formed. The mean number of offspring may not be sufficient to define the fitness of a genotype: the distribution of the number of its offspring may also be of importance. Thus, Gillespie has shown that genotypes having the same mean number of progeny, but differing in variances, have a different evolutionary fate. Everything else being equal, an increase in variance of the number of progeny (from generation to generation, or spatially, or developmentally) is, in the long run, disadvantageous. From the actual mean expected number of progeny we should subtract a quantity equal to $1/2s^2$ for the case of temporal variation (where s^2 is the variance in offspring number) and $1/Ns^2$ for the case of developmental variation to arrive at an estimate of fitness. However, with the exception explained above, concerning the one-generation delayed effects of a certain genotype affecting reproduction,

we will restrict fitness definition to only one generation, (where N is the effective population size) thus avoiding the temporal variation. (The long-term evolutionary fate addressed by Thoday and Cooper will not be considered here.) Furthermore, we will restrict the definition of fitness to a certain homogeneous selective environment, thus avoiding complications such as those described by Brandon (1990), where a genotype having two different fitnesses in two environments, both lower than the respective two fitnesses of a different genotype, may end up having a higher mean fitness due to an unequal distribution of individuals of the two respective genotypes in these two environments]. The reason for these restrictions is that fitness values serve to put some flesh onto the models describing allelic frequency changes from generation to generation and thus allowing short-term genetic predictions. Fitness is a useful device in quantifying the kinetics of a genetic change; it is otherwise devoid of any other independent meaning and cannot serve as a substitute to the nebulous concept of adaptation. Of course in some models, one may consider complex fitness functions, e.g., the weighted mean fitness in two environments.

Medawar (in Krimbas, 1984) expresses this view in the following statement:

The genetical usage of 'fitness' is an extreme attenuation of the ordinary usage: it is, in effect, a system of pricing the endowments of organisms in the currency of offspring; i.e. in terms of net reproductive performance. It is a genetic valuation of goods, not a statement about their nature or quality.

Historical Overview

The first use of 'fitness' with a loosely similar meaning is found in Darwin's *On the Origin of Species*. From the first to the sixth edition, Darwin employed the verb 'fit' and the adjective 'fitted' as synonyms for 'adapt' and 'adapted,' respectively. The noun first appears in 1859:

Nor ought we to marvel if all the contrivances in nature be not, as far as we can judge, absolutely perfect; and if some of them be abhorrent to our idea of fitness.

(Paul, 1992)

Of course, Darwin inherited the concept of a fitness between the organism and its environment from natural theology and from the concept of adaptation. In 1864, Herbert Spencer used the expression "survival of the fittest" as a synonym for natural selection, which was later used by Darwin. Thus, from the beginning, fit and fitness were seen to be semantically closely related to the process of natural selection and to 'adaptation.' Even today, Brandon (1990) equates fitness with adaptedness.

In 1798, Malthus compared the rates of increase of population size with the amount of food produced. According to Tort (1996), the ratio λ of the number of individuals of one generation (N_{t+1}) to that of its parental generation (N_t) is the Darwinian fitness or the Malthusian fitness. No differences among individuals are considered in this formulation, which describes a geometric or rather an exponential increase of population size, if λ is constant. In 1838, Verhulst gave another formulation, taking into consideration the change in ratio as the population reaches its carrying capacity, K. Thus Verhulst distinguishes w, the biological fitness (the Verhulstian fitness is the number of offspring produced by an individual at its sexual maturity) and population fitness, which varies also according to K and to the present population size, N_t:

$$N_{t+1} = wN_t - \frac{(w-1)}{K}(N_t)^2$$

Thus, the relation between a nonconstant Malthusian fitness and a Verhulstian fitness is:

$$\lambda_t = w - \frac{(w-1)}{K}N_t$$

Let b be the percentage of the individuals in a population that during a small time interval Δt give birth to one individual ($b\Delta t$) and d is the percentage of individuals dying at the same time interval ($d\Delta t$), the net change in individuals at the same time interval will be:

$$N_t = (b-d)N_t\Delta t$$

By substituting $b-d$ with m (where m is, according to Fisher, the Malthusian parameter) and integrating we get the form of increase of population size:

$$N_t = N_0 e^{mt}$$

Lotka, as well as Fisher, used mortality and fertility tables for the different biological ages to estimate fitness from m. The Darwinian fitness is related to the Malthusian parameter in the following:

$$\lambda = e^m$$

Furthermore, Fisher considered that Malthusian parameters, and thus fitnesses, are inherited, different genotypes having different fitnesses. The course of evolution is to maximize population fitness, that is the (weighted) mean value of the individual fitnesses of a population.

The rate of increase in fitness in any organism at any time is equal to its [additive] genetic variance in fitness at that time.

Since variances are always positive the change will always be in the direction of an increase of this quantity. Fisher considered this 'fundamental theorem of natural selection' as a general law, equivalent to the second law of thermodynamics, which stipulates always an increase of a physical quantity, i.e., entropy. The generality of Fisher's law was questioned and, in some cases, it was shown not to hold true. Furthermore, as Crow and Kimura remarked,

One interpretation of the theorem is to say that it measures the rate of increase in fitness that would occur if the gene frequency changes took place, but nothing else changed.

Thus an environmental deterioration that would affect fitness values, and thus decrease mean population fitness, is not considered by Fisher.

Wright used the population fitness as varying according to the gene frequencies in the population. Excluding competition among individuals, Wright states that every genotype is characterized by a fitness value and each individual belonging to that genotype has an expected number of progeny, which is the fitness of that genotype. The population fitness, $\bar{W}$, is the expected mean number of progeny of every individual of the parental generation. $\bar{W}$ is a composite function, the sum total of the products of all genotype frequencies by their specific fitnesses (or adaptive values). Contrary to Fisher, Wright, in his shifting balance theory, envisages most of the species to consist of many small and more or less isolated populations, each with its specific gene frequencies. Populations occupy the peaks of an adaptive surface, formed by the values of $\bar{W}$ (population fitnesses), for every point corresponding to certain gene frequencies. These peaks are positions of stable local equilibria. Due to drift, gene frequencies may change and, thus, populations may cross a valley of the adaptive surface and be attracted by another peak. Equilibrium points are local highest points of population fitness values.

Components of Fitness: Inclusive Fitness

It is often stated that selection acts on survival and reproduction. This is not an exact phrasing: fitness is the mean number of progeny left; therefore viability components (survival, longevity) are important as far as they affect the net reproductive effect. Longevity may be important only in those cases where it may affect the net reproductive effect. Selection is blind to longevity at a postreproductive age. This is the reason why the inherited pathological syndrome of Huntington disease, which appears after the reproductive years, seems not to be selected against.

According to Hartl and Clark (1989), starting from the stage of the zygote, the components of fitness are as follows: viability; subsequently sexual selection operates favoring or prohibiting a genotype to find mates; in the general case every combination of genotypes of the mating pair may correspond to a specific fecundity. Thus, fecundity depends on the genetic constitution of both partners. For a simple one gene/two alleles case, nine different fecundity values are defined. Before the formation of the zygote, gametic selection (one aspect of it being meiotic drive) may take place, and sometimes counteract the direction of selection exercised at the diploid phase. Fitness is estimated by counting zygotes produced by a zygote. A proposal to overcome the difficulty of counting zygotes in many animals was to start from another well-recognized stage of the biological cycle and complete this cycle to the same stage of the progeny. This proposal is, however, mistaken, since the progeny of a genotype do not necessarily have the same genotype as their parent. As a result this estimates fitness components corresponding to different genotypes.

Developmental time is an important, but generally neglected or ignored, component of fitness in populations of overlapping generations at the phase of increase of their size (e.g., at the beginning of colonization of a new unoccupied territory; r-selection). Lewontin (1965) examined the case of insects that follow a triangle schedule of oviposition (a triangular egg productivity function is characterized by three points: the age of first production, that of peak production, and that of the last production reported in a time coordinate and the number of eggs produced at the other). In his specific model, a shortening of developmental time may be equivalent to a doubling of total net fecundity. This shortening is equal to a 1.55 day decrease of the entire egg production program (what Lewontin calls a transposition of the triangle to an earlier age), or to a 2.20-day decrease only of the age of sexual maturity (the age at which the first egg is produced), leaving the other ages unchanged as well as the total number of eggs deposited. It is also equivalent to a 5.55-day decrease of the age of the highest egg production only (the peak of the triangle), other things remaining unchanged or, finally, to a 21-day decrease of the age at which the last egg is deposited, other variables remaining the same.

Hamilton's concept of 'inclusive fitness' was formulated to provide a Darwinian explanation for altruistic actions that may endanger the life of the individual performing such acts. An individual may multiply its genes in two different ways: directly by its progeny, and indirectly by protecting the life of other individuals of a similar-to-it genetic constitution. If the danger encountered is outweighed by the gain (all calculated in genes) then the performance of such acts may be fixed by natural selection. Estimations of inclusive fitness do not take into account only the individual's fitness but also that of its relatives (of similar genetic constitution): it is the sum total of two selective processes, individual selection and kin selection. In this case, in fact, the counting tends to change from the number of individuals in the progeny to the number of genes preserved by altruistic acts in addition to those transmitted directly through its progeny.

Adaptation, Adaptedness and the Propensity Interpretation of Fitness

Natural selection acts on phenotypes; certain traits of these phenotypes are the targets of selection. The individuals bearing some traits are said to be adapted. However, no common and general property may characterize adaptation. A search through the literature of all the important neo-Darwinists reveals that, in spite of the suggestion that adaptation has an autonomous meaning, it is used in fact as an alternative to selection. Van Valen seems to differ from all other authors because he equates adaptation with the maximization of energy appropriation, both for multiplying and for increasing biomass, thus solving the problem of lianas and other clone organisms. The concept of adaptation was shown to be completely dependent on that of selection (Krimbas, 1984). Brandon provided an argument proving the impossibility of establishing an independent of selection criterion or trait for adaptation. He argued that we may be able to select in the laboratory against any character except for one, fitness. There is no reason to exclude from natural selection the selection experiments performed in the laboratory, since the laboratory is also part of nature. Thus there is no character or trait in the diploid organism that could be taken in advance as an indication of adaptation independently of selection. Fitness is a variable substantiating and quantifying the selective process.

While one would expect adaptation to disappear from the evolutionary vocabulary, it is still used for describing the selective process that changes or establishes a phenotypic trait as well as the trait itself. Sometimes the engineering approach is used:

adaptation, it is argued, is in every case the optimal solution to an environmental problem. The difficulties with such an approach are twofold. First, we are often unable to define precisely the problem that the organism faces (it might be a composite problem) in order to determine in advance the optimal solution and, as a result, we tend to adapt the 'solution' encountered to the nature of the problem the organism faces. Second, it is evident that several selection products are not necessarily the optimal solutions, the evolutionary change resembling more a process of tinkering rather than an application of an engineering design.

Recently, several authors (Brandon, Mills and Beatty, Burian, and Sober; see Brandon, 1990) have supported the propensity interpretation of fitness (or adaptedness). In so doing they try first to disentangle 'individual fitness' (something we are not considering here; as mentioned earlier we have taken into consideration only the fitnesses of a certain category or group of individuals) from the fitness that is expected from its genetic constitution. Indeed, all kinds of accidents may drastically modify the number of progeny one individual leaves behind. A sudden death may zero an individual's contribution to the next generation. But selection is a systematic process in the sense that in similar situations similar outcomes are expected. Thus, in order to pass from the individual or actual fitness to the expected one, these authors are obliged to consider two different interpretations of 'probability.' The first interpretation considers probability as the limit of a relative frequency of an event in an infinite series of trials, but since this series is never achieved, the observed frequency in a finite series of trials might be used instead. The second interpretation is that of propensity, where the very constitution, i.e., the physical properties, of the individual underlies the propensity for performing in a given way. This may be a dispositional property, i.e., it might be displayed in a certain way in some situations and in another way in others. The propensity interpretation of fitness attributes to physical causes, linked to the very structure of the individual, the tendency to produce a specific number of offspring in a particular selective environment. This is another way of reifying fitness, and via fitness relative adaptedness, and finally adaptation. It is reminiscent of the Aristotelian *potentia et actu*, where the propensity is 'potentia' and the actual mean number of offspring corresponds to the 'actu.' In some situations of viability selection this interpretation seems quite satisfactory (e.g., in mice resistant to warfarin). No one would deny that the selection process depends most of the time on the properties of a genotype performing in a certain environment. But this may not be as general as one may think. There are situations in which the contribution to fitness from the part of the organism is not clear or does not seem preponderant. Thus, it is more difficult and much less satisfactory to attribute to a certain genetic constitution the mating advantage of the males when they are rare and the mating disadvantage when they are frequent. It is a case of frequency-dependent selection.

On the other hand, the definition of genotypic fitness might also suffer from some disadvantages. Let us consider the case of dextral and sinistral coiling in shells of certain species of snails. The direction of coiling is genetic, due to one gene with two alleles. The allele d (for dextral coiling) is dominant to the *l* allele (recessive). But the phenotype of the individual is exclusively determined by the genotype (not the phenotype) of its mother and not by its own genotype. Thus, there is a delay of one generation in phenotypic expression. Selection operates on phenotypes (the interactors of D. Hull). In the case of selection for dextral or sinistral direction of coiling, the phenotypic fitnesses may be clearly understood and simple but useless, while the genotypic fitnesses would be a complicated function depending on the frequency of the alleles in the population and the mating system.

Thus, it seems better to consider genotypic fitness as a useful device in performing some kinetic studies regarding changes in gene frequencies or attraction to an equilibrium point. It is useless to attribute other qualities or properties to this device. Modern evolutionary theory is basically of a historical nature (although some processes may be repeated). A complete and satisfactory explanation of a specific case should comprise a historical narrative including information of the phenotypic trait being the target of selection, the ecological, natural history, or other reason driving the selective process (why this trait is being selected), the genetics of the trait, the subsequent change to selection of the genetic structure of the population, and the corresponding change in the phenotypes. In natural history, generality and the search for hidden and nonexisting entities and properties may only contribute to an increase in the metaphysical component of evolutionary theory inherited from natural theology.

On Population Fitness

It is much more difficult to define population fitness: population geneticists use to calculate the mean adaptive value or the mean individual fitness in a population. But this exercise is quite futile when comparing two different populations. A group of adapted organisms is not necessarily an adapted group of organisms. Demographers earlier equated size (or increase in size) with population fitness. However, as Lewontin once remarked, it is not certain that a greater or denser population is better adapted, since it may suffer from parasites and epidemics; on the other hand, a population depleted of individuals may suffer collapse and extinction. I have argued (Krimbas, 1984) that according to the 'Red Queen hypothesis' of Van Valen, all populations (at least of the same species) seem to have, a priori, the same probability of extinction, and thus possess, a priori, the same long-term population fitness.

In addition, it is not clear enough how we should consider a group: a group is not an organism that survives and reproduces. Although individuals of the group interact in complex ways and thus provide some image of cohesion, the 'individuality' of the groups seems most of the time to be quite a loose subject. Should we consider group extinction per unit of time to determine group fitness? What about group multiplication? In order to achieve a model in group selection cases, one may resort to different population selective coefficients, or population adaptive coefficients (something related to the population fitness). In these cases, the search for the nature of population fitness becomes even more elusive. As a result, population fitness is a parameter useful exclusively for its expediency; no search for its hidden nature is justified.

References

Brandon RN (1990) *Adaptation and Environment.* Princeton, NJ: Princeton University Press.

Gillois M (1996) Fitness. In Tort P (ed.) *Dictionnaire du Darwinisme et de l'Evolution,* vol. 2, 1676–1688. Paris: Presses Universitaires de France.

Hartl DL and Clark AG (1989) *Principles of Population Genetics,* 2nd edn. Sunderland, MA: Sinauer Associates.

Krimbas CB (1984) On adaptation, neo-Darwinian tautology, and population fitness. *Evolutionary Biology* 17: 1–57.

Lewontin RC (1965) Selection for colonizing ability. In: Baker HG and Stebbins GL (eds) *The Genetics of Colonizing Species,* 77–94. New York: Columbia University Press.

Paul D (1992) Fitness: historical perspectives. In: Fox Keller E and Lloyd EA (eds) *Keywords in Evolutionary Biology,* 112–114. Cambridge, MA: Harvard University Press.

See also: **Adaptive Landscapes; Darwin, Charles; Fitness Landscape; Natural Selection**

Fitness Landscape

J F Crow

Copyright © 2001 Academic Press
doi: 10.1006/rwgn.2001.0461

A fitness landscape, or adaptive surface, is a geometrical construct in which the fitness or adaptive value of a genotype is the ordinate and the two abscissae

are gene (or sometimes genotype) frequencies. It is usually pictured in three dimensions, but conceptually can involve a larger number. In some models it has an exact mathematical meaning, in others it is employed as a metaphor.

The idea of an adaptive surface was introduced by Sewall Wright. He thought of a surface on which each point on the surface corresponded to a combination of allele frequencies on the abscissae. **Figure 1** shows a simple two-locus example. Random mating proportions and linkage equilibrium are assumed. The two abscissae are the frequencies of the dominant A and B alleles. The relative genotype fitnesses of $aa\ bb$, $A-\ bb$, $aa\ B-$, and $A-B-$ are 1, $1-s$, $1-s$, and $1+t$, where s and t are both positive and $A-$ (or $B-$) indicates that the second allele can be either A or a (or B or b). The ordinate represents the average fitness of a population with particular allele frequencies. There are two peaks, one when the genotype $AA\ BB$ is fixed, the other a lower peak for the genotype $aa\ bb$. Genotypes $AA\ bb$ and $aa\ BB$ are at the other two corners and are least fit. Ordinarily a population, located at a point on the surface, climbs the nearest peak, but not necessarily in a straight line. The complications of mutation, linkage, and epistasis may cause the path upward to be circuitous. And, as these complications are introduced, along with more loci, the mathematics becomes more difficult.

This is the situation envisioned by Sewall Wright. A population cannot change from the lower peak to the higher one, because it has to pass through a less fit region. It was this dilemma that led Wright to propose his shifting-balance theory whereby a combination of random drift and differential migration make it possible to cross the valley and reach a higher peak. Wright regarded the fitness surface more as a

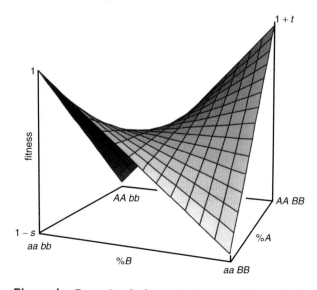

Figure 1 Example of a fitness landscape, with two loci.

metaphor than as a mathematical model. As a result his papers present different, often confusing concepts. Sometimes the abscissae are allele frequencies, sometimes they are genotype frequencies, and sometimes phenotypes. How rugged the fitness surface is has been a matter of continual discussion since Wright first introduced his ideas in the 1930s. Wright thought of the multidimensional surface as quite rugged, with numerous peaks and valleys. Others, R. A. Fisher in particular, have suggested that the surface is more like an ocean with undulating wave patterns. Furthermore, as the number of dimensions increases, only a small fraction of the stationary points are maxima. A population is much more likely to be on a ridge than on a peak. The debate was not settled while Wright was alive and still continues. Wright summarized his lifetime view of the subject in a paper entitled "Surfaces of selective value revisited," published in 1988, shortly before his death (Wright, 1988).

Although Wright, more than anyone else, was responsible for introducing random processes into population genetic theory, he never attempted to model the whole shifting-balance process stochastically. Recently there has been considerable mathematical work in this area, partly as a way of developing and testing Wright's theory. The entire process has been treated stochastically, something that was missing in Wright's formulations.

The landscape idea has been extended to concepts other than fitness, such as developmental morphology and protein structure. The ruggedness of the landscape determines whether orderly change is possible or whether alternatives, such as stasis or chaos, emerge. In evolution, the lower the peaks and the higher the valleys, the more likely it is that selection can carry a population, if not to the highest peak, at least to one that has a respectable fitness. Similar considerations apply to the study of morphological development in the presence of various constraints. The ruggedness of the landscape can be deduced from parameters, such as the number of factors involved, and especially the degree to which they are coupled, or in genetic terms, the degree of epistasis.

Further Reading

Kaufmann SA (1993) *The Origins of Order.* Oxford: Oxford University Press.

Provine WB (1986) *Sewall Wright and Evolutionary Biology.* Chicago, IL: University of Chicago Press.

Reference

Wright S (1988) Surfaces of selective value revisited. *American Naturalist* 131: 115–123.

See also: Fisher, R.A.; Fitness; Wright, Sewall

Fix Genes

J Vanderleyden, A Van Dommelen, and J Michiels

Copyright © 2001 Academic Press
doi: 10.1006/rwgn.2001.1637

The process of biological nitrogen fixation is an extremely energy-demanding process requiring, under ideal conditions, approximately 16 moles of ATP per mole of N_2 fixed. The property of reducing atmospheric dinitrogen to ammonia is found among a wide variety of free living, associative, and strictly symbiotic bacteria. The genetics of nitrogen fixation were initiated in the free-living diazotroph *Klebsiella pneumoniae*. This analysis led to the identification of 20 *nif* (for nitrogen fixation) genes. The identification of these *nif* genes has substantially facilitated the study of nitrogen fixation in other prokaryotes such as *Sinorhizobium meliloti* (formerly *Rhizobium meliloti*) and *Bradyrhizobium japonicum* in identifying genes that are both structurally and functionally equivalent to *K. pneumoniae nif* genes, including *nifHDK, nifA, nifB, nifE, nifN, nifS, nifW,* and *nifX.* In addition, in these organisms, genes essential for nitrogen fixation were identified for which no homologs are present in *K. pneumoniae.* These were named *fix* genes and are often clustered with *nif* genes or regulated coordinately. **Table 1** summarizes the properties and functions of *fix* genes.

Regulation of Nitrogen Fixation

Owing to the extreme oxygen sensitivity of the nitrogenase enzyme, a major trigger for *nif* and *fix* gene expression in all systems studied so far is low oxygen tension. For instance, in the legume nodule, the dissolved oxygen concentration is 10–30 mmol l^{-1}, creating a hypoxic environment. Conversely, all nitrogen-fixing bacteria deploy a complex regulatory cascade preventing aerobic expression of *nif* and *fix* genes. In addition, the deprivation of fixed nitrogen also controls the process of nitrogen fixation in free-living fixers but not in symbiotic bacteria (except *Azorhizobium caulinodans*). Many, but not all, *nif* and *fix* genes including the nitrogenase structural genes and accessory functions are preceded by a characteristic type of promoter, the $-24/-12$ promoter, recognized by the alternative sigma factor σ^{54} (or RpoN). Activation of this promoter requires the presence of an activator protein, i.e., the nitrogen regulatory protein NtrC or the nitrogen fixation regulatory protein NifA. While NtrC regulates gene expression in response to the nitrogen status, NifA

Table I Function or putative function of *fix* genes

Gene or operon	Homology and/or function	Renamed
fixABCX	Required for nitrogen fixation, function unknown; FixX shows similarity to ferredoxins	
fixD	Transcription activator of *nif*, *fix*, and additional genes	*nifA*
fixF	Codes for a polypeptide homologous to the *nifK* gene product	*nifN*
fixGHIS	Required for the formation of the high-affinity cbb_3-type cytochrome oxidase; FixI shows similarity to the Cu-transporter CopA	
fixLJ	Regulatory two-component system involved in oxygen regulation of *fixK* and *nifA* (*S. meliloti*) transcription	
fixK	Regulatory protein, belongs to the Crp/Fnr family of prokaryotic transcriptional activators	
fixNOQP	Microaerobically induced, membrane bound high-affinity cbb_3 cytochrome oxidase	*cytNOQP*
fixR	Sequence similarity to NAD-dependent dehydrogenases, not essential for symbiotic nitrogen fixation in *B. japonicum*	
fixT	Negative regulator of FixL	
fixU	Function unknown	
fixW	Function unknown	
fixY	Deduced amino acid sequence of the sequenced part of *fixY* shows similarity to the regulatory NifA protein from *K. pneumoniae*	
fixZ	May contain an iron–sulfur cluster; the sequence of FixZ is very similar to *K. pneumoniae* NifB	

senses the oxygen concentration either directly, as in *S. meliloti*, or indirectly through the activity of the NifL regulatory protein, as observed in *K. pneumoniae*. In addition to the basic NifA-mediated regulatory mechanism, most symbiotic diazotrophs have evolved additional control mechanisms of *nif* and *fix* gene expression.

The *FixL–FixJ* Regulatory Cascade

The activation of nitrogen fixation genes in *S. meliloti* involves a regulatory cascade of which the *fixLJ* genes are the primary controllers (**Figure 1**). The FixL and FixJ proteins are members of the ubiquitous two-component family of regulatory systems in which the sensor (*in casu* FixL), a histidine kinase, activates the response regulator (FixJ) by phosphorylation in response to a specific environmental signal. The *S. meliloti* FixL protein is a membrane-anchored hemoprotein that acts as an oxygen sensor. Oxygen binds to a heme group joined to a histidine residue that is located within a PAS structural motif. These motifs are found in a wide variety of protein modules that sense diverse stimuli such as the redox potential, light, and oxygen. Under hypoxic conditions, FixL autophosphorylates on a conserved histidine residue with a γ-phosphate from ATP. In the absence of bound

oxygen, the kinase activity of FixL is turned on and the phosphate group is subsequently transferred to an aspartate residue in the cognate receiver protein, FixJ. Upon phosphorylation, FixJ is turned into a transcriptional activator of two regulatory genes, *nifA* and *fixK*. In addition, FixL also has phosphatase activity, reducing effectively the amount of FixJ-phosphate under aerobic conditions.

Besides *nifA*, FixJ-phosphate activates the expression of the *fixK* gene. FixK is homologous to members of the Crp/Fnr family of prokaryotic transcriptional activator proteins. FixK acts as an activator of the *fixNOQP* genes, coding for a high-affinity respiratory oxidase complex, *fixGHIS*, and *fixT* genes. This function can be taken over in other rhizobia such as *B. japonicum*, *Rhizobium leguminosarum* biovar. *viciae*, and *Rhizobium etli* by the FnrN protein. The latter protein possesses a distinct cysteine signature believed to play a role in redox sensing as does a similar motif in Fnr. In contrast, the FixK protein does not show conserved cysteines and the activity of this protein is not subject to oxygen control. FixK and FnrN bind to conserved DNA motifs called anacroboxes in the promoter of their target genes.

In *S. meliloti*, a repressor of nitrogen fixation gene expression was also identified. The *fixT* gene codes for a small protein that modulates the activity of the

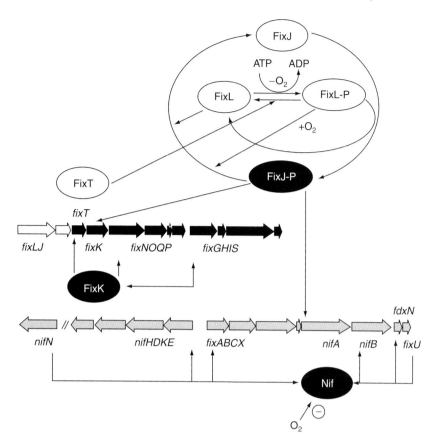

Figure I Model of the FixL–FixJ regulatory cascade in *Sinorhizobium meliloti*. Low oxygen conditions stimulate the autophosphorylation activity of FixL and repress the phosphatase activity of FixL–P. In contrast, the phosphatase activity of the unphosphorylated FixL as well as phosphoryl transfer from FixL–P to FixJ protein are independent of the oxygen concentration. Activity of the transcriptional regulator NifA is repressed by oxygen (see text for details). Genes regulated by NifA and FixK are marked in grey and black respectively. Proteins marked in a black oval are transcriptional activators. Not all known open reading frames (ORFs) associated with *nif* and *fix* genes are shown.

two-component system FixLJ. The target of FixT is the C-terminal domain of the FixL protein and the interaction of both proteins leads to inhibition of FixL-phosphate synthesis and consequently a decrease in *nifA* and *fixK* transcription.

Conclusion

Up to now many nitrogen fixation genes, different from the previously characterized *nif* genes, have been identified. These *fix* genes are involved either in basic cellular functions in nitrogen fixing conditions (e.g., respiration), or in processes more directly linked to the nitrogen fixation process (e.g., electron transport to the nitrogenase enzyme), or have a regulatory role (oxygen sensing). *fix* genes are present not only in symbiotic bacteria but also in free-living nitrogen fixers. Genes homologous to *fix* genes are even found in non-nitrogen-fixing bacteria (e.g., *fixABCX* homologs in *Escherichia coli*). It is to be expected that ongoing sequencing projects and the associated gene expression and functional analyses of prokaryotic

organisms will ultimately lead to complete models describing processes as complex as biological nitrogen fixation.

See also: **Symbionts, Genetics of**

Fixation of Alleles

See: **Fixation Probability**

Fixation Probability

T Ohta

Copyright © 2001 Academic Press
doi: 10.1006/rwgn.2001.0463

The fixation probability of a mutant allele is the probability that it becomes fixed and substitutes for the original allele in a population. The probability depends on the initial frequency and selective value of the mutant as well as on the effective population size. Fixation probability becomes larger as the advantageous effect of the mutant increases. However, selectively neutral as well as very slightly deleterious mutations have a finite probability of fixation. Many mutants at the molecular level have very small effects, and neutral and slightly deleterious mutations are prevalent. Fixation probability is a most basic quantity for discussions of evolution, particularly at the molecular level.

In the following, let us consider the simple case of genic selection. Let A and a be the original and mutant alleles, and s be the selection coefficient of the allele, a, such that the relative fitness of the genotypes, AA, Aa, and aa are 1, $1+s$ and $1+2s$, respectively. N_e and p denote the effective population size and the initial frequency of a. M. Kimura has shown that the fixation probability of a, $u(p)$, becomes as follows:

$$u(p) = \frac{1 - e^{-4N_e sp}}{1 - e^{-4N_e s}} \qquad (1)$$

Most mutations are unique, and the initial frequency, p, is the reciprocal of the actual size of the population, $1/(2N)$, where N is the actual size:

$$u\left(\frac{1}{2N}\right) = \frac{1 - e^{-2(N_e/N)s}}{1 - e^{-4N_e s}} \qquad (2)$$

For a neutral mutant, i.e., $s = 0$, the fixation probability is equal to the initial frequency, p. Equation (1) tells us that $u(p)$ simply depends on the product, $N_e s$. When there is dominance, the formula becomes more complicated.

The relationship between $u(p)$ and $N_e s$ is given in **Figure 1**. As seen from the figure, the fixation probability is a monotonically increasing function of $N_e s$. In other words, for advantageous mutants ($s > 0$), $u(p)$ increases as N_e and/or s get larger. If the selective advantage of a mutant remains the same in varying population size, the mutant has a greater chance of becoming fixed in large populations than in small ones. For slightly deleterious mutations ($s < 0$), $u(p)$ decreases as N_e and/or the absolute s value get larger. Therefore, slightly deleterious mutations have less chance of survival in large populations than in small ones.

When $4 N_e s \gg 1$, mutants are said to be definitely advantageous, and equation (2) reduces to $2s$, provided $N = N_e$. In other words, the fixation probability of a

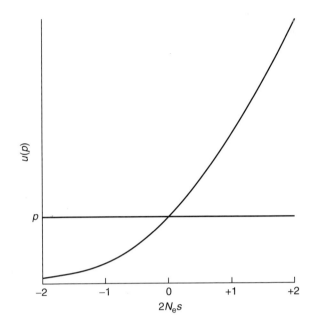

Figure 1   Fixation probability, $u(p)$, of mutant genes as a function of $2 N_e s$.

definitely advantageous mutant is twice its selective advantage, as found by J. B. S. Haldane a long time ago.

In natural populations, the values of N_e and s are not constant, but varies in space and in time and the above simplified treatment is an approximation. The application of the theory is mostly on molecular evolution. The data on the rate of gene substitution are available on many proteins and DNA sequences. By examining such data, one may infer various selective forces in relation to fixation probability. The rate of substitution may be obtained in terms of fixation probability. In one generation, $2Nv$ new mutations appear in the population, if v is the rate of occurrence of mutations per gamete per generation. The rate of substitution (k) is the product of $2Nv$ and fixation probability:

$$k = 2Nv \times u\left(\frac{1}{2N}\right) \qquad (3)$$

This equation is useful in interpreting data.

Further Reading

Gillespie JH (1998) *Population Genetics: A Concise Guide*. Baltimore, MD: Johns Hopkins University Press.

See also: Gene Substitution; Genetic Drift; Natural Selection; Nearly Neutral Theory; Neutral Theory

Flagella

S-I Aizawa

Copyright © 2001 Academic Press
doi: 10.1006/rwgn.2001.0464

The flagellum is a molecular machine whose function is to generate motility, a common characteristic of many species of bacteria. Motility gives cells the freedom to move into a wider world, but, in order to confer a survival advantage, movement must be coupled to some form of sensory machinery that allows movement toward a favorable environment. The sensory machinery is a protein complex consisting of various receptors and signal-transducing enzymes. The behavior imparted by a harmonious combination of motility and chemical sensing is called chemotaxis.

Of the 50 genes required for chemotactic behavior in *Salmonella enterica* serovar *typhimurium*, 10 genes encode the sensory complex (four genes for chemoreceptors, six genes for components of signal-transducing enzymes). The remaining 40 genes are required for the biogenesis of the flagellum (**Figure 1**). Phenotypes caused by mutations in each gene are divided into three groups: Che⁻ (chemotaxis-deficient), Mot⁻ (motility-deficient), Fla⁻ and (flagella-deficient).

Defects in genes of the chemosensory system give rise to Che⁻ mutants, which show either smooth (no change of direction) or tumbling (continuous changes of direction) swimming. There are two authentic genes (*mot*A and *mot*B) and three pseudogenes (*fli*G, *fli*M, and *fli*N) necessary for torque generation. Most of the remaining genes are responsible for flagellar construction and were originally called *fla* genes. When the number of *fla* genes surpassed 26, in 1988, *fla* genes were assigned to four groups according to the gene clusters on the chromosome: *flg*, flag (A–N; 23 min); *flh*, fluh (A–E; 40 min); *fli*, fly (A–T, Y, Z; 42 min); and *flj*, flaj (A, B; 56 min). This unified nomenclature was proposed for *Escherichia coli* and *Salmonella enterica* serovar *typhimurium* and is now widely applied to many other bacterial species.

In *Salmonella* and related species, there is a three-tier regulatory hierarchy that governs the transcription of the flagellar genes. The master operon (*flh*D, *flh*C, or *flh*DC), the only operon in class 1, activates the class 2 genes (37 genes in 8 operons) that mostly encode structural proteins of the hook-basal body (HBB). The class 2 level contains two regulatory genes, *fli*A and *flg*M: FliA is a sigma factor (σ²⁸) for initiating class 3 gene transcription, and FlgM is an anti-sigma factor that binds FliA to halt its action. FlgM is secreted through the central channel of the

complete HBB, resulting in release of FliA, which can then freely interact with RNA polymerase and direct transcription of the class 3 operons. FlgM is transcribed from class 2 as well as class 3 operons. The amount of FlgM expressed at class 3 is much higher than that at class 2, indicating an autogenous regulation of the class 3 operons. Therefore, flagellar gene regulation is strictly coupled with the flagellar construction, preventing unnecessary production of abundant proteins of class 3. Upon initiation of class 3 gene expression, flagellar filaments are formed by flagellin (FliC) export and polymerization, and the sensory system (composed of 10 Che proteins) is organized. At the same time, the MotA/B complex is assembled on the periplasmic side of the membrane to rotate the motor, thereby completing a functional flagellum.

There are three multifunctional genes (*fli*G, *fli*M, and *fli*N) that show three different phenotypes depending on the mutational sites. These gene products, FliG, FliM, and FliN, form a cup-shaped complex (called a C ring) at the cytoplasmic side of the MS ring complex see (**Figure 1**). In fact, the C ring is multifunctional: It works as a part of the export apparatus, it generates torque by the interaction with Mot complexes, and it switches the rotational direction of the motor by binding to a signal protein, the phosphorylated form of CheY.

Salmonella species have two sets of flagellin genes: *fli*C and *flj*B. The *hin* gene upstream of the *flj*B gene flip-flops, allowing *flj*B gene expression in only one direction. When the *flj*B gene is expressed, the *flj*A gene downstream of the *flj*B produces a repressor of the *fli*C gene, inhibiting a concomitant expression of the latter. By switching these flagellins (a property known as phase variation), cells can evade the immune system of the host.

Flagellar construction is not independent from the cell division cycle. The number of flagella on a peritrichously flagellated cell must stay constant after each cell division, otherwise the number will quickly become either zero or infinite after several generations. Hence, it is reasonable to assume that the flagellar system is under global regulation, occurring synchronously with cell division. The gene(s) directly controlling the flagellar master genes has (have) not yet been identified.

Although as many as 70% of the bacterial species so far studied show flagellar motility, flagella are not necessarily expressed at all time points throughout the life cycle. Some *E. coli* cells do not grow flagella in rich medium because of catabolite repression. Many species living in water grow flagella only at lower temperatures. Some soil bacteria and freshwater bacteria show flagella only during early log-phase. The master operon *flh*DC is transcribed with the help of the

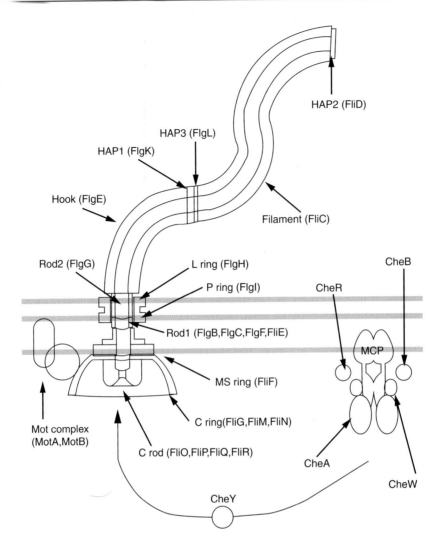

Figure I Schematic drawing of the flagellar system. Stimuli go through the receptor, MCP, that triggers signal transduction to send signal proteun, CheY, to the flagellar base. Name of each substructure of the flagellum is indicated by arrows. Their component protiens are shown in parenthesis.

housekeeping sigma factor σ^{70}. The sequences upstream of the *flh*DC operon are fairly diverse among bacterial species. In *E. coli*, several genes and physiological factors affecting the *flh*DC expression have been known: the heatshock proteins (DnaK, DnaJ, and GrpE), the pleiotropic response regulator (OmpR) activated by acetyl phosphate, and the DNA-binding protein H-NS. More such genes and factors have been discovered in other species, and their mechanisms are under investigation.

Further Reading

Aizawa S-I and Kubori T (1998) Bacterial flagellation and cell division. *Genes to Cells* 3: 1–10.

Aizawa S-I, Harwood CS and Kadner RJ (2000) Signaling components in bacterial locomotion and sensory reception. *Journal of Bacteriology* 182: 1459–1471.

See also: Escherichia coli; Salmonella

Flagellar Phase Variation (Biology)

See also: Phase Variation

FLI1 Oncogene

C S Cooper

Copyright © 2001 Academic Press
doi: 10.1006/rwgn.2001.1570

FLI1 (Friend leukemia virus integration 1), identified as a common mouse viral integration site, encodes a protein that is a member of the Ets family of transcriptional regulators. As a consequence of the t(11;22)(q24;q12) translocation found in Ewing's sarcoma (EWS) and primitive neuroectodermal tumours, *FLI1* at 11q24 becomes joined to the *EWS* gene producing a fusion protein in which N-terminal EWS protein sequences become fused to the C-terminal *FLI1* DNA-binding domain.

***See also:* Ets Family; Ewing's Tumor**

Flower Development, Genetics of

G Theissen

Copyright © 2001 Academic Press
doi: 10.1006/rwgn.2001.1674

Essentials of Flower Developmental Genetics

Flowers are the well-known reproductive structures of flowering plants (angiosperms) which are by far the largest group of extant plants. Flowers are composed of up to four different types of specialized floral organs: green, leaf-like sepals; showy petals which may attract pollinators; stamens, being the male reproductive organs which produce the pollen; and carpels, being the female reproductive organs inside which the ovules and seeds develop (**Figure 1**). The number, arrangement and morphology of these organs is diverse, but species-specific, since flower development is under strict genetic control. This guarantees that flower development is initiated only under conditions favorable for reproduction, but that once started it proceeds in a highly standardized way. Flower development can be subdivided into several major steps, such as floral induction, floral meristem formation, and floral organ development. Accurate genetic control of the different steps of flower development is achieved by a hierarchy of interacting regulatory genes, most of which encode transcription factors (**Figure 1**). Close to the top of that hierarchy are 'flowering time genes' which are triggered by developmental cues and environmental factors such as plant age, day length, and temperature. 'Flowering time genes' mediate the switch from vegetative to reproductive development by activating meristem identity genes. 'Meristem identity genes' control the transition from vegetative to inflorescence and floral meristems and work as upstream regulators of 'floral organ identity genes.' Combinatorial interactions of these genes specify the identity of the different floral organs by activating organ-specific 'realizator genes.' Most of the genes controlling flower development belong to highly conserved gene families, such as the MADS-box, *FLO*-like, and *AP2/EREBP*-like genes, which are assumed to encode transcription factors.

Our current knowledge about the genetics of flower development has been mainly worked out in two model plants, thale cress (*Arabidopsis thaliana*) and snapdragon (*Antirrhinum majus*). While *Arabidopsis* has been of great importance for studies on all different kinds of genes involved in flower formation, *Antirrhinum* was of special importance during cloning of the first floral meristem and organ identity genes. Therefore, the descriptions outlined below focus on these predominant model systems, unless stated otherwise

Floral Induction

When flowering plants have reached a critical age, environmental signals may trigger a switch to floral development. The shoot apical meristem, a small group of progenitor cells, ceases production of leaf primordia and switches to the production of floral meristems which develop into flowers. Since flowering at the wrong time may seriously hamper reproductive success, the angiosperms have evolved multiple genetic pathways to regulate the timing of the floral transition in response to environmental stimuli and developmental cues. Since plants live under very different environmental conditions and follow diverse life strategies, the mechanisms controlling the transition to flowering vary a lot, often even within single species.

The analysis of natural variants (ecotypes) and of mutants that flower later or earlier than wild-type has revealed more than 80 gene loci that affect flowering time in *Arabidopsis*. These flowering time genes may contribute to two different components of the floral transition: the production of flowering signals and the competence of the shoot apical meristem to respond to these signals. The flowering time mutants can be grouped into different classes defining different pathways of floral induction. *Arabidopsis* is a facultative long-day plant which responds to long days (indicating spring and summer) by flowering earlier than when grown in short days. One class of mutants displays a reduced response to changes in photoperiod (day length) when compared to wild-type. The corresponding genes, therefore, may participate in a photoperiod promotion pathway. A second class of late-flowering mutants are unaffected in their response to photoperiod. The corresponding genes thus may be involved in an autonomous promotion pathway. This pathway monitors the signals of an internal developmental clock that measures plant age. A third pathway, termed vernalization promotion pathway, confers susceptibility to vernalization, i.e., an extensive

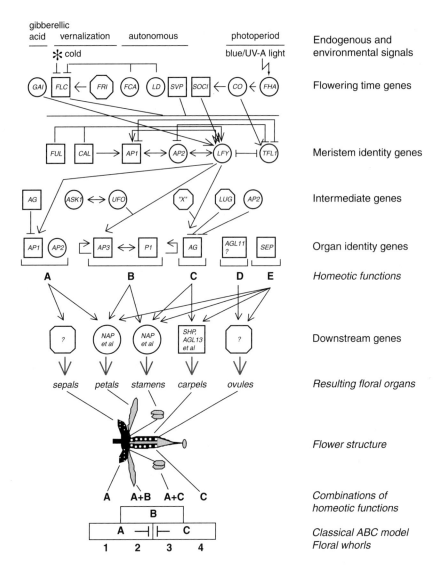

Figure 1 A simplified and preliminary depiction of the genetic hierarchy that controls flower development in *Arabidopsis thaliana*. Examples for the different types of genes within each hierarchy level are shown. 'Gibberellic acid,' 'vernalization,' 'autonomous,' and 'photoperiod' refer to the different promotion pathways of floral induction. 'Intermediate genes' summarizes a functionally diverse class of genes including 'cadastral genes.' MADS-box genes are shown as squares, non-MADS-box genes as circles, and genes whose sequence has not been reported up to now as octagons. Some regulatory interactions between the genes are symbolized by arrows (activation), double arrows (synergistic interaction), or barred lines (inhibition, antagonistic interaction). For a better overview, by far not all of the known genes and interactions involved in flower development are shown. In case of the downstream genes, just one symbol is shown for every type of floral organ, though whole cascades of many direct target genes and further downstream genes are probably activated in each organ of the flower. A flower structure is shown in the lower region of the figure. At the bottom of the figure, the classical 'ABC model' of flower organ identity is depicted. According to this model, flower organ identity is specified by three classes of 'floral organ identity genes' providing 'homeotic functions' A, B, and C, which are each active in two adjacent whorls. A alone specifies sepals in whorl 1; the combined activities of A + B specify petals in whorl 2; B + C specify stamens in whorl 3; and C alone specifies carpels in whorl 4. The activities A and C are mutually antagonistic, as indicated by barred lines: A prevents the activity of C in whorls 1 and 2, and C prevents the activity of A in whorls 3 and 4.

Abbreviations of gene names used: *AG, AGAMOUS; AGL, AGAMOUS-LIKE GENE; AP, APETALA; ASK1, ARABIDOPSIS SKP1-LIKE1; CAL, CAULIFLOWER; CO, CONSTANS; FLC, FLOWERING LOCUS C; FRI, FRIGIDA; FUL, FRUITFULL; LD, LUMINIDEPENDENS; LFY, LEAFY; LUG, LEUNIG; NAP, NAC-LIKE, ACTIVATED BY AP3/PI; PI, PISTILLATA; SEP, SEPALLATA; SHP, SHATTERPROOF; SOC1, SUPPRESSOR OF OVEREXPRESSION OF CO1; SVP, SHORT VEGETATIVE PHASE; UFO, UNUSUAL FLORAL ORGANS; TFL1, TERMINAL FLOWER1.*

exposure to cold signaling the passage of winter and the onset of spring. A fourth pathway that mediates floral induction, the gibberellic acid promotion pathway, depends on the plant hormone gibberellic acid (**Figure 1**).

Quite a number of flowering time genes have already been cloned, among them *GA1*, *LUMINIDEPENDENS (LD)*, *CONSTANS (CO)*, *FCA*, *FHA*, *FPA*, *FLOWERING LOCUS C (FLC)*, and *SHORT VEGETATIVE PHASE (SVP)*. Mutations in the first six genes mentioned confer late-flowering phenotypes – hence they are termed 'late flowering genes' indicating that they normally function to promote the floral transition. While *CO* and *FHA* belong to the photoperiod promotion pathway, *FCA*, *FPA* and *LD* are involved in the autonomous flowering pathway. *GA1* is a key gene of the gibberellic acid promotion pathway, which eventually may activate the floral meristem identity gene *LEAFY* (see below).

The late flowering genes encode proteins with very diverse biochemical or biophysical properties. *GA1* encodes *ent*-kaurene synthetase, a key enzyme of gibberellin biosynthesis. The *FHA* gene encodes the blue light receptor CRYPTOCHROME2 (CRY2) which is probably involved in photoperiod perception. The FCA and FPA gene products show similarity to RNA-binding proteins, suggesting that they promote flowering via a posttranscriptional mechanism. *CO* and *LD* encode putative transcription factors which promote flowering by activating early target genes such as *SUPPRESSOR OF OVEREXPRESSION OF CO 1 (SOC1)*, and *FLOWERING LOCUS T (FT)* in case of CO. *FT* encodes a protein with similarity to Raf kinase inhibitor protein. *SOC1*, like many other genes involved in flower development (**Figure 1**), is a member of the MADS-box gene family encoding transcription factors. MADS-box genes share a highly conserved, approximately 180 bp long DNA sequence, termed the MADS-box, which encodes the DNA-binding domain of the respective MADS-domain proteins.

In contrast to the late-flowering phenotypes of the genes mentioned above, *flc* and *svp* null mutations result in early flowering, indicating that *FLC* and *SVP* are repressors of flowering. Reduction of *FLC* expression is an important component of the vernalization response. Both *FLC* and *SVP* are MADS-box genes.

Floral Meristem Formation

As a consequence of floral induction, shoot meristems become committed to flowering. In *Arabidopsis* and *Antirrhinum*, floral meristems arise at the flanks of the inflorescence meristems at the shoot apices. Two key genes ('floral meristem identity genes') are responsible for the transition from inflorescence to floral meristems and the specification of floral meristem identity in *Antirrhinum*, *FLORICAULA (FLO)* and *SQUAMOSA (SQUA)*. The putative orthologs and functional equivalents from *Arabidopsis* are *LEAFY (LFY)* and *APETALA1 (AP1)*, respectively. The function of these genes is indicated by the phenotype of loss-of-function mutants. In these mutants, floral meristems often fail to form, and secondary inflorescences form instead, indicating that the transition from inflorescence to floral meristems does not take place. Three other floral meristem identity genes, *APETALA2 (AP2)*, *FRUITFULL (FUL)*, and *CAULIFLOWER (CAL)*, have little effect on meristem identity as single mutations, but the *ap2*, *ful*, and *cal* mutations enhance effects of *lfy* and *ap1* mutations on floral meristem identity.

In the apical inflorescence meristem of *Antirrhinum* the action of the floral meristem identity genes is antagonized by the *CENTRORADIALIS (CEN)* gene. Therefore, loss-of-function of this gene results in ectopic expression of *FLO* and *SQUA* in the inflorescence meristem, thus transforming it into a floral meristem which generates a terminal flower. The putative ortholog and functional equivalent of *CEN* from *Arabidopsis* is *TERMINAL FLOWER1 (TFL1)*. TFL1 and *CEN* encode putative membrane-associated proteins which may be involved in a signal transduction chain required to repress the expression of the floral meristem identity genes in the inflorescence meristem. In contrast, all known meristem identity genes that promote floral fate encode putative transcription factors. *AP1*, *CAL*, *FUL*, and *SQUA* belong to the family of MADS-box genes. *LFY* and *FLO* are members of a small family termed *FLO*-like genes. *AP2* is a founder member of a large gene family called *AP2/EREBP*-like genes.

Flower Formation and Floral Organ Development

When the transition from inflorescence to floral meristems has taken place, floral organs arise at defined positions from within these meristems under the control of different types of genes. In *Arabidopsis*, 'floral meristem size genes' such as *CLAVATA1 (CLV1)*, *CLV2*, *CLV3*, and *WIGGUM (WIG = ERA1)* regulate the size of the floral meristem and also influence floral organ number. *CLV1* encodes a receptor protein kinase, *CLV3* encodes the presumed extracellular protein ligand for CLV1, and *CLV2* encodes a receptor-like protein that may form a heterodimer with CLV1. *WIG* encodes a farnesyltransferase β-subunit involved in numerous aspects of plant development. 'Cadastral genes' like *LEUNIG (LUG)*, *AP2*, and *AG* are involved in setting the boundaries of floral organ identity gene functions. 'Floral organ pattern genes' such as *PERIANTHIA (PAN)*, which encodes a bZIP-type transcription factor, act to establish floral organ primordia in specific numbers and positions. These primordia develop into the different types of floral organs under the control of specific homeotic selector genes, termed 'floral organ identity genes.'

The function of floral organ identity genes was recognized during the study of homeotic mutants in which the identity of floral organs is changed. In *Arabidopsis* and *Antirrhinum* such mutants come in three classes, A, B, and C. Ideal class A mutants have carpels in the first whorl instead of sepals, and stamens in the second whorl instead of petals. Class B mutants have sepals rather than petals in the second whorl, and carpels rather than stamens in the third whorl. Class C mutants have petals instead of stamens in the third whorl, and replacement of the carpels in the fourth whorl by sepals. In addition, these mutants are indeterminate, i.e., there is continued production of mutant floral organs inside the fourth whorl.

Based on these classes of mutants and all combinations of double and triple mutants the 'ABC model' proposes three classes of combinatorially acting floral organ identity genes, called A, B, and C, with A specifying sepals in the first floral whorl, A + B petals in the second whorl, B + C stamens in the third whorl, and C carpels in the fourth whorl (**Figure 1**). The model also maintains that the class A and class C genes negatively regulate each other. Based on studies in petunia (*Petunia hybrida*), the ABC model was later extended by class D genes, specifying ovules. Meanwhile it has been demonstrated by a reverse genetic approach that yet another class of floral organ identity genes, tentatively termed class E genes here, is involved in specifying petals, stamens and carpels. The floral organ identity genes can be interpreted as acting as major developmental switches that activate the entire genetic program for a particular organ.

In *Arabidopsis*, class A genes comprise *APETALA1 (AP1)* and *APETALA2 (AP2)*. The class B genes are represented by *APETALA3 (AP3)* and *PISTILLATA (PI)*, and the class C gene is *AGAMOUS (AG)*. In *Antirrhinum*, the class B genes comprise *DEFICIENS (DEF)* and *GLOBOSA (GLO)*, and the class C gene is *PLENA (PLE)*. Class D genes have been recognized only in petunia so far, where they have been termed *FLORAL BINDING PROTEIN7 (FBP7)* and *FBP11*. The class E genes in *Arabidopsis* comprise *SEPALLATA1 (SEP1)*, *SEP2*, and *SEP3*, which have highly redundant functions. All these genes have been cloned, which revealed that they all encode putative transcription factors. Thus the products of the floral organ identity genes probably all control the transcription of other genes ('target genes') whose products are involved in the formation or function of the different floral organs. Except for *AP2*, all floral organ identity genes are MADS-box genes.

Among the regulators of the floral organ identity genes in *Arabidopsis* is the transcription factor LFY. LFY alone can induce expression of the class A gene *AP1*, i.e., other, flower- or region-specific coregulators are not needed. In contrast, the class B gene *AP3* and the class C gene *AG* are activated by LFY in region-specific patterns within flowers, depending on other factors such as the F-box gene *UNUSUAL FLORAL ORGANS (UFO)* in case of *AP3* and an unknown factor 'X' in case of *AG* (**Figure 1**). Recently, it was shown that *AP1* and *AG* are direct downstream targets of LFY.

Not much is known so far about the downstream target of the 'floral homeotic genes' itself. How floral organ identity is realized at the molecular level is, therefore, not well understood. The first proven direct target gene of a floral homeotic gene (*AP3*), termed *NAP*, was identified just recently. It may play a role in the transition between growth by cell division and cell expansion in stamens and petals.

In contrast to the actinomorphic (polysymmetric) flowers of most angiosperms, including *Arabidopsis*, the flowers of *Antirrhinum* and many other species are zygomorphic, meaning that they have only one plane of reflectional symmetry. Genetic analyses revealed that the development of zygomorphic *Antirrhinum* flowers requires the interaction between several genes that affect the upper (dorsal) region (*CYCLOIDEA*, *RADIALIS*, *DICHOTOMA*) or the lower (ventral) region (*DIVARICATA*) of the flower. *CYCLOIDEA (CYC)* and *DICHOTOMA (DICH)* have been cloned and were shown to encode quite similar and functionally partially redundant transcription factors that are expressed in dorsal regions of the flower. CYC and DICH are founder members of a small group of transcription factors termed the TCP family.

The formation of seeds, which are just ripened ovules, could be considered as the final goal of any flower development. Quite a number of genes involved in different stages of ovule development have been identified. Some of these genes have already been cloned, among them several encoding transcription factors such as the AP2-like gene *ANTITEGUMENTA (ANT)* and the homeobox gene *BELL1 (BEL1)*.

Future Prospects

In the future, the genes involved in flower development will be studied less and less individually, but rather more and more as components of complex gene networks. Since most human food is derived from flower parts or products, such as fruits and grains, there will be intensive attempts to apply the knowledge obtained with the model plants (which are higher eudicots) to commercially important crop plants (which are predominantly monocots). The goal will be to design these plants according to our desires with respect to traits such as time to flowering, and inflorescence, flower, and fruit structure. Comparative studies on genes controlling reproductive development in a diverse range of phylogenetically informative taxa, including monocotyledonous and basal angiosperms, but also nonflowering plants, will provide a better understanding of flower evolution and the origin of biodiversity.

Further Reading

Coen ES (1996) Floral symmetry. *EMBO Journal* 15: 6777–6788.

Gasser CS, Broadhvest J and Hauser BA (1998) Genetic analysis of ovule development. *Annual Review of Plant Physiology and Plant Molecular Biology* 49: 1–24.

Meyerowitz EM (1994) The genetics of flower development. *Scientific American* 271(5): 40–47.

Simpson GG, Gendall AR and Dean C (1999) When to switch to flowering. *Annual Review of Cell and Developmental Biology* 99: 519–550.

Theissen G, Becker A, Di Rosa A *et al.* (2000) A short history of MADS-box genes in plants. *Plant Molecular Biology* 42: 115–149.

See also: Arabidopsis thaliana: The Premier Model Plant; Plant Development, Genetics of; Plant Embryogenesis, Genetics of; Seed Development, Genetics of

Flp Recombinase-Mediated DNA Inversion

M Jayaram

Copyright © 2001 Academic Press
Doi: 10.1006/rwgn.2001.0466

Site-Specific DNA Recombination

Recombination is a universal strategy employed by life forms to reshuffle and reorganize their genetic information from time to time. Recombination can be classified broadly into two types: homologous and site-specific. The former is dependent on rather long stretches of homology between the participant DNA substrates (as in mitotic or meiotic recombination between chromosomes in eukaryotic cells). By contrast, the latter utilizes much shorter segments of homology embedded within sequence-specific DNA targets. In a more extreme form of site-specific recombination, DNA transposition, for example, the recombination partners often share little or no homology between them.

Families of Site-Specific Recombinases

Two families of site-specific recombinases have been well characterized: the resolvase/invertase family and the integrase family (see Site-Specific Recombination). Members of these two families bring about recombination by breaking specific phosphodiester bonds within their DNA targets, and reforming them across substrate partners. The reaction does not require an exogenous energy source such as ATP, and proceeds without degradation or synthesis of DNA. Hence these recombinases have been classified as 'conservative' site-specific recombinases. While the resolvase/invertase family appears to be confined to the prokaryotic world, the integrase family (named after the Int protein of phage lambda) includes members from bacteriophage, bacteria, and yeasts. The Flp recombinase, the subject of this article, is an Int family member from the yeast *Saccharomyces cerevisiae*.

Recombination mediated by the Int family proteins can lead to DNA fusions or dissociation, DNA deletions or inversions, and DNA translocations. A particualr outcome depends on whether the DNA substrates are circular or linear, whether the two sites partaking in a recombination event are present on a single DNA molecule or two separate DNA molecules, and, for the intramolecular case, whether they are in the same (head-to-tail) or opposite (head-to-head) orientations. The DNA rearrangements resulting from recombination have profound genetic and physiological consequences: ranging from phage integration into and excision from bacterial genomes to developmental regulation of gene expression in specific cell types; stable segregation of unit copy or low copy circular genomes by the resolution of dimers and higher oligomers into monomers; and (as will be discussed here for the Flp system) copy number amplification of yeast plasmids.

2-Micron Plasmid and Flp Site-Specific Recombination

The 2-micron plasmid is a circular, multicopy extrachromosomal element present in most strains of *Saccharomyces* yeasts (**Figure 1**). The steady-state copy number of the plasmid is approximately 60 per yeast cell. Under normal growth conditions, the plasmid does not appear to confer any advantage to its host cell; nor is it a burden on the cellular metabolic machinery. The plasmid may be regarded as a typical 'benign parasite genome' that has optimized functions for its stable inheritance and its copy number maintenance.

The 2-micron circle molecules exist in the yeast nucleus as minichromosomes, and are replicated during the S phase of the cell cycle by the same replication apparatus that duplicates the chromosomes. Replication is initiated at the origin (*ORI*; **Figure I**), and the replication forks proceed bidirectionally along the circular contour of the plasmid genome. Normally, each plasmid molecule is restricted to one round of replication per cell cycle. Equal partitioning of the duplicated circles is achieved by the Rep1 and Rep2 proteins (coded for by the *REP1* and *REP2* plasmid genes) acting in concert with the partitioning locus *STB* (**Figure I**).

The 2-micron circle contains a duplicated sequence, 599 bp long, and arranged in a head-to-head orientation (indicated by the parallel lines in **Figure I**). These inverted repeats divide the plasmid into two unique regions, represented by the circular arcs in **Figure I**. The Flp site-specific recombinase is the product of the *FLP* locus, and acts on the *FRT* (*Flp Recombination Target*) sites located within the inverted repeats. The result of the recombination reaction is an inversion of the left unique region with respect to the right unique region. As a consequence, the plasmid population within the yeast cell consists of an equilibrium mixture of the two forms A and B, present in roughly equimolar amounts (**Figure I**). The relative flipping of the DNA by recombination is what gives the recombinase its name Flp (pronounced either as 'flip' or as the letters F-L-P).

Mechanism of Flp Recombination

The *FRT* site consists of three 13 bp Flp-binding elements (1a, 1′a, and 1′b) and an 8 bp strand exchange region (or spacer) arranged as shown in **Figure I**

(bottom). The two phosphodiester bonds that take part in recombination at the la-spacer junction and at the 1′a-spacer junction are indicated in **Figure I**. Note that la and 1′a bordering the spacer at the left and right ends, respectively, are oriented in a head-to-head fashion. The third element 1′b is not directly involved in the recombination reaction, although it may modulate the reaction efficiency *in vivo* in yeast. For simplicity, the mechanism of Flp recombination will be described for the 'minimal' 34 bp *FRT* site consisting of the 1a-1′a Flp-binding elements and the included spacer sequence.

The Flp recombination reaction follows the typical Int family recombination pathway (**Figure 2A**). The reaction is initiated by the synapsis of two DNA substrates, each bound by two Flp monomers. In order to appreciate the geometry of the recombination complex, it is useful to divide each substrate into a left DNA arm (corresponding to 1a) and a right DNA arm (corresponding to 1′a). The reuslts from a number of studies are most easily accommodated by arranging the two substrates, L1R1 and L2R2, in the antiparallel configuration: L1 and L2 (also R1 and R2) being placed at opposite ends of the synaptic structure. The bend introduced into each substrate, L1R1 and L2R2, results from the interaction of the bound Flp monomers. This left-to-right dimeric interaction is essential for assembling the Flp active site. During this functional interaction, one Flp monomer orients the scissile phosphate using an active site cleft that includes three invariant Int family residues, Arg191, His305, and Arg308. This phosphate is then attacked by Tyr343, the fourth invariant family residue, from the second Flp monomer to break the DNA strand (see **Figure 2B**). The result of the cleavage reaction is the formation of the 3′-phosphotyrosine bond and a 5′-hydroxyl group in each substrate at one end of the spacer (the left end in **Figure 2A**). This transesterification mechanism, as opposed to a hydrolytic cleavage mechanism, conserves the energy of the

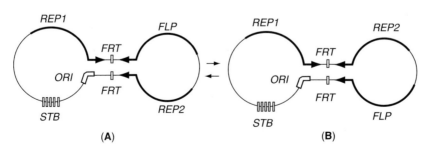

(A)   (B)

Figure I In a schematic representation of the 2-micron plasmid, the 599 bp inverted repeat (shown by the parallel lines) divides the circular genome into two unique regions (circular arcs at the left and at the right). Flp-mediated recombination at the *FRT* sites is responsible for interconversion between forms (A) and (B) by DNA inversion. The products of the *REP* genes, Rep1 and Rep2 proteins, together with the *STB* locus are responsible for plasmid partitioning at cell division. *ORI* is the plasmid replication origin.

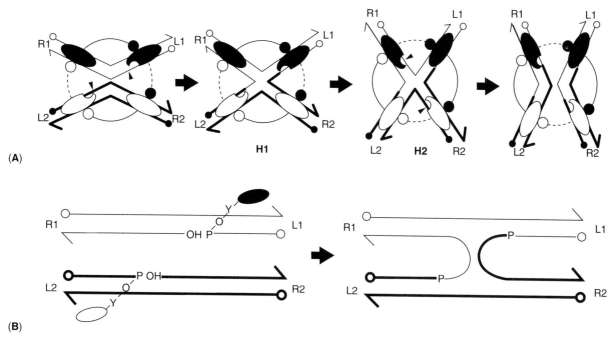

Figure 2 (A) Recombination between two DNA substrates, L I R I and L2R2, is initiated by strand cleavage and exchange at one end of the spacer (at the left end in the scheme shown here). The interactions between recombinase monomers bound to each substrate (the two shaded monomers in substrate I, and the two unshaded monomers in substrate 2) are responsible for the first strand cleavage and exchange reaction. The resulting Holliday intermediate is resolved into the recombinants, L I R I and L2R2, by cleavage and exchange at the right end of the spacer. During the resolution step, the catalytic dimers are formed between Flp monomers bound on the left and right arms of partner substrates. Each 'active dimer' is constituted by a darkly shaded and a lightly shaded monomer. Note that, throughout the reaction pathway, a cyclic peptide connectivity is maintained among the four Flp monomers. The catalytically active and inactive associations between pairs of Flp monomers are indicated by the solid and dashed arcs, respectively. The switch in the configuration of the active Flp dimers involves the isomerization of the Holliday junction from H1 to H2. These junctions have an approximate fourfold symmetry, but are strictly only twofold symmetric. The circles and split arrowheads indicate the 5' and 3' ends, respectively, of DNA strands. (B) The strand exchange reaction at the initiation and termination steps of recombination involves the formation of a covalent protein–DNA intermediate in which the 3'-phosphate end of a cleaved strand is linked to the active site tyrosine of Flp.

phosphodiester bond for the strand-joining reaction. Attack by the 5'-hydroxyl groups on the phospho-tyrosine bonds across substrates (**Figure 2B**) results in the first exchange of strands and the formation of the Holliday intermediate (H1; **Figure 2A**). The junction rearranges (isomerizes) to the H2 form in preparation for its resolution, and thus the termination of recombination. In H2, the two Flp dimers constituted by the R2 and L1 arms and the R1 and L2 arms are in the proper geometric configuration for strand cleavage and exchange at the right end of the spacer. The outcome of Holliday resolution is the formation of the two reciprocally recombinant products, L1R2 and L2R1. The mechanism of the reaction as outlined in **Figure 2A** is supported by the X-ray structure of a Flp–DNA complex solved recently by

P. Rice and colleagues at the University of Chicago (**Figure 3**).

In addition to the Arg–His–Arg triad and the tyrosine nucleophile, two other amino acids in Flp (Lys223 and Trp330) are thought to be active site residues that assist or participate directly in catalysis. Amino acid sequence comparisons indicate that the conserved residue corresponding to Trp330 of Flp is a histidine in most Int family members. The lysine corresponding to Lys223 of Flp is located in a three β-sheet region in solved X-ray structures of the Int type recombinases. An equivalent lysine is also seen in the crystal structures of human and vaccinia topoisomerases. The overall similarity between the Int family recombinases and type IB topoisomerases in their active site architecture is consistent with the

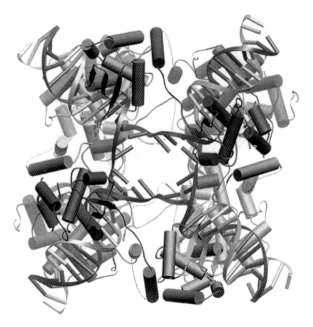

Figure 3 In the structure of a Flp–DNA complex, two Flp monomers execute strand cleavage by providing the tyrosine nucleophile, while the other two monomers assist cleavage by orienting the target phosphodiester bonds. Note that recombination is completed in two steps, each step exchanging one pair of strands between the DNA partners. Hence the 'cleavage-active' and 'cleavage-assisting' monomers switch roles during a recombination event. The Flp–DNA structure has a roughly fourfold symmetry and is consistent with the reaction pathway drawn in **Figure 2A**. (From P. Rice, University of Chicago.)

common chemical mechanism they employ for strand cutting.

Relevance of Flp Recombination to Plasmid Physiology

The Flp recombination reaction serves an important function in the physiology of the 2-micron plasmid. In the event of a stochastic drop in copy number, caused, for example, by a missegregation event, the amplification system constituted by the Flp protein and the *FRT* sites is brought into play to restore it quickly to the steady-state value. Thus, the Flp recombination system, together with the plasmid stability system constituted by the Rep proteins and *STB* (see **Figure 1**), provides a dual strategy to ensure the persistence of the plasmid as a benign parasite genome.

A clever model for how the recombination reaction can be utilized to mediate plasmid amplification has been proposed by Bruce Futcher. The essential features of the 'Futcher model' are illustrated in **Figure 4**. The model is critically dependent on the asymmetric

location of the plasmid replication origin (*ORI*) with respect to the *FRT* sites (see **Figure 1**), and the bidirectional replication mode by which plasmid molecules are duplicated during the yeast cell cycle. One of the two replication forks initiated at the *ORI* sequence will traverse the proximal *FRT* site well before the second fork crosses the distal *FRT* site. Imagine a Flp recombination reaction to occur (as illustrated in **Figure 2**) within a replicating plasmid when only the proximal *FRT* site has been duplicated. The result is the inversion of one fork with respect to the other. Instead of meeting head-on and terminating replication, as they do during a normal cell cycle, the forks now chase each other around the plasmid contour, spinning out multiple copies of it. The tandemly linked copies can be reduced to the monomeric units, also by Flp-mediated recombination. This reductional recombination will occur between alternate (as opposed to adjacent) *FRT* sites, which are in direct (head-to-tail) orientation. Thus, the recombinational inversion of a bidirectional replication fork allows a single initiation event (dictated by the cell cycle control of replication) to be transformed into a multiple plasmid copying mechanism. Note that amplification can be terminated when a second recombination even reinverts the forks, thereby restoring their bidirectional movement.

Although the Futcher model has not been exhaustively verified, it has been clearly demonstrated that the act of recombination *per se* is essential for amplification. When the Flp protein is mutated to a catalytically inactive variant, or when the FRT site is altered to a recombination-incompetent state, a plasmid substrate, which is present at an initial low copy state, fails to amplify.

Control of Copy Number Amplification

Under steady-state growth conditions, when the plasmid is at its normal copy number, the amplification system is unnecessary, and may even be disadvantageous to the 2-micron plasmid. A runaway increase in plasmid copy number by unregulated expression of Flp would be harmful to the host, and hence indirectly so for a benign parasite that it harbors. Hence, it is logical to suppose that the amplification system would be tightly controlled, either at the level of Flp expression, or at the level of the recombination reaction, or both. For the system to act beneficially and efficiently, it must not only be silenced at normal copy number, but also should be rapidly commissioned into action when there is a downward fluctuation in copy number. Preliminary genetic evidence suggests that the 2-micron circle Rep proteins may provide an indirect readout of the plasmid levels in a cell, and act as

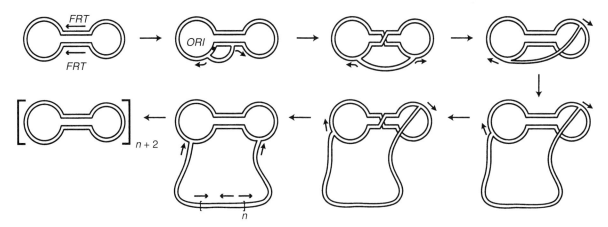

Figure 4   The 2-micron circle replication is initiated at the replication origin (*ORI*; located close to one *FRT* site and away from the other) and proceeds bidirectionally. A plasmid molecule is restricted to one round of replication during a normal cell cycle. During the amplification mode by the Futcher model, a Flp-mediated recombination event (indicated by the DNA crossover) inverts one fork with respect to the other. As the two forks chase each other around the circular template, multiple tandem copies of the plasmid are made from the single replication event initiated at *ORI*. Amplification can be terminated by a second recombination event that now redirects the forks toward each other. In the example shown, $n + 1$ copies of the plasmid are made before replication is terminated. A single plasmid unit in the 'amplicon' is indicated by the square brackets, with the arrows representing the 2-micron circle inverted repeats. After resolution to individual copies, there would be a total of $n + 2$ plasmid molecules.

negative regulators of the *FLP* gene expression in a concentration-dependent manner. However, the details of this regulatory circuit remain to be resolved.

Site-Specific Recombination in Evolution: the Means to Many Ends

The circular geometry of the 2-micron plasmid, its structural organization, and its genetic potential are all part of the elegant biological design of a successful selfish DNA element. One central outcome from this molecular architecture is that a carefully controlled site-specific recombination event can be exploited to promote replicative amplification of the genome. It is not surprising therefore that circular plasmids found in yeasts that are rather distantly related to *Saccharomyces* are structurally similar to the 2-micron plasmid (despite their large diversity in nucleotide sequences), and harbor their own individual site-specific recombination systems. Furthermore, the observed kinship among site-specific recombination systems found in phage, bacteria, and yeasts attests to the axiom that evolution is adept at reutilizing or retooling the same basic biochemical strategy to bring about widely varied end results under distinct biological contexts.

Further Reading

Broach JR and Volkert FC (1991) In: *The Molecular and Cellular Biology of the Yeast Saccharomyces*, 297–331. Plainview, NY: Cold Spring Harbor Laboratory Press.

Futcher AB (1988) The 2 μm circle plasmid of *Saccharomyces cerevisiae*. *Yeast* 4: 27–40.

Landy A (1993) Mechanistic and structural complexity in the site-specific recombination pathways of Int and FLP. *Current Opinions in Genetics and Development* 3: 699–707.

Nash HA (1996) In: *Escherichia coli* and *Salmonella Cellular and Molecular Biology*, vol. 2, 2363–2376. Washington, DC: ASM Press.

Sherratt DJ, Arciszewska LK, Blakely G *et al.* (1995) *Philosophical Transactions of the Royal Society of London* 347: 37–42.

See also: Hin/Gin-Mediated Site-Specific DNA Inversion; Integrase Family of Site-Specific Recombinases; *Ori* Sequences; Resolvase-Mediated Deletion; Site-Specific Recombination

FMS Oncogene

R A Padua

Copyright © 2001 Academic Press
doi: 10.1006/rwgn.2001.1571

The *FMS* oncogene was identified on the basis of its homology to the *v-fms* gene, transduced from the Susan McDonough strain of feline sarcoma virus (SM-FeSV) (McDonough *et al.*, 1971). The gene was sequenced and found to code for a transmembrane glycoprotein and the C-terminal region was found to be homologous to protein tyrosine kinases. Human

FMS was isolated in 1983 and was localized to chromosome 5. *FMS* was shown to code for the colony monocytic stimulating factor-1 receptor (CSF-1R) (Sherr *et al.*, 1985). It is expressed in monocytes produced by the bone marrow, where it is required for monocytic differentiation and survival of macrophages, and is also expressed in the spleen, liver, brain, and placenta. *v-fms* exhibits constitutive tyrosine kinase activity in the absence of ligand and transforms cells. The differences between oncogenic *v-fms* and normal cellular *FMS* are a number of scattered point mutations and the replacement of 50 amino acids at the C-terminus of the human gene with 11 unrelated amino acids in the viral gene. The effects of *FMS* are cell type dependent. In NIH3T3 mouse fibroblast cells the sequence responsible for transformation was localized to amino acid 301 in the extracellular domain. Regulatory sequences at position 969, when mutated, enhance transformation mediated by mutations in codon 301. However, in hematopoietic FDCP-1 cells the 969 mutations transform these cells, rendering them anchorage- independent and tumorigenic in nude mice, whereas the 301 mutant construct is not transforming. Cells infected with the 969 mutant construct cannot be saturated with concentrations of CSF-1 observed to saturate the wild-type receptor (McGlynn *et al.*, 1998). Screening myeloid (pre)leukemia patients for these mutations revealed that mutations at codon 969 were more frequent than those at codon 301 (Ridge *et al.*, 1990; Tobal *et al.*, 1990), suggesting that the *FMS* oncogene may be involved in the pathogenesis of this disease (see Gallagher *et al.*, 1997 and references therein).

References

Gallagher A, Darley RL and Padua R (1997) The molecular basis of myelodysplastic syndromes. *Haematologica* 82: 191–204.

McDonough SK, Larsen S, Brodey RS, Stock ND and Hardy WD, Jr (1971) A transmissible feline fibrosarcoma of viral origin. *Cancer Research* 31: 953–956.

McGlynn H, Baker AH and Padua RA (1998) Biological consequences of a point mutation at codon 969 of the *FMS* gene. *Leukemia Research* 22: 365–372.

Ridge SA, Worwood M, Oscier D, Jacobs A and Padua RA (1990) FMS mutations in myelodysplasia, leukaemia and normal subjects. *Proceedings of the National Academy of Sciences, USA* 87: 1377–1380.

Sherr CJ, Rettenmier CW, Sacca R *et al.* (1985) The c-fms protooncogene product is related to the receptor for the mononuclear phagocyte growth factor, CSF-1. *Cell* 41: 665–676.

Tobal K, Pagliuca A, Bhatt B *et al.* (1990) Mutations of the human FMS gene (M-CSF receptor) in myelodysplastic syndromes and acute myeloid leukemia. *Leukemia* 4: 486–494.

See also: Cancer Susceptibility; Leukemia

Foldback DNA

J H Miller

Copyright © 2001 Academic Press
doi: 10.1006/rwgn.2001.0467

Single-stranded DNA with sequences that permit it to make stable secondary structures by folding back upon itself and forming hydrogen bonds.

Follicular Lymphoma

P G Isaacson

Copyright © 2001 Academic Press
doi: 10.1006/rwgn.2001.1572

Follicular lymphoma is a neoplasm of germinal center B cells that recapitulates the histology of reactive B-cell follicles. It is one of the commonest nonHodgkin's lymphomas in Western countries. Follicular lymphoma is characterized by t(14;18)(q32;q21) that leads to overexpression of the apoptosis inhibitory bcl-2 protein. It is clinically indolent but ultimately incurable.

See also: Cancer Susceptibility

Footprinting

Copyright © 2001 Academic Press
doi: 10.1006/rwgn.2001.1843

Footprinting is a technique used to identify the binding site of, for example, a protein in a nucleic acid sequence by virtue of the protection given by the binding site against nuclease attack.

See also: Nuclease

Ford, Charles

E P Evans

Copyright © 2001 Academic Press
doi: 10.1006/rwgn.2001.0471

The first of the many significant contributions of Charles Ford (1912–1999) to mammalian cytogenetics was an involvement in the 1956 correction of the human diploid chromosome number. For over 30 years there had been debate as to whether it was 47 or 48 before Ford and Hamerton (1956) unequivocally showed the presence of 23 pairs of chromosomes at meiosis in direct preparations obtained from the germ cells of normal men and so corroborated the mitotic counts of 46 obtained by Tijo and Levan in the same year. Ford, with others, went on to show correlations between aberrant chromosome numbers and phenotype in known human syndromes such as Turner (XO) and Klinefelter (XXY). These revelations led to a worldwide surge of interest in human cytogenetics but also gave rise to increasing conflict in the reporting of observations. To resolve the disparities, Ford was instrumental in convening a study group to decide on an acceptable international nomenclature system. Their recommendations were published in 1960 as the Denver Report (after the venue of the meeting) and this has served as a model for nomenclature and further updates to this day.

Although widely recognized as one of the initiators of a golden era of mammalian cytogenetics, before 1956 Ford was exclusively involved with plant material. After graduating in botany from King's College, London, he studied the chromosome translocation complexes in the genus *Oenothera* before departing in 1938 for what became a war-interrupted seven-year period as the geneticist at the Rubber Research Scheme in the then Ceylon (Sri Lanka). An increasing postwar concern about the genetic damaging effect of radiation and radiomimetic chemicals saw his recruitment to work with one of the classic tools of chromosome breakage study, the root tips of *Vicia faba*. He started at the Atomic Energy Laboratory at Chalk River, Canada and then returned to the UK to head the Cytogenetic Section at the newly founded Medical Research Council Radiobiology Unit at Harwell. Here a failure of root-tip growth (subsequently found to result from the toxic effect of copper leaching from the new pipework) played a significant role in his destiny. To await new pipework, he experimented with the more readily available supply of animal tissue and perfected the technical methods that were used to correct the human chromosome number. At the same time he became aware of the potential value to radiobiology of combining the use of these new methods with the expertise of other scientists both within and outside the Unit. Mice with induced chromosome aberrations were produced by the geneticists and these yielded valuable information in assessing genetic risk and also in the study of the effects of gross genome imbalance on survival. One of these aberrations, an unequal reciprocal translocation, proved additionally useful in that it presented a derived chromosome much smaller than the smallest normal chromosome. In the then absence of any convenient cell marker,

Ford realized its value in tracking donor cell contributions in the ongoing experiments by the immunologists to 'rescue' lethally irradiated mice by bone marrow injection. The small chromosome, named T6, was and is still used worldwide as a convenient cell marker and the early experiments laid the foundations of the basic principles of immunosuppression and tissue transplantation such as for human bone marrow replacement.

In over 20 years of involvement with animal cytogenetics, first at Harwell and then at the University of Oxford, Ford worked with the chromosomes of innumerable species in a variety of situations. Only a few examples can be cited. The chromosome marker studies were continued in analyzing cellular contributions in mouse chimeras 'created' by morula fusion or blastocyst injection. The chimeras also produced insights into the masculinizing effect of the mammalian Y chromosome in XX:XY combinations, an interest that was extended into studies of the natural secondary chimeras found in cattle (freemartins) and marmoset monkeys. An earlier interest in the Robertsonian translocation systems discovered in the common shrew broadened with the discovery of similar systems in feral mice and their property to induce high levels of nondisjunction and zygotic imbalance when crossed to laboratory mice. At the same time, human cytogenetics was not neglected with such studies as meiosis in XYY males and the chromosomal screening of cultured blood from athletes competing in the Mexico City Olympic Games.

Ford was renowned for his inspirational enthusiasm in all branches of cytogenetics and his many contributions were acknowledged by his election to a Fellowship of the Royal Society of London in 1965 and in the compilation in 1978 of a special issue of an international journal in honor of his 65th birthday. The contents, by friends and associates, reflect many of his interests and the esteem in which he was held.

References

Ford CE and Hamerton JL (1956) The chromosomes of man. *Nature* 178: 1020–1023

Cytogenetics and Cell Genetics (1978) vol. 20.

See also: Aneuploid; Blastocyst; Chimera; Diploidy; Feral; Germ Cell; Human Chromosomes; Klinefelter Syndrome; Levan, Albert; Mammalian Genetics (Mouse Genetics); Meiosis; Mitosis; Network; Robertsonian Translocation; Sex Chromosomes; Tjio, Joe-Hin; Translocation; Turner Syndrome; Y Chromosome (Human)

Forward Mutations

Forward mutations are those that inactivate a wild-type gene.

See also: **Wild-Type (WT)**

Fosmid

J Hodgkin

A fosmid is a low-copy-number cosmid vector based on the *Escherichia coli* F factor, which is present in only a few copies in each bacterial cell. Eukaryotic DNA cloned into vectors that are present in many copies per cell is sometimes unstable, tending to undergo deletion or rearrangement. Unstable inserts of this type can often be stably propagated as fosmid clones.

See also: **F Factor**

Founder Effect

L Peltonen

The founder effect implies that a small number of individuals have a significant and lasting effect on the gene pool of a population. Since the genes can migrate only when carried in or out of a location by individuals, the founder effect is linked to the history of a population. Typically, the genes in the current population originated from a well-defined, restricted group of individuals that became separated from a larger initial population and migrated to a new location. The gene pool of the migrating population represents a small sample from the original population, since only a small number of the original population migrated. The migration event is an example of what is called a bottleneck in population genetics. The selection of particular alleles of the genes that moved to the new population is entirely a matter of chance.

Identification of evidence of the founder effect at the gene level does not necessarily require DNA analyses of the population. The founder effect can become evident through the observation of some diseases. Some populations show an exceptionally high prevalence of recessive diseases, which are rare elsewhere. The frequency of a recessive disease allele might have been very low in the initial population, but in a small subset comprising the new migrating population this allele might have a relatively high frequency due to the small number of founders. Its frequency thus becomes markedly higher than in the initial population. (The recessive genes are a good example, since there is less selection pressure on these genes that 'remain silent' in the population and their prevalence reflects the history of the founders of the population better than the prevalence of dominant disease genes, which can be selected against since they express themselves in an individual's disease phenotype.)

Good examples of populations exhibiting founder effect are small, isolated, or remote populations, such as the Sardinians or Finns, which exhibit a uniquely high prevalence of some disease genes and a very low prevalence of others. Some 30 recessive diseases are more common in Finland than elsewhere in the world and diseases like cystic fibrosis (CF) and phenylketonuria (PKU), which are common in other Caucasian populations, are extremely rare. Characterization of the molecular background of Finnish diseases that are enriched in the population showed that they exhibit striking locus and allelic homogeneity. Although some of the diseases enriched in the Finnish population, such as Meckel syndrome (early lethal malformation syndrome) or PLOSL (early adulthood-onset progressive dementia), show a feature called locus heterogeneity (the occurrence of multiple genes causing the same clinical phenotype), globally, all Finnish patients share the same chromosomal locus. Furthermore, one major mutation has been systematically identified in the vast majority of diseases, the prevalence of one mutation being as high as 98 % (**Table 1**). These findings strongly support the hypothesis that one founder mutation was brought to this population in the genome of a single immigrant generations ago, and Finnish patients living today originate from one common ancestor. A similar founder effect has been demonstrated in the French Canadian population. One mutation resulting in tyrosinemia I (caused by the deficiency of an enzyme, fumarylacetoacetate hydrolase) was found in 90 % of disease alleles. In contrast this mutation is found only in 28 % of the tyrosinemia alleles in the rest of the world.

The founder effect can be further exemplified by the fact that some Finnish disease alleles show major regional variations in their population frequencies, as well as in the number of affected individuals. This is

Table I Examples of disease mutations demonstrating the founder effect in the Finnish population

Disease (OMIM number)	Defective protein	Major mutation occurrence in Finland (%)
APECED (240300)	Novel nuclear protein	82
Aspartylglucosaminuria (AGU, 208400)	Aspartylglucosaminidase	98
Congenital chloride diarrhea (CCH, 214700)	Product of the gene downregulated in adenoma	100
Congenital nephrosis (CNF, 256300)	Nephrin	78
Diastrophic dysplasia (DTD, 222600)	Sulfate transporter	90
Familial amyloidosis, Finnish type (FAF, 105120)	Gelsolin	100
Gyrate atrophy of choroid and retina (HOGA, 258870)	Ornithine-aminotransferase	85
Hypergonadotrophic ovarial dysgenesis (ODG1, 2333300)	Follicle-stimulating hormone receptor	100
Infantile neuronal ceroid lipofuscinosis (INCL, 256730)	Palmitoyl protein thioesterase	98
Lysinuric protein intolerance (LPI, 222700)	L-Amino acid transporter	100
Nonketotic hyperglycinemia (NKH, 238300)	Glycine cleavage system; protein P	70
Progressive myoclonus epilepsy (PME, 254800)	Cystatin B	96
Retinoschisis (RS, 312700)	XLRSI	70
Sialic acid storage disease (SIASD, 268740)	Novel transporter	94
Finnish variant of late infantile neuronal ceroid lipofuscinosis (vLINCL, 256731)	Novel membrane protein	94

the result of an internal migration after initial settlement in the country. Some 2000 years or 100 generations ago, small immigrant groups inhabited Finland. Later, small subgroups of this initial population moved to still more remote regions of Finland and established small population subisolates. Perhaps only 20–40 families moved to remote areas 200–300 years ago, and the founder effect and chance (genetic drift) resulted in the enrichment of some disease genes in these subisolates.

The founder effect in one ancestral mutation makes the mapping and identification of disease genes a straightforward task. Genome-wide searches for disease genes are based on the identification of a chromosomal region containing genetic markers which co-segregate with the disease, due to the close vicinity of the marker and the mutated gene. Families with multiple affected children are needed to reveal this co-segregation. In the presence of the founder effect, mapping strategies based on the analyses of only diseased individuals can be applied. Monitoring of shared marker alleles among the affected individuals has been highly successful in the identification of genes and alleles causing inherited diseases in genetic isolates. The shared chromosomal regions indicate that the alleles are identical by descent (IBD), since they share a common ancestor. In the case of recessive diseases, this strategy has been called homozygosity mapping. Typically for disease alleles showing a founder effect, linkage disequilibrium or the nonrandom association of alleles is seen over a long genetic interval

flanking the disease gene. The length of this interval is negatively correlated with the number of generations that have passed since the founder effect took place and with the expansion rate of the population.

The founder effect has been invoked to explain the exceptionally high prevalence of some worldwide genetic disorders in specific populations. Good examples are cystic fibrosis in Northern Europeans and Tay–Sachs disease in Eastern European Jewish populations. Furthermore, in some genetic isolates, such as in Sardinia, the prevalence of some common diseases like type I diabetes is exceptionally high. One hypothesis for this phenomenon is a founder effect. This concept of limited variation in the genetic background caused by a founder effect has raised significant interest in those projects designed to map genes contributing to complex diseases using population isolates. Examples are studies of asthma in Tristan da Cunha or schizophrenia in Palau, Micronesia.

The founder effect has some practical consequences for DNA testing and disease diagnostics. If one mutation is found in 90 % of the disease alleles, diagnostic DNA tests providing high specificity and reliability are easy to develop. This is different from tests for mutations in other, more heterogeneous populations, in which the value of DNA diagnostics has remained limited due to the high number of disease mutations.

Further Reading

Jorde L (1996) Linkage disequilibrium as a gene mapping tool. *American Journal of Human Genetics* 56: 11–14.

Kaplan NL, Hill W and Weir B (1995) Likelihood methods for locating genes in nonequilibrium populations. *American Journal of Human Genetics* 56: 18–32.

Lander ES and Botstein D (1987) Homozygosity mapping: A way to map human recessive traits with the DNA of inbred children. *Science* 236: 1567–1570.

Online Mendelian Inheritance in Man (OMIM) http://www.ncbi.nlm.nih. gov/Omim/

Peltonen L, Jalanko A and Varilo T (1999) Molecular genetics of Finnish disease heritage. *Human Molecular Genetics* 8: 1913–1923.

See also: **Alleles; Genetic Drift; Locus; Marker**

Founder Principle

See: **Bottleneck Effect**

Fragile Chromosome Site

G R Sutherland

Fragile sites are specific points on chromosomes that show nonrandom gaps on breaks when the cells from which the chromosomes were prepared have been exposed to a specific chemical agent or condition of tissue culture. The fragile site is an area of chromatin that is not compacted when seen at mitosis.

Fragile sites are classified as rare (on less than 1 in 40 chromosomes) or common (on all chromosomes) and by the conditions under which they are seen. There are more than 120 recognized fragile sites in the human genome (Sutherland *et al.*, 1996).

Reference

Sutherland GR, Baker E and Richards RI (1996) Fragile sites. In: Meyers RA (ed.) *Encyclopaedia of Molecular Biology and Molecular Medicine*, vol. 2, pp. 313–318.

See also: **Fragile X Syndrome**

Fragile X Syndrome

G R Sutherland and R I Richards

Fragile X syndrome is the most common form of familial mental retardation. It is so-called because it is associated with a fragile site (*FRAXA*) on the end of the long arm of the X chromosome. The most prominent feature of the condition is moderate to severe mental retardation in most affected males and milder intellectual deficits in a proportion of females. In addition to the mental retardation, there is a syndrome of minor subtle malformations, again more evident in males than females.

The syndrome was first described with the fragile X chromosome in 1969 but its relatively common occurrence (about 1 in 4000 boys and 1 in 6000 girls) was not recognized until the early 1980s. This was largely because in 1977 it was discovered that chromosome studies needed to be performed in a specific way for the fragile X chromosome to be seen. It was recognized in the mid-1980s that the fragile X syndrome had anomalous inheritance patterns and was not a simple X-linked recessive disorder. The reasons for this were unknown until the molecular basis of the disease was elucidated in 1991. The fragile site was shown to be due to expansion of a naturally occurring polymorphic CCG trinucleotide repeat in the $5'$ untranslated region of the *FMR1* gene. The number of copies of the repeat can change on transmission from parent to child and when the number exceeds about 230 the expression of the *FMR1* gene is extinguished and this is the molecular cause of fragile X syndrome.

Clinical Features

There are many physical and behavioral features of fragile X syndrome. Those which occur in more than 50% of males are listed in **Table 1**.

These features are shown by those with a full mutation. Individuals with a premutation are intellectually and physically normal. The only significant exception

Table 1 Clinical signs present in more than 50% of Fragile X males[a]

Physical signs	Behavioral signs
Long face	Hand flapping
Prominent ears	Hand biting
High arched palate	Hyperactivity
Hyperextensible fingers	Perseveration
Double-jointed thumbs	Aggression
Flat feet	Shyness
Macroorchidism	Anxiety
Strabismus	Poor eye contact
Soft smooth skin	Tactile defensiveness
Mitral valve prolapse	
Tall as children, short as adults	
Large heads as children, small as adults	

[a]From Hagerman and Cronister (1996).

to this is that females with premutations appear to be prone to premature ovarian failure, which can occur at the age of 30 years onwards, although most premutation carriers do not have premature menopause.

Treatment

Cure of fragile X syndrome is not possible. A number of the behavioral difficulties exhibited by fragile X syndrome are amenable to both pharmaceutical and behavior treatments. Integrated approaches to treatment will maximize the potential of affected individuals and minimize the disruption to family life that this condition can produce (Hagerman and Cronister, 1996).

Cytogenetics

The appearance of the fragile X chromosome is shown in **Figure 1**. This is most easily seen in chromosomes prepared from lymphocyte cultures. The lymphocytes need to be cultured in media which have a relative deficiency of thymidine or deoxycytidine. This can be achieved by using special commercially available media, using medium TC199, or by adding a variety of inducing agents such as the antifolate aminopterin, the thymidylate synthetase inhibitor fluorodeoxyuridine, or high concentrations of thymidine which inhibit the availability of deoxycytidine (Sutherland, 1991). Cytogenetic testing for fragile X syndrome has largely been replaced by DNA testing.

Molecular Genetics

The molecular basis of fragile X syndrome is lack of FMRP, the protein encoded by the *FMR1* gene. Within the 5′ untranslated region of the *FMR1* gene there is a polymorphic CCG repeat, which on normal chromosomes varies in size from about 5 to 55 copies. Once this repeat exceeds about 55 copies the chromosome is said to have a fragile X premutation. Beyond about 230 copies of the repeat a full mutation is present (Warren and Nelson, 1994). The full mutation results in CpG methylation of the DNA in both the promoter region of the *FMR1* gene, and of the expanded repeat, and this results in transcriptional silencing of this gene. Males with the full mutation have fragile X syndrome.

The fragile X chromosome is subject to random X inactivation. This, and possibly other factors, influences the clinical picture in females with a full mutation on one of their X chromosomes. About 60% of such females will be mildly mentally impaired or worse. This presents a difficulty at prenatal diagnosis as the phenotype of a female fetus with a full mutation cannot be accurately predicted.

Some individuals (males and females) show somatic instability of the expanded repeat and can be termed 'mosaics.' This means that there are populations of cells in which the number of copies of the CCG repeat are different. In extreme cases the one patient may have normal, premutation, and full mutation cells. Full mutations are inherited via the ovum and apparently exhibit somatic instability ('breakdown') very early in embryonic development.

More than 99% of fragile X syndrome mutations are due to expansion of the CCG repeat by a mechanism known as dynamic mutation (see Dynamic Mutations). The other 1% or so are due to a variety of mutations, primarily deletions of various sizes, but point mutations have been recorded. The function of the FMR1 protein is not fully understood, but it is an RNA-binding protein (Oostra, 1996). The protein is widely expressed during development and, later on, in brain, testis, and uterus. There appears to be extensive

Figure 1   Sex chromosome complements from individuals expressing FRAXA. A female (left) showing the fragile X and normal X chromosome, and a male (right) showing the fragile X and a normal Y chromosome.

alternative splicing of the FMR1 mRNA and some forms of the protein locate in the cytoplasm, and others in the nucleus of the cell.

Diagnosis of fragile X syndrome now is primarily by measuring the number of CCG repeats in the *FMR1* gene. This is usually performed by Southern blot analysis to estimate the size of a DNA restriction fragment, with increases in the size being due to additional copies of the CCG repeat. This can be performed either postnatally or on DNA extracted from chorionic villus samples for prenatal diagnosis.

Genetics

The paradoxical nature of the fragile X chromosome was documented by Sherman *et al.* (1985). They showed that normal males could 'carry' the condition, an anomalous situation for an X-linked disease but now known to be because of the premutations being clinically harmless. They showed that the mothers and daughters of normal fragile X carrier males had different risks of having children with fragile X syndrome ('the Sherman paradox').

It is now recognized that when women transmit the fragile X mutation it usually increases in size and the risk of going from a premutation to a full mutation depends upon the size of the premutation (Fisch *et al.*, 1995). When a male with a premutation transmits it, the size of the premutation changes little. When a male with a full mutation transmits his fragile X chromosome (to a daughter) she always receives it as a premutation.

It is worth noting that whenever a child is identified with fragile X syndrome, the mother is always a carrier (either pre- or full mutation) as is one of the maternal grandparents.

Conclusion

Fragile X syndrome is a common disorder. Its genetics are reasonably well understood but much remains to be learned about the molecular pathway from genotype to phenotype. Diagnosis by DNA analysis is very reliable, and prenatal diagnosis is appropriate and available to women who are carriers of this disorder.

References

Fisch GS, Snow K, Thibodeau SN *et al.* (1995) The fragile X premutation in carriers and its effect on mutation size in offspring. *American Journal of Human Genetics* 56: 1147–1155.

Hagerman RJ and Cronister A (eds) (1996) *Fragile X Syndrome: Diagnosis, Treatment and Research*, 2nd edn. Baltimore, MD: Johns Hopkins University Press.

Oostra BA (1996) FMR1 protein studies and animal model for fragile X syndrome In: Hagerman RJ and Cronister A (eds) *Fragile X Syndrome: Diagnosis, Treatment and Research*, 2nd edn, 193–209. Baltimore, MD: Johns Hopkins University Press.

Sherman SL, Jacobs PA, Morton NE *et al.* (1985) Further segregation analysis of the fragile X syndrome with special reference to transmitting males. *Human Genetics* 69: 289–299.

Sutherland GR (1991) The detection of fragile sites on human chromosomes. In: Adolph KW (ed.) *Advanced Techniques in Chromosome Research*, 203–222. New York: Marcel Dekker.

Warren ST and Nelson DL (1994) Advances in molecular analysis of fragile X syndrome. *Journal of the American Medical Association* 271: 536–542.

See also: **Fragile Chromosome Site; Genetic Diseases**

Frameshift Mutation

B S Guttman

Copyright © 2001 Academic Press
doi: 10.1006/rwgn.2001.0478

A nucleic acid sequence is translated into the protein it encodes by means of transfer RNAs (see Transfer RNA (tRNA)) interacting with the ribosomal apparatus. Transfer RNAs bind to three nucleotides at a time and thus divide the nucleic acid sequence into codons, each specifying one amino acid. However, depending on the point at which division into codons begins, the nucleic acid can be read in three distinct phases (three distinct reading frames) and, aside from the signal for initiation of translation, the sequence does not contain 'punctuation signals' to indicate which frame should be used. A frameshift mutation is an alteration in the nucleic acid sequence, generally an addition or deletion, that shifts the translation mechanism from one reading frame to another.

In hypothesizing possible coding mechanisms, Crick and his colleagues suggested in 1961 that the code might be commaless; in other words, that there are no intrinsic 'commas' to show the proper reading by marking off groups of three nucleotides as being the correct codons. In this case, they suggested, a short insertion or deletion might act as a frameshift mutation, and it might be corrected by a nearby suppressor mutation that would shift the reading frame back into the proper phase. Suppose a gene encoding a certain protein is properly divided into codons as shown by the following spaces (which do not exist in reality):

CAT CAT CAT CAT CAT CAT CAT CAT CAT...

A deletion of one nucleotide would shift the reading frame one space to the left and encode the wrong peptide after a certain point:

CAT CAT CAC ATC ATC ATC ATC ATC ATC...

However, a nearby insertion of one nucleotide would shift the reading frame back into its proper phase:

CAT CAT CAC ATC ATX CAT CAT CAT CAT...

Although a few codons still specify the wrong amino acids, in many proteins this would make little difference and the double mutant will still exhibit the normal phenotype.

Crick *et al.* (1961) tested this hypothesis by collecting *rII* mutants of phage T4 caused by the mutagen proflavine, which was known to produce insertions and deletions. (T4 *rII* mutants are particularly suited for this study because wild-type phage multiply in bacteria that are lysogenic for phage lambda but mutants do not.) They started with one mutant, which we may designate arbitrarily as having a phase shift to the left (L). Proflavine-induced suppressors of this mutation must therefore have a phase shift to the right (R). In turn, suppressors of these R mutants must be L mutants. After collecting several mutants, arbitrarily designated L or R, they showed that in general a phage will have a wild-type phenotype if it bears an L and an R mutation that are quite close together. Furthermore, they confirmed that a phage with three L mutations or three R mutations close together also has the wild-type phenotype, as expected if the code is triplet, since three frameshifts in one direction will then restore the proper reading frame.

Reference

Crick FHC, Barnett L, Brenner S and Watts-Tobin RJ (1961) General nature of the genetic code for proteins. *Nature* 192: 1227–1232.

See also: Commaless Code; Genetic Code; Transfer RNA (tRNA)

Freedom, Degrees of

T P Speed

Copyright © 2001 Academic Press
doi: 10.1006/rwgn.2001.0479

Degrees of freedom are part of the specification of χ^2 and of certain other statistical distributions such as the t and the F. We only discuss the χ^2 test here.

In practice, degrees of freedom (abbreviated df) need to be known when carrying out a χ^2 test, in order to identify the appropriate column of a table of critical values to consult or to calculate the appropriate p value.

If the df are not known a priori, then the question arises of determining their correct value in a given context. In some cases there are straightforward rules that can be followed, but in general this is not a simple question to answer. The general determination of the df of a χ^2 test is embedded in the statistical theory underpinning a particular test in a given context, and is thus only accessible to those familiar with this theory. Many computer programs calculate the df automatically using rules, not always correctly.

The most common example of a χ^2 test and its associated single degree of freedom (equivalently, 1 df) comes with the 2×2 contingency table. This can arise when comparing two binomial proportions or when cross-classifying units according to two binary characteristics. A familiar genetic example is as follows. Suppose that we have a random sample of individuals who are classified as affected or not in relation to some disease, and that we also classify them as *aa* or not *aa* (i.e., *Aa* or *AA*) at a biallelic locus. A statistical test of the null hypothesis of no association between disease status and this particular genetic dichotomy can be carried out by organizing the data in a 2×2 table, and computing a χ^2 test statistic. As indicated above, this test will have 1 df, and this is used in the assessment of significance. If we did not collapse the genotypes as described, but kept all three separate, we would have a 2×3 classification: 2 disease states (affected, unaffected) and 3 genotypes (*aa*, *Aa*, and *AA*). A χ^2 test of the null hypothesis of no association could still be carried out, but in this case the df would typically be 2, failing to be so only if one of the rows or columns had no entries. More generally, a χ^2 test of no association based on data from a table with r rows and c columns normally has $(r-1)(c-1)$ df, though different df can be appropriate if not all cells have positive counts. The two-way contingency table is an example of the situation in which the calculation of the df is usually but not always by a simple rule.

Another such example arises with the χ^2 test of goodness-of-fit. Here the χ^2 statistic might be the familiar sum over all cells of observed minus expected cell count squared, divided by expected cell count. If no unknown parameters need to be estimated to calculate the expected cell counts, then the df are the number of cells minus 1. When k parameters have to be estimated to calculate the expected cell counts, the df are typically the number of cells minus $k+1$. This rule is not universally true, for there are conditions that need to apply, but they are beyond the scope of this entry.

In summary, the degrees of freedom of a χ^2 distribution will usually be determined in a particular context by a simple rule. The rule will cover most but not all cases that arise in practice.

See also: Null Hypothesis

Frequency-Dependent Fitness

See: Frequency-Dependent Selection

Frequency-Dependent Selection

T Prout

Frequency-dependent selection means that the fitness of a genotype is a function of its rarity or commoness relative to other genotypes. Several types of such selection have been reported, including rare male mating advantage, rare male fertility advantage in the histocompatibility system in mammals, rare type predator resisitance, rare type survival advantage, rare type allele advantage in self-incompatibility systems in plants, a similar system of sex determination in bees and some other hymenoptera, and mimicry. These cases will be discussed in turn.

Rare male mating advantage was first reported for laboratory experiments with *Drosophila* by Petit (1954) and additional cases by Ehrman and others (reviewed by Ehrman and Probber, 1978). The competing genotypes were visible mutants, inversions, and strains from different locations. Usually the alternative competitors had an advantage when rare, but equal mating success when common, which would result in a polymorphic equilibrium. Rare male advantage has also been observed in the wasp *Nasonia*, the beetle *Tribolium*, the ladybird beetle *Adalia*, the guppy *Poeciliopsis* and the mosquito fish *Gambusia*. These last two were observed under natural and seminatural conditions. The laboratory experiments with *Drosophila* have been criticized with respect to experimental conditions and statistics (Bryant *et al.*, 1980; Merrell, 1983; Knoppien, 1985; Partridge, 1988) so, at this point, these *Drosophila* results can be considered controversial.

There have been many experiments showing that competition for resources other than mates is frequency dependent. Such experiments have been done with *Drosophila* and different strains of crop plants. In most of the *Drosophila* experiments different allozyme genotypes or inversion karyotypes were placed as young larvae in the medium at different frequencies and the survival to adulthood recorded. In many cases the rare type had higher survival than the common type surrounding it.

Many such experiments have been done with different strains of crop plants (reviewed by Donald and Hamblin, 1983). For seed crops, at least, the yield is fitness – survival × fertility. In many cases the strain performs much better when surrounded by competitors than by pure stands, and sometimes better than competitors showing rare type advantage. In some cases the data have been simulated showing a stable polymorphic equilibrium (Allard and Adams, 1969). However, theory shows that yield is not maximized at equilibrium.

The self-incompatibility system in plants is another case of rare type advantage. There is a single incompatibility locus where if the pollen and style have the same allele the cross is sterile. It is evident that a rare new mutation would have an advantage. Plant species with this system have a large number of alleles. One clover (*Trifoliume*) species has 100 alleles. A similar system occurs in bees and some other hymenoptera in which there is a 'sex locus' where a homozygote is a male that dies. Normal males (drones) are haploid and heterozygotes are workers or a queen; this system favors rare alleles. There are 14–20 sex locus alleles in bees. There is an analogous system in the major histocompatibility complex (MHC) in mammals, which in humans is called human leukocyte antigen (HLA). It has been shown that if the embryo has the same genotype as the mother abortion results, which favors fathers with different genotypes. There are 100 alleles in the HLA system. In mice females prefer males with a different MHC genotype, which the female recognizes by odor.

Frequency-dependent selection has been proposed in predator–prey interactions, where the predator is conditioned to favor the most common prey phenotype. The resulting rare type prey advantage is termed 'apostatic advantage.' A number of experiments have been carried out in which the prey population is contrived, usually but not always with artificial prey, and bird predation has been measured. Both apostatic selection and preference for the rare type has been observed (anti-apostatic selection) in these experiments.

Finally, mimicry shows frequency-dependent selection. In the case of Batesian mimicry, the mimetic morph resembles a bad-tasting model which predators are conditioned to avoid. Here there is a rarity advantage so that the predators will not experience the good-tasting mimic. This results in a low frequency of the mimic equilibrium. If the abundance of the model varies then there should be a positive correlation with the frequency of the mimetic morph within the mimetic species. This relation has been shown by Edmunds (1966) in Africa for temporal variation in frequencies of the butterfly model and mimic. *Danaus chrysippus* and *Hypolymnas misippus*, respectively, and by Brower and Brower (1962) in North America for spatial variation in frequencies of the butterfly model and mimic *Battus philenor* and *Papilio glaucus*, respectively.

Müllerian mimicry is where the distasteful or harmful species conspicuously resemble each other. The poisonous coral snakes of South America are all striped – black, white, and red. There are 50 species which all appear the same. Another case of Müllerian mimicry is the *Heliconius* butterfly complex in South America. In this case there are several different warning designs in different places. Some species are polymorphic for different warning designs in different places with steep clines between. Because they belong to the same species the design genetics has been studied. Mallet and Barton (1989) performed a field study in which they released one member of a design group into a different design group and showed a rare type disadvantage. Those individuals with a locally rare design were conspicuous to bird predators that were conditioned to avoid a different design. This case of frequency dependent selection is the opposite to that discussed up to this point, being a rare type disadvantage. The steep clines between these design regions have been studied and successfully modeled by Mallet *et al.* (1990).

References

Allard RW and Adams J (1969) Population studies in predominantly self-pollinating species. XIII. Intergenotypic competition and population structure in barley and wheat. *American Nature* 103:934, 621–645.

Brower LP and Brower JV (1962) The relative abundance of model and mimic butterflies in natural populations of the *Battus philenor* mimicry complex. *Ecology* 43(1): 154–158.

Bryant EH, Kance A and Kimball KT (1980) A rare male advantage in the house fly induced by wing clipping and some general considerations for *Drosophila*. *Genetics*. 96: 975–993.

Donald CM and Hamblin J (1983) The convergent evolution of annual seed crops in agriculture. *Nature* 212: 1478.

Edmunds M (1966) Natural selection in the mimetic butterfly *Hypolimnas misippus* in Ghana. *Nature* 212: 154–158.

Ehrman L and Probber J (1978) Rare *Drosophila* males: the mysterious matter of choice. *American Scientist* 66: 216–222.

Knoppien P (1985) The number of rare males stored per vial: a possible source of bias in rare male experiments. *Drosophila Information Service* 62: 101.

Mallet J and Barton NH (1989) Strong natural selection in a warning-color hybrid zone *Evolution* 43(2): 421–431.

Mallet J, Barton NH, Lamas GM et al. (1990) Estimates of selection and gene flow, from measures of Cline width and linkage disequilibrium in *Heliconius* hybrid zones. *Genetics* 124(4): 921–936.

Merrell DJ (1983) Frequency dependent mating? *Evolution* 37(2): 413–414.

Partridge L (1988) The rare male effect: what is its evolutionary significance? *Philosophical Transactions of the Royal Society of London, Series B, Biological Sciences* 319: 525–539.

Petit C (1954) L'isolement sexuel chez *Drosophila melanogaster*: étude du mutant white et de son allélomorph sauvage. *Bulletin of Biology* 88: 435–443.

See also: **Frequency-Dependent Selection as Expressed in Rare Male Mating Advantages; Major Histocompatibility Complex (MHC); Predator–Prey and Parasite–Host Interactions**

Frequency-Dependent Selection as Expressed in Rare Male Mating Advantages

L Ehrman

The rare male mating advantage, representing frequency-dependent selection, has long fascinated population geneticists. The term implies that the fitness of a given genotype depends on its proportions in a population. Frequency dependence may be positive (in favor of the common type) or negative (in favor of the rare type). A situation is conceivable in which the advantage, or disadvantage, holds only for one type when rare. In that case it is called one-sided frequency dependence. When the rare type has a higher fitness than the common type, selection is balancing, because as soon as the rarer type becomes more common its advantage disappears. Models implying some kind of balancing selection can explain high levels of genetic variability, routinely maintained in natural populations. (For definitions of assorted types of natural selection, see Natural Selection.) The model most commonly employed for this purpose is the overdominance model, implying that the heterozygote has a higher fitness than the homozygote.

(However, this model implies the occurrence of genetic load. Two-sided negative frequency dependence, on the other hand, can maintain genetic variation even without any genetic polymorphism.)

As a consequence of frequency-dependent fitness values, the frequency of the rare type will increase until an equilibrium value is reached, wherein all genotypes have equal fitnesses. For this reason frequency-dependent selection, with advantage for the rare type, is proposed as a possible mechanism for the maintenance of genetic variation in nature. It is claimed that there is strong evidence for frequency-dependent selection with an advantage for the rare type among prey as a result of predation, as an aspect of mimicry, among hosts as a result of parasitism, and also due to competition. Consideration of the rare male advantage from the viewpoint of population genetics leads to the hypothesis that an initially rare genotype will increase in frequency if there are no other selective forces operating against it. As the rare type becomes more common, its advantage diminishes, leading to equilibrium (see **Figure I** where this has been recorded as happening in competitions between epistatic eye color mutants).

A successful experimental approach employed to detect frequency dependence of mating success in *Drosophila* in the laboratory has involved two types of flies in mating chambers, the frequency of the types of flies being varied in different replicas (Ehrman and Parsons, 1981). For an excellent and comprehensive review see Knoppien (1985) and the references therein. We also recommend articles by Ehrman *et al.* (1991) and Lofdahl *et al.* (1992), which deal with toxic media and with newer approaches to geotaxis, respectively.

A number of gene and chromosomal polymorphisms have been documented as maintained by such frequency-dependent equilibria. The magnitude and reproducibility of the effect appears to depend on the species, but it has been observed in insects other than *Drosophila* (house flies), as well as in a vertebrate (the guppy). Because these are polymorphisms for which minimal fitness differentials between competing component genotypes are expected at equilibrium, a different sort of selection would prevail from that of the heterozygote advantage model. Therefore, frequency dependence may represent a way of maintaining a high level of genetic variability without obviously associated fitness differentials. This could be of considerable evolutionary significance, since it has been argued that there is a limit to the amount of variability a population can maintain under the classic heterozygote fitness advantage model (see Dobzhansky, 1970; Dobzhansky *et al.*, 1977; Ehrman and Parsons, 1981).

References

Dobzhansky Th (1970) *Genetics of the Evolutionary Process.* New York: Columbia University Press.

Dobzhansky Th, Ayala F, Stebbins G and Valentine J (1977) *Evolution.* San Francisco, CA: WH Freeman.

Ehrman L and Parsons PA (1981) *Behavior Genetics and Evolution,* ch. 8. New York: McGraw-Hill.

Ehrman L, White M and Wallace B (1991) A long-term study involving *Drosophila melanogaster* and toxic media. *Evolutionary Biology* 25: 175–209.

Knoppien P (1985) Rare-male advantage: a review. *Biological Reviews* 60: 81–117.

Lofdahl K, Hu D, Ehrman L, Hirsch J and Skoog L (1992) Incipient reproductive isolation and evolution in laboratory *Drosophila melanogaster* selected for geotaxis. *Animal Behaviour* 44: 783–786.

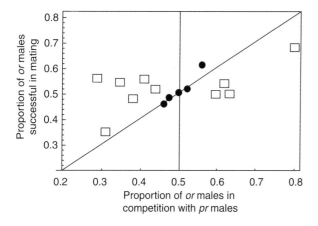

Figure I Distribution of matings in mass cultures of *Drosophila pseudoobscura* in which orange-eyed (*or*) and purple-eyed (*pr*) females had a choice of *or* and *pr* males, showing that as the minority male becomes more common, its advantage diminishes, leading to equilibrium (similar results were obtained in reciprocal experiments reversing rarity). These unlinked marker genes are useful in determining paternity because: *or/or +/+* or *or/or +/pr* = orange-eyed; *+/+ pr/pr* or *+/or pr/pr* = purple-eyed; *or/or pr/pr* = white-eyed; and *+/or +/pr* or *+/+ +/pr* or *+/or +/+* or *+/+ +/+* = wild-type red-eyed. ☐ indicates rare male advantage.

See also: Balanced Polymorphism; Disruptive Selection; Frequency-Dependent Selection; Genetic Load; Overdominance; Polymorphism; Selection Intensity

Fruit Fly

See: Drosophila melanogaster

Functional Genomics

D Seemungal and D Carr

Copyright © 2001 Academic Press
doi: 10.1006/rwgn.2001.1862

Functional genomics is the development and implementation of techniques to examine – both in time and space – the global patterns by which genes and their protein products act in concert to effect function.

The human genome consists of a DNA complement of approximately 3 billion nucleotide pairs in length, divided into 24 distinct chromosomal segments (autosomes 1–22 plus X and Y). Contained within this linear array of nucleotides is anything between 30 000 and 40 000 genes. The protein products of genes interact in complex pathways to effect cellular function. The combination of genes which is expressed in a cell at any particular point in time determines the protein complement of the cell and hence its functional mechanics.

During the process of development, subsets of cells in the body contain different subpopulations of activated genes and, as a result of the different combinations of proteins that result, differentiate to form distinct tissues with their own specialized physiology. Proteins, as mobile functional elements, allow communication between cells, both locally and remotely, and are hence responsible for the integration of various cell types into the coordinated, highly complex physiological systems that comprise a living organism.

Differences in the genetic complement between individuals can cause differential expression of genes or the production of proteins which function in slightly different ways. In extreme cases, some individuals possess harmful variants which cause disease directly, but this variation also underpins disease processes in more subtle ways, influencing an individual's susceptibility to disease and varying responses to therapeutic interventions. In addition, intracellular protein systems allow cells to respond to changes in their environment. Certain environmental stimuli will perturb the normal cellular functions of proteins and cause changes in gene expression. These kinds of environmental factors can also lead to the pathology of disease. Often the development of a disease will be the result of a complex mix of factors including inherent genetic susceptibility and a series of environmental changes or challenges.

The determination of the full sequence of the human genome provides a tremendous opportunity to the international research community to begin to get the "full picture" of the ways in which cells work and the mechanisms underlying disease. Hidden within this DNA sequence is the information that underpins the biochemistry by which our cells function, interact, and differentiate during development to effect the complex physiology which makes the human body function. However, although the characterization of the sequence is a critical first step, determination of the primary sequence itself leaves us a long way from characterization of the complex mechanisms by which genes interact to impart this function. Knowledge of this sequence, for instance, does not in itself elucidate the mechanisms governing the control of gene expression, so that the correct proteins are present in our cells at the correct time during development, nor the ways in which gene expression changes during the development of certain types of disease. Further studies are also needed to shed light on the ways in which the protein products of genes interact both temporally and spatially within the cell to form the complex pathways which effect cellular process. Until we find ways of dissecting these processes and making sense of these mechanisms, we will never truly understand how genes act and the factors which underpin the development of complex diseases.

The new discipline of functional genomics aims to develop and apply technologies to use the information generated from the characterization of human and other genomes to dissect the complexity of function.

See also: **Gene Expression; Genetic Diseases; Genome Organization; Human Genome Project**

Fundamental Theorem of Natural Selection

W J Ewens

Copyright © 2001 Academic Press
doi: 10.1006/rwgn.2001.0483

Ever since it was first put forward by Fisher in 1930, the 'fundamental theorem of natural selection' (henceforth referred to as FTNS) has provoked as much controversy, and caused as much misunderstanding, as perhaps any other result in evolutionary population genetics. The reasons for the misunderstandings arise from Fisher's cryptic writing style, the fact that the precise statement of the theorem was never clear, the existence of typing errors in almost every account he gave of the theorem, and the leaps of faith apparently made in the mathematical derivations. The position was not helped by the appearance of Fisher's 1958 book, in which none of these problems was remedied, and in which various further printing errors added to the confusion.

Fisher (1958) gives the following statement for the FTNS: "The rate of increase in fitness of any organism at any time is equal to its genetic variance in fitness at that time." The contemporary statement of the classical version of the theorem would be, approximately: "The rate of increase of mean fitness of any population at any time is equal to its additive genetic variance in fitness at that time." This statement does not imply any change to the content of the theorem, but is intended to clarify three points. First, the result relates to a population, not some given organism in that population. Second, it relates to the mean fitness of that population. Finally, the contemporary expression 'additive genetic variance,' denoted here V_A, clarifies the meaning of the perhaps ambiguous term 'genetic variance.'

Why did Fisher place so much weight on this theorem, claiming that it holds the "supreme position among the biological sciences?" Fisher's central aim was to restate the Darwinian theory – that evolution by natural selection requires variation, and that evolution by natural selection is a process of 'improvement' – in Mendelian terms. Because a parent passes on a gene at each locus to an offspring, not his/her genotype at that locus, and because entire genome genotypes are regularly broken up over successive generations by recombination, he focused on the gene as the fundamental unit of transmission and thus as the natural entity for describing evolution as a Mendelian process. This is why V_A, being that component of overall genotypic variation in fitness ascribable to genes, was relevant to him. To show that this component is equal to the increase in mean fitness must surely have appeared to him as encapsulating the restatement that he desired.

'Classical' Interpretation of the Theorem

Since its foundation in the 1920s by Fisher, Wright, and Haldane, population genetics theory has consisted in large part of results that assume random mating in the population considered; that is, that the choice of one's mate is made at random, independent of the genetic constitution of the mate. A second assumption often made is that, in studying the evolution of gene frequencies at any locus through the effects of mutation and selection, all other loci can be ignored and the locus of interest treated in isolation. A third assumption, often made in connection with the second, is that the fitness of an individual of any given one-locus genotype is a fixed quantity, independent of the

genes in the remainder of the genome. All three assumptions were initially made in large part to simplify the theory, which otherwise would have encountered almost insuperable mathematical obstacles. It was, however, recognized from the start that a complete theory would eventually relax these assumptions.

This set of assumptions led to the following 'classical' interpretation of the FTNS: if an arbitrary number of different allelic types is allowed at some gene locus, if the fitness of any individual depends only on its genotype defined by these alleles, and if these genotype fitnesses are fixed constants, then assuming mating is random, the population mean fitness will increase from one generation to the next, or at least remain constant, with the increase in mean fitness from one generation to the next being approximately equal to the additive genetic variance at that locus. A proof of the classical version of the theorem, under these assumptions, appears in almost every textbook in population genetics, the formal result being:

$$\Delta \bar{w} \approx V_A \qquad (1)$$

where $\Delta \bar{w}$ is the change in the mean fitness $\bar{w}$ between parental and offspring generations and V_A is the parental generation additive genetic variance. Further, it is also easy to show, if the various single-locus genotype fitnesses differ from each other by a small term of order δ, that the actual increase in mean fitness differs from V_A by a term of order δ^3. The reason for the random-mating requirement is that it is easy to find examples, when random mating is not the case, for which mean fitness decreases between parental and offspring generations.

The fact that $\Delta \bar{w}$ is not exactly equal to V_A appears to contradict the claims by Fisher that "the rate of increase in fitness ... is exactly equal to the genetic variance" and that "the theorem is exact," and throws doubt on "the rigor of the demonstration ..." and the use of the word "theorem" to describe the result. This observation prompted some involved in the exegesis of the theorem to doubt the correctness of Fisher's calculations, a view apparently supported by the "failure" of the theorem in the multiple locus, as described below. Others took the view that at best Fisher intended his result to be approximate, a view hard to reconcile with his words quoted above. It is shown below that two modern interpretations of the theorem claim that the FTNS as correctly understood is an exact statement, involving no approximations.

The classical version of the theorem implies that mean fitness is a potential function, that is a mathematically defined time-dependent quantity that, in a dynamic process, increases steadily (or at worst remains constant) as time goes on. Yet Fisher, who was well aware of the properties of potential functions, steadfastly disclaimed any such interpretation to the FTNS. Thus in a (1956) letter to Kimura (Bennett, 1983) he said:

…I preferred to develop the theory without (the) assumption [of a potential function], which…is a restriction.…I should like to be clear that the expression I have obtained…does not depend on the existence of any potential function.

This claim, of course, merely adds to the mystery: what can Fisher have claimed to be discussing, and what can he claim to have proved?

Strict mathematical proofs that mean fitness does increase (under the above assumptions) was accomplished independently by various authors around 1960. The most direct proof was given by Kingman, who showed further that mean fitness strictly increases unless gene frequencies are at equilibrium values.

Equation (1) is found under the assumption of a nonoverlapping generation model – there is a distinct parental generation, giving rise to a distinct offspring generation, and so on. Continuous-time models, in which generation membership does not arise, have also been studied, with conclusions similar to those given. The sex-linked case has also been analyzed. Further, a generalization of the theorem, referring to any character, not only fitness – sometimes called (Robertson's) 'secondary theorem' of natural selection – has been made. This generalization states, roughly, that the between-generation increase in any character is equal to the parental generation covariance between additive effects of that character and fitness.

Of course, the assumptions made for the proof of (1) describe a situation that often is far from biological reality. Mating might not be random, fitnesses will usually involve fertility as well as viability, will depend on all genes in the genome, can change for extrinsic ecological reasons, and will often not be fixed constants, but rather be frequency-dependent. Later versions of equation (1) were devised, incorporating several of these factors, leading to formulas of increasing degrees of complexity. All versions, however, had much the same flavor as that encapsulated in equation (1), differing from equation (1) in details but not in fundamentals.

The classical version was, for many years, the accepted statement of the theorem. What influence has the classical interpretation had in evolutionary thinking? The classical version of the FTNS is attractive in that it appears to quantify in Mendelian terms the two prime themes of the Darwinian theory, namely that variation is needed for evolution by natural selection, and that evolution by natural selection is a process of steady improvement in the population.

A variety of views exists about the biological value of the classical version of the theorem. Whatever its biological value might be, the theorem received continual attention from the purely mathematical point of view. The most interesting discussion concerned the 'multiple-locus case.' Fisher's statement of the FTNS clearly claimed that it was derived assuming that the fitness of any individual depends on one's complete genetic make-up. The Kingman analysis showing that mean fitness does increase, assumes, however, that fitness depends on the genotype at one locus only. Thus, immediately after this result was firmly established, attempts were made in the literature to remove the 'one locus' assumption and to derive a mathematical theorem for the case where fitness depends on an individual's genotype at two loci, the first step in moving to a multiple-locus result and thus coming closer to Fisher's claimed general statement.

It is not possible to obtain the multiple-locus generalization of the approximation (1) by summing both sides over all loci in the genome, since when epistasis exists the total additive genetic variance is not the sum of the single locus marginal values. Nor is it possible to obtain the desired result using only gene frequencies in the analysis. Even under random mating the vehicle needed to study the evolution of a randomly mating Mendelian population, when fitness depends on the genotype at many loci, is the set of gametic frequencies in the population.

Eventually an analysis of the theorem was carried out using these gametic frequencies. When this analysis was done, it was found that in the multilocus case the population mean fitness can decrease from one generation to the next. The change in mean fitness, being in such cases negative, could not then be equated with any form of variance, so that the classical version of the FTNS fails in the multiple-locus case. This reinforced the views of those who had claimed that Fisher's calculations were always at best approximate.

The reason why mean fitness can decrease in the two-locus case, even under random mating, derives from the existence of recombination. Recombination can cause an offspring chromosome to differ from either parental chromosome, and to this extent the offspring does not resemble the parent. It is thus not unexpected that the FTNS, in its classical form, will fail in the multiple-locus case.

Despite these comments, cases where mean fitness decreases have the nature of comparatively rare oddities. When fitness differentials are small and linkage is loose, mean fitness 'usually' increases, and is 'usually' approximately equal to V_A. Several important results of this type are given, for example, by Nagylaki (1991, 1992).

Thus in the multiple-locus case the classical version of the theorem is often 'almost true.' However, a mathematical theorem as an exact statement must always be true. Further, from the biological and evolutionary point of view, the fact that mean fitness could decrease is disturbing. The position of the theorem as an exact mathematical statement was thus still unresolved.

Recent Versions of FTNS

The 'classical' version of the FTNS, described in detail above, and about which so much has been written, cannot have been what Fisher meant by the theorem. This is most clearly seen in the case of nonrandom mating. Fisher emphasized frequently that mating (in human populations in particular) is not random, and claimed that the FTNS is true even for nonrandom-mating populations. For example, in an acerbic comment on Wright's evolutionary work he said Wright's formulas are "foredoomed to failure just as soon as the simplifying, but unrealistic, assumption of random mating is abandoned." It is easy to find cases where the population mean fitness decreases when mating is not at random, and Fisher was well aware of these, so that the theorem in its classical version cannot have been what he had in mind. However, his writings unfortunately do not make clear, with any degree of certainty, what he did have in mind, and sometimes seem to state results that cannot be what he meant. Thus the focus changes from problems of exegesis of Fisher's written work to the more dangerous undertaking of reading Fisher's mind, and finding what must have been his interpretation of the FTNS, camouflaged though it may be in his writings. This change of direction has led to two recent interpretations of the theorem, both quite different from the classical interpretation.

The breakthrough in this direction came with a little-appreciated paper by Price (1972). Price claimed that Fisher was not interested in the actual change of mean fitness, but rather only in that part of the change "due to natural selection [rather than] due to environmental change, [where we regard] dominance and epistasis as environmental effects." This 'natural selection' change was also thought of as the change in mean fitness due to changes in gene frequencies. Difficult though it is to make immediately concrete the concept of "change due to natural selection and gene frequencies," this insight nevertheless led to both modern interpretations of the theorem. The "change due to natural selection" has been called the "partial change" in mean fitness, and the interpretation of this change is clarified by considering the case where fitness values depend on the genotype at one gene locus only.

Suppose then that the fitness of any individual depends entirely on his genotype at a single locus 'A' at which may occur genes (here and elsewhere, more exactly 'alleles') $A_1, A_2, \ldots, A_k$. Denote the frequency of the genotype $A_i A_j$, at the time of conception of the parental generation, as P_{ii} (when $i = j$) and $2P_{ij}$ (when $i \neq j$). This notation implies that the frequency p_i of the gene A_i is $\sum_j P_{ij}$. If the fitness of an individual of genotype $A_i A_j$ is w_{ij}, the mean fitness $\bar{w}$ of the parental generation, at this time, is then given by:

$$\bar{w} = \sum_i \sum_j P_{ij} w_{ij} \qquad (2)$$

As noted above, Fisher's main evolutionary focus was on the genes at any locus, not the genotypes, and a key concept for Fisher was the average effect in fitness of any gene. The average effects $\alpha_1, \alpha_2, \ldots, \alpha_k$ of the genes $A_1, A_2, \ldots, A_k$ are defined as the values that minimize the quadratic function $\sum_i \sum_j P_{ij}(w_{ij} - \bar{w} - \alpha_i - \alpha_j)^2$, subject to the constraint $\sum_i p_i \alpha_i = 0$. These average effects may be thought of as roughly the 'fitnesses' of the various genes, and $\bar{w} + \alpha_i + \alpha_j$ as the best additive approximation to the fitness w_{ij} using these average effects. The additive genetic variance V_A is the amount removed from the above quadratic function by fitting these α_i values.

The next step in the argument is to note that, for evolutionary analyses, Fisher appears to have conceived of the fitness of the typical genotype $A_i A_j$ not as the actual fitness w_{ij}, but rather as the additive approximation $\bar{w} + \alpha_i + \alpha_j$. This interpretation is justified from the excerpts such as the following from his 1958 book:

…for any specific gene combination we build up an 'expected value'…by adding [to the mean] appropriate [α values] according to the…genes present. This expected value will not necessarily represent the real [fitness]…but its statistical properties will be more intimately involved in the inheritance of real [fitness] than [fitness] itself.

This additive approximation is called the 'breeding value' in animal breeding programs.

This change of viewpoint implies that Fisher thought of the mean fitness not as in equation (2), but rather as:

$$\sum_i \sum_j P_{ij}(\bar{w} + \alpha_i + \alpha_j) \qquad (3)$$

This change of viewpoint is purely conceptual, since the expression (3) is numerically identical to the mean fitness defined in (2). Despite this identity, this new conceptualization leads to the concept of the partial change in mean fitness as the change, over one generation, of the expression (3) brought about by changes in the genotype frequencies P_{ij}, with the quantities $\bar{w}$, α_i, and α_j being unchanged, remaining at the parental generation values. This partial change in mean fitness is then:

$$\sum_i \sum_j (P'_{ij} - P_{ij})(\bar{w} + \alpha_i + \alpha_j) \qquad (4)$$

where P'_{ij} is the daughter generation frequency of the genotype A_iA_j, defined as for the parental generation value. It is straightforward to show, with minimal evolutionary assumptions, that:

$$\text{partial increase in mean fitness} = V_A/\bar{w} \qquad (5)$$

whether or not random mating occurs. The additive genetic variance V_A in this expression may be computed as:

$$V_A = 2\bar{w}\sum_i \alpha_i(\Delta p_i) \qquad (6)$$

This exact single-locus discrete-time result involving no approximations, and together with its analogous continuous-time version, is in reasonable agreement with Fisher's wording.

The parallel multiple-locus statement of the theorem, namely that if fitness depends in an arbitrary way on an arbitrary number of genes at an arbitrary number of loci, with an arbitrary recombination structure, and with no assumption made about random mating, is:

$$\text{partial increase in mean fitness} = V_A^{(m)}/\bar{w} \qquad (7)$$

where $V_A^{(m)}$ is the full multiple-locus additive genetic variance. This equation, again exact and embodying no approximations, is the statement of one of the modern interpretations version of the FTNS.

Equation (7) is not achieved by simply summing both sides of equation (5) over all loci. Despite this, a summation result of a different form does hold. An expression parallel to that in (6) is that the multiple-locus additive genetic variance may be written as:

$$V_A^{(m)} = 2\bar{w} \sum_i \sum_j \alpha_{ij}(\Delta p_{ij}) \qquad (8)$$

where α_{ij} is the multiple-locus average effect of gene A_i at locus j, Δp_{ij} is the one-generation change in the frequency of that gene, and the sum is over all genes at all loci in the genome.

Equations (7) and (8) imply that this version of the FTNS can be restated in the form:

$$\text{partial increase in mean fitness} = 2 \sum_j \sum_i \alpha_{ij}(\Delta p_{ij}) \quad (9)$$

the double sum being over all alleles at all loci. The expression on the right-hand side of (9) derives from the following argument. All multiple-locus genotypes are thought of as being listed in order, the typical such genotype being described as genotype g. The partial change in mean fitness is then

$$\Sigma_g \Delta P(g)w(g)_\alpha, \qquad (10)$$

where $\Delta P(g)$ is the between-generation change in frequency of the multiple-locus genotype g and $w(g)_\alpha$ is the sum of the average effects of all genes at all loci in the genotype g, any average effect being counted in twice in the sum if the corresponding gene occurs twice in genotype g. The expression (10) may be shown to be identical to the expression $2\omega\Sigma_i\Sigma_j\alpha_{ij}(\Delta P_{ij})\bar{w}$ arising on the right-hand side of (9), so that the above interpretation of the FTNS can be written as

$$\Sigma_g \Delta P(g)w(g)_\alpha = V_A \qquad (11)$$

A second modern interpretation of the FTNS, due to Lessard (1997), appears initially to be similar to (11), but is arrived at by a quite different analysis than that leading to (11), and differs from (11) in several important ways. Lessard's equation is

$$\Sigma_g \Delta P(g)_\alpha w(g) = V_A \qquad (12)$$

The difference between the two expressions (11) and (12) is the following. In (11), $\Delta P(g)$ is the actual change in frequency of genotype g over one generation, and $w(g)_\alpha$ can be thought of as the best estimate of the fitness of genotype g, given the genes in this genotype. In (12), $w(g)$ is the actual fitness of genotype g and $\Delta P(g)_\alpha$, defined as $P(g)w(g)_\alpha - P(g)$, may be thought of as the best estimate in the change in the frequency of genotype g, given the genes in this genotype.

Lessard's interpretation of the theorem appears to agree more closely with Fisher's words than does the interpretation deriving from (9) and (10), and may very well be the correct interpretation of the theorem. If so, a final resolution of the interpretation of the FTNS has been reached. A full discussion of this point is given in Lessard (1997).

The above discussion in terms of a discrete time model with viability fitnesses only. Lessard (1997) and Ewens (1989) show that the two modern interpretations hold, with appropriate changes, for continuous time models, and Lessard discusses models with age structure and fitness defined as the mean number of offspring produced. Lessard and Castilloux (1995) show that the modern interpretations hold also when fitnesses relate to fertility differences among couples. Frank (1997) discusses the relation of the FTNS with Price's equation.

References

Bennett JH (1983) *Natural Selection, Heredity, and Eugenics.* Oxford: Clarendon Press.

Ewens WJ (1989) An interpretation and proof of the Fundamental Theorem of Natural Selection. *Theoretical Population Biology* 36: 167–180.

Fisher RA (1958) *The Genetical Theory of Natural Selection.* New York: Dover.

Frank SA (1997) The Price equation, Fisher's Fundamental Theorem, kin selection, and casual analysis. *Evolution* 51(6): 1712–1729.

Lessard S (1997) Fisher's fundamental theorem of natural selection revisited. *Theoretical Population Biology* 52: 119–136.

Lessard S and Castilloux AM (1995) The fundamental theorem of natural selection in Ewens' sense: case of fertility selection. *Genetics* 141: 733–742.

Nagylaki T (1991) Error bounds for the primary and secondary theorems of natural selection. *Proceedings of the National Academy of Sciences, USA* 88: 2402–2406.

Nagylaki T (1992) *Introduction to Theoretical Population Genetics.* New York: Springer-Verlag.

Price GR (1972) Fisher's 'Fundamental Theorem' made clear. *Annals of Human Genetics* 36: 129–140.

See also: Additive Genetic Variance; Fisher, R.A.; Fitness; Wright, Sewall

Fungal Genetics

D Stadler

Copyright © 2001 Academic Press
doi: 10.1006/rwgn.2001.0484

Fungal genetics is the experimental study of the properties of genes and chromosomes carried out with filamentous fungi (such as *Neurospora*, *Aspergillus*, and *Ascobolus*) or with yeasts (such as *Saccharomyces*, *Schizosaccharomyces*, and *Candida*). These organisms have been important in basic genetics because they are eukaryotes but are also amenable to the elegant methods of bacteriology.

See also: Ascobolus; Aspergillus nidulans; Neurospora crassa; Saccharomyces cerevisiae (Brewer's Yeast); Schizosaccharomyces pombe, the Principal Subject of Fission Yeast Genetics

Fungi

D Stadler

Copyright © 2001 Academic Press
doi: 006/rwgn.2001.0485

A group of simple, nongreen plants that includes molds, mushrooms, rusts and smuts, and sometimes yeasts.

See also: Ascobolus; Aspergillus nidulans; Neurospora crassa

FUS-CHOP Fusion

See: Myxoid Liposarcoma and *FUS/TLS-CHOP* Fusion Genes

Fusion Gene

P Riggs

Copyright © 2001 Academic Press
doi: 10.1006/rwgn.2001.0486

A gene fusion is defined as two genes that are joined so that they are transcribed and translated as a single unit. Gene fusions can occur *in vivo*, both naturally and as a result of genetic manipulations, and can be constructed *in vitro* using recombinant DNA techniques. They occur in nature over the course of evolution, for example, where two genes whose products are part of a metabolic pathway fuse, giving rise to a fusion protein that carries out both steps of the pathway.

History

The first gene fusions created by design were between the rIIA and rIIB genes of phage T4, studied by Champe and Benzer. They used the effects of missense, nonsense and frameshift mutations in the rIIA gene on RIIB activity to elucidate the properties of the genetic code. Subsequently, fusions were created in *Escherichia coli* using *in vivo* genetic techniques to join various genes to the *lacZ* gene, which codes for the easily assayed enzyme β-galactosidase. These fusions were used as a way to examine the expression level and regulation of the gene fused to *lacZ*. Fusions were originally limited to genes that were located near the β-galactosidase gene, but later Casadaban and coworkers pioneered *in vivo* and *in vitro* techniques that allowed fusion to virtually any gene.

Current Uses

The major current use of gene fusions is still the study of gene expression, including levels of expression and location of gene products. Both gene fusions and reporter constructs (where the gene of interest is replaced by a 'reporter' gene instead of being fused to it) are used for this purpose. Fusions to *lacZ* are common, but any gene whose product is active as a fusion and can be assayed is suitable for this purpose. In this method, an extract of a cell or tissue containing a gene fusion is prepared and the level of gene expression is measured by assaying the fusion. Gene fusions can also be used to study the differential expression of a gene in different tissues of an organism, by histochemical staining for the fused gene in sections, tissues, or the whole organism. Two genes commonly used for this technique are the *lacZ* and *gfp* genes. The *lacZ* gene has been used primarily because of the vast experience researchers have with β-galactosidase fusions, and the many substrates available for this enzyme. One of these substrates, X-gal, produces a dark-blue insoluble product when cleaved by β-galactosidase. Thus, the blue color does not diffuse away from the site of cleavage, and one can infer the location and level of expression from the intensity of the blue color. The *gfp* gene codes for green fluorescent protein, which fluoresces green when excited by blue or UV light. This allows visualization, and in many cases can be used on intact, live organisms.

Further Reading

Casadaban M J, Martinez-Avias A, Shapiro D K and Chou J (1983) β-galactosidase gene fusion for analyzing gene expression in *Escherichia coli* and yeast. *Methods Enzymology* 100: 293–307.

Champe S P and Benzer S (1962) An active cistron fragment. *Journal of Molecular Biology* 4: 288–292.

See also: **Beta (β)-Galactosidase; Fusion Proteins**

Fusion Proteins

P Riggs

Copyright © 2001 Academic Press
doi: 10.1006/rwgn.2001.0487

A fusion protein is a protein consisting of at least two domains that are encoded by separate genes that have been joined so that they are transcribed and translated as a single unit, producing a single polypeptide. Fusion proteins can be created *in vivo*, but are usually created using recombinant DNA techniques. The fusion often consists of the protein that is being studied joined to one of a small number of proteins that have useful properties to aid in the study.

History

Some of the first fusion proteins were created in *Escherichia coli* using *in vivo* genetic techniques to join various proteins to the β-galactosidase enzyme. These fusions were used initially as a way to assay the expression level of the protein of interest. Fusions were originally limited to proteins whose genes were located near the β-galactosidase gene, but later, Casadaban and coworkers pioneered *in vivo* and *in vitro* techniques that allowed fusion to virtually any protein. Researchers were originally surprised that some of the fusions were bifunctional, i.e., when the C-terminus of a protein was fused to the amino terminus of β-galactosidase, both the proteins retained activity. As more and more fusions to β-galactosidase were obtained and found to have activity, researchers began to make fusions to other proteins besides β-galactosidase and found that they could be bifunctional as well.

Uses of Fusion Proteins

The technique of creating fusion proteins has been extended to other fusion partners, and additional uses have been developed for the fusion partner. Three of the most important uses of fusion proteins are: as aids in the purification of cloned genes, as reporters of expression level, and as histochemical tags to enable visualization of the location of proteins in a cell, tissue, or organism.

For purification, a protein that can be easily and conveniently purified by affinity chromatography is fused to a protein that the researcher wishes to study. A number of proteins and peptides have been used for this purpose, including staphylococcus protein A, glutathione-S-transferase, maltose-binding protein, cellulose-binding protein, chitin-binding domain, thioredoxin, strepavidin, RNaseI, polyhistidine, human growth hormone, ubiquitin, and antibody epitopes.

The proteins used most often as fusion partners for reporter constructs are β-galactosidase, luciferase, and green fluorescent protein (GFP). β-galactosidase has the advantage of numerous commercially available substrates, including some that produce a colored product and some that lead to the production of light. Luciferase and GFP both produce light, and can be visualized directly or quantitated using a luminometer or a fluorometer, respectively. GFP has an advantage in that it does not require a substrate, whereas luciferase requires its substrate, luciferin, as well as ATP, O_2, and Mg^{2+}. GFP emits green light when excited by blue or UV light, and in many cases can be used on live, intact cells and organisms.

A useful extension of fusion proteins as reporters is the two-hybrid system. In this method, two separate fusions are employed to test for interaction between two proteins, where binding of the two proteins brings together their fusion partners and results in activated transcription of a reporter gene.

See also: **Beta (β)-Galactosidase; Fusion Gene**

G

G1

The G1 phase of the eukaryotic cell cycle is that between the end of cell division and the start of DNA synthesis. G1 refers to the first gap phase.

See also: **Cell Cycle**

G2

The G2 phase of the eukaryotic cell cycle is that between the end of DNA synthesis and the start of cell division. G1 refers to the second gap phase.

See also: **Cell Cycle**

Galactosemia

S Segal

Galactosemia is the most common form of abnormal galactose metabolism and is a recessively inherited disorder with an incidence of 1:20 000 to 1:60 000 live births. Although the deficient enzyme is known, the etiology of the clinical syndrome is enigmatic. The clinical picture evolves in two phases. The first occurs after birth when the feeding of milk and other formulas containing lactose produces a galactose toxicity syndrome manifested by hyperbilirubinemia, failure to thrive, vomiting, cataract formation, blood coagulation defects, and renal tubule dysfunction. With a galactose-restricted diet these abnormalities regress. The second phase occurs despite the diet therapy with later development of speech abnormalities, mental retardation, neurological ataxias, and ovarian failure.

The diagnosis is suggested by the presence of abnormally high galactose levels in blood and urine and elevated red blood cell galactose-1-phosphate. It is confirmed by quantitation of red blood cell enzyme activity of less than 7% of normal as well as determination of the genotype. Heterozygotes express about 50% of normal enzyme activity. Subjects who are carriers of the defective galactosemia gene compounded with the Duarte gene express about 25% of normal red cell activity.

The normal disposition of dietary galactose involves the conversion of the sugar to glucose via a series of three enzymes known as the Leloir pathway: (1) galactokinase catalyzes the phosphorlylation of galactose with ATP to form galactose-1-phosphate; (2) galactose-1-phosphate uridyltransferase reacts the sugar phosphate with UDPglucose to form UDPgalactose and glucose-1-phosphate; and (3) UDP-galactose-4-epimerase converts the UDPgalactose to UDPglucose. The net result of the series is the conversion of galactose-1-phosphate to glucose-1-phosphate. Inherited deficiencies of galactokinase and UDPgalactose-4-epimerase are known but occur much less frequently than transferase deficiency galactosemia. The main manifestation of galactokinase deficiency is cataract formation when a galactose-containing diet is ingested. Epimerase occurs in two forms: one benign with reduced red blood cell enzyme activity, and the other, which is extremely rare, exhibits a toxicity syndrome similar to transferase deficiency.

With a block in the pathway due to absence of transferase, galactose-1-phosphate and galactose accumulate. As a consequence two alternative routes of galactose disposal are activated. The first is reduction of the sugar by aldose-reductase to galactitol, which is not further metabolized. The second involves oxidation to galactonate, which can be further metabolized to CO_2 and xylulose. Both galactitol and galactonate are excreted in urine in large quantities, and red blood cell galactose-1-phosphate remains elevated despite a galactose-restricted diet. The explanation for the elevation of these abnormal metabolites appears to be a large endogenous synthesis of galactose, presumably from turnover of galactose-containing complex glyco-conjugates. Short-term 2-h oxidation of isotopic galactose to CO_2 is very slow in most patients, but 24-h oxidative capacity is similar to that measured in normal patients after 5 h. The oxidative pathways involved have not been completely defined. This capacity to oxidize the sugar plus the urinary metabolite excretion maintain the patient in a steady-state with plasma galactose levels in the low micromolar range.

The human galactose-1-phosphate uridyltransferase gene of 4 kb has been cloned and sequenced and consists of 11 introns and exons on chromosome 9. The cDNA codes for a 374 amino acid protein, with about 49% conservation between human and *Escherichia coli* enzymes. The active enzyme is a dimer with a molecular mass of 96 kDa. There are over 100 mutations known to occur in galactosemic patients. Most are missense mutations with a single base change, but stop mutations, splice site changes, frameshifts, and large deletions are found. The most common mutation, accounting for over 60% of mutant alleles, is Q188R, in which arginine is substituted for glutamine in the highly conserved region of exon 6. About 45% of patients are homozygous for Q188R. A number of Q188R alleles are compounded with other mutations. In African American and South African black galactosemics the prevalent mutation is S135L. The Q188R mutant is believed to be devoid of enzyme activity, while the S135L mutation results in residual liver enzyme activity. Heterodimer formation may be a significant determinant of enzyme activity. The N314D mutation with an asparagine to aspartic change is prevalent and the basis of the Duarte variant. It results in diminished but not absent erythrocyte enzyme activity and is itself benign.

There appears to be no clear genotype–phenotype correlation. However, the ability to oxidize administered 1-^{13}C galactose to $^{13}CO_2$ of less then 2% in 2 h appears to indicate a more severe disorder as observed in many Q188R homozygotes than that present in compound heterozygotes.

Neither the pathobiochemical basis of galactose toxicity in the newborn period, nor the late onset long-term diet-independent complications are known. Accumulation of galactose-1-phosphate and galactitol are believed to be responsible, but the mechanism of multiorgan involvement is unclear. Cataract formation is associated with galactitol accumulation. A knockout mouse with absent transferase activity shows no manifestations of the human phenotype, suggesting that absence of transferase is necessary but not sufficient to cause disease. This points to epigenetic factors and abnormal alternative pathway metabolites as the possible basis of the human disease.

The only known treatment of galactosemia has been restriction of lactose and other galactose-containing foods. Although the postnatal toxicity is alleviated, the long-term complications have not been averted. Speech therapy, special schooling, and hormonal therapy of ovarian failure are indicated and may be helpful. The disorder remains an enigma requiring the search for new therapeutic strategies.

See *also*: Lactose

Galton, Francis

R Olby

Copyright © 2001 Academic Press
doi: 10.1006/rwgn.2001.0490

The Victorian intellectual Francis Galton (1822–1911) was one of the chief founders of the science of biometry, or the statistical and quantitative study of living things. He described its chief objective to be "to afford material that shall be exact enough for the discovery of incipient stages in evolution," stages that are "too small to be otherwise apparent." His goal was to establish the foundations upon which to base a policy to control and direct the future of mankind. In 1883 he coined the word 'eugenics' for the science of improving stock, a term, he remarked,

that is not confined to questions of judicious mating, but which especially in the case of man, takes cognizance of all influences that tend in however remote a degree to give to the more suitable races or strains of blood a better chance of prevailing speedily over the less suitable than they otherwise would have.

Galton was particularly concerned to show that the behavioral as well as the physical traits of mankind are inherited, that what is acquired in life cannot be passed to the offspring, and that nature (that which is inherited) has much more influence on the individual than has nurture (that which is gained from experience and education). He was, in other words, an 'hereditarian.' But he did not simply make hereditarian claims. He developed the statistical techniques of regression and correlation to analyze the biometric data he collected from sampling human populations. His passion for quantitative treatment he also applied to establish weather patterns, to test the effectiveness of prayer, and to explore sensory perceptions, imagery, and memory.

The zoologist Raphael Weldon and the mathematician Karl Pearson became devoted followers of

Galton. From the 1890s they vigorously developed the science of biometry, and in the early years of the twentieth century they opposed the newly rediscovered science of Mendelian heredity. Nearly two decades passed before biometry and Mendelism were effectively united to form what we call population genetics. Meanwhile Galton's science of eugenics passed through phases of popular approval and disapproval. The intensive study of the chemical sequence of the genetic material of our genes that has been ongoing since the 1980s has once more brought the subject of eugenics to popular attention. Could new and powerful techniques now make possible a kind of eugenics "by the backdoor"? Not, in other words, public legislation enforcing eugenic policies, but covert pressures of the market through discrimination, and limited access to resources.

Francis Galton's Life

Galton came of a wealthy and well-connected family, his mother being the daughter of Charles Darwin's grandfather, Dr Erasmus Darwin. As a boy and the youngest in the family, much affection was bestowed upon Francis, especially by his three sisters. Adèle, the youngest, acted as his tutor, and those around him soon considered him an infant prodigy. However, formal education, neither in France where he was sent at the age of eight, nor in England from the age of 14, proved to be to his liking. At 16 he began to study medicine, but two years later he turned to mathematics and moved to Cambridge. Four years later he gained a BA without honors and prepared to return to his medical studies. Then his father died, and he came into an inheritance that permitted him to forget medicine and indulge in his love of exploration.

The resourceful and courageous young Galton traveled through Egypt, Syria, and South West Africa, where he covered some 1700 miles of uncharted country and came to know the Damara, Namaqua, and Ovampo tribes. He was struck by their distinctive behavioral and physical characteristics, and those of their domesticated animals. On his return to England in 1852 the Royal Geographical Society awarded him their gold medal for his achievement. During the 1850s he worked to promote geographical exploration, published his guide *The Art of Travel* (1855), introduced weather maps, discovered anticyclones, and worked for the British Association for the Advancement of Science. When in 1859 his cousin Charles Darwin published his book *On the Origin of Species*, Francis read it and was greatly impressed. If all life is the product of evolution, we should be able, given sufficient knowledge, to control our own evolution – our future. But, like his cousin, he realized there

was a weak spot in the theory, the lack of sound knowledge of the nature of heredity. Accordingly, he turned in the 1860s to address this subject. Between 1865 and 1889 he worked often obsessively on gathering material. Publications on the subject flowed from his pen, the most important being his two books: *Hereditary Genius* (1869) and *Natural Inheritance* (1889). He lived on to 1911 – long enough to seize the opportunity that the changing political climate of the new century afforded him to publicly appeal for the establishment and support of the science he called eugenics.

Human Heredity

Conceptual

During the nineteenth century Herbert Spencer and Francis Galton gave up the term 'inheritance' and following the French they substituted the term 'heredity' (*hérédité*). This signaled Galton's conception of heredity as based upon the continuity from generation to generation through an unbroken line not of persons but of the elements in the fertilized eggs from which they came. The term 'inheritance' suggested the legal concept of the transmission of a person's estate to his descendants. Here the link is between the visible characteristics of the grown person (the parent) and the corresponding features of the offspring. But heredity is often indirect. The offspring bear similarities to many ancestors, not just to the parents. Moreover Galton was convinced that nothing we acquire in our organic constitution can thus be passed on. If we behave more virtuously, will our children do likewise? Do the sons of old soldiers, he asked in 1865, learn their drill more quickly than others, or the sons of fishermen escape sea-sickness? And if acquired characters are inherited, why have the many tribes of American Indians, though scattered over the vast range of different climates and situations of the Americas, remained much the same? Yet, if heredity is so unyielding, why is it that a father's characters are sometimes revealed in the son, sometimes in the daughter, or the child may bear the character seen only in a grandparent or more distant ancestor? How can so hard a process be so fickle? Galton saw that the answer lay in the statistical study of large numbers of ancestors and descendants, and in making an analysis of their statistical relations one to another. This is the heart of his project for what was later to be called biometry.

Observational

First he wanted to gather evidence that behavioral as well as physical characters are inherited. He chose to study what he called 'genius,' or as he defined it, an ability that is "exceptionally high, and at the same time inborn." It excludes any ability that can be attributed to the effects of education, but it includes an energetic disposition. Brilliance without application, persistance, and stamina, is of little use. Then he made the questionable assumption that ability correlates with eminence in public and professional life. Noting that great ability seemed to cling, as it were, to particular families like his cousin's, the Darwins, or the Bachs with its musicians and the Bernoullis with its mathematicians, he turned to the legal profession and extracted the names of 109 judges sufficiently eminent to be mentioned in Foss's *Lives of the Judges* (1865). Then he tracked the 85 families involved to establish how many relatives of these judges also achieved eminence in the legal or other professions. He found that one in every nine of these judges was either father, son, or brother to another judge, not to mention the relations of judges that attained higher legal office. He set out his results in tabular form (**Table 1**). The table illustrates how fewer and fewer relations of the most gifted member of a family attain to eminence the more distant is their kinship to that member. The percentages, wrote Galton, "are quartered at each successive remove." He concluded that the data show "in the most unmistakable manner the enormous odds that a near kinsman has over one that is remote, in the chance of inheriting ability." To consolidate this claim he turned to another eight professions, and to oarsmen and wrestlers. Most of the data were supportive of his claim, though he noted that some sons of very pious parents occasionally turn out extremely badly!

Methods in Population Studies

Pedigrees

To the objection that Galton was ignoring the effects of nepotism, and the advantages of privileged upbringing and expensive education, he replied with the names of great men who, despite their lowly origin, had become eminent. This criticism, of course, struck at Galton's assumption that public eminence is a measure of native ability. He was aware of another problem, i.e., the underrepresentation of family data. This is the Achilles heel of the pedigree method, i.e., the use of family pedigrees for genetic data collection. Have some of the 'failures' in life been left out? Are more representatives of the male kin included than those of the female? And how does one assess the contribution to ability coming from the females in the line using professional achievement at a time when the professions studied by Galton were not open to them? By the 1890s whole-population studies were being undertaken in Germany to escape such criticisms in the debate over the supposed inheritance of tuberculosis,

Table 1 The judges from 1660 to 1865. (From data in Galton (1889) *Ancestral Inheritance*.)

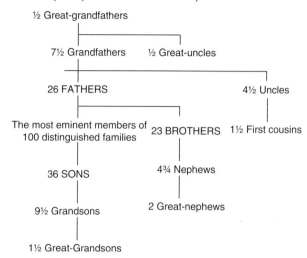

and in the 1870s Galton developed his famous method of twin studies in his effort to gather reliable evidence concerning the relative power of heredity and environment upon the shaping of the offspring. This has become one of the classic approaches whenever dealing with human traits, since the experimental approach is excluded.

Twin Studies

For his study of what he called "The history of twins" Galton used the questionnaire method. Darwin had circulated a questionnaire in his study of heredity and variation in the 1830s, and Galton had followed his example in his investigation into the upbringing and personal characteristics of Fellows of the Royal Society. His appeal for information about twins resulted in 35 adequately answered responses from parents of 'closely similar' twins, and 20 from parents of 'exceedingly unlike' twins. These allowed him to distinguish between what we call identical and non-identical twins. From this comparison of the two groups he concluded that "nature prevails enormously over nurture when the differences of nurture do not exceed what is commonly to be found among persons of the same rank of society and in the same country." This was a wise qualification because he did not have data on twins reared apart either in identical or nonidentical environments. Subsequent researches by Galton's successors did extend the data collection in this way, but it is questionable how different were the environments of the separate homes in which the two members of each pair of twins grew up. In the 1970s the most extensive collection of twin data, that of the British psychologist, the late Sir Cyril Burt, was exposed as fraudulent. On a subject as politically sensitive as the heredity–environment equation, this revelation had a damaging impact upon the field, but careful twin studies continue, particularly as a tool in the study of hereditary predispositions to diseases, including mental illnesses.

Regression

Another method of central importance in the study of populations is that of the statistical distribution of traits. Galton was aware of the curve of 'normal' distribution, also known as the Gaussian or error curve after the mathematician Gauss who applied it to the study of errors in astronomical measurement. Following Gauss, the Belgian Adolph Quetelet found that the measurement of the chests of 5738 Scottish soldiers and the stature of 100 000 French conscripts, when compared with the expectation from Gaussian curves, showed a "marvelous concordance." The graph is bell-shaped, its top or plateau representing the median of the data (**Figure 1**), the median being that value which divides the data on either side equally and symmetrically. As an error curve the sides of the bell represent the 'population' of error measurements, and the top itself is hopefully the 'true' measurement. As a representation of the distribution of the soldiers' heights, Quetelet envisioned the top as marking the height of the 'average man.' Those taller or shorter than this measure were 'errors' as it were in attempts to copy the ideal of the race. The fact that these data fitted the error curve demonstrated, in his view, that they were homogeneous.

Galton focused his attention less on the homogeneity of the population than on its variability. How, in spite of variability, did its median remain the same in successive generations, for of this he was already convinced? Therefore, he wanted to dissect the curve into its parts and follow the progeny of those parts. So he devised an exploratory study in which he got his friends to help by asking them to grow sets of sweet pea seeds (*Lathyrus odoratus*), which he had divided into seven classes by weight. They returned the crop to him and he was then able to plot the progeny seed weights against the weights of their respective parents. The result revealed the presence of a tendency of the progeny of heavy seeds to be lighter than their parents and those of lighter seeds to be heavier. There was a 'reversion' toward the ancestral mean. Since the aggregate mean remained the same and because his helpers all lived in different parts of the British Isles he was confident that the data did not reflect the effects of environment. This tendency to counteract the extremes of individual variation by 'shrinking' the excesses whether dwarfs or gaints in their progeny he called 'reversion,' and later more wisely, 'regression,' since reversion was the term already in use to refer to the return of the progeny of hybrids to their originating

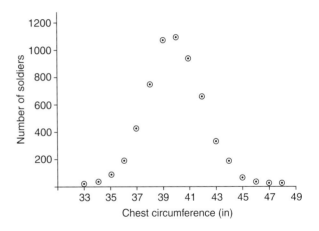

Figure 1 Graph of the distribution of chest circumferences of 5732 Scottish militia men. The figure approximates to the bell-shaped curve of a 'normal distribution.' (Redrawn from data in Quetelet (1817) *Edinburgh Medical and Surgical Journal*: 260–264.)

species. Now he could understand why variability does not change the median or 'center of gravity' of the population.

Having established in a rough manner evidence for this regression in plants, Galton cast around for data on human characteristics, but in vain. However, when he advertized prizes of £500 for those who best filled in the elaborate set of questions that he prepared concerning them, their grandparents, parents, sisters, brothers, children, and other relatives, he was rewarded with a good response. These family records included stature of family members, so he was able to plot the statures of parents against offspring from which he calculated the regression (**Figure 2**). He expressed what he called the 'coefficient of regression' as the ratio between the deviation of the offspring and that of the mid-parents from the population mean. This is measured on the graph by the distances AB and AC or EF and EG. Since the data fall approximately on a straight line, the ratio is constant throughout its length, giving a coefficient of regression of two-thirds. Now he had measured a statistical relation between two generations.

Correlation

Initially he considered regression in one direction only, but later realized that the regression of the parent on the child is the reciprocal of the child on the parent. Then in 1885 he hit upon the concept of 'correlation,' namely that where there is a relation between the variation of one entity and that of another, they can be considered causally related. This important conception was developed more fully later by Karl Pearson.

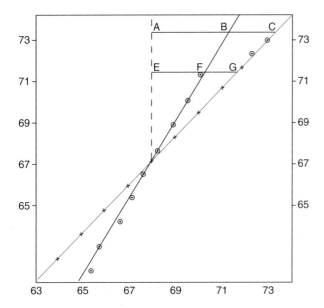

Figure 2 Graph of the relation between the mid-parental heights of parents and the mean height of their children. The diagonal line represent all points on the graph corresponding to hypothetical parents and their children where the means of the children's heights are identical with their parents. The steeper line is plotted from Galton's data. The ratios AB:AC or EF:EG give the coefficient of regression. (Redrawn from Galton (1869) *Hereditary Genius: An Inquiry into its Laws and Consequences*, pp. 83. London: Macmillan.)

Natural Selection

Stabilizing Selection

Darwin had focused upon the slight individual differences that constitute the variation to be found among members of a family and of a species. "They afford material," he said, "for natural selection to accumulate, in the same manner as man can accumulate in any given direction individual differences in his domesticated productions." But Galton was convinced that natural selection cannot work effectively against heredity, and he considered that his work was showing heredity maintaining the racial mean. According to Galton, the mean is the adapted form and deviations from it will be less adapted to the conditions of life. Therefore, natural selection will be aiding heredity in preserving it. In other words, he granted natural selection its 'stabilizing' role, but not its 'creative role.' As the reason for this state of affairs he turned to the physiology of reproduction. The fertilized egg is composed of hereditary material from two parents, so that each time a new generation is produced there is a bringing together of two such materials. Inevitably the contributions of each parent and each ancestor will be diluted. So he accepted the long-held tradition

of fractional inheritance, according to which the parents collectively contribute one-half, the grandparents one-quarter, and the great-grandparents one-eighth to the hereditary constitution of the offspring. These ancestral contributions exert their influence and tend to bring back the progeny of deviants toward the mean of the ancestral population as a whole.

Discontinuous Variation

Having thus restricted the role of natural selection, he turned to 'sports' of nature, those marked deviations that possess a stability shown by the absence of regression to the existing type among their progeny. These deviations create a new mean toward which any progeny will tend to regress instead of regressing toward the mean of the original population. Hence, he explained, these sports may give rise to a new race with but little help from natural selection. He was thus opposed to cousin Darwin who in his *On the Origin of Species* had stressed how unlikely it was that such sports could serve as the starting point for new species. Granted they were strongly inherited, but most sports were closer to monstrosities than to newly adapted forms. In any case, thought Darwin, their rarity would result in the dilution of their type in successive generations of breeding with other members of the species. Darwin proved to be largely right on his first objection, but wrong on his second.

It should not be assumed that Galton's apostasy over natural selection was unusual for the nineteenth century. The consensus was in favor of evolution by descent, but not under the principal agency of natural selection. Therefore, it is ironic that those who most strongly supported the pre-eminence of natural selection considered themselves Galton's successors. Thus Karl Pearson and Raphael Weldon developed Galton's statistical techniques and corrected his errors. But when Weldon sought to demonstrate natural selection in its creative role, shifting the mean of a population, he only succeeded in demonstrating its stabilizing role. Pearson exposed some of the confusions in the several differing representations of the ancestral law offered by Galton. He corrected the figure of two-thirds for the regression of offspring on parents, explained why regression was not the barrier to selection that Galton claimed, and explored the effects of 'assortative' mating, i.e., the choice of mates based on similarities of ability and background, which promotes the shifting of the mean of the resulting offspring further and further away from that of the general population.

Finger Prints

The minute and distinctive patterns of ridges on the skin were used to make finger prints before Galton took up the subject. But it was he who conducted a systematic study leading to his classification of the differing types and it was he who persuaded the police to adopt the practice of fingerprinting for personal identification of criminals. For Galton the subject had a compelling theoretical interest because the trait appeared to have no function such that natural selection could act upon it. Marriage selection does not depend on it; the different patterns are not confined to particular classes or races. Therefore, there is complete 'promiscuity' with respect to this trait. Yet the varieties remain distinct. Here, then, we have a trait whose varieties do not blend and are not subject to selection. This, he believed, was an example of the existence and persistence of distinctive types independent of selection. Of course he could not really have known whether the trait was connected to some other trait that is subject to natural selection.

Eugenics

Controlling our Evolution
The driving force behind Galton's extensive and long-continued research was not just his curiosity, great as that was, but his vision of a future in which mankind would attain to greater energy and coadaptation. But he realized how easy it was to follow the wrong course. To accept the evolutionary process passively would be to surrender to "blind and wasteful processes" in which raw material is produced extravagantly and all that is superfluous is rejected "through the blundering steps of trial and error." He favored the alternative that we should take control of our evolution, for it may be that we are the "only executives on earth." Hence the importance of eugenics in providing the proper scientific basis for action. To support such work he settled an endowment on University College London so that in 1905 a Research Fellow in eugenics could be appointed. Further expansion led to the creation of the Eugenics Laboratory. In 1911, again through Galton's munificence, a chair of eugenics was established at the College, the first occupant being Pearson.

Victorian Attitudes
Galton's attitude to racial differences, to women, and to the indigent was typical of a wealthy Victorian. Mild of manner and gentle in his disposition, yet his attitude to the less fortunate was unquestionably harsh. Many at the time endorsed the policy of negative eugenics, i.e., to discourage the marriage and procreation of offspring by the exceptionally unfit, but Galton went further and wanted to favor those families that were "exceptionally fit for citizenship." He argued that since there was substantial giving to the poor and destitute, could not support be forthcoming

to promote "the natural gifts and the national efficiency of future generations"? His concern, like Pearson's, was over the differential between the reproductive rates of the upper middle and lower classes in favor of the latter.

Publicizing Eugenics
In 1904 the time was judged opportune for Galton to address the Sociological Society on eugenics. Here he argued for the maintenance of diversity, but "each class or sect" represented "by its best specimens," and then to leave them to "work out their common civilization in their own way." The best were the healthy, the energetic, the able, the manly (!), the courteous, but he advised leaving out the cranks and refusing the criminals. Eugenics should study the conditions that cause families to thrive and leave more descendants, so that the most useful members of society could be encouraged to adopt such conditions. The main task ahead was to establish eugenics as an academic question, then to bring about consideration of its practical development, and third to introduce eugenics "into the national conscience, like a new religion." He ended by cautioning his audience against too much zeal which could lead to hasty action. A golden age is not round the corner, he warned. Such expectations would lead to discrediting of the science. In the event it took Hitler's treatment of the Jews in World War II to achieve that.

Conclusion
Galton was the confident English gentleman, well aware of the superiority of his nation and his class, condescending to the former colonies, and dedicated to turning back the degeneration of his countrymen. But he disparaged the institution of the aristocracy, rejected the Christian religion, and considered many of our behavioral characteristics as outworn relics from a primitive stage in our social evolution. Although his mathematical skills were limited, his imagination, insight, and inventiveness were remarkable. Allied to his incessant curiosity, these talents made him one of the founders of the statistical revolution that occurred in his lifetime. His book *Natural Inheritance* (1889) proved an inspiration and a turning point in the lives of several of those who became important contributors to the development of biometry, statistics, and evolutionary biology. The imaginative psychological studies that he published in *Inquiries into Human Faculty and its Development* (1883), proved an important influence among psychologists.

Further Reading
Forrest DW (1974) *Francis Galton: The Life and Work of a Victorian Genius*. London: Paul Elek.

Gayon J (1998) *Darwin's Struggle for Survival*, ch. 4. Cambridge: Cambridge University Press.

Pearson K (1914–1930)*The Life, Letters and Labours of Francis Galton*, 3 vols. Cambridge: Cambridge University Press.

Stigler S (1986)*The History of Statistics: The Measurement of Uncertainty before 1900*. Cambridge: Cambridge University Press.

See also: **Darwin, Charles**

Gametes

J R S Fincham

Copyright © 2001 Academic Press
doi: 10.1006/rwgn.2001.0492

Gametes are the haploid cells that fuse in the sexual life cycle to form the diploid zygote. Not all sexual organisms have gametes in the sense of specialized uninucleate cells, but they nevertheless contrive to bring haploid nuclei together for fusion (karyogamy), and we can refer to these as gamete nuclei. Not all gametes and gamete nuclei are differentiated into male and female, and not all are the immediate products of meiosis. In the majority of sexually reproducing organisms other than animals there is usually a haploid phase of mitotic division interposed between meiosis and sexual nuclear fusion, and in ferns, mosses, and liverworts, and most fungi and algae, the haploid phase is free-living.

This article briefly reviews the variations found in different groups of sexually reproducing organisms.

Animals

All groups of animals, other than the unicellular forms (Protozoa), have differentiated female eggs and male spermatozoa, the former contributing both nucleus and cytoplasm to the zygote and the latter little more than the nucleus. Both are the immediate products of meiosis, but whereas all four spermatozoa formed in a sperm mother cell (spermatocyte) are potentially viable, only one nucleus from meiosis in the oocyte survives in the egg, which is generally released from the ovary for fertilization as a free cell. The spermatozoa are each propelled by a single flagellum.

Variations in the Form of Gametes in Other Organisms

All Gametes Motile in some (not all) Algae

In algae we see all the stages in the hypothetical evolution of male and female gametes from the supposed primordial state of gametes of similar form and size. In some green algae, including the unicellular, motile *Chlamydomonas reinhardtii*, the gametes are motile biflagellate cells all the same size, though of two different mating types. In some related species, sexual fusion is between larger and smaller motile cells, which may be called male and female; the female gametes may, as in the colonial genus *Volvox*, lose their flagella and become nonmotile, so becoming more like eggs.

Among the brown algae, some forms, such as the filamentous *Ectocarpus*, have equal-sized biflagellate motile gametes. The large brown seaweeds, exemplified by the genera *Laminaria* and *Fucus* and their allies, have nonmotile free-floating ova and motile sperm (called antherozoids). These genera are predominantly diploid, and the gametes in *Fucus* are the immediate products of meiosis.

In the red algae there is another variation, with the male gametes not motile at all but rather nonmotile spermatia, which are released into the water in great numbers with the object of fusing with female receptive filaments which connect to the ova, which are retained within the female organ rather than allowed to drift.

One well-studied group of Fungi, the Blastocladiales (e.g., *Blastocladiella*, *Allomyces*) can be mentioned here because of their remarkably alga-like mode of reproduction, with motile uniflagellated 'male' and 'female' gametes of different size.

Gametes and Vegetative Cells Interchangeable

In the budding yeasts such as *Saccharomyces cerevisiae*, the haploid products of meiosis are ready to function as gametes immediately, provided that different mating types come together (as they always do in strains with mating-type switching), but if restricted to one mating type they can bud indefinitely as vegetative haploid cells.

Ferns, Mosses, etc.: Male Gametes Motile

In both ferns and mosses, the female gametes are eggs held within female receptive structures (archegonia), while the male gametes are motile sperms, biflagellate in mosses but with many fine cilia in ferns. Two orders of much larger plants, the Cycadales and Ginkgoales, sometimes classified as distantly related to the gymnosperms (pine trees, etc.), also have multiciliated motile male gametes.

Seed Plants: Female Eggs and Male Gamete Nuclei

The two main groups of seed plants, the angiosperms (flowering plants) and gymnosperms have egg cells within their respective female reproductive structures, but, strictly speaking do not have male gametes, in the sense of separate cells, but only gamete nuclei.

In the angiosperms the product of meiosis on the female side is a megaspore, which undergoes haploid mitosis to produce eight nuclei, one of which becomes the nucleus of the egg. On the male side, the pollen grain (microspore) germinates to give a pollen tube which, after two mitotic divisions, contains three haploid nuclei. The pollen tube grows down the style of the flower to the embryo sac, where one of its nuclei fuses with the egg nucleus while another fuses with two other embryo sac nuclei to found the (usually) triploid tissue of the seed endosperm, which has a nutritive function. The pollen tube nucleus that fertilizes the egg can certainly be called a gamete nucleus, and the one that contributes to the endosperm, which has no genetic future, is a gamete nucleus in a more special sense.

The gymnosperms are different in that the haploid tissue derived from the megaspore is much more extensive than in the angiosperms. It includes the endosperm of the seed, which here is purely maternal, and, embedded within it, the archegonia that contain the eggs. The pollen grains make no genetic contribution to the endosperm and, of the few haploid nuclei (usually four) in the pollen tube, only the one that fuses with the egg can be called a gamete nucleus.

Fungi: Gamete Nuclei in Gametangia and Dikaryons

In fungi of the important group Mucorales, which include the bread-mold genus *Mucor*, and *Phycomyces blakesleeanus* (much worked on by Max Delbrück for its response to light) the cells which fuse sexually are called gametangia, borne as club-shaped branches on the filamentous mycelium. They appear to be multinucleate and generally similar in size, though often of different mating types. Cell fusion is followed by nuclear fusion (karyogamy), but whether of one or several pairs of nuclei is not completely clear. The multinucleate gametangia cannot be described as gametes in themselves, but the nuclei that they contain can be termed gamete nuclei.

Most of the fungi that have been used for genetics belong either to the Ascomycetes or the agaric (mushroom) division of the Basidiomycetes. In both of these groups the growth that gives rise to the diploid cells, within which meiosis occurs, is haploid and except in the yeasts, dikaryotic – that is to say consisting of binucleate cells, with the pairs of nuclei dividing in synchrony. In the mushrooms the dikaryon is the major proliferative phase of the life cycle. Haploid basidiospores, the immediate products of meiosis, germinate to give mycelia which remain monokaryotic only so long as it takes them to find another monokaryon of compatible mating type with which they can fuse to form a dikaryon. The dikaryon produces the mushrooms which bear the basidia – specialized cells within which fusion of the mutually compatible nuclei finally takes place, with meiosis following immediately. There are no gametes in the life cycle, but the nuclei fusing within the basidium might be called gamete nuclei, though they are not usually so termed.

In the filamentous Ascomycetes (which include such genetically important genera as *Aspergillus*, *Neurospora*, *Sordaria*, *Podospora*, and *Ascobolus*) dikaryon formation follows the fertilization of female structures (ascogonia), which have receptive filamentous outgrowths called trichogynes. The male fertilizing elements come in various forms: as specialized fertilizing spores (microconidia), as conidia of the same kind as propagate the fungus vegetatively, or as vegetative hyphal tips. Following fusion with a trichogyne, a male nucleus migrates into the ascogonium to establish a dikaryon in partnership with the ascogonial nucleus. The dikaryon proliferates briefly within the developing fruit body (ascogenous hyphae) but soon form ascus initials within each of which a pair of nuclei, the descendants of the original pair, undergo fusion. Meiosis in the ascus follows immediately, with the formation of haploid ascospores. In this system the term gamete, if used at all, should be reserved for the nuclei within the dikaryon which finally fuse, rather than the cells which initiate the dikaryon.

Contrasting Styles in the Protozoa

The Protozoa – single-celled animals – are a vast and diverse group. The ciliates *Paramecium* and *Tetrahymena* are probably the most extensively studied from the genetic point of view.

Paramecium and *Tetrahymena* spp. are diploid organisms, and a cell about to enter into sexual fusion (conjugation) undergoes meiosis; three of the four haploid nuclei degenerate, and the survivor divides once mitotically to give two haploid nuclei. Conjugating pairs of cells remain joined for long enough for one haploid nucleus from each cell to pass into the other, where it fuses with the resident nucleus. This is another example of gamete nuclei, rather than gamete cells.

In a very different protozoan, *Plasmodium falciparum*, the mosquito-transmitted cause of malaria, differentiated male and female gametes are formed in the mosquito; the female gametes are nonmotile spherical cells and the motile male gametes are whip-like, not with distinct head and flagellum as in higher animals.

See also: **Meiotic Product**

Gametes, Mammalian

L Silver

Copyright © 2001 Academic Press
doi: 10.1006/rwgn.2001.0491

Gamete is the general term used to describe the reproductive cells of animals or plants. Thus, in animals, sperm and eggs are both considered gametes.

See also: **Meiosis**

Gametic Disequilibrium

G Thomson

Copyright © 2001 Academic Press
doi: 10.1006/rwgn.2001.0493

Definition and Relationship to Recombination under Neutral Model

The description of genetic variation at the population level usually begins with consideration of allelic variation at a single gene locus. The next step is to consider genetic variation at two or more loci simultaneously, including nonrandom associations. Gametic disequilibrium (also referred to as linkage disequilibrium) describes the nonrandom association of alleles at different genetic loci. The pairwise gametic disequilibrium parameter, usually deCell Cyclenoted D, is given by the difference between the observed frequency of a gametic type and the frequency expected on the basis of random association of alleles in gametes. Gametic disequilibrium can occur in populations as a consequence of mutation, selection, migration or admixture, and random genetic drift. The amount of gametic disequilibrium observed in a population is affected by recombination, selection, nonrandom mating, and the demographics of the population.

Consider two genes (denoted A and B), with two alleles each (A, a and B, b), and four gametic types (also referred to as haplotypes): AB, Ab, aB, and ab. The frequencies of the four gametes (denoted $f(AB)$, etc.) can be described in terms of the allele frequencies p_A ($p_a = 1 - p_A$) and p_B ($p_b = 1 - p_B$) at the two loci, and the gametic disequilibrium parameter D, as follows:

$$f(AB) = p_A p_B + D, \qquad f(Ab) = p_A p_b - D,$$
$$f(aB) = p_a p_B - D, \qquad f(ab) = p_a p_b + D$$

where $D = f(AB) - p_A p_B = f(AB)f(ab) - f(Ab)$ $f(aB)$. If $D = 0$ (a state referred to as gametic or linkage equilibrium) then the alleles at the two loci are randomly associated. If $D > 0$ the allele A occurs more often with allele B than expected by chance (and hence a with b), while if $D < 0$ alleles A and b (and hence a and B) are preferentially associated. The possible value of D in a two locus system is constrained by the fact that the haplotype frequencies must be ≥ 0. The normalized gametic disequilibrium $D' = D/D_{max}$ is often considered, where D_{max} is equal to either the lesser of $p_A p_b$ or $p_a p_B$ if D is positive, and the lesser of $p_A p_B$ or $p_a p_b$ if D is negative. The advantage of this measure over D is that it has a range from -1 to $+1$, regardless of the allele frequencies. The correlation coefficient: $r = D/(p_A p_a p_B p_b)^{1/2}$, with a range from -1 to $+1$, is also often used, as well as r^2 with a range from 0 to $+1$.

It is possible, although relatively rare, that unlinked loci can be in significant gametic disequilibrium, and that very closely linked loci may be in gametic equilibrium. The value of gametic disequilibrium is often expected to change each generation. Further, a population at genetic equilibrium can have significant gametic disequilibrium (e.g., under various selection schemes). The term linkage disequilibrium is thus an unfortunate choice and the term gametic disequilibrium is preferable, although also not perfect; however, the term linkage disequilibrium is commonly used.

There is a relationship between D and the recombination fraction c between two loci. Changes in gamete frequencies over generations only occurs by recombination in individuals heterozygous at both loci under consideration. In fact, the value of D decreases by a fraction $(1-c)$ each generation under random mating and a neutral model. Thus the gametic disequilibrium in this case converges to zero (random association of alleles) with time as $(1-c)^n$, where n is the number of generations. The more loosely linked are two loci, the faster the decay of gametic disequilibrium. On the other hand, for very tightly linked loci gametic disequilibrium may exist for a very long time.

The definition of gametic disequilibrium is easily extended to accommodate more than two alleles at a locus. When considering three or more loci, higher order disequilibrium terms are also needed. For example, in a three-locus system a gametic frequency can be expressed in terms of the three-allele frequencies, the three pairwise gametic disequilibria, and a single measure of third-order gametic disequilibrium.

Evolutionary Forces Creating Gametic Disequilibrium

Historical

When a new mutant arises it occurs in one individual and is in gametic disequilibrium with all polymorphic loci in the population. For example, if the A locus is monomorphic initially (allele A), and the B locus is polymorphic (alleles B and b), then when a new mutant (allele a) arises it will occur on a chromosome carrying either B or b, but not both, so the alleles are nonrandomly associated with $D \neq 0$. Although the absolute value of D is relatively small in this case, the normalized disequilibrium D' is $+1$ or -1. The new allele may increase in frequency due to, for example, genetic drift or selection and although recombination will break down this nonrandom association, significant gametic disequilibrium may be maintained for a long time between very closely linked loci.

Selection (Direct and Hitchhiking)

Selection for different combinations of alleles can produce $D \neq 0$. If the selection is acting directly on the two loci being considered gametic disequilibrium can be maintained in an equilibrium state. While this can apply to unlinked loci, it is expected more often for closely linked loci. Transient gametic disequilibrium can also be created with neutral loci via a hitchhiking event. If an allele, say b, at a neutral locus is in gametic disequilibrium with an allele favored by selection at another locus, say a, then changes in the frequency of the allele b will occur due to this nonrandom association. Such hitchhiking can noticeably increase the absolute value of the gametic disequilibrium, if selection in favor of the new mutant is greater than the recombination rate between the neutral and selected loci. Further, gametic disequilibrium can be generated between two neutral loci via hitchhiking at a third closely linked locus. Gametic associations built up via hitchhiking events are expected to decline in strength as recombination breaks up haplotypes bearing the selected allele.

Migration or Admixture

Mixing of two genetically different populations can create gametic disequilibrium. As an extreme consider two populations – one monomorphic for alleles A and B, the other monomorphic for alleles a and b, so initially when the populations are mixed there are only two gametic types AB and ab, hence $D \neq 0$. For pairwise gametic disequilibrium to be generated by migration or admixture, the allelic frequencies of both loci in the two populations must be different, and the difference in allele frequencies must be substantial in order to generate very much gametic disequilibrium. Again recombination will break down this association over time.

Finite Population Size

Genetic drift can cause nonrandom associations between alleles at different loci. While the expected value of pairwise gametic disequilibrium due to drift over many generations is zero, the variance is large for closely linked loci in small populations. The demographic structure of a population will affect the amount of gametic disequilibrium observed. A small founder population or a bottleneck in the recent past can cause significant gametic disequilibrium for closely linked loci. While less gametic disequilibrium will be generated by genetic drift in a rapidly growing population, gametic disequilibrium present before or during the early phase of the expansion will persist.

Nonrandom Mating

The mating or reproductive system can retard the rate of approach to random allelic association. For example, a high level of self-fertilization leads to a reduction in the proportion of double heterozygotes, from which recombinants are subsequently formed, and hence retards the decay to gametic equilibrium.

Population Level Observations

Significant nonrandom association (gametic disequilibrium) between alleles at two loci can be tested using the chi-square (χ^2) test and the Fisher's exact test on the contingency table of gametic types. Algorithms are available to perform Fisher's exact test and Monte Carlo methods for approximating the results for the exact test so that examples where the expected numbers of some gametic types are small can be considered. While genetic drift and demographic effects should be randomly distributed over the genome, the effects of natural selection are expected to be nonrandomly distributed.

General observations are that there is an overall proportionality between gametic disequilibrium and the inverse of the recombination distance, although this breaks down in very closely linked regions. Gametic disequilibrium is nonrandomly distributed throughout the genome. Some regions, such as the immune response human leukocyte antigen (HLA) system on chromosome 6, show strong evidence of selection and significant gametic disequilibrium which may span 3 centimorgans (cM) or more.

Disequilibrium (Association) Mapping of Disease Loci

The existence of gametic disequilibrium has been a very powerful tool in mapping over 200 diseases to the HLA region. An increased frequency of an HLA antigen (allele) in patients over that in an ethnically matched control population is inferred to be due either to the direct effect of the HLA antigen itself on disease, or to gametic disequilibrium (association) of the HLA allele with the actual disease-causing allele at a separate locus. Stratification analyses can be used to distinguish between these two possibilities.

For monogenic traits, most disease genes mapped to date show gametic disequilibrium with markers sufficiently close to the disease gene, 0.5 cM or more, e.g., cystic fibrosis, Huntington disease, Wilson disease, Batten disease, Friedreich ataxia, myotonic dystrophy, torsion dystonia, hemochromatosis, diastrophic dysplasia, adult onset polycystic kidney disease, and many others. The familial breast cancer gene *BRCA1* is an exception to this rule; gametic disequilibrium is not seen with closely linked markers since each family usually has a unique mutation.

For complex diseases involving multiple loci, incomplete penetrance, and genetic heterogeneity, association mapping has been successfully applied in the study of candidate regions. The greater number of markers needed for an association genome scan to detect disease-predisposing genes compared with standard linkage analysis techniques (LOD score analysis and affected sib pair methods) has prevented their wide-scale implementation to date. However, the use of DNA pooling for the study of microsatellite variation in patients and controls, and the current development of DNA chip technology for the study of single nucleotide polymorphisms (SNPs), has opened the way for future routine disequilibrium mapping of disease genes.

See also: Linkage Disequilibrium; Linkage Map

Gametogenesis

J Hodgkin

doi: 10.1006/rwgn.2001.0494

Gametogenesis is the process leading to the production of specialized reproductive cell types, either eggs or sperm, collectively known as gametes. It entails meiotic division, to generate haploid cells, together with maturation into the appropriate functional gamete.

See also: Oogenesis in *Caenorhabditis elegans*; Oogenesis, Mouse; Spermatogenesis in *Caenorhabditis elegans*; Spermatogenesis, Mouse

Gamma Distribution

N Saitou

doi: 10.1006/rwgn.2001.0495

Gamma distribution is based on gamma function $\Gamma(a)$, so first this function must be explained. Gamma function $\Gamma(a)$ is defined as $\int_0^\infty e^{-t} t^{a-1} dt$. This function has various interesting properties. For example, $\Gamma(a + 1) = a\Gamma(a)$. Therefore, when variable a is integer $\Gamma(n+1) = n_!$. Hence, gamma function is also called 'factorial' function.

Gamma distribution $f(r)$ is defined as $[b^a/\Gamma(a)]e^{-br}r^{a-1}$, where $a = \text{mean}(r)^2/\text{var}(r)$ and $b = \text{mean}(r)/\text{var}(r)$. Mean($r$) and var($r$) are mean and variance of variable r, respectively. Shape of the gamma distribution $f(r)$ is determined by a, while b is a scaling factor. The gamma distribution is known to be very flexible and takes various shapes depending on the value of a. Therefore, variable a is often called a 'shape parameter' of gamma distribution. When a is small, its distribution is skewed to the left. When a is infinite, r takes only one value (Dirac's delta function).

In molecular evolutionary studies, this gamma distribution is sometimes used when a certain distribution is empirically known to be quite heterogeneous. For example, evolutionary rate of amino acid or nucleotide substitution is often assumed to be constant for every site in simple evolutionary models. However, the rate varies greatly in reality. In this case, the gamma distribution may be used. If we can estimate the value of shape parameter a, application of the gamma distribution is possible. Evolutionary distance thus estimated is often called 'gamma distance.'

See also: Evolutionary Rate

GAP (*RAS* GTPase Activating Protein)

A C Lloyd

doi: 10.1006/rwgn.2001.1574

Ras-GAP is a 120-kDa, ubiquitously expressed cytosolic protein. It was initially identified as an activity in cell extracts able to stimulate the intrinsic GTPase activity of p21Ras (Trahey and McCormick, 1987).

The catalytic GAP activity resides in the carboxy region of the protein and acts as a negative regulator of Ras signaling, modulating the levels of Ras-GTP (van der Geer *et al.*, 1997). There is increasing evidence for functions independent of GAP activity via the N-terminus that consists of two SH2 domains flanking an SH3 domain which mediate interactions with other cellular proteins such as p190 and p62 (Kulkarni *et al.*, 2000). A central region contains a plekstrin homology (PH) domain, and a CaLB domain thought to be important in regulating membrane interactions. Mice homozygous for a mutant GAP allele die at about day 10.5 of embryogenesis displaying a variety of defects including vascular abnormalities and increased apoptosis in the nervous system.

References

Kulkarni SV, Gish G, van der Geer P, Henkemeyer M and Pawson T (2000) Role of p120 Ras-GAP in directed cell movement. *Journal of Cell Biology* 149: 457–470.

Trahey M and McCormick F (1987) A cytoplasmic protein stimulates normal N-ras p21 GTPase, but does not effect oncogoenic mutants. *Science* 238: 542–545.

van der Geer P, Henkemeyer M, Jacks T and Pawson T (1997) Aberrant Ras regulation and reduced p190 tyrosine phosphorylation in cell lacking p120-Gap. *Molecular and Cell Biology* 17: 1840–1847.

See also: **Ras Gene Family**

Gastrulation

See: **Developmental Genetics**

Gaucher's Disease

T M Cox

Copyright © 2001 Academic Press
doi: 10.1006/rwgn.2001.0498

Frequency of Gaucher's Disease

The overall frequency of lysosomal diseases in the general population worldwide is estimated to be about 1 in 5000 live births of which Gaucher's disease is the most common having an estimated frequency of 1 in 50 000 to 60 000 live births. In selected populations the frequency appears to be much greater and although the predicted frequency of Gaucher's disease in the Ashkenazi population is unknown, homozygotes for the N370S mutation and compound heterozygotes with the N370S 84GG genotype would occur in an overall frequency of about 1 per 855 individuals in the population at large.

Definition

Gaucher's disease is a multisystem disorder principally affecting macrophages and classified in the Online Mendelian Inheritance in Man (OMIM) website as OMIM 23080, 23091, and 23100. It is a prototype of the glycosphingolipidoses, an important group of lysosomal disorders characterized by deficiency of specific acid hydrolases responsible for the degradation of complex membrane glycolipids.

Gaucher's disease is caused by a recessively inherited deficiency of an acid β-glucosidase, glucocerebrosidase (EC.3.2.1.45). As a consequence of this deficiency N-acyl-sphingosyl-1-0-β-D-glucoside and other minor glycolipid metabolites such as glucosylsphingosine accumulate. All the accumulated glycolipids represent metabolic intermediates derived from the cellular turnover of membrane lipid macromolecules of the ganglioside and globoside classes.

Genetics

The human glucocerebrosidase locus has been mapped to chromosome 1q21 where it is found in close proximity to a nonprocessed cognate pseudogene which is absent in several other vertebrates. The human acid-β-glucosidase gene is also found in proximity to two other genes, metaxin and thrombospondin 3. Expression of mRNA encoding human acid-β-glucosidase is constitutive in nearly all cells but varies in abundance. In the 5′ untranslated region of the functional acid-β-glucosidase gene in humans there are two CAAT boxes and two sequences encoding putative CAAT boxes. Promoter/expression studies have revealed several transcription factors, octamer-binding transcription factor 1 (OCT binding protein), oncogene Jun activator protein-1 (AP-1), ets-related transcription factors: Polyomavirus Enhanced Activator-3 (PEA3), and the CAAT binding protein). These and other transcription factors yet to be characterized indicate that the expression acid-β-glucosidase may be regulated by transcriptional activation associated with proliferative cell responses.

Clinical Spectrum of Gaucher's Disease

Gaucher's disease may be associated solely with systemic (nonneuronopathic) or with neuronopathic (neurological) features. In the nonneuronopathic (type 1) form of Gaucher's disease partial enzymatic deficiency of acid-β-glucosidase is associated with the accumulation of glycolipids in macrophages that belong to the mononuclear phagocyte system located principally in the liver, bone marrow, and spleen. Pathological macrophages containing excess lysosomal stored lipid may also be found within the lung and, on rare occasions, pericardium and kidney. In the neuronopathic forms of Gaucher's disease (type 2 and type 3) severe deficiency of glucocerebrosidase caused by disabling or inactivating mutations is additionally associated with disease of the nervous system. The pathogenesis of neuronopathic Gaucher's disease is complex but in many instances failure to degrade endogenous glycosphingolipids present in brain tissue is a contributing factor although the accumulation of Gaucher's cells around adventitial spaces in cerebral blood vessels as a result of uptake of circulating glucosylceramide present in plasma may be a contributory factor.

Genotype–Phenotype Correlations

In a particular variant of neuronopathic Gaucher's disease described in Arabic, Japanese, and Spanish populations, an intermediate phenotype associated with corneal opacities and endocardial thickening with mitral and aortic valve disease of the heart has been identified and related to homozygosity for a particular missense mutation D409H in the glucocerebrosidase gene. Other genotype/phenotype correlations are less close but of the common widespread mutations L444P is associated with disease severity and has been described in all three phenotypes of Gaucher's disease. Homozygosity for L444P is a significant cause of neuronopathic Gaucher's disease in the so-called Swedish or Norrbottnian variant with slowly progressive neurological symptoms associated with survival to adult life. In type 2 Gaucher's disease the rapid onset of bulbar paresis, spastic paraparesis, and opisthotonus with swallowing difficulties is noted within the first few months of life and survival beyond the first few years of life is very unusual.

A recently recognized and rare variant of Gaucher's disease is associated with premature fetal loss and stillbirth as well as infants with a desquamating and dehydrating skin lesion that die shortly after birth of dehydration. This condition associated frequently with an abnormal appearance of the dermis 'collodion' is associated with severely inactivating lesions in the human glucocerebrosidase gene and has parallels with the short-lived lethal murine glucocerebrosidase deficiency state generated by targeted disruption of the glucocerebrosidase gene in embryonic stem cells. Homozygous animals die within 24 h of birth and although scant Gaucher's cells are present within systemic organs and excess glucosyl ceramide accumulates, the principal cause of death appears to be skin desquamation and dehydration. Ceramides released by the action of acid-β-glucosidase appear to be essential for the maintenance of dermal integrity and the prevention of water loss.

Several mutations in the glucocerebrosidase gene appear to result from genetic rearrangements between the functional and human pseudogenes as a result of gene conversion or recombination events. This leads to the transfer of multiple missense or point mutations in the presence of the closely related glucocerebrosidase pseudogene that may create difficulties for facile detection and more precise identification of causal mutations. Definitive genomic sequencing, cDNA sequencing procedures are recommended. Apart from the widely distributed L444P allele, the N370S allele harboring a missense mutation is widespread in several populations and may have occurred on a background of several haplotypes. This mutation appears to be associated with the presence of only a mild catalytic impairment of the cognate enzyme polypeptide.

The presence of at least one copy of the N370S allele militates against the occurrence of neuronopathic Gaucher's disease. Several mutations, and in particular the N370S mutation, as well as the 84GG mutation, occur with particular frequency in the Ashkenazi Jewish population. Population studies reveal diverse phenotypes associated with homozygosity for N370S but the N370S mutation is widespread in populations throughout the world including South America, Spain, and Portugal and in patients with no known Ashkenazi ancestry. The high gene frequency for N370S and 84GG have been estimated to be approximately 0.03 and 0.002 in the Ashkenazi population, respectively. The basis for this high allele frequency has not been fully explained. The operation of selective evolutionary pressure has been postulated. It has been suggested that homozygotes or heterozygotes for the N370S mutation may have constitutive activation of macrophages in target organs such as the spleen that would confer resistance against infection with pathogenic microorganisms – particularly tuberculosis. No experimental evidence to support this speculation has been yet provided.

Clinical Presentation

Symptoms of type 1 Gaucher's disease usually result from the presence of splenic enlargement and either the enlarged viscera are noted by the patient or the consequences of hypersplenism (anemia, thrombocytopenia, or leukopenia) declare themselves by the occurrence of spontaneous bruising or unexplained sepsis. Abnormal blood counts combined with enlargement of the liver and spleen (hepatosplenomegaly) may lead ultimately to bone marrow examination or tissue biopsy that may reveal the presence of the characteristic Gaucher's cells. Bone marrow biopsy or tissue biopsy is no longer necessary for the diagnosis, however, which can be easily made by enzymatic assay of circulating leucocytes using fluorescent substrates to reveal a profound deficiency of acid-β-glucosidase in affected homozygotes. Retrospective enquiry may reveal a history of bone pains attributable to the so-called bone infarction crises resulting from marrow infiltration particularly in regions and occurring particularly at the growing ends of long bones (epiphyses). A prior diagnosis of Perthe's disease is common. Pallor, fatigue, and palpitations often presage the diagnosis. Occasionally, massive enlargement of the liver and spleen occurs in infancy. Patients with neurological disease may present at any age. In the more indolent type 3 neuronopathic forms natural gaze, disturbances of vertical gaze, result from neuronophagia and other localized injury within the nuclei of the brain stem that are key to control of conjugate eye movements. Later, ataxia, mild spasticity, myoclonic or complex epilepsy, and slowly progressive dementia may become clear. These patients have a degree of systemic involvement with hepatosplenomegaly and bone marrow infiltration that is very variable ranging from massive hepatosplenomegaly with a bleeding tendency and gross abdominal swelling to only subtle enlargement of the liver and spleen detectable by ultrasonic examination.

In the acute neuronopathic variant, type 2, difficulty in swallowing, paralytic squint, and persistent hyperextension of the head is common followed by spasm of the jaw (trismus), generalized spasticity, and psychomotor retardation; respiratory obstruction due to laryngospasm also occurs with aspiration pneumonia, myoclonus, and generalized seizures in the late stages of the illness.

Treatment of Gaucher's Disease

As emphasized earlier, Gaucher's disease is in many respects the prototypic lysosomal disorder. It affects all ages and as the most common lysosomal disorder has been subject to intensive investigation of its biochemistry and genetics – and of definitive methods for therapy.

Marrow Transplantation

Because macrophages, the principal focus of Gaucher's disease, are derived from granulocyte–monocyte progenitor cells in the bone marrow, it was likely that bone-marrow transplantation would provide a population of cells competent in the degradation of glycosphingolipids and thereby correct the

defect. Marrow transplantation has been successfully carried out in infants, children, and young adults with Gaucher's disease. The donors have been HLA-matched sibling donors either normal or heterozygous for the glucocerebrosidase defect. Successful engraftment of the bone marrow stem cells has been associated with clinical regression of the disease with catch-up growth in stunted children and, ultimately, almost complete disappearance of the pathological storage cells in the tissues including the liver. Although only a minority of patients with Gaucher's disease will be suitable candidates for bone-marrow transplantation, particularly with the emergence of enzyme replacement therapy (see below), its success in eradicating the disease demonstrates that a complement of tissue macrophages derived from the bone marrow with at least 50% of normal β-glucosidase activity is sufficient to correct the nonneuronopathic manifestations of this systemic disease.

Enzyme Replacement Therapy

Early studies using preparations of human glucocerebrosidase prepared from placental tissue were conducted at the National Institutes of Health by Roscoe Brady and colleagues. Infusion of the native protein was associated with a reduction of erythrocyte and plasma glucocerebroside over a few days. No convincing clinical improvement was demonstrated. However, since the pioneering discovery of the lysosome and its access to the aqueous phase by Christian de Duve and contemporaneous studies on the uptake of glycoproteins by parenchymal and nonparenchymal hepatic cells, it was considered that native human glucocerebrosidase may lack the critical recognition signals for uptake and delivery to the disease macrophage of Gaucher's tissue. With the identification of a mannose receptor on the surface of macrophages and the preferential uptake of mannosylated proteins by human alveolar macrophages, experiments were undertaken to modify the terminal carbohydrate residues of placental glucocerebrosidase by sequential enzymatic deglycosylation. Mannose-terminated preparations of human glucocerebrosidase were then shown to be taken up preferentially by nonparenchymal (Kupffer-cell-rich) rather than parenchymal hepatic cells in rats and prompted further studies of enzyme replacement therapy in patients with Gaucher's disease.

Early clinical trials of mannosylated human placental glucocerebrosidase (alglucerase) showed rapid regression of symptoms and visceromegaly with improvement and blood counts and other parameters of Gaucher's disease activity. The preparation secured approval as an Orphan Drug under the Food and Drug Administration of the USA in 1990.

With advances in genetics and the cloning of the human glucocerebrosidase gene by several groups the development of a recombinant enzyme replacement strategy was a key element of pharmaceutical investment by the Genzyme Company, the commercial partners in this pioneering work. Recombinant human glucocerebrosidase, imiglucerase (Cerezyme™), is now produced by Genzyme as a recombinant product purified from Chinese hamster ovary cells transfected with the human glucocerebrosidase gene. Many thousands of patients worldwide with Gaucher's disease are now able to receive this agent, which also appear to relieve some aspects of the mild neuronopathic (type 3) forms of Gaucher's disease. Immunological and sensitivity reactions to the infusions are rare and this is accounted for by the observation that most patients with Gaucher's disease harbor mutations that allow expression of residual glucocerebrosidase polypeptide antigens.

With the commercial success of enzyme therapy in Gaucher's disease preparations to treat other lysosomal disorders such as Fabry's disease (an X-linked endothelial disorder with principal effects on the heart, peripheral nerves, and kidneys) and MPS-1 (Hurler's and Scheie's diseases) are now available and have emerged as effective agents from clinical trials. An interesting further development has been that of recombinant human acid-glucosidase, the enzyme deficient in glycogenosis type II (Pompe's disease) which has been prepared as a recombinant product with appropriate mannose-6-phosphate terminal residues for uptake by skeletal and heart muscle in this classical lysosomal disorder – the first so to be recognized (by Henri-Gery Hers).

Other Therapeutic Opportunities

Although enzyme therapy has proved to be effective, clearly the generation of recombinant protein with appropriate safety and stability profiles for human use is expensive, and research continues for alternative methods for controlling lysosomal disease such as Gaucher's disease by other means. Recently, the concept of substrate depletion to prevent the accumulation of glycosphingolipids by inhibiting their biosynthesis that was first suggested by Norman Radin has been developed by Frances Platt and Terry Butters in the Glycobiology Institute in Oxford. This proposal followed their discovery that a derivative of N-butyl deoxynojirimycin and N-butyl deoxygalactonojirimycin selectively inhibit the glucosyl transferase step in the biosynthesis of glycolipids without affecting glucocerebrosidase and other acid glucosidases. The administration of these iminosugars to genetically modified animals that represent experimental models of the debilitating glycosphingolipidoses such as Tay–Sachs disease and Sandhoff disease have shown reduced glycolipid storage with partial delay or arrest of the ineluctable progression of these lysosomal storage diseases affecting brain tissue.

Since N-butyl deoxynojirimycin had been previously used in clinical trials in an attempt to arrest the proliferation of human immunodeficiency virus (HIV), and was shown not to have major human toxicity, a clinical trial of substrate depletion with this agent was undertaken in Gaucher's disease. This open-labeled clinical trial demonstrated slow regression of the major disease parameters of Gaucher's disease including organomegaly and surrogate disease markers of Gaucher's activity together with slow but steady and statistically significant improvement in blood counts. Further trials are in progress to determine whether or not substrate depletion therapy can synergize with enzyme replacement therapy and whether the medication will have any therapeutic value for patients suffering from the otherwise intractable neuronopathic forms of this disorder.

Gene Therapy

Since Gaucher's disease can be corrected by transplantation of allogenic bone marrow providing a source of granulocyte–monocyte progenitor cells, the possibility of gene therapy directed toward hematopoietic stem cells is raised. Several trials have been approved in the USA for the genetic transduction of CD34[+] hemopoietic stem cells that have been therapeutically corrected by transfer of the human glucocerebrosidase gene in retroviral vectors. This approach has already been successful in normal mice where prolonged expression of human glucocerebrosidase at a high level has been achieved in the macrophages of mice that have received primary and secondary marrow transplants. At present it is not clear how in humans transfected cells would have a selective advantage for survival and to populate the entire bone marrow that is diseased in Gaucher's disease thereby providing long-term remission of glycolipid storage by the metabolism of endogenous glucocerebroside. However, high efficiency vectors for long-term expression in grafted autologous cells are currently being studied to secure corrective expression with the wild-type glucocerebrosidase gene. The means to secure a selective advantage within the marrow population continues to be explored actively.

Genetic Studies of Pathophysiology

The Gaucher's cell, a pathological macrophage, is a striking feature of Gaucher's disease but the connection between the pathological storage of glycosphingolipid and the diverse manifestations of the disease remain unexplained. Gaucher's disease is accompanied by weight loss, fatigue, increased metabolic rate, sustained acute inflammatory reaction with B-cell proliferative responses as well as massive enlargement of the spleen and liver and in tissue destruction in the bone, lung, liver, and brain stem. Although the visceral organs may enlarge 50–80-fold pathological lipid that accumulates within the tissues accounts for less than 2% of the additional tissue mass. Thus the link between the macrophage abnormality and the complex phenotype that characterizes Gaucher's disease and other lysosomal disorders due to glycolipid activation remains unknown. Studies are under way to understand better the pathogenesis of Gaucher's disease and related glycolipid disorders. And clearly cDNA microarray analysis would offer the chance of a cluster analysis of genes upregulated and downregulated as part of the cellular response to the presence of stored glycolipid. Recent studies have been reported by the author's group based on the polymerase chain reaction to identify genes whose transcriptional products are increased in Gaucher's disease tissue as a first step toward understanding the pathogenesis of this condition and opening up new avenues of therapy. Several genes including those encoding for chemokine and three lysosomal cysteine proteinases which are known to participate in tissue modeling antigen presentation and bone matrix destruction, respectively, were shown to be upregulated in Gaucher's disease tissue. The proteinases were present also in excess in the plasma and serum of affected patients. Expression of several proteinases appear to be correlated with Gaucher's disease activity and severity score indices and serum levels of the cysteine proteases decreased upon reduction of Gaucher's disease activity with enzyme replacement treatment.

Thus the study of the secondary genetic abnormalities in the lysosomal disease, such as Gaucher's disease, may prove to be revealing to identify the pathological cascades that are activated as a result of abnormal lipid storage and may ultimately provide avenues for additional therapy. Proinflammatory cytokine pathways such as that mediated by interleukin 6 have been implicated in Gaucher's disease. Since this cytokine influences gene expression of several cathepsins and has been shown to be increased in the serum of patients with Gaucher's disease it may thus represent one critical triggering factor for disease activation. The identification of increased expression of the cathepsin K proteinase with preferential activity against collagen 1, the principal bone matrix protein, is also of significance and provides an example of how a new candidate for therapeutic attack can emerge from the genetic study of Gaucher's disease. Specific inhibitors of cathepsin K have been developed for

pharmaceutical use for the treatment of metabolic bone diseases including osteoporosis: enhanced cathepsin K expression associated with active Gaucher's disease and lytic bone lesions immediately suggest the potential for the use of selective cathepsin K inhibitors for those patients afflicted.

With the introduction of cluster analysis of pathological gene expression profiling and systematic proteome analysis further opportunities for studying the pathogenesis of Gaucher's disease and related glycosphingolipid disorders will undoubtedly come to light. From every aspect, therefore, Gaucher's disease represents a landmark condition as a prototype for the glycosphingolipid storage disorders and provides a vivid example of many productive interactions between clinical, biochemical, and genetical research.

Further Reading
Online Mendelian Inheritance in Man (OMIM) http://www3.ncbi.nlm.nih.gov/Omim/

See also: Fabry Disease (α-Galactosidase A Deficiency); Hurler Syndrome; Tay–Sachs Disease

G-Banding

G-banding (Giemsa banding) is a technique that generates a banded pattern in metaphase chromosomes, thus allowing identification of the separate chromosomes. It involves brief treatment with protease and staining with Giemsa.

See also: Giemsa Banding, Mouse Chromosomes

Gel Electrophoresis
B A Roe

Electrophoresis, initially described by Arne Tiselius in 1937, is the process by which charged particles move through a media in the presence of an electric field at a given pH. The charged particles move at a constant velocity. The electric force (Eq) is equal to the frictional force or viscous drag (fv), as defined by the relationship:

$$Eq = fv$$

where E is electric field strength (volts per centimeter), q is net charge of the particle (electrostatic units), f is frictional coefficient (a function of the size and shape of the particle), and v is velocity of the particle (centimeters per second).

Ohm's Law states that voltage (V) and current (I) are related by the relationship:

$$V = IR$$

The electric field strength (E) is defined either by the voltage (V) or the current (I), one of which typically is held constant. Therefore, the velocity (v) of a particle in an electric field is defined by the relationship:

$$v = Eq/f$$

If the electric field strength (E) is kept constant, then the velocity (v) of the particle depends on its net charge (q) and its frictional coefficient (f). The frictional coefficient is directly proportional to the particle's Stokes radius, i.e., the radius of a spherical particle with equivalent hydrodynamic properties.

In practical terms, if a mixture of different proteins or varying-sized nucleic acids are electrophoresed, generally the higher molecular weight proteins or nucleic acids will have larger effective diameters, a higher frictional coefficient, and travel slower than lower molecular weight proteins or nucleic acids. The distance traveled then is approximately inversely proportional to the log of the molecular weight of the particle. There are exceptions, especially in the case of very small polypeptides and oligonucleotides or very large proteins or nucleic acids. Also, if a protein is rich in proline or charged amino acids the solution structure will be distorted. Such proteins will have an electrophoretic mobility different than that predicted by its molecular weight. Similarly, double-stranded DNA also can have a supercoiled circle, a relaxed circle, or a linear structure. Although all three forms would have the same molecular weight, each species would have a different electrophoretic mobility on the same agarose gel.

The major electrophoretic media and apparatus used today for protein separations is polyacrylamide gel electrophoresis (PAGE) on 'slab' gels. If the separation voltage for the PAGE is kept constant, based on the preceding discussion of Stokes radius and frictional coefficient, the major factors that affect the separation of proteins with varying molecular weights are the pore size and amount of cross-linking of the gel media, but typically a 6% PAGE gel will resolve almost all proteins in the 10 000 to 100 000 molecular weight range. There are two types of PAGE gels, a 'native' gel and a 'denaturing' gel. A native gel will separate proteins that are monomers from dimers, from tetramers, etc. For example, hemoglobin, which has two copies of two identical subunits and the structure $\alpha_2\beta_2$, will be resolved as a single band on a native gel. However, a denaturing gel typically contains the detergent sodium dodecyl sulfate (SDA). Prior to loading, the protein mixture usually is heated in the presence of a reducing agent such as b-mercaptoethanol and a chelating agent such as EDTA to disrupt any subunits. Thus, hemoglobin, with its $\alpha_2\beta_2$ structure, will be resolved into two bands on a denaturing gel, where the faster moving band is the smaller α-subunit, and the slower moving band is the larger β-subunit. Proteins are detected on both native and denaturing gels by staining either with methylene blue or the more sensitive silver stain.

Nucleic acid electrophoretic media typically is either a polyacrylamide gel for nucleic acids that are shorter than 1000 bases, or an agarose gel for nucleic acids that contain more than several hundred or several thousand bases but less than a few hundred thousand bases. In both instances, the nucleic acid can be either single-stranded or double-stranded. The pore sizes for the electrophoretic media is adjusted by varying the percentage of media and the amount of cross-linking. For example, a 0.8% or 1% agarose gel, which can separate larger nucleic acids, would have much larger pore size and less cross-linking than a 4% or 6% polyacrylamide gel used to resolve smaller nucleic acids. Extremely large nucleic acids that contain more than several hundred thousand bases can be resolved on pulse-field agarose gels. Large nucleic acids such as plasmids, cosmids, and restriction endonuclease-digested DNA usually are separated on agarose gels, and the DNA bands are detected by ethidium bromide staining. Mixtures of nucleic acids shorter than 1000 bases such as DNA-sequencing reaction nested fragment sets or multiple restriction endonuclease-digested DNA often are separated on polyacrylamide gels. Here the nucleic acids are either fluorescently or radioactively labeled. The fluorescent-labeled nucleic acids can be detected by a photomultiplier tube or CCD camera after laser activation of the associated fluorescent dye. Alternatively, radioactivity-labeled nucleic acids can be detected by direct exposure to X-ray film, or the gel can be sliced and the radioactivity measured in a liquid scintillation counter.

More recently, slab gels have begun to give way to capillary electrophoretic gels, which are much thinner, contain less media, require protein or nucleic acid samples that are several orders of magnitude lower, and resolve the samples in minutes rather than hours.

The capillary electrophoresis instrument also is coupled with automated detection equipment and an associated computer, on which the results can be stored for further analysis. Various media have been described which can provide single-base resolution of either single- or double-stranded nucleic acids as large as 500–1000 bases. These media include linear polyacrylamide, methyl cellulose, hydroxyethyl cellulose either alone or mixed with polyethylene oxide, and hydroxypropyl cellulose either alone or mixed with polyethylene oxide. The capillaries are either coated with a siliconizing reagent to reduce the charges on the glass capillary or used directly without coating. Capillary electrophoretic-based instrumentation now is quite robust and has gained wide acceptance in both the gene-mapping and DNA-sequencing communities. In the case of protein separation by capillary gel electrophoresis, media similar to that used in nucleic acid separations have been described. However, coated capillaries almost always are required because of the greater tendency for proteins to bind to the capillaries and thereby cause altered observed electrophoretic mobility and irreproducible quantitation of any resolved samples.

There are numerous manufacturers and resellers of electrophoretic equipment that range from simple but effective Plexiglas acrylamide apparatus and power supplies to self-contained PAGE gel or capillary electrophoresis instruments. These suppliers also include detailed protocols for optimal use of their apparatus or instrumentation that are easy to follow and typically yield reproducible results. Finally, the electrophoresis literature is extensive and provides many detailed procedures. However, the reader is encouraged to investigate the following two books: *Molecular Cloning: A Laboratory Manual* (Sambrook and Russell, 2000); and *Proteins* (Walker, 1984), as they provide an almost complete review of the existing literature.

References

Deyl Z, Chramhach A, Everaeı ls EM and Prusik Z (eds) (1983) Electrophoresis, a survey of techniques and application. *Journal of Chromatography A* 18: 390.

Deyl Z, Miksik I, Tagliaro F and Tesarova E (1998) Advanced chromatographic and electromigration methods in biosciences. *Journal of Chromatography A* 60: 1091.

Heftman E. (ed.) (1983) *Chromatography: fundamentals and application of chromatographic and electrophoretic methods. Journal of Chromatography A* 22: 331.

Sambrook J and Russell D (2000) *Molecular Cloning: A Laboratory Manual*, 3rd edn. Plainview, NY: Cold Spring Harbor Laboratory Press.

Walker JM (ed.) (1984) *Methods in Molecular Biology*, vol. 1, *Proteins*. Clifton, NJ: Humana Press.

***See also:* Proteins and Protein Structure; Pulsed Field Gel Electrophoresis (PFGE)**

Gene

J Merriam

Copyright © 2001 Academic Press
doi: 10.1006/rwgn.2001.0500

The gene is the unit of heredity. While this definition may seem to stand on its own and not need further explanation, in fact it represents a continuing evolution in the way we view the biological process of inheritance. Mendel's view of the gene (the mechanism of inheritance) was an important conceptual change from the established view in the nineteenth century. Darwin, the establishment figure, promulgated the idea, first voiced by Hippocrates (400 BCE), that inheritance derived from miniature body parts or characters transmitted through copulation. Darwin's theory of 'pangenesis' saw semen as being replenished by 'gemmules' derived from all the somatic tissues of the body. Mendel's explanation instead makes it clear that it is information about the characters rather than the characters themselves that are transmitted.

Mendel recognized the impossibility of a pangenesis-like model to explain his experimental observations as well as those of earlier plant hybridizers. They saw that recessive traits could be carried unchanged through several generations, that the traits could reappear by the F_2 generation, and that the recessive homozygotes extracted from such crosses could form pure breeding stocks indistinguishable from the original parental strains. Mendel's explanation, with genes present in pairs that segregate during gamete formation, is the basis of the present science of genetics. The coupling of the idea of single gene inheritance with differences for individual traits was the breakthrough that unified biology in its disciplines ranging from evolution to physiological function.

After the rediscovery of Mendel's work in 1900, Bateson introduced many of the terms in current usage: genetic, zygote, homozygote, heterozygote, allelomorph (later shortened to allele), and F_1 and F_2 generations. Mendel's term 'Merkmal' was translated as either character, unit character, or factor (Bateson's choice). Johannsen proposed using the word gene in 1909.

Mendel introduced a convenient symbolism to describe genetic relations between parents and their offspring. One character could abstractly be referred to as A, another as B, and so forth. He used the upper case letter to indicate which was the dominant trait and the lower-case letter for the recessive trait. Subsequently, workers extended this to name the gene after the character, initially describing the dominant trait but subsequently shifting to usually naming genes after recessive traits. The reason for the shift is that most genes were identified by recessive mutations that departed from the 'wild-type' standard appearance. It was understood that each gene so identified had two alleles, one mutant and the other representing the wild-type.

The discovery of multiple alleles called into question what is meant by the gene. Two eminent geneticists, A. H. Sturtevant and G. W. Beadle, who had coauthored a textbook on genetics, discovered that each had used the term 'gene' differently. The white gene to Sturtevant was the specific white mutant but to Beadle it represented the group of white alleles including the wild-type allele. While most geneticists today follow Beadle's usage, medical geneticists frequently refer to those mutations associated with genetic diseases as genes, e.g., the Duchenne muscular dystrophy gene or the Huntington disease gene, without mentioning the normal alleles. Many geneticists, however, are adopting the compromise language of referring to the Duchenne muscular dystrophy gene 'mutation' or the Huntington disease gene 'mutation' to distinguish between gene and allele.

By this definition it is alleles, not genes, that are observed to be the units of segregation. Allelism is also conferred from the shared properties of similar mutant phenotypes and the failure to complement other mutant alleles. Alleles will almost invariably segregate from each other in *trans* heterozygotes, at least in multicellular organisms having low levels of recombination. Rarely is a wild-type recombinant progeny observed to indicate that those parental alleles cannot be mutant in the same location. Instead they are described as 'pseudoalleles,' at different sites but still regarded as marking the same gene. Other examples pose more of a challenge to the idea of one gene, however. In the phage T4, mutants of the rII class, named after their phenotype, map to one section of the linkage group in a cluster that is over seven map units long! Do rII mutants mark one gene? Employing what he called the *cis–trans* test, Seymour Benzer described the mutants in the cluster as forming two complementation groups, or cistrons. He showed that mutants in one cistron localized by recombination studies to one side of the cluster and mutants of the other cistron localized to the other side. Moreover,

Benzer showed that the T4 linkage group is linear both inside and outside the rII cluster. Mutation positions are continuously distributed along this line with no obvious demarcations between the flanking adjacent genes or the rII cistrons. This means that genes cannot be separated solely by mutant position or by mutant phenotype. The remaining alternative, defining a gene by complementation testing, implies that genes can be separated on the basis of biochemical function.

The function of genes, known since the work of Beadle and Tatum in 1945 and Linus Pauling in 1949, is to code for the structure of proteins (more accurately, polypeptide chains). The disciplines of biochemistry and genetics are united in the DNA nucleotide sequence of a gene coding for the amino acid sequence of a polypeptide chain. The latter sequence determines not only how a protein folds into three dimensions but also the specific enzymatic reaction(s) or other chemical role(s) of the protein in the organism. A considerable body of effort over the past 50 years consisted of finding which proteins were coded by which genes. Starting with an isolated protein, the amino acid sequence was used to predict the nucleotide coding sequence. Eventually an oligonucleotide with this sequence could be made and used by hybridization to isolate or locate the chromosomal gene. Starting with a mutant gene in an organism, biochemical assays are used to determine which protein is aberrant, or mapping studies are used to determine which candidate nucleotide sequence is mutated. Automation in sequencing techniques has led to great advances in linking genes and proteins. The goal of organism genome projects is to place every gene/protein-coding unit on the complete sequence of that organism's genome.

Predicting genes from long nucleotide sequence tracts is a matter of identifying open reading frames (ORFs). Of the six possible reading frames in any interval, a genuine protein-coding region is expected to have one reading frame consisting of sense or amino acid codons, that is open long enough to specify a candidate polypeptide chain. Reading frames that are interrupted by stop codons do not qualify. With three stop codons out of 64 possible triplet codons, random noncoding sequence is expected to be interrupted by stop codons every 21 codons on average. The complete sequence of the *Saccharomyces cerevisiae* genome yields 6023 predicted ORFs. About a third could be connected initially with known mutant genes or known biochemical products. About a third were experimentally verified to be genes by showing a mutant phenotype after knocking the gene out. In *S. cerevisiae* homologous recombination is used to replace the normal allele with a nonfunctional allele or 'knockout' in order to test for function. The remaining third of the predicted ORFs cannot be confirmed as being genes by these tests. It is more difficult still to apply this approach to the genome sequences of multicellular plants and animals because of the complications introduced by introns interrupting ORFs. Computer-generated predictions taking into account species-specific codon usage preferences and the preferred splice donor and splice acceptor sequences (to recognize the ends of introns) still do not recognize all of the known genes. The race to guess the number of genes from the human sequence has assumed the status of a TV game show, with lotteries and prizes promised to the winners. The estimates range between 20 000 and 120 000 genes.

There are other complications that get in the way of precisely identifying genes on the basis of DNA sequence alone. Equating genes with biochemical functions has problems because of multiple gene families and/or multiple gene products. Are two ORFs that code for the same protein counted as one gene or two? One example among many is the two α-globin-coding regions on human chromosome 16. Many genes, perhaps most, produce more than one mRNA product through alternative transcription starts or alternative splicing. Awareness is growing that proteins with some shared domains and some different domains play an important role in fine-tuning tissue-specific development. Still other polypeptide chains are subsequently cleaved or modified to yield different kinds of products. Mutation locations can vary to include hits in the regulatory regions outside the recognized coding intervals. What should be counted as a gene at the DNA level: the code for each function (chemical reaction), the template for each mRNA (cDNA) transcript, the code for each polypeptide chain, or each mutation location? Epigenetic regulation by imprinting means that some aspects of gene expression are above the sequence; the implication, reminiscent of Goldschmidt's 1946 argument, is that individual genes cannot be separated from functioning of the larger chromosomal unit. A modern synthesis might state that reproductive success, natural selection, and evolution value only what works for the organism. There is no design to biology, just history in the form of inheritance and tinkering through the noise of mutations and environmental variations to result in the individual, for better or worse.

If gene as a concept is not as well founded as say, atoms or molecules, does this mean the term should be discarded? Probably not. Gene still has heuristic value. We mean by it the awareness of the origin of specific molecules that are well founded, as in the gene for telomerase, or any other enzyme under active investigation. We mean also the importance of inheritance over environment for individual differences, not to close discussion but to stimulate investigation of the mechanisms influencing biology. And gene codifies what is known about the mechanism of inheritance, as the statements "it's in the genes" or "genes run in families," cannot be imagined with gemmules replacing genes. Ultimately it is the responsibility of authors to make clear how they are using the term in order to convey their message.

See also: Alleles; Benzer, Seymour; Linkage Group; Mutation; Nomenclature of Genetics

Gene Action

C Yanofsky

Copyright © 2001 Academic Press
doi: 10.1006/rwgn.2001.0501

Gene action is the consequence(s) of the presence and activities of the product of a gene.

See also: Operon

Gene Amplification

T D Tlsty

Copyright © 2001 Academic Press
doi: 10.1006/rwgn.2001.0502

A fundamental property of living cells is their orderly transmission of genetic information from generation to generation. One aspect of this property involves a mechanism which controls replication and ensures one complete doubling of each replicon during each cell generation. Another aspect of this property is the placement of genes such that they are expressed properly and that each daughter cell receives the genes in the appropriate configuration. It is now appreciated that deviations from this principle occur commonly and that amplification of DNA sequences as well as rearrangement of sequences occurs often.

Amplification with of DNA sequences, the differential increase in a specific portion of the genome in comparison with the remainder, occurs during development as well as during the vegetative growth of cells. The processes of polyploidization and endoreduplication, where the entire chromosome complement is multiplied in one nucleus, will not be discussed in this section. Additionally, aneuploidy (or trisomy) can also result in a differential increase in a portion of the genome, but this is distinct from DNA amplification and will not be further discussed. Developmental amplification has been documented extensively in both germline cells and somatic cells of many organisms. Clearly these changes in the DNA content of the nucleus of cells are carefully regulated and lead to the appearance of the extra DNA at a predetermined time and, in many cases, the dissipation of this DNA on cue. Another type of DNA amplification (sporadic) is detected when cells are overcoming adverse environmental conditions. Amplification of this type, usually visualized by selection for a desired phenotype, has been found in bacteria, yeast, insects, and vertebrates. Salient features of this amplification process are being studied in several laboratories and several general characteristics have emerged. These include: (1) a multistep process leading to the generation of highly amplified sequences; (2) a karyotype characterized by chromosomal abnormalities; (3) a genetic instability of the resistant (amplified) phenotype which may extend into a marked clonal variation among cells; (4) a spontaneous rate of detection which varies; and (5) a spontaneous rate of detection which may be increased by manipulations of the cellular growth conditions (among these, treatment with carcinogens). Several excellent reviews of amplification have appeared in the literature (Hamlin *et al.*, 1984; Stark and Wahl, 1984; Stark, 1986; Schimke, 1988).

Development

In Germline Cells

The earliest suggestion of developmental DNA amplification was made by King in 1908 when she described the extra 'chromatin' (associated with forming nucleoli) which arose in the oocytes of the toad, *Bufo* sp., during pachytene (King, 1908). Later studies show that these 'masses' contain DNA, and hybridization studies demonstrate a great excess of sequences coding for rRNA (Gall, 1968). This differential synthesis of rRNA genes is correlated with the appearance of hundreds of nucleolar organizers during the pachytene stage of meiosis. Similar examples of this phenomenon are seen in other amphibians: *Xenopus*, *Rana*, *Eleutherodactylus*, *Triturus* (Gall, 1968), as well as in an echiuroid worm, the surf clam (Brown and David, 1968), and many insects (Gall *et al.*, 1969). Documentation of rDNA amplification is particularly amenable, because the rDNA forms nucleoli which are distinctive in appearance and because most rDNA sequences have a relatively high guanosine–cytosine content, which allows their separation from bulk DNA on CsCl gradients. These two aspects of rDNA have been important in the recognition of this and other phenomena (magnification and compensation) in

which the rDNA copy number changes under various genetic conditions.

The most complete study of oogenic rRNA gene amplification has been made in *Xenopus laevis*. In this animal the early primordial germ cells do not contain amplified rDNA. Amplification is initiated in both oogonia and spermatogonia of the tadpole during sexual differentiation and germ cell mitosis and results in a 10- to 40-fold increase in rDNA genes. The premeiotic amplification is lost at the onset of meiotic prophase. This loss seems to be permanent in male germ cells but temporary in female germ cells. Early during meiotic prophase, the oocyte nucleus undergoes a second burst of rDNA amplification which results in a 1000-fold increase in ribosomal genes (Kalt and Gall, 1974). The mechanism by which the first extrachromosomal rDNA copies are produced in the premeiotic stage is unknown, although circular structures have been observed (Bird, 1978). Since the number and placement of ribosomal cistrons does not change, a mechanism involving disproportionate replication is favored. The second burst of rDNA amplification probably involves a rolling-circle intermediate. Such structures have been visualized by Hourcade *et al.* (1973) and could account for the increase in rDNA in the oocyte during the given period of time (Rochaix *et al.*, 1974). While the rDNA content is constant, the size and number of the rings are variable, suggesting fission and fusion of nucleoli during oogenesis (Thiebaud, 1979). The circular molecules exhibit sizes which are integral multiples of a basic unit. Using molecular techniques, this basic unit was found to be the DNA segment coding for one precursor rRNA molecule plus the accompanying nontranscribed spacer region. Extrachromosomal, circular molecules containing rDNA are also found at a low frequency (0.05–0.15% of total number of molecules) in *Xenopus* tissue culture cells and in *Xenopus* blood cells (Rochaix and Bird, 1975). It has been suggested that the difference in the state of rDNA between a somatic cell and an amplified germ cell may only be one of degree (Bird, 1978).

The cytological literature contains many other references to extrachromosomal DNA in oocytes, of which one of the most striking examples is found in dytiscid water beetles such as *Rhyncosciara*. The nucleus of each oocyte contains a large chromatin mass, termed 'Giardina's body,' in addition to the chromosomal complement. In older oocytes this DNA is associated with multitudes of nucleoli and has been shown to contain an increase in rDNA sequences (Gall *et al.*, 1969). Hence, amplification of rDNA occurs during the maturation of the oocyte. However, hybridization studies indicate that only a fraction of the extrachromosomal DNA is made up of rDNA sequences, indicating that other DNA sequences (of unknown function) are also amplified (Gall *et al.,* 1969).

The phenomenon of gene amplification in *Tetrahymena*, while also resulting in a considerable increase in rDNA sequences, is somewhat different from that in oocytes described above. *Tetrahymena* contain two types of nuclei in each cell: a transcriptionally quiescent micronucleus, which is responsible for genetic continuity; and a macronucleus, which is derived from the micronucleus in a developmentally regulated process whereby micronuclear sequences are eliminated, rearranged, and amplified (Yao and Gall, 1974; Yao and Gorovsky, 1974; Yao *et al.,* 1978). Considerable amplification of many micronuclear sequences occurs, which results in the macronucleus containing 45 times the haploid amount of DNA. Some of this increase is accounted for by amplification of rDNA sequences which are present as a single integrated copy in the micronucleus but are present in 200 copies in the macronucleus (Gall and Rochaix, 1974; Yao and Gall, 1974; Yao *et al.,* 1978). These amplified rDNA sequences are present as linear, extrachromosomal, palindromic molecules (Yao *et al.,* 1978; Yao, 1981). The present data favors a model whereby excision of the single rDNA copy is followed by amplification (Yao *et al.,* 1978).

The amplification, rearrangement, and elimination of sequences is also developmentally regulated in the slime molds *Physarum* and *Dictyostelium*, as well as another ciliated protozoan, *Stylonychia*, to varying extents. Studies concerning the molecular structure of these amplified DNA sequences have been undertaken and have revealed interesting structures at the termini.

In Somatic Cells

Gene amplification during the differentiation of somatic cells also occurs and was again first detected by morphological criteria. Several regions of the polytene chromosomes found in the larval salivary glands of the fly *Rhynchosciara americana* showed a puffing response to hormone treatment which was carefully defined both temporally and spatially. These puffs were found to contain greater amounts of DNA than surrounding region (Breuer and Pavan, 1955) and code for peptides utilized in the synthesis of the cocoon. Amplification and puff formation are dependent on the developmental stage of the cell as well as the cell's position within the gland (Glover *et al.,* 1982). Puff formation is not peculiar to the salivary gland chromosomes of *Rhynchosciara*, because it has also been observed in the cells of Malpighian tubules and intestinal cells of the same insect. In addition, puff formation has also been observed in different tissues of Chironomidae, *Drosophila*, *Sciara*, *Hybosciara*, and other Diptera (Breuer and Pavan, 1955), in some cases thought to be under hormonal control (Pavan and da Cunha, 1969; Bostock and Sumner, 1978).

In many cases of differentiation, somatic cells acquire adequate amounts of mRNA for production of abundant proteins by accumulation of stable mRNA molecules over a period of days. Examples of such are the silk fibroin genes, the ovalbumin genes, and the β-globin-chain genes. This is thought to be controlled at the transcriptional or posttranscriptional level. However, in other cases, such as during the synthesis of the insect eggshell by the ovarian follicle cells of *Drosophila*, little time is allotted for the production of the specific mRNAs which are needed in large quantities. In these latter cases, the rates of transcription and translation do not seem to be high enough for the production of adequate amounts of protein. Spradling and Mahowald (1980) found that this need is met by differential amplification of the chorion gene sequences in the ovarian follicle cells. Spradling and Mahowald (1981) found that the chorion genes that are located on the X chromosome are in two clusters, s36 and s38, and are amplified 15-fold. The s15 and s18 loci are on the third chromosome and are amplified 60-fold. The genes in both clusters are amplified at the same time and both homologs are amplified equally. Sequences which flank these genes are also disproportionately replicated but not to as great an extent. This results in a gradient of amplification which spans 90 kb of DNA, is maximal in the center, and does not evidence any discrete termination sites.

Changes in the DNA content of the nucleus of germline cells and somatic cells during development is common. Amplification of rRNA sequences, as well as others coding for proteins which are needed in large amounts during that particular phase of development, have been extensively documented. In still other cases, the nature of the amplified DNA cannot be totally accounted for by known sequences and probably contains amplified sequences of unknown function. Such sequences have been implicated in the differentiation of the orchid *Cynidium* sp. (Nagl *et al.,* 1972), the differentiation of peas (Van'T Hof and Bjerknes, 1982), and in the flowering of the tobacco plant, *Nicotiana* sp. (Wardell, 1977).

Acquisition of a Selected Phenotype

Duplication and amplification of genetic material in cells has long been documented as a means for overcoming deleterious growth conditions. Unlike amplification events that are specifically regulated during development, amplification as a means of survival is a more sporadic event, whose frequency is often detected at much lower levels than that seen in developmental amplification.

Bacteria

The first well-documented instance of gene duplications in bacteria which provided an adaptive advantage was seen by Novick and coworkers (Novick and Horiuchi, 1961; Horiuchi et al., 1962, 1963). Bacterial strains were grown for long periods of time in limiting concentrations of lactose in the chemostat. The bacterial strains which emerged were able to synthesize four times the maximal normal amount of β-galactosidase. The ability to produce large amounts of enzyme was unstable and could be transferred by conjugation at a time when the lactose operon genes were expected to be transferred. It was concluded that the ability to overproduce β-galactosidase was due to extra copies of the lactose genes present in these strains. The spontaneous rate of duplication was estimated to be 10^{-3} (Horiuchi et al., 1963) and, in a similar system, 10^{-4} (Langridge, 1969). Overproducers with similar characteristics were subsequently reported for other enzymes such as ribitol dehydrogenase, β-lactamase (Normark et al., 1977), and several others (for an excellent review, see Anderson and Roth, 1977). Roth and coworkers have demonstrated that duplications of up to a quarter of the *Salmonella typhimurium* chromosome occur at large homologous segments such as the rRNA genes (Anderson and Roth, 1977, 1981). Such large duplications are dependent on the *recA* system. On the other hand, duplications in the range of 10–30 kb appear to be independent of *recA* and do not involve very large homologous segments of DNA (Emmons et al., 1975; Anderson and Roth, 1977; Emmons and Thomas, 1981). Edlund and Normark (1981) have detected tandem duplications of 10–20 kb at the *E. coli* chromosomal *ampC* locus, which codes for β-lactamase and confers resistance to ampicillin. After a step-wise selection in increasing concentrations of ampicillin, 30–50 copies of the duplication events were analyzed at the restriction fragment level and were found to have different endpoints. The junction point of the amplified unit in one case was sequenced and it was found that the original duplication event occurred at a sequence of 12 bp, which was repeated on each side of the *ampC* locus, 10 kb apart. Tlsty et al. (1984a), have detected amplification events as revertants of certain leaky *Lac* mutants. The DNA sequences, which were amplified, contained the lactose operon and anywhere from 7 to 32 kb of flanking sequences. These regions were amplified 100-fold. Virtually all of the duplications, which have been detected and subsequently studied in bacteria, have been tandem in nature.

Several bacteriophages are also known to amplify genetic markers. The phage lambda (Edlund and Normark, 1981), T4 (Kozinski et al., 1980), and P1 phage (Meyer and Lida, 1979) may use mechanisms that parallel those used by viral sequences in mammalian cells. At the present time, the mechanism is unknown.

Yeast

Resistance to the toxic effects of copper in *Saccharomyces cerevisiae* is mediated by tandem gene amplification of the CUP1 locus (Fogel et al., 1983) but is different from sporadic amplification events characterized in other organisms. The CUP1 locus of yeast codes for a small molecular-weight copper-binding protein. Copper-sensitive strains contain one copy of this locus and, when grown in elevated concentrations of copper, failed to produce resistant derivatives with a higher gene copy number of CUP1 genes on one chromosome. Infrequently, copper-resistant strains were isolated in the laboratory and were found to carry up to 10 tandem duplications of the region, which is 2 kb in size. Further amplification could be achieved by growing the copper-resistant strains in elevated copper concentrations. The authors postulate that the mechanism of amplification in this instance proceeds through the formation of a disomy for chromosome VIII (which carries the CUP1 locus). Copper-resistant mutants of the sensitive strain were found to be disomics for chromosome VIII. The amplification or random iteration could then result primarily for subsequent unequal chromosome or sister chromatid exchanges. The formation of a disomy may constitute an initial event in the process.

A second example of gene amplification in yeast exhibits molecular structures that are more similar to those described in other organisms. A yeast strain resistant to antimycin A, an alcohol dehydrogenase inhibitor, has been found to contain multiple copies of a nuclear gene, ADH4, an isoenzyme of alcohol dehydrogenase. The amplified copies are 42 kb in length, display a linear, extrachromosomal, palindromic structure and contain telomeric sequences. Their structure resembles that of the amplified rDNA genes in the macronucleus of *Tetrahymena* and related cilliated protozoa, except that the nuclear copy remains within the chromosome in this situation. In contrast to what is often observed in mammalian amplification, the extrachromosomal copies of this gene were stable during mitotic growth. Amplification of the ADH4 gene is a relatively rare event ($\sim 10^{-10}$ mutations/cell/generation); alternative mutations compose the majority of antimycin A-resistant events.

Protozoans

Drug resistance in protozoan parasites is a common occurrence and presents a serious problem for the chemotherapy of diseases caused by such pathogens as *Trypanosoma*, *Leishmania*, and *Plasmodium* (Browning, 1954; Peters, 1974; Rollo, 1980). Recently, *Leishmania* strains resistant to the well-known chemotherapeutic agent methotrexate (MTX), have been isolated and analyzed as to their mechanism of resistance. Organisms which were resistant to high concentrations of MTX (1 mM) had a 40-fold increase in dihydrofolate reductase (DHFR), which in this organism is associated with thymidylate synthetase (Beverley *et al.*, 1984).

Recent studies have shown that *Plasmodium falciparum* contains genes that are analogous to the multidrug resistance genes in mammalian cells. Parasites that become resistant to chloroquine have also proven to be resistant to other antimalarial drugs; similar to the phenomenon of multi-drug resistance seen in human tumors. Wilson *et al.* (1989) found that sequences that were similar to the mammalian P-glycoprotein existed in *P. falciparum*. Their studies showed that drug-resistant parasites contained amplified copies of these specific DNA sequences when compared with their drug-sensitive siblings.

Invertebrates

Amplification of rRNA genes in drosophila during oogenesis does not occur as has been described in *Xenopus* (see previous section, "Development"). However, sporadic amplification of the rRNA genes during one generation has been observed under specific genetic conditions. Amplification of the rDNA genes in this situation results in a reversion from a mutant to a wild-type phenotype. Each sex chromosome carries approximately 130–150 rRNA genes (Ritossa *et al.*, 1966). The phenotype is wild if the diploid cell carries at least one normal locus (~130 genes) while the phenotype is altered (bobbed) if the genome carries less than 130 genes (Ritossa and Scala, 1964; Ritossa *et al.*, 1966). The intensity of the bobbed phenotype (slow development, thin chitinous cuticle, reduced body traits, and short bristles) is inversely proportional to the number of genes for rRNA. Ribosomal DNA magnification, the increase in rDNA copy number, is observed in the progeny of phenotypically bobbed males. It involves rapid accumulation of rDNA by unknown mechanisms at either nucleolus organizer. The rDNA, which is accumulated during the first generation, does not have a noticeable effect on the phenotype of the fly. The phenotypic effects of

the magnified rDNA become evident in the F_2 progeny if the rDNA has been transmitted by males and if the genotype of the F_2 generation is again characterized by rDNA deficiencies. In other words, the extra copies of rDNA are eliminated if the magnified fly is crossed with a normal (bb^+) female. The phenotypic inheritance of magnified rDNA requires its integration and inheritance through the male germline. Two hypotheses have been proposed to explain rDNA magnification: disproportionate replication of rDNA (Ritossa and Scala, 1964; Ritossa *et al.*, 1971) or unequal sister chromatid exchange (Tartof, 1974). Recent experiments demonstrating the decrease in magnification frequency in organisms which carry the rDNA on a ring chromosome strongly suggest unequal crossover as the mechanism.

Another type of amplification occurs in *Drosophila melanogaster*, which differs from rDNA magnification in several characteristics. This amplification, called 'compensation,' occurs when one nucleolus organizer of the two homologs is completely deleted (X/O or X/X-no females). In such mutants, the remaining organizer 'compensates' for the deletion of the rDNA sequences by a disproportionate replication of the remaining sequences on the intact homolog. Compensation may only occur on the X chromosomal nucleolus organizer, and the extra rDNA is not inherited in subsequent generations. Utilizing various deficiencies for the X-chromosomal heterochromatin, evidence has been presented for the existence of a genetic locus that regulates rDNA compensation (Procunier and Tartof, 1978). This locus, called the 'compensatory response' (cr), is located outside the ribosomal cluster and in the X-chromosomal heterochromatin. The locus acts in *trans* to sense the presence or absence of its partner locus on the opposite homolog. If only one cr locus is present, it acts in *cis* by driving compensation (disproportionate replication) of adjacent rRNA genes. Not all embryos with the proper genotype undergo compensatory amplification to emerge with an increased number of rRNA genes. Only a small fraction undergoes the putative compensatory amplification. In this respect the amplification behaves like a mutagenic reversion event to restore the functional phenotype.

Resistance to environmental agents such as pesticides and toxic chemical waste has been documented in laboratory stocks and natural populations of invertebrates. Selection of *Drosophila* larvae in increasing concentrations of cadmium yields strains that contain duplications of the metallothionien gene (Otto *et al.*, 1986). The duplication is stably inherited in the absence of selection pressure and produces a corresponding increase in metallothionien messenger RNA. A survey of natural populations found that this event is common (Maroni *et al.*, 1987) and may signal the early stages of the evolution of a gene family. The mosquito *Culex quinquefasciatus* develops resistance to various organophosphorus insecticides by overproducing the enzyme esterase B1. Molecular studies have demonstrated that the overproduction of the enzyme is the result of amplification of the esterase B1 gene some 250-fold (Mouches *et al.*, 1986). The resistant mosquito was described as normally developed and had reproductive capacity. This observation raises questions of evolutionary significance for the duplication and amplification event at least in invertebrates.

Plants

DNA changes in plants during response to environmental stress have been reported for flax (see Cullis, 1977, 1979, 1983). The suggestion that these changes in DNA content are induced by the environment awaits further studies for verification.

Vertebrates

DNA amplification in mammalian cells was first detected when murine tumor cell populations became resistant to chemotherapeutic drugs. MTX, an oft-used chemotherapeutic drug, inhibits the action of dihydrofolate reductase, which is required for the biosynthesis of thymidylate, glycine, and purines. Step-wise selection of cells in increasing concentrations of MTX generated highly resistant cells (Hakala *et al.*, 1961; Alt *et al.*, 1976; Flintoff *et al.*, 1976a,b; Haber *et al.*, 1981): Beidler and Spengler (1976) detected chromosomal abnormalities in the cells, which overproduced dhfr and suggested that they reflected an increase in gene dosage. Schimke and coworkers obtained a cDNA for the dhfr sequence and were able to show that the overproduction of dhfr enzyme was the result of amplification of the DHFR DNA sequence (Alt *et al.*, 1978). It is now known that amplification of the DNA sequence coding for the target enzyme of a metabolic inhibitor is a common mechanism for overcoming growth restriction (Dolnick *et al.*, 1979; Melera *et al.*, 1980; Tyler-Smith and Bostock, 1981; Flintoff *et al.*, 1983; Stark and Wahl, 1984 for review).

Other examples of this phenomenon were subsequently found, the best studied being amplification of the CAD gene. The CAD gene codes for a multifunctional protein which catalyses the first three steps in the synthesis of pyrimidines. The asparate transcarbamylase activity can be inhibited by the transition state analogue, *N*-phosphoacetyl-L-aspartate (PALA). PALA-resistant cells overproduce not only the aspartate transcarbamylase but the other two enzymes as well (carbamyl synthetase and dihydroorotase; Kempe *et al.*, 1976). Wahl *et al.* (1979) have shown that overproduction of these enzymes is the direct result of amplification of DNA coding for these proteins.

Numerous other instances of DNA amplification have now been described (for a comprehensive list and references, see Stark and Wahl, 1984). In all cases the growth of cells is inhibited either by metabolic inhibitors, toxic agents, or altered enzymes with reduced efficiency. Of clinical importance has been the discovery that multidrug resistance in cancer chemotherapy is, in some cases, mediated by amplification of the *mdr* locus (Roninson *et al.*, 1984a,b). Selection pressure can also lead to the amplification of sequences with initially unknown functions.

The acquisition of a selected phenotype may often result from selection pressures, which are unknown at that time. In these cases a certain phenotype may be accompanied by the manifestations of gene amplification for sequences unknown. Such an instance has been described in studying the sequences, which are carried on the double-minute chromosomes (DMs) and found in the homogeneously staining regions (HSRs) of neuroblastoma cells, where these structures were first described. The sequences, which are amplified in these lines, are cellular *onc* genes, the N-*myc* gene (Schwab *et al.*, 1983). Amplification of oncogenes has now been found in several tumor types containing DMs and HSRs.

Mammalian Gene Amplification

There are several characteristics of amplification which many of the systems above share in common. To illustrate these characteristics, general properties of MTX-resistant cells, which result from amplification of the DHFR gene, will be described.

Classically, mammalian cells containing amplified DHFR genes were obtained by a stepwise selection for cells, which were highly resistant to MTX. High MTX resistance (by virtue of amplification of the DHFR gene) cannot be obtained by a large, single-step selection protocol; it is a multistep process. The initial step seems to be rate-limiting, since cells with a low copy number can be rapidly stepped up to a high level of resistance and a high copy number. When the initial increase in gene copy was examined more closely, Brown *et al.* (1983a,b) and Tlsty *et al.* (1984b) found that the stringency of selection is critical in obtaining cells, which have amplified DHFR. It was found that incremental increases in drug concentration not only promote the rapid emergence of resistance but also specifically promote the rapid amplification of the DHFR gene (Rath *et al.*, 1984).

The second property of MTX-resistant cells, which have amplified their DHFR gene, is the frequent presence of karyotypic abnormalities in the cells. As indicated previously, abnormal chromosomal structures were associated with overproduction of the DHFR in the early studies of Beidler and Spengler (1976). They described a marker chromosome in over producing cells which contained an elongated chromosomal arm. The term 'homogeneously staining region' (HSR) was coined to describe a region of this chromosome which banded abnormally when stained with Giemsa and which was subsequently shown to be the site of the amplified DHFR sequences (Alt et al., 1978). This structure was associated with stable resistance to MTX; that is, retention of the resistant phenotype even after subsequent growth in the absence of selection pressure. This is in contrast to the karyotype of cells, which were unstably resistant to MTX; i.e., with extended growth in nonselective medium, the resistant phenotype (amplification) diminished rapidly and disappeared. HSR structures were not found in unstably resistant cells. Close examination of the karyotype of unstably resistant cells did, however, bring to light the presence of small chromosomal fragments known as DMs. These structures (as well as HSRs) had been described by Balaban-Malenbaum and Gilbert (1980) in cell lines obtained from human neuroblastoma. Subsequent work demonstrated that unstably resistant cells contained the amplified copies of DHFR on the DMs. The lack of centromeric structure in these fragments leads to their random (unequal) segregation at mitosis and a diminution in their number if selection pressure is no longer exerted on the cells (see Kaufman and Schimke, 1981).

The molecular structure of HSRs and DMs has been studied. The first obstacle in characterizing the amplified unit derives from its large size. The DHFR gene, which is amplified to confer MTX resistance, is large: 31 kb including introns. The size of the amplified region is greater still; gross estimates vary from 120 to 1000 kb as the unit of DNA, which is amplified. Analysis of the end points of the amplified units provided information on the structure of the amplified unit. Although the sequence of DNA which needs to be characterized is long, cloning of neighboring fragments ('chromosomal walking') has been accomplished by several laboratories in both mouse and hamster model systems (Zeig et al., 1983; Federspiel et al., 1984; Giulotto et al., 1989). The information derived from these endeavors has not provided the desired portrait of the amplified unit because of another obstacle. The amplified structure, at the molecular level, seems to be continually changing, as evidenced in the chromosomal walking studies. On each amplified cell studied, the amplified sequences correlate with the cloned map only up to a certain point and then diverge. Rearrangements of DNA

accompany the amplification of genes. The basic molecular event of DNA amplification is obscured by the dynamic aspect of the process. HSRs, DMs, and translocations are karyotypic abnormalities which have been detected in cells which are highly resistant to a given metabolic inhibitor (i.e., cells which have already progressed through much of the multistep process). Contrasting results have been obtained by Hamlin and Montoya-Zavala (1985) in their study of DHFR gene amplification in Chinese hamster ovary (CHO) cells. They find the amplified unit to be uniform in size and exist in head-to-head and head-to-tail tandem repeats.

A third characteristic of MTX-resistant cells, which have amplified the DHFR gene, is the initial genetic instability of the resistant (amplified) phenotype, which is accompanied by a marked heterogeneity in the population. Cells newly selected for MTX resistance are unstable with respect to DHFR levels, and the loss of the elevated DHFR levels was variable in the progeny of different cloned cells. The initial instability of the amplified DHFR genes in emerging, resistant CHO cells is consistent with the hypothesis that they are present as extrachromosomal pieces of DNA. Stabilization of the resistant phenotype could be the result of integration of these sequences into the chromosome, either at the site of amplification or elsewhere in the genome, or could be the result of processes that are unknown at the present time.

A final characteristic is that the frequency of DNA amplification can be manipulated. Several agents have been found that increase the incidence of DNA amplification. Pretreatment with hydroxyurea, ultraviolet light, or MTX itself increases the incidence of the initial amplification of the DHFR sequences (Brown et al., 1983a,b; Tlsty et al., 1984b). Similar observations have been made using an SV40-transformed cell system to detect amplification of SV40 sequences (Lavi, 1981). Viral sequences undergo an amplification process that is enhanced by pretreatment with carcinogenic agents. Lavi and coworkers (Lavi and Etkin, 1981) observed dramatic increases in viral sequences after the cells were treated with agents such as benzopyrene, aflatoxin, methylmethane sulfonate, and a host of other carcinogens. The extent of the enhancement of amplification can be as little as a few-fold or exceed a 1000-fold. The basis for the enhancement of gene amplification by carcinogen pretreatment is not known at the present time.

Frequency of Sporadic Amplification in Mammalian Cells

In the last few years, it has become obvious that the frequency of gene amplification in different cells can vary dramatically. Initially, gene amplification was measured in the model systems that were used to study the phenomenon: established rodent cell lines such as S180, BHK cells, CHO cells, and 3T6 cells. Reported values for the rodent model systems were incidences of 10^{-6} or 10^{-4} or rates that approached 10^{-3} events/cell/generation. Several laboratories have begun examining the incidence of gene amplification in different cell populations. Early results suggested that tumorigenic cells could amplify more frequently than nonturmorigenic cells (Sager *et al.,* 1985; Otto *et al.,* 1989). In general, highly tumorigenic cells amplify at a greater frequency than nontumorigenic cells. Earlier studies of gene amplification used immortalized cell lines and biopsied tumor samples. However, in two studies, the amplification potentials of primary diploid cells, both human and rodent, were examined and quantitatively compared with the amplification potentials of their transformed counterparts. Strikingly, the difference in amplification incidence between 'normal' cells and their transformed counterparts (in some cases tumorigenic) is immense (Tlsty, 1990; Wright *et al.,* 1990). Amplification potential was measured at two loci, the CAD gene and the DHFR gene. Comparatively quantitative data for both normal ($<2\times10^{-8}$) and transformed cell lines (10^{-4}) indicated a difference in frequency which is greater than four orders of magnitude (Tlsty, 1990). These studies suggest that there is some fundamental difference between normal cells and transformed cells that affects their ability to amplify; diploid cells lack a detectable frequency of gene amplification, while tumorigenic cells readily amplify DNA sequences (at least a fourth order of magnitude difference). Subsequent studies have identified the p53 tumor suppressor gene as a regulator for the amplification event in mammalian cells (Livingstone *et al.,* 1992; Yin *et al.,* 1992).

Experiments with tissue culture cells have shown us that a wide variety of loci may be amplified in mammalian cells. The amplification is usually manifested as an overproduction of the protein product that is targeted by the chemotherapeutic agent. Luria–Delbrück fluctuation analysis has demonstrated that the amplification events are occurring spontaneously at a constant rate; it is the selective environment that allows them to be visualized. A recent study has compared the amplification rate in nontumorigenic and tumorigenic cells and found that the tumorigenic cells amplified the endogenous locus 100 times more than the nontumorigenic cell line (Tlsty *et al.,* 1989). Restrictions on the loci that can spontaneously amplify have not been encountered. Studies have also shown that more than one locus can be amplified at the same time (Giulotto *et al.,* 1989).

Summary

The literature suggests that when gene amplification does occur in normal tissues it is developmentally regulated. This evidence is mostly compiled from studies on *Xenopus* and *Drosophilia* (see the section "Development"). In higher organisms, the documentation of gene amplification as a developmental event is lacking. At the present time we do not know if gene amplification can be developmentally programmed in mammalian cells.

Sporadic amplification can occur in unicellular organisms such as bacteria and yeast, but seems to be lacking in the normal somatic tissues of higher eukaryotes. Several reports of sporadic amplification in the germline cells of several organisms have been reported and have been shown to be heritable. In all of these cases, the phenotype demonstrated an increased resistance to an environmental toxin (Mouches *et al.,* 1986; Maroni *et al.,* 1987; Prody *et al.,* 1989). The extensive documentation of sporadic amplification in neoplastic tissues raises questions of when the neoplastic cell acquires the ability to amplify and if the manipulation of this event can aid in the treatment of cancer.

References

Alt FW, Kellems RE and Schimke RT (1976) Synthesis and degradation of folate reductase in sensitive and methotrexate-resistant lines of S-180 cells. *Journal of Biological Chemistry* 251: 3063–3074.

Alt FW, Kellems RE, Bertino JR and Schimke RT (1978) Selective multiplication of dihydrofolate reductase genes in methotrexate-resistant varients of cultured murine cells. *Journal of Biological Chemistry* 253: 1357–1361.

Anderson P and Roth JR (1977) Tandem genetic duplications in phage and bacteria. *Annual Review of Microbiology* 31: 473–505.

Anderson P and Roth J (1981) Spontaneous tandem genetic duplications in *Salmonella typhimurium* arise by unequal recombination between rRNA (rrn) cistrons. *Proceedings of the National Academy of Sciences, USA* 78: 3113–3117.

Balaban-Malenbaum G and Gilbert F (1980) The proposed orgin of double minutes from homogeneously staining region (HSR)-marker chromosomes in human neuroblastoma hybrid cell lines. *Cancer Genetics and Cytogenetics* 2: 339–348.

Beverley SM, Coderre JA, Santi DV and Schimke RT (1984) DNA amplification in methotrexate-resistant *Leishmania*: extra chromosomal circles and relocation in to chromosomal DNA. *Cell* 38: 431–433.

Beidler JL and Spengler BA (1976) Metaphase chromosome anomaly: association with drug resistance and cell-specific products. *Science* 191: 185–187.

Bird AP (1978) A study of early events in ribosomal gene amplification. *Cold Spring Harbor Symposia on Quantitative Biology* 38: 1179–1183.

Bostock CJ and Sumner AT (1978) *The Eukaryotic Chromosome*, pp. 256–259. Amsterdam: North Holland Publishers.

Breuer ME and Pavan C (1955) Behavior of polytene chromosomes of *Rhynchosciara angelae* at different stages of larval development. *Chromosoma* 7: 371–386.

Brown DD and David IB (1968) Specific gene amplification in oocytes. *Science* 160: 272–280.

Brown PC, Johnson RN and Schimke RT (1983a) Approaches to the study of mechanisms of selective gene amplification in cultured mammalian cells. In: *Gene Structure and Regulation in Development*, pp. 197–212.

Brown PC, Tlsty TD and Schimke RT (1983b) Enhancement of methotrexate resistance and dihydrofolate reductase gene amplification by treatment of mouse 3T6 cells with hydroxyurea. *Molecular and Cellular Biology* 3: 1097–1107.

Browning CH (1954) The chemotherapy of trypanosomic infections. *Annals of the New York Academy of Sciences* 59: 198–213.

Cullis CA (1977) Molecular aspects of the environmental induction of heritable changes in flax. *Heredity* 38: 129–154.

Cullis CA (1979) Quantitative variations of ribosomal RNA genes in flax genotrophs. *Heredity* 42: 237–246.

Cullis CA (1983) Environmentally induced DNA changes in plants. *CRC Critical Reviews in Plant Sciences* 1: 117–131.

Dolnick BJ, Berenson RJ, Bertino JR, Kaufman RJ, Nunberg JH and Schimke RT (1979) Correlation of dihydrofolate reductase elevation with gene amplification in a homogeneously staining chromosomal region in L5178Y Cells. *Journal of Cell Biology* 83: 394–402.

Edlund T and Normark S (1981) Recombination between short DNA homologies causes tandem duplication. *Nature* 292: 269–271.

Emmons SW and Thomas JO (1981) Tandem genetic duplications in phage lambda. *Journal of Molecular Biology* 91: 147–152.

Emmons SW, MacCosham V and Baldwin RL (1975) Tandem genetic duplications in phage lambda. *Journal of Molecular Biology* 91: 133–146.

Federspiel NA, Beverley SM, Schilling JW and Schimke RT (1984) Novel DNA rearrangements are associated with dihydrofolate reductase gene amplification. *Journal of Biological Chemistry* 259: 9127–9140.

Flintoff WF, Davidson SV and Siminovich L (1976a) Isolation and partial characterization of three methotrexate-resistant phenotypes from Chinese hamster ovary cells. *Somatic Cell Genetics* 2: 245–261.

Flintoff WF, Spindler SM and Siminovitch L (1976b) Genetic characterization of methotrexate-resistant Chinese hamster ovary cells. *In Vitro* 12: 749–757.

Flintoff WF, Weber MK, Nagainis CR, Essani AK, Tobertson D and Salser W (1983) Overproduction of dihydrofolate reductase and gene amplification in methotrexate-resistant Chinese hamster ovary cells. *Molecular and Cellular Biology* 2: 275–285.

Fogel S, Welch JW, Cathala G and Karin M (1983) Gene amplification in yeast: CUPO copy number regulates copper resistance. *Current Genetics* 7: 347–355.

Gall JG (1968) Differential synthesis of the genes for ribosomal RNA during amphibian oogenesis. *Proceedings of the National Academy of Sciences, USA* 60: 553–560.

Gall JG and Rochaix JD (1974) The amplified ribosomal DNA of dytiscid beetles. *Proceedings of the National Academy of Sciences, USA* 71: 1819–1823.

Gall JG, MacGregor HC and Kidston ME (1969) Gene amplification in the oocytes of dytiscid water beetles. *Chromosoma* 26: 169–187.

Giulotto E, Saito I and Stark GR (1989) Structure of DNA formed in the first step CAD gene amplification. *EMBO Journal* 5: 2115–2951.

Glover DM, Zaha A, Stocker AJ, Santelli RV, Pueyo MT, De Toledo SM and Lara FJS (1982) Gene amplification in *Rhynchosciara* salivary gland chromosomes. *Proceedings of the National Academy of Sciences, USA* 79: 2947–2951.

Haber DA, Beverly SM, Kiely ML and Schimke RT (1981) Properties of an altered dihydrofolate reductase encoded by amplified gene in cultured mouse fibroblasts. *Journal of Biological Chemistry* 256: 9501–9510.

Hakala MT, Zakrzewski SF and Nichol CA (1961) Relation of folic acid reductase to amethopterin resistance in cultured mammalian cells. *Journal of Biological Chemistry* 236: 952–958.

Hamlin JL, Milbrandt JD, Heintz NH and Azizkhan JC (1984) DNA sequence amplification in mammalian cells. *International Review of Cytology* 90: 31–82.

Horiuchi T, Tomizawa JI and Novick A (1962) Isolation and properties of bacteria capable of high rates of β-galactosidase synthesis. *Biochemical and Biophysical Research Communications* 55: 152–163.

Horiuchi T, Horiuchi S and Novick A (1963) The genetic basis of hyper-synthesis of β-galactosidase *Genetics* 48: 157–169.

Hourcade D, Dressler D and Wolfson J (1973) The nucleolus and the rolling circle. *Cold Spring Harbor Symposia on Quantitative Biology* 38: 537–550.

Kalt MR and Gall JG (1974) Observations on early germ cell development and premeiotic ribosomal DNA amplification in *Xenopus laevis*. *Journal of Cell Biology* 62: 460–472.

Kaufman RJ and Schimke RT (1981) Amplification and loss of dihydrofolate reductase genes in a Chinese hamster ovary cell line. *Molecular and Cellular Biology* 1: 1069–1076.

Kempe TD, Swyryd EA, Bruist M and Stark GR (1976) Stable mutants of mammalian cells that overproduce the first three enzymes of pyrimidine nucleotide biosynthesis. *Cell* 9: 541–550.

King HD (1908) The oogenesis of *Bufo lentiginosus*. *Journal of Morphology* 19: 369–438.

Kozinski A, Ling S, Hutchingson N, Halpern M and Mattson T (1980) Differential amplification of specific areas of phage T4 genome as revealed by hybridization to cloned genetic segments. *Proceedings of the National Academy of Sciences, USA* 77: 5064–5068.

Langridge J (1969) Mutations conferring quantitative and qualitative increase in β-galactosidase activity in *Escherichia coli*. *Molecular and General Genetics* 105: 74–83.

Lavi S (1981) Carcinogen-mediated amplification of viral DNA sequences in SV40-transformed Chinese hamster embryo cells. *Proceedings of the National Academy of Sciences, USA* 78: 6144–6148.

Lavi S and Etkin S (1981) Carcinogen-mediated induction of SV40 DNA synthesis in SV40 transformed Chinese hamster embryo cells. *Carcinogenesis* 2: 417–423.

Livingstone LR, White A, Sprouse J, Livanos E, Jacks T and Tlsty TD (1992) Altered cell cycle arrest and gene amplification potential accompany loss of wild-type p 53. *Cell* 70: 923–935.

Maroni G, Wise J, Young JE and Otto E (1987) Metallothionein gene duplication and metal tolerance in natural populations of *Drosophila melanogaster. Genetics* 117: 739–744.

Melera PW, Lewis JA, Biedler JL and Hession C (1980) Antifolate-resistant Chinese hamster cells. Evidence for dihydrofolate reductase gene amplification among independently derived sublines overproducing different dihydrofolate reductases. *Journal of Biological Chemistry* 255: 7024–7082.

Meyer J and Lida S (1979) Amplification of chloromphenicol resistance transposons carried by phage PICm in *Escherichia coli. Molecular and General Genetics* 176: 209–219.

Mouches C, Pasteur N, Berge JB, Hyrien O, Raymond M, Vincent BR, DeSilvestri M and Georghiou GP (1986) Amplification of an esterase gene is responsible for insecticide resistance in a California *Culex* mosquito. *Science* 233: 778–780.

Nagl W, Hendon J and Rucker W (1972) DNA Amplification in cymbidium protocorms *in vitro* as it relates to cytodifferentiation and hormone treatment. *Cell Differentiation* 1: 229–237.

Normark S, Edlund T, Grundstrom T, Bergstrom S and Wolf-Watts H (1977) *Escherichia coli* K-12 mutants hyperproducing chromosomal β-lactamase by gene repititions. *Journal of Bacteriology* 132: 912–922.

Novick A and Horiuchi T (1961) Hyper-production of β-galactosidase by *Escherichia coli* bacteria. *Cold Spring Harbor Symposia on Quantitative Biology* 21: 239–245.

Otto E, Young JE and Maroni G (1986) Structure and expression of a tandem duplication of the *Drosophila* metallothionien gene. *Proceedings of the National Academy of Sciences, USA* 83: 6025–6029.

Otto E, McCord S and Tlsty TD (1989) Increase incidence of CAD gene amplification of tumorigenic rat lines as an indicator of genomic instability of neoplastic cells. *Journal of Biological Chemistry* 264: 3390–3396.

Pavan C and da Cunha B (1969) Gene amplification in ontogeny and phylogeny of animals. *Genetics* 61 (supplement): 289–304.

Peters W (1974) Chagas' diseases. *Ciba Foundation Symposium* 20: 309–334.

Procunier JD and Tartof KD (1978) A genetic locus having *trans* and contiguous *cis* functions that control the disproportionate replication of ribosomal RNA genes in *Drosophila melanogaster. Genetics* 88: 67–79.

Rath H, Tlsty TD and Schimke RT (1984) Rapid emergence of methotrexate resistance in cultured mouse cells. *Cancer Research* 44: 3303–3306.

Ritossa FM and Scala G (1964) Equilibrium variations in the redundancy of rDNA in *Drosophila melanogaster. Genetics* 61: 305–37.

Ritossa FM, Atwood KC and Spiegelman S (1966) A molecular explanation of the bobbed mutants of drosophila as partial deficiencies of ribosomal DNA. *Genetics* 54: 819–834.

Ritossa FM, Boncinelli C, Graziani F and Polita L (1971) The first steps of magnification of DNA complementary to ribosomal RNA in *Drosophila melanogaster. Proceedings of the National Academy of Sciences, USA* 68: 1580–1584.

Rochaix JD and Bird AP (1975) Circular ribosomal DNA and ribosomal DNA: replication in somatic amphibian cells. *Chromosoma* 52: 317–327.

Rochaix JD, Bird AP and Bakken A (1974) Ribosomal RNA gene amplification by rolling circles. *Journal of Molecular Biology* 87: 473–487.

Rollo EM (1980) Drugs used in the chemotherapy of malaria. In: Gilman AG, Goodman LS and Gilman A (eds) *The Pharmacological Basis of Therapeutics*, 6th edn, pp. 1038–1069. New York: Macmillan.

Roninson IB, Abelson HT, Housman DE, Howell N and Varshavsky A (1984a) Amplification of specific DNA sequences correlates with multi-drug resistance in Chinese hamster cells. *Nature* 309: 626–628.

Roninson IB, Chin JE, Choi K, Gros P, Housman DE, Fojo A, Shen D, Gottesman MM and Pastan I (1984b) Isolation of human mdr DNA sequences amplified in multidrug resistant KB carcinoma cells. *Proceedings of the National Academy of Sciences, USA* 83: 4538–4542.

Sager R, Gadi I, Stephens L and Grabowy C (1985) Gene amplification: an example of accelerated evolution in tumorigenic cells. *Proceedings of the National Academy of Sciences, USA* 82: 7015–7019.

Schimke RT (1988) Gene amplification in cultured cells. *Journal of Biological Chemistry* 263: 5989–5992.

Schwab M, Alitalo K, Klempnauer KH *et al.* (1983) Amplified DNA with limited homology to *myc* cellular oncogene is shared by human neuroblastoma cell lines and a neuroblastoma tumor. *Nature* 305: 245–248.

Spradling AC and Mahowald AP (1980) Amplification of genes for chorion proteins during oogenesis in *Drosophila melanogaster. Proceedings of the National Academy of Sciences, USA* 77: 1096–1100.

Spradling AC and Mahowald AP (1981) A chromosome inversion alters the pattern of specific DNA replication in *Drosophila* follicle cells. *Cell* 27: 203–209.

Stark GR (1986) DNA amplification in drug-resistant cells and in tumours. *Cancer Surveys* 5: 1–23

Stark GR and Wahl GM (1984) Gene amplification. *Annual Review of Biochemistry* 53: 447–491.

Tartof KD (1974) Increasing the multiplicity of ribosomal RNA genes in *Drosophila melanogaster. Science* 171: 294–297.

Thiebaud CH (1979) Quantitative determination of amplified rDNA and its distribution during oogenesis in *Xenopus laevis*. *Chromosoma* 73: 37–44.

Tlsty TD (1990) Normal diploid human and rodent cells lack a detectable frequency of gene amplification. *Proceedings of the National Academy of Sciences, USA* 87: 3132–3136.

Tlsty TD, Albertini AM and Miller JH (1984a) Gene amplification in the *lac* region of *E. coli*. *Cell* 37: 217–224.

Tlsty TD, Brown PC and Schimke RT (1984b) UV radiation facilitates methotrexate resistance and amplification of the dihydrofolate reductase gene in cultured 3T6 mouse cells. *Molecular and Cellular Biology* 4: 1050–1056.

Tlsty TD, Margolin B and Lum K (1989) Differences in the rates of gene amplification in non-tumorigenic and tumorigenic cell lines as measured by Luria-Delbrück fluctuation analysis. *Proceedings of the National Academy of Sciences, USA* 86: 9441–9445.

Tyler-Smith C and Bostock CJ (1981) Gene amplification in methotrexate resistant mouse cells. II. Rearrangement and amplification of non-dihydrofolate reductase gene sequences accompany chromosomal changes. *Journal of Molecular Biology* 153: 219–236.

Van'T Hof J and Bjerknes CA (1982) Cells of pea (*Pisum sativum*) that differentiate from G2 phase have extrachromosomal DNA. *Molecular and Cellular Biology* 2: 339–345.

Wahl GM, Padgett RA and Stark GR (1979) Gene amplification causes overproduction of the first three enzymes of UMP synthesis in *N*-(phosphoacetyl-l-aspartate) resistant hamster cells. *Journal of Biological Chemistry* 254: 8679–8689.

Wardell WL (1977) Floral induction of vegetative plants supplied a purified fraction of deoxyribonucleic acid from stems of flowering plants. *Plant Physiology* 60: 885–891.

Wright J, Smith H, Hancock M, Hudson D and Stark G (1990) DNA amplification is rare in normal human cells. *Proceedings of the National Academy of Sciences, USA* 87: 1791–1795.

Yao MC (1981) Ribosomal RNA gene amplification in tetrahymena may be associated with chromosome breakage and DNA elimination. *Cell* 24: 765–774.

Yao MC and Gall JG (1974) A single integrated gene for ribosomal RNA in a eukaryote, *Tetrahymena pyriformis*. *Cell* 12: 121–132.

Yao MC and Gorovsky MA (1974) Comparison of the sequences of macro- and micronuclear DNA *Tetrahymena pyriformis*. *Chromosoma* 48: 1–18.

Yao MC, Blackburn E and Gall JG (1978) Amplification of the rRNA genes in *Tetrahymena*. *Cold Spring Harbor Symposia on Quantitative Biology* 38: 1293–1296.

Yin Y Tainsky MA Bischoff FZ Strong LC and Wahl GM (1992) Wild-type p53 restores cell cycle control and inhibits gene amplification in cells with mutant p53 alleles. *Cell* 70: 937–948.

Zeig J, Clayton C, Ardeshir F, Giulotto E, Swyryd E and Stark G (1983) Properties of single-step mutants of Syrian hamster cell lines resistant to *N*-(phosphonacetyl)-l-aspartate. *Molecular and Cellular Biology* 3: 2089–2098.

See also: Gene Expression; Polytene Chromosomes; RecA Protein and Homology; Rolling Circle Replication; Transformation

Gene Cassettes

R M Hall

Copyright © 2001 Academic Press
doi: 10.1006/rwgn.2001.1720

Gene cassettes are small, discrete mobile elements. A gene cassette generally comprises a single gene and a downstream 59-be (59-base element) which is a recombination site. Cassettes differ from most other known mobile elements in that they do not encode the enzymatic machinery responsible for their movement; this is supplied by a companion element called an integron (see Integrons). Cassette integration involves a site-specific recombination reaction between the 59-be and the *attI* site in an integron, which is catalyzed by an integron-encoded IntI-type integrase. Excision of cassettes occurs via both *attI* × 59-be and 59-be × 59-be reactions. The mobility of gene cassettes is thus dependent on the presence of an integron in the same cell but, as the most common location for gene cassettes is within an integron, this condition is normally satisfied.

Structure of Gene Cassettes

The organization of gene cassettes is very compact. They generally include only a single gene (or open reading frame) and a downstream recombination site called a 59-be (59-base element) and any gene can, in theory, be part of a cassette. Occasionally, two open reading frames (ORFs) are found in a single cassette. Gene cassettes are normally found in a linear form integrated at the *attI* site of an integron, but can also exist transiently in a free, closed-circular form (**Figure 1A**) which is created as a product of excision of a cassette from an integron. Circular cassettes can be reincorporated at the *attI* site of an integron, and IntI1-catalyzed integration of gene cassettes into the *attI1* site of a class 1 integron has been demonstrated experimentally using the IntI1 integrase (see Integrons). As the recombination crossover has been localized to a unique position between the conserved G and TT in the 1R site of 59-be, integrated cassettes begin with TT and end with G (**Figure 1B**).

As few as 7 bp separate the first inframe initiation codon of the gene from the start of the linear form of the cassette, and the termination codon normally lies very close to the 59-be or even within it. Thus, there is usually no space for transcription signals and such cassettes rely on the presence of an upstream promoter for expression of their genes. This promoter is normally supplied by the integron (see Integrons) and the correct orientation of the gene in a cassette with respect to the promoter in the integron is essential if the gene is to be expressed. This is achieved only when the 59-be is located downstream of the gene, as is the general rule. In rare cases, a promoter is located within the cassette. Both a promoter and translational attenuation signals have also been found upstream of cmlA genes that confer resistance to chloramphenicol, and production of the protein is induced by chloramphenicol. The presence of a promoter within a cassette will permit expression of genes in cassettes that are situated too far from the integron's promoter to permit expression from it.

Cassette-Associated Genes

It is presumed that any gene can become packaged in cassette form, though how and where this happens remains a matter for speculation. Many (over 60) of the known cassettes contain an antibiotic resistance gene, and these genes determine resistance to various antimicrobial agents (β-lactams, aminoglycosides, trimethoprim, chloramphenicol, erythromycin, rifampicin, and antiseptic quaternary ammonium compounds) using a variety of mechanisms. Over 150 further cassettes have been found in the Vibrio cholerae small chromosome. Only a few of the ORFs (potential genes) contained in these cassettes have been identified. They include genes for a toxin, a virulence determinant, and a lipoprotein as well as a few potential antibiotic resistance genes. Restriction and modification enzymes have also been found to be encoded in cassettes.

Cassette-Associated Recombination Sites

The 59-be recombination sites found in cassettes provide the signal that permits cassettes to be mobilized. They were originally recognized as a consensus of 59 bp found downstream of several different genes and were subsequently shown to be recombination sites recognized by the integron-encoded IntI integrases. Each cassettes includes a unique 59-be and members of the 59-be family were later found to vary considerably in sequence and length; the shortest

are 57 bp and the longest 141 bp. However, all 59-be share a set of identifiable features. The term 59-be has been retained because it has been widely used and is generally understood. The VCR (Vibrio cholerae repeat) found in cassettes from the V. cholerae integron region also share these features and are thus 59-be. Each 59-be is made up of two regions of 25–30 bp located at the outer ends (labeled LH and RH simple sites in Figure I) that each have an organization equivalent to that of the simple sites of other integrase-type recombinases. Each simple site includes a pair of inversely oriented core sites of 7 bp (boxed in Figure I) that are part of somewhat longer IntI binding domains. Both simple site regions are needed for the 59-be to be an effective recombination site, though one simple site can participate in recombination at greatly reduced efficiency. The overall organization of 59-be is unusual as the sites recognized by other integrases (tyrosine recombinases) include only one simple site.

The sequences of the LH and RH simple sites are only moderately conserved and the consensus regions in 59-be are confined to them. Indeed, the variation between the sequences of individual 59-be is such that only eight bases, four in each simple site region, are completely conserved in known 59-be. The sequences of the two consensus or simple site regions are imperfect inverted repeats of one another and, in any individual 59-be, complementarity between key bases in the two simple site regions appears to be preserved in preference to conformity to the consensus. The length of the central region of 59-be between the two simple sites is highly variable and this accounts for differences in the lengths of 59-be. The central sequence is also variable, but it commonly includes an inverted repeat. The importance of these features in recognition of 59-be-type sites remains to be established. However, differences between the LH and RH simple site regions, such as the extra residue in 2L (× in Figure I), may play a role in ensuring that the RH simple site is the location for strand exchange and hence that the cassette gene is correctly oriented with respect to the promoter in the integron.

A small number of examples of variants of known gene cassettes that have lost most of the 59-be have been found. In each of these, only one simple site remains. This simple site is made up of the 1L and 1R core sites of the original 59-be separated by a spacer. Cases where the spacer is derived from the spacer of either the RH or LH simple site of the original 59-be or from the attI1 simple site have been found. While it is probable that cassettes containing only a simple site can move, this has not yet been demonstrated.

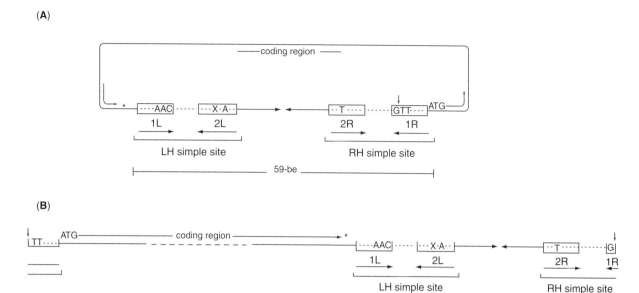

Figure 1 Structure of a circular gene cassette. A generalized typical cassette in (A) its free, circular form and (B) its linear integrated form showing the coding region of the gene and the 59-be recombination site. The extent of the coding region (not to scale) is delineated by start (ATG) and stop(*) codons. 7-bp core site sequences, related to the consensus GTTRRRY, that lie within putative IntI binding sites are boxed, and arrows indicate their relative orientations. Only bases found in all 59-be are shown, while other consensus bases are represented by dots. An extra base in 2L is marked by an ×. In any individual 59-be, sites numbered 1 are closely related, as are sites numbered 2, but the bases between them are not. The left hand (LH) and right hand (RH) simple sites consist of pairs of core sites (1L and 2L; 2R and 1R, respectively) together with flanking sequences. The region located between the LH and RH simple sites is an inverted repeat (represented by a pair of arrows) which has a variable sequence and length. A vertical arrow indicates the recombination crossover point. On integration of the cassette into the *attI* site of an integron, the 1R core site is split at the recombination crossover point so that the last six bases of 1R in the circular cassette become the first six bases of the integrated cassette.

Cassettes usually Congregate in Integrons

The normal location for gene cassettes is within the *attI* site of an integron (see Integrons). Indeed, cassettes were first identified as discrete entities because they constituted variable regions found in the surrounding conserved integron structure, and only subsequently was their mobility established experimentally. Integrons can capture one or many gene cassettes to form arrays of cassettes. These arrays can include one, a few, or many gene cassettes. The arrays can readily be lengthened by incorporation of new cassettes, shortened by excision of one or more cassettes, or reshuffled to create new orders. All of these events can be effected by IntI1-mediated recombination between *attI* and a 59-be or between two 59-be. Several different classes of integron that include different *intI/attI* modules have been found and identical cassettes have been found in the cassette arrays of the integrons belonging to different classes (1, 2, 3, etc.). This indicates that integrons share gene cassettes and that the 59-be sites are recognized by all IntI1 integrases.

However, cassettes are not always found associated with an integron. The IntI1 integrase and possibly other IntI integrases can, at low frequency, catalyze recombination between a 59-be site (a primary recombination site) and a secondary site, and this reaction can lead to the integration of a gene cassette at a location other than an *attI* site. The secondary sites conform to a simple consensus (Ga/tT) and this potentially permits incorporation of new genes at many different positions. Though this reaction is much less efficient than recombination between two primary sites, it may be quite important in the evolution of bacterial chromosomes. However, when a cassette is incorporated at a secondary site, the gene it contains can only be expressed if the cassette includes a promoter or an appropriately oriented promoter is located upstream.

Further Reading

Hall RM and Collis CM (1995) Mobile gene cassettes and integrons: capture and spread of genes by site-specific recombination. *Molecular Microbiology* 15: 593–600.

Recchia ED and Hall RM (1995) Gene Cassettes: a new class of module element. *Microbiology* 141: 3015–3027.

Recchia GD and Hall RM (1997) Origins of the mobile gene cassettes found in integrons. *Trends in Microbiology* 389: 389–394.

Stokes HW, O'Gorman DB, Recchia GD, Parsekhian M and Hall RM (1997) Structure and function of 59-base element recombination sites associated with mobile gene cassettes. *Molecular Microbiology* 26: 731–745.

See also: **Integrase Family of Site-Specific Recombinases; Integrons; Site-Specific Recombination**

Gene Conversion

F W Stahl

Copyright © 2001 Academic Press
doi: 10.1006/rwgn.2001.0503

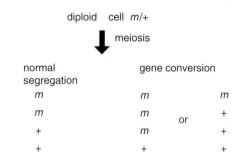

Figure 1   Gene conversion in meiosis. When a heterozygous diploid cell (genotype *m*/+) undergoes meiosis, the usual outcome is two haploid cells of genotype *m* and two of genotype +. Occasionally, however, this normal, Mendelian segregation is disturbed, with three of the haploid cells being of one genotype and one of the other. This aberrant segregation is a manifestation of meiotic gene conversion.

Gene conversion is an event in which a gene in a heterozygous diploid appears to have taken on the identity of its allele. It is distinguished conceptually and operationally from crossing-over by its nonreciprocal nature. This nonreciprocality can be convincingly demonstrated for conversion that occurs during meioses in which all four products of individual acts of meiosis can be recovered, as in many fungi. When the heterozygous diploid *A*/*a* undergoes meiosis, most of the resulting tetrads are composed of two *A* cells and two *a* cells (normal 2:2 segregation). Gene conversion is manifest as occasional violations of this Mendelian rule, in which ratios 3*A*:1*a* or 3*a*:1*A* (3:1 segregation) occur (**Figure 1**).

Related Aberrant Segregation Ratios

Other variations in the segregation ratio can be seen in fungi that have eight spores due to a postmeiotic mitosis. These variations can be seen also in four-spored fungi when care is taken to determine separately the genotypes of the chromosomes carried in the daughter cells of the first mitosis following meiosis. The common variations are 5:3 (5*A*:3*a* or 5*a*:3*A*) or aberrant 4:4. In 5*A*:3*a* tetrads two of the haploid cells are *A* on both strands, one is *a* on both strands, and one is heteroduplex *A*/*a*. These tetrads are 'half-conversion' tetrads by virtue of having an allele ratio which is half-way between the normal 2:2 ratio and the 3:1 ratio of (full) conversion. In aberrant 4:4 tetrads, one haploid cell is *A* on both strands, one is *a* on both strands, and two are *A*/*a* heteroduplexes. In both 5:3 and aberrant 4:4 tetrads, segregation of alleles is completed only at the first postmeiotic mitosis. Accordingly, such tetrads are often called

postmeiotic segregation (PMS) tetrads. Tetrads with segregation other than normal 2:2 are collectively called aberrant segregation tetrads. The rarity of tetrads whose ratios are more extreme than 6:2 implies that two of the four meiotic chromatids are uninvolved in any given interaction that leads to aberrant segregation.

Conversion Results in Nonreciprocal Recombination

In two-factor crosses, aberrant segregation at one site can occur separately from that at the other site – aberrant segregation is local. When the aberrant segregation is 3:1, such tetrads produced from the diploid *AB*/*ab* will be of four kinds depending on which site converts and in which direction: (*AB AB Ab ab*) and (*AB aB ab ab*) have both been converted at the site marked by the alternatives *A* and *a*; (*AB AB aB ab*) and (*AB Ab ab ab*) have been converted at the site marked by the alternatives *B* and *b*. Each tetrad contains a recombinant spore, *Ab* or *aB*, but not a pair of complementary recombinants. Thus, conversion produces recombinants nonreciprocally.

In two-factor crosses, the other kinds of aberrant tetrads contain recombinants, too. In a 5*a*:3*A* tetrad, the eight strands of DNA, paired as they would be in a tetrad, are *AB AB, AB aB, ab ab, ab ab*. In a tetrad that is aberrant 4:4 at the *A* site, the eight strands are *AB AB, AB aB, Ab ab, ab ab*. Thus, 5:3 tetrads are recombinant (on one polynucleotide strand) nonreciprocally; aberrant 4:4 are recombinant (on two strands) reciprocally. Coaberrant segregation in PMS tetrads is common for markers within a few hundred base pairs of each other.

Conversion Gradient

The frequency of conversion (half and/or full) for markers within a given gene varies with the position of the marker. These rates may vary monotonically from one end of the gene to the other (conversion gradient).

Aberrant Segregation and Crossing-Over

In three-factor crosses with linked markers, aberrant segregation at the central site is accompanied by crossing-over of the flanking markers (as long as they lie outside the aberrant segregation tract) about half the time. Conversion gradients and the correlation of aberrant segregation with crossing-over have motivated models for meiotic recombination.

The Double-Strand-Break-Repair Model for Conversion and Crossing-Over

A double-strand-break-repair (DSBR) model has enjoyed support both from genetic analysis of tetrads and from physical analysis of meiotic DNA. In the DSBR model as currently understood (**Figure 2**), one chromatid is cut at a place that, for reasons of chromatin structure, is sensitive to a meiosis-specific endonuclease. In *Saccharomyces cerevisiae*, these places are often at promoters of transcription. The 5′-ended strands on each side of the break are resected by an exonuclease. The resulting 3′-ended single strands bind a protein related to RecA of *Escherichia coli*, which enables them to invade a chromatid of the homolog. The invading ends form hybrid DNA with the complementary strand of the intact homolog displacing the resident strand. Heteroduplexes (hybrid DNA with point(s) of noncomplementarity between the two strands) may activate the mismatch repair system, resulting in degradation of some of the invading strand. DNA synthesis, primed by the invading 3′ ends and using the intact homolog as template, replaces DNA lost by the initial resection and by mismatch repair. These rounds of DNA destruction and replacement can result in 5:3 and 6:2 segregations (**Figure 2**). As a result of synthesis and covalent completion of both invading strands, the two participants are held together in a joint molecule by two Holliday junctions. The junctions may be moved outwards by the action of proteins like RuvA and RuvB of *E. coli*. Resolution of the joint molecule to give two duplexes can result in either crossing-over or noncrossing-over. Crossing-over will occur when one Holliday junction is cut (by an enzyme, resolvase) on one pair of strands and the other is cut on the other strands ('vertically' and 'horizontally' in **Figure 2**, left). If both junctions

are cut on the same strands (both 'vertically' or both 'horizontally'), noncrossing-over will result (**Figure 2**, right). Noncrossovers will result, also, if only one junction is cut by resolvase and the other junction slides to the still open site of the first. A topoisomerase could also effect this alternative noncrossover resolution of the joint molecule. Strand interruptions introduced by resolvase may direct a second round of mismatch repair.

In *S. cerevisiae*, genetic support for the DSBR model includes the following: (1) in diploids heterozygous for an endonuclease-sensitive site, the chromatid that carries the active site loses markers near that site; (2) a conversion gradient is demonstrable on both sides of the initiating site; (3) when conversion is accompanied by crossing-over of flanking markers, the exchange effecting the crossing-over cannot be located uniquely to one side or the other of the converted site; (4) in the absence of the major mismatch repair system, or when the markers used escape detection by that system, the frequency of 5:3 tetrads rises at the expense of 6:2 tetrads; and (5) aberrant 4:4 tetrads are seen for markers at the low end of the conversion gradient.

In the *ARG4* gene of *S. cerevisiae* the noncrossover resolution of the joint molecule intermediate appears to occur only rarely by cutting of the two Holliday junctions. In tetrads that segregate 5:3 for markers close to and on opposite sides of the initiation site, most observed double heteroduplexes are in the same chromatid, in the configuration shown in **Figure 2** for the alternative resolution (Gilbertson and Stahl, 1996). Furthermore, 5:3 tetrads of the type shown on the right in **Figure 2** are so rare in yeast as to be called 'aberrant 5:3s.' They are recognized as noncrossovers manifesting quasi-reciprocal exchange of a short segment between the two participating chromatids.

Physical analyses of isolated meiotic yeast DNA have supported the DSBR model: (1) double-strand breaks occur at hot spots for meiotic recombination at rates commensurate with the rates of aberrant segregation of markers near those hot spots; (2) the 5′-ended strands on either side of a double-strand break are eroded; (3) joint molecules, in which homologous duplexes are held together by a pair of Holliday junctions, arise near recombination hotspots (Schwacha and Kleckner, 1995); and (4) mutations that block the progression of physically monitored events block meiotic recombination.

The DSBR model offers the opportunity for full conversion that is independent of mismatch repair of heteroduplexes. If the 3′-ending as well as the 5′-ending strand should be resected, a double-strand gap arises. The repair of this gap using the homolog as template will result in a full conversion tetrad for

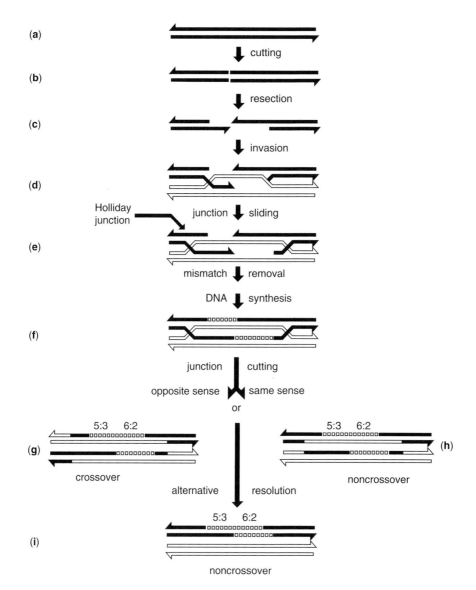

Figure 2 Double-strand-break-repair model. A duplex (a) is cut on both strands at a hot spot (b). Resection of 5′ ends creates 3′ overhangs (c). These single strands of DNA bind proteins like RecA of *Escherichia coli* and invade the homolog, creating regions of hybrid DNA (d). (Alternatively, invasion by one end, followed by DNA synthesis primed by that end, displaces a strand from the intact chromatid, which can then anneal with the resected end on the other side of the initiating break. The resulting double Holliday junction intermediate is the same for either scenario.) The Holliday junctions (where the strands swap partners) may be pushed outward (branch migration) (e). Mismatch repair of heteroduplexes (shown only on the right side of the initial break) removes invading DNA from the break site to a mismatch. DNA lost by resection is resynthesized (broken lines) using the intact homolog as template, creating a joint molecule (f). Resolution of the joint molecule results in crossing-over if one junction is cut vertically and the other horizontally (g). Noncrossovers result if both junctions are resolved in the same way (h) or if the two participating duplexes are separated from each other by an alternative route that could involve a topoisomerase or could result from the cutting of one junction followed by sliding of the other (i). In the *ARG4* gene of *Saccharomyces cerevisiae*, the rarity of 5:3 segregations of the type shown on the bottom-right (h) and the occurrence of tetrads like those shown on the bottom-middle (i) imply that cutting of the two Holliday junctions is rarely the route to the noncrossover resolution of joint molecules. (Modified from Szostak *et al.*, 1983.)

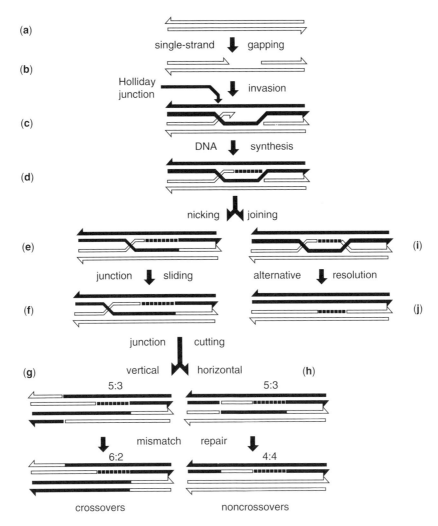

Figure 3   Single-strand-gap-repair model. Premeiotic replication of chromosomes (a) results in occasional single strand gaps in one daughter or the other (b). RecA-like protein binds to this single-stranded DNA and catalyzes interaction with a chromatid of the homolog (c). The invading 3′ end primes DNA synthesis using the intact homolog as template (D). The resulting joint molecule may contain two Holliday junctions (i), or only one (e). Resolution of the joint molecule by cutting the single Holliday junction (f) may yield either crossover (g) or noncrossover products (h). Examples of segregation ratios prior to and consequent to mismatch repair are shown. If two Holliday junctions are formed, alternative resolution to form noncrossovers, effected by topoisomerase or by cutting of only one junction, would preserve the homolog genetically intact (j), in keeping with most observations of noncrossovers in *Saccharomyces cervisiae*. (Modified from Kuzminov, 1996.)

any marker that was in the gap. The failure to replace all 6:2 tetrads with 5:3 tetrads consequent to the removal of mismatch-repair systems suggests such gap repair. When the marker examined is a deletion of the initiation site, the only conversions seen are full conversions, which favor the deletion and which occur independently of mismatch repair.

Conversion Initiated by Single-Strand Gaps?

Hot spot recombination is demonstrably due to hot spots for meiosis-specific double-strand cuts and

accounts for a major fraction of meiotic recombination in *S. cerevisiae*. The possibility of other routes to conversion, perhaps with attendant crossing-over, remains open. A prime candidate for another route is single-strand gap repair. When DNA is damaged on one strand, replication can skip across the impediment, producing two duplexes, one of which is gapped in its daughter strand. In mitotic cells, such a gap can be repaired, and the impediment removed, with help from the sister duplex (West *et al.*, 1981). In meiotic cells, a chromatid left gapped on one strand following premeiotic DNA replication may be repaired with the aid of the homolog, rather than the sister chromatid

(**Figure 3**). Such events could be responsible for crossing-over and conversion that may be unaccounted for by hot-spot-initiated recombination.

Meiotic Conversion Separable from Crossing-Over

Factors that alter the fraction of conversions that are accompanied by crossing-over, without altering the total rate of conversion, are understood to be operating on the resolution of joint molecules. However, the existence of factors that alter the frequency of meiotic conversion without changing the frequency of crossing-over suggests that some conversions are formed by a route that does not lead to crossing-over. Single-strand gap repair (**Figure 3**) could be such a route if resolution of joint molecules so formed were constrained to noncrossover modes. It is plausible, however, that some treatments alter conversion rates, but not crossover rates, by altering the lengths of heteroduplex DNA and/or the probabilities of mismatch repair in joint molecule intermediates like those of **Figure 2** without altering the rate of formation of such joint molecules or the mode of their resolution.

Common Misuse of 'Conversion'

'Conversion' is often used to denote just those meiotic conversion events that are not accompanied by crossing-over of flanking DNA. This use tends to create the false impression that meiotic conversion and crossing-over are mutually exclusive events.

Other Occurrences of Conversion

'Conversion' is widely used to denote any recombination event that appears to involve nonreciprocal exchange of a segment of DNA, especially when those events are unaccompanied by exchange of flanking markers. Transformation of cells by introduced fragments of genomic DNA qualify for this use of 'conversion,' as does phage-mediated transduction. In vegetative or somatic cells that contain reverse transcriptase, a DNA copy of an mRNA molecule can 'convert' a homolog. Conversion (the nonreciprocal change of a gene by its homolog) may be responsible for maintaining sequence identity between multicopy genes. Some of these conversions may occur by the same mechanisms as does meiotic conversion.

Some conversions serve to alter gene expression. Among such events are mating-type switching in some yeasts, surface antigen changes in trypanasomes, and functional diverse immunoglobulin gene formation in chickens. The first two systems involve the nonreciprocal transfer of information from a silent locus to an expression locus and are not normally accompanied by crossing-over.

Further Reading

Fogel S, Mortimer R, Lusnak K and Tavares F (1979) Meiotic gene conversion: a signal of the basic recombination event in yeast. *Cold Spring Harbor Symposia on Quantitative Biology* 43: 1325–1341.

Kuzminov A (1996) *Recombinational Repair of DNA Damage.* Austin, TX: RG Landes.

Stahl FW (1996) Meiotic recombination in yeast: coronation of the double-strand-break repair model. *Cell* 87: 965–968.

Szostak JW, Orr-Weaver TL, Rothstein RJ and Stahl FW (1983) The double-strand-break repair model for recombination. *Cell* 33: 25–35.

Whitehouse HLK (1982) *Genetic Recombination.* Chichester, UK: John Wiley.

References

Gilbertson LA and Stahl FW (1996) A test of the double-strand break repair model for meiotic recombination in *Saccharomyces cerevisiae*. *Genetics* 144: 27–41.

Schwacha A and Kleckner N (1995) Identification of double Holliday junctions as intermediates in meiotic recombination. *Cell* 83: 783–791.

West SC, Cassuto E and Howard-Flanders P (1981) Mechanism of *E. coli* RecA protein directed strand exchange in post-replication repair of DNA. *Nature* 294: 659–662.

See also: Double-Strand Break Repair Model; Genetic Recombination; Hot Spot of Recombination; Non-Mendelian Inheritance; Recombination, Models of

Gene Dosage

Copyright © 2001 Academic Press
doi: 10.1006/rwgn.2001.1848

Gene dosage is the number of copies of a particular gene locus in the chromosome. In most cells, this is either one or two.

See also: Dosage Compensation

Gene Duplication

D Carroll

Copyright © 2001 Academic Press
doi: 10.1006/rwgn.2001.0505

Gene duplication is a process that occurs periodically (usually rarely) within genomes of all types of organisms. As the name implies, one or more additional copies of a preexisting gene are generated. The new copy may reside adjacent to the original (tandem duplication) or be inserted at a novel chromosomal location (dispersed duplication). The duplication process may be reiterated a number of times, leading to the production of gene families; the history of various family members can often be deduced by sequence comparisons.

Tandem duplications, i.e., the creation of a new copy of a gene right next to the old copy on a chromosome, probably occur by unequal crossing-over between homologous chromosomes or sister chromatids (**Figure 1**). If the whole gene and the regulatory sequences that control its expression are duplicated, the new copy will be expressed in the same way as the old one. If one of the two copies accumulates mutations that inactivate the gene product, this will have no consequence for the organism, since the other copy will provide the necessary function. Inactivated gene copies are called pseudogenes. In rare circumstances, mutations in a gene copy will lead to a new function or a new pattern of expression for the gene product. In

this fashion, the gene repertoire is augmented without the loss of preexisting functions.

In the genomes of higher organisms, there are many examples of gene families that have arisen by gene duplication. For instance, the five human genes for the various β chains of hemoglobin that are expressed at different times during development are located in a cluster on chromosome 11 (**Figure 2**). There is also one pseudogene in this cluster. The four genes for the α chains of hemoglobin reside in a separate cluster on chromosome 16, where there are also three pseudogenes. Examining the sequence relationships among these genes, we can deduce that an ancient duplication event created separate α and β genes, and they were subsequently dispersed. Then, each of these was amplified by several tandem duplication events. Accumulated mutations created the current distinctions among the family members.

Dispersed duplications are sometimes the result of making a DNA copy of a messenger RNA and

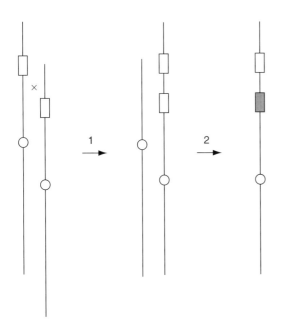

Figure 1 A hypothetical gene duplication event. In step 1, two homologous or sister chromatids undergo unequal crossing-over, as indicated by the x in the left diagram. This creates one chromosome with a deletion of the gene indicated by the open rectangle, and another with a duplication of that gene. In step 2, one of the copies of the duplicated gene is modified by mutation, indicated by the shading. This modification may inactivate the second copy, or it may alter its function or pattern of expression. The circle represents a centromere.

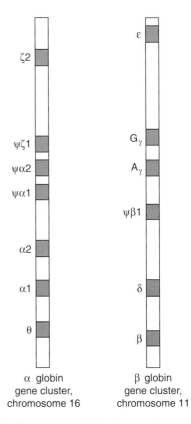

α globin gene cluster, chromosome 16

β globin gene cluster, chromosome 11

Figure 2 Depiction of the human hemoglobin gene family. In the α globin cluster, α1 and α2 are expressed in fetal and adult stages, ζ2 in early embryos. φζ1, φα2 and φα1 are pseudogenes that are no longer functional. The role of the θ gene is not known. In the β globin cluster ε is expressed in embryos, ^{G}γ and ^{A}γ in the fetal stage, while β and δ are the major and minor adult forms, respectively. φβ1 is a pseudogene.

inserting that copy at a novel chromosomal site. Typically these copies will not be expressed because they do not have the appropriate regulatory sequences around them at the new location. Because the organism does not rely on the new copy for a functional gene product, mutations that accumulate in it will be neutral, i.e., there is no selection against them, and most such duplicates exist as pseudogenes. They can be recognized because, like the mRNAs that are their progenitors, they lack introns, and the sequences surrounding them in the chromosome bear no resemblance to sequences around the real gene from which they were derived.

Sometimes very large segments of a chromosome are duplicated at once. This has been observed in bacteria, where as much as 25% of the chromosome may be duplicated in a single event. Such large tandem duplications are relatively unstable because the entire duplicated segment is a target for elimination by homologous recombination.

In some instances, gene amplification by tandem duplication can give cells a growth advantage. This has been observed in some human tumors, where duplication of a gene whose product is involved in promoting cell proliferation can overcome normal cell cycle regulation and lead to uncontrolled growth. An example is amplification of the *N-myc* gene in some cancers of the nervous system. The amplified copies are arranged in tandem and they may be located at the normal *N-myc* chromosomal site or spun off as extrachromosomal elements called double-minute chromosomes. These amplification events usually occur in somatic cells during the life of the organism and they are not passed on to succeeding generations as stable gene families.

Further Reading

Lewin B (1997) *Genes VI*. New York: Oxford University Press.

Romero D and Palacios R (1997) Gene amplification and genome plasticity in prokaryotes. *Annual Review of Genetics* 31: 91–111.

See also: Double-Minute Chromosomes; Evolution of Gene Families; Gene Amplification

Gene Expression

J Parker

Copyright © 2001 Academic Press
doi: 10.1006/rwgn.2001.0506

The genetic material contains the information necessary for an organism to develop, function, and reproduce, but it is necessary that this information be expressed in order for any activity, even maintenance, to be carried out. Therefore, one can consider gene expression as encompassing all the processes which are necessary to produce a gene product from a gene. One can also include all regulatory steps, those necessary to synthesize the gene product in appropriate amounts at an appropriate time and those involved in regulating the activity of the gene product. Therefore, the following description is not meant to be an exhaustive account of gene expression, but an overview of some of the processes involved. Many of these processes are explored in more detail in articles elsewhere in this volume.

Transcription

The genes of all cellular organisms are composed of double-stranded DNA (some viruses have single-stranded DNA genomes and others even RNA genomes) and the first step in their expression is transcription (see Transcription). Transcription involves using one of the two strands of DNA as a template to make an RNA copy by an enzyme called RNA polymerase (see RNA Polymerase). All RNA polymerases synthesize an RNA chain from the 5′ end to the 3′ end while reading the template strand of the DNA in the 3′ to 5′ direction. The RNA molecules are synthesized from specific starting sites on the DNA and also terminate at specific sites. The sites where RNA polymerase (using accessory factors) recognizes the beginning of a transcriptional unit are termed promoters (see Promoters). In higher organisms, the unit of transcription is almost always a single gene. However, in prokaryotes the transcriptional unit may contain several contiguous genes. These genes are often related in function and/or belong to one pathway.

Transcription is a target of several regulatory mechanisms. These can serve to repress or activate transcription, or lead to premature termination. One common mechanism in bacteria is the binding of a repressor protein to a specific region of the DNA near the promoter which then blocks transcription (see Repressor). The sequence to which the repressor protein binds is termed an 'operator' (see Operators), a term which has given its name to the transcriptional unit called an 'operon' (see Operon). In bacteria an operon may contain one or more genes, all under the control of the single operator. Another mechanism for regulating gene expression is the binding of a regulatory protein to the DNA which activates transcription. Such positive control is widespread in eukaryotic genes. It is not uncommon for genes to be under more than one form of regulation, nor is it uncommon, in bacteria, for some regulatory proteins to be both repressors and activators for different genes. Attenuation is another form of transcriptional regulation, but in this case the transcript is terminated early in elongation (see Attenuation). The mechanism by which attenuation takes place can vary quite dramatically between different organisms. Also not all regulatory molecules are proteins; regulatory RNA can also play a role (see Regulatory RNA).

The majority of genes encode proteins, and the RNA transcript must then be used as (or processed to become) a messenger RNA (mRNA). As mentioned above, eukaryotic transcriptional units are almost always single genes, but some transcripts from protein-encoding genes (particularly from animals) can be very long (more than one million bases). The great length of these transcripts results from the fact that the protein-encoding genes of eukaryotes often have several introns (noncoding sequences) interspersed within the coding sequences (exons), and these are transcribed as a unit. Such genes are sometimes referred to as 'split genes' (see Introns and Exons; Split Genes). In genes containing introns, then, one part of gene expression is the processing of the transcript to remove these introns. Indeed, in eukaryotes most transcripts from protein-encoding genes need three distinct processing steps to be converted into mRNA: capping, splicing, and tailing. Capping involves adding a modified guanosine to the 5′ end of the pre-mRNA. It is this cap that allows the RNA to be recognized by the translational machinery of the cell as an mRNA. The RNA splicing process removes introns and joins the exons together. Tailing involves cutting the transcript at a specific site downstream of the region encoding the protein and polyadenylating the newly created 3′ end.

These processing events are coupled to transcription. Capping takes place very soon after transcription has started. At least in the higher eukaryotes, where genes may have, in the extreme, many large introns, splicing is also coupled to transcription. The splicing process in eukaryotic pre-mRNA is complex and involves ribonucleoprotein particles called 'spliceosomes' that contain various protein factors and small nuclear RNA molecules (snRNPs or 'snurps'; see Pre-mRNA Splicing). Splicing involves recognition of specific sites on the RNA and very precise cleavage and ligation of the RNA (since an error of a single nucleotide will result in a frameshifted message). Splicing is also regulated, and some genes have transcripts that can be spliced in more than one way (alternative splicing) to yield more than one protein from a single gene. Alternative splicing pathways are particularly prevalent in the transcripts from genomes of small animal viruses but occur in other genomes also.

The transcripts of protein-encoding genes from prokaryotes do not require processing to be functional;

therefore, the transcripts of these genes are mRNAs. Also, as mentioned above, some transcriptional units in prokaryotes contain information from several contiguous genes. The mRNAs produced from such units are said to be 'polycistronic,' in contrast to 'monocistronic' mRNA, which carries information for only one gene product (see Polycistronic mRNA). In *Escherichia coli* over 70% of the mRNA is monocistronic and about 30% is polycistronic (with about 6% containing the information from four or more genes).

For some genes the final product is an RNA molecule, but even here processing is involved, and in this case processing occurs in both prokaryotes and eukaryotes. (Therefore, the only major class of RNA that can be used directly as transcribed is mRNA from prokaryotes.) The only genes we shall discuss here whose final product is RNA are genes encoding transfer RNA (tRNA) and genes encoding ribosomal RNA (rRNA). In both prokaryotes and eukaryotes, some of both types of genes may contain introns. Although the process by which these introns are removed involves excising the intron and ligating the exons, and is called 'splicing,' the machinery which performs these reactions is not related to that which splices eukaryotic mRNA (see Introns and Exons). Some of the introns in rRNA and tRNA are self-splicing (and self-splicing introns are also known in a few bacteriophage mRNAs). Self-splicing introns (a particular kind of self-splicing intron) are widely found in nature and they are the only type found in bacteria and bacteriophages. In both eukaryotes and prokaryotes, tRNAs and rRNAs are made initially as longer precursors and all must be cut to their final size. In addition, tRNAs contain many modified bases (and in some cases the final conserved CCA sequence at the 3′ end must be added enzymatically; see Transfer RNA (tRNA)). Modification of rRNAs is less extensive (see Ribosomal RNA (rRNA)).

All these RNAs, whether they are informational intermediates like mRNA or final products of gene expression like tRNA and rRNA, are used in the next step of gene expression: translation.

Translation

In prokaryotes, transcription and translation are coupled, that is, the translation of a mRNA begins before its synthesis is complete. There are even some regulatory mechanisms which take advantage of this coupling (see Attenuation). In eukaryotes, however, transcription (and processing) occurs in the nucleus and translation occurs in the cytoplasm. Therefore, in eukaryotes the mature mRNA must be transported to the cytoplasm. In all organisms, most mRNA is reasonably unstable (as contrasted to tRNA and rRNA), but the stability of mRNAs from different genes can vary widely, and mRNA stability is another area where gene expression can be regulated.

Translation itself is the process whereby protein is synthesized using the information in the mRNA as a template (see Translation). This process takes place on large ribonucleoprotein particles called 'ribosomes' (see Ribosomes) which contain many different proteins and one copy of each of the different rRNAs that the cells make. A large number of different protein factors as well as the cell's tRNAs are involved in the overall process. However, it has been demonstrated that peptide bond formation (the linkage together of the amino acid residues) is catalyzed by the large sub-unit rRNA.

In translation the ribosome (and attendant factors) must first recognize the start site of the information encoding the protein and then proceed down the mRNA (in the 5′ to 3′ direction) until a stop codon is reached and chain growth is terminated. The protein synthesized will have an amino acid sequence corresponding in identity and order to the three base codons of the genetic code (see Genetic Code). Prokaryotic ribosomes bind to mRNA at a ribosome-binding site, which is a larger sequence than just the start codon (see Ribosome Binding Site). Eukaryotic ribosomes typically bind to the cap at the 5′ end of the mRNA and travel down the ribosome, initiating protein synthesis at the first possible start codon (AUG, methionine). The differences in the signal for binding of the ribosome to the mRNA and initiating the synthesis of a protein allow prokaryotic ribosomes to use polycistronic mRNA, since downstream cistrons will require initiation of protein synthesis from some genes toward the middle of such an mRNA molecule.

The codons on the mRNA are 'read' by anticodons on aminoacylated tRNAs, and peptide bonds are formed between consecutive amino acid residues carried by adjacent tRNAs. The protein being synthesized is typically folding during synthesis and, when a stop codon is reached, the completed protein is hydrolyzed from the last tRNA and released; its tertiary structure may be nearly formed. Translation is also a step at which regulation can occur, and mechanisms of translational regulation are known that involve both regulatory RNA and regulatory protein.

Posttranslation Steps in Gene Expression

There are several possible steps that can take place after translation which alter the activity of a protein (and therefore alter gene expression). Of course, many enzymes can be inhibited or activated by a number of noncovalent interactions with small molecules. However, many proteins are subject to covalent modifications which also affect their normal activity, location, or stability. Indeed the majority of proteins undergo at least some modification as the initiating methionine (or N-formyl-methionine) is removed. Some proteins require more extensive processing. For instance, trypsin is cleaved from an inactive precursor, and many peptide hormones such as insulin are cleaved in a more complicated pattern from larger molecules. There are also examples known where the protien must be 'spliced.' Protein splicing involves cutting out intervening amino acid residues (called 'inteins') and ligating together those portions of the protein required for activity ('exteins'). Although not as common as RNA splicing, protein splicing has been found in a number of organisms, both prokaryotic and eukaryotic (see Protein Splicing).

One other type of cleavage that may occur relates to proteins that are specifically transported into various membrane-bound cellular compartments or exported from the cell. Such proteins have a signal sequence, or leader peptide, at their N-terminus which is cleaved off by the cellular machinery during transport of the protein across the membranes (see Leader Peptide).

Finally there are many examples known of small molecules being specifically covalently attached to proteins and at least some of these have regulatory significance. Proteins from higher eukaryotes are often extensively glycosylated, but other modifications also occur and modifications can also occur in prokaryotes. Some modifications such as the protein phosphorylations involved in signal transduction (a type of transcriptional control; see Signal Transduction) and the adenylation of glutamine synthetase are reversible. Some posttranslational covalent modifications convert a 'standard' amino acid inserted translationally into a modified amino acid, such as the iodotyrosine in the thyroxin hormones. Although all of these process are considered 'posttranslational,' at least some can occur cotranslationally.

See also: Attenuation; Autoregulation; Cistron; Derepression; Enhancers; Genetic Code; Induction of Transcription; Introns and Exons; Leader Peptide; Messenger RNA (mRNA); Operators; Operon; Polycistronic mRNA; Pre-mRNA Splicing; Promoters; Protein Splicing; Regulatory Genes; Regulatory RNA; Repressor; Ribosomal RNA (rRNA); Ribosome Binding Site; Ribosomes; RNA Polymerase; Signal Transduction; Split Genes; Transcription; Transfer RNA (tRNA); Translation; Translational Control

Gene Family

L Silver

Copyright © 2001 Academic Press
doi: 10.1006/rwgn.2001.0507

Origins and Examples

Much of the functional DNA in the genome is organized within gene families and hierarchies of gene superfamilies. The superfamily term was coined to describe relationships of common ancestry that exist between and among two or more gene families, each of which contains more closely related members. As more and more genes are cloned, sequenced, and analyzed by computer, deeper and older relationships among superfamilies have unfolded. Complex relationships can be visualized within context of branches upon branches in evolutionary trees. All of these superfamilies have evolved out of combinations of unequal crossover events that expanded the size of gene clusters and transposition events that acted to seed distant genomic regions with new genes or clusters.

A prototypical small-size gene superfamily is represented by the very well-studied globin genes. All functional members of this superfamily play a role in oxygen transport. The superfamily has three main families (or branches) represented by the β-like genes, the α-like genes, and the single myoglobin gene. The duplication and divergence of these three main branches occurred early during the evolution of

vertebrates and, as such, all three are a common feature of all mammals. The products encoded by genes within two of these branches – α-globin and β-globin – come together (with heme cofactors) to form a tetramer which is the functional hemoglobin protein that acts to transport oxygen through the bloodstream. The product encoded by the third branch of this superfamily – myoglobin – acts to transport oxygen in muscle tissue.

The β-like branch of this gene superfamily has duplicated by multiple unequal crossing over events and diverged into five functional genes and two β-like pseudogenes that are all present in a single cluster on mouse chromosome 7 as shown in **Figure 1**. Each of the β-like chains codes for a similar polypeptide which has been selected for optimal functionality at a specific stage of mouse development: one functions during early embryogenesis, one during a later stage of embryogenesis, and two in the adult. The α-like branch has also expanded by unequal crossing-over into a cluster of three genes – one functional during embryogenesis and two functional in the adult – on mouse chromosome 11. The two adult α genes are virtually identical at the DNA sequence level, which is indicative of a very recent duplication event (on the evolutionary time scale).

In addition to the primary α-like cluster are two isolated α-like genes (now nonfunctional) that have transposed to dispersed locations on chromosomes 15 and 17. When pseudogenes are found as single copies in isolation from their parental families, they are called 'orphons.' Interestingly, one of the α-globin orphons (*Hba-ps3* on Chr 15) is intronless and

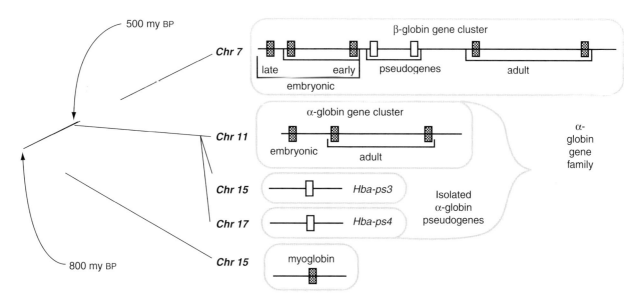

Figure 1   Mouse chromosome 7.

would appear to have been derived through a retro-transposition event, whereas the other orphon (*Hba-ps4* on chromosome 17) contains introns and may have been derived by a direct DNA-mediated transposition. Finally, the single myoglobin gene on chromosome 15 does not have any close relatives either nearby or far away. Thus, the globin gene superfamily provides a view of the many different mechanisms that can be employed by the genome to evolve structural and functional complexity.

The *Hox* gene superfamily provides an alternative prototype for the expansion of gene number. In this case, the earliest duplication events (which predate the divergence of vertebrates and insects) led to a cluster of related genes that encoded DNA-binding proteins used to encode spatial information in the developing embryo. The original gene cluster has been duplicated *en masse* and dispersed to a total of four chromosomal locations (on chromosomes 2, 6, 11, and 15) each of which contains nine to twelve genes. Interestingly, because of the order in which the duplication events occurred – unequal crossing-over to expand the cluster size first, transposition *en masse* second – an evolutionary tree would show that a single 'gene family' within this superfamily is actually splayed out physically across all of the different gene clusters. Some gene additions and subtractions within individual clusters have occurred by unequal crossing-over since the *en masse* duplication so that differences in gene number and type can be seen within a basic framework of homology among the different whole clusters.

A final example of a gene super-superfamily is the very large set of genes that contain immunoglobulin-like (Ig) domains and function as cell surface or soluble receptors involved in immune function or other aspects of cell–cell interaction. This set includes the immunoglobulin gene families themselves, the major histocompatibility genes (called *H2* in mice), the T cell receptor genes, and many more. There are dispersed genes and gene families, small clusters, large clusters, and clusters within clusters, tandem and interspersed. Dispersion has occurred with the transposition of single genes that later formed clusters and with the dispersion of whole clusters *en masse*. Furthermore, the original Ig domain can occur as a single unit in some genes, but it has also been duplicated intragenically to produce gene products that contain two, three, or four domains linked together in a single polypeptide. The Ig superfamily, which contains hundreds (perhaps thousands) of genes, illustrates the manner in which the initial emergence of a versatile genetic element can be exploited by the forces of genomic evolution with a consequential enormous growth in genomic and organismal complexity.

Tandem Families of Identical Genetic Elements

A limited number of multicopy gene families have evolved under a very special form of selective pressure that requires all members of the gene family to maintain essentially the same sequence. In these cases, the purpose of high copy number is not to effect different variations on a common theme, but rather to supply the cell with a sufficient amount of an identical product within a short period of time. The set of gene families with identical elements includes those that produce RNA components of the cell's machinery within ribosomes and as transfer RNA. It also includes the histone genes which must rapidly produce sufficient levels of protein to coat the new copy of the whole genome that is replicated during the S-phase of every cell cycle.

Each of these gene families is contained within one or more clusters of tandem repeats of identical elements. In each case, there is strong selective pressure to maintain the same sequence across all members of the gene family because all are used to produce the same product. In other words, optimal functioning of the cell requires that the products from any one individual gene are directly interchangeable in structure and function with the products from all other individual members of the same family. How is this accomplished? The problem is that once sequences are duplicated, their natural tendency is to drift apart over time. How does the genome counteract this natural tendency?

When ribosomal RNA genes and other gene families in this class were first compared both between and within species, a remarkable picture emerged: between species, there was clear evidence of genetic drift with rates of change that appeared to follow the molecular clock hypothesis. However, within a species, all sequences were essentially equivalent. Thus, it is not simply the case that mutational changes in these gene families are suppressed. Rather, there appears to be an ongoing process of 'concerted evolution' which allows changes in single genetic elements to spread across a complete set of genes in a particular family. So the question posed previously can now be narrowed down further: how does concerted evolution occur?

Concerted evolution appears to occur through two different processes. The first is based on the expansion and contraction of gene family size through sequential rounds of unequal crossing-over between homologous sequences. Selection acts to maintain the absolute size of the gene family within a small range around an optimal mean. As the gene family becomes too large, the shorter of the unequal crossover products will be selected; as the family becomes too small, the longer products will be selected. This cyclic process will cause a continuous oscillation around a mean in size. However, each contraction will result in the loss of divergent genes, whereas each expansion will result in the indirect 'replacement' of these lost genes with identical copies of other genes in the family. With unequal crossovers occurring at random positions throughout the cluster and with selection acting in favor of the least divergence among family members, this process can act to slow down dramatically the continuous process of genetic drift between family members.

The second process responsible for concerted evolution is intergenic gene conversion between 'nonallelic' family members. It is easy to see that different tandem elements of nearly identical sequence can take part in the formation of Holliday intermediates which can resolve into either unequal crossing over products or gene conversion between nonallelic sequences. Although the direction of information transfer from one gene copy to the next will be random in each case, selection will act upon this molecular process to ensure an increase in homogeneity among different gene family members. As discussed above, information transfer – presumably by means of gene conversion – can also occur across gene clusters that belong to the same family but are distributed to different chromosomes.

Thus, with unequal crossing-over and interallelic gene conversion (which are actually two alternative outcomes of the same initial process) along with selection for homogeneity, all of the members of a gene family can be maintained with nearly the same DNA sequence. Nevertheless, concerted evolution will still lead to increasing divergence between whole gene families present in different species.

See also: **Concerted Evolution; Gene Conversion; Globin Genes, Human; Immunoglobulin Gene Superfamily; Molecular Clock; Unequal Crossing Over**

Gene Flow

J B Mitton

Copyright © 2001 Academic Press
doi: 10.1006/rwgn.2001.0508

Gene flow is defined as the movement of genes among populations. The rate of gene flow, m, is the proportion of the gene copies in a population that have been carried into that population by immigrants. Gene flow can be mediated by the dispersal of either gametes or individuals. But gene flow is not equivalent to dispersal, for gametes or individuals that move among populations but fail to incorporate genes into the gene pool have not mediated gene flow.

Population structure, the pattern of genetic variation among populations, is produced by the joint action of gene flow, genetic drift, and natural selection. Genetic drift is change in allelic frequencies produced by accidents of sampling and chance variation in survival, mating success, and family size. Natural selection is defined as the differential reproduction of genotypes.

Elaboration

Genetic Drift Differentiates Populations

If populations are not connected by gene flow, stochastic changes will cause them to diverge in time. Imagine a large, genetically diverse, randomly mating population that is suddenly broken apart into two perfectly isolated populations, each considerably smaller than the initial population. Initially, these populations might share the same alleles at similar frequencies. The Hardy–Weinberg Law demonstrates that, in the absence of selection, mutation, and migration, allelic frequencies will not change in an infinitely large population with random mating. But in finite populations, allelic frequencies drift over time, with stochastic variation in survival and reproduction. The stochastic loss of alleles may differ between populations, increasing the genetic distance between them. In addition, mutations may introduce new alleles into populations, further distinguishing them. With sufficient time, perfectly isolated populations will become completely differentiated, so that they do not share any alleles.

The rate of genetic drift in a population is dependent on the number of breeding adults in the population. Consider a gene segregating two alleles, A and a, at frequencies p and q, respectively, so that:

$$p + q = 1.0$$

The standard error (SE) of the allelic frequency is a measure of the magnitude of drift of the frequency of an allele in a single generation. The standard error of the frequency of an allele is:

$$\mathrm{SE} = \sqrt{\frac{pq}{2N}}$$

where N is the number of breeding adults. Most of the time (95%), the change in frequency will be less than two standard errors. For example, in a population with 1000 breeding adults, and p and q equal to 0.5, the standard error of allelic frequency is 0.01. Thus, in the next generation, 95% of the time, p will be greater than 0.48 but less than 0.52. However, in a population with the same frequencies but only 10 breeding adults, the standard error is 0.11, so the likely range of p in the next generation will be from 0.38 to 0.72. Thus, the rate of genetic drift increases with diminishing population size.

Gene Flow Tends to Homogenize Populations

Gene flow among populations makes them more similar. This point can be made intuitively by considering an exercise with two glasses of wine, one red and one white. Imagine pouring a small amount from the glass of white wine into the other glass, then swirling it. The red wine will still be red, but careful inspection would reveal that the intensity of the color has diminished. Now pour some of the red wine into the other glass. A few drops of red wine bring a tinge of red to the white wine. Now imagine repeating the exchanges many times. Ultimately, the colors in the two glasses will be identical. Similarly, some gene flow between populations will make them more similar, and high gene flow will make them indistinguishable.

The impact of gene flow on population structure can be illustrated quantitatively by modeling genetic variation at a single gene. Now consider gene flow into a population from populations that have different allelic frequencies. If the proportion of migrants into a population is m, and the frequency of A in the migrants is $\bar{p}$, then p', the new frequency of A in the population, will be:

$$p' = \bar{p}m + p(1 - m)$$

If gene flow were unopposed by other forces, the populations connected by gene flow would ultimately share the same alleles, at the same frequencies.

Natural Selection Can Overcome Gene Flow

Natural selection can oppose the homogenizing effect of gene flow, sustaining genetic differences among populations linked by gene flow. For example, the blue mussel, *Mytilus edulis* (**Figure 1**), exhibits an abrupt genetic boundary despite high gene flow. Blue mussels are native to the North Atlantic, and are common in the rocky intertidal. They are dioecious, i.e., an individual is either male or female, and they release their gametes into the water. The gametes unite to form veliger larvae, which are carried by currents for at least 3 weeks. Studies of coastal currents suggest that larvae could be carried more than 100 km, and an estimate of gene flow from genetic data (see below) indicates that blue mussels exchange many individuals among populations each generation. Despite high

levels of gene flow, the mussels in Long Island Sound remain distinctly differentiated from other populations.

Long Island Sound receives water from several major rivers (Housatonic, Quinnipiac, Connecticut, Thames), which dilutes the salinity of the Sound to about one half of the salinity of the open ocean. Thus, the Sound is a distinct environment for mussels, which must make physiological adjustment to retain osmotic cell pressure. Variation at the gene coding for leucine aminopeptidase (*Lap*) plays an important role in the maintenance of cell pressure; some genotypes are most efficient at high salinity, while other genotypes are most efficient at low salinity. Each spring, millions of larvae are carried into Long Island Sound by currents sweeping west along the coast of Rhode Island and Connecticut. But each fall, mortality in the young mussels creates a sharp genetic cline in *Lap* frequencies near Guilford, Connecticut, where salinity changes abruptly. Although the veliger are capable of dispersing more than 100 km, the genetic cline is only 20 km wide.

In addition, studies of both ribbed mussels, *Geukensia demissa*, and acorn barnacles, *Semibalanus balanoides*, have reported significant differentiation between the samples taken from the upper and lower portions of the intertidal zone – distances of one or two meters. These species, like the blue mussel, also have pelagic larvae, and consequently gene flow would homogenize the frequencies of neutral or unselected genes within the intertidal zone. Both cases of differentiation were produced by selection differing among habitats in a heterogeneous environment.

Some Generalizations concerning Gene Flow

Dispersal

Although gene flow is not synonymous with dispersal, it is certainly true that long-distance dispersal

Figure 1   (See Plate 17) The blue mussel, *Mytilus edulis*.

provides the opportunity for long-distance gene flow, and hence for high levels of gene flow among populations. The larvae of some marine mollusks have been documented to be carried by equatorial currents from the coast of Africa to the Caribbean Sea, and we would expect those species to have high levels of gene flow among populations in Africa or in the Caribbean. On the other hand, some marine mollusks brood their young, or attach egg cases to the substrate, severely limiting the opportunity for dispersal, and restricting gene flow. Species that are philopatric with respect to breeding sites, such as salamanders and some species of birds, are characterized by very low gene flow.

Mating System

The mating system can have a profound impact on gene flow. For example, the mating systems of plants can be characterized as predominantly selfing, or predominantly outcrossing, or a mixed system, employing an intermediate balance of selfing and outcrossing. Many species of plants, such as wheat, barley, oaks, and pines, are monoecious, meaning that an individual produces both male and female gametes. Wheat and barley produce their seeds predominantly (> 99%) by selfing. This mating system is characterized by very low gene flow, for there is no gene flow in the fertilization of selfed seed, and the seeds typically disperse less than 2 m. Gene flow is much higher in oaks and pines, which are typically outcrossed and wind-pollinated. Outcrossed seeds have separate maternal and paternal parents, and the wind pollination provides the possibility that the parents are distant from one another. How far can oak or pine pollen travel? Pollen traps on ships 150 km from shore have captured pine pollen, confirming long-distance dispersal, and providing the opportunity for long-distance gene flow.

Behavior can have a major impact on gene flow. Plants with animal pollinators will have gene flow determined by the behavior of their pollinators. Plants pollinated by bees that visit many flowers on a plant before visiting an adjacent plant will have low gene flow. Gene flow mediated by various species of hummingbirds can be low or high, depending on whether the birds defend small territories or are 'trapliners,' flying substantial distances between sequential pollinations.

Pods of killer whales around the San Juan Islands, Washington State, have distinct feeding behaviors that constrain their social systems and limit gene flow among pods. Some of the pods prey predominantly on marine mammals, such as seals and sea lions, while other pods prey almost exclusively on salmon. Long-term studies of the behaviors of the pods revealed that the pods defend their territories, and are stealthy when they trespass into the territories defended by neighboring pods. Studies of mitochondrial DNA identified diagnostic differences between mammal-eating and fish-eating pods, and suggest that gene flow between these pods had not occurred for 2000 years.

Direct Measurement of Gene Flow

The most direct measure of gene flow is to tag permanently an individual at or near its natal site, and then record where it breeds. For example, bird bands, which are amulets or rings placed on a bird's leg, have been used to study gene flow in many species of birds. Tags have been attached to the fins of fish, and tiny bar code signs have been glued on insects. Radio beacons fashioned into collars have revealed the movements of wolves and lynx. Fluorescent dyes prepared as a fine dust have been used to mark birds and small mammals for short periods of time. Radio transmitters have been placed in the stomachs of snakes and beneath the skin of sharks. Mammals have had their coats numbered with bleach or paint. These marking techniques have the advantage of providing clear evidence of dispersal and, if the animal breeds at its destination, evidence of gene flow. They have the disadvantage that they are often labor-intensive, and some of the tags, such as radio transmitters, are both expensive and short-lived. But these techniques cannot be used in all species and, in addition, they provide just a single estimate of gene flow. Because animal behavior is flexible, and can vary among years and generations, tagging studies may not reflect the average gene flow. Finally, population structure may be predominantly determined by historical events rather than the current rate of gene flow.

Inference from Genetic Data

F_{st} Measures the Differentiation of Populations

F_{st} is a quantitative estimate of the degree of differentiation of populations. Consider a gene segregating two alleles, A and a, at frequencies p and q, respectively. F_{st}, a standardized variance of allelic frequencies, is defined as:

$$F_{st} = \frac{S_p^2}{\bar{p}\bar{q}}$$

where the numerator is the variance of p among populations, and the denominator is the product of the means of the allelic frequencies. The variance of allelic frequencies among populations is calculated as:

$$S_p^2 = \frac{1}{d}\sum_i (p_i - \bar{p})^2$$

where d is the number of populations, p_i are the frequencies of the A allele in the populations, and $\bar{p}$ is the mean of the frequencies. F_{st} is zero if all populations have the same alleles at the same frequencies, and 1.0 for two populations fixed for different alleles. F_{st} will increase over time between isolated populations, and because genetic drift increases with decreasing population size, the rate of divergence increases with decreasing population size.

The degree of differentiation among populations will come to an equilibrium that reflects a balance between genetic drift and gene flow. The relationship between differentiation and gene flow is:

$$F_{st} = 1/(4Nm + 1) \quad \text{or, equivalently,}$$
$$Nm = (1/F_{st} - 1)/4$$

where N is the number of breeding individuals in a population. Thus, if populations are completely isolated for a long time, F_{st} will decline to zero, but if just one member of a population is a new immigrant (e.g., $Nm = 1$) then the rate of gene flow is:

$$m = \frac{1}{N}$$

and the equilibrium value of F_{st} will be 0.20. Higher rates of gene flow will make the populations even more similar. For example, if the number of immigrants is 5 per generation, then F_{st} will be less than 0.05, and the populations will be, for all practical purposes, very similar.

An important threshold is placed at the rate of gene flow of $Nm = 1.0$. Effectively, when $Nm < 1$, gene flow is not sufficient to offset the effects of genetic drift. So populations connected by $Nm < 1$ will diverge in time, while for populations connected by $Nm > 1$, gene flow will prevent differentiation by genetic drift.

Inference of Gene Flow in Limber Pine

The organellar genomes of pines are ideal for measuring gene flow, as mitochondrial DNA (mtDNA) has maternal inheritance and chloroplast DNA (cpDNA) has paternal inheritance in pines. These different modes of inheritance allow us to explicitly identify gene flow mediated by pollen and by seeds. In addition, pollen and seeds have disparate potentials for dispersal. The wind-borne pollen have the potential to travel great distances, but in contrast, the seeds of pines usually fall within a circle that has a radius equal to the height of the tree.

Limber pine, *Pinus flexilis* (**Figure 2**), is native to western North America, where it is primarily restricted to windy ridges and scree slopes from the

Figure 2   (See Plate 18) The limber pine, *Pinus flexilis*.

Sierra Madre of Mexico to the Canadian Rockies, from Mt Pinos in southern California to the Black Hills of South Dakota. The seeds of limber pine are dispersed and planted by Clark's nutcracker, *Nucifraga columbiana*. The bird and pine are engaged in a mutualism sculpted by evolution. Limber pine relies on the bird to harvest, disperse, and plant its seeds. Clark's nutcracker relies on limber pine seeds to get through the winter. Both the bird and the pine have evolved morphological traits (a sublingual pouch, wingless seeds) to better serve and exploit their partner. The birds usually cache seeds on windy or south-facing slopes that will be free of snow in winter, and this explains the curious distribution of limber pine. A bird can carry approximately 30 limber pine seeds in its sublingual pouch. When its pouch is full, the bird flies to a propitious site for caching and harvesting seed. The flight distances are highly variable; although the record flight exceeds 20 km, most flights are very short, a few meters to a few hundred meters.

The potentials for dispersal of pollen and seed lead biologists to expect high gene flow in genes dispersed by pollen (nuclear genes, cpDNA) and low gene flow for genes dispersed solely by seed. This hypothesis was tested with a study of gene flow among populations of limber pine in the Front Range of Colorado. The populations were distributed from tree line at the Continental Divide to an isolated stand of trees 100 miles to the east, on an escarpment on the Great Plains. Haplotype frequencies were used to calculate F_{st} for both cpDNA and mtDNA, and gene flow was inferred from F_{st} with the equation directly above. F_{st}s were 0.02 and 0.68 for cpDNA and mtDNA, respectively, suggesting that the number of migrants among populations per year are 12.25 for pollen and 0.12 for seeds. The gene flow of cpDNA is high, and should tend to homogenize the frequencies of cpDNA haplotypes and nuclear genes among populations within distances of approximately 100 miles. In contrast, the gene flow of mtDNA is below the threshold at which the influence of genetic drift predominates. So mtDNA is expected to vary more among populations than nuclear genes and cpDNA, and genetic drift will cause populations to diverge with respect to mtDNA haplotypes.

Private Alleles Estimate Gene Flow

Private alleles, or alleles found only in a single population, can also be used to infer rates of gene flow among populations. The private alleles can be from markers from mtDNA, cpDNA, nuclear DNA, or allozyme markers, and they are usually taken from surveys of geographical variation within a species. For example, a survey of allozyme variation throughout the range might reveal several or many private alleles. The average frequency of the private alleles, $\bar{p}$, is plotted on a regression line on a plot of $\ln(\bar{p})$ on the ordinate versus $\ln(Nm)$ on the abscissa. The regression line was estimated from a computer simulation study examining the relationship between genetic drift and gene flow in the determination of the geographical distribution of new mutations. Consider a species that has very low gene flow among populations. When mutation produces a novel allele in a single population, it could drift to moderate or even high frequencies before an individual bearing that allele migrated to another population and reproduced. However, if gene flow in the species was very high, then it is likely that the new mutation would still be at a low frequency when it was successfully introduced to another population. Thus, low gene flow allows private alleles to drift to higher frequencies while high gene flow holds private alleles to low frequencies.

Estimates of gene flow from private alleles are usually, but not always, consistent with estimates from F_{st}. In a compilation of estimates of gene flow from private alleles, Slatkin (**Table 1**) found the very highest rate of gene flow in the blue mussel, *M. edulis* ($Nm = 42$) (**Figure 1**). This estimate is probably realistic, for the mussels have pelagic larvae that ride ocean currents for weeks. At the other end of the scale were four species of salamanders, all with values of Nm considerably below 1.0. Once again, this estimate of gene flow seems reasonable given our knowledge of salamanders. Salamanders forage only short distances, and they usually breed in their natal ponds. Consequently, movement of individuals among populations is rare.

Caveats concerning the Relationship of F_{st} to Nm

The relationships between Nm and F_{st} and between Nm and the frequency of private alleles are both dependent on assumptions that may be frequently violated in the data collected in range-wide surveys of genetic variation.

Assumption of 'Evolutionary Equilibrium'

The inference of rates of gene flow from either F_{st} or private alleles depends on the assumption that there has been sufficient time for population structure to come to an evolutionary equilibrium determined by the joint action of gene flow and genetic drift. Conformation to this assumption is rarely considered, but some biologists believe that very few species have reached equilibrium. For example, limber pines were

Table 1 Estimates of the number of migrants moving among populations (Nm) from the average frequency of private alleles ($\bar{p}(1)$)

Common name	Formal name	$\bar{p}(1)$	Nm
Blue mussel	*Mytilus edulis*	0.008	42.0
Fruit fly	*Drosophila willistoni*	0.014	9.9
Milkfish	*Chanos chanos*	0.030	4.2
Desert lizard	*Lacerta melisellensis*	0.066	1.9
[Annual plant]	*Stephanomeria exigua*	0.054	1.4
Pacific treefrog	*Hyla regilla*	0.081	1.4
Valley pocket gopher	*Thomomys bottae*	0.087	0.86
Pacific slender salamander	*Batrachoseps pacifica*	0.117	0.64
Red back salamander	*Plethodon cinereus*	0.200	0.22
Oldfield mouse	*Peromyscus polionotus*	0.158	0.31
Camp's slender salamander	*Batrachoseps campi*	0.338	0.16
Zigzag salamander	*Plethodon dorsalis*	0.294	0.10

Note: values of Nm have been adjusted for the sample sizes, so there is not a perfect rank-order correlation between $\bar{p}(1)$ and Nm.
(Adapted from Slatkin, 1985.)

displaced from high elevations by the glaciers that reached their most recent glacial maximum 18 000 years ago. Once the glaciers subsided, limber pine were able to colonize numerous sites above 10 000 feet in the Rocky Mountains, where limber pines commonly attain ages in excess of 1000 years. The populations with ancient trees are certainly not at an evolutionary equilibrium between drift and gene flow, for very few of their generations have passed since they recolonized high elevations. Similar scenarios apply to the plants and animals that moved northward in North America and Europe since the last glacial maximum.

Heterogeneity among Estimates

In studies of gene flow based on F_{st}, the values of F_{st} are commonly heterogeneous. This should not be the case for neutral characters, for migration and drift should influence all loci in similar ways. The relationship between F_{st} and Nm is appropriate only for neutral genes; selection on a subset of the loci can produce heterogeneous estimates of F_{st}. One of the most striking cases of heterogeneity of estimates of gene flow comes from a series of studies of the American oyster, *Crassostrea virginica*. Estimates of gene flow from allozyme markers suggest that the larvae move great distances, homogenizing allelic frequencies from Massachusetts to Texas. However, both mtDNA and several nuclear DNA markers reveal a picture of limited gene flow, with a major barrier to gene flow in the vicinity of Cape Canaveral, Florida. The authors attribute the heterogeneity of estimates of gene flow to balancing selection on the allozyme loci.

Heterogeneity of estimates of gene flow frequently involve lower estimates of F_{st} from microsatellite loci than from other nuclear markers. The differences are particularly pronounced when the populations are well differentiated, and gene flow between them is low. This heterogeneity is attributable to heterogeneous mutation rates. While the mutation rates for nuclear loci are typically 10^{-6}–10^{-8}, mutation rates for microsatellite loci are much higher, often around 10^{-3}, but reaching 1/20. High mutation rates at microsatellite loci are due to the nature of the variation at these loci. Microsatellite alleles differ in their numbers of tandem repeats, and the different sizes of the alleles produces chromosomal rearrangements when chromosomes are unable to synapse perfectly in the first division of meiosis. The high rates of mutation generate many size variants in each population. For microsatellite loci, the sharing of alleles among populations may be due to independent mutations, rather than gene flow.

Biologists using genetic data to infer rates of migration are obliged to be cognizant of the assumptions underlying their methods. If there are egregious violations of the assumptions, estimates of gene flow may be unreliable.

Further Reading

Avise JC (1994) *Molecular Markers, Natural History and Evolution.* New York: Chapman & Hall.

Endler JA (1977) *Geographic Variation, Speciation, and Clines.* Princeton, NJ: Princeton University Press.

Futuyma DJ (1998) *Evolutionary Biology*, 3rd edn. Sunderland, MA: Sinauer Associates.

Latta RG and Mitton JB (1997) A comparison of population differentiation across four classes of gene marker in limber pine (*Pinus flexilis* James). *Genetics* 146: 1153–1163.

Mitton JB (1997) *Selection in Natural Populations.* New York: Oxford University Press.

Slatkin M (1985) Gene flow in natural populations. *Annual Review of Ecological Systems* 16: 393–430.

Reference

Slatkin M (1985) Rare alleles as indicators of gene flow. *Evolution* 39: 53–65

See also: Genetic Colonization; Genetic Drift; Genetic Migration; Hybrid Zone, Mouse; Phylogeography; Population Genetics; Population Substructure

Gene Frequency

C F Aquadro

doi: 10.1006/rwgn.2001.0509

Gene frequency refers to the proportion of a population that carries one type of variant, or allele, at a locus. More appropriately referred to as 'allele frequency,' gene frequency ranges from 0 (where the particular variant is absent from the population) to 1 (where the variant type is the only allele present). In the latter case, the population is said to be 'fixed' for this particular allele. While often defined in terms of a locus or gene and, in the early days of genetics, assessed by phenotype of the corresponding genotype, the gene frequency is now applied to the frequency of any alternative form found segregating in a population, e.g., alternative nucleotides at a single site in a sequence, whether it be in coding, intron, or intragenic regions, as well as insertion/deletion variants and even alternative gene rearrangements such as inversion types.

Gene frequency is estimated by taking a random sample of individuals from what might be considered a population of the species of interest (e.g., from a geographic locale). By random, we simply mean that the individuals are chosen without regard for their genotype or phenotype associated with the locus of interest. For a haploid organism like *E. coli*, we could estimate the frequency of a particular nucleotide polymorphism (A versus T for example) by sampling 1000 bacterial cells and assaying them for the sequence variant of interest. We might find that 15 of the 1000 are A. Our estimate of the gene frequency of the A allele is thus $15/1000 = 0.015$ or 1.5 %. The frequency of the alternative allele in this case, T, would be $985/1000 = 0.985$, or 98.5 %. We could also calculate the frequency of T as simply $1 - p$ (where p is the frequency of the A allele), or $1 - 0.015 = 0.985$. The larger the sample, the more precise our estimate of the gene frequency. The sampling variance of this estimate is $p(1 - p)/n$, where p is the frequency of the allele of interest and n is the number of alleles sampled (also equal to the number of haploid individuals sampled since each individual cell has only one copy of the genome and thus can only carry one allele). In this case, the sampling variance is 1.4775×10^{-5}, and the standard error (SE) of the estimate is the square root of the variance or 3.8438×10^{-3}. We can thus be 95% sure that the true population frequency is within the interval of $p \pm 1.96(SE)$, or (0.00747, 0.02253).

It is important to distinguish between gene and genotype frequency for organisms other than haploids. Consider a diploid like ourselves. At a particular site in our DNA, some chromosomes carry a C (in frequency p) and some a G (frequency q). Some individuals will have two Cs, some a C and a G, and some two Gs. Individuals carrying two copies of the same allele are called homozygotes (e.g., C/C or G/G) and individuals with two different alleles are called heterozygotes (e.g., C/G). Genotype frequencies represent the proportion of each type of genotype in the population sample. For one set of gene frequencies, there can be many different genotype frequencies: e.g., for $p = q = 0.5$, we could have 0.5 C/C, 0.0 C/G, and 0.5 G/G; or we could have 0.0 C/C, 1.0 C/G, 0.0 G/G; or we could have 0.25 C/C, 0.50 C/G, 0.25 G/G; the latter is what is expected with random mating and no selection, drift, mutation, or migration – the Hardy–Weinberg equilibrium genotype frequencies for these gene frequencies.

For a dipoid organism like ourselves, we estimate gene frequency and the associated sampling variance as we did for haploids, but we must take account of the fact that if we are examining an autosomal gene, then each individual carries two copies of each gene. Thus, allele frequency is calculated as twice the number of homoygotes (for example, A/A individuals) plus the number of heterozygotes (A/T individuals), all divided by twice the total number of individuals sampled. For X-linked genes, the heterogametic sex (males in humans) only carries one gene copy, while females carry two copies. Similar logic applies to the estimation of gene frequencies in polyploid species, or in haplodiploid species.

What do gene frequencies tell us about the evolutionary forces shaping variation? Variation ultimately has its origin as mutation. Mutation introduces alleles into populations. The alleles can be spread to other populations by gene flow (migration). If the population of interest is infinitely large, and the variants do not confer any advantage or disadvantage to their carriers (so called selectively neutral), then from one generation to the next there will be no chance sampling 'genetic drift' of gene frequencies; they will change in frequency only by additional mutation. However, all real populations are finite, and thus genetic drift is a process that contributes to allele frequency change in all populations.

If drift is the only factor influencing gene frequencies, then the higher the frequency of a particular variant, the older that variant is likely to be. That is, a new variant in a diploid population of size N individuals (and thus $2N$ copies of each locus) starts at a frequency of $1/2N$ and increases or decreases by drift. The probability that an allele is eventually fixed in the population by drift alone turns out to be simply its frequency ($1/2N$ for a new allele), but it will take on average $4N$ generations for this to occur. This is the average time it takes for all alleles in a population to share the same single common ancestor allele, that is, all alleles present now are descendant copies of a single allele present in the population on average $4N$ generations ago. The probability that a new mutation is ultimately lost from the population by drift alone (barring new mutation) is $1 - 1/2N$. For a large population, this is very high and most new mutations are destined to be lost. The continual introduction of alleles into a population and their inexorable march to fixation or loss, leads to an steady-state ('equilibrium') distribution of gene frequencies expected in a population of size N and with a given mutation rate. Most alleles will be of low or high frequency, with relatively few of intermediate frequency.

Some variants do affect the contribution of their carriers to the next generation (e.g., influence survival, number of progeny produced, etc.). For those new mutants that are favored by selection, they will increase in frequency to fixation, provided their selective advantage is large enough to overcome drift. In some cases, natural selection favors individuals with multiple different allelic types (for example human heterozygotes for normal and sickle-cell β-chain hemoglobin in regions of the world with malaria). Here, selection maintains a stable 'equilibrium' frequency of both allele types. Even harmful (deleterious) mutants will exist in an equilibrium frequency in populations, due to the balance of the introduction of the allele type into the population by mutation and its elimination by natural selection. Alleles that are phenotypically recessive, often due to a loss of function, can reach moderate mutation–selection balance frequencies, since they are 'hidden' as heterozygotes and only selected out as homozygotes. Knowledge of the selective disadvantage of such alleles allows the estimation of mutation rates for these types of alleles and has been widely used to do so, particularly in human genetics.

See also: **Balanced Polymorphism; Gene Flow; Genetic Drift; Genetic Equilibrium; Hardy–Weinberg Law**

Gene Insertion

J H Miller

Copyright © 2001 Academic Press
doi: 10.1006/rwgn.2001.0510

Gene insertion is the term for a gene that has been altered by the insertion of extra DNA within it. In most cases this leads to loss of function.

Gene Interaction

J Hodgkin

Copyright © 2001 Academic Press
doi: 10.1006/rwgn.2001.1530

The term gene interaction is regrettably ambiguous, being used with many different meanings in the scientific literature. Generally, it describes situations where the presence of two mutations in an organism leads to a phenotype that is different from what might be expected from either mutant phenotype alone. The mutations may affect the same gene (allelic interactions) or different genes.

For allelic interactions (those affecting a single gene), one allele may be recessive, dominant, incompletely dominant or codominant to the other. If a gene has a wild-type allele + and a mutant allele x, with a mutant phenotype X in x/x homozygotes, then if the phenotype of the heterozygote $x/+$ is wild-type, x is recessive to wild-type. If the heterozygote $x/+$ has the phenotype X, then x is dominant to wild-type. If the phenotype is intermediate, then x is incompletely

dominant (also known as semidominant). If two alleles x and y confer two distinguishable phenotypes X and Y, then they are said to be codominant if both phenotypes are seen in the heterozygote x/y. Other forms of interaction include overdominance, or heterozygote superiority. In the case of overdominance the fitness of a heterozygote, x/y, is higher than that of either homozygote, x/x or y/y. The reciprocal situation, when the fitness of x/y is lower than that of either homozygote, is called underdominance.

For interactions involving two genes, a variety of possibilities exist. These include suppression, epistasis and hypostasis, and synergy. In the case of suppression, mutation of a second gene results in the amelioration of the phenotypic effects of mutation in the first gene, either partly or wholly to a wildtype phenotype. In the case of epistasis, if two genes have distinct mutant phenotypes, then the double mutant exhibits only one of these phenotypes and the other is masked. If A is the gene with the masking phenotype, and B the gene with the masked phenotype, then in this situation gene A is said to be epistatic to gene B. The different term hypostasis, which is less frequently used, has the opposite meaning, so in this example gene B is hypostatic to gene A. The distinction between epistasis and suppression is that the suppressing mutation restores the wild-type phenotype, and may have no other phenotype of its own.

Synergistic interactions are observed when the combination of two mutant genes results in a much more severe phenotype than either mutant alone. For example, the combination of two viable mutations may result in lethality. Synergy is often caused by redundancy in gene action, so that loss of function in either gene alone has little or no effect on a given process, but loss of both blocks the process completely, and therefore has much more drastic consequences.

The term gene interaction has also been used with reference to direct physical interactions between gene products, most usually protein–protein interactions, though protein–RNA, protein–DNA, RNA–RNA, and RNA–DNA interactions may also be encountered. These physical interactions often involve high-affinity, stable binding, which can be easily detected by biochemical methods, but may alternatively involve transient phenomena, such as protein modification or cleavage. Weak or transient interactions may still be biologically important, and can often only be detected by genetic methods. Genetic approaches, however, cannot usually distinguish between direct and indirect interactions. It may be that the genetic analysis of two genes suggests strongly that their products interact, but this interaction may in fact be mediated by some additional factor or factors, so that the two products never come into actual physical contact. The genetic data can nevertheless provide evidence for involvement in the same pathway or process.

See also: **Alleles; Dominance; Epistasis; Recessive Inheritance; Suppression; Suppressor Mutations**

Gene Library

I Schildkraut

Copyright © 2001 Academic Press
doi: 10.1006/rwgn.2001.0512

A gene library is a collection of single cells (usually bacteria), each of which has received a single segment of DNA usually carried by a plasmid, bacteriophage, or viral vector, the DNA segments having been derived from genomic DNA or cDNA.

See also: **Genomic Library**

Gene Mapping

J R S Fincham

Copyright © 2001 Academic Press
doi: 10.1006/rwgn.2001.0513

The Detection of Recombination within Genes

In the sense intended here, gene mapping means the ordering of sites within genes. In the earlier history of genetics, the gene was considered to be a single indivisible unit of mutation and recombination, but from the early 1950s onwards it became apparent that different mutations in the same gene were nearly always at different sites and able to recombine to yield wild-type and, where they were looked for, doubly mutant genes. In eukaryotic organisms (e.g., fungi and *Drosophila*) the recombinants were generated in meiosis following a sexual cross between mutants. In bacteria, recombination occurred in the course of conjugation, transduction, or transformation, as the donor genomic fragment was integrated into the whole genome of the recipient cell. In bacteriophage it occurred during mixed infection of bacterial cells.

The frequency of recombination within genes is low compared with that between genes. In the budding yeast *Saccharomyces cerevisiae* it may amount to a few percent of the meiotic products, but in most other eukaryotes favored by geneticists it is very much

lower. Consequently, the use of recombination for mapping within genes depends on some method for selecting wild-type recombinants from a large excess of nonrecombinant mutants. This is straightforward when the mutants have a growth handicap not shared by the wild-type, as when they have special nutritional requirements (i.e., are auxotrophs) or are sensitive to higher temperatures, or (in bacteriophage genetics) unable to grow in a particular host.

Mapping by Recombination Frequency

It is a general principle of linkage mapping that the frequency of recombination between sites of mutation increases with their distance apart. But the use of recombination frequency for mapping within genes is complicated by the fact that the sites being recombined are close to the recombination event, which is a complex process probably always involving local nonreciprocality. In fungi and *Drosophila*, which are the eukaryotic organisms most studied in this regard, much or most recombination between mutant sites in the same gene (usually detected as production of wild-types from intermutant crosses) is due to the nonreciprocal conversion of one mutant site to wild-type. The contribution of reciprocal crossing-over is greater when the mutant sites are relatively widely spaced.

However, even if most recombination within genes is due to conversion and not to crossing-over, we still expect, and generally find, a strong correlation between recombination frequency and distance. The reason is that conversion involves tracts of DNA rather than single base pairs, and recombination between two sites by conversion will occur only when the conversion tract covers one but not the other, and this will obviously be less likely the closer the spacing of the sites.

In practice, recombination frequency is a good general guide to gene mapping but sometimes gives ambiguous results. One source of confusion is that the nature of the mutational site may strongly influence its probability of conversion (see Gene Conversion and Mismatch Repair (Long/Short Patch)). This is called a marker effect.

The Use of Flanking Markers

To the extent that recombination between mutant sites within a gene is due to reciprocal crossing-over, it should result in recombination of genetic markers placed on either side of the gene. Thus, wild-type recombinants should be associated with one new flanking marker combination, and double-mutant recombinants (if they are recoverable) with the reciprocal

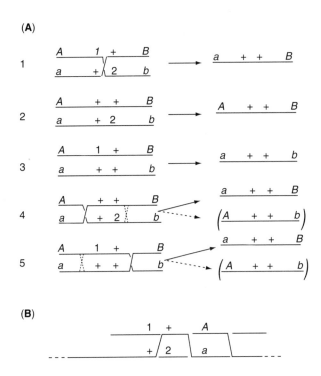

Figure 1 The use of flanking markers to determine the order of mutant sites within a gene. (A) The use of flanking markers in the study of recombination within a gene. The two mutant strains being crossed (with mutational sites 1 and 2, with corresponding wild-type sites shown as +) are also distinguished by allelic differences A/a and B/b at closely placed flanking loci. Wild-type (++) recombinants from the A 1 B × a 2 b cross can arise in the following ways: (1) Reciprocal crossing-over between the sites: all ++ recombinants will be a B if the 1 − 2 order is as shown; (2) and (3) conversion of either 1 or 2 to wild-type without crossing-over, conveying no information about the order of the sites; (4) and (5) conversion of 1 or 2 to wild-type, with crossing-over in the interval adjacent to the conversion event: all ++ recombinants will be a B if the 1 − 2 order is as shown. If the conversion-associated crossover is on the other side of the gene (shown as a dotted cross), the outcome will be A ++ b recombinants; this is usually a less common event. Overall, therefore, the order A–1–2–B will be indicated by a predominance of a ++ B over A ++ b products. (B) The use of one flanking marker in a transduction cross in bacteria. Mutant sites 1 and 2 are present in the donor and recipient, respectively; A is a flanking marker, not subject to selection, present in the donor, as opposed to a in the recipient. If the sites are arranged 1–2–A, integration of a donor phage-borne fragment to give a ++ recombinant requires one exchange and another either (i) between 2 and A or (ii) to the right of A. If the sites are the other way round, the exchange between 1 and 2 will always exclude A unless there are more than two exchanges.

flanking marker combination (**Figure 1**). If the intragenic recombination is due to conversion at one site without crossing-over, the flanking markers will retain their parental combinations and give no information about order within the gene.

However, tetrad analysis in fungi, particularly in the budding yeast *S. cerevisiae*, shows that even though much, or sometimes nearly all, recombination within genes is due to conversion and not to reciprocal crossing-over, it is still, with a frequency which is often about 40% or 45%, associated with crossing-over between flanking markers. Most of the conversion-associated crossovers were found to be on the side of the gene where the conversion event had occurred and could have been immediately adjacent to the conversion

tract (Fogel and Hurst, 1967). This result is consistent with the hypothesis that conversions and crossovers have a common origin in hybrid DNA structures, formed by interaction between chromatids, that always involve local unilateral transfer of DNA (and hence gene conversion if the transferred segment happens to carry a distinguishing marker), but lead to crossing-over only in a certain proportion of cases (see Recombination, Models of). To the extent that conversion-associated crossovers really are adjacent to the conversion tracts, the wild-type intragene recombination will be associated with the same crossover combination of flanking markers as they would have been had they originated by reciprocal crossing-over (**Figure 1**). In this case the relative frequencies of the two flanking marker recombinant classes will reveal the order of the sites within the gene, even if most or all intragenic recombination is due to gene conversion.

In practice, flanking markers usually give a clear order of sites, though in the fungus *Neurospora* the data are often complicated by the occurrence of a substantial minority of wild-type recombinants with the 'wrong way round' flanking marker recombination. In some cases, these 'exceptions' are too numerous for an unambiguous ordering of sites within the gene (**Figure 1A**).

This complication does not appear to arise in *Drosophila*, where in the best-analyzed case – the mapping of sites within the rosy (*ry*) gene which encodes the enzyme xanthine dehydrogenase – only one of the two flanking marker crossover combinations occurred in the ry^+ recombinants (Chovnick *et al.*, 1971). In this study the rare ry^+ recombinants were selected through their ability to survive on purine-containing medium, which kills *ry* mutants.

The flanking marker principle has also been used in transduction experiments to order sites within bacterial genes, though here the gene being mapped has usually been flanked by only one marker. If the transductants are selected for intragenic recombination, the probability of the donor flanking marker being included will depend on whether it is more closely linked to the selected or to the excluded site (**Figure 1B**).

Deletion Mapping

Among any large collection of mutations within a particular gene some are likely to be due to deletions of gene sequence rather than to changes of single base pairs (point mutations). Deletion mutations can in general be distinguished from point mutations through their inability to back-mutate to wild-type. More definitively, they fail to give wild-type recombinants in crosses to sets of point mutations that are able to recombine with each other. Deletions provide the most unambiguous method of mapping within genes. The analysis proceeds in two steps.

First, the deletions are arranged in a linear order defined by their overlaps. Nonoverlapping deletions can give wild-type recombinants when crossed, whereas overlapping deletions can not. Then, when the map of deletions has been established, the point mutations can all be placed in one or other of the segments defined by the deletion overlaps and nonoverlaps. The ability to recombine with a deletion shows that the point mutation falls outside the deleted segment; conversely, the failure to recombine with a deletion shows that the point mutation falls within, or at least very close to, the deleted segment. The principle is explained in **Figure 2**

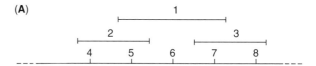

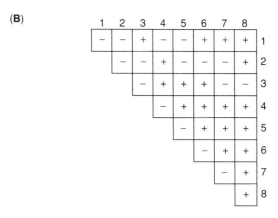

Figure 2 The principle of deletion mapping. (A) Of eight mutations within a gene, 1 to 3 are deletions and 4–8 are 'point' mutations. Crosses between them in all combinations yield either some wild-type recombinants (+) or none (−). (B) The results determine the order of the point mutations (above).

The deletion method was first used for intragene mapping by Seymour Benzer (1959), whose fine-structure map of the bacteriophage T4 *rII* gene was a major factor in the demise of the doctrine of gene indivisibility. In yeast (*S. cerevisiae*) probably the best example has been the mapping by Sherman *et al.* (1975) of the *CYC1* gene, which encodes the major cytochrome *c* protein.

The principle of collinearity

As soon as it became clear that genes determined protein structure, it was an obvious hypothesis that the gene was a linear code for the sequence of amino acids in the protein polypeptide chain. The sequence of mutational sites within the gene, determined by one of the methods outlined above, should correspond to the order in the polypeptide chain of the amino acids changed by the mutations (the principle of collinearity – the double-l is optional). This prediction was confirmed wherever it was tested, firstly for the *Escherichia coli* gene encoding the A subunit of tryptophan synthetase (Yanofsky *et al.*, 1964), and later in several other cases, including the yeast example mentioned above.

Today the ordering of mutational sites by genetic crosses has been largely superseded by the direct determination of the DNA sequences in wild-type and mutants. The principle of collinearity has been confirmed by molecular methods in countless cases.

References

Benzer S (1959) On the topology of genetic fine structure. *Proceedings of the National Academy of Sciences, USA* 45: 1607–1620.

Chovnick A, Ballantyne GH and Holm DG (1971) Studies on gene conversion and its relationship to linked exchange in *Drosophila melanogaster. Genetics* 69: 179–209.

Fogel S and Hurst DD (1967) Meiotic recombination in yeast tetrads and the theory of recombination. *Genetics* 57: 455–481.

Sherman F, Jackson M, Liebman SW, Schweingruber M and Stewart JW (1975) A deletion map of *cyc1* mutants and its correspondence to mutationally altered iso-1 cyctochrome c of yeast. *Genetics* 81: 51–73.

Yanofsky C, Carlton BC, Guest JR, Helinski DR and Henning U (1964) On the colinearity of gene structure and protein structure. *Proceedings of the National Academy of Sciences, USA* 51: 266–272.

See also: Bacteriophage Recombination; Colinearity; Deletion; Deletion Mapping; Gene Conversion; Mismatch Repair (Long/Short Patch); Mutant Allele; Recombination, Models of; Transduction

Gene Number

J Hodgkin

Copyright © 2001 Academic Press
doi: 10.1006/rwgn.2001.0515

Gene numbers of free-living organisms range from about 1500, for the simplest bacteria, to probably more than 100 000 in some higher eukaryotes, although the true upper limit is impossible to determine at present. Parasitic organisms can survive with much smaller numbers of genes, so bacteriophage genomes may contain as few as four genes, in the case of some RNA bacteriophages, and mycoplasma (bacteria that can live only as intracellular parasites) have fewer than 500 genes.

Gene numbers can only be known accurately for species with completely sequenced genomes, and even then good numbers can be hard to come up with. At the time of writing a rough draft of the human genome sequence has been completed, and estimates of the number of genes included in this total sequence still range from 30 000 to 150 000. Undoubtedly improvements in sequencing and gene prediction will rapidly refine the estimates, but it will probably be a long time before the number is known to better than ± 5%.

There are many sources of difficulty in attempting to count genes in raw sequence data obtained from large eukaryotic genomes. A major problem is that exon prediction becomes more and more difficult as the size and number of introns increases. Similarly, knowing where one gene ends and another begins may be very difficult. Genes may be embedded within the introns of other genes, or overlapped with them. Small genes may be missed, especially those that encode RNAs rather than proteins. All these factors will tend to lead to underestimates; conversely, failing to distinguish between functional genes and pseudogenes will lead to overestimates.

Gene counting in prokaryotes is much easier, since introns are usually absent and signals for translational initiation and termination are well defined. Some bacteria with large genomes, such as streptomycetes and myxobacteria, must have more genes than lower eukaryotes such as fungi. Complete sequences for the budding yeast *Saccharomyces cerevisiae* and the pathogenic bacterium *Pseudomonas aeruginosa* show that both have about 6000 protein-coding genes, so there is clearly overlap between the prokaryotic and eukaryotic worlds in this respect. The minimal eukaryotic gene set may contain as few as 4000 genes, though this is still much larger than the minimal prokaryotic set.

A conspicuous failing in the classical genetic analysis of eukaryotes has been the consistent underestimates of true gene number. For the two best-studied examples, the fruit fly *Drosophila melanogaster* and the nematode *Caenorhabditis elegans*, predictions of gene number derived from genetic studies were low by factors of at least two. Observations on banding patterns on the polytene chromosomes of *Drosophila* created a longstanding bias. Saturation mutagenesis of some regions of the fly genome suggested an exact correspondence between the number of polytene bands and the number of essential genes (as defined by lethal mutations), and led to a prediction of about 5000 genes in all. In hindsight, it is clear that the correspondence between bands and essential genes is no more than an unfortunate coincidence, and the current gene number inferred from genome sequencing is much higher, about 13 600. Surprisingly, this is a lower number than the estimate for *C. elegans* (about 19 000 protein-coding genes), and lower yet than the estimate for *Arabidopsis* (about 25 000 genes), although the apparent organismal complexity of *Drosophila*, in terms of cell types and anatomical detail, is higher in fly than in worm or weed. The apparent paradox can be explained by larger gene families and more extensive gene duplication in *C. elegans* and *Arabidopsis*. Also, genes in *Drosophila* may be more complex, undergoing more alternative splicing and therefore generating a greater variety of final proteins. Total gene number therefore should not be regarded as a very useful or informative genomic property.

See also: **Genome Organization**

Gene Pool

K E Holsinger

Copyright © 2001 Academic Press
doi: 10.1006/rwgn.2001.0516

The genetic information encoded in all the genes of a population or species at one time comprises the gene pool from which the genes of the next generation are derived. The gene pool determines the genetic characteristics that future generations will have, except to the extent that the genetic information currently present is altered by mutation in the production of gametes. The characteristics of the gene pool determine how readily a population can respond to natural selection.

According to the classical view of population structure, as exemplified in the writings of H. J. Muller, individuals in a population are homozygous for a single wild-type allele at almost every locus. Rare alleles are maintained only by continual mutation, because they are unconditionally deleterious. Because the population is nearly uniform genetically, the power of natural selection to provoke a response is quite limited. The rate at which adaptation can occur is limited by the rate at which favorable mutations arise.

According to the balance view of population structure, as exemplified in work by Th. Dobzhansky, the gene pool consists of several to many alleles at many loci. Balancing selection, either in the form of heterozygote advantage or negative frequency-dependent selection, is presumed to be responsible for maintaining large amounts of genetic variability at many loci. Because the population is highly variable, natural selection can provoke a dramatic response, and the rate at which adaptation occurs is not limited by the rate at which favorable mutations accumulate, at least in the short run.

The concept of a gene pool is not restricted to populations that are panmictic. It will often take more generations for particular gene combinations to be formed in a population that is inbreeding or divided into geographically distinct subpopulations than in one that is panmictic. Nonetheless, those gene combinations will eventually be formed. Once formed they will persist longer in a population with inbreeding or geographical structure than in one that is panmictic.

In a panmictic population, genotypes at each locus will be found in approximately Hardy–Weinberg proportions, unless genotypes differ substantially in their abilities to survive and reproduce. In an inbred population, heterozygotes will be less common and homozygotes will be more common than in a panmictic population with the same allele frequencies. If inbred and outbred populations have the same amount of genetic diversity in terms of the numbers and types of alleles at each locus, the allelic composition of the two gene pools is equivalent. The genotypic composition of the two gene pools will, however, be different.

See also: **Balanced Polymorphism; Demes; Dobzhansky, Theodosius; Hardy–Weinberg Law; Heterozygote and Heterozygosis; Natural Selection; Panmixis**

Gene Product

J Parker

Copyright © 2001 Academic Press
doi: 10.1006/rwgn.2001.0517

The product of a gene is the protein or the RNA which it encodes. The vast majority of genes encode proteins. For instance, the bacterium *Escherichia coli* has 4288 possible protein-encoding genes, representing 87.8% of the chromosome, while only 0.8% of the genome encodes RNA as its final product. Messenger RNAs (mRNAs), or in eukaryotes the RNAs which are precursors to messenger RNAs, are informational intermediates in protein synthesis (translation). Since they are not the ultimate product of the gene, mRNAs are not included in lists of gene products. The RNAs that are included are the ribosomal RNAs, the transfer RNAs, and other stable RNAs. In prokaryotes these include 4.5S RNA, 10S RNA, and the RNA component of RNase P, while in eukaryotes there are large numbers of such small RNAs.

Although they can be considered the products of transcription and/or translation, most gene products undergo one or more processing steps before they reach their final form. This is almost universally true for the stable RNAs, which are cut from longer precursors and/or require modifications of one or more bases. Posttranslational processing of proteins is also common, from rather minor changes such as the removal of the initiating methionine to much more complex processing and modification steps.

See also: **Coding Sequences; Transcription; Translation**

Gene Rearrangement in Eukaryotic Organisms

K L Hill and B C Coughlin

Copyright © 2001 Academic Press
doi: 10.1006/rwgn.2001.0518

The entire pool of genetic information (DNA) in an organism is referred to as the organism's genome. This genetic information is organized into units called genes and the physical location of a gene within a genome is called a locus. In general, the position of a gene within a genome is fixed. However, in some cases a gene may be moved from one physical location to another. Such gene rearrangements can contribute to several important processes, including the regulation of gene expression, generation of diversity in a population, generation of diversity in proteins, and cellular differentiation. Sometimes gene rearrangements can be harmful and may lead to inherited disease. On an evolutionary time scale, DNA rearrangements can produce gene duplications, giving rise to repetitive DNA elements, pseudogenes, and gene superfamilies.

Transposable Genetic Elements

Transposons are small pieces of DNA (500–1500 bp long) capable of moving themselves from one place to another within a genome. These mobile genetic elements were first recognized in maize (corn), but are now known to be present in essentially all organisms. In the fruit fly, *Drosophila melanogaster*, transposons may constitute as much as 10% of the entire genome! Transposons usually have repetitive DNA sequences at each end to facilitate their excision from the genome, and include a gene for the enzyme (transposase) that catalyzes excision. Once excised, transposons reenter the genome at random positions and usually do not disrupt the general architecture of the genome. However, transposons often have dramatic effects on gene expression and may cause deleterious gene rearrangements if their integration disrupts important regulatory or protein coding sequences, or if pieces of the genome surrounding the transposon are inadvertently deleted during transposon excision.

Regulation of Mating Type in Fungi

Haploid cells of the budding yeast *Saccharomyces cerevisiae* are able to repeatedly switch between two alternate mating types, **a** and α. The choice between these two mating types is determined by the identity of the gene in the mating type (MAT) locus. Cells with a *MATa* gene in the MAT locus become mating type **a**, while those with a *MATα* gene in the MAT locus become mating type α. Each yeast cell harbors an unexpressed ("silent") copy of the *a* gene and the α gene. These silent genes are located at the HMR and HML loci, 100–200 kb away from the MAT locus. The silent *a* and α genes are never expressed in wild-type cells. When a yeast cell switches between mating types, the active gene at the MAT locus is removed and replaced with a duplicate version of one of the silent mating type genes from either the HMR or HML locus. Once placed into the MAT locus, the newly duplicated mating-type gene is turned on and cellular differentiation proceeds, generating a cell of the new mating type. Rearrangement of the yeast mating type genes occurs over a distance of 100–200 kb of DNA and is controlled by stringent regulatory mechanisms that are coordinated with cell division. Similar gene rearrangements control mating-type switching in other fungi.

Antigenic Variation in African Trypanosomes

African trypanosomes are eukaryotic pathogens that infect a wide variety of mammals, including humans. In order to disguise themselves from the mammalian immune system, these single-celled parasites periodically change the identity of their major surface glycoprotein antigen, a process known as 'antigenic variation.' Although trypanosomes contain several hundred genes for these variant surface glycoproteins (VSGs) scattered throughout their genome, only one VSG gene is expressed at any given time. Expression of the active VSG gene occurs exclusively at a telomere-linked 'expression site,' while silent VSG genes can be located internally in the genome, or at inactive telomere expression sites. Three types of gene rearrangements are generally associated with the activation of a silent VSG gene. First, a duplicated copy of a silent VSG gene may be transposed into the active expression site, displacing the previously active VSG gene. This duplicative transposition may include all or part of the VSG gene. A variation of this type of gene rearrangement is a telomere conversion whereby the active telomeric region containing the expressed VSG gene is completely replaced with a duplicated copy of a silent telomeric region. Finally, two telomeres may undergo a reciprocal exchange, activating one gene and inactivating the other. These dramatic gene rearrangements occur over a distance of 100 kb or more, and may even occur between different

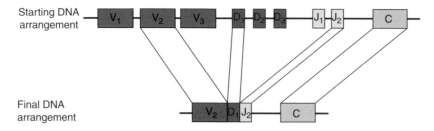

Figure 1 Simplified diagram of antibody gene rearrangements. The boxes labeled 'V,' 'D,' 'J,' and 'C' represent the four DNA segments of the gene for an antibody heavy-chain subunit. Rearrangements of these gene segments is necessary to produce a functional gene, as shown in the final DNA arrangement.

chromosomes. Similar gene rearrangements are responsible for antigenic variation in other microbial pathogens.

Generation of Diversity in the Vertebrate Immune System

Perhaps the most elaborate example of gene rearrangement in eukaryotes occurs during assembly of the genes for antigen-recognizing molecules of the vertebrate immune system. Invading pathogens are recognized as foreign by the immune system on the basis of the structures of their pathogen-specific macromolecules (proteins, carbohydrates, and lipids). This recognition is mediated by two groups of specialized proteins of the immune system called 'antibodies' and 'T-cell antigen receptors.' Pathogen molecules that are recognized by antibodies and T-cell receptors are collectively referred to as 'antigens,' since antibodies are generated in response to them. Any given pathogen is composed of its own unique set of tens to thousands of antigens. Hence, in order to recognize and respond to all potential pathogens, the immune system must have an extremely large repertoire of antibodies and T-cell receptors. Indeed, it is estimated that the human immune system has the capacity to produce as many as 100 billion different antibody molecules! There is a similarly large pool of variant T-cell receptor molecules. Since the human genome is estimated to be 30 000–40 000 protein coding genes, there is not enough genetic material for every different antibody and T-cell receptor to be derived from its own, individual gene. How then, is this great diversity generated? It turns out that vertebrate cells employ a complex and highly regulated series of gene rearrangements to generate variant antibodies and T-cell receptors from a relatively small number of variant gene segments.

Antibodies are dimeric proteins, composed of one heavy-chain subunit and one light-chain subunit. The genes for each antibody subunit are not present in the genome as single, contiguous units. Rather, each subunit gene is arranged as linear array of fragmented segments (V, D, J, and C in **Figure 1**), each encoding a different part of the antibody subunit molecule. In this initial DNA arrangement, a functional antibody is not produced. Instead, the fragmented gene segments are first repositioned to generate, in the final DNA arrangement, a single gene that encodes one complete subunit of the antibody. This shuffling of gene segments is accomplished through an ordered series of highly regulated gene rearrangements.

For each gene segment, there are multiple segments that can all be mixed and matched with other segments to construct a complete antibody subunit gene. Each antibody-producing cell undergoes antibody gene rearrangements independently of other antibody-producing cells. This process results in a pool of antibody genes with an overall diversity that is several million times greater than the diversity of the original pool of variant gene segments, and is referred to as 'combinatorial diversity.' Antibody gene rearrangements are primarily considered in the context of their ability to generate diversity in antibody proteins. However, these gene rearrangements also cause activation of important regulatory elements (called 'promoters' and 'enhancers') that control antibody gene expression. Similar gene rearrangements are responsible for generating diversity in variant gene segments that encode T-cell receptor subunits. Thus, regulated gene rearrangements are used by the immune system to generate immense diversity in antibodies and T-cell receptors, which in turn are necessary to effectively combat infection by microbial pathogens.

Further Reading

Bennetzen JL (2000) Transposable element contributions to plant gene and genome evolution. *Plant Molecular Biology* 42: 251–269.

Donelson JE and Turner MJ (1985) How the trypanosome changes its coat. *Scientific American* 252: 44–51.

Haber JE (1998) A locus control region regulates yeast recombination. *Trends in Genetics* 14: 317–321.

Weill JC and Reynaud CA (1996) Rearrangement/hypermuta-
tion/gene conversion: when, where and why? *Immunology
Today* 17: 92–97.

**See also: Antibody; Antigenic Variation; Gene
Expression; Immunoglobulin Gene Superfamily;
Mating-Type Genes and their Switching in Yeasts;
T Cell Receptor Gene Family; Transposable
Elements**

Gene Rearrangements, Prokaryotic

R Haselkorn

doi: 10.1006/rwgn.2001.1451

Gene rearrangements have occurred in prokaryotes
since the dawn of unicellular life. They play a critical
role in bacterial evolution. The remarkable plasticity
of bacterial genomes has been revealed by the comparison of restriction maps and, more recently, the comparison of whole genome DNA sequences. We find, in
the latter, evidence for lateral gene transfer between
species as well as long- and short-range rearrangements within strains of the same species. The recombination events leading to these rearrangements are
relatively rare, so that most genetic experiments conducted on the time-scale of years result in a unique
physical map for a given strain. However, there are
several examples of rearrangements of bacterial genes
that occur on a much shorter time-scale, minutes or
seconds; these are described below.

Some contemporaneous rearrangements are
stochastic: they occur at a given frequency throughout
vegetative growth, providing variant phenotypes
available for selection when the need arises. Others
are developmentally regulated, responding to environmental cues to provide new proteins for the developmental program.

Stochastic Rearrangements

The general idea of a stochastic rearrangement is to
provide a new promoter for a gene encoding a structural protein or an enzyme. A classical example is
the phenomenon of phase variation in *Salmonella*,
observed originally in the 1920s and studied further
in the 1950s. Certain strains of *Salmonella* can switch
from a form expressing one flagellar antigen (H1) to a
form expressing a different flagellar antigen (H2) and

then back. This switch, or phase variation, corresponds at the molecular level to the inversion of a DNA
segment by a site-specific recombination enzyme. The
inverted segment carries a promoter such that, in one
orientation, it drives the transcription of a gene encoding the H2 flagellar protein and a repressor of
transcription of the distant H1 antigen gene. In the
opposite orientation, the promoter points the 'wrong'
way, preventing transcription of both the H2 gene and
the repressor of H1. Thus, transcription of H1 occurs
and the phase is switched. Flipping of the promoter
segment is accomplished by a site-specific recombinase operating on short inverted repeat sequences at the
ends of the segment, which also encodes the recombinase. This antigenic variation occurs in about one cell per
thousand per generation, allowing the population as a
whole to survive antibody directed against one or the
other flagellar antigen.

A variation of this theme in *Escherichia coli* has the
promoter for the *fimA* gene, encoding the structural
protein for type 1 fimbriae, alone on the invertable
segment. Two recombinases (FimB and FimE) are
each encoded nearby. Inversion of the promoter-
containing segment by FimB results in transcription
of the *fimA* gene, while the FimE recombinase flips
the promoter, shutting off transcription of *fimA*.
Other proteins, such as IHF, play a role in this inversion. Fimbriae are important in virulence, mediating the attachment of *E. coli* to epithelial and other
human cells.

Other rearrangements occur as a consequence of recombination between repeated sequences in the genome, catalyzed by the general recombination system,
or by the movement of elements that encode site-
specific recombinases that catalyze their own transposition. Any pair of repeated sequences can be found
in two different relative orientations: direct or inverted.
Recombination between two identical sequences in inverted orientation results in inversion of the entire DNA
segment between the recombining repeated elements.
One such event involves the genes encoding ribosomal
RNA in *E. coli*, two of which flank the origin of
chromosome replication (ORI), oriented away from
the ORI. Transcription seems to be more efficient,
for any gene, if it is oriented in the same direction
as DNA replication. Recombination between the inverted rRNA operons flanking ori does not, of course,
change the direction of transcription relative to ORI,
since DNA replication is bidirectional, but nevertheless growth of cells having one arrangement is slightly
faster than growth of cells with the other arrangement.
General recombination between the rRNA operons
flips the ORI in a few cells per thousand in each generation. Under normal circumstances, the rearranged
chromosomes are lost because the cells containing

them grow more slowly than their predecessors. However, if a selectable gene is inserted such that it can be transcribed only in the less preferred orientation of the rRNA operons, that orientation can be selected and maintained. Relaxation of selection results in repopulation of the culture with the more preferred orientation.

Recombination between two identical sequences in direct orientation results in deletion, rather than inversion, of the intervening sequences. Most bacterial genomes (*Bacillus subtilis* is a notable exception) contain substantial numbers of genetic elements called 'insertion sequences,' usually about 1 kb long, containing an open reading frame encoding a site-specific recombinase (transposase) flanked by short (up to 40 bp) sequences themselves in inverted repeat orientation. Expression of the encoded transposase results in flipping of the entire insertion sequence at the same locus. But general recombination can occur between two copies of the same insertion sequence at different chromosomal locations, again leading to either long-range inversions or deletions. If the deleted DNA segment contains an essential gene, the cell in which that rearrangement occurred will die. Insertion sequences are also responsible for recombination events between chromosomes and plasmids, such as the insertion of the F plasmid into the chromosome of F⁺ *E. coli*, generating the high-frequency conjugating strains called Hfr. Plasmid–plasmid recombination to form cointegrates also occurs via insertion sequences. Finally, DNA segments flanked by two identical or nearly identical insertion sequences, in direct or inverted orientation, are capable of transposition from one chromosomal location to another as a unit. Such 'transposons' are also responsible for large-scale genome reorganization in bacteria.

Developmentally Regulated Gene Rearrangements in Bacteria

Recombination between two directly repeated DNA elements leads to deletion of the DNA between the elements. Several such events have been described in connection with specific developmental programs in bacteria: induction of bacteriophage lysogens, differentiation of nitrogen-fixing heterocysts in cyanobacteria, and sporulation in bacilli.

The *E. coli* bacteriophage lambda has two alternative life styles. Upon infection of a naive cell, the linear viral DNA is circularized. It then chooses between replication, leading to lysis of its host accompanied by release of several hundred progeny virus particles, or integration by site-specific recombination between a special site (attP) on the viral chromosome and a corresponding site (attB) on the host chromosome.

There it is content to rest and be replicated once each generation by the host's DNA replication machinery. If the host is endangered by any one of several insults, the viral DNA is excised from the chromosome by reversal of the recombination events that inserted it originally. It then replicates, eventually yielding several hundred virus particles, as in the lytic cycle. The excision of lambda DNA from the chromosome of a lysogen is perhaps the best-studied example of a developmentally regulated gene rearrangement in bacteria. All of the enzymes and participating protein factors have been purified and the role of each nucleotide in the insertion and excision sites has been determined *in vitro*.

Regulation of this rearrangement is essentially negative. That is, the inserted viral chromosome expresses one gene, yielding a repressor protein that effectively blocks transcription of every viral gene except its own. Insults to the host cell result in activation of a protease that cleaves the repressor protein, leading to expression of viral genes encoding the excision recombinase and DNA replication proteins; the pathway to virus production and cell lysis described above is then followed. Induction of this lytic pathway requires relief of repression in the lysogen.

Such negative regulation has not been detected yet in the case of cyanobacterial heterocyst differentiation. Cyanobacteria are oxytrophic photosynthetic bacteria; that is, they carry out green plant photosynthesis, evolving oxygen in the light. Although some cyanobacteria can utilize fructose or glucose as a carbon source, most known species cannot do so, but rather are obligate phototrophs dependent upon light and the fixation of CO_2 for their reduced carbon. Some species, such as *Anabaena*, grow in filaments of several hundred cells, indistinguishable from one another as long as a good source of reduced nitrogen (ammonia or nitrate) is available. Deprived of such a source, *Anabaena* differentiates cells specialized for nitrogen fixation along each filament, usually spaced about ten cells apart (**Figure 1**). The undifferentiated vegetative cells continue to fix CO_2 and to generate O_2. The specialized cells, called heterocysts, are anaerobic factories for nitrogen fixation, the reduction of atmospheric nitrogen gas to ammonia.

The patterned conversion of a dividing, oxygen-evolving vegetative cell to an anaerobic, nitrogen-fixing heterocyst requires the orderly expression of many genes, up to 20% of the 7300 genes in the *Anabaena* genome. Among the genes needed in the heterocyst are those encoding the machinery for nitrogen fixation, including the polypeptides of the nitrogenase complex. These are organized in an operon, nifHDK, encoding the protein called dinitrogenase reductase (NifH) and the two subunits of dinitrogenase (NifD

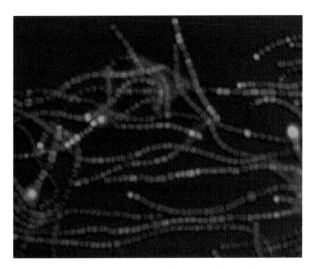

Figure 1 (See Plate 19) Filaments of the cyanobacterium *Anabaena* 77 h after transfer to nitrogen-free medium. Nitrogen-fixing heterocysts have differentiated at regular intervals along each filament. The image shown is a composite of a fluorescence image showing the location of green fluorescent protein expressed from the promoter of the *hetR* gene and a DIC image that outlines the cells. The HetR protein is required for heterocyst differentiation. It is expressed early in the differentiation of only those cells destined to develop.

5'*nifH*...*nifD*..GGCA----T-C---**GCCTCATTAGG**-----CAC—AA----C..*nifD*....*nifK*.

5'*nifB*...*fdxN*..T-G-----A-T—**TATTC**—AGAA-TTT-C---A..*fdxN*....*nifS*...*nifU*.

5'*hupL*..G----**CACAGCAGTTATATGG**-------T---G—A..*hupL*.

Figure 2 Nucleotide sequences involved in the excisions that occur during cyanobacterial heterocyst differentiation. In each case, the sequences shown are repeated directly, separated by, from top to bottom, 11 kb, 55 kb, and 10.5 kb, respectively. Recombination, catalyzed by a recombinase encoded within the excised element, occurs within the bold-faced sequence, resulting in the excision of circular elements of the size mentioned. Dashes represent nucleotides that differ in the two copies of the sequence prior to excision. Plain capitals represent nucleotides that are conserved around both copies of the repeated sequences. The recombinases that catalyse the *nifD* and *hupL* rearrangements are related proteins. The *fdxN* recombinase is unrelated to these but is related to the enzyme that excises the *skin* element during *Bacillus subtilis* sporulation.

and NifK). Most strains of *Anabaena* contain an 11-kb DNA element interrupting the *nifD* gene in vegetative cell DNA. During heterocyst differentiation, and only during differentiation, the 11-kb element is excised by a site-specific recombinase, acting on directly repeated sequences at the ends of the element. The resulting circular element does not replicate or reinsert during the limited life of the heterocyst. These nitrogen-fixing cells do not divide. Eventually they die or are diluted out by growth of the vegetative cells if a new supply of reduced nitrogen is found. Under nitrogen-fixing conditions, the vegetative cells can grow by virtue of amino acids supplied directly to them by the heterocysts. Continued differentiation of heterocysts, halfway between existing heterocysts once each vegetative cell generation, maintains the spacing pattern.

Excision of the 11-kb element in differentiating heterocysts is catalyzed by a site-specific recombinase encoded by the *xisA* gene, located within the 11-kb element. The repeated sequences at the ends of the 11-kb element at which recombination occurs have the same feature as those of the bacteriophage lambda attachment site: a fully conserved core flanked on both sides by regions of partial sequence identity. In these respects, the 11-kb element looks like a remnant of a bacterial virus chromosome, but lacking genes for

head and tail components. The element seems not to contribute materially to *Anabaena* vegetative cell life, because cells cured of the element grow as well as wild-type cells in medium containing ammonia and they differentiate and fix nitrogen normally.

The 11-kb element interrupting the *nifD* gene is only one of three such elements that interrupt *Anabaena* genes whose products are involved in nitrogen fixation. A 55-kb element interrupts a nearby operon that includes the *nifB*, *nifS*, and *nifU* genes. Just as the 11-kb element prevents transcription through the nifHDK operon, the 55-kb element prevents transcription of *nifU* and *nifS*, genes whose products are required for formation of iron–sulfur clusters and their insertion into dinitrogenase. The 55-kb element is excised precisely during heterocyst differentiation, using a site-specific recombinase encoded by the element, acting on directly repeated sequences at the ends of the element (**Figure 2**). Both the amino acid sequence of the recombinase and the DNA sequences at the excision sites differ from those of the 11-kb element.

These two elements appear to provide ultimate examples of selfish DNA. They provide no known advantage to the cells carrying them, but they are clever enough to get out of the way when the genes they invade are necessary for survival. At the time of their discovery, each of the excision enzyme sequences defined new families of recombination enzymes. Subsequently, another small element was discovered interrupting a gene encoding hydrogenase in *Anabaena*. Like the first two, its excision occurs only during heterocyst differentiation. In this case, the sequence of the excisase encoded by the new element puts it in the same family as the excisase of the 11-kb element. The excisase of the 55-kb element remained an orphan until the discovery of another element, described below, that interrupts a gene required for sporulation in *B. subtilis*.

Many gram-positive bacteria, such as the soil inhabitant *B. subtilis*, produce heat-stable spores when conditions become unfavorable for vegetative growth. The process of sporulation involves the regulated expression of a very large number of genes, ending with the lysis of the mother cell within which the spore develops. In response to environmental signals such as carbon or nitrogen starvation, a cascade of two-component regulators is brought into play. The final step in this cascade of phosphorylations is the activation of a sigma factor that permits transcription of the earliest acting sporulation-related genes. A septum forms asymmetrically and one bacterial chromosome partitions into each of the daughter cells. The smaller of these cells pinches off within the intact mother cell and is supplied with several layers of protein to provide the characteristic tough coat of the spore. This program is managed by the differential expression and use of sigma factors in the two cell compartments, the developing spore and the mother cell. One of the last events is the expression in the mother cell of a gene encoding the sigma factor that directs transcription of the major spore coat protein gene. The sigma factor gene is interrupted by a 42-kb element that must be excised for the functional sigma factor to be made. As might now be expected, the excision is carried out by a site-specific recombinase acting on directly repeated sequences at the ends of the element. Since this event occurs only in the mother cell, which will die, it has to be repeated whenever a vegetative cell sporulates. Finally, the amino acid sequence of this excisase puts it in the same family as the enzyme that excises the 55-kb element in *Anabaena*.

The similarities between excision of bacteriophage lambda DNA from a lysogen and the excision of these elements interrupting genes in *Anabaena* and *Bacillus* suggest that the latter elements entered their respective host chromosomes as viral DNA. In the case of *Anabaena*, many strains from different parts of the world have one or more of these elements in their chromosomes, so some comparative sequencing might permit analysis of their age and evolution. There are also parallels between these bacterial gene rearrangements and the transactions in developing lymphocytes that generate the reorganized genes responsible for antibody diversity.

Further Reading

Carrascc CD, Buettner JA and Golden JD (1995) Programmed DNA rearrangement of a cyanobacterial *hupL* gene in heterocysts. *Proceedings of the National Academy of Sciences, USA* 92: 791–795.

Glasgow AC, Hughes KT and Simon MI (1989) Bacterial DNA insertion systems. In: Berg DE and Howe MM (eds) *Mobile DNA*. pp. 637–659. Washington, DC: ASM Press.

Haselkorn R (1989) Excision of elements interrupting nitrogen fixation operons. In: Berg DE and Howe MM (eds) *Mobile DNA*. pp. 735–742. Washington, DC: ASM Press.

Haselkorn R (1992) Developmentally regulated gene rearrangements in prokaryotes. *Annual Review of Genetics* 26: 113–130.

Stragier P and Losick R (1996) Molecular genetics of sporulation in *Bacillus subtilis*. *Annual Review of Genetics* 30: 297–341.

***See also:* Alternation of Gene Expression; *Bacillus subtilis*; Hin/Gin-Mediated Site-Specific DNA Inversion; Insertion Sequence; Site-Specific Recombination**

Gene Regulation

P Laybourn

Copyright © 2001 Academic Press
doi: 10.1006/rwgn.2001.0520

Gene Regulation Occurs Primarily at the Level of Transcription

Gene regulation is the highly controlled turning on and off of gene expression. In single celled organisms it directs the efficient use of cellular resources in response to the cell's environment. In multicellular organisms gene regulation defines the cell, its structure and function, and ultimately the whole organism. Aberrant gene regulation results in cancer, birth defects, and even death. The first step in gene expression is transcription. Therefore, transcription is the primary point of regulation in the process of gene expression. However, one must keep in mind that before a gene can be transcribed the chromatin packaging must be opened up to allow the transcriptional machinery access to that gene.

RNA Polymerase II General Transcription Factors and Basal Transcription Mechanism

Eukaryotes versus Prokaryotes

In the most basic sense, the mechanism of transcription in eukaryotes is very similar to that of prokaryotes. The promoter and RNA start site must be recognized and the RNA transcript must be initiated, elongated, and terminated. In addition, several RNA polymerase subunits have conserved structures and functions indicating that they have a common ancestry.

In eukaryotes, however, the DNA is in a form called chromatin, and only a small proportion of the genome is expressed (in higher eukaryotes). The expressed regions of chromosomes were shown over a decade ago to be more open or accessible than repressed regions, for example, to nuclease cleavage. Other differences between the expressed and unexpressed regions include the presence of an RNA polymerase, the nonhistone proteins, modified histones, and undermethylated DNA (mammals).

Another major difference between prokaryotic and eukaryotic transcription results from the sequestration of chromosomes in the nucleus, or compartmentalization. Transcription and translation occur in separate compartments, the nucleus and the cytoplasm. In the nucleus the RNA (hnRNA) is transcribed and processed to mRNA. It is then transported out of the nucleus to the cytoplasm (and on the endoplasmic reticulum, ER) to be translated by the ribosomes (mRNA) or to participate in the process of translation (rRNA and tRNA).

Finally, there are differences in genome complexity and gene structure. Eukaryotic cells have 10^3 to 10^5 single-copy genes. In addition, their genes are discontinuous, containing introns and exons.

Eukaryotes have Three Classes of Genes, Each Transcribed by a Separate RNA Polymerase

These three RNA polymerases were originally identified by Robert Roeder while in William Rutter's laboratory in the early 1970s by fractionation of nuclear extracts from cells on a DEAE-Sephadex column and are found in all eukaryotes (yeast, plants, insects, mammals). These enzymes were named RNA polymerase I, II, and III (Pol I, Pol II, and Pol III, respectively). Subsequently, the polymerases have been shown to transcribe pre-rRNA genes (class I), hnRNA genes (class II), and pre-tRNA and 5S RNA genes (class III), respectively. The subunit compositions of the three eukaryotic polymerases are similar (**Figure 1**). In addition, the two largest subunits of each eukaryotic polymerase have structural and functional conservation with *Escherichia coli* subunits β and β'. None of the subunits seem to correspond to sigma factor (in *E. coli*). The corresponding activities associated with sigma factor are found in TFIIF and TBP, thus these functions are divided between proteins.

The RNA polymerase II largest subunit contains an unusual domain on its C-terminus referred to as the C-terminal domain (CTD). The CTD is not found in Pol I and Pol III largest subunit or β' of *E. coli* RNA polymerase. The CTD consists of 26–52 repeats of a 7-amino acid sequence: Tyr-Ser-Pro-Thr-Ser-Pro-Ser. This domain is highly phosphorylated on the Ser, Thr, and Tyr residues. The less phosphorylated form is called IIa and highly phosphorylated form is called

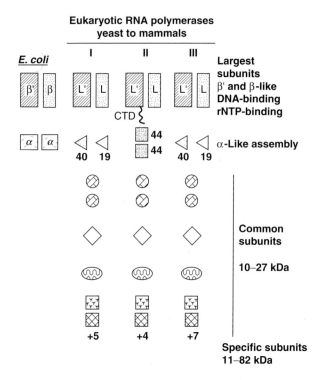

Eukaryotic RNA polymerases yeast to mammals

Figure 1 RNA polymerase I, II, and III subunits and their functions.

IIo. Generally, the IIA form of RNA polymerase II, containing subunit IIa, is associated with inactive genes, whereas the IIO form, containing IIo, is associated with active genes.

Each RNA polymerase is regulated independently and has a different promoter structure. All three have more complex promoter structures than prokaryotic promoters do. Each polymerase has multiple subunits (8–14) and has a molecular weight on the order of 5×10^5. In each case, the RNA polymerases themselves are not sufficient for promoter recognition and promoter specific transcription. Alone, they will initiate transcription essentially randomly, primarily at the ends and nicks in DNA. To initiate transcription promoter specifically they require additional protein factors or 'accessory factors' that have been identified by fractionation from cell extracts. These are also called transcription factors. These transcription factors can be further divided into two groups: the basal or general transcription factors (GTFs), which are absolutely required for promoter-dependent transcription, and the regulatory or promoter-specific transcription factors.

RNA Polymerase Subunits and Ancillary Factors were Identified by Purification from Cell-Free Extracts

The three types of cell extracts that have been used are whole cell, nuclear, and cytoplasmic. These extracts were the starting material for fractionation, purification, and identification of these protein factors. One factor, TBP, which stands for T̲ATA-b̲inding protein, is required by all three RNA polymerases. The TATA box is a DNA motif found in the minimal promoter of many class II genes. TBP is a component of the general transcription factors TFIID (pol II), TFIIIB (pol III), and TIF-IB (pol I). Class II genes are more numerous and are regulated differently from class I and III genes. Class I and III genes have simpler promoter structures and require fewer accessory factors. Class II genes have much more complex promoter structures and require a much larger number and variety of general transcription factors and transcription regulatory factors.

RNA Polymerase II Basal Transcription Factors and the Basic Mechanism of RNA Polymerase II Transcription

Fractionation of crude cell-free extracts has identified seven general (basal, minimal) transcription factors (GTFs). These have been designated as transcription factors (TF) IIA, IIB, IID, IIE, IIF, and IIH. Initially these proteins or protein complexes were separated by chromatography. Their functions were identified using *in vitro* transcription assays and electrophoretic mobility shift assays.

The minimal eukaryotic promoter consists of an RNA start site and the TATA-element (**Figure 2**). The RNA start site is often an A surrounded by pyrimidines, and is called an initiator (Inr). The consensus TATA-element sequence is TATA(A/T)A(A/T) that tends to be surrounded by GC-rich sequences. The TATA-element is found at −25 to −35 in higher eukaryotes and −40 to −90 in yeast. However, this promoter element is absent on many constitutively expressed genes, sometimes called housekeeping genes. In addition, there is some variation in TATA box sequences recognized by TBP, as well.

TBP is a 38-kDa protein that binds the TATA box in the DNA minor groove and has a modular structure

Figure 2 The minimal or core promoter sequences.

(seen in many transcription factors). The C-terminus of this protein has homology to *E. coli* sigma factor. When TBP binds the TATA-element it bends the DNA 80° (shown by TBP–DNA co-crystal). In addition, it puts a kink in the DNA and unwinds DNA 110°, opening up the minor groove (**Figure 3**).

The 'saddle' is actually perpendicular to the main DNA long axis, but is parallel with it through the 8 bp of the TATA box. The 80° bend and the 110° unwinding of the DNA nearly compensate; the net result is no measurable change in the supercoiling or any measurable bend in the DNA associated with TBP. When TBP binds in the minor groove, it opens it up and molds it to the underside of the TBP saddle.

TFIID is the only GTF that makes a sequence-specific contact with the DNA template. TFIID has TAFs (TBP-associated factors). TBP plus the TAFs make up TFIID (total MW of 750 000). There are eight TAFs with molecular weights of 250, 150, 110, 80, 60, 40, 30-α, and 30-β. TBP is a part of pol I and III transcription factors and has pol I and III specific TAFs. Hence TBP is sometimes referred to as the universal transcription factor. TFIID TAFs are designated by, for example, $TAF_{II}250$. The TAF proteins are thought to function as adapters or surfaces for interaction with other protein factors or DNA sequences. TFIIA contacts TAF_{II} 110, 250 and TBP. $TAF_{II}150$ binds the Inr sequence and $TAF_{II}150$ and TBP alone cover the same length of DNA as native TFIID (TBP plus all TAFs, **Figure 4**). TBP alone has an approximately 20-bp footprint right over the TATA box, whereas TFIID has a 75-bp footprint centered on the RNA start site.

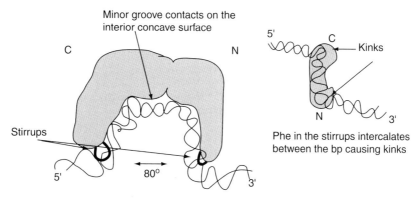

Figure 3 The interactions between TBP and the TATA element DNA.

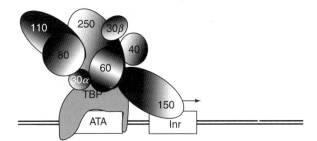

Figure 4 The complete native or holo-TFIID bound to the TATA element and the Inr.

Role of the RNA Polymerase II GTFs in Transcription

The 'preinitiation complex' that is formed is very large, having a molecular weight of greater than 2 million Da and consisting of as many as 40 polypeptides (**Figure 5**). On a supercoiled template only a minimal set of TFs, TBP, TFIIB, TFIIF, and RNA pol II, are required for transcription; the free energy of the supercoiled template may promote open complex formation. On a TATA-less promoter there is evidence for Inr-binding protein, TAF$_{II}$150, and RNA polymerase II-mediated complex formation on the Inr site. In addition, TATA-less promoters have multiple RNA start sites, while TATA plus Inr containing promoters often have a single RNA start site.

The following is a summary of RNA polymerase II and its GTFs by function and by order of appearance. TBP (or TFIID) binds to the TATA element, forming a stable or template committed complex. TFIIA stabilizes IID binding and counteracts negative factors, but is not required with purified factors. TFIIB also stabilizes IID binding. In addition, TFIIB functions in RNA polymerase II docking during preinitiation complex formation and in measuring the distance to the RNA start site(s) from the TATA element. TFIIF tightly associates with RNA polymerase II and functions to repress nonspecific DNA binding by polymerase II and in RNA polymerase II docking on the preinitiation complex. RNA polymerase II binds DNA and is the catalytic component, functioning to synthesize RNA from the DNA template. TFIIE has a regulatory role, functioning to recruit TFIIH and to stimulate the CTD kinase activity and to inhibit the helicase activity. TFIIH contains DNA duplex melting and CTD phosphorylation activity and functions in promoter clearance.

The intermediate complexes formed by the sequential binding of the transcription factors were identified by EMSA (gel shift), footprinting, and order-of-addition *in vitro* transcription assays. Interactions between transcription factors were determined by affinity chromatography, glycerol gradient. This multistep, multifactor process provides many opportunities

First step: recognition and binding of TATA box by TBP (or TFIID)

Second step: binding of TFIIA

Third step: binding of TFIIB

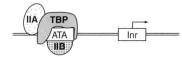

Fourth step: RNA polymerase II and TFIIF bind

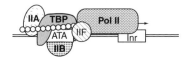

Fifth step: TFIIE and IIH bind sequentially to form the complete preinitiation complex

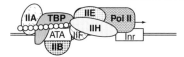

Sixth step: Energy-dependent step(s) (ATP bond hydrolysis)

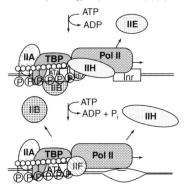

Seventh step: Initiation and elongation or RNA synthesis

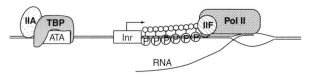

Figure 5 A summary of the basic mechanism of RNA polymerase II transcription initiation.

for regulation. Protein–protein interactions between GTFs and Pol II make up most of the interactions that hold the preinitiation complex together. Elongation is carried out by RNA polymerase IIO and is regulated and stimulated by TFIIF and TFIIS (elongation factors). Termination is difficult to study owing to the rapid processing of the RNA transcript. Termination sites are not well defined, but there is some evidence that termination does occur and requires 3′ cleavage.

Regulatory Elements and Factors

Basic RNA Polymerase II Promoter Structure and the Proteins that Bind the Promoter Motifs

It could be said that there are five steps in gene expression (protein coding genes): (1) activation of gene chromatin structure; (2) initiation of transcription; (3) RNA processing; (4) transport to the cytoplasm; and (5) translation. Transcription initiation is an early step, and is therefore an important control point.

Basic RNA polymerase II promoter structure involves *cis*-acting sequences that are bound by *trans*-acting protein factors (**Figure 6**). The *cis*-acting sequences are often identified by deletion and 'linker-scanning' mutagenesis.

The proximal, minimal, or core promoter region consists of TATA and Inr/Start site (**Figure 2**). Upstream promoter elements often act constitutively or in an unregulated manner (**Figure 6**). These elements are often found on the promoters of 'housekeeping' genes. Some examples of constitutive promoter elements and the factors that bind them are the GC box and the CCAAT box. GC boxes (GGGCGG) are bound by Sp1, which is expressed in all cell types in humans. There are often multiple GC boxes (G/C islands) found in gene promoters. These may function in concert with regulated factors to increase their effect on transcription. In addition, GC boxes are often seen upstream of TATA-less and Inr-less promoters. CCAAT boxes are bound by several factors including CTF/NF1. CTF/NF1 is present in all tissues, as well.

Regulatory promoter elements and the factors that bind them respond to environmental stimuli or are cell-type specific. Inducible element or response elements and transcription factors include those induced by stress (for example heat shock). The HSE (heat shock element) is bound by the heat shock transcription factor (HSTF). These elements can also be hormone inducible, for example, the hormone response elements, which are bound by the hormone receptors. The hormones include steroid hormones, derived from cholesterol, and thyroid hormone, derived from tyrosine. A well-studied example is the glucocorticoid receptor (GR), which is bound by glucocorticoid in the cytoplasm causing it to move into the nucleus. The GR–hormone complex then binds to the GRE (glucocorticoid response element). In contrast, membrane-bound receptors act through second messengers to activate transcription regulatory factors bound to their cognate sequences (e.g., CREB/ATF on CRE).

Cell type-specific regulatory elements and transcription factors regulate cell type-specific gene expression. The transcription factors are expressed or active only in particular cell types. For example, in the B-cell-specific expression of immunoglobulin genes the promoter is bound and activated by Oct-2, which is only expressed in B cells.

Enhancers are made of many of the same DNA elements, GC boxes, CCAAT boxes, response elements, cell-specific elements, and are bound by the same factors. However, enhancers can function over very large distances. Enhancers have been described as "a promoter element that might have been designed by an overenthusiastic graduate student" (Gary Felsenfeld). These elements can function downstream as well as upstream (3′ and 5′ from proximal promoter), and they can function in either orientation (can be inverted 180°). Enhancers are composed of different combinations and often of redundant regulatory elements, including constitutive elements, providing a wide range of regulatory possibilities. They can be cell type-specific or respond to external factors. These elements provide cell type-specific or factor-regulated expression to heterologous genes.

Transcription factors interact with the nucleotide bases on chemical groups outside of those participating in H-bonding between base pairs (bp) in the major and minor grooves. Each bp has a set of H-bond donors and acceptors, and hydrophobic surfaces. This is true of both the major and minor grooves. However, only the major groove has a unique pattern for each bp. Therefore, most sequence-specific factors bind in the major groove.

Structural Families of Regulatory Transcription Factors

Transcription factors have modular structures, a feature common to many eukaryotic proteins. Nearly all regulatory transcription factors are DNA-binding proteins. One of the first transcription factors to be purified and identified was Sp1, which binds many promoters at GC boxes. Sp1 can be purified by GC box double-stranded oligonucleotide affinity chromatography.

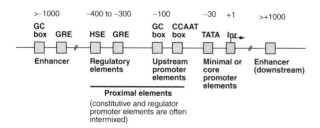

Figure 6   A generic example of an RNA polymerase II promoter structure.

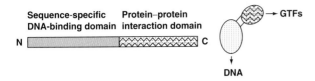

Figure 7 The 'modular' structure of transcription regulatory proteins.

These factors must bind to promoters containing their cognate binding site to activate transcription. They have two to four modules or domains. Nearly every transcription factor contains a sequence recognition domain, which binds DNA, and a protein–protein interaction domain, which binds the general transcription factors or RNA polymerase II (**Figure 7**). Some transcription factors have dimerization domains and some have regulatory domains, where they are modified or bound by regulatory factors. They also often have flexible connector regions between domains. The 'modular' or domain structure of these proteins has been demonstrated by the ability to form hybrid proteins consisting of domains from two different proteins that will function in transcriptional activation.

Activation domains interact with basal transcription factors and RNA polymerase II via protein–protein interactions. A common example is the acidic activation domain containing several asp, glu residues. GAL4, GCN4, VP16 (binds Oct-1), and the glucocorticoid receptor (GR) contain acidic activation domains. These activation domains have no specific sequence homology. However, all these proteins have a net negative charge. Another example is the glutamine-rich domain, found in Sp1, Antennapedia, Oct-1, Oct-2 N-terminus, and homeobox proteins. These domains contain approximately 25% glutamines and few negatively charged residues. A third example is the proline-rich domain found in CTF/NF1, Jun, AP2, Oct-2 C-terminus, which consists of 25% proline residues. Activation domains are thought to be somewhat unstructured and to contain hydrophobic amino acids and variously placed characteristic side chains. The interaction of activation domains with their targets is driven by hydrophobic forces (similar to protein folding). The order or periodicity of amino acid side chains in the cohesive surfaces determines the specificity of the interaction. Activation domains appear to be essentially unstructured and may only adopt a specific structure upon binding to their targets, in other words undergo an induced fit. This model accounts for both the specificity and the flexibility in the activator–target interactions seen in transcription.

DNA-binding (sequence-specific) domains are generally made up of α-helices and bind in the DNA

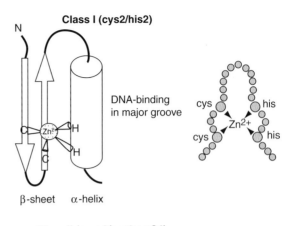

Figure 8 The structure of class I and II zinc finger domains.

major groove. One of the first DNA-binding domains identified is the zinc finger domain of which there are three types (**Figure 8**). Class I zinc finger proteins TFIIIA, Sp1, Krüppel, and steroid hormone receptors have a group of conserved amino acids that bind a zinc ion (Zn^{2+}) to form a particular structure. Class I zinc fingers have a single finger consensus (cys2/his2) that forms a tetrahedral structure with Zn^{2+} and that contains 23 amino acids. Class I zinc fingers have seven to eight amino acids between fingers. TFIIIA has nine zinc fingers and Sp1 has three zinc fingers. Members of this class are usually monomeric and have multiple fingers. Class II zinc finger proteins include the steroid hormone receptors and have cys2/cys2 (C4) zinc fingers. These zinc finger proteins have a region on the first zinc finger that determines DNA-binding specificity and bind as a dimer, for example, the GR. Class III zinc finger domains are typified by the GAL4 DNA-binding domain and have cys6 (C6) zinc fingers. They contain dimerization α-helices that form a coiled coil. The cys6 zinc finger domains are compact globular domains.

Other DNA-binding domains include the homeodomain proteins (or basic helix–turn–helix), the helix–loop–helix domain, the leucine zipper domain (**Figure 9**), and the POU (Pit-Oct-Unc) domain. Homeodomain proteins include the homeotic gene

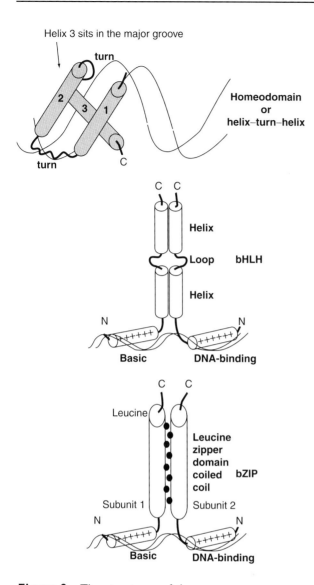

Figure 9 The structures of three common sequence-specific DNA-binding domains.

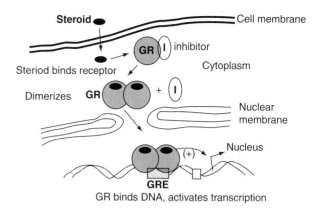

Figure 10 Regulation of steroid hormone receptor transcription activation activity.

products (Antp, en), Oct-1, Oct-2, and α2. These proteins are well conserved (80–90% similarity among *Drosophila* factors) in a 60-amino acid domain. They are made up of three α-helical regions and are related to the CAP protein, lambda repressor, and the Lac repressor in structure but are monomeric. Basic helix–loop–helix (bHLH) proteins include E12 and E47, myoD, c-myc, and *Drosophila* neuronal development factors. The bHLH domain is a 40–50-amino acid domain made up of two amphipathic α-helices. These transcription factors form homodimers and heterodimers through interactions between the hydrophobic face of the helices. The bHLH proteins have a basic region just N-terminal to the HLH domain that is required for DNA binding. Whether they are homodimers or heterodimers and what their partner is determines whether they will bind DNA at their cognate site. Dimerization is required for stable DNA binding, which is why they have double DNA-binding motifs. Non-basic HLH proteins, when dimerized with bHLH proteins, render them unable to bind DNA. Leucine zipper-containing proteins include C/EBP, Jun, Fos (or AP1), and Gcn4p. These proteins also have a basic region required for DNA-binding. A leucine zipper is an amphipathic helix in which every seventh amino acid is a leucine protruding from the hydrophobic face, with four to five repeats of this motif per protein (called bZIP proteins). These leucines interdigitate with those on a second bZIP molecule, and the two helices wind around each other. The DNA-binding site consists of two inverted repeats, with no separation. The bZIP transcription factors are often heterodimers.

Transcription Regulatory Mechanisms

Regulating the Regulators

Transcription factors can be regulated at the level of gene expression. They can also be regulated through covalent modification (phosphorylation, etc.). For example, CREB (cAMP response element binding factor) is activated by phosphorylation and AP1 is inactivated by phosphorylation. Some transcription factors are regulated through ligand-binding. The binding of these lipid-soluble hormones regulates the steroid hormone receptors (**Figure 10**). Lipid-soluble hormones include cortisol, retinoic acid, and thyroxinine. The hormone-binding domain of glucocorticoid receptor (GR) inhibits transcription activation in the absence of hormone. GR is thought to bind an inhibitor that anchors the GR in the cytoplasm in the absence of bound hormone, as well. Thyroid hormone receptor (THR) binds DNA and represses transcription in the absence of hormone. THR becomes a transcriptional activator on hormone binding. Other factors, such as NF-κB, are regulated, by protein inhibitors

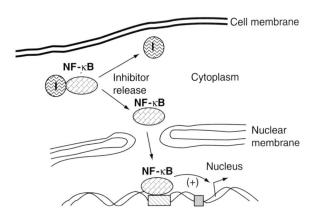

Figure 11 Regulation of transcription factors through protein inhibitor-binding.

(**Figure 11**). The released NF-κB enters the nucleus, binds DNA, and activates transcription. Transcription factors that form heterodimers, such as MyoD/ID and MCM1/α2, can be regulated through a change of dimerization partner. Homodimers of E12 and MyoD bind poorly or dimerize poorly, respectively. Finally, an important regulatory mechanism is the accessibility of binding sites, which can be determined through changes in chromatin structure.

Peptide hormones often function via posttranslational modification. These hormones first activate a membrane-bound receptor that sends a signal through a signal transduction pathway or a second messenger (small molecule). The end result is a modification, such as phosphorylation, of the transcription regulatory factor. This modification can affect nuclear localization, DNA binding, or transcription activation. A classic example is G-protein mediated signaling through cAMP. cAMP activates PKA, which in turn phosphorylates and activates CREB.

Regulation of Transcription; Regulation of the Function of GTFs

Transcription regulatory factors can act in at least four ways. First, they can act through stabilizing or increasing the rate of general transcription factor-binding or association with the DNA or preinitiation complex. Second, they can act by activating (increasing the catalytic rate) of the activity of a factor (e.g., the CTD kinase activity of TFIIH). Third, they can function by inducing a conformational change in basal transcription factors. There are several steps and GTFs to serve as targets. Finally, activators may function to counteract negative factors, for example, those that are part of or associated with TFIID, nucleosomes, and histone H1.

The protein–protein interactions between the transcription regulatory factors and the GTFs are well conserved. Activation domains from yeast work in *Drosophila*, plants, and mammals, although in most of these experiments acidic activators were used.

Synergistic activation is observed when there are multiple factors bound to a promoter. Transcription factors may interact simultaneously with the same or different targets in the complex. Synergism may be the result of there being many factors and steps that act as targets. A related phenomenon, 'squelching' (repression that occurs from high concentrations of a transcriptional activator), indicates that GTFs are targets for activators. Squelching is thought to result from high concentrations of an activating transcription factor titrating out a GTF and inhibiting transcription.

While the process of transcription initiation is often discussed as though there are several individual steps, many of the GTFs may be associated before promoter binding. This pre-assembled transcription complex is referred to as the holoenzyme. The holoenzyme model has important implications for how transcription regulatory factors work.

Finally, RNA polymerase II transcription forms a cycle of initiation, elongation, termination, and reinitiation. This cycle indicates that transcription activators can stimulate multiple rounds of transcription. In addition, postinitiation steps like promoter clearance and elongation are also regulated.

Adapters, Coactivators, or Mediators

The TAF$_{II}$40 is required for activation by GAL4-VP16 (an acidic activator) and the TAF$_{II}$110 is required for Sp1 (glutamine-rich activator). TBP alone is not sufficient for activated transcription by these transcription factors.

Highly purified TFIID (includes all the TAF$_{II}$s) is not sufficient for activated transcription by certain transcript factors, as well. These transcription factors require adapter proteins or coactivators. These adapter proteins may also be titrated out in 'squelching.'

Another type of coactivator protein are the architectural transcription factors. These proteins mediate protein–protein interactions and bend the DNA to promote interactions between transcription factors bound to an enhancer.

Promoter–Proximal Attenuation

Attenuation sites are located 20–30 bp downstream of the RNA start site and are found in *c-myc*, *hsp70*, *hsp26*, *hsp27*, α- and β-tubulin, polyubiquitin, and GAPDH genes. The *Drosophila hsp70* gene is a model for gene regulation through this mechanism. RNA polymerase II is paused on the uninduced *hsp70* promoter *in vivo* with a ~25 nt transcript, and is distributed between −17 to +37 bp on the promoter (**Figure 12**). Gene activation by stress (heat

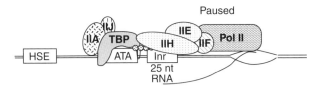

Figure 12 The paused transcription complex on the uninduced *hsp70* promoter.

shock, etc.) releases the pause with a concomitant phosphorylation of the CTD.

Gene Regulation through Chromatin Structure

Most of what we know about the role of chromatin structure in the regulation of specific genes is at the nucleosome level. A growing number of gene promoters have been shown to have positioned nucleosomes that play an important role in transcriptional repression and activation. In addition, some transcription factors function, at least in part, to counteract nucleosomal (core histone, H1) repression.

If cells are depleted of one of the core histones many genes are deregulated, apparently by the loss of nucleosomal repression. This is accomplished by shutting off the expression of one of the histone genes. When these cells go through S-phase and replicate their DNA, they end up with half the amount of one of the core histones and so have only half the nucleosomes needed for the assembly of two copies of the genome into chromatin. Several genes are derepressed, indicating that histones in the form of nucleosomes are required to maintain these genes in the inactive state. A similar situation is seen when trying to reconstitute transcription *in vitro*. Mutations in specific domains in the core histones disrupts repression and activation. This suggests a direct interaction between the transcription regulatory machinery and the core histones. In addition, mutations in the histones that affect their stability can suppress the effect of defective promoter elements, further supporting the idea that transcription factors can function to counteract nucleosomal repression.

Examples of Positioned Nucleosomes in Repression and Activation of Specific Genes

Nucleosomes can be positioned such that key promoter elements are wrapped around a nucleosome. Some regulatory transcription factors can bind their recognition sequences when they are wrapped around the nucleosome. Alternatively, nucleosomes can be positioned such that a key promoter element is placed in the linker DNA between nucleosomes and constitutively available.

The mouse mammary tumor virus-long terminal repeat (MMTV-LTR) promoter is regulated by the GR (glucocortrioid receptor). The MMTV-LTR promoter is incorporated into six positioned nucleosomes (**Figure 13**). When this promoter is activated the glucocorticoid hormone binds and activates the GR. The activated GR can bind the GREs (glucocortrioid response elements) when wrapped around nucleosome B, indicating that some transcription factors can recognize their binding sites on the surface of a nucleosome. GR-binding appears to displace histone H1 and recruit the Swi/Snf complex, which disrupts or reconfigures nucleosome B. This allows transcription factors NF1 and Oct-1 to bind their cognate sequences. All three transcription factors then displace nucleosome A or help the basal transcriptional machinery to displace nucleosome A. The preinitiation transcription complex is formed and the promoter is transcribed.

The α2/MCM1 transcription factor complex functions to inhibit a-cell-specific genes in α-cells (yeast *Saccharomyces cerevisiae*). Transcription factor α2 is absent in a-cells so no inhibition of these genes occurs. The α2/MCM1 complex binds the α2 operator in the promoter of these genes (e.g., *STE6*) and positions a nucleosome next to the operator over the TATA element, repressing transcription. This complex is thought to recruit Tup1p and Ssn6p (transcriptional repressors) to the *STE6* promoter. The histone H4 tail is required for nucleosome positioning and transcriptional repression. In addition, Tup1p and Ssn6p have been shown to bind the histone H4 tail. Insertion of 75 bp between the α2 operator and the TATA element in the linker DNA between nucleosomes does not relieve the repression. This may explain the fact that on the *STE6* promoter an array of nucleosomes is formed. This array of positioned nucleosomes is thought to be stabilized by a backbone of Tup1p/Ssn6p molecules. The transcription factors Mig1p and Rox1p also recruit the Tup1p/Ssn6p complex and repress transcription through nucleosome positioning on the promoters of the metabolic genes *SUC2* and *GAL1–GAL10*.

Role of Core Histone Acetylation in Transcriptional Regulation

The core histone N-terminal tails are unstructured and highly positively charged, containing several lysines and a few arginines. The ε-amino groups on the lysines are posttranslationally modified by acetylation, which removes the positive charge. The unmodified, positively charged core histone tails may interact with the linker DNA and with negatively charged patches in the core histones on the exposed surface of adjacent nucleosomes. These interactions

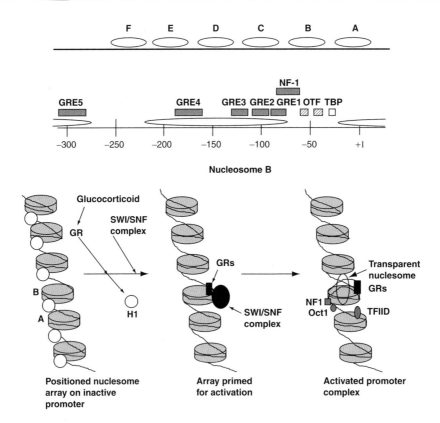

Figure 13   The modulation of chromatin structure in MMTV-LTR promoter activation.

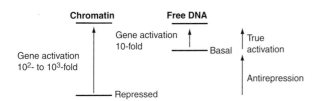

Figure 14   Transcriptional activation on chromatin vs. free DNA templates.

stabilize the folding of the chromatin fiber into higher order structures that repress transcription. Acetylation and removal of the positive charges disrupts these interactions and tends to derepress transcription. Sequence-specific DNA-binding transcription regulatory proteins can recruit the activities (histone deacetylases and histone acetyltransferases) responsible for maintaining the histone acetylation state.

Biochemical Analysis of the Mechanism of Transcription Regulation with Chromatin Templates

Most biochemical studies on the mechanism of transcription and its regulatory factors have used naked DNA templates for purposes of simplicity. The level of activation observed with these templates typically has only been in the range of 5- to 10-fold, possibly 20-fold at the outside. Reconstitution of the DNA template into nucleosomes results in a general repression of transcription (**Figure 14**). If TFIID or sequence-specific DNA binding activators are bound prior to nucleosome formation then these templates are activated. The net result is a much greater-fold activation (10^2 to 10^3) on chromatin templates than seen with free DNA templates, a level similar to that seen *in vivo*. This finding has led to the hypothesis that some transcriptional activators function at least in part to counteract chromatin-mediated repression ('antirepression'). This antirepression occurs on top of 'true activation,' which is the result of recruitment and stimulation of the basal transcription factors.

Further Reading

Elgin SCR (ed.) (1995) *Chromatin Structure and Gene Expression*. New York: Oxford University Press.

Hames DB and Higgins SJU (1994) *Gene Transcription: A Practical Approach*. Oxford: Oxford University Press.

Latchman DS (1995) *Eukaryotic Transcription Factors*, 2nd edn. San Diego, CA: Academic Press.

Latchman DS (1995) *Gene Regulation: A Eukaryotic Perspective*, 2nd edn. London: Chapman & Hall.

Wolffe AP (1998) *Chromatin: Structure and Function*, 3rd edn. San Diego, CA: Academic Press.

See also: **Chromatin; Enhancers; Gene Expression; Operators; Promoters; Repressor; Transcription**

Gene Replacement

See: **Gene Targeting**

Gene Sequencing

See: **DNA Sequencing**

Gene Silencing

W Filipowicz and J Paszkowski

Gene silencing is defined as an epigenetic modification of gene expression leading to inactivation of previously active genes. Epigenetic modification does not alter the DNA sequence and, although it is heritable, variable frequencies of reversions to expression are observed. Gene silencing is used in the course of normal development and differentiation to repress genes whose products are not required in specific cell types or tissues. This may apply to individual genes or larger chromosome regions. In some special situations, such as chromosome dosage compensation in mammals, one of the two female X chromosomes is almost completely repressed. Mechanisms responsible for repression of genes involve changes in chromatin structure and levels of DNA methylation, or destabilization of mRNA. Modifications of chromatin and DNA template make genes inaccessible to the transcription machinery. Mechanisms of RNA destabilization are still largely unknown. Aberrant silencing of genes may lead to disease in mammals and generate developmental variants in plants. For example, methylation of tumor suppressor genes contributes to the onset and progression of cancer, while methylation of genes controlling flower development results in heritable changes of flower morphology.

Gene silencing can act at the transcriptional or posttranscriptional level; the two phenomena being referred to as transcriptional gene silencing (TGS), and posttranscriptional gene silencing (PTGS). Genes affected by TGS are either not transcribed at all, or transcripts are produced at very low levels. TGS has been observed in fungi, plants, and animals. It is probably triggered by redundancies of genetic information since its occurrence correlates well with the presence of repeated genes or subgenomic fragments. As shown in plants, increased levels of ploidy may likewise act as a trigger of TGS. In organisms that are able to methylate their DNA, levels of DNA methylation are significantly increased in genes silenced by TGS. In the fungus *Neurospora crassa*, DNA methylation of redundant sequences is followed by a modification of their nucleotide sequences in a process referred to as repeat induced point mutation (RIP). In the fungus *Ascobolus nidulans*, methylation and inactivation of redundant genes occurs in a specific phase of the life cycle, in a process called MIP (methylation induced premeiotically). TGS has been well studied genetically in yeast, and more recently also in plants. These studies revealed a number of genes that are required for silencing. Their protein products are either chromatin components or posttranslational modifiers of chromatin proteins. In organisms that are able to methylate DNA, TGS regulators also include DNA methyltransferases and proteins recognizing methylated DNA. The biological role of TGS is still under debate. One postulated function is to extinguish transcription of transposable elements in order to prevent their movement and propagation in chromosomal DNA. In plants, TGS is also able to affect single copy genes, giving rise to semistable epigenetic variants. Creation of reversible epialleles adds to the phenotypic variability important in evolving plant populations.

In PTGS, also referred to as cosuppression in plants, or quelling in *N. crassa*, the affected gene is transcriptionally active but its transcripts undergo rapid degradation, resulting in the absence of translatable mRNA. PTGS is frequently observed in transgenic organisms, in particular when multiple copies of the transgene are present. Transcripts of both the transgene and host genes having 80% or more sequence identity with the transgene, are subject to the degradation. In plants, infection with RNA viruses engineered to express sequences homologous to host genes, will likewise result in specific degradation of host and viral RNAs. Available evidence indicates that small antisense or double-stranded (ds) RNAs are responsible for specific RNA degradation. Such aberrant RNAs may be formed as a result of the artifactual bidirectional transcription from the transgene loci, or may be produced from endogenous genes modified by ectopic interactions with homologous transgenes. The strongest support for the role of dsRNA in PTGS comes from RNA interference (RNAi) experiments. Injection of dsRNA into the nematode *Caenorhabditis elegans* or into the eggs of *Drosophila melanogaster* leads to potent and sequence-specific PTGS. Injected dsRNA is fragmented into ~23-nt-long RNA pieces which appear to act as guides hybridizing to endogenous mRNAs and targeting them for degradation. In both plants and *C. elegans*, the PTGS/RNAi effect spreads across cellular and tissue boundaries and small RNA fragments are the best candidates for the diffusible silencing signals. Genetic screens have identified several genes essential for establishing and/or maintaining PTGS. Some of them are also required for RNAi, indicating that the two phenomena are mechanisticaly related. Like TGS, PTGS may also represent a mechanism to defend the organism and its genome against invasive nucleic acids such as transposons, retroelements, and viruses. Certain forms of PTGS, in particular RNAi, offer a targeted and efficient way of inactivating genes, providing a powerful tool for investigating gene function.

Recent experiments point to links between TGS and PTGS. In plants, dsRNA which acts as a trigger of PTGS can also direct methylation of the homologous sequences in DNA, leading to transcriptional inactivation of the gene.

See also: Epigenetics; Transposable Elements; X-Chromosome Inactivation

Gene Splicing

See: Recombinant DNA

Gene Substitution

T Ohta

Copyright © 2001 Academic Press
doi: 10.1006/rwgn.2001.0524

Gene substitution is the process in which a mutant allele replaces the original allele in a population. Many mutants arise in natural populations, but the majority of them are lost within a few generations by chance. Those lucky mutants that survive the first few generations are tested by natural selection, i.e., selectively advantageous mutations increase their frequencies in the population, and disadvantageous ones are eliminated from the population. For selectively neutral mutants, their rise and fall in the population is governed by random genetic drift. When the frequency of a mutant gene in the population becomes one, it is said to have fixed in the population, and a gene substitution has occurred.

The process of gene frequency change in a population has been studied by population geneticists. Two approaches are deterministic and stochastic. When the population size is large and random drift is negligible, the deterministic model is applicable, i.e., the change of gene frequency by natural selection can be predicted by simple formulas. However, when the population size is not large, the chance effect becomes significant, and stochastic treatments are needed.

Behavior of molecular mutants is often influenced by random genetic drift even in a large population because of their minute effect. In other words, many molecular mutants are selectively neutral or nearly neutral, and their behavior depends on random drift. The dynamics of a completely neutral mutant has been theoretically described, i.e., the average course of the substitution process is known. For nearly neutral mutants, interaction of selection and random drift is important.

The number of gene substitutions at a locus is estimated by comparing gene sequences at this locus between species. From such comparative studies of gene sequences, the rates of gene substitutions at various protein loci and noncoding regions have been obtained. The rate is defined as the number of substitutions per unit time. As an example, the rate of substitution of the α hemoglobin gene is about 0.5 per site for amino acid replacement sites, and about 4 per site for synonymous sites (rates are given per 10^9 years). In general, unimportant sites are evolving rapidly and important sites are evolving slowly.

See also: Population Genetics

Gene Targeting

See: Genetic Recombination

Gene Therapy, Human

T Friedmann

Copyright © 2001 Academic Press
doi: 10.1006/rwgn.2001.0526

From the beginning of medical history, attempts to treat human disease have necessarily been aimed at ameliorating symptoms and easing suffering rather than correcting the underlying causes of most human disease. The reasons for this are quite clear. In most cases, until late into the twentieth century, other than for attributing human disease and misfortune to irate gods, healers simply did not recognize or understand the true causes of most human afflication and could envision no alternatives to simply bringing relief and comfort to death and suffering. An exception to this general rule might be represented by even the most ancient forms of surgery, in which root causes of disease, as imaginary and wrong as they may often have been, were identified and invasive surgical procedures developed to rid the afflicted patient of the offense.

The emergence of modern medical science over the past few centuries, and particularly the last few decades of the twentieth century, was to change that approach. Epochal advances first slowly and then with increasing speed:

- The development of the art/science of human anatomy.
- The discovery of blood circulation by the English physician William Harvey in 1628.
- The invention of the compound light microscope by the Dutch cloth merchant Anton van Leeuwenhoek in 1674 and the identification by Robert Hooke in 1665 of 'cells' as the structural basis of life.
- The development of the science of cell biology by Theodor Schwann, Matthias Schleiden, and Rudolph Virchow in the early 1800s.
- The first public demonstration of anesthesia by William Morton in 1846 and the consequent birth of modern surgery.
- The discovery by Gregor Mendel in 1865 of the laws of genetic inheritance.
- The development of the germ theory by Louis Pasteur and others during the same period.
- The revelation of the concepts of chemical pathology by Archibald Garrod at the beginning of the twentieth century (the concept of 'inborn' errors of metabolism – the principle that genetic errors lead to disruptions of normal metabolic processes to produce disease).
- The discovery of antibiotics by Alexander Fleming in London and a group of chemists at Oxford in the 1920s and the 1930s.
- The invention of experimental genetics in *Drosophila* by Thomas Hunt Morgan in the early 1900s.
- The discovery in the 1940s and 1950s by Oswald Avery, Colin McLeod, MacLyn McCarty, Alfred Hershey, and Martha Chase that genetic information is carried by deoxyribonucleic acid (DNA).
- The discovery of the chemical rules by which DNA stores and transmits its genetic information during the full flowering of molecular biology in the 1960s and 1970s, under the leadership of the giants of the era – Francis Crick, James Watson, Sydney Brenner, François Jacob, Jacques Monod, Fred Sanger, Max Perutz, and many others.

The result of this explosion of knowledge of the physical and chemical basis of life became applied very quickly to an understanding of the nature of the genetic errors that lead to disease, making the design of drugs ever more rational and effective. An important product of all these developments was an understanding that most human disease results from a combination of inborn genetic factors and environmental influences, with predominating genetic factors in some disorders (such as cystic fibrosis, sickle-cell anemia, Tay–Sachs disease, Huntington disease, etc.), and a combination of genetic and environmental influences in most of the common and severe diseases (cancer, heart disease, degenerative disorders, neurological diseases such as Parkinson and Alzheimer diseases, and even infectious disease). Yet still, until the late 1960s and early 1970s, even with all this new understanding of the causes of disease, the predominant treatment model was still one in which the target for therapy was the drug treatment of abnormal cellular processes that resulted from the underlying defect. Treatment was still not aimed at the defect itself – what was being fixed was not what was broken.

A new and more definitive approach to therapy began to surface in the mid to late twentieth century, one that is destined to provide a rational attack not only on the results of the causative defects but also, for the first time, more directly on the causes themselves. In 1944, Oswald Avery, Colin McLeod, and their colleagues at the Rockefeller Institute in New York first demonstrated that purified DNA from one strain of bacteria could be introduced into another strain to produce transfer genetic traits to the recipient bacteria. This process came to be called 'genetic transformation.' Inevitably, as the science of mammalian cell biology matured in the 1950s and 1960s, scientists would begin to try the same sort of experiment in mammalian cells rather than bacterial cells. Could normal human and other mammalian cells be changed permanently 'transformed' by exposure to DNA from another mammalian cell? Could cells carrying disease traits be changed to normal cells 'cured' by exposure to DNA from normal cells?

The answer was that the genetic modification was found to be much, much more difficult and far less efficient to correct errors in mammalian cells than it was in bacteria. There were a number of early experiments indicating that, after exposure to foreign normal DNA, genetically altered cells could indeed be found among defective cells, but only at a frequency of one in a million or less – certainly not efficient enough to imagine correcting a disease by such an approach. But by the mid-1960s, a number of investigators came to realize that there were agents all around us in nature that were able to do the job of introducing foreign DNA into human and other mammalian cells with very great efficiency. These agents are called viruses. Their life cycles depend on their ability to insert their genetic material, whether it is DNA or RNA, into target cells and to express their genes for varying lengths of time in those cells, thereby both reproducing themselves and imparting new genetic traits to those cells. They have therefore, had to learn to carry out such gene transfer with great efficiency. Viruses are essentially packages of DNA or RNA surrounded by protein, sugars, and fat molecules. The functions of these viral 'coats' is not only to package and protect the viral genetic material but also to help the virus identify specific molecules on the surface of their target host cells (virus receptors) that serve to attach the virus to the cells and promote its entry into the cell. It is because of this interaction of viruses with their specific cell-surface receptors that viruses are so much more efficient at transferring their genes into cells than other nonviral methods of gene transfer into mammalian cells. Unfortunately, at least for the infected cell, the cell often becomes nothing more than a factory more or less single-mindedly devoted to reproducing the virus and subverting all other cell functions necessary for cell survival, thereby killing the cells. However, some viruses have come to a very happy accommodation with their host cells and are able to exist for long periods in the infected cell without producing any apparent damage to the cell. In such cells, the foreign piece of new genetic information can become integrated into the genetic information of the cells, thereby providing the cell stably and permanently with new genetic functions without killing the cells. Unfortunately, as Renato Dulbecco and his colleagues at the Salk Institute in California showed in the mid-1960s, those new genes can have the effect of causing the cell to forget how to stop growing in its usual controlled fashion, thereby producing a cell that grows out of control – a cancer cell.

For someone interested in human disease and, by good fortune, exposed to the environment of such a laboratory, the leap from inefficient gene transfer into defective human cells with purified ('naked') DNA to the use of viruses as agents to carry foreign and potentially therapeutic DNA into cells seemed obvious to several of us in the Dulbecco laboratory. In 1972, my colleague and I proposed that such viruses might be genetically modified to make them incapable of replicating and also abrogate their pathogenicity, while at the same time using them as vehicles to transfer therapeutic genes into defective cells (Friedmann and Roblin, 1972). We envisioned two general approaches to the therapeutic applications:

- The *ex vivo* approach in which the genetic correction would be accomplished by removing target cells from a patient, introducing a therapeutic viral vector into them *in vitro*, and then returning the genetically corrected cells to the patient.
- The *in vivo* approach in which the gene transfer vector is introduced directly into the target defective cells in the patient.

While very attractive in principle, these concepts could not be put into practice at the time because no efficient viral gene transfer vectors existed and the recombinant DNA techniques needed to produce them had not yet been developed. Fortunately, over the next few years, methods of recombinant DNA manipulation were developed and refined and allowed, in the early 1980s, the design and production of the first truly efficient viral vectors for gene transfer into mammalian cells. These vectors were derived from mouse viruses that used RNA as their genetic material and that were associated with several kinds of cancer in laboratory mice. These original vectors were derived from viruses that were called retroviruses because they have the property of converting their RNA into DNA after infection. They are able to integrate the DNA copies of their genomes into the genome of the host cell, thereby allowing them to express some of the viral genes in a stable and heritable way in the cell for the lifetime of the cell. Recombinant DNA methods allowed investigators to remove the potentially deleterious genes from the viruses and replace them with other genes that could also be expressed permanently in the infected cells. These methods provided a proof of principle that such retrovirus vectors could carry out the functions required of a gene therapy vector for human disease. Very quickly thereafter, our laboratory showed for the first time that a retrovirus vector carrying a normal copy of a human disease-related gene could correct the abnormal properties of cells derived from patients. We transferred a cDNA corresponding to the normal allele of the hypoxanthine guanine phosphoribosyl transferase (HPRT) via a retrovirus vector into cultured cells from patients with the rare but devastating Lesch–Nyhan disease, and found that we could identify modified cells that not only demonstrated restored expression of the normal gene but also correction of some of the secondary metabolic defects resulting from their HPRT deficiency (Willis *et al.*, 1984). It was the development of the retrovirus vectors that represented the single most important early technical advance that opened the door to the subsequent explosion of gene transfer with many additional disease-related genes.

One of those other disease-related genes that was applied early to gene transfer studies with retrovirus vectors was the gene encoding adenosine deaminase (ADA), a defect of which is responsible for a severe immunological defect in human patients. Model studies with the normal ADA gene similar to those with HPRT demonstrated correction of the enzyme defect in cells from ADA patients and it was this disease model that was eventually to become the subject of the first potentially therapeutic human gene therapy study.

Retrovirus vectors have many advantages for potential gene therapy applications, but they were also quickly found to demonstrate a number of disadvantage. They are unable to infect nonreplicating cells such as neurons or hepatocytes, both important potential target cells for gene therapy. They are also relatively unstable *in vivo* and they cannot be made to sufficiently high titers to make gene delivery efficient *in vivo*. For these and other reasons, vectors have been developed from a number of other parent viruses, including human retroviruses such as HIV-1 and HIV-2 and other lentiviruses, adenoviruses, herpes viruses, and adeno-associated viruses. This growing collection of vectors now allows gene transfer into virtually any and every possible human or other mammalian cell, either *in vitro* or *in vivo*. Furthermore, gene transfer methods using nonviral vectors such as liposomes and naked DNA have become increasingly efficient and useful in a wide variety of disease models. Some of the important properties of the more common of these vectors are summarized in **Table I**. These represent the properties of the most commonly used versions of each of the major vector systems. It must be kept in mind that major improvements are rapidly being made in each of the systems, which will significantly improve their properties and make some of the disadvantages described in the list above out of date quite quickly. For example, methods are emerging to permit the targeting of some of these vectors to specific cells *in vivo*, allowing efficient delivery by the bloodstream. Titers and vector concentrations are improving, and

cytotoxic and immunogenic properties are being reduced. Methods are emerging for the integrating viruses (retroviruses, lentiviruses, AAV) that will eventually allow insertion into specified sites in the host cell genome, thus abrogating the possibility of insertional mutations in the cell.

Even though the initial disease models were those of the single gene defects, the 'inborn errors of metabolism' such as Lesch–Nyhan disease, adenosine deaminase deficiency described above, cystic fibrosis, familial hypercholesterolemia, and others, other more complex diseases also became targets for gene therapy studies. Cancer quickly became one of the most attractive targets for gene therapy studies because of the enormous importance of the public health problem posed by cancer, and because of the identification of a variety of cancer-causing genes (oncogenes, tumor suppressor genes, apoptosis and cell death genes, cell-cycle-regulating genes, immune-modulatory genes, and others) that presented appealing targets for genetic manipulation and disease intervention. Other complex disease also came to be identified more and more as potential gene therapy targets, including degenerative diseases such as atherosclerosis and many forms of cardiovascular disease, arthritis, diabetes mellitus, familial and sporadic forms of neurological degenerative disorders such as Parkinson and Alzheimer diseases, and others. Even infectious diseases such as AIDS became potential targets for genetic intervention, and genetic approaches toward the control of agents responsible for other infectious diseases such as malaria have also become active areas of research. In many of the direct human disease models, laboratory studies have shown that foreign genes introduced into affected cells or into animals subjects by one or another of the gene transfer techniques could modify or even prevent the disease phenotype.

This plethora of gene transfer techniques techniques and the availability of a growing number of

Table I Some important properties of the more common vectors

Vector	Advantages	Disadvantages
Retrovirus	Noncytotoxic, integrates, stable expression	Requires replicating cells, low titers, unstable *in vivo*, insertional mutations
Lentivirus (HIV, FIV, etc.)	Noncytotoxic, infects nonreplicating cells, stable expression	Low titers, unstable *in vivo*
Adenovirus	High titers, efficient expression	Usually transient expression, cytotoxic, immunogenic
Herpes simplex	High titers, latency in some cells, prolonged expression	Cytotoxic
Liposomes	Noncytotoxic	Inefficient
Naked DNA	Noncytotoxic, stable in some cells, vaccination uses	Inefficient

convincing disease models made it appear in the late 1980s that the road to successful gene therapy in human patients was going to be relatively smooth and uncomplicated. Beginning in 1989 and 1990, proposals for human application of promising laboratory gene transfer results began to pour in to the federal regulatory bodies empowered to evaluate human gene therapy trials – the Gene Therapy Subcommittee of the Office of Recombinant DNA Advisory Committee (RAC) at the National Institutes of Health and the Food and Drug Administration (FDA). By the mid-1990s, several hundred clinical studies had been reviewed and approved by the RAC and FDA in the US and by their equivalent agencies in Britain, France, Japan, Italy, Germany, and a number of other countries. Clinical gene therapy trials were undertaken in many forms of cancer, ADA deficiency, cystic fibrosis, and hypercholesterolemia, involving several thousand patients. Despite high levels of expectation for some evidence of therapeutic efficacy even in the early phase I studies, the results of this first rigorous set of clinical studies published in 1985 were disappointing, since they failed to provide definitive proof for clinical benefit to any patients. However, the studies did demonstrate clearly that foreign genes could be introduced into humans without any apparent deleterious effects and that such genes could be expressed for prolonged periods (up to several years) and even that they produced physiological effects that were relevant to the disease processes. But no convincing evidence was presented by any of these studies for a cure, reversal, stabilization, or cessation of a disease process or for improved quality of life for any of the patients.

These early experiments should not be seen as outright failures but rather as experiments that were carried out in an atmosphere of unrealistically exaggerated expectations and overstated claims by some scientists, by their institutions (including universities) and even the National Institutes of Health, and by both the lay and scientific media. Several investigators, as well as the director of the NIH, became concerned that the general field of human gene therapy was promising more than it could deliver and began to call for more through basic and clinical research and more restraint in public statements from all parties regarding immediacy of clinical benefit from a field so obviously in its infancy (Friedmann, 1994).

In the few years since those studies and the criticisms that followed, all aspects of the basic and clinical science of human gene transfer have improved markedly. New and vastly improved vectors have become available and many new disease-related genes have been described and their role in disease better understood. Gene transfer studies in tissue culture systems

and in the growing number of faithful animal model systems for human disease have provided very convincing evidence for continuously improving efficiency and stability of gene transfer and expression. But most exciting of all is the clear evidence for some clinical benefit to patients that is beginning to percolate to the surface through the layer of uncertainty and doubt from so many previous inconclusive clinical studies. Only the most pessimistic could fail to see or believe that the clinical promise of human gene therapy is about to be delivered, slowly at first but with increasing speed and efficiency as our techniques and tools improve.

Gene therapy is actually two things. It is the concept that much of human disease can and should be treated at the level of the underlying genetic mechanisms. That part of the revolution of human gene therapy is over. Gene therapy is now a widely accepted and even a central driving force in modern medicine. It will not vanish or fail in the long run. Gene therapy is also the implementation to clinical reality. That part of the revolution is now occurring. Within the coming several years, patients will survive who would have died without genetic intervention, suffering will be eased that could not have been ameliorated by traditional means, and quality of life will improve for many people because of the power of genetic modification.

This newly justified optimism does not mean that the road ahead for gene therapy in humans will be entirely smooth. There will many technical and conceptual obstacles to the treatment of disease, and, inexorably, public policy and ethical problems posed by the inevitable extension of disease management to manipulations of traits not so clearly disease-related, e.g., physical stature, memory, cognitive, and even some personality traits. As the technology of gene transfer into human somatic cells becomes more and more efficient, predictable, and error-free, extension of genetic manipulation to the human germline to reduce the expression of disease not only in a treated patient but also in the patient's progeny will become more and more irresistible. The debates surrounding human gene therapy will be far from over with the imminent demonstration of therapeutic success in current clinical studies. Nevertheless, it is clear that medicine is on the verge of being able finally to deliver truly definitive therapy for so many diseases that have been otherwise intractable scourges since the beginning of medical history (Friedmann, 1996, 1997). It is a truly remarkable time for medicine.

References

Friedmann T (1994) The promise and overpromise of human gene therapy. *Gene Therapy* 1(4): Editorial.

Friedmann T (1996) Gene therapy: an immature genie but certainly out of the bottle. *Nature Medicine* 2: 144–147.

Friedmann T (1997) Overcoming the obstacles to gene therapy. *Scientific American* 276: 95–101.

Friedmann T and Roblin R (1972) Gene therapy for human genetic disease? *Science* 175: 949–955.

Willis RC, Jolly DJ, Miller AD *et al.* (1984) Partial phenotypic correction of human Lesch–Nyhan (HPRT-deficient) lymphoblasts with a transmissible retroviral vector. *Journal of Biological Chemistry* 259: 7842–7849.

See also: **Cancer Susceptibility; Ethics and Genetics; Genetic Diseases; Metabolic Disorders, Mutants**

Gene Transfer

See: **Horizontal Transfer**

Gene Trapping

L Silver

One side product of many transgenic experiments is the generation of mice in which a transgene insertion has disrupted an endogenous gene with a consequent effect on phenotype. Unlike spontaneous or mutagen-induced mutations, 'insertional mutations' of this type are directly amenable to molecular analysis because the disrupted locus is tagged with the transgene construct. Unexpected insertional mutations have provided instant molecular handles not only for interesting new loci but for classical loci, as well, that had not been cloned previously.

When insertional mutagenesis, rather than the analysis of a particular transgene construct, is the goal of an experiment, one can use alternative experimental protocols that are geared directly toward gene disruption. The main strategies currently in use are based on the introduction into embryonic stem (ES) cells of β-galactosidase reporter constructs that either lack a promoter or are disrupted by an intron. The constructs can be introduced by DNA transfection or within the context of a retrovirus. It is only when a construct integrates into a gene undergoing transcriptional activity that functional β-galactosidase is produced, and producing cells can be easily recognized by a color assay. Of course, the production of

β-galactosidase will usually mean that the normal product of the disrupted gene can not be made and thus, this protocol provides a means for the direct isolation of ES cells with tagged mutations in genes that function in embryonic cells. Mutant cells can be incorporated into chimeric embryos for the ultimate production of homozygous mutant animals that will display the phenotype caused by the absence of the disrupted locus. This entire technology, referred to as 'gene trapping,' is clearly superior to traditional methods for the production of mutations at novel loci that use chemical mutagens or irradiation.

See also: **Beta (β)-Galactosidase; Embryonic Stem Cells**

Gene Trees

N Saitou

The phylogenetic trees of genes are called 'gene trees.' Reconstruction of gene trees is quite important for evolutionary studies, because replication of nucleotide sequences automatically produces a bifurcating tree of genes. It should be emphasized that the phylogenetic relationship of genes is different from the mutation process. The former always exists, while mutations may or may not happen within a certain time period and DNA region. Therefore, even if several nucleotide sequences happen to be identical, there must be a genealogical relationship for those sequences. However, it is impossible to reconstruct that genealogical relationship without the occurrence of mutational events. In this respect, the extraction of mutations from genes and their products is important for reconstructing phylogenetic trees of genes. The advancement of molecular biotechnology has made it possible routinely to produce nucleotide sequences.

Phylogenetic trees of genes and species are called 'gene trees' and 'species trees,' respectively, and there are several important differences between them. One such difference is illustrated in **Figure 1**. Because a gene duplication occurred before speciation of species A and B in **Figure 1A**, both species have two homologous genes in their genomes. In this situation, we should distinguish 'orthology,' which is homology of genes reflecting the phylogenetic relationship of species, from 'paralogy,' which is homology of genes caused by gene duplication(s). Thus, genes 1 and 3 (and 2 and 4) are 'orthologous,' while genes 1 and 4 (and 2

(A)

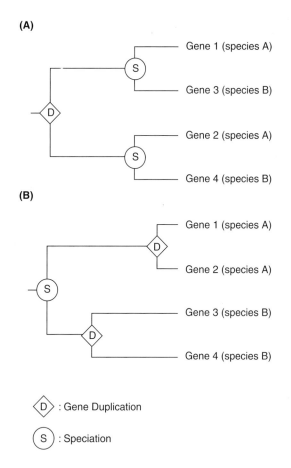

(B)

◇D◇ : Gene Duplication

Ⓢ : Speciation

Figure 1 Two possible relationships of four homologous genes sampled from two species. (A) When gene duplication preceded separation. (B) When two independent gene duplications occurred after speciation.

and 3) are 'paralogous,' as well as homologous genes in the same genome (gene pairs 1–2 and 3–4). If one is not aware of the gene duplication event, the gene tree for 1 and 4 may be misrepresented as the species tree of A and B, and thus a gross overestimation of the divergence time may occur. Note also that the divergence time between genes 1 and 3 is identical to that between genes 2 and 4, since both times correspond to the same speciation event.

When two homologous gene copies are found in species A and B, another situation is possible, as shown in **Figure 1B**. Now two gene duplications have occurred after the speciation of species A and B, and two gene copies in the genome of each species are more closely related with each other than the corresponding homologous genes at different species. Because two duplication events occurred independently, the divergence time between genes 1 and 2 is different from that between genes 3 and 4.

(A)

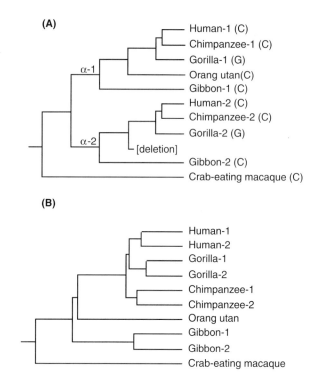

(B)

Figure 2 Effect of gene conversions to tandem duplicated IgA genes. (A) Plausible gene trees. (B) Spurious gene tree adopted from Kawamura S, Saitou, N and Veda S (1992) *Journal of Biological Chemistry* **267**: 7359–7367.

When gene conversion and/or recombination has occurred within the gene region under consideration, a gene tree may be different from the species tree. **Figure 2A** shows the plausible gene tree for primate immunoglobulin α genes 1 and 2, and the gene duplication clearly preceded speciation of hominoids, followed by deletion of the α-2 gene from the orang utan genome. However, there are many nucleotide sites that possibly experienced gene conversion. One example is shown in **Figure 2A**: two gorilla genes were both G at a particular nucleotide site, while the remaining genes were C. This suggests either parallel substitution in the gorilla lineage or gene conversion between two gorilla genes occurred. If this kind of nucleotide configuration occurs multiple times close to each other, gene conversion is suspected. The resulting 'spurious' gene tree (**Figure 2B**) is distorted from the tree of **Figure 2A** because of the strong effect of gene conversion.

When closely related genes, such as genes sampled from the same species or same population, are compared, the resulting gene trees are often called 'gene genealogies.' Although basic characteristics do not change from gene trees in which remotely related gene are compared, a somewhat different approach may be necessary. This short-term evolution has been central to population genetics theories, where allele frequency changes were considered. When the overall divergence time of a gene genealogy is small, the total number of mutations occurring in that genealogy may be quite small. In this case, detailed reconstruction of a gene genealogy is not easy, especially when only a short nucleotide segment is examined. Therefore, allele frequency change can be more powerful to delineate short-term evolution.

See also: **Homology; Orthology; Paralogy; Phylogeny; Species Trees; Trees**

Genetic Code

S Brenner

Copyright © 2001 Academic Press
doi: 10.1006/rwgn.2001.0528

One of the main outcomes of the elucidation of the structure of DNA was that the gene could be considered as a one-dimensional sequence of the four bases, adenine, guanine, cytosine, and thymine. It was known from the work of Frederick Sanger and the protein chemists who followed him that proteins were folded versions of linear polypeptide chains, i.e., one-dimensional sequences of the 20 different amino acids. How the sequence of four bases in DNA determined the sequence of 20 amino acids in proteins came to be known as the coding problem. The physicist G. Gamow proposed a special code in which he supposed that three bases were used to specify one amino acid (a triplet code) and that the triplets overlapped; this was a degenerate code in which each base was used three times in successive triplets. By choosing a particular rule to classify the triplets he showed that the 64 triplets could code for exactly 20 amino acids. The fact that the magic number 20 could be derived in what seemed to be a natural way lent encouragement to the idea that the code could be deduced theoretically. In due course, Gamow's code was shown to be wrong and, in fact, the theory of overlapping triplet codes could be eliminated simply by showing that there were more dipeptide sequences than the 256 to which overlapping codes were limited. It became clear that the code would have to be determined experimentally.

In the 1960s it became known that proteins were translated in special particles called ribosomes and that the messenger RNA was read not directly by amino acids but by special transfer RNAs to which the amino acids had become linked (see Adaptor Hypothesis). This provided a way of studying which triplets corresponded to which amino acids and in this way the code was determined experimentally. Thus GCA, GCU, GCG, and GCC all specify alanine while histidine is coded by two triplets, CAC and CAU. Three of the triplets, UAA, UAG, and UGA, are reserved as chain termination signals while AUG, which normally codes for methonine, also has a special tRNA for initiating translation of the sequence (see **Tables 1** and **2**).

Table 1 The genetic code

UUU phenylalanine	UCU serine	UAU tyrosine	UGU cysteine
UUC phenylalanine	UCC serine	UAC tyrosine	UGC cysteine
UUA leucine	UCA serine	UAA stop (ochre)	UGA stop
UUG leucine	UCG serine	UAG stop (amber)	UGG tryptophan
CUU leucine	CCU proline	CAU histidine	CGU arginine
CUC leucine	CCC proline	CAC histidine	CGC arginine
CUA leucine	CCA proline	CAA glutamine	CGA arginine
CUG leucine	CCG proline	CAG glutamine	CGG arginine
AUU isoleucine	ACU threonine	AAU asparagine	AGU serine
AUC isoleucine	ACC threonine	AAC asparagine	AGC serine
AUA isoleucine	ACA threonine	AAA lysine	AGA arginine
AUG methionine	ACG threonine	AAG lysine	AGG arginine
GUU valine	GCU alanine	GAU aspartic acid	GGU glycine
GUC valine	GCC alanine	GAC aspartic acid	GGC glycine
GUA valine	GCA alanine	GAA glutamic acid	GGA glycine
GUG valine	GCG alanine	GAG glutamic acid	GGG glycine

Table 2 Variations on the genetic code

Organism[1]	Genes	Codon[2]	Universal meaning	Actual meaning	References
Prokaryotes					
Various	Selenoproteins[3]	UGA	Stop	SeCys	Low and Berry, 1996, *Trends Biochem. Sci.* 21: 203
Mycoplasma sp.	All genes	UGA	Stop	Trp	Yamao *et al.*, 1985, *Proc. Natl Acad. Sci., USA* 82: 2306
Organellar genomes					
Mammals	All mitochondrial	UGA	Stop	Trp	Anderson *et al.* (1981) *Nature* 290: 457
		AGA, AGG	Arg	Stop	
		AUA	Ile	Met	
Drosophila	All mitochondrial	UGA	Stop	Trp	Clary *et al.* (1984) *Nucl. Acids Res.* 12: 3747.
		AGA	Arg	Ser	
		AUA	Ile	Met	
Saccharomyces cerevisiae	All mitochondrial	UGA	Stop	Trp	Sibler *et al.* (1981) *FEBS Letters* 132: 344
		CUN	Leu	Thr	
		AUA	Ile	Met	
Fungi	All mitochondrial (?)	UGA	Stop	Trp	Waring *et al.* (1981) *Cell* 27: 4
Maize[4]	All mitochondrial (?)	CGG	Arg	Trp	Fox and Leaver (1981) *Cell* 26: 315
Eukaryotic nuclear genomes					
Protozoa	All nuclear	UAA, UAG	Stop	Gln	Caron and Meyer (1985) *Nature* 314: 185; Preer *et al.* (1985) *Nature* 314: 188; Horowitz and Gorovsky (1985) *Proc. Natl Acad. Sci., USA* 82: 2452; Kuchino *et al.* (1985) *Proc. Natl Acad. Sci., USA* 82: 4758
Candida cylindracea	All nuclear	CUG	Leu	Ser	
Various mammals	Selenoproteins[3]	UGA	Stop	SeCys	Low and Berry (1996) *Trends Biochem. Sci.* 21: 203

[1]Where a single species is given it is possible that related organisms also display the same code modifications.
[2]N = any nucleotide.
[3]The following are known to be selenoproteins: formate dehydrogenase (*Escherichia coli, Enterobacter aerogenes, Clostridium thermoaceticum, C. thermoautotrophicum, Methanococcus vannielii*), NiFeSe hydrogenase (*Desulphomicrobium baculatum, M. voltae*), glycine reductase (*C. sticklandii, C. purinolyticum*), cellular glutathione peroxidase (human, cow, rat, mouse), plasma glutathione peroxidase (human), phospholipid hydroperoxide glutathione peroxidase (pig, rat), selenoprotein P (human, cow, rat), selenoprotein W (rat), type 1 deiodinase (human, rat, mouse, dog), type 2 deiodinase (*Rana catesbiana*), type 3 deiodinase (human, rat, *R. catesbiana*). See http://www.tigr.org/tdb/at/at.html
[4]In maize and other plants the CGG codon is probably converted into UGG (the correct codon for tryptophan) by RNA editing. (Reproduced with permission from *Molecular Biology Labfax 1: Recombinant DNA*. London: Academic Press.)

For some time it was thought that the code was universal, that is, identical for all living organisms from viruses to humans. However, in certain protozoa and in the mitochondrial organelles of higher organisms there are differences. For example, a codon that normally signifies chain termination can encode an amino acid, or a codon that codes for a particular amino acid in one organism can code for a different amino acid in another.

See also: **Adaptor Hypothesis; Codon Usage Bias; Codons; Universal Genetic Code; Variable Codons**

Genetic Colonization

J B Mitton

Copyright © 2001 Academic Press
doi: 10.1006/rwgn.2001.0529

Genetic colonization refers to the establishment of new breeding populations. This process is more than just the arrival of individuals at an unpopulated site; 'genetic colonization' indicates that the colonists breed and establish a self-sustaining population.

Each spring, the pelagic larvae of the blue mussel, *Mytilus edulis*, colonize the Outer Banks of North Carolina from breeding populations further north. However, the summer temperatures on the Outer Banks exceed the tolerance level of the mussels, so the populations go extinct before they have a chance to breed. In contrast, the mussel native to the Mediterranean, *Mytilus galloprovincialis*, successfully colonized sites in southern Africa and Australia during the last glacial maximum.

Genetic colonizations are, in a historic sense, quite common, and they can be organized into three general groups: changes in geographical range as climates shift, contemporary invasions facilitated by man, and the normal flux of establishment and extinction of local populations in species with metapopulations.

Genetic Colonizations Associated with Climate Change

The waxing and waning of glaciers modifies the distributions of species in both temperate terrestrial and marine environments. As the glaciers grow, they displace species from high elevations and high latitudes into glacial refugia at lower elevations and latitudes. The accumulation of glacial ice lowers sea levels, exposing land bridges that connect continents and islands. For example, during the last glaciation, Native Americans colonized North America by crossing the land bridge between Siberia and Alaska.

At the height of the most recent glaciation, 18 000 years ago, Scandinavia and parts of the British Isles were covered with ice, and tundra and permafrost covered central Europe. So it comes as no suprise that the modern ranges of European plants and animals were colonized from glacial refugia further south. Plants and animals occupied at least four glacial refugia, in the Iberian Peninsula, and areas in Italy, Greece, and Turkey. The plants and animals migrated to the north and into the mountains as the glaciers receded.

Genetic Colonizations by Invading Species

Biological invasions are genetic colonizations of an environment by a non-native species. Man's activities have produced numerous genetic colonizations, often with disastrous results. For example, domesticated cats introduce to New Zealand and Australia have caused the extinction of ground-dwelling birds and the local extinctions of some marsupials. When ocean-going freighters do not have a full load, they pump sea water into their tanks to adjust their buoyancy. The next time they take on cargo, they pump water out of the tanks, often introducing marine pelagic larvae into new environments. There are more than 60 marine species that have successfully colonized San Francisco Bay in this way.

Genetic Colonizations in Species with Metapopulations

Some species, such as several species of songbirds in the British Isles, do not live in large, continuous populations, but in a metapopulation, i.e., a series of small populations linked by occasional gene flow. The small populations sometimes die out, but migrants from nearby populations have the opportunity to recolognize the site.

Founder Effect

Genetic variability in a population is a function of the number of breeding colonizers. Reduction in the genetic variability of a population due to a small number of breeding colonists is called founder effect. Genetic variability in areas once covered by glaciers is often reported to be low, probably as a consequence of repeated founder effects in successive genetic colonizations.

See also: **Founder Effect; Gene Flow; Phylogeography**

Genetic Correlation
W G Hill

Copyright © 2001 Academic Press
doi: 10.1006/rwgn.2001.1423

The phenotypic correlation (r_P) is a measure of association between the observed performance (phenotypic value) of individuals for a pair of quantitative traits, for example, stature and body weight of man. The genetic correlation is the corresponding measure of association between the genotypes of individuals, formally their genotypic or breeding values. It is important in describing how traits are associated at the genetic level and in predicting the effect of selection on one trait on changes in other traits.

For a single trait, the phenotypic variance can be partitioned into genetic and environmental components, and the genetic variance into further components. In the same way the phenotypic covariance (but not correlation) cov_P can be expressed as a sum of covariance components, $cov_P = cov_A + cov_D + cov_I + cov_E$ of which only the phenotypic, (additive) genetic (cov_A) and environmental (cov_E) are much used, the dominance (cov_D) and epistatic covariances (cov_I) typically being subsumed into cov_E. Unless otherwise qualified, the genetic correlation is usually defined as the correlation, r_A, of breeding values (or sums of average effects), A_X and A_Y for traits X and Y, rather than as the correlation of genotypic values, because r_A can be estimated from the correlation between relatives and is useful in predicting selection response. Thus $r_A = cov(A_X, A_Y)/\sqrt{[V_{AX} V_{AY}]}$. The genetic correlation is visualized most simply for individuals with large numbers of progeny, such as dairy sires used in artificial insemination, where it becomes approximately equal to the correlation of progeny group means for the two traits.

The correlation may be caused by the pleiotropic effects of individual genes on the two traits or by linkage disequilibrium between genes each affecting only one of the traits. Although pleiotropy is likely to lead to essentially stable correlations, for example, genes influencing appetite may affect size and obesity, correlations due to disequilibrium are likely to be transient, perhaps following the crossing or introgression between populations. For example, the cross between a line with large body size and high prolificacy and a line with small size and low prolificacy will induce a positive genetic correlation between the traits, which may be sustained by disequilibrium.

As pointed out by Falconer in 1952, the genetic correlation can also be defined where the two traits specify performance in two different environments, or indeed in two sexes. As an individual is reared in only one environment, an equivalent phenotypic correlation can not be defined, however. A high genetic correlation between environments then specifies a lack of genotype × environment interaction, and shows, for example, that selection in one environment will lead to genetic change in another.

The genetic correlation can be estimated from resemblance among relatives using the same designs and similar methods as used to estimate heritability, including offspring–parent and sib correlations, and maximum-likelihood methods using all relationships in the data. The covariances (directly, or scaled as correlations or regressions) are now computed between the performance of individuals for one trait with that of their relatives for another. For example, if $c(X, Y)$ is the sample covariance between trait X on the parent and trait Y on the offspring, $2c(X, Y)$ is an estimate of the (additive) genetic covariance and $\sqrt{\{[c(X, Y)c(Y, X)]/[c(X, X)c(Y, Y)]\}}$ an estimate of the genetic correlation. Unless the data set is large, the

estimate of the genetic correlation typically has a high standard error.

Estimates of genetic and phenotypic correlations have been obtained for many traits and populations. Because of real differences among populations and species and because of sampling errors they are not all consistent, but some patterns emerge:

1. Genetic correlations between repeat records such as milk yield of cattle in different lactations or number of bristles on two abdominal segments of *Drosophila* typically show genetic correlations close to 1, although the phenotypic correlations (the repeatability of the record) may be much lower, say 0.5.
2. Correlations among general size and among conformation traits are high (over 0.4); correlations between growth rates and fatness are generally quite small; further, for such characteristics, genetic and phenotypic correlations tend to be very similar.
3. Genetic and phenotypic correlations between production traits such as milk yield and the concentration of its components, e.g., fat%, are negative (say −0.3).
4. Genetic and phenotypic correlations between traits of growth and reproduction are usually small, but signs are not consistent. Typically, however, there is a positive correlation between body size and offspring number (litter size).

The genetic change or correlated response (CR_Y) in a trait (Y) to selection on another trait (X) is proportional to the genetic covariance or correlation. If $S_X = i\sigma_x$ is the selection differential on X, then $CR_Y = (cov_A/V_{PX})S_X = ih_X h_Y r_A \sigma_{PY}$. The genetic correlation can therefore also be estimated from the correlated response to selection if one of a pair of lines is selected for X and the other for Y. Correlations often change substantially over generations in selection experiments, however, presumably as a consequence of gene frequency change at pleiotropic loci. If selection intensities are unaffected, the correlated response in Y from selection in X compared to the direct response from selecting on Y alone is given by $r_A h_X/h_Y$. This specifies the relative effectiveness of indirect selection, for example on growth rate to improve feed conversion efficiency.

The magnitude of genetic correlations reflect both the pleiotropic nature of genes present and arising in the population from mutation, and the evolutionary forces to which the species or population has been exposed. Thus it would be surprising to find very strong correlations between body size or conformation traits and reproduction traits, on the assumption that the latter were exposed to natural selection, and if there were any associations they would be nonlinear, i.e., intermediates at an optimum. Negative associations are to be expected among traits that individually contribute to fitness, because positive variants will have been removed by selection. The magnitude of such correlations among life history traits is a subject of active research.

Further Reading

Falconer DS and Mackay TFC (1996) *Introduction to Quantitative Genetics*, 4th edn. Harlow, UK: Longman.

Kearsey MJ and Pooni HS (1996) *The Genetical Analysis of Quantitative Traits*. London: Chapman & Hall.

Lynch M and Walsh B (1998) *Genetics and Analysis of Quantitative Traits*. Sunderland, MA: Sinauer Associates.

Roff DA (1997) *Evolutionary Quantitative Genetics*. New York: Chapman & Hall.

See also: Artificial Selection; Genetic Variation; Heritability; Selection Index

Genetic Counseling

R Harris

Copyright © 2001 Academic Press
doi: 10.1006/rwgn.2001.0530

Kelly (1986) has defined genetic counseling as an individual- and family-based "educational process that seeks to assist affected and/or at-risk individuals to understand the nature of the genetic disorder, its transmission and the options open to them in management and family planning." A comprehensive review can be found in *Practical Genetic Counselling* (Harper, 1998).

Specialist Genetic Clinic

In the specialist genetic clinic, counseling frequently involves risks of recurrence of genetic disorders and reproductive options rather than treatment and has five components (see **Table 1**). A family tree and a precise genetic diagnosis are important for accurate risk estimation. All available clinical records of the patient and family and appropriate special investigations, including the latest DNA methods, will be used. Risk estimation may be a simple process if the diagnosis and family history are known and the genetic disorder is consistent in its manifestations, but in

many cases there is a need to combine such data with new findings to arrive at a final risk estimation. The use that the client makes of information, including decisions about reproductive options, may be influenced by the way it is communicated. Effective communication requires training in counseling as well as knowledge of medical genetics. The counselor needs to be aware of the client's attitudes, level of emotional involvement, perception of the facts, and religious or other precepts. Counselors are consequently required to identify the individual client's 'agenda,' knowledge, and needs rather than to deploy standardized explanations or advice. Continuing support may be highly desirable during, for example, prenatal diagnosis, termination of pregnancy, or the consequences of adverse results of predictive tests.

Ethical Issues

The aim is to be nondirective and to supply clients with the facts, understanding, and confidence to make reproductive or other decisions that are best for them. Trained medical geneticists and counselors, especially when dealing with reproductive decisions, will always attempt to adhere to the gold standard of nondirectiveness. This means that counselors provide the five components of genetic counseling (see **Table 1**) but rarely advise a 'correct' or 'incorrect' course of action for the client to follow. This is not always true of other specialists, because they are accustomed to advising patients to accept various forms of treatment for physical illness. There are shades of opinion amongst health professionals (and their patients) about giving advice rather than only information, but the rule is to reject coercion aimed at the subjection of an individual patient's wishes to the public good (Ethics and Genetics).

Who Else Does Genetic Counseling?

The preceding description is based on counseling developed in specialist centers dealing with indi-

Table 1   The five components of genetic counseling as developed in specialist genetic clinics

No.	Components
1.	Taking a family history
2.	Making a diagnosis
3.	Estimating risk
4.	Empathic communication of facts to the client
5.	Follow-up and support

viduals and families who request counseling because of the birth of an affected infant or a preexisting genetic disorder. However, there are many other circumstances requiring genetic counseling but where the full process is not available because of time restraints and the absence of fully trained geneticists. For example, family studies and population screening identify individuals who may not have sought genetic counseling themselves but are at increased risk or are shown to be carriers or to have genetic susceptibility factors. Genetic counseling is also needed in many specialities as part of routine practice including treatment and prevention. Common diseases of complex etiology, diabetes mellitus, coronary heart disease, cancer, etc. may require quantitative and probabilistic risks to currently healthy individuals. Here similar ethical principles apply respecting individual autonomy, but counseling is increasingly likely to be provided by physicians and others concerned with the management of common disease and more familiar with therapy than counseling.

Audit of Genetic Counseling

Maintaining overall quality, accuracy, respect for patient autonomy, and nondirectiveness requires continuous audit, especially as medical and nursing undergraduate and postgraduate genetic education have tended to lag behind scientific advances. Appropriate clinical management of genetic disorders must include records of timely counseling so that the avoidance of genetic disease can always be seen to result from informed patient choice. Only when there are records to document that counseling was accurate and empathic can we be confident that rejection or *acceptance* of screening, prenatal diagnosis, or termination of pregnancy are autonomous decisions made by adequately informed patients (Harris *et al.*, 1999).

References

Harper PS (1998) *Practical Genetic Counselling*, 5th edn. Oxford: Butterworth-Heinemann.

Harris R, Lane B, Harris HJ et al. (1999) National confidential enquiry into counselling for genetic disorders by non-geneticists: general recommendation and specific standards for improving care. *Journal of the Royal College of Obstetricians and Gynaecologists* 106: 658–663.

Kelly TE (1986) *Clinical Genetics and Genetic Counseling*. Chicago, IL: Year Book.

See also: Ethics and Genetics; Genetic Diseases

Genetic Covariance

E Pollak

Copyright © 2001 Academic Press
doi: 10.1006/rwgn. 2001.1424

Consider a quantitative trait. The phenotype of an individual X, which is its measurement with respect to this trait, may be written as:

$$P_X = G_X + e_X + (Ge)_X$$

where G_X is the mean of all individuals with the same genotype as X, e_X is the effect of the environment, and $(Ge)_X$ is the effect of the genotype–environment interaction. Then if it is assumed that $(Ge)_X$ is equal to 0 and that genotypes are randomly distributed among environments, the covariance between the phenotypes of pairs of individuals X and Z with a particular pattern of relationship is

$$\text{cov}(P_X, P_Z) = \text{cov}(G_X, G_Z)$$

where the right side of the equation is the genetic covariance between X and Z. The genetic covariance is the average of cross-products of deviations of G_X and G_Z from the mean of the population when all pairs of individuals X and Z with the same particular pattern of relationship are considered.

Let us assume that there is an infinite random mating population and independent assortment. Let the pairs of parents of X and Z be respectively (P, Q) and (R, S) and f_{AB} be the probability that independently chosen random copies of a gene from individuals A and B are identical by descent. Then

$$\text{cov}(G_X, G_Z) = 2f_{XZ}\sigma_A^2 + u_{XZ}\sigma_D^2 \\ + \sum_{r+s \geq 2}(2f_{XZ})^r(u_{XZ})^s\sigma_{A^rD^s}^2$$

where $u_{XZ} = f_{PR}f_{QS} + f_{PS}f_{QR}$. The variance components $\sigma_A^2, \sigma_D^2,$ and $\sigma_{A^rD^s}^2$ are, respectively, the additive genetic variance, the dominance variance, and the variance associated with all interactions of single alleles at r loci and genotypes at s other loci. This general expression for the genetic covariance was independently derived by Cockerham and Kempthorne.

If there is no epistasis but loci are not in gametic phase equilibrium, a general expression for $\text{cov}(G_X G_Z)$ is obtainable, but it has a very complicated form, as shown by Weir, Cockerham, and Reynolds. Genetic covariances can also be calculated if there is inbreeding, but the resulting expressions contain covariances that are not present when there is random mating. A thorough analysis of this problem if two loci are involved was presented by Weir and Cockerham.

Other models in the literature allow for sex-linked loci, maternal effects, effects of cytoplasmic genes that are maternally inherited, polyploidy, and covariances between relatives when one relative of a pair is measured in trait Y_1 and the other in trait Y_2. There is also some theory that applies when there is assortative mating. Discussions of these topics and references to the original papers mentioned above can be found in the books listed below.

Further Reading

Cockerham CC (1954) An extension of the concept of partitioning heredity variance for analysis of covariances among relatives when epistasis is present. *Genetics* 114: 859–882.

Falconer DS and Mackay TFC (1996) *Introduction to Quantitative Genetics*, 4th edn. Harlow, UK: Longman.

Kempthorne O (1954) The correlation between relatives in a random mating population. *Proceedings of the Royal Society of London B* 143: 103–113.

Kempthorne O (1957) *An Introduction to Genetic Statistics*. New York: John Wiley.

Lynch M and Walsh B (1998) *Genetics and Analysis of Quantitative Traits*. Sunderland, MA: Sinauer Associates.

Weir BS, Cockerham CC and Reynolds J (1980) The effects of linkage and linkage disequilibrium on the covariances of noninbred relatives. *Heredity* 45: 351–359.

See also: Complex Traits; QTL (Quantitative Trait Locus)

Genetic Diseases

K M Beckingham

Copyright © 2001 Academic Press
doi: 10.1006/rwgn.2001.0531

In contrast to an infectious disease, which is acquired adventitiously during an individual's lifetime as a result of invasion by a foreign organism, a genetic disease results from a defect (mutation) within an individual's own genetic material (DNA) that causes detectable malfunction of certain tissues and organs.

The first molecular genetic change underlying an inherited genetic disease was identified in 1957, when Ingram demonstrated that sickle-cell hemoglobin differs from normal hemoglobin by a single amino acid substitution. Since the advent of recombinant DNA technology in the 1970s, the mutated genes responsible for many genetic disorders have been identified and the precise molecular lesions to these genes that are produced by individual mutations have been established.

For a genetic disease caused by a recessive mutation, both parents must be heterozygous for the mutation and there will be a one in four chance of producing a homozygous affected child. However, given that heterozygous individuals (carriers) are asymptomatic, in the case of relatively rare genetic diseases, carriers can be completely unaware of their status and thus unprepared for the birth of a genetically compromised child. Some recessive genetic disease, such as Tay–Sachs disease, show dramatically increased prevalence in certain ethnic groups. Considerable effort has been devoted to identifying and counseling potential carriers within these groups. For X-linked recessive diseases such as hemophilia, where half the sons of a heterozygous mother are affected, and for dominant mutation diseases such as achondroplasia (see Achondroplasia), individuals carrying a single mutant allele will, in general, be aware of their problem and thus able to make informed choices about parenthood. Unfortunately some diseases caused by dominant mutations, such as Huntington disease, do not usually manifest until after the reproductive years.

Two complementary approaches to the eradication of genetic disorders are ongoing, both of which rely on preliminary identification of the affected gene. One approach is preventative and involves genetic testing of asymptomatic potential carriers or potentially affected individuals so that those carrying a mutant allele may know their status and, if necessary, avoid passing on the mutation in question. In combination with genetic testing of embryos produced by *in vitro* fertilization and selection of embryos carrying non-mutant genes as potential progeny, this approach can permit carriers to produce their own children while simultaneously eradicating the disease mutation from their family lineage. Some in the medical community advocate neonatal genetic testing of all individuals for all possible genetic diseases.

The second approach is a therapeutic one for affected individuals and involves supplying a correctly functioning version of the mutated gene to the malfunctioning tissue(s). The first experiments using this approach, termed gene therapy, were initiated in 1990. Introducing a gene into the cells of certain body (somatic) tissues within an individual still leaves the germline sperm or egg progenitor cells mutant and thus does not eliminate the potential for disease in any offspring. However, current methods for permanent gene integration into germline cells carry a potential for causing further genetic damage and thus are ethically unacceptable. The option of genetically screening embryos and selecting unaffected embryos for implantation (as discussed above) is also possible for genetic disease patients. Currently, harmless, modified versions of viruses are proving to be the best vectors for introducing genes into somatic tissues of individuals with genetic diseases. However, to date gene therapy has had few clinical trials, and very limited success. At least one individual has died as a direct result of this type of treatment.

All forms of cancer involve mutations to genes within an individual. In contrast to genetic disease mutations these defects are not initially present in the genome but arise from damage to DNA in a particular tissue during the individual's lifetime.

See also: **Cancer Susceptibility; Clinical Genetics; Gene Therapy, Human; Genetic Counseling**

Genetic Distance

M Nei

Copyright © 2001 Academic Press
doi: 10.006/rwgn.2001.0532

Genetic distance is the degree of genetic difference (genomic difference) between species or populations that is measured by some numerical method. Thus, the average number of codon or nucleotide differences per gene is a measure of genetic distance. There are various molecular data that can be used for measuring genetic distance. When the two species to be compared are distantly related, data on amino acid or nucleotide sequences are used Nei and Kumar (2000). In the comparison of closely related species or populations, however, the effect of polymorphism cannot be neglected, and one has to examine many proteins or genes. For this reason, it is customary to measure the genetic distance between populations in terms of a function of allele frequencies for many genetic loci.

Genetic distances are useful for constructing phylogenetic trees of populations as well as for estimating times of divergence between populations. In the past, many investigators have used allele frequency data obtained by protein electrophoresis and immunological methods. In recent years, many different types of molecular data such as microsatellite DNA and RAPD data are used, but the basic principle of computing genetic distances and constructing phylogenetic trees remains essentially the same. Here only the basic methods for computing genetic distances are discussed. The reader who is interested in more detailed information should refer to Nei and Kumar (1983). Some results from recent studies of the evolution of human populations will also be presented.

Commonly Used Distance Measures

Rogers' Distance

Suppose that there are q alleles at a locus, and let x_i and y_i be the frequencies of the ith allele in populations X and Y, respectively. Each allele frequency may take a value between 0 and 1. Therefore, it is possible to represent populations X and Y in a q-dimensional space. The distance between the two populations in the space is then given by

$$d_R = \left[\sum_{i=1}^{q} (x_i - y_i)^2 \right]^{1/2} \qquad (1)$$

This distance takes a value between 0 and $\sqrt{2}$, the latter value being obtained when the two populations are fixed for different alleles. This property is not very desirable. So, Rogers (1972) proposed the following measure, which takes a value between 0 and 1:

$$D_R = \left[\frac{1}{2} \sum_{i=1}^{q} (x_i - y_i)^2 \right]^{1/2} \qquad (2)$$

When allele frequency data are available for many loci, the average of this value is used. Note, however, that this measure has one deficiency. When the two populations are both polymorphic but share no common alleles, D_R is given by $[(\sum x_i^2 + \sum y_i^2)/2]^{1/2}$. This value can be much smaller than 1 even if the populations have entirely different sets of alleles. For example, when there are five nonshared alleles in each population and all allele frequencies are equal ($x_i = 1/5$; $y_i = 1/5$), we have $D_R = 0.45$. This property is clearly undesirable.

Bhattacharyya's Distance and its Modifications

Representing two populations on the surface of a multidimensional hypersphere, Bhattacharyya (1946) suggested that the extent of differentiation of populations be measured in terms of the angle (θ) between the two lines projecting from the origin to the two populations (X and Y) on the hypersphere (**Figure 1**). When there are q alleles, we consider a q-dimensional hypersphere with radius 1 and let each axis represent the square root of the allele frequency, i.e., $\xi_i = \sqrt{(x_i)}$ and $\eta_i = \sqrt{(y_i)}$. Therefore, $\sum \xi_i^2 = \sum \eta_i^2 = 1$. When there are only two alleles, populations X and Y can be represented on a circle, as shown in **Figure 1**. Elementary geometry shows that in the case of q alleles the angle θ is given by

$$\cos \theta = \sum_{i=1}^{q} \xi_i \eta_i = \sum_{i=1}^{q} \sqrt{x_i y_i} \qquad (3)$$

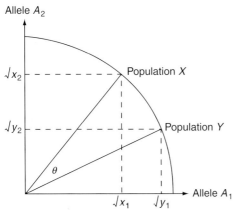

Figure I Bhattacharyya's geometric representation of populations X and Y for the case of two alleles.

Bhattacharyya proposed that the distance between two populations be measured by

$$\theta^2 = \left[\arccos\left(\sum_i \sqrt{x_i y_i}\right)\right]^2$$
$$\approx \frac{1}{2}\sum_{i=1}^{q}\frac{(x_i - y_i)^2}{(x_i + y_i)} \qquad (4)$$

This measure takes a value between 0 and 1. When there are allele frequency data for many loci, the average of this quantity is used as a genetic distance measure as in the case of D_R.

In a computer simulation, Nei *et al.* (1983) noted that the following distance measure is quite efficient in recovering the true topology of an evolutionary tree when it is reconstructed from allele frequency data.

$$D_A = \sum_{k=1}^{L}\left(1 - \sum_{i=1}^{q_k}\sqrt{x_{ik} y_{ik}}\right)\Big/ L \qquad (5)$$

where q_k and L are the number of alleles at the kth locus and the number of loci examined, respectively, and the subscript ik refers to the ith allele at the kth locus. This measure takes a value between 0 and 1, the latter value being obtained when the two populations share no common alleles. Since the maximum value of D_A is 1, D_A is nonlinearly related to the number of gene substitutions. When D_A is small, however, it increases approximately linearly with evolutionary time.

The standard error of D_A or the difference in D_A between two pairs of populations can be computed by the bootstrap method if it is based on many loci. In this case, a bootstrap sample will represent a different set of loci, which have been chosen at random with replacement (Nei and Kumar, 2000). Similarly, the standard errors of average D_R, θ^2, and d_C can be computed by the bootstrap.

F_{ST}^{*} Distance

The allele frequencies of different populations may differentiate by genetic drift alone without any selection. When a population splits into many populations of effective size N in a generation, the extent of differentiation of allele frequencies in subsequent generations can be measured by Wright's F_{ST} Nei and Kumar, 2000. When there are only two populations but allele frequency data are available from many different loci, it is possible to develop a statistic whose expectation is equal to F_{ST}. One such statistic is given by

$$F_{ST}^{*} = [(\hat{J}_X + \hat{J}_Y)/2 - \hat{J}_{XY}]/(1 - \hat{J}_{XY}) \qquad (6)$$

where $\hat{J}_X$, $\hat{J}_Y$, and $\hat{J}_{XY}$ are unbiased estimators of the means (J_X, J_Y, and J_{XY}) of $\sum x_i^2$, $\sum y_i^2$, and $\sum x_i y_i$ over all loci, respectively. For a single locus, unbiased estimates of $\sum x_i^2$, $\sum y_i^2$, and $\sum x_i y_i$ are given by

$$\hat{j}_X = \left(2m_X \sum \hat{x}_i^2 - 1\right)\Big/(2m_X - 1) \qquad (7)$$

$$\hat{j}_Y = \left(2m_Y \sum \hat{y}_i^2 - 1\right)\Big/(2m_Y - 1) \qquad (8)$$

$$\hat{j}_{XY} = \sum \hat{x}_i \hat{y}_i \qquad (9)$$

where m_X and m_Y are the numbers of diploid individuals sampled from populations X and Y, respectively, and $\hat{x}_i$ and $\hat{y}_i$ are the sample frequencies of allele A_i in populations X and Y. Therefore, $\hat{J}_X, \hat{J}_Y$, and $\hat{J}_{XY}$ are the means of $\hat{j}_X$, $\hat{j}_Y$, and $\hat{j}_{XY}$ over all loci, respectively. The expectation of F_{ST}^{*} is given by

$$E(F_{ST}^{*}) = 1 - e^{-t/(2N)} \qquad (10)$$

where t is the number of generations after population splitting. Therefore, we have

$$D_L = -\ln(1 - F_{ST}^{*}) \qquad (11)$$

which is expected to be proportional to t when the number of loci used is large [$E(D_L) = t/(2N)$]. This indicates that when evolutionary time is short and new mutations are negligible, one can estimate t by $2ND_L$ if N is known. In practice, however, new mutations always occur, and this will disturb the linear relationship between D_L and t when a relatively long evolutionary time is considered. N is also usually unknown.

Standard Genetic Distance

Nei (1972) developed a genetic distance measure called the standard genetic distance, whose expected value is proportional to evolutionary time when both effects of mutation and genetic drift are taken into account. It is estimated by

$$D = -\ln I \tag{12}$$

where

$$I = \hat{J}_{XY} / \sqrt{\hat{J}_X \hat{J}_Y} \tag{13}$$

The variances of I and D can be computed by the bootstrap method.

When the populations are in mutation–drift balance throughout the evolutionary process and all mutations result in new alleles following the infinite-allele model (Nei and Kumar, 2000), the expectation of D increases in proportion to the time after divergence between two populations. That is,

$$E(D) = 2\alpha T \tag{14}$$

where α is the rate of mutation or gene substitution per year and T is the number of years after divergence of the two populations. Therefore, if we know α, we can estimate divergence time from D.

The α value varies with genetic locus and the type of data used. For the genetic loci that are commonly used in protein electrophoresis, it has been suggested that α is approximately 10^{-7} per locus per year. If this is the case, the time after divergence between two populations is estimated by

$$T = 5 \times 10^6 D \tag{15}$$

This formula is based on the assumption that all loci have the same rate of gene substitution. In practice, the α value varies from locus to locus approximately following the gamma distribution. In this case, the average value of I over loci is given by

$$I_A = \left[\frac{a}{a + 2\bar{\alpha}t} \right]^a \tag{16}$$

where a is the shape parameter of the gamma distribution and $\bar{\alpha}$ is the mean of α over loci. Therefore, the number of gene substitutions per locus is given by

$$D_\nu = 2\alpha t = a[(1 - I_A)^{-1/a} - 1] \tag{17}$$

When $a = 1$, this becomes

$$D_\nu = (1 - I_A)/I_A \tag{18}$$

Here I_A is estimated by equation (14). In the case of $a > 0$, T can be estimated by replacing D in equation

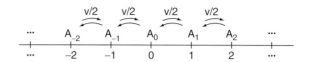

Figure 2 Stepwise mutation model.

(16) by D_ν. Note that D_ν is nearly equal to D when $I_A \geq 8$ and $a = 1$.

$(\delta\mu)^2$ Distance

Microsatellite DNA loci are segments of repeated DNA with a short repeat length, usually two to six nucleotides. Thus, an allele for a CA repeat locus may be represented by CACACACACACACA, where the dinucleotide CA is repeated seven times. Microsatellite loci are believed to be subject to a mutational change following the slippage model of duplication or deletion of repeat units. Therefore, new alleles are supposed to be generated by following the stepwise mutation model given in **Figure 2**. Microsatellite loci are usually highly polymorphic with respect to the number of repeats, and therefore they are useful for studying phylogenetic relationships of populations. Goldstein et al. (1995) proposed that the following distance measure be used for microsatellite DNA data.

$$(\delta\mu)^2 = \sum_k^L (\mu_{Xk} - \mu_{Yk})^2 / L \tag{19}$$

where $\mu_{Xk} (= \sum i x_{ik})$ and $\mu_{Yk} (= \sum i y_{jk})$ are the mean numbers of repeats at the kth locus in populations X and Y, respectively. The expectation of $(\delta\mu)^2$ is given by $E(\delta\mu)^2 = 2\alpha T$, where α is the mutation rate per year. Therefore, T can be estimated by $(\delta\mu)^2/(2\alpha)$.

In practice, however, there are a number of problems with this method. First, the α value apparently varies considerably with locus and organism, and it is not a simple matter to estimate α for each locus. Second, the variance or the coefficient of variation of $(\delta\mu)^2$ is very large compared to that of other distance measures such as d_C and D_A. Therefore, a large number of loci must be used to obtain a reliable estimate of T even if α is known. Third, there is evidence that the actual mutational pattern is irregular and deviates considerably from the stepwise mutation model on which this distance measure is based.

Genetic Distance and Phylogenetic Trees

A linear relationship of a distance measure with evolutionary time is important for estimating the time of divergence between two populations. It is also a nice property for constructing phylogenetic trees, other things being equal. In practice, however, different distance measures have different variances, and for this

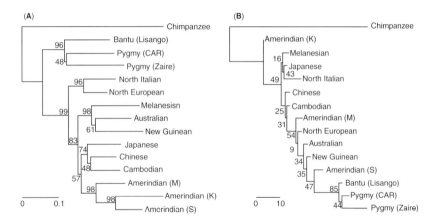

Figure 3 Neighbor-joining trees of human populations obtained by using Bowcock *et al.*'s (1994) data of 25 microsatellite loci. (A) Tree obtained by D_A distance. (B) Tree obtained by $(\delta\mu)^2$ distance. The number for each interior branch is the bootstrap value from 1000 replications. M, Maya; K, Karitiana; S, Surui; CAR, Central African Republic.

reason a distance measure that is linear with time is not necessarily better than a nonlinear distance in obtaining true trees (topologies).

A number of authors have studied this problem by using computer simulation. The general conclusions obtained from these studies are as follows:

1. For all distance measures, the probability of obtaining the true topology (P_T) is very low when the number of loci used is less than ten but gradually increases with increasing number of loci. In general, P_T is lower for the stepwise mutation model than for the infinite-allele model. This indicates that a larger number of loci should be used for microsatellite DNA data than for electrophoretic data when the level of average heterozygosity is the same.
2. Distance measures D_A and d_C are generally more efficient in obtaining the true topology than other distance measures under many different conditions.
3. When the total number of individuals to be studied is fixed, it is generally better to examine more loci with a smaller number of individuals per locus rather than fewer loci with a large number of individuals in order to have a high P_T value, as long as the number of individuals per locus is greater than about 25. When average heterozygosity is as high as 0.8, however, a larger number of individuals per locus need to be studied.

Evolutionary Relationships of Human Populations

Bowcock *et al.* (1994) examined microsatellite DNA (mostly CA repeats) polymorphisms for 25 loci from 14 human populations and one chimpanzee species. **Figure 3A, B** show the phylogenetic trees obtained by using the D_A and $(\delta\mu)^2$ distances, respectively, from allele frequency data for the 25 loci.

The tree obtained by D_A distances (**Figure 3A**) shows that Africans (Pygmies and Bantu) first separated from the rest of the human groups and that the bootstrap values for the interior branches connecting Africans and chimpanzees and non-Africans and chimpanzees are both very high. (A bootstrap value for an interior branch is an indicator of the accuracy of distinction of the two population groups separated by the interior branch.) This result supports the currently popular view that modern humans originated in Africa. The same tree shows that Europeans first diverged from the other non-African people and then the group of New Guineans and native Australians separated from the remaining group. The first separation of Europeans from the rest of non-Africans is well supported by a high bootstrap value, but the next separation of New Guineans and Australians is less clear, because the bootstrap value for one of the two interior branches involved is only 53%. In fact, a similar study using classical markers (blood group and allozyme data) has suggested that New Guineans and Australians are genetically close to southeastern Asians (Indonesians, Filipinos, Thais). To clarify this aspect of evolutionary relationships, it seems necessary to examine many more loci.

Figure 3B shows the tree obtained by $(\delta\mu)^2$ distances. The topology of this tree is very different from that for D_A distances and is poorly supported by the bootstrap test. This unreliable tree was obtained mainly because the sampling error of $(\delta\mu)^2$ is very large, as mentioned earlier.

Other Genetic Markers

In recent years, a number of other genetic markers have been used for studying the phylogenetic relationships of populations. They are restriction fragment length polymorphism (RFLP), amplified fragment length polymorphism (AFLP), and random amplification of polymorphic DNA (RAPD) data. The allele frequency data obtained by these markers can be analyzed by the same methods as mentioned earlier. They can also be used to estimate the average number of nucleotide differences per site between two populations. In the latter analysis somewhat sophisticated statistical methods are required, and they are presented in Nei and Kumar (2000).

During the last two decades, RFLP data for mitochondrial and chloroplast DNA have been used extensively to study the extent of genetic differentiation of closely related species or populations. RFLP data can be obtained inexpensively and give sufficiently accurate results for studying closely related populations. In recent years, however, many authors sequence polymorphic alleles to obtain more accurate results. Statistical methods for analyzing these polymorphic DNA sequences are described in Nei and Kumar (2000).

References

Bhattacharyya A (1946) On a measure of divergence between two multinomial populations. *Sankhya* 7: 401–406.

Bowcock AM, Ruiz-Linares A, Tomfohrde J et al. (1994) High resolution of human evolutionary trees with polymorphic microsatellites. *Nature* 368: 455–457.

Goldstein DB, Ruiz-Linares A, Cavalli-Sforza LL and Feldman MW (1995) Genetics absolute dating based on microsatellites and origin of modern humans. *Proceedings of the National Academy of Sciences, USA* 92: 6723–6727.

Nei M (1972) Genetic distance between populations. *American Naturalist* 106: 283–292.

Nei M and Kumar S (2000) *Molecular Evolution and Phylogenetics.* Oxford: Oxford University Press.

Nei M, Tajima F and Tateno Y (1983) Accuracy of estimated phylogenetic trees from molecular data. II. Gene frequency data. *Journal of Molecular Evolution* 19: 153–170.

Rogers JS (1972) Measures of genetic similarity and genetic distance. In *Studies in Genetics*, vol. 7, University of Texas Publication 7213, pp. 145–153. Austin, TX: University of Texas.

See also: Evolutionary Rate; Gene Substitution; Microsatellite; Phylogeny; Trees

Genetic Drift

J Arnold

Copyright © 2001 Academic Press
doi: 10.1006/rwgn.2001.0533

Genetic drift is the random variation in gene frequencies due to sampling. Since populations are finite in size, those individuals contributing genes to the next generation constitute a sample from the population.

As an example consider a single genetic locus with two alleles, A and a, with frequencies p and q in the gene pool ($p + q = 1$). If there were 50 A alleles ($p = 50/100 = 0.5$) and 50 a alleles ($q = 50/100 = 0.5$) among 50 ($N_e = 50$) diploid parents that interbreed, the sampling process might yield 48 A alleles ($p = 48/100 = 0.48$ and 52 a alleles ($q = 52/100 = 0.52$) among offspring of the next generation. Buri (1956) gives an example of this sampling process in 107 populations of *Drosophila melanogaster* segregating for two alleles both initially with frequencies of $p = q = 0.5$ and breeding size of $N_e = 16$. As can be seen in **Figure 1A** gradually the empirical density (histogram) of allele frequency (p) of each replicate population drifts or spreads out from 0.5.

Wright (1931) and Fisher (1930) calculated the predicted effects of genetic drift on natural and experimental populations. If $f(p;t)$ is the theoretical density (histogram) of an allele frequency under drift at time t, Wright calculated this density of allele frequencies over time in a population undergoing genetic drift as shown in **Figure 1B**.

The theoretical density of an allele frequency $f(p;t)$ flattens out with time (t) at $1/2N_e$ (with all allele frequencies being equally likely) after about $2N_e$ generations. Eventually a population is expected to drift to fixation or loss of the A allele under genetic drift. Thus, if genetic drift goes on long enough, a consequence is reduction in genetic variation within a population.

Under random drift the variance in the current allele frequency per generation is approximately $pq/2N$, where N_e is the breeding size of the population and p is the allele frequency in the current generation. When the breeding size of the population is small, then there is more variation in an allele frequency from generation to generation. As shown in **Figure 1** in both panels, the eventual outcome of genetic drift is that the allele frequency p does a random walk to $p = 0$ or $p = 1$ so that that the population becomes fixed for either the A allele or a allele. Another consequence of drift is that as the gene pool becomes fixed for A or a, the heterozygosity (H_t) in the population (the frequency of heterozygotes) in generation t is expected to decline each generation from the initial heterozygosity (H_0) according to the rule:

$$H_t = (1 - 1/2N_e)^t H_0$$

Genetic drift is one of four factors (mutation, migration, genetic drift, and natural selection) causing gene pools to change over time, and genetic drift is at the heart of several recent theories of evolution. In the shifting-balance theory of evolution (Wright, 1931) genetic drift is part of a two-phase process of adaptation

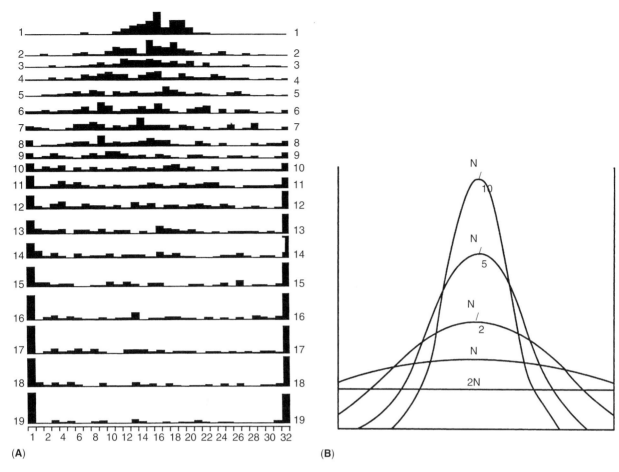

Figure 1 (A) Empirical histogram of allele frequency in 107 experimental populations of *Drosophila melanogaster* all started with an allele frequency of $p = 0.5$. (B) Theoretical histogram of allele frequency $f(p;t)$ after multiple of N generations and all started with an allele frequency of $p = 0.50$. (Redrawn from Buri, 1956 and Wright, 1969.)

of a subdivided population. In the first phase genetic drift causes each subdivision to undergo a random walk in allele frequencies to explore new combinations of genes. In the second phase a new favorable combination of alleles is fixed in the subpopulation by natural selection and is exported to other demes by factors like migration between demes. Much of the basic theory of genetic drift was developed in the context of understanding the shifting balance theory of evolution.

Genetic drift has also played a fundamental role in the neutral theory of molecular evolution. In this theory most of the genetic variation in DNA and protein sequences is explained by a balance between mutation and genetic drift. Mutation slowly creates new allelic variation in DNA and proteins, and genetic drift slowly eliminates this variability, thereby achieving a steady state.

Consider for example a new mutation arising in a gamete's DNA with probability u each generation. If there are $2N_e$ alleles in the gene pool, then the number of new mutations per gamete per generation is $u \times 2N_e$. As the distribution flattens out in **Figure 1**, the chance that a new neutral allele becomes fixed is $1/2N_e$, in that this copy is equally likely to be fixed. The rate of new substitutions becoming fixed per generation is then = (number of new mutations) (probability of fixation) = $(u \times 2N_e) (1/2N_e) = u$. A fundamental prediction of genetic drift theory is then the substitution rate λ in genes or replacement rate in proteins is constant and equal to the mutation rate. This prediction amounts to the prediction that there is a molecular clock in DNA and protein sequences, a prediction for which there is now considerable supporting data (Kimura, 1983).

References

Buri P (1956) Gene frequency in small populations of mutant *Drosophila*. *Evolution* 10: 367–402.

Fisher RA (1930) The distribution of gene ratios for rare mutations. *Proceedings of the Royal Society of Edinburgh* 50: 205–220.

Kimura M (1983) *The Neutral Theory of Molecular Evolution*. New York: Cambridge University Press.

Wright S (1931) Evolution in Mendelian populations. *Genetics* 16: 97–159.

Wright S (1969) *Evolution and the Genetics of Populations*, vol. 2, *The Theory of Gene Frequencies*. Chicago, IL: University of Chicago Press.

See also: Gene Substitution; Neutral Theory; Shifting Balance Theory of Evolution

Genetic Engineering

I Schildkraut

doi: 10.1006/rwgn.2001.0534

Genetic engineering is the manipulation of genetic material by either molecular biological techniques or by selective breeding. While selective breeding has been practiced for thousands of years (domestication of the dog; farming corn; brewer's yeast) the manipulation of genetic material *in vitro* was developed in the 1970s. The DNA is manipulated within a test tube and subsequently introduced back into a cell in order to change the processes of a cell or organism. In its simplest conception a molecular biologist can combine molecules of DNA from different organisms encoding different properties. Most typically manipulating DNA *in vitro* requires first, isolating DNA from cells, cleaving the DNA with sequence specific restriction endonucleases, mixing two independently isolated DNAs and joining the DNA molecules with DNA ligase. Lastly, reintroducing the DNA into cells and identifying the cells which carry the newly joined DNA molecules. For example, an antibiotic resistance gene is isolated from one bacteria and combined *in vitro* with a plasmid (vector) that is capable of replicating in another bacteria. This engineered plasmid is introduced into the bacterium where it confers the antibiotic resistance to the newly transformed bacteria.

The term 'genetic engineering' is also used to refer to the process of altering the expression level of a protein, for example a protein may be overexpressed for purposes of purifying large amounts of the protein by changing its promoter. 'Genetic engineering' can also be used as a term synonymous with 'protein engineering' where the biochemical characteristics of a protein are altered by mutating the gene which encodes the protein.

See also: Biotechnology; Breeding of Animals; Recombinant DNA; Recombinant DNA Guidelines

Genetic Equilibrium

M Tracey

doi: 10.1006/rwgn.2001.0535

Equilibrium is a state in which opposing forces balance to create a steady state. This steady state may be stable, in which case perturbation away from the steady state is followed by a return to that state. Or the equilibrium state may be unstable in which case changes in the equilibrium lead either to the establishment of a new equilibrium value or loss of equilibrium. The most well-known genetic equilibrium is the Hardy–Weinberg equilibrium.

Hardy–Weinberg Equilibrium

In a sexually reproducing, diploid population of infinite size in which there is no mutation or migration, no natural selection, and where mating takes place at random, the frequencies of alleles and genotypes will remain unchanged in Hardy–Weinberg equilibrium. If this hypothetical population is not at equilibrium at the outset, it will take only a single generation to establish equilibrium under the conditions defined above. Imagine a population like the one defined above in which we focus on a gene with two alleles, h1 and h2. Let us assume that the frequency of h1 is defined as 0.1 and the frequency of h2 equals 0.9, since h2 is always $1 - f(h1)$ in a two-allele system, because the sum of allele frequencies must always equal one. Since h1 cannot change into h2 by mutation, nor can h2 change into h1 and there is no natural selection, we may represent random mating by multiplying the frequencies of male gametes by the frequencies of female gametes:

$$(f(h1) + (f(h2)) \times (f(h1) + (f(h2))$$
$$= h1^2 + 2h1h2 + h2^2$$

For this h1 = 0.1 and h2 = 0.9 example this is

$$(0.1 + 0.9) \times (0.1 + 0.9)$$
$$= 0.01h1h1 + 0.18h1h2 + 0.81h2h2$$

and the allele frequencies which will produce the next generation are ascertained by collecting alleles from the diploid individuals on the right side of the equation:

$$f(h1) = 0.01 + (1/2)0.18 = 0.1 \quad \text{and}$$
$$f(h2) = (1/2)0.18 + 0.81 = 0.9$$

Neither the allele frequencies nor the genotype frequencies will change under the conditions defined; the single locus, two-allele genetic system is at equilibrium. This is an inherently unstable equilibrium, because there are no active forces balancing the equilibrium state.

If, for example, we relax the requirement that the population size is infinite and instead reduce the population size to 505, allele and genotype frequencies will change simply because the population is of finite size (see Genetic Drift). If we go from the infinite population to the population of 505 at the diploid stage we would expect to see:

$$0.01(505)h1h1 + 0.18(505)h1h2 + 0.81(505)h2h2$$
$$= 5.05h1h1 + 90.9h1h2 + 409.05h2h2$$

The total number of individuals must equal 505, but organisms come only as whole units, so one or more of the genotypic classes will gain and others will lose by chance. There may, for example, be 6 h1h1, 90 h1h2, and 409 h2h2 for a total of 505 progeny. When these reproduce the allele frequencies which will produce the next generation will be:

$$f(h1) = ((12 + 90)/1010) = 0.10099 \quad \text{and}$$
$$f(h2) = ((90 + 818)/1010) = 0.89901$$

Using these allele frequencies it is obvious that the genotype frequencies will be different and if the infinite population assumption is restored alleles will remain at these new frequencies until another force acts to disturb the equilibrium. This Hardy–Weinberg equilibrium is, thus, an unstable equilibrium. However, stable equilibria are seen in genetic systems. For example some dominant alleles are lethal or produce sterility in the heterozygous state. In this case, natural selection acts to remove these dominant alleles from the population in a single generation, because the heterozygotes die or fail to reproduce. The equilibrium state is represented by the balance of elimination by death or sterility and production by mutation.

Mutation–Selection, a Balanced Equilibrium

Dominant alleles which confer prereproductive death or sterility on their carrier are eliminated from the population in a single generation. All the alleles seen in a population must be new mutations and the equilibrium is simply a balance between elimination by death or sterility and new mutations.

Overdominance

Perhaps the most well-known case of stable equilibrium in human genetics is sickle-cell anemia, where the equilbrium is determined by the balanced loss of the normal genotype that is susceptible to malaria and the sickle-cell homozygote that is lost to anemia. The heterozygote is resistant to malaria and does not suffer anemia; thus both normal and sickle-cell alleles are maintained in the population at an equilibrium value determined by severity of the anemia and prevalence of malaria in specific populations (Sickle Cell Anemia and Overdominance).

See also: **Balanced Polymorphism; Genetic Drift; Hardy–Weinberg Law; Overdominance; Sickle Cell Anemia**

Genetic Homeostasis

J Phelan

Copyright © 2001 Academic Press
doi: 10.1006/rwgn.2001.0536

In environments that fluctuate and change unpredictably, the adaptability of a population is critically dependent upon the maintenance of reserves of genetic variability. Genetic homeostasis describes this property of populations that emerges from stabilizing selection which operates on individuals. In such environments, the adaptedness of individuals is increased by the enhanced buffering against developmental instability that comes from heterozygosity. Such buffering enables individuals to produce the proper adaptive phenotype despite the inevitable environmental fluctuations that occur during development.

What is Homeostasis?

From an organism's perspective, the ability to maintain normal physiological functioning despite inconstant environmental variables – both internally and externally – is a central aspect of their fitness. Consequently, homeostasis is one of the fundamental adaptations. The ability of an animal embryo to develop normally even in the face of large insulin fluctuations, for example, or of a mammal to maintain a constant body temperature despite changing weather, confer a survival advantage.

Populations as well as Individuals Can Be Homeostatic

Just as individuals exhibit physiological or developmental homeostasis, populations, too, may exhibit homeostatic devices. That is, the genetic composition of populations may have properties that confer on that population a greater or lesser likelihood of persisting in the face of environmental change. Among populations of sexually reproducing organisms, balanced polymorphism is a likely suspect in achieving population homeostasis. In nature, heterozygous genotypes are often more fit than homozygotes. For example, in a population where all individuals share a single genotype, e.g., A_1A_1, individuals may be fertile only within the temperature range of 20–30 °C. In another population, all individuals may be A_2A_2 and fertile only between 15 and 25 °C. A population containing individuals of both genotypes would, of course, have fertile individuals throughout the larger range of 15 to 30 °C and, consequently, in the face of varying environments would be more likely to persist than either monomorphic population.

The converse of this situation is true as well. To the extent that balanced polymorphism is best for population homeostasis under conditions of environmental fluctuation, long-term environmental stability results in populations that exhibit decreased polymorphism. In a clever experiment utilizing a naturally occurring inversion system in *Drosophila*, Richard Lewontin demonstrated this, documenting the destruction of a polymorphic system by natural selection when it was maintained under extremely constant conditions for more than 40 generations.

Additional evidence of genetic homeostasis comes from artificial selection experiments. From these we can observe that natural selection tends to resist changes in allele frequencies. In an equilibrium population, selection on one particular feature (say abdominal bristle number in fruit flies or toe morphology in chickens) predictably decreases one or more of the major components of fitness as a correlated response. Moreover, when the artificial selection is suspended before too much of the genetic variance is lost due to fixation, the frequencies of those genes for fitness components return to equilibrium and the mean value of the selected character reverts towards its original value. Given this strong property of populations to resist change, genetic homeostasis is sometimes referred to as genetic inertia.

In similarly revealing experiments on heterozygosity and homeostasis, 178 strains of *Drosophila pseudoobscura* were created from a balanced polymorphism population (Lewontin, 1956). Within each of these strains, all of the flies were completely homozygous for the second chromosome. Larval viability was then measured for each strain and compared with that found among the heterozygotes. Somewhat surprisingly, more than a dozen of the homozygous strains showed greater larval viability than the average

of the heterozygotes. This prompted the question of why the original population would be polymorphic at all when there were clearly some homozygotes with higher fitness.

The key is that when the environment (either temperature or food composition) was altered just slightly, the homozygous strains that had exhibited high larval viability could no longer match the viability of the heterozygotes, which barely changed at all. Occasionally, under a narrow and specific set of environmental conditions, homozygotes may be more fit than heterozygotes. But like a 100 m sprinter versus a decathlete, these superspecialists just cannot compete when the playing field varies over time. Hence, the polymorphic populations persist.

Organisms May Achieve Homeostasis via Heterozygosity

An important feature of populations in which balanced polymorphisms are maintained is that they inevitably produce more heterozygous individuals than less genetically variable populations. It turns out that just as the populations that these heterozygous individuals come from exhibit greater genetic homeostasis, the heterozygous individuals themselves have greater developmental homeostasis in the face of varying environments.

Imagine a gene for a generalized enzyme. Suppose that alternative alleles for this enzyme code for slightly different forms of the protein, each with a slightly different range of conditions (temperature, pH, or salt concentration) of optimal activity, perhaps operating through slightly different synthetic pathways. Heterozygous individuals, with two different forms of the enzyme, would better be able to accommodate the vagaries of the environment. Now multiply this homeostatic effect across hundreds or even thousands of genes. The greater the number of heterozygous loci, the greater the biochemical diversity and the stronger the potential buffering, homeostatic effect.

In the absence of knowledge about a trait's adaptive significance, of course, simple measures of variability do not necessarily represent an index of homeostasis. For some traits, evolution may actually favor high variability rather than uniformity within individuals. On the other hand, there are traits such as histone structure, for which selection may favor minimal variability across changing environments. Because there is, unfortunately, no consistent relationship between homeostasis and variability, it is not always possible to estimate homeostasis by observing phenotypic variability for just a single trait.

Inbreeding Leads to Genetic Uniformity but not Phenotypic Uniformity

The fitness benefits of maintaining some heterozygosity in populations is most clearly demonstrated via inbreeding. The process of inbreeding has two effects: First, it creates populations with no genetic variance, where all individuals are genetically identical; and second, by reducing the number of unique alleles occurring at each locus to one, it produces individuals which all are completely homozygous. The first effect is usually the desired goal of inbreeding. The second effect is an unavoidable by product of inbreeding. In an inbred strain, after 20 or more generations of brother–sister mating there is no genetic variability and individuals can be presumed homozygous at every locus. In a population of F_1 hybrid animals, on the other hand, there also is no genetic variability, but all of the individuals are heterozygous at every locus for which the parent strains had different alleles. Populations of random-bred and wild-caught animals, by comparison, usually have some level of genetic variability and individuals have some intermediate level of heterozygosity.

Phenotypic variability can be compared between populations by measuring coefficients of variation (CVs) for a variety of physical, biochemical, and behavioral characters. The CV is simply the ratio of the standard deviation to the mean for the group. It serves to make the variance measure independent of the mean.

For many characters, in many organisms, inbreeding increases the developmental instability so much that it overrides any decrease in phenotypic variance that may have been achieved by the decreased genetic variance. Somewhat surprisingly to many researchers, in these cases, using inbred strains of animals makes it *more* difficult to detect significant differences between treatment groups than if F_1 hybrids or random-bred animals were used. In one study utilizing data from fourteen species, including invertebrate and vertebrate animals as well as plants, the CVs for 172 characters were calculated and compared between inbred strains and F_1 hybrids. In more than 80% of the cases, the F_1 hybrids exhibited significantly less phenotypic variability, sometimes several-fold less. The characters analyzed spanned a broad range and included life history characters such as rate of development, reproductive output, and longevity, physical traits, and behavioral traits such as learning, wheel running, and open-field activity.

Fluctuating Asymmetry is Another Indicator of Poor Homeostasis

In order to test whether heterozygosity enhances homeostasis, it is useful to have a single, standard measure which can serve as an index of developmental stability, rather than having to compare the CV for multiple traits. One such measure is fluctuating asymmetry. Fluctuating asymmetry (FA) is defined as random deviations in the expression of normally bilateral characters and is generally ascribed to 'developmental accidents' or noise.

The empirical calculation of fluctuating asymmetry is straightforward. First, the asymmetry of a character in an individual is measured by noting the difference in the measure of that character between the right and left side of the individual as a proportion of the mean value of the character. This is then repeated for six to ten additional characters for that individual, and by summing these values a single composite 'symmetry index' for the individual is computed. This makes it possible to compare the symmetry index between different populations of individuals, such as those with high versus low average heterozygosity.

Because they have virtually no heritability, deviations from bilateral symmetry do not appear to represent genetic differences. Similarly, it seems unlikely that any aspect of the external environment differs systematically or consistently between the left and right side of an organism. Instead, such deviations seem to indicate a breakdown in normally well-buffered developmental pathways or a lack of homeostasis. Thus, the greater the average FA among the individuals in a population, the lower their homeostasis.

Observations of FA and its relationship to heterozygosity have been made for many traits in a wide variety of taxa. For instance, in *D. melanogaster*, the length of the right and left wing, as well as the number of bristles on the left and right sides of the body within an individual, vary significantly more among individuals as homozygosity increases. Similar patterns are seen for structural features in fish, mammals, and molluscs. Generally, it appears that: (1) populations and individuals with higher heterozygosity generally exhibit lower frequencies of FA, and (2) the frequency of FA increases with the degree of inbreeding. Because of the apparent link between FA and homeostasis, researchers have used it to assess exposure to environmental stress in humans and other animals. Interestingly, it has even been noted that some animals preferentially choose mates that exhibit greater symmetry than the population average and in humans, ratings of attractiveness, too, are significantly correlated with measures of physical symmetry.

Increased Variability among Inbreds has Practical Implications

The evolutionary origin of developmental stability is important in its own right, but it also has significant practical implications for experimental biologists. Because it is critically important to obtain research animals that offer maximum likelihood of detecting real differences between treatment and control groups, it may be unwise to rely on a small number of inbred strains of animals. Increased developmental instability among inbred organisms may obscure true relationships between biological characters and the effects of experimental manipulations. Testing the response of an experimental treatment in an inbred strain is the equivalent of repeated testing on a single individual, since an inbred strain is a single genotype. Thus, studies utilizing a single strain (or a small number of strains) may produce results which do not necessarily characterize the general pattern of response to the treatment.

This problem is manifest, for example, as significant differences among strains of mice and rats in rates of occurrence of common lesions. An investigator using Fischer 344 rats, for instance, might conclude that most rat mortality and morbidity is due to adenomas of Leydig cells, bile duct hyperplasia, and hepatic microabscesses, since they are observed in 51% (of males), 56%, and 33% of the animals, respectively. An investigator using Brown Norway rats, on the other hand, might not observe a single incidence of any of these lesions and instead might conclude that the pathologies of greatest concern are testicular atrophy, chronic dacryoadenitis of the harderian gland, and nodular vacuolation of adrenal cortical cells (observed in 57% (of males), 52%, and 31%, respectively).

Such unique patterns of pathology make the study of individual diseases easier by utilizing a single inbred strain. Inbred animals may not, however, be the best tools for dissecting a multifactorial process such as aging or development. Researchers may gain experimental power by using F_1 hybrids in place of any specific inbred strain. The F_1 hybrid genotypes are equally replicable and their inter individual phenotypic variability may be significantly lower.

Further Reading

Lerner IM (1954) *Genetic Homeostasis*. Edinburgh: Oliver & Boyd.

Lewontin RC (1956) Studies on homeostasis and heterozygosity. I. General considerations: abdominal bristle number in second chromosome homozygotes of *Drosophila melanogaster*. American Naturalist 90: 237–255.

Lewontin RC (1958) Studies on homeostasis and heterozygosity. II. Loss of heterosis in a constant environment. *Evolution* 12: 494–503

Markow TA (1993) *Developmental Stability: Its Origins and Evolutionary Implications*. Boston, MA: Kluwer Academic Publishers.

Palmer AR and Strobeck C (1986) Fluctuating asymmetry: measurement, analysis, patterns. *Annual Review of Ecology and Systematics* 17: 391–421.

See also: **Adaptive Landscapes; Heterosis**

Genetic Load

J F Crow

Copyright © 2001 Academic Press
doi: 10.1006/rwgn.2001.0537

Genetic load is a measure of the extent to which the average fitness, viability, or other favorable attribute of a population is decreased by the factor under consideration. Thus there are the following types of load: a mutation load, caused by deleterious mutations; a segregation (or balanced) load, caused by segregation of poor homozygotes at loci where the heterozygote is favored; a recombination load, caused by the breakup of favorable gene combinations by recombination; a load due to meiotic drive or gamete selection in which these processes produce less favored genotypes; an incompatibility load, cause by maternal–fetal incompatibility, as in the Rh blood groups; a drift load, caused by unfavorable alleles increasing in frequency by random processes in small populations; and a migration load, caused by immigrants adapted to a different environment.

The word 'load' was introduced in 1950 by H.J. Muller in an article entitled "Our load of mutations" (Muller, 1950). His purpose was to quantify the reduction in mean fitness caused by recurrent mutation using the Haldane–Muller principle, which says that the effect of mutation on fitness is to reduce it by the total mutation rate per zygote. The word was then extended to include all the fitness-reducing processes mentioned above.

The choice of word, load, is unfortunate in its implication that a load is necessarily bad. A genetic load may be a reflection of the opportunity of the species to undergo further evolution. For example, a variable natural population has a lower average fitness than one consisting entirely of the genotype of maximum fitness. Yet the uniformly high-fit population lacks the genetic variability necessary for evolution by natural selection. Likewise, mutation is a requisite for evolution.

The load can be the 'expressed' load, i.e., that which occurs in a natural, usually randomly mating population. There is also a 'total' load, which includes the 'hidden' load, i.e., that which is brought out by special circumstances, such as inbreeding. Separating these loads, usually by studies of inbreeding, has revealed a great deal about the amount of hidden variability in natural populations.

The loads that have been the most extensively researched and discussed are the mutation and segregation loads. Beginning in the 1950s, the hidden mutation load, as revealed by inbreeding, was used as a way to estimate the genomic mutation rate in organisms such as the human, where experimental measures were not feasible. Load principles were also invoked in an attempt to assess the impact on the population of an increased mutation rate, such as might be caused by radiation or environmental mutagens.

In the 1960s there was controversy between those, especially H.J. Muller, who favored the 'classical' hypothesis of population structure, and those, especially Th. Dobzhansky, who favored the 'balance' hypothesis. According to the classical hypothesis, most genetic variability in a sexually reproducing population is caused by recurrent mutation and overdominant loci are rare. The balance hypothesis assumes that most loci are overdominant and that most variability is caused by segregation from superior heterozygotes. The genetic load is much larger under the balance hypothesis. The reason is that under the classical hypothesis deleterious mutants are kept at low frequency by natural selection, whereas with overdominance deleterious homozygotes are relatively common. Some argued that the balance hypothesis entails a large segregation load, perhaps too large to be realistic. Others countered that with rank order selection, the load could readily be accommodated. Although the issue was strongly debated and many experimental studies were undertaken, often with useful by-product information, they failed to settle the issue as to how much genetic variability a natural population contains. The answer came later with the discovery of molecular methods, first protein polymorphisms and later direct measurements of DNA. The answer, curiously, is between what the two hypotheses predict. The amount of protein heterozygosity, about 5–10%, is less than the balance school would have predicted, but higher than had been argued by the classical school. Now other causes of genetic variability have been discovered and this, along with the realization that much molecular variability may be neutral, has caused the debate to subside. Genetic load is now a part of population genetics theory and not a matter of controversy. With the evidence that the mutation rate may be higher than was earlier suspected and the number of overdominant loci fewer, the question of current interest is not a large segregation load but a large mutation load.

The other kinds of loads mentioned above have had much less theoretical treatment, but all are factors in the structure and evolution of natural populations. Current research on these subjects emphasizes direct measurements of allele frequencies rather than indirect assessments from load theory.

References
Muller HJ (1950) Our load of mutations. *American Journal of Human Genetics* 2: 111–176.

See also: **Fitness Landscape; Haldane–Muller Principle; Muller, Hermann J; Mutation Load**

Genetic Mapping
See: **Chromosome Mapping, Gene Mapping**

Genetic Marker
Copyright © 2001 Academic Press
doi: 10.1006/rwgn.2001.1849

A genetic marker is any identifiable allele of interest in an experiment.

See also: **Marker; Marker Effect; Marker Rescue**

Genetic Material
J Merriam
Copyright © 2001 Academic Press
doi: 10.1006/rwgn.2001.0539

The term genetic material describes the physical substance that is inherited from parents by offspring. Generally this refers to DNA. This physical substance is a constant connection for all cells in a body or colony, all individuals in a species, or ultimately, all organisms. The physical substance carries the information specifying the enzymes, structural proteins, and other gene products that are characteristic of life. More abstractly, genetic material refers to that information, or code, that directs life processes.

See also: **DNA; Universal Genetic Code**

Genetic Migration
J B Mitton
Copyright © 2001 Academic Press
doi: 10.1006/rwgn.2001.1422

Genetic migration is the expansion of the geographical distribution of a species by expansion of populations and the founding of new populations in a previously unoccupied area. Many of the best-documented cases involve the movement out of a glacial refugium into an area from which the species had been excluded by the glacier.

Fourteen to 17 000 years ago, when sea levels were lowered by the accumulation of glacial ice, Native Americans migrated from Asia to North America across the land bridge between Siberia and present-day Alaska. They continued their migration east to the Atlantic Ocean and south to the tip of South America.

The most recent cycle of the Wisconsin glaciation forced plant species occupying either high latitudes or elevations to migrate substantial distances. For example, during the Wisconsin glaciation, ponderosa pine retracted to refugia in two general areas, northern Mexico and the Pacific Coast. At the end of the ice age, ponderosa pines began to spread north from Mexico, reaching the San Andres Mountains of southern New Mexico 14 920 years ago (ya), the Santa Catalina Mountains of southern Arizona a few centuries later, and the Grand Canyon about 10 000 ya. They arrived in eastern Nevada 6100 ya, and in northern Colorado 5090 ya. Ponderosa pines reached northeastern Wyoming about 4000 ya, and continued north around the northern edge of the Great Basin where they formed a narrow transition zone in eastern Montana with the ponderosa pines spreading east from their Pacific refugia.

At the height of the most recent glaciation, 18 000 years ago, Scandinavia and parts of the British Isles were covered with ice, and tundra and permafrost covered central Europe. So it comes as no surprise that the modern ranges of European plants and animals were colonized from glacial refugia further south. Phylogeographical studies of eight animals (including a newt, *Triturus cristatus*, a grasshopper, *Chorthippus parallelus*, hedgehogs, *Erinaceus* spp., and bear, *Ursus arctos*) and four plants (alder, *Alnus glutinosa*, oaks, *Quercus* spp., beech, *Fagus sylvatica*, and fir, *Abies alba*) identified four refugia: the Iberian Peninsula, and areas in Italy, Greece, and Turkey. The northern areas of modern distributions generally exhibit less genetic diversity than the southern areas, almost certainly as a consequence of successive population bottlenecks as populations spread to the north, trickling over high passes and dispersing across inhospitable terrain.

Concordant plant and animal phylogeographies have revealed genetic migrations from a previously unappreciated glacial refugium in western North America. A comparison of cpDNA phylogenies of plants in the Pacific Northwest revealed similar geographical patterns of cpDNA variation in six of seven species analyzed. These species include three herbaceous perennials (*Tolmiea menziesii*, *Tellima grandiflora*, *Tiarella trifoliata*), a shrub (*Ribes bracteosum*), a tree (*Alnus rubra*), and a fern (*Polystichum munitum*).

Similar phylogeographical patterns were found in mtDNA of black bear (*Ursus americanus*), brown bear (*U. arctos*), marten (*Martes americana*), and short-tailed weasel (*Mustela erminea*). Deep clefts in the intraspecific phylogenies separate populations north and south of the border between Oregon and Washington. The concordance of these phylogeographies is attributable to isolation of both the plants and animals in the Haida Gwaii refugium, in the present Queen Charlotte Islands of British Columbia. The clefts in the phylogeographies reveal the differentiation that evolved between populations in the Haida Gwaii refugium, surrounded by ice, and the populations occupying the ice-free area south of Washington.

Further Reading

Byun SA, Koop BF and Reimcher TE (1997) North American black bear mtDNA phylogeography: implications for morphology and the Haida Gwaii glacial refugium controversy. *Evolution* 51: 1647–1653.

Hewitt GM (1999) Post-glacial re-colonization of European biota. *Biological Journal of the Linnean Society* 68: 87–112.

Soltis DE, Gitzendanner MA, Strenge DD and Soltis PS (1997) Chloroplast DNA intraspecific phylogeography of plants from the Pacific Northwest of North America. *Plant Systematics and Evolution* 206: 353–373.

See also: **Allopatric; Phylogeography; Speciation**

Genetic Polarity

See: **Polaron**

Genetic Ratios

L Silver

Copyright © 2001 Academic Press
doi: 10.1006/rwgn.2001.0542

In Mendel's original experiments on the transmission of different alleles from crosses that segregated various mutations, certain ratios of offspring were observed consistently. Thus, with a cross between two parents both heterozygous for a recessive mutation, the offspring appeared in a 3:1 ratio of wild-type to mutant. In a cross between one parent heterozygous for a dominant mutation and a second wild-type parent, the offspring appeared in a 1:1 ratio of wild-type to mutant. More complicated ratios were obtained by Mendel in crosses that involved mutations at more than one locus.

See also: **Mendel's Laws; Punnett Square**

(A)

(B)

Plate 13 Ehlers–Danlos Syndrome. (A) face and (B) hands of acrogeric EDS IV patients. The large eyes and lobeless ears are typical, whilst the hands show pulp atrophy from terminal phalyngeal erosions.

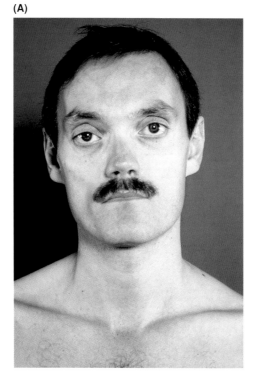

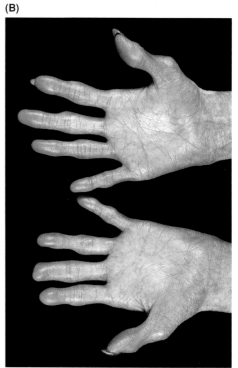

Plate 14 Ehlers–Danlos Syndrome. Typical late facial scarring of EDS I/II.

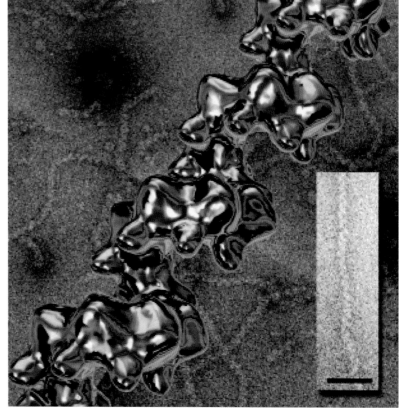

Plate 15 Electron Microscopy. The surface of a three-dimensional reconstruction of a helical filament of human Rad51 protein on DNA is shown in gold in the foreground. In the background is an electron micrograph of the actual filaments that Rad51 protein forms on single-stranded DNA in the presence of ATP. (The Rad51 protein is from the laboratory of Dr Steve West, ICRF, UK.) The inset (right), a portion of such a Rad51-DNA filament (scale bar represents 400 Å) shows the very poor signal-to-noise ratio present in such images. To surmount this problem, the reconstruction has been generated using an algorithm for processing such images (Egelman (2000) *Ultramicroscopy* 85: 225-234) and involved averaging images of 7620 segments. The reconstruction shows that the filaments contain ~6.4 subunits per turn of a 99 Å pitch helix.

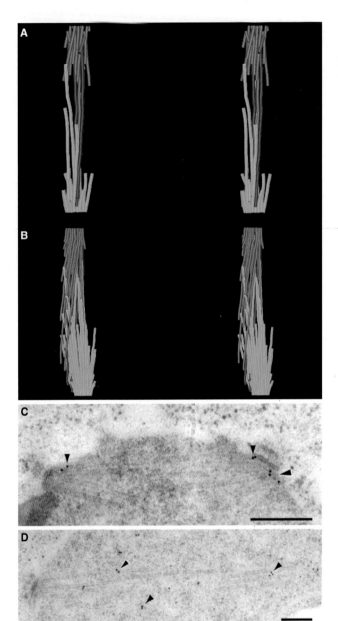

Plate 17 Gene Flow. The blue mussel, *Mytilus edulis.*

Plate 18 Gene Flow. The limber pine, *Pinus flexilis.*

Plate 16 Electron Microscopy. Mitotic spindles in the yeast *Saccharomyces cerevisiae*, analyzed using thin sections. Stereo three-dimensional reconstructions of mitotic spindles from wild-type (A) and a *cdc20* mutant (B). Light gray and dark gray lines represent microtubules; red lines represent microtubules that are continuous between the two poles. The *cdc20* cell division cycle mutant was grown at the nonpermissive temperature (36°C) for 4h, where these cells arrest in mitosis with an average spindle length of ~2.5 μm and contain many more micretubules than wild-type spindles of comparable lengths (see Winey *et al.*, 1995 and O'Toole *et al.*, 1997). Immunoelectron microscopy localization of Kar3-GFP (C) and Slk19-GFP (D) fusion proteins (arrowheads). Spindle microtubules appear as straight structures emanating into the nucleus from dense spindle pole bodies which are embedded in the nuclear envelope. Kar3-GFP is a motor enzyme of the kinesin family that localizes close to the spindle poles, whereas slk19-GFP localizes to kinetochores and the spindle midzone (see Zeng *et al.*, 1994) The scale bars represent 250nm.

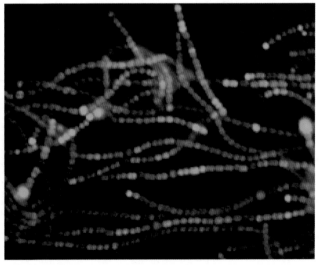

Plate 19 Gene Rearrangements, Prokaryotic. Filaments of the cyanobacterium *Anabaena* 77 h after transfer to nitrogen-free medium. Nitrogen-fixing heterocysts have differentiated at regular intervals along each filament. The image shown is a composite of a fluorescence image showing the location of green fluorescent protein expressed from the promoter of the *hetR* gene and a DIC image that outlines the cells. The HetR protein is required for heterocyst differentiation. It is expressed early in the differentiation of only those cells destined to develop.

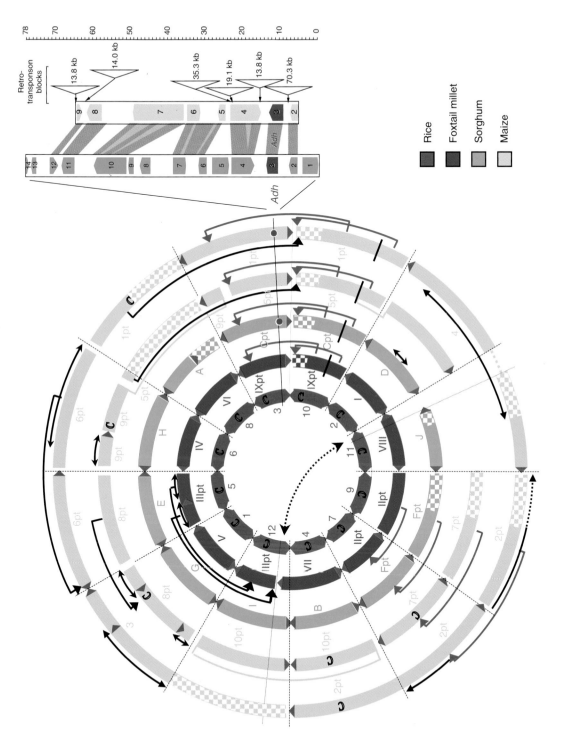

Plate 20 Genome Relationships: Maize and the Grass Model. Consensus map showing the relationship at the map level between the rice, foxtail millet, sorghum, and maize genomes. Arrows indicate rearrangements; **Red** arrows indicate rearrangements that are common to species within a taxonomic group; C, centromere positions; pink circles, location of orthologous *Adh* genes in sorghum and maize. In the detailed comparison of the orthologous *Adh* regions of sorghum (green) and maize (yellow) at the DNA sequence level, arrows indicate the location and predicted transcriptional orientation of the identified genes; conserved sequences in sorghum and maize are connected by gray shading.

Rice

Foxtail millet

Sorghum

Maize

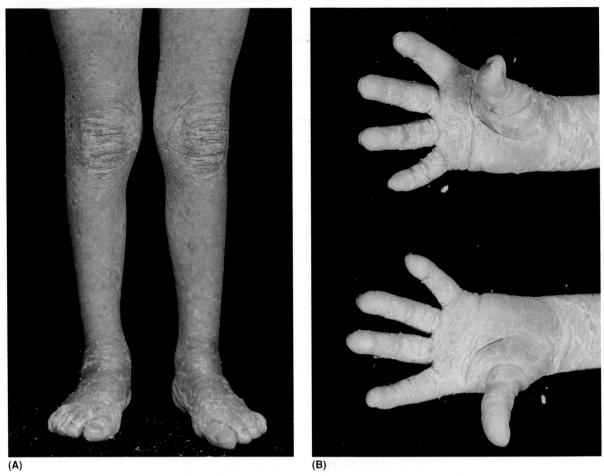

(A)

(B)

Plate 21 Ichthyosis. (A) Generalized hyperkeratosis with background erythema of the lower limbs; (B) palmoplantar hyperkeratosis extending proximally, typical of epidermolytic hyperkeratosis.

Plate 22 Ichthyosis.
Generalized cutaneous features
of a harlequin fetus.

Genetic Recombination

D Carroll

Copyright © 2001 Academic Press
doi: 10.1006/rwgn.2001.0543

Definitions

Genetic recombination refers to the rearrangement of DNA sequences by some combination of the breakage, rejoining, and copying of chromosomes or chromosome segments. It also describes the consequences of such rearrangements, i.e., the inheritance of novel combinations of alleles in the offspring that carry recombinant chromosomes. Genetic recombination is a programmed feature of meiosis in most sexual organisms, where it ensures the proper segregation of chromosomes. Because the frequency of recombination is approximately proportional to the physical distance between markers, it provides the basis for genetic mapping. Recombination also serves as a mechanism to repair some types of potentially lethal damage to chromosomes.

Genetic recombination is often used as a general term that includes many types of DNA rearrangements and underlying molecular processes. Meiotic recombination is an example of a reaction that involves DNA sequences that are paired and homologous over very extended lengths. This type of process, which is illustrated in **Figure 1**, is termed general, legitimate, or homologous recombination. Recombination of this type is reciprocal, because each participating chromosome receives information comparable to what it donates to the other partner. The event shown in **Figure 1** is also designated as a crossover, since all the information on both sides of the effective break has been exchanged.

Gene conversion is a form of homologous recombination that is nonreciprocal. This is recognized by the recovery of unequal numbers of the parental markers at a particular locus, and a simple example is shown in **Figure 2**. Conversion events can be accompanied by a crossover, or not (as shown in **Figure 2**). In the latter case, conversion looks like a very localized double crossover, but it is nonreciprocal and is likely the result of a single event.

Homologous recombination can occur between homologous chromosomes or sister chromatids in mitotic cells as well. In addition, essentially analogous events may take place between homologous sequences that are present at different locations on nonhomologous chromosomes; this is often called ectopic recombination. Recombination that involves very limited or no homology between the interacting DNA sequences

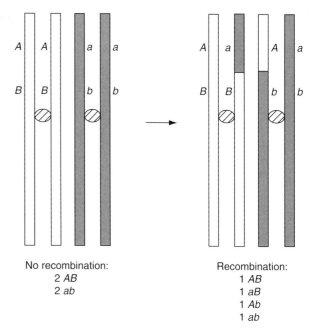

No recombination:
2 *AB*
2 *ab*

Recombination:
1 *AB*
1 *aB*
1 *Ab*
1 *ab*

Figure 1 Simplified diagram of a meiotic recombination event. Vertical bars indicate individual chromatids (i.e., double-stranded DNA molecules); shaded ovals are centromeres. We imagine two pairs of sister chromatids after premeiotic DNA synthesis that are distinguished by color and by genetic markers at locations *A/a* and *B/b*. If meiosis were to proceed without recombination, the markers would segregate 2:2 in linked pairs in the resulting gametes or spores, as indicated below the left diagram. If one reciprocal recombination event takes place between the two markers, the linkage relationships are changed, yielding two new chromatids as shown and ultimately four distinct haploid products.

is termed illegitimate or nonhomologous recombination. Sometimes a few matched base pairs are seen precisely at illegitimate recombination junctions, and these are called microhomologies. An event supported by homologies of 100 bp or more would typically be classified as homologous, a match of 10 bp or fewer would be nonhomologous, and there is evidently a gray area in between.

In conservative recombination events, the number of copies of the interacting chromosomes or DNA sequences is maintained throughout the process, while in nonconservative events, two original copies are reduced to one in the product. This distinction can be made for both homologous and nonhomologous recombination.

Site-specific recombination events are mediated by sequence-specific recombination enzymes often encoded by viruses or transposable elements. The molecular processes they catalyze may rely on very short stretches of homology between the interacting DNAs, or they may be entirely nonhomologous.

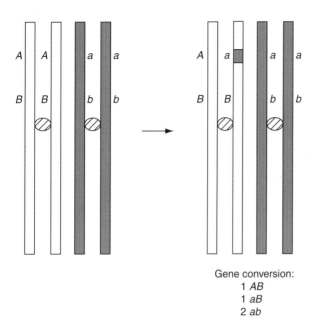

Gene conversion:
1 *AB*
1 *aB*
2 *ab*

Figure 2 Illustration of a gene conversion event. Unlike the reciprocal recombination shown in **Figure 1**, information has been transferred only from one parent to the other, and the extent of the information exchanged is smaller.

Genetic Mapping

To a first approximation, the probability that a genetic recombination event will occur during meiosis is distributed equally along the length of each chromosome, and in most organisms the number of crossovers in each chromosome arm is limited to one or a few. This means that it is quite unlikely that an event will occur between two genes that are very close to each other on a chromosome, but much more likely between distant genes. The closer genes *A* and *B* are along the DNA, the less likely an exchange that rearranges the alleles of these genes, as shown in **Figure 1**. This forms the basis of genetic mapping.

The frequency of recombination is defined as the fraction of all cases in which two genetic markers that came from the same parent are found separated in the offspring. If two markers are on different chromosomes, they will not be linked as they pass through meiosis, and their recombination frequency will be 0.5, i.e., they will segregate into the same gamete by chance half the time and into different gametes half the time. Markers that are very close to each other on the same chromosome arm will be separated very rarely and will have a recombination frequency close to zero. Markers more distant from each other on the same chromosome will show recombination frequencies between zero and 0.5.

Now imagine a situation in which a third marker, *c*, is added to the same chromosome. When the three

markers are monitored in pairwise combinations, the measured recombination frequencies (if they are not too high) are essentially additive, and the numbers are consistent with the physical order of the corresponding genes on the chromosome. For example, if *b* lies between *a* and *c*, the recombination frequencies for the *ab* and *bc* pairs will be smaller than that for *ac*, and the latter will be approximately the sum of the two smaller numbers. In this way, measured recombination frequencies are used to determine the order of genes along chromosomes and the relative distances between them.

We now know that recombination frequencies are not uniform throughout the length of a chromosome. When examined very closely, there are hot spots with elevated frequencies and relatively cold spots with reduced frequencies. This reflects the interaction of the recombination machinery with specific DNA sequences and chromosomal configurations. Nonetheless, since genetic recombination measures genetic, not physical, distances, distant markers usually obey the additivity rules.

Recombination and DNA Structure

Each cellular chromosome usually consists of a single molecule of double-stranded DNA. Genetic recombination may begin with the exchange of only one of the two DNA strands, and recombination outcomes often reflect this fact. Some examples are shown in **Figure 3**, which illustrates gene conversion, postmeiotic segregation, and DNA repair. We imagine two replicated homologous chromosomes (indicated by different shading in **Figure 3**) undergoing homologous recombination. At any particular location along the chromosomes, one or both strands of DNA may be exchanged. When only one strand is exchanged, a heteroduplex is formed that contains one strand from each parent. If the parents differ in sequence in this region, the heteroduplex will be subject to correction by the mismatch repair machinery of the cell. Mismatch repair is frequently responsible for gene conversion, as shown in products 4 and 7 in **Figure 3**.

When meiosis is completed, each haploid gamete will receive one of the four DNA duplexes shown in each of the diagrams in **Figure 3**. Homoduplex will be inherited at all sites that were not directly involved in the recombination event and at all sites where a heteroduplex was repaired. In cases where mismatches escape repair, the heteroduplex will be transmitted to a gamete, and information from both parents will be present at the unrepaired site (*B/b* in diagrams 3 and 6). When the heteroduplex DNA is replicated, after fertilization or germination (depending on the type of organism), homoduplexes of the two parental types

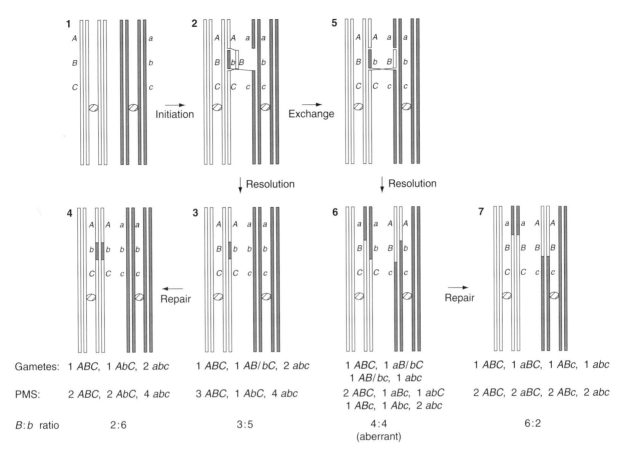

Figure 3 Illustration of some events that occur following the initial exchange of one strand of DNA between homologous chromosomes. Each DNA single strand is shown as a vertical bar, and the parental chromosomes are imagined to differ in allelic markers in three genes: *A/a*, *B/b*, and *C/c*. In diagram 1, all white strands carry *ABC* information, while all shaded strands are *abc*. In subsequent diagrams, the information in the strands of the interacting chromosomes is noted explicitly. After the initiation event (2), in which a segment of one strand invades corresponding sequences in a homologous chromosome, several outcomes are possible. The initial patch, carrying marker *b*, can be incorporated into the recipient chromosome, and the gap it left behind can be filled by DNA synthesis (3). Mismatches in the resulting heteroduplex at *B/b* can be repaired (4). An alternative fate of intermediate 2 is the reciprocal exchange of a single strand from the invaded chromosome (5). One possible way this intermediate can be resolved is by cleavage and religation of the nonexchanged strands; this leads to a crossover, with heteroduplexes remaining at site *B/b* (6). If these heteroduplexes are both repaired to the *B* allele, the result will be that shown in 7. Below the diagrams of products 3, 4, 6, and 7, the genetic outcomes are indicated. The line labeled 'Gametes' shows the status of the double-stranded DNAs, with *B/b* indicating the persistence of heteroduplexes. The results of postmeiotic segregation (PMS) are indicated, as are the recoveries of the parental alleles at *B/b* for each of the outcomes. Product 4 would be scored as a gene conversion, while product 7 is a gene conversion associated with a crossover.

will be produced. This phenomenon is referred to as postmeiotic segregation, or PMS.

As shown in **Figure 1**, the usual outcome of meiosis would be the recovery of equal numbers of parental alleles, in either parental or recombinant configuration. The processes illustrated in **Figure 3** can alter this distribution for markers that are close to the site of the recombination event itself. The number of alleles of each parental type that are present in each of the recombination products is tabulated in the figure. The

2:6 and 6:2 segregation (products 4 and 7) represent gene conversion. The 3:5 (product 5) and aberrant 4:4 (product 6) segregation patterns are revealed as PMS.

DNA Repair by Recombination

So far we have emphasized recombination events that occur in a programmed fashion during meiosis. In mitotic cells, the principal function of recombination appears to be the repair of double-strand breaks (DSBs)

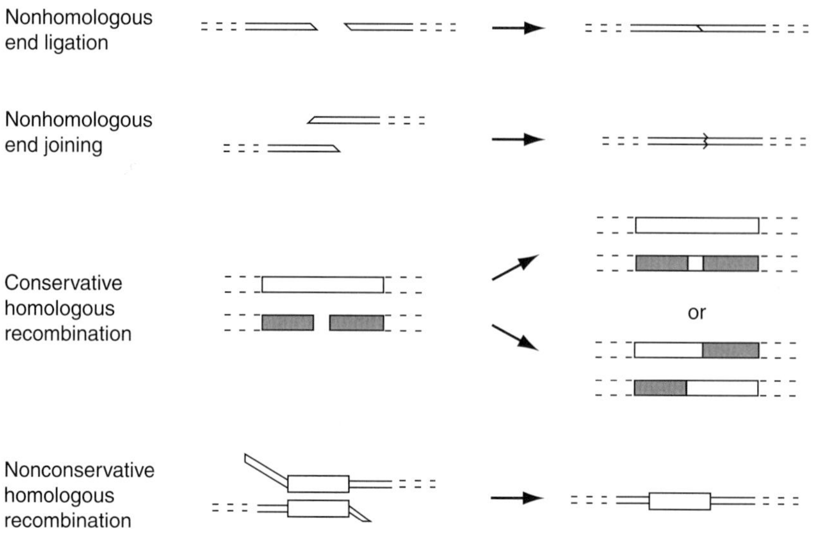

Figure 4 Modes of double-strand break repair by recombination. Each horizontal bar represents a double-stranded DNA molecule. Thin bars indicate no homology, while thick bars denote homologous sequences. In nonhomologous end ligation, broken ends are rejoined precisely without gain or loss of information. Nonhomologous end joining is typically accompanied by deletion (as shown) or insertion of DNA. In conservative homologous recombination, the break is repaired by copying information from homologous sequences elsewhere in the genome, often from a homologous chromosome or sister chromatid, and the repair may or may not be accompanied by crossing over, as shown in the two alternative outcomes. Nonconservative homologous recombination relies on repeated sequences near the break, and repair is accompanied by deletion of one copy of the repeat and all sequences between the two copies.

in DNA. By definition, this type of damage must be repaired by recombination: what came apart must be put back together. DSBs can be generated by external agents, like ionizing radiation and some types of chemicals, or during normal cellular processes, like generation of reactive oxygen species and problems in DNA replication. Essentially all types of recombination play a role in DSB repair in some organisms or cell types: homologous and nonhomologous, conservative and nonconservative, reciprocal and nonreciprocal. Some examples are illustrated in **Figure 4**.

Some types of breaks in DNA can be rejoined by the simple action of DNA ligase without a need for extensive sequence homology. Examples would be the short, complementary single-stranded tails generated by many restriction endonucleases. More frequently, however, the broken DNA ends are not ligatable and end joining occurs between novel sequences, often with concomitant deletions or insertions. Examination of the junctions produced in these illegitimate events frequently reveals microhomologies ($\sim$1–5 nucleotides) between the parental sequences.

Two forms of homologous recombination are involved in DSB repair. If an unbroken copy of the same sequence is available on a homologous chromosome or sister chromatid, the most reliable way to restore the integrity of the broken DNA is to copy that information in repairing the break. Conservative

homologous recombination of this sort may result in a gene conversion, i.e., replacement with information from another allele, close to the break, and it may or may not be accompanied by a crossover (**Figure 4**). If the donor and recipient sequences were not on homologous chromosomes, a crossover would lead to a reciprocal chromosome translocation. The conservative mechanism operates very efficiently in fungi, where it shows considerable similarity to meiotic recombination, both in mechanism and in genetic requirements. In mammalian cells, a nonconservative homology-dependent mechanism seems to predominate. As shown in **Figure 4**, repeated sequences flanking the break can recombine with each other. All the DNA between these two interacting copies is deleted in the process.

Genetic Engineering by Recombination

Genetic recombination is a natural process that plays critical roles in DNA metabolism. In the research laboratory it is sometimes possible to make use of cellular recombination machinery to produce specific genetic alterations. For example, a yeast researcher may want to replace the normal version of a gene with a mutant copy, then examine the effects of the mutation on the life of the organism. The mutant version of the gene can be produced and verified using

DNA cloning and sequencing techniques. If it is then introduced into living yeast, some fraction of the cells will incorporate it at the homologous chromosomal site, using the normal recombination apparatus.

This type of experiment, called gene targeting, works rather well in fungi, but is less efficient in multicellular organisms. The frequency of homologous recombination events between an introduced DNA molecule and the corresponding chromosomal target can be improved if both the target and the introduced DNA are broken. Apparently, the cell sees the DSBs as damage that needs to be repaired. Picture a variant of the conservative homologous DSB repair illustrated in **Figure 4**, in which the broken chromosome is shaded and the white DNA is a linear fragment introduced into the cells. The non-crossover product carries the targeted insertion. As described in the preceding section, cells have multiple pathways of DSB repair, and in practice both homologous and nonhomologous events occur at the broken ends.

In addition to adding to the arsenal of the experimental geneticist, gene targeting holds promise for human gene therapy. In principle, the disease-causing version of a gene could be replaced by the normal allele using this same procedure. At present, however, the efficiency of gene targeting in human cells is too low to make this approach practical.

Further Reading

Kucherlapati R and Smith GR (eds) (1988) *Genetic Recombination.* Washington, DC: American Society for Microbiology.

Low KB (ed.) (1988) *The Recombination of Genetic Material.* San Diego, CA: Academic Press.

Stahl FW (1979) *Genetic Recombination: Thinking about It in Phage and Fungi.* San Francisco, CA: W.H. Freeman.

Whitehouse HLK (1982) *Genetic Recombination: Understanding the Mechanisms.* New York: John Wiley.

See also: **Crossing-Over; DNA Recombination; Gene Conversion; Gene Therapy, Human; Recombination, Models of**

Genetic Redundancy

D C Krakauer and M A Nowak

Copyright © 2001 Academic Press
doi: 10.1006/rwgn.2001.0519

Following the genetic knockout of particular genes there is often no detectable or 'scoreable' change in the phenotype of the organism. Such studies have alerted biologists to the presence of genes with overlapping or redundant functions. These include, among numerous others, the *Drosophila* genes gooseberry and sloppy paired, the mouse tanascin and Hox genes, and yeast myosin genes. In each of these cases, removal of the gene does not result in a quantifiable change to the phenotype in the laboratory.

There are two distinct sets of questions relating to genetic redundancy. One set involves determining the origins of correlations among gene functions, whereas the other relates to the preservation or persistence of these correlations through evolutionary time. There are two principal theories for the origin of redundancy, functional shift and genetic duplication. In functional shift, two independent genes evolve toward some degree of overlap in relation to their current functions or a third, novel function. By contrast, after a random gene duplication event there are two identical genes performing the same function.

Once a degree of redundancy has emerged how is it maintained? This presents a problem because random inactivation of one gene from a redundant pair of genes is not expected to produce any selectable consequences. Several mechanisms for preserving redundancy have been proposed. These include a cumulative benefit from gene copy number (dosage effects), increased fidelity from overlapping functions (error buffering), structural constraints such that genes with independent activities and overlapping function remain redundant through selection on their independent functions (pleiotropy), and convergent functions emerging from common structures.

It can be seen that the problem of functional shift is to provide theories for why genes should evolve toward correlated function, whereas the problem of gene duplication is how to preserve the redundant function following the random mutation. Thus theories for the origin of functional shift are effectively equivalent to theories for the persistence of duplicated genes. In the discussion that follows, we shall concentrate on the problem of persistence.

Cumulative Benefit Theories

When increasing the quantity of a gene product increases fitness, genetic redundancy is easy to understand. Thus, each eukaryotic cell harbors multiple copies of the mitochondrial genome, which enables cells to metabolize efficiently, and multiple copies of tRNA and mRNA genes for efficient translation. Eliminating copies would reduce net fitness and hence redundancy is maintained by stabilizing selection acting on the ensemble of identical genes.

Mutational Error Buffering Theories

Genetic error buffering is defined as any mechanism that reduces mutational load. Consider two identical genes with slightly different mutation rates. One observes that the gene with the higher mutation rate is eliminated from the population – it becomes a pseudogene. If mutation rates are equal, one of two genes will eventually become silenced by drift. Only by allowing a slight asymmetry in each of the gene's abilities to perform their correlated function, might redundancy be preserved. The essential asymmetry is that the less efficient gene also possesses the lower rate of mutation. If the more efficient gene had the lower rate of mutation, there would be no selective reason to maintain redundancy.

Developmental Error Buffering Theories

Developmental error buffering is defined as any mechanism that reduces the deleterious effects of nonheritable perturbations of the phenotype during ontogeny. An appropriate analogy is that of the duplicate flight systems employed in aircraft design or an external storage device used to back up data from a personal computer hard disk. The recurrent risk of error during the lifetime of a device can select for noise-buffering. Once again, considering two genes it can be shown that to preserve a sizeable frequency of both genes within the population, we require that one gene from the pair must mutate less frequently than its duplicate experiences an ontogenetic defect. In other words, a gene that acts as a developmental buffer must have a high mutation rate and be in support of a developmentally unstable gene with a low genetic mutation rate.

Pleiotropic Theories

Some forms of pleiotropy can also ensure the conservation of redundant function. Recall that pleiotropy refers to cases where a single gene experiences selection in more than one context. Consider again two genes with two independent functions. Furthermore, assume that one gene can, in addition to its own function, perform the function of the other gene, but less efficiently. Redundancy is partial and measured in terms of correlated function. When mutations to the pleiotropic gene can either eliminate its unique function or both functions, then redundancy is preserved whenever the rate of elimination of the pleotropic function is lower than elimination of its unique function. In other words, when one gene's pleiotropic function is more robust than the other gene's unique function, the correlated function can be preserved.

Genetic Regulatory Element Theories

We should consider not only redundancy among coding regions, but also redundancy among regulatory elements associated with these genes. For example, if two duplicated genes are each accompanied by duplicates of subsets of regulatory elements (where each subset overlaps to some degree through shared elements), the redundant genes can be maintained by selection acting through their unique patterns of expression and the shared regulatory element. The shared element is assumed to control the correlated function. If one assumes that the shared regulatory element is a smaller mutational target than the coding region, this will prolong the half-life of the redundant function. It is not sufficient, however, to prevent one or more shared elements from becoming silenced in the long term. To preserve redundancy indefinitely we would require, as with the pleiotropic model, some asymmetry in mutation and/or efficacy of the regulators.

Summary

In summary the problem of redundancy is the problem of its preservation. Evolutionary stability of non-trivial redundancy requires asymmetries in mutation and functional efficiency. However, trivial redundancy such as dosage effects could provide the most parsimonious explanation for the observed data. Assuming that weak selection acting in large populations over long time scales has played an important role in genome evolution, cumulative benefit appears to be refuted simply because experimental assays are not sufficiently sensitive.

See also: **Gene Regulation; Mutation Load; Pleiotropy**

Genetic Screening

See: **Gene Mapping; Gene Therapy, Human; Genetic Counseling; Pedigree Analysis; Prenatal Diagnosis**

Genetic Stock Collections and Centers

M K B Berlyn

Copyright © 2001 Academic Press
doi: 10.1006/rwgn.2001.0300

A stock center is a repository for strains or varieties or species of organisms for the purpose of preservation and distribution of a wide range of useful organisms. Genetic stock centers collect primarily mutant derivatives of one or more founder strains, along with related nonmutant strains for long-term preservation and distribution to researchers. This contrasts with collections of germplasm and type cultures, which are more heterogeneous holdings of a wide range of species and varieties (see, for example, Comprehensive Centers for Microbes, in Berlyn, 2000). A few such comprehensive collections are cited in this entry because they have incorporated one or more genetic stock collections within their aggregate of accessions. Many of these general collections in Europe have a history predating genetic stock centers, having been founded before the rediscovery of Mendel.

Early Genetic Stock Centers

In most cases, genetic stock centers originated when geneticists working with the species recognized that the stocks they were making and using were valuable for contemporary and future research in other laboratories and made accommodations for preserving and distributing these stocks to colleagues. The recognition of the need for the stocks was often accompanied by the realization that a means for disseminating new scientific results in a rapid, informal way was also required for advancing scientific progress with the organism. The information function for stock centers has made a natural progression from species-specific newsletters and published genotypes and linkage maps to comprehensive on-line databases (**Table 1**). Many of the genetic stock center databases are part of or are linked to genome databases for their species. The earliest genetic stock centers include research collections of *Drosophila* and maize in the 1920s, the Jackson Laboratory mouse collection in 1929, and, for microbes, the Fungal Genetics Stock Center in 1960 and the *E. coli* Genetic Stock Center (CGSC) in 1970.

Contributions of stocks and strains to the stock centers have been for the most part from the academic research community. Support for public stock center operations in the US has primarily come from the National Science Foundation (NSF) Living Stock Collections Program (http://www.nsf.gov/pubs/1997/nsf9780/nsf9780.htm), the US Department of Agriculture (http://www.ars-grin.gov), the National Institutes of Health (NIH) National Center for Research Resources (http://www.ncrr.nih.gov), and, for some centers, in part from industry.

Drosophila

T. H. Morgan and his students at Columbia, as part of their studies of *Drosophila melanogaster* mutants that began in 1913, preserved their stocks in a collection maintained by C. Bridges. These stocks were provided to anyone requesting them. In 1928, Morgan, Bridges,

A.H. Sturtevant, and the collection moved to the California Institute of Technology. A stock list was first published in the *Drosophila* Information Service in 1934 and it consisted of 572 stocks. E.B. Lewis directed the stock center from 1948 until his retirement, with an approximate tripling of the number of stocks. In 1987, it moved to Indiana University, under the direction of T.C. Kaufman and K.A. Matthews, growing from approximately 4500 stocks in 1995, with the merger of some of the stocks from the *Drosophila* Mid-America Center (*Drosophila* Species Collection) at Bowling Green, KY, in 1997, and subsequent acquisitions, to more than 7700 stocks in the year 2000. It is supported by the NSF and NIH. History, information about stocks, mutations, and nomenclature, and procedures for ordering and culturing the stocks, are given on the web site for the Bloomington *Drosophila* Stock Collection, http://flystocks.bio.indiana.edu. Links to the stock information as well as many other kinds of molecular, genetic, morphological, and mapping information on *Drosophila* are found in the comprehensive *Drosophila* database, FlyBase, at http://fly.ebi.ac.uk:7081 or http://flybase.bio.indiana.edu or http://www.grs.nig.ac.jp:7081.

Collaborative European *Drosophila* stock centers in Umea, Sweden, and Szeged, Hungary, have been supported by the European Union. The P Insertion Mutant Stock Centre in Szeged has recessive lethal P insertion mutants on chromosome 2 and 3 and a collection of mobile element insertions causing altered-expression phenotypes, the EP element lines (http://www.bio.u-szeged.hu/genetika/stock). The European *Drosophila* Stock Center in Umea included general stocks, maternal-effect lethals, zygotic lethals, non-*melanogaster* species and wild-type *D. melanogaster*. It closed at the end of February 2001, with stocks to be transferred to a new center in Kyoto (http://www.grs.nig.acup/.data/doecs/reefman-B.hcml).

The National Institute of Genetics in Mishima, Japan, also maintains about 700 mutant stocks of *D. melanogaster* and 400 stocks of several other *Drosophila* species in various locations and distributes them upon the request of researchers (http://www.shigen.nig.ac.jp/fly/nighayashi.html). They also maintain a data depository of information and documents for the Japanese-speaking community, http://jfly.nibb.ac.jp.

There are also regional stock centers: the *Drosophila* Stock Center, Mexico (http://hp.fciencias.unam.mx/Drosophila/LOSHTML/portada.html), the Moscow Regional *Drosophila melanogaster* Stock Center, and the Indian *Drosophila* Stock Centre at Devi Ahilya University (see addresses at http://flystocks.bio.indiana.edu/other-centers.html).

Stocks from the *Drosophila* Species Stock Center at Bowling Green that were not incorporated into the Bloomington Stock Center have been moved to the University of Arizona (http://stockcenter.arl.arizona.edu).

Maize

At the 1928 Winter Science meetings in New York, a work on the maize linkage maps, and the idea of an group of maize geneticists discussed organized Maize

Table 1   Table of internet resources for Genetic Stock Centers

Type	Organism	Address (Dated March 2001)
Microbial	*Bacillus subtilis*	http://bacillus.biosci.ohio-state.edu
	Chlamydomonas	http://www.biology.duke.edu/chlamy
	Escherichia coli	http://cgsc.biology.yale.edu; http://shigen.lab.nig.ac.jp/ecoli/strain; http://www.shigen.nig.ac.jp/cvector/cvector.html
	Filamentous fungi	http://www.fgsc.net; http://www.hgmp.mrc.ac.uk/research/fgsc/intro.html
	Pseudomonas	http://www.pseudomonas.med.ecu.edu
	Salmonella	http://www.acs.ucalgary.ca/~kesander
	Yeast	http://phage.atcc.org/searchengine/ygsc.html or http://www.atcc.org (Berkeley); http://panizzi.shef.ac.uk/msdn/peter (Peterhof); http://www.ifrn.bbsrc.ac.uk/NCYC (UK); see also http://genome-www.stanford.edu/Saccharomyces
	Miscellaneous (having subcollections of *Agrobacterium, Escherichia coli*, yeast, etc.)	http://www.cabi.org/; http://www.belspo.be/-bccm; http://www.dsmz.de; http://www.pasteur.fr/applications/CIP; http://www.ukncc.co.uk/ http://www.jcm.riken.go.jp/JCM/ aboutJCM.html; http://wdcm.nig.ac.jp; http://www.atcc.org; http://mgd.nacse.org/ocid/prospect3.html
Plant	*Arabidopsis*	http://aims.cps.msu.edu/aims; http://nasc.nott.ac.uk/home.html; see also http://www.arabidopsis.org
	Barley	http://www.ars-grin.gov/ars/PacWest/Aberdeen/hang.html
	Maize	http://w3.ag.uiuc.edu/ maize-coop/mgc-info.html; see also http://www.agron.missouri.edu/
	Pea	http://www.ars-grin.gov/ars/-PacWest/Pullman/GenStock/pea/MyHome.html
	Rice	http://shigen.lab.nig.ac.jp/rice/oryzabase; see also http://ars-genome.cornell.edu
	Tomato	http://tgrc.ucdavis.edu
	Wheat	http://www.ars-grin.gov/ars/PacWest/Aberdeen/hang.html
Animal	Axolotl	http://www.indiana.edu/~axolotl
	Chicken	http://danr013.ucdavis.edu/publications/indexa.htm
	Drosophila	http://flystocks.bio.indiana.edu (Bloomington); http://www.bio.u-szeged.hu/stock (Szeged); http://www.grs.nig.ac.jp:7081/.data/doc/refman/refmanB.html (see Section B.11.2.2.) http://www.shigen.nig.ac.jp/fly/nighayashi.html (Japan); http://flystocks.bio.indiana.edu/other-centers.html (Moscow and India); http://stockcenter.arl.arizona.edu http://hp.fciencias.unam.mx/Drosophila/LOSHTML/portada.html (Mexico); see also: http://jfly.nibb.ac.jp, http://fly.ebi.ac.uk:7081 or http://flybase.bio.indiana.edu, or http://www.grs.nig.ac.jp:7081
	Mouse	http://www.jax.orgnd http://jaxmice.jax.org (Jackson Lab); http://imsr.har.mrc.ac.uk (MRC); http://lsd.ornl.gov/htmouse; http://www.nih.gov/science/models/mouse/resources/ornl.html (Oak Ridge); http://stkctr.biol.sc.edu (*Peromyscus*); see also http://www.informatics.jax.org
	Caenorhabditis elegans	http://biosci.umn.edu/CGC
	Zebrafish	http://zfin.org/zf_info/stckctr/stckctr.html; see also http://zfin.org/index.html

Genetics Cooperation originated in that discussion. It was formalized in 1932 at the 6th International Genetics Congress. It included provision for the Maize Genetics Cooperation Newsletter and the Maize Genetics Cooperation Stock Center. M. M. Rhoades served as first secretary of the Newsletter and first director of the stock center. The responsibility for these activities rotated among several prominent maize geneticists during 1936–1952, while the Stock Center was located at Cornell University, and then again after 1953, when the collection moved to the University of Illinois, Urbana. It was supported by grants from the NSF (1953–1981) and then by the USDA Agricultural Research Service and Plant Genetic Resources Program. Currently, the Maize Cooperation Stock Center is at the University of Illinois under the direction of M.M. Sachs, USDA/ARS and University of Illinois, and the Maize Genetics Cooperation Newsletter secretary is E.H. Coe of the USDA/ARS and the University of Missouri and current co-secretaries are M. Polocco and J. Birchler, also at the University of Missouri. The collection includes nearly 80 000 pedigreed samples, including alleles of several hundred genes, combinations of such alleles, chromosome aberrations, ploidy variants, and other variations. Details about the collection, its history, available stocks, and request forms can be found at http://w3.ag.uiuc.edu/maize-coop/mgc-info.html. The Stock Center database is an integral part of the Maize Genome Database (Maize DB), which links stock center data to the Maize DB information on alleles, genes, molecular markers, maps, probes, etc. The newsletter is also found at this site. http://www.agron.missouri.edu/mnl.

Mouse

The Jackson Laboratory is a nonprofit, independent research institution which was founded in 1929 by C. C. Little to conduct basic genetic and biomedical research and to provide training and genetic resources to the scientific community. It has since that time supplied inbred and mutant strains of mice to the research community. Its current resource includes over 2500 strains of genetically defined mice, both live stocks and frozen embryos, and a transgenic mouse resource and DNA resources (http://www.jax.org and http://www.jaxmice.jax.org). Mouse Genome Informatics is served from (http://www.informatics.jax.org).

The MRC Mammalian Genetics Unit, Harwell, UK (http://www.mgu.har.mrc.ac.uk) maintains a Frozen Embryo and Sperm Archive of almost 1000 stocks and live mouse stocks of 200 mutant, chromosomal anomaly, and inbred lines available on request. The Archive provides free cryopreservation and storage to researchers, with charges for withdrawals from the Embryo Bank.

Oak Ridge National Laboratory Mutant Mouse Collection has several hundred mouse stocks, propagating mutations induced by radiation or chemical mutagenesis, plus several standard inbred strains. Stocks include live mice, frozen embryos or sperm, and frozen tissues (http://www.nih.gov/science/models/mouse/resources/ornl.html and http://lsd.ornl.gov/htmouse).

The *Peromyscus* (deer mouse) Genetic Stock Center originated in 1985 at the University of South Carolina, under the direction of W.D. Dawson, with support from the NSF, and in 1998 contained 35 mutant lines, stocks of wild-type animals of 7 species, and 2 inbred lines. Planning for a comprehensive database, PeroBase, began in 1997. The Stock Center also has received support from the NIH to develop strains to serve as animal models for disease (http://stkctr.biol.sc.edu)

The Fungal Genetic Stock Center

The prominent role of *Neurospora* and other ascomycetes in the study of the genetics of nutritional, biochemical mutants in the 1940s resulted in the isolation of many important strains for research in biochemical and molecular genetics. The Fungal Genetic Stock Center (FGSC) was organized as a result of recommendations by the Genetics Society of America in 1960 and has been funded continuously by the NSF. It was originally located at Dartmouth College, directed by R. Barratt, then moved to California State University at Humboldt and, in 1985, to the University of Kansas Medical Center, directed by J.A. Kinsey and K. McCluskey. Its holdings include nearly 9000 strains of filamentous fungi, mostly genetic derivatives of *Neurospora crassa* and *Aspergillus nidulans*, but also strains of *Aspergillus niger*, *Neurospora tetrasperma*, and isolates of other *Neurospora* and *Aspergillus* species. The collection also contains *Fusarium* species and mutants, *Nectria*, and *Sordaria* mutants and species. The stock center publishes the Fungal Genetics Newsletter, originally a mailed publication, now on-line, as well as meeting abstracts and announcements, and a bibliography available on its web site. The website includes information on genes, alleles, and maps, and on plasmids, clones, and gene libraries for *N. crassa* and *A. nidulans* that the center supplies (http://www.fgsc.net and http://www.hgmp.mrc.ac.uk/research/fgsc/intro.html).

Escherichia coli

Sexuality and the ability to make genetic crosses between mutant strains of bacteria were discovered in *E. coli* only in the 1940s. Individual laboratories studying biochemical and molecular genetics then accumulated large numbers of *E. coli* mutants. It soon became apparent that a national repository would greatly aid the free exchange of strains and the advance of molecular genetics. In the US, the NSF supported a proposal to begin with E.A. Adelberg's Yale University collection of stocks and add important strains and sets of strains from laboratories worldwide. This became the *E. coli* Genetic Stock Center (rGSC) at Yale, curated and directed for 25 years by B.J. Bachmann (until her retirement in 1993, then succeeded by M. Berlyn) and supported continuously since 1971 by the NSF. The collection holds over 7800 strains and a plasmid library encompassing cloned segments of nearly all of the *E. coli* genome. Unlike other early stock centers, it did not establish a newsletter as an integral part of its activities, but it soon assumed the functions of registering gene names and allele numbers, as set forth in the widely accepted guidelines for bacterial nomenclature by Demerec *et al.* (1966), and for registry of designations for deletions, insertions, and F′ plasmids. It also took on responsibility for periodic publishing of the linkage map for *E. coli*. These information functions provided a natural progression to the development of an online database, established in 1989 and also supported by NSF, covering gene names, functions, map locations, strain genotypes, mutation information, and supporting documentation (http://cgsc.biology.yale.edu).

In Japan, the National Institute of Genetics in Mishima established a Genetic Stock Center in 1976 which has a collection of about 4000 genetic derivatives of *E. coli* and 400 cloning vectors. Its reorganization in 1997 created the Genetic Strains Research Center, the Microbial Genetics Center, and the Center for Genetic Research Information (http://shigen.lab.nig.ac.jp/ecoli/strain and http://www.shigen.nig.ac.jp/cvector/cvector.html).

In Europe, many of the broader collections, such as those cited at the end of the next section, carry large numbers of *E. coli* genetic stocks. Phabagen in particular was an early collection of *E. coli* strains and bacteriophage, which has broadened its range of bacteria and also merged with other collections (see BCCM in **Table 1**). The American Type Culture Collection (ATCC) in the US also has many *E. coli* strains and is cited in the next section ("Yeast").

Other Bacteria, Yeast, and *Chlamydomonas*

Salmonella

The *Salmonella* Genetic Stock Center (SGSC) at the University of Calgary, Alberta, Canada, originated in

the laboratory of M. Demerec at Cold Spring Harbor Laboratory and Brookhaven National Laboratory, Long Island, NY, in the 1950s and 1960s, as derivatives primarily of *Salmonella typhimurium* (aka *Salmonella enterica* subspecies *enterica* serovar *typhimurium*) strain LT2. After Demerec's death, the collection was moved and expanded at the University of Calgary by K. Sanderson. It currently has several thousand strains, cosmid and phage libraries, and a set of cloned genes. Many of the mutant strains are organized into special-purpose kits, useful for specific genetic techniques or analyses. In addition to the mutants, it has the *Salmonella* Reference Collection (SARC) representing all subgenera of *Salmonella*. The Center is supported by the Natural Sciences and Engineering Research Council of Canada (http://www.acs.ucalgary.ca/~kesander).

Agrobacterium, Escherichia coli, and Bacteriophages

Phabagen, the Phage and Bacterial Genetics Collection, includes 3500 mutant bacterial strains, 450 cloning vectors, 800 other plasmids, 2 plasmid-containing gene banks of *E. coli*, and over 100 phages. It was established in the early 1960s with deposits of bacterial mutants from researchers of the Working Community Phabagen. Since 1990 it has been part of the Centraal Bureau voor Schimmelcultures (CBS), and has merged with the Laboratory for Microbiology at Delft (LMD) Collections to become the National Culture Collection of Bacteria of the Netherlands, which includes mutant derivatives of *E. coli* K-12 and B and also mutants of *Agrobacterium tumefaciens*, wild-type and reference strains of other bacteria, plasmids, and phages (http://www.cbs.knaw.nl/nb).

Bacillus subtilis

Bacillus subtilis is a spore-forming bacterium that has been used particularly to study that process in prokaryotes. The *Bacillus* Genetic Stock Center at Ohio State University was established in 1978. It is supported by the NSF under the direction and management of D.H. Dean and D.R. Ziegler. The collection includes 1000 genetically characterized *B. subtilis* strains and 300 strains of other *Bacillus* species, as well as a bacterial artificial chromosome (BAC) library, cloned DNA, and shuttle plasmids in *E. coli* strains. The Center publishes a newsletter and a genetic map for *B. subtilis* (http://bacillus.biosci.ohio-state.edu).

Pseudomonas aeruginosa

The *Pseudomonas* Genetic Stock Center is a collection of genetic derivatives of the prototrophic *Pseudomonas aeruginosa* strain PAO1. The collection was originally created at Monash University in Australia by

B. Holloway and is currently located at the Brody School of Medicine, East Carolina University (ECU), Greenville, NC, under P.V. Phibbs. The Center maintains and distributes, in addition to these strains, generalized transducing phages for *P. aeruginosa*, some *P. putida* strains from the J. Sokatch and R. Gunsalus laboratories, and the Holloway cosmid library. It is supported by the Department of Microbiology and Immunology of the Brody School of Medicine at ECU (http://www.pseudomonas.med.ecu.edu).

Yeast

The Yeast Genetic Stock Center originated at the University of California at Berkeley, in 1960, founded and administered by R.K. Mortimer. It included 1200 strains of *Saccharomyces cerevisiae*, primarily derivatives of the stocks of C.C. Lindegren at Southern Illinois University. Professor Mortimer and the stock center annually published updated linkage summaries and linkage maps. After his retirement, the collection moved, in 1998, to the ATCC, where it is maintained as a separate collection (http://phage.atcc.org/search-engine/ygsc.html). The ATCC also has a number of the mutant lines of *Schizosaccharomyces pombe*. In addition it will serve as a repository for a complete set of deletion strains (http://www.atcc.org). Maps and sequence are now presented on-line as part of the *Saccharomyces* Genome Database (http://genome-www.stanford.edu/Saccharomyces).

The Peterhof Genetic Collection of Yeasts (PGC) is part of the Biotechnology Center at St. Petersburg State University in Russia. It has over 1000 genetically marked yeast strains, with an origin distinct from the Carbondale/Berkeley collection. The Peterhof lines originated from a diploid cell of an inbred strain of *Saccharomyces cerevisiae*. The Collection includes mutants derived from this line, other genetically marked yeast strains, and segregants of crosses between the Peterhof-derived and other strains. (panizzi.shef.ac.uk/msdn/peter).

The National Collection of Yeast Cultures (NCYC), Institute of Food Research, in Norwich, UK, includes brewing yeast strains, genetically defined strains of *Saccharomyces cerevisiae* and *Schizosaccharomyces pombe*, and general yeast strains, totalling over 2700 nonpathogenic yeasts. In addition to supplying cultures, it is a patent and safe repository and it performs yeast identification services. A searchable database for the NCYC is found at http: //www.ifrn.bbsvc.ac.uk/ncyc.

Chlamydomonas

The *Chlamydomonas* Genetics Center (CGC) at Duke University was founded in 1984. It collects, describes, and distributes nuclear and cytoplasmic mutant strains, and genomic and cDNA clones of *Chlamydomonas reinhardtii*. The web and gopher sites provide, in addition, information on genetic and molecular maps of *Chlamydomonas*, plasmids, sequences, and bibliographic citations (http://www.biology.duke.edu/chlamy www.biology.date.edu/chlamy).

Large Diverse Collections that Include Microbial Genetic Stocks

A number of the large national collections of microorganisms include genetic stocks. These are described in more detail in the Stock Centers entry in the *Encyclopedia of Microbiology* (Berlyn, 2000).

For example:

1. The International Mycological Institute (IMI) for Culture Collections, which was founded in 1920 as an organization supported by 32 governments and has over 16 500 strains of filamentous fungi, yeasts, and bacteria. It is part of Commonwealth Agricultural Bureaux (CAB) International, a nonprofit intergovernmental organization (http://www.cabi.bioscience.grc.htm).
2. The Belgian Coordinated Collections of Microorganisms (BCCM), a consortium of four research-based collections, include 50 000 documented strains of bacteria, filamentous fungi, and yeasts and over 1500 plasmids, supported by the Belgian Federal Office for Scientific, Technical, and Cultural Affairs. They provide patent and safe-deposit services, as well as fingerprinting/biotyping and identification services, contract research, and training (http://www.belspo.be/bccm).
3. The Deutsche Sammlung von Mikroorganismen und Zellkulturen (DSMZ) is the national culture collection in Germany, founded in 1969 and supported by the Federal Ministry of Research and Technology and the State Ministries. It includes genetic stocks of bacteria, filamentous fungi, and yeast. It also has plant and animal cell cultures. In addition to supplying scientists and institutions with its cultures, it acts as a patent and safe repository (http://www.dsmz.de).
4. The Collection of the Institut Pasteur (CIP) traces its origin to Dr. Binot's collection of microbial strains in 1891. It now includes genetic stocks of *E. coli* (http://www.pasteur.fr/applications/CIP).
5. The NCYC has been cited for its yeast collections. The National Collection of Type Cultures in London (NCTC) is another of the UK National Culture Collections (http://www.ukncc.co.uk) and it has genetic derivatives as well as natural isolates and pathogenic strains of *E. coli* in its large collection, which emphasizes pathogenic bacteria and mycoplasmas. It is a patent and safe depository

and, jointly with the DSMZ, is the resource centre for plasmid-bearing bacteria for Europe. It is supported by the UK Public Health Laboratory Service and is part of the Central Public Health Laboratory, Colindale (http://www.phls.co.uk/services/nctc/index.htm).

6. The Japan Collection of Microorganisms (JCM), in the Institute of Physical and Chemical Research (RIKEN), has over 6000 strains of bacteria, filamentous fungi, yeast, and archea (http://www.jcm.riken.go.jp/JCM/aboutJCM.html) and http://wdcm.nig.ac.jp).

7. The Cloning Vector Collection at the National Institute of Genetics provides vectors of *E. coli* as purified DNA (gillnet.lab.nig.ac.jp/~cvector/NIG_cvector/aboute.html).

8. The ATCC, in the US, has already been mentioned for its Yeast Genetic Stock Collection; it also has genetic stocks of *E. coli* and other bacteria and is a patent and safe depository (http://www.atcc.org).

9. The Microbial Germplasm Database (MGD), at Oregon State University, is not a physical collection, but a database that contains information on collections maintained for research purposes in laboratories of universities, industry, and government and on NSF-supported collections, including contact information for researchers holding these collections. The MGD provides a newsletter and maintains a website where queries can be made (http://mgd.nacse.org/ocid/prospect3.html).

Arabidopsis and Crop Plant Genetic Stock Centers

Arabidopsis

Research using *Arabidopsis thaliana* as a model organism for flowering plants increased dramatically from the mid-1980s through the 1990s. Resource centers that included genetic stocks, genomic libraries, and cloned DNA were established in response to recommendations from a series of workshops sponsored by NSF and culminating in a long-range plan for a multi-national-coordinated *Arabidopsis* genome project presented in 1990. The *Arabidopsis* Biological Resource Center (ABRC) at Ohio State University includes seeds, restriction fragment length polymorphism (RFLP) markers, and yeast artificial chromosome (YAC) libraries. The seed collection and distribution activities in Europe are performed by the *Arabidopsis* Centre at Nottingham, in England, and there is a clone center in Germany. The centers manage the same (mirrored) collection of seed stocks and collaborate and coordinate their efforts to meet the needs of the world *Arabidopsis* research community. The US center, directed by R. Scholl, receives funding from the

NSF, and the UK center, directed originally by M. Anderson and then S. May, is funded by the Biotechnology and Biological Sciences Research Council and the European Union, in addition to user and local institutional support (http://aims.cse.msu.edu/aims and http://nasc.nott.ac.uk/home.html).

Genetic Stocks within Plant Germplasm Collections

The largest and best-known crop plant collections are primarily germplasm repositories for cultivars, landraces, and plant breeding stocks, rather than genetic derivatives of specific stocks. For example, the *ex situ* conservation efforts administered by the USDA are centered at the National Seed Storage Laboratory at Fort Collins, CO, with a base collection that includes over 232 000 accessions of nearly 400 genera and over 1800 species. It preserves valuable germplasm for the US and, by agreement with the International Board for Plant Genetic Resources (http://www.ipgri.cgiar.org), for the global network of genetic resources centers called the Consultative Group for International Agricultural Research (CGIAR) (http://www.sgrp.cgiar.org) and provides these seeds to researchers worldwide. In addition, the USDA National Genetic Resources Collections include a number of collections of genetically defined mutant strains. Besides the USDA/ARS Maize Cooperation Stock Center previously described, there are genetic stock collections for tomato, wheat, barley, and pea. Rice genetic stocks are available through the international Rice Genetic Cooperative.

Tomato

The tomato collection was started by C.M. Ricks in the Department of Vegetable Crops, the University of California at Davis, with collections he made of wild species and mutant marker and cytogenetic stocks created in the laboratory. Others then contributed both germplasm and mutant stocks. It has *c.* 3000 accessions. The C.M. Ricks Tomato Genetic Resource Center has become part of the USDA National Plant Germplasm System, the NPGS (http://www.ars-grin.gov/npgs) and is supported by them, by the University of California, and by industry-sponsored endowments and grants. Seeds are stored in Davis and also, for long-term storage and backup, at the National Seed Storage Laboratory (NSSL) in Fort Collins and are provided to researchers. The annual Tomato Genetics Cooperative Report includes a list of stocks, which is also available through the website. History, query capability, gene and allele descriptions, and links to related sites are provided at the web site, http://tgrc.ucdavis.edu.

Wheat
The E.R. Sears Wheat Genetic Stock Collection origin ated with the cytogenetic and breeding work of E.R. Sears at the University of Missouri in Columbia and includes aneuploids of Chinese Spring wheat – monosomic, trisomic, tetrasomic, nullisomic, and more complicated variations – as well as addition, subtraction, and translocation lines. There are 334 accessions of *Triticum aestivum* subsp. *aestivum* and a total of *c.* 600 accessions from the Columbia collection. Data are available from the GRIN system (http://www.ars-grin.gov/ars/PacWest/Aberdeen/hang.html).

Rice
The international Rice Genetic Cooperative (RGC) was founded in 1985 for the purposes of maintaining genetic stocks, enhancing rice genetics, publishing a Rice Genetics Newsletter that includes gene symbol coordination and linkage map information, and holding periodic symposia. A Japanese committee constructed a network of the genetic stock centers that were located at universities and research stations in Japan, and the information is available through the National Institute of Genetics. Like the Maize Genetic Cooperation stocks, rice stocks include mutant lines, polyploids, trisomics, translocation lines, landraces, varieties, and wild species (http://shigen.lab.nig.ac.jp/rice/oryzabase/Strain.html). The Oryzabase web site includes information about the stock centers, strains, alleles, linkage maps, genes, and other information. Oryzabase was established in 2000 to bring together information ranging from classical genetics to genomics and basic descriptions of rice biology (http://shigen.lab.nig.ac.jp/rice/oryzabase).

RiceGenes at Cornell University (http://arsgenome.cornell.edu/rice) is also a database of the rice molecular marker map and genomic information, particularly quantitative trait loci. Its sister databases, SolGenes for the Solanaceae (including a periodic downloading from the Tomato Stock Center website, see above) and GrainGenes, for wheat and relatives, as well as other crop and animal genome databases, can be reached from the USDA–ARS Center for Bioinformatics and Comparative Genomics, http://arsgenome.cornell.edu.

Barley
The Barley Genetic Stock Center previously housed at Colorado State University and the NSSL moved in 1993 to the USDA–ARS National Small Grains Germplasm Research Facility in Aberdeen, Idaho. It includes over 2500 accessions of *Hordeum vulgare* subsp. *vulgare*. The database of information on the collection are part of the GRIN system. The collection includes aneuploids (primary trisomics) and desynaptic mutants (http://www.ars-grin.gov/ars/PacWest/Aberdeen/hang.html).

Pea
G. Marx, at Cornell University, collected pea germplasm and mutants, and upon his retirement the G.A. Marx Collection became part of the NPGS, and the collection was moved to Washington State University, with accessions numbering *c.* 3000. It includes mutations affecting foliage, flowers, seeds, pods, productivity, and photoperiodism, and a special subset tagged 'Mendel's Genes' (http://www.ars-grin.gov/ars/PacWest/Pullman/GenStock/pea/MyHome.html).

Caenorhabditis, Zebrafish, and Other Animal Stock Centers

Nematode *Caenorhabditis elegans*
Use of this model organism for the genetics of development, behavior, and neurobiology began in S. Brenner's laboratory in the early 1970s and encompasses mutant isolation and analysis, documentation of cell lineage and development, and the complete genome sequence (see, for example, Cell Division in *Caenorhabditis elegans*). The *Caenorhabditis* Genetic Center (CGC) at the University of Minnesota keeps genetic stocks of *C. elegans*, approximately 3500, and a database linked to the *C. elegans* genomic database, http://biosci.umn.edu/CGC.

Zebrafish Resource Center
A more recently developed model system, the zebrafish, for study of vertebrate development and genetics, has a repository for strains at the University of Oregon, supported by funds from the NIH and the state of Oregon.

The International Resource Center for Zebrafish preserves sperm samples, embryos, and live stocks of zebrafish wild-type and mutant stocks submitted by researchers and available for distribution to the research community, maintains the genetic map and information on genetic markers, publishes information on methods for maintenance and use use of zebrafish in research, and studies disease and health of zebrafish strains. The Center maintains the ZFIN, the Zebrafish Information Network database for disseminating information on genetics, genomics, and development of the organism and community information. The database project was founded in 1994, with initial support from the NSF and the Keck Foundation and current support from the NIH (http://zfin.org/zfinfo/stckctr/stckctr.htmland http://zfin.org/index.html).

Domestic Chicken Genetic Stocks

An Avian Genetic Stock Collection for mutants of the domestic chicken at the University of California, Davis, was funded by NSF in 1997 for preservation of existing stocks and planning for future long-term preservation of the collection (http://danr013.ucdavis.edu/publications/indexa.htm).

The Axolotl Colony

The Axolotl Colony, a colony of the Mexican axolotl (*Ambystoma mexicanum*) was founded at Indiana University in 1957 by R.R. Humphrey and has been supported since 1957 by the NSF. It serves as a genetic stock center with mutant lines that affect coloration, organs, limbs, development, and isozymic variation. It has approximately 80 000 axolotls. Embryos, larvae, and adults are sent to research scientists and to classrooms. Information on axolotls and methods of care and a newsletter, as well as mutant descriptions, are found on their web site, http://www.indiana.edu/~axolotl.

Further Reading

Knutson L and Stoner AK. (1998) *Biotic Diversity and Germplasm Preservation*, Beltsville Symposia in Agricultural Research. Boston, MA: Kluwer.

Letovsky SI (1999) *Bioinformatics: Databases and Systems*. Boston, MA: Kluwer.

World Federation of Culture Collections (WFCC) publications: http://wdcm.nig.ac.jp/wfcc/ publications.html.

References

Berlyn MKB (2000) Stock culture collections and their databases. In: Lederberg J (ed.) *Encyclopedia of Microbiology*, vol. 4, pp. 404–427. London: Academic Press.

Demerec M, Adelberg EA, Clark AJ *et al.* (1966) A proposal for a uniform nomenclature in bacterial genetics. *Genetics* 54: 61–76.

See also: *Agrobacterium*; *Arabidopsis thaliana*: The Premier Model Plant; *Aspergillus nidulans*; *Bacillus subtilis*; *Caenorhabditis elegans*; Cell Division in *Caenorhabditis elegans*; *Chlamydomonas reinhardtii*; *Drosophila melanogaster*; *Escherichia coli*; Genome Relationships: Maize and the Grass Model; Grasses, Synteny, Evolution, and Molecular Systematics; *Hordeum* Species; Inbred Strain; *Neurospora crassa*; *Oryza sativa* (Rice); *Pisum sativum* (Garden Pea); *Salmonella*; *Schizosaccharomyces pombe*, the Principal Subject of Fission Yeast Genetics; *Triticum* Species (Wheat)

Genetic Transformation

See: **Bacterial Transformation**

Genetic Translation

B E Schoner

Copyright © 2001 Academic Press
doi: 10.1006/rwgn.2001.0551

Genetic translation refers to the process whereby messenger RNA (mRNA) serves as a template for ribosome-mediated protein synthesis. The process of translation occurs in the cytoplasm of a cell and can be divided into three distinct phases: translation initiation, polypeptide chain elongation, and chain termination. For translation initiation, the small ribosomal subunit must bind to the mRNA to form, along with initiation factors, an initiation complex. The subsequent formation of a polypeptide chain starts with a methionine, which is donated by a unique initiator transfer RNA (met-tRNA$_i$). A different met-tRNA functions in chain elongation. Once the initiation process is completed, the initiation factors are released from the initiation complex and the large ribosomal subunit binds. Additional amino acids are then added to the growing polypeptide chain, in a stepwise manner, where the choice of amino acid is determined by consecutive triplets (codons) along the mRNA. Chain elongation is terminated when one of the three translational stop codons, UAA, UGA, or UAG is encountered.

There are many aspects of translation that are common among both prokaryotic and eukaryotic organisms. For example, the existence of two ribosomal subunits with similar overall structure and similar biochemical steps involved in peptide bond formation. However, there are significant differences in the structure of the mRNAs that prokaryotic and eukaryotic organisms produce, requiring a different process for translation initiation. Bacterial mRNAs are typically polycistronic, which means that more than one gene is contained on a single mRNA. Since these genes are often functionally related or part of a common biosynthetic or degradative pathway, this organizational arrangement has the advantage of allowing coordinate expression of these genes. Translation initiation of bacterial messages is dependent on two elements: an initiator codon and a purine-rich sequence that must be located approximately 10 bases upstream from the initiator codon. The most common initiator codon is AUG, but others such as GUG, AUU, or UUG are being used. The purine-rich sequence, also referred to as 'Shine-Dalgarno' sequence, is complementary to the 3′ end of the 16S ribosomal RNA and is found not only upstream of the initator codon but also in intercistronic regions, or in some cases, within the 3′ end of the preceding gene. Abolishing this sequence, either by mutation or deletion, will result in premature termination of translation.

By contrast, eukaryotic messages are strictly monocistronic. Large precursors, synthesized in the nucleus, are processed (spliced) during their transport into the cytoplasm, where they are further post-transcriptionally modified. These modifications include addition of a methylated cap to the 5′-terminus of the message and addition of a poly(A) tail to the 3′-terminus. In eukaryotes, the smaller ribosomal subunit binds to the capped 5′-terminus, and according to the scanning model proposed by Kozak (1989), migrates linearly until it encounters the first AUG codon. At this point, the larger ribosomal subunit binds and translation begins. The sequence context in which the AUG resides determines the efficiency with which translation initiation takes place. In certain instances, where the AUG is in an unfavorable sequence context, ribosomes can bypass the first AUG and proceed to the next one. However, this is more the exception than the rule.

Regulation at the translational level is less well understood than regulation at the transcriptional level. However, it is clear that the sequences within the 5′ end of both prokayotic and eukaryotic messages have a profound impact on its ability to be translated. In general, high G+C content that promotes secondary structure formation causes poor translational efficiency. This is an important consideration for generating engineered cell lines for the purpose of maximizing gene expression.

Further Reading

Kozak M (1983) Comparison of initiation of protein synthesis in procaryotes, eucaryotes, and organelles. *Microbiological Reviews* 47: 1– 45.

Schoner B, Belagaje RM and Schoner RG (1987) Expression of eukaryotic genes in *Escherichia coli* with a synthetic two-cistron system. *Methods in Enzymology* 153: 401– 416.

References

Kozak M (1989) The scanning model for translation: an update. *Journal of Cell Biology* 108: 229– 241.

See also: Gene Expression; Messenger RNA (mRNA); Protein Synthesis; Ribosomal RNA (rRNA); Ribosomes

Genetic Variation

W J Ewens

Copyright © 2001 Academic Press
doi: 10.1006/rwgn.2001.0552

Evolution by natural selection in a population can occur only if genetic variation exists within that population. Genetic variation is however important not only in evolution but also in all areas where genetics is involved, and many empirical and theoretical studies have been made into the nature and extent of genetic variation and the reasons for its existence and maintenance. The subject is indeed a vast one and here we can only touch on a few aspects of this important topic.

Perhaps the most important fact concerning the maintenance of genetic variation is that the Mendelian hereditary system itself is a 'variation-preserving' one: if there are no selective forces, then genetic variation in any population is maintained (except for random sampling effects in small populations) from one generation to another. A hereditary scheme in which the character of any offspring is a kind of average, or blend, of the values of the character in the two parents rapidly extinguishes variation in the character. In Darwin's time the hereditary mechanism was assumed to be some form of 'blending,' and the loss of variation in such a scheme was recognized by Darwin as an important argument against his theory. The discovery of the Mendelian hereditary mechanism immediately removed this problem.

The amount of genetic variation at any gene locus is usually measured by the degree of heterozygosity at that locus, although other measures (for example the number of alleles present) are sometimes more appropriate. In subdivided populations, the degree of variation both within and between populations can be measured in various ways, the most frequently used measured of such variation being Wright's F-statistics.

Some characters are determined by the genes at a single locus and thus exhibit classical Mendelian segregation. Other characters are determined by a small number of major loci, together with minor effects from other loci. In other cases a character is determined by a large number of loci, with no one locus being predominant in the determination of the character.

The latter case includes many examples of a measurable character such as height or weight. Characters are often also determined in part by environmental factors. The attempt to apportion variation in a measurable character to genetic and environmental effects has long fascinated scientists and laymen alike. Artificial selection on any character depends on the variation in that characters' having in part a genetic basis. For characters depending on many gene loci, the store of genetic variation in a population is such that artificial selection can bring about substantial changes to the values of many characters, often well outside presently observed limits.

A quantitative measure of variation of some measurable character within a population is naturally provided by the statistical concept of a variance. This variance can be estimated from measurements taken from a sample of individuals from a population. The similarities between two characters, either two different characters (for example, height and weight) in the same individual, or the same character in two related individuals, are measured by the covariance, and from this by the correlation, between these characters.

The simplest possible variance calculations arise where the character measurement of any individual depends on its genetic constitution at one single gene locus, with no environmental component, with only two alleles, A_1 and A_2, possible at the locus. Suppose that in diploids (the only case we consider) individuals of the three possible genotypes, A_1A_1, A_1A_2, and A_2A_2 have measurement values m_{11}, m_{12}, and m_{22}, respectively. Let the population frequencies of these three genotypes be P_{11}, $2P_{12}$, and P_{22}. Then the population mean for this measurement is $\bar{m} = P_{11}\,m_{11} + 2P_{12}\,m_{12} + P_{22}\,m_{22}$ and the population variance in the character is $\sigma^2 = P_{11}(m_{11} - \bar{m})^2 + 2P_{12}(m_{12} - \bar{m}^2 + P_{22}(m_{22} - \bar{m})^2$.

In statistical terminology, this variance has two degrees of freedom and can thus be split up into two components, each describing some significant component to the population variation in the measurement. By far the most useful subdivision of this type is the partition of σ^2 into the additive genetic variance (see Additive Genetic Variance) and the dominance variance. Roughly speaking, the additive genetic variance is the variance due to genes within genotypes and the dominance component is the variance not explainable by genes. The former is important in evolution and artificial breeding programs because a parent passes on a gene, and not an entire genotype, to an offspring. In the two-allele case, genetic variation is preserved when the fitness of the heterozygote exceeds that of both homozygotes. When many alleles are possible, a complicated mathematical criterion is needed to assess whether genetic variation is preserved.

Generalizations of these ideas to the case where the character depends on the genes at many loci are also possible. The criteria for the maintenance of genetic variation are now far more complicated than in the single-locus case. It is also interesting to ask how many loci influence the variation in a particular character. In the case of genetic diseases, this is associated with the distinction between 'simple Mendelian' and 'multifactorial' diseases.

Genetic variation is also preserved when a population is divided into small subpopulations, with selective forces acting in different directions in the subpopulations, provided that there is a small migration rate between them. Another agency preserving genetic variation is a selective force acting in different directions between the sexes.

Genetic variation is lost, in small populations, by random sampling effects. Quantitative expressions for the rate of loss of variation through this agency are available in simple cases, particularly those where genes are not subject to natural selection. Generalizations of these expressions in cases where whole subpopulations are subject to extinction are also available.

See also: Additive Genetic Variance; QTL (Quantitative Trait Locus)

Genetics

A Campbell

Copyright © 2001 Academic Press
doi: 10.1006/rwgn.2001.0545

Genetics has been defined as the scientific study of heredity. It has three major subdivisions: transmission genetics, physiological genetics, and population genetics.

Transmission Genetics

Transmission genetics concerns the germinal substance (deoxyribonucleic acid or DNA) and its mode of transmission from parent to progeny. DNAs are distinguished from one another by the sequence of nucleotides along their length. Linear molecules of DNA constitute the core of microscopically visible structures (the chromosomes) that divide and segregate at cell division so that each cell of a multicellular organism generally has the same chromosome complement. Sexually reproducing eukaryotes are typically diploid: an individual has one chromosome set from each of his or her parents. At reproduction, a meiotic division produces gametes (sperm or egg), each of which has a single chromosome set.

The discipline of genetics follows the work of Gregor Mendel, who discovered in 1865 the regular pattern of transmission of units (later called genes) that affect visible properties (later called phenotypes) of organisms. Mendel's units are segments of linear DNA molecules. Most of the basic rules of genetics were deduced before 1951 (when the germinal substance was shown to be DNA) and long before DNA sequencing. Among the processes fundamental to deducing the rules are mutation, recombination, and the meiotic behavior of structurally aberrant chromosomes. A mutation is a heritable change, almost always a change in DNA sequence. Mutations can be used to mark the chromosome entering a cross from the parents and recovered in the progeny. Genetic recombination of such marked chromosomes allows the construction of linkage maps. All these processes aid in equating genetic determinants with specific chromosomal segments – a goal that is superseded by complete genome sequencing, when that is available.

Transmission genetics includes the study of DNA transfer between individuals by means other than sexual reproduction, and its incorporation into the recipient genome. This process is conspicuous in prokaryotes, which lack a meiotic cycle and frequently have circular rather than linear chromosomes. Transmission genetics also include the study of organelles such as mitochondria or plastids that contain DNA but are not distributed in a regular manner at cell division, and of viruses and related elements that contain RNA rather than DNA (some of which are transmitted vertically from parent to progeny cell).

Physiological Genetics

Physiological genetics concerns the mechanisms whereby genes affect organismal properties through transcription of DNA to RNA, translation of RNA to protein, and their regulation. Some subdivisions are biochemical genetics, developmental genetics, and cell genetics. A major tool of physiological genetics has been the characterization of mutant organisms. Such studies have identified various regulatory elements including activators and repressors of transcription and translation and nucleases and proteases that affect protein concentrations. Mutant studies frequently reveal complex pathways leading from primary gene functions to visible traits, sometimes allowing genes to be classified into regulatory hierarchies that define temporal patterns of gene expression during such processes as metazoan development and cell cycle progression.

Population Genetics

The subject matter of population genetics is the distribution of heritable variation among the members of an interbreeding population. It includes both the genesis of such variation (through mutation and selection) and its maintenance from one generation to the next. Much of the variation in natural populations has no detected phenotypic consequences and is observed only at the level of nucleotide sequence. Development and application of appropriate mathematical theory has been central to the discipline of population genetics. The basic rules of population genetics were put forward by R.A. Fisher, J.B.S. Haldane, and S. Wright, from about 1920 onward. One of their principal goals was to explain how Darwinian selection should affect diploid populations.

See also: **Developmental Genetics; Mendel, Gregor; Mouse, Classical Genetics; Population Genetics**

Genome

F Ruddle

Copyright © 2001 Academic Press
doi: 10.1006/rwgn.2001.0560

The term genome has been used traditionally to define the haploid set of chromosomes in the nuclei of multicellular organisms. Hence, one sees reference to the 'human genome,' the 'mouse genome,' and the 'fly genome.' Today the term is used more generally, as for example, to define the chromosomes in cytoplasmic organelles such as mitochondria and chloroplasts, and the chromosomes of prokaryotes and viruses. Hence one sees reference to the 'mitochondrial genome,' the 'yeast genome,' the '*Salmonella* genome,' and the 'SV40 genome.'

Genome is a noun as formerly used, but today it is also used as an adjective. One sees reference to 'genomic variability' or 'genomic size.' The adjectival form is also used as a noun as in the journal dealing with genomic matters, entitled, *Genomics*. The study of genomes is referred as 'genomics.' Researchers investigating genomes are referred to as 'genomasists.'

Genomes vary greatly in size as measured by their DNA content. In general there is a positive correlation between size and developmental complexity. This correlation is imperfect, because developmental complexity cannot be defined in strictly quantitative terms. The SV40 virus genome contains approximately 5000 base pairs. The *Escherichia coli* bacterial genome has 4.6 million bp. The yeast *Saccharomyces cerevisiae* genome has been measured at 12 million bp. The multicellular worm *Caenorhabditis elegans* has a genome size of 100 million bp, while the simple flowering plant *Arabidopsis thaliana* has a nuclear genome of comparable size. The fruit fly *Drosophila melanogaster*, an important experimental organism, has a genome size of 140 million bp. *Homo sapiens* has a genome size of 3 billion bp and the genome of the laboratory mouse, *Mus musculus*, is only slightly larger at 3.3 billion bp.

The genomes of nucleated organisms, eukaryotes, are generally organized into nuclear organelles termed chromosomes. In the somatic cells, the chromosomes exist as two sets, one of maternal, the other of paternal origin. This is known as the diploid condition. In the germ cells, sperm and ova, the nuclei contain a single or haploid set of chromosomes. In humans, the haploid number of chromosomes is 23. Irrespective of cell type, the genome always refers to the haploid set of chromosomes. The nuclear genomes of male and female are slightly different, since the male possesses a Y + X sex chromosome pair, while the female is characterized by a XX condition.

Eukaryotic chromosomes generally consist of a linear DNA duplex complexed with histone proteins plus a variety of minor proteins. The complex of DNA plus associated proteins is termed chromatin. Chromatin has the capacity to alter the compaction of the chromosomes over many orders of magnitude, to replicate the chromosome, and to appropriately regulate the expression of genes encoded in the DNA. The ends of chromosomes terminate in structures termed telomeres that stabilize the DNA strand terminus. Centromeres are structures located at positions between the telomeres and serve as attachment points to the mitotic spindle and serve to distribute replicated chromosomes to daughter cells. Chromosomes can be identified morphologically on the basis of their overall length and position of the centromere. An average size human chromosome contains approximately 130 million bp.

The chromosomes of cytoplasmic organelles, bacteria, and viruses are generally organized as circular structures obviating telomeres, but frequently containing structures analogous to centromeres. These chromosomes are generally much smaller than eukaryotic nuclear chromosomes. The human mitochrondrial chromosome contains 17 000 bp, while the rice *Oryza sativa* chroloplast genome contains 136 000 bp. The *Escherichia coli* genome contains 4.6 million bp. The SV40 virus has a genome size of only 5000 bp.

The number of genes residing in a genome can be most accurately determined by DNA sequencing and sequence analysis. Modern DNA sequencing methods are now producing data on the gene content of even very large genomes. Currently, the large genomes of important eukaryotic research organisms such as the yeast *S. cerevisiae*, the worm *C. elegans*, and the fly *D. melanogaster* have been completely sequenced. The human genome will be completely sequenced in the very near future. The complete sequences of prokaryotic genomes and organellar cytoplasmic genomes are also known.

The number of genes within a genome does not necessarily correlate directly with DNA content. This is because of the complex organization of the genome into coding and noncoding components. Coding components specify the amino acid composition and sequence of proteins. The coding elements of the genome define the total collection of proteins of a cell or organism, termed the proteome. The noncoding elements of DNA fall into a number of categories. One consists of control elements that are essential for the proper expression of coding regions. Principal control elements are promoters which reside proximal to the coding elements and initiate their transcription and enhancers that may reside at a distance from the coding elements and regulate the spatiotemporal expression of coding regions. Genes may be defined as coding elements producing a particular protein product in association with their noncoding control elements. Additional noncoding elements are satellite DNA consisting of long (1–10 kb) tandemly arranged repetitive elements usually concentrated near centromeres, microsatellite DNA made up of short repeats of about 20 bp generally distributed throughout the chromosome and serving as useful genetic markers, and transposable elements that have the capacity to remodel genomes by recombination and additional repetitive and nonrepetitive DNA elements that have no known function. Higher organisms with large genomes may have large noncoding components compared to coding elements. The human genome contains only 3.0% coding DNA. Smaller genomes have a relatively higher content of coding DNA and are said to be more 'compact.' Organellar genomes are highly compact with little noncoding DNA.

Complete DNA sequencing of genomes allows an accurate estimate of gene number. *Hemophilis influenzae*, a pathogenic bacterium, has 1709 predicted genes, while *S. cerevisiae* has 6241, *C. elegans* 18 424, *D. melaogaster* 13 601, and *Homo sapiens* not yet fully sequenced with an estimated gene number of approximately 30 000. It is interesting that as genome size increases, the gene number increases correspondingly less. For example, the human genome is approximately 25 times larger than that of the worm and fly genomes, but the increase in gene number is only twofold. One possible explanation for this discontinuity is that genes interact combinatorially so that fewer genes may by interaction accomplish more complex functions.

Genes increase in number by several mechanisms. One such is by unequal crossing-over whereby a gene undergoes lateral duplication to give rise to two daughter genes residing initially side by side on a chromosome. A second mechanism is by whole genome duplication whereby the initial gene will give rise to daughter genes residing initially on separate chromosomes. Duplicated genes within a genome are termed paralogs and constitute gene families the members of which are related both structurally and functionally. The large genomes of higher organisms are characterized by numerous gene families frequently of large size. For example, the homeobox genes concerned with developmental regulation exist as large gene families in the worm with 88 members and in the fly with 113 members.

Recent progress in our understanding of genome organization in a variety of organisms promises advances in a number of important areas. These include an understanding of evolution and its associated mechanisms, developmental control and the design of body plan, mechanisms associated with the aging process, and practical advances in medicine, agriculture, and biotechnology. An expanding genomic knowledge base will also generate ethical and legal problems which will require political solution and cultural adjustment.

Further Reading

Alberts B, Bray D, Lewis J et al. (1994) *The Cell,* 3rd edn. New York: Garland.

Brown TA (1999) *Genomes.* New York: Wiley-Liss.

Ridley M (1999) *Genome.* London: Harper Collins.

Science (2000) 287: 2105–2364.

See also: C-Value Paradox; Evolution of Gene Families; Genome Organization

Genome Organization

G L Gabor Miklos

Copyright © 2001 Academic Press
doi: 10.1006/rwgn.2001.0556

This review highlights important aspects of the genome architectures of humans, a number of mammals, fish, invertebrates, fungi, plants, protoctists, and bacteria. Many of the differences in genome size and organization among organisms at the same morphological grade are due to variations in the amounts of tandemly repetitious DNA sequences located around centromeres and telomeres, in the amounts of active and degenerate transposable elements, and in the sizes of introns and the spacing between genes. Distantly related genomes differ more by whole and partial genome duplications, the piecemeal amplification and

contraction of gene families, and the evolution of more complex multidomain proteins.

Whole Genome DNA Sequencing

The collective understanding of global genome organizations was accelerated as a result of the industrialization of whole genome sequencing technologies. Different consortia have completed the genomes of 31 bacteria, the megabase (Mb)-sized nuclear genomes of baker's yeast *Saccharomyces cerevisiae* (12 Mb), the nematode worm *Caenorhabditis elegans* (100 Mb), and the fly *Drosophila melanogaster* (180 Mb). The genomes of humans (*Homo sapiens*; 3300 Mb) and two plants, the wall cress (*Arabidopsis thaliana*; 125 Mb) and rice (*Oryza sativa*; 430 Mb), have been completed.

Genomes of Bacteria

The variation in bacterial genome size and gene number is large. *Mycoplasma genitalium* (0.58 Mb) has approximately 470 protein coding genes, whereas *Myxococcus xanthus* (9.5 Mb) probably has in excess of 8000 genes. Genome sizes also vary within a taxonomic group, e.g., from 2.7 to 6.5 Mb in cyanobacteria and from 6.5 to 8 Mb in different strains of *Streptomyces ambofaciens*. Furthermore, the genome organizations of *Mycoplasma genitalium* (470 genes), *Haemophilus influenzae* (1709 genes), *Synechocystis* ssp. (3200 genes), *Bacillus subtilis* (4000 genes), and *Escherichia coli* (4300 genes) reveal that gene order is not conserved, and that there is no absolute functional requirement for specific gene juxtapositions. Bacterial genomes are organized as linear and circular structures. Linear chromosomes occur in *Borrelia burgdorferi*, various species of *Streptomyces*, *Agrobacterium tumefaciens*, and in *Rhodococcus fasciens*. Circular chromosomes are found or inferred in other species: *Mycoplasma genitalium*, *Haemophilus influenzae*, *Escherichia coli*, *Deinococcus radiodurans*, *Leptospira interrogans*, and *Rhizobium meliloti*.

Genomes of Placental Mammals

The variation in genome organization and size is striking. The Indian barking deer, *Muntiacus muntjac*, has only three pairs of chromosomes, whereas the black rhinoceros, *Diceros bicornis*, has 67 pairs. Genome size varies from 1650 Mb in the Italian bat *Miniopterus schreibersi* to 5500 Mb in the South African aardvark, *Orycteropus afer*. In short evolutionary time spans, these differences in genome size have little effect on embryological development, morphology, or physiology, as revealed by comparisons of the Indian muntjac, *Muntiacus muntjac* (2400 Mb), with its three pairs of chromosomes, and the Chinese muntjac, *Muntiacus reevesi* (2900 Mb), with its 23 pairs of chromosomes. Despite these different genome architectures and genome sizes, the species are morphologically similar and yield viable hybrids.

Genomes of Fish, Plants, Yeasts, Ciliates, Crustaceans, and an Ant Species

The African lungfish, *Protopterus aethiopicus*, has a genome of 130 000 Mb (40 times that of humans), whereas the puffer fish, *Fugu rubripes*, has a genome of only 400 Mb, yet both are osteichthyian fish. The lily, *Lilium henryii* has a genome of 33 000 Mb, whereas the wall cress, *Arabidopsis thaliana*, has a genome of only 125 Mb. The similarly sized genomes of the yeasts *Saccharomyces cerevisiae* and *Schizosaccharomyces pombe* are organized in 17 and three chromosomes, respectively. In protoctists, the genes in the macronucleus of ciliates occur either in large chromosomes, as in *Tetrahymena pyriformis*, or as ten thousand or so individual gene-sized pieces in *Oxytricha similis*. Finally, the ultimate reductionist is the ant *Myrmecia pilosula*; its genome consists of just one pair of chromosomes.

Localized Repetitive DNA Sequences in Rats, Humans, and Flies

In genomes of the morphologically similar American kangaroo rats, *Dipodomys ordii monoensis* (5300 Mb) and *Dipodomys heermani tularensis* (3400 Mb), the difference of 1900 Mb is largely accounted for by 700 million copies of just three simple DNA sequences, (AAG, TTAGGG, and ACACAGCGGG), located in the centromeric heterochromatin of *D.ordii monoensis*. By contrast, *D. heermani tularensis* has a small amount of centromeric heterochromatin and a minimal investment in such sequences. These three nontranscribed sequences constitute an amount of DNA equivalent to over half the human genome and are obviously dispensable for centromeric and cellular functions and make no significant contribution to morphology. In humans, there can be differences of many megabases in the size of the Y chromosome among different individuals, the differences being due to varying amounts of two tandemly repetitious DNA sequences. Polymorphisms involving many megabases of centromeric heterochromatin occur on other human chromosomes, particularly chromosome 9, and none of these inherited polymorphisms has any known clinical manifestation. A similar situation is found in *Drosophila melanogaster*, where the satellite DNA-rich centromeric heterochromatin is polymorphic, with differences of many megabases among different strains of flies. In addition, deletion analysis of the satellite DNA-rich X chromosome heterochromatin reveals that at least 12 Mb can be deleted and viability is maintained, attesting to this DNA being devoid of essential genes. In populations of the grasshopper, *Atractomorpha similis*, there are extensive polymorphisms in telomeric heterochromatin, with differences of the order of tens of megabases between individuals in the same population. Finally, in the crustacean *Cyclops strenuus*, 600 Mb of centromeric, telomeric, and interstitial heterochromatic DNA is excised from the chromosomes during the early cleavage divisions of embryogenesis and degraded. The remaining DNAs are spliced together to leave a somatic genome of 400 Mb. The DNAs of these disposable heterochromatic segments are clearly not critical for embryogenesis or cellular functions.

Organization of Centromeres in Fungi, Worms, Flies, and Humans

The localized centromeres of the budding yeast, *Saccharomyces cerevisiae*, consist of a 125-bp region of DNA, while those of the fission yeast, *Schizosaccharomyces pombe*, occupy 40 000 to 100 000 bp. In contrast, those of the fungus *Neurospora crassa* are made up of degenerate transposable elements. The one centromere characterized in *Drosophila melanogaster* is a 0.42 Mb region consisting of tens of thousands of copies of two simple sequence DNAs. The centromeric regions of human chromosomes typically consist of 2–4 megabases of different combinations of the four satellite DNAs and various other repetitive elements. However, stable human chromosomes exist in which direct sequencing reveals that their neocentromeres totally lack all repetitive sequences. The worm *Caenorhabditis elegans* does not have localized centromeres at all; its chromosomes are holocentric. Thus, while the centromeres of humans, flies, and some fungi are embedded in blocks of repetitious sequences, there is no common underlying sequence organization between them, and yeast, and some human centromeres, are totally devoid of repetitive sequences.

Organization of Telomeres in Humans, Ciliates, Yeasts, and Flies

The ends of human chromosomes consist of thousands of tandemly repeated copies of the simple sequence TTAGGG, internal to which are a heterogeneous group of 93 bp repetitive sequences found at the telomeres of chromosomes 5, 7, 17, 19, 20, 21, and 22. The telomeres of trypanosomes also have these TTAGGG repeats, whereas the ciliates *Tetrahymena* and *Euplotes* have variants of these; TTGGGG and TTTTGGGG, respectively. The telomeres of *Saccharomyces cerevisiae* consist of repetitive sequences based on T(G) 2–3 (TG) 1–6 and in, addition, the *Y'* family of conserved repetitive sequences is found at 19 of the yeast telomeres. In contrast to these G-rich sequences, the telomeres of *Drosophila melanogaster* lack the characteristic simple TTAGGG-rich repeats of humans and other organisms. Instead, fly telomeres are composed of a tandem array of elements related to non-LTR retrotransposons of the HeT-A and TART families, which are related to the LINE families of vertebrate transposons.

Dispersed Transposable Sequences in Humans, Flies, Plants, Worms, Yeasts, and Bacteria

The second major component of eukaryotic genomes are the dispersed repetitive sequences, the bulk of which originate from the activities of transposable elements. The sequencing of the human genome reveals that the euchromatic regions of human chromosomes contain a heterogeneous array of transposable elements that were once mobile, but now mostly degenerate and sessile. These elements are finely interspersed with protein coding genes. The bulk of these elements, of the retrotransposon and DNA transposon types (the Alu, MIR, LINE1, LINE2, HERV, Ma1R, mariner, and other miscellaneous transposons), account for approximately 1300 Mb of the human genome. One group, the Alu family, has over a million members dispersed between genes and within intronic regions.

The genes of *Drosophila melanogaster* are interspersed with the members of at least 90 different families of transposable elements. In addition, *D. melanogaster* has approximately eight times as many dispersed transposable elements as its sibling species *D. simulans*, and this accounts for the 20 Mb difference between these two genomes. This huge imbalance in the amount of dispersed repetitive DNA has not manifested itself in significant morphological change, as the two species are near identical and viable hybrids can be produced.

The interdigitation of transposons and other sundry repetitive elements with protein coding genes is a general feature of all genomes, the main difference being the types and amounts of sequences involved. In the lily *Lilium henryii* (33 000 Mb), there are 13 000 copies of just one family of transposons, while in *Arabidopsis thaliana* there are smaller memberships of many different transposable element families, of the LTR, non-LTR retrotransposons, En-like, TNP2-like and MuDR families. In *C. elegans* there

are at least 40 families of dispersed repetitive elements, most of which probably arose from transposition events. The *Saccharomyces cerevisiae* genome is more modest in this regard, with the Ty elements and some solo LTRs together constituting about 3 % of the genome. In bacterial genomes, dispersed repetitive sequences constitute no more than 2 % of genomes. In *E.coli*, 18 repetitive families make up a heterogeneous mixture of autonomously transposable elements, cryptic prophage and phages, and short DNA sequences (such as the 40 bp elements termed REP/BIME/PU). Family memberships vary from a few to approximately 600 and they are dispersed throughout the chromosome.

Gene Numbers in Bacteria, Yeasts, Worms, Flies, Plants, Fish, and Humans

The number of protein-coding genes in fully sequenced bacterial genomes varies enormously from 470 genes in *Mycoplasma genitalium* to a number estimated to be in excess of 8000 in *Myxococcus xanthus*. In free-living eukaryotes, the variation is from 6200 in *Saccharomyces cerevisiae*, 18 000 in *Caenorhabditis elegans*, 14 000 in *Drosophila melanogaster* to 26 000 in *Arabidopsis thaliana*. Gene numbers in human beings, and in mammals in general, are still controversial, with estimates varying from below 40 000 to well over this figure.

Gene Families

All genomes contain different-sized gene families, the members of which have arisen by duplicative processes. Thus in *Haemophilus influenzae*, while there are 1709 genes in total, 284 of these are duplicated products, or paralogs. Thus, there are only 1425 distinct families some of which have more than one family member. In *E. coli*, nearly 50 % of the genes are duplicated, a figure not very different from the percentage of paralogs in the genomes of the worm (49 %) and the fly (41 %). Thus, independently of their grade of evolutionary organization, genomes have undergone a significant degree of duplication of their genes. These duplications can be local and form a cluster, such as a tandem array of 10 glutathione S-transferase genes in the fly and the cluster of 10 kallikrein serine proteases in the rat. Alternatively, the duplicated family members can be dispersed, such as the G-protein-coupled receptor genes (GPCRs), which are distributed throughout the entire fly genome. Furthermore, the extent of these duplication events is different in each evolutionary lineage. In the case of the trypsin-like (S1) proteases, yeast has one gene, the worm has seven, and the fly has 199.

In the case of the GPCRs, there are 160 in the fly, 1100 in the worm, and an estimated 700 in the human genome. In the case of neurotransmitter-gated ion channels, there are 27 genes in the fly, 81 in the worm, and none in yeast. In summary, hundreds of gene families, with very different membership sizes, characterize the various evolutionary lineages.

Pseudogenes

In addition to duplicated gene products that are functional, many metazoan genomes are littered with pseudogenes, duplicated copies that have become inactivated. For example, while there is only a single functional copy of the glyceraldehyde 3-phosphate dehydrogenase (GAPDH) gene in humans, mice, and rats, there are 10 to 30 nonfunctional GAPDH pseudogenes in humans and more than 200 in mice and rats. In the completely sequenced human chromosome 22, there are estimated to be 545 genes and 134 pseudogenes. Furthermore, at least half of the human olfactory GPCRs are pseudogenes. In the worm, at least 300 of the 1100 GPCRs are pseudogenes. In yeast and bacteria, on the other hand, pseudogenes are rare (usually less than 1%).

Orphan Genes

The most surprising result that has emerged from all completely sequenced genomes, be they from bacteria, eukaryotes, or metazoans, is that irrespective of their genic content, at least 20% of the genes (and sometimes much more) are orphans. ORFans are genes whose protein products have no clear sequence similarities to proteins encoded from their own genome or to any other protein in existing public databases. For example, even in *Mycoplasma genitalium* with only 470 genes, 120 are ORFans of totally unknown origin or function. In yeast and the fly, the figures are in excess of 25%. Whether ORFans constitute an irreducible core of genes that evolve rapidly and whose protein products can maintain old functions, or acquire new ones, is not yet clear. They remain a mystery.

Gene Sizes

A comparison of the partially characterized 400 Mb genome of the puffer fish with that of the 3300 Mb human genome is intriguing, particularly since these two organisms are believed to contain the same number of genes. When the human dystrophin, utrophin, and Huntington genes are compared with their puffer fish homologs, it is found that the human genes are 2500, 1000, and 170 kb in length, respectively, whereas their puffer fish homologs are only 200, 100, and

23 kb, respectively. The number of exons and their sizes are near identical between homologs, but the intron sizes in the human genes are almost eight times larger on average than those in the puffer fish. In addition, the puffer fish has less DNA between its contiguous genes than the human genes. When one takes into consideration that the human genome has 2000 megabases of localized and dispersed repetitive sequences, as well as much larger introns and between gene distances than the puffer fish, then the initial eightfold difference in genome size becomes much less mysterious.

Summary

Eukaryote genome organization of the have been dominated by a mixture of whole genome as well as piecemeal genome duplications. Thus the genes presently constituting the mammalian lineage likely stem from a combination of whole genomic amplifications and subsequent reductions of a much smaller genome. Layered on top of this are the local and dispersed duplicative processes that have resulted in the expansion and contraction of individual protein coding families, as well as the expansion and contraction of noncoding tandemly repetitious and transposable families. Layered on top of this again are the molecular processes that generate larger proteins with a greater combinatorial complexity of protein domains. Finally, it is clear that much of the variation that is seen in present-day genomes, particularly in the localized heterochromatic and transposable element compartments of the genome, is essentially the flotsam and jetsam of genomic turnover events. Most of these processes have little effect on phenotype in the short term.

Further Reading

Adams MD, Celniker SE, Holt RA et al. (2000) The genome sequence of *Drosophila melanogaster*. *Science* 287: 2185–2195.

Bendich AJ and Drlica K (2000) Prokaryotic and eukaryotic chromosomes: what's the difference? *BioEssays* 22: 481–486.

Burge CB and Karlin S (1998) Finding the genes in genomic DNA. *Current Opinion in Structural Biology* 8: 346–354.

International Human Genome Sequencing Consortium (2001) Initial sequencing and analysis of the human genome. *Nature* 409: 860–921.

John B and Miklos GLG (1988) *The Eukaryote Genome in Development and Evolution*. London: Allen & Unwin.

Miklos GLG (1985) Localized highly repetitive DNA sequences in vertebrate and invertebrate genomes. In: MacIntyre RJ (ed.) *Molecular Evolutionary Genetics*. New York: Plenum Press.

Miklos GLG and Rubin GM (1996) The role of the genome project in determining gene function: insights from model organisms. *Cell* 86: 521–529.

Rubin GM, Yandell MD, Wortman JR, Miklos GLG *et al.* (2000) Comparative genomics of the eukaryotes. *Science* 287: 2204–2215.

Venter JC, Adams MD, Myers EW, Li PW *et al.* (2001) The sequence of the human genome. *Science* 291: 1304–1351

See also: **Genome**

Genome Relationships: Maize and the Grass Model

K M Devos and J L Bennetzen

Copyright © 2001 Academic Press
doi: 10.1006/rwgn.2001.1668

Gene sequences have remained highly conserved during evolution. Hence, a single set of complementary DNAs (cDNAs) i.e., sequences derived from transcribed genes, can be used as hybridization probes across a range of related species to construct comparative genetic maps, outlining their genome relationships. Comparative mapping within the grass family (Poaceae), including the major cereals rice, maize, and wheat, has demonstrated that both gene content and orders have remained highly conserved during 60 million years of evolution. Thus, it is possible to describe each grass genome, irrespective of its genome size or chromosome number, by its relationship to a single reference genome, rice. These relationships can be depicted by a series of concentric circles with the inner and outer circles representing the smallest and largest genomes in the comparison, respectively (**Figure 1**). Within a genome, chromosomes are ordered so that a minimum number of rearrangements are needed in the overall comparison. Corresponding genes across the species can be found on the radii.

Maize ($2n = 20$; C = 2.5 pg), a species belonging to the subfamily Panicoideae, originated about 16 to 11 million years (My) ago through the hybridization of two diploid ancestors and subsequent diploidization (Gaut and Doebley, 1997). The ancient tetraploid origin of the maize genome is revealed in the comparative maps. Each of two sets of five maize chromosomes (1, 2, 3, 4, 6 and 5, 7, 8, 9, 10) corresponds to a complete rice genome, albeit with a different order of the rice linkage blocks (**Figure 1**). Some rearrangements relative to the rice genome are common to both genomes (indicated by red arrows in **Figure 1**), and also extend to other Panicoideae species. These chromosomal mutations provide information on species' relationships and evolution. However, the rate at which rearrangements occur and are fixed may be species-specific, and thus dependent on the genome structure

rather than on evolutionary divergence time (Zhang *et al.*, 1998; Devos *et al.*, 2000).

Once the comparative maps have identified corresponding, or orthologous, regions across species, DNA sequencing can provide more detailed information on the extent to which gene orders have remained conserved. DNA sequence analysis of orthologous *Adh* regions in maize and sorghum, which diverged 16–20 My ago, showed that nine genes were present in the same order and orientation, while three had apparently been deleted in maize (Tikhonov *et al.*, 1999) (**Figure 1**). The difference in physical length of the region (78 kb in sorghum and 225 kb in maize) was mainly due to the presence of nonconserved retroelements, which inserted within this maize region over the last 6 My (SanMiguel *et al.*, 1998). The region identified by the most conserved *Adh* gene in rice displayed no colinearity with the maize and sorghum *Adh* regions (Tikhonov *et al.*, 1999; Tarchini *et al.*, 2000). This indicated that the *Adh* region had undergone rearrangements in either rice or the Panicoideae lineage since their divergence from a common ancestor. Similar studies across the grass family have indicated that single gene and small segmental duplications and transpositions within otherwise colinear regions may be common events in genome evolution (Bennetzen, 2000; Devos and Gale, 2000).

The main application of the integration of genomic data is the transfer of knowledge across species and the exploitation of common resources including marker sets, mutant collections, and ever-increasing rice genomic sequence data. For example, a maize dwarf mutant that maps in a region orthologous to a plant height QTL in sorghum may be the homolog of the gene underlying the sorghum trait. If rice genomic sequence data are available for this region, it may even be possible to readily identify a candidate gene. The small sorghum genome may also be used as a tool for the isolation of genes in its large-genome relative, maize. Following the identification of the region in sorghum that is orthologous to the target region in maize, chromosome walking and gene isolation can be carried out in the threefold smaller sorghum genome. Although this approach circumvents many of the problems associated with the presence of highly repetitive DNA elements in maize, any disruption of colinearity in the orthologous regions may affect successful isolation of the target gene.

The high level of conserved colinearity within the grass family is in stark contrast with the almost complete lack of gene order conservation between the grass and *Arabidopsis* genomes (Devos *et al.*, 1999; Tikhonov *et al.*, 1999; van Dodeweerd *et al.*, 1999). Although the eudicot and monocot species diverged some 130–240 My ago, this large erosion of colinearity

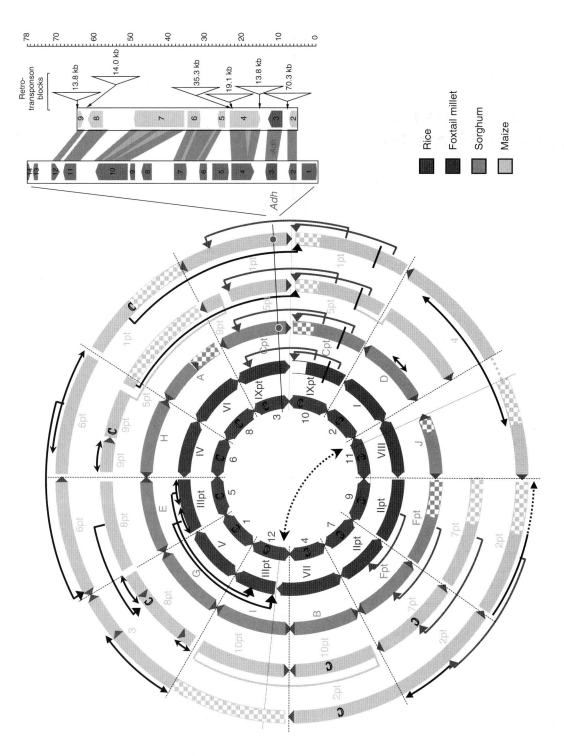

Figure I (See Plate 20) Consensus map showing the relationship at the map level between the rice, foxtail millet, sorghum, and maize genomes. Arrows indicate rearrangements; **Red** arrows indicate rearrangements that are common to species within a taxonomic group; C, centromere positions; pink circles, location of orthologous *Adh* genes in sorghum and maize. In the detailed comparison of the orthologous *Adh* regions of sorghum (green) and maize (yellow) at the DNA sequence level, arrows indicate the location and predicted transcriptional orientation of the identified genes; conserved sequences in sorghum and maize are connected by gray shading.

was unexpected and suggests a high rate of genome rearrangements in the lineage leading to *Arabidopsis*. Although this needs to be confirmed by further comparative data, one can speculate that the extensive duplication of the *Arabidopsis* genome (Bancroft, 2000; Blanc *et al.*, 2000) may have contributed to its faster evolution. Gene duplication, and subsequent divergence of the two copies may also be an important mechanism through which species acquire new gene functions.

In conclusion, comparative genome analyses have demonstrated that gene orders have remained conserved during 60 My of evolution, both at the map and at the DNA sequence level. The wealth of information provided by the integrated grass maps can now be exploited to enhance our knowledge of both well-studied major cereals and under-resourced orphan crops.

References

Bancroft I (2000) Insights into the structural and functional evolution of plant genomes afforded by the nucleotide sequence. *Yeast* 17: 1–5.

Bennetzen JL (2000) Comparative sequence analysis of plant nuclear genomes: microcolinearity and its many exceptions. *Plant Cell* 12: 1021–1029.

Blanc G, Barakat A, Guyot R, Cooke R and Delseny M (2000) Extensive duplication and reshuffling in the *Arabidopsis* genome. *Plant Cell* 12: 1093–1101.

Devos KM and Gale MD (2000) Genome relationships: the grass model in current research. *Plant Cell* 12: 637–646.

Devos KM, Beales J, Nagamura Y and Sasaki T (1999) *Arabidopsis*–rice: will colinearity allow gene prediction across the eudicot–monocot divide? *Genome Research* 9: 825–829.

Devos KM, Pittaway TS, Reynolds A and Gale MD (2000) Comparative mapping reveals a complex relationship between the pearl millet genome and those of foxtail millet and rice. *Theoretical and Applied Genetics* 100: 190–198.

Gaut BS and Doebley JF (1997) DNA sequence evidence for the segmental allotetraploid origin of maize. *Proceedings of the National Academy of Sciences, USA* 94: 6809–6814.

SanMiguel P, Gaut BS, Tikhonov A, Nakajima Y and Bennetzen JL (1998) The paleontology of intergene retrotransposons of maize: dating the strata. *Nature Genetics* 20: 43–45.

Tarchini R, Biddle P, Wineland R, Tingey S and Rafalski S (2000) The complete sequence of 340 kb of DNA around the rice *Adh1–Adh2* region reveals interrupted colinearity with maize chromosome 4. *Plant Cell* 12: 381–391.

Tikhonov AP, SanMiguel PJ, Nakajima Y, Gorenstein NM, Bennetzen JL and Avramova Z (1999) Colinearity and its exceptions in orthologous *adh* regions of maize and sorghum. *Proceedings of the National Academy of Sciences, USA* 96: 7409–7414.

van Dodeweerd AM, Hall CR, Bent EG, Johnson SJ, Bevan MW and Bancroft I (1999) Identification and analysis of homologous segments of the genomes of rice and *Arabidopsis thaliana*. *Genome* 42: 887–892.

Zhang H, Jia J, Gale MD and Devos KM (1998) Relationship between the chromosomes of *Aegilops umbellulata* and *wheat*. *Theoretical and Applied Genetics* 96: 69–75.

See also: *Arabidopsis thaliana*: The Premier Model Plant; Colinearity; Grasses, Synteny, Evolution, and Molecular Systematics

Genome Size

J Hodgkin

Copyright © 2001 Academic Press
doi: 10.1006/rwgn.2001.0557

Genome sizes are usually expressed in terms of the number of base pairs in the haploid genome, either in kilobases (1 kb = 1000 bp) or megabases (1 Mb = 1 000 000 bp). Kilobases are related to other units by the useful 1-2-3 mnemonic: 1 µm of linear duplex DNA has an approximate molecular weight of 2 million daltons and contains approximately 3 kb of DNA. One megabase of duplex DNA has a mass of 1 fg (10^{-15} g). Genome sizes of bacteriophages and viruses range from a few thousand bases to several hundred kilobases. Bacterial genomes range from 0.5 Mb to 10 Mb. Eukaryotic genomes are diverse, from approximately 10 Mb in some fungi to more than 100 000 Mb in certain plants. Genome size in eukaryotes is poorly correlated with organismal complexity. For example, the largest genome known is that of the protozoan *Amoeba dubia*, at 670 000 Mb. The Database of Genome Sizes contains convenient listings of genome sizes for a large number of organisms.

References

Database of Genome Sizes, http://www.cbs.dtu.dk/databases/DOGS/index.html

See also: C-Value Paradox; Genome Organization

Genomic Library

W C Nierman and T V Feldblyum

Copyright © 2001 Academic Press
doi: 10.1006/rwgn.2001.0559

A genomic library is a set of DNA clones that ideally contains the entire DNA content of a genome from which the library was derived. A DNA clone is a DNA construct that is propagated by replication in a microorganism. The clone is composed of two parts that are fused into a single continuous DNA molecule. One part is the vector, which at a minimum contains genes coding for the proteins and other DNA elements necessary for the propagation and selection of the clone in the host microorganism. The other part of the clone is the insert DNA. This is the DNA that is isolated from the organism under study and inserted into the vector.

History

Genomic libraries were constructed in the early days of the development of recombinant DNA technology in the mid to late 1970s. The libraries were the source of clones for the analysis of genes of interest. The first libraries were constructed using partial restriction digests as the means for fragmenting the genomic DNA in a way that generated overlapping fragments of length suitable for cloning. Such fragments were cloned into plasmid vectors by Clark and Carbon (genomic libraries of *Escherichia coli* and *Saccharomyces cerevisiae* DNA). Maniatis constructed genomic libraries of *Drosophila*, rabbit, and human using a bacteriophage lambda vector.

Procedures were developed for the rapid screening of these libraries for sequences of interest based on sequence similarity to a labeled nucleic acid probe. Colony hybridization for screening plasmid libraries and plaque hybridization for screening bacteriophage lambda libraries revealed which clones contained DNA sequences with identity or very high similarity to the sequence of the probe. In these procedures DNA from the high-density plates of colonies or plaques was transferred to solid hybridization membranes, initially nitrocellulose and subsequently various formulations of modified nylon. The initial pattern of colonies or plaques on the plates was preserved on the membrane. A DNA fragment serving as a probe was typically labeled with the ^{32}P isotope of phosphorus. The membrane was hybridized to the probe in solution and after washing away the unhybridized probe, it was exposed to X-ray film to reveal those colonies or plaques that contained DNA identical or similar to the probe. Based on the location of the hybridizing signal on the membrane, the corresponding colonies or plaques were recovered for further analysis.

Recombinant DNA technology lead to the explosive development of molecular genetics as numerous genes of biological and medical importance were isolated from genomic libraries, characterized, and their products expressed in *E. coli*, other bacterial species, *S. cerevisiae*, and insect and mammalian cultured cells.

The plan to map and sequence the human genome emerged as the Human Genome Project in the late 1980s, bringing genomic libraries into this new application. Up to this point the libraries were the source of clones for studying individual genes or sequences. For whole genome scale analysis, the properties required of genomic libraries were more rigorous. Three characteristics emerged as the requirement for use in libraries for genome projects. These characteristics (large cloning capacity, stable propagation of insert DNA, and curtailing of chimeric inserts) are critical features of vectors and library construction protocols for genome mapping and sequencing. For these applications cosmids, yeast artificial chromosome vectors (YACs), bacteriophage P1 vectors, P1 artificial chromosome vectors (PACs), and bacterial artificial chromosome vectors (BACs) have been developed and used.

The decade of the 1990s has brought a focus on high throughput genomic DNA sequencing of many species including the human, the laboratory mouse, the roundworm *Caenorhabditis elegans*, plants, including *Arabidopsis thaliana*, rice and potato, and numerous species of bacteria. A critical component of all of these projects is the construction of genomic libraries from either the entire genome or from a large insert genomic clone such as a BAC. These libraries are constructed by shearing the genomic DNA to randomly generate overlapping fragments of the appropriate size. A fraction of the sheared DNA is then selected by size for construction of the library. The resultant library has a narrow range of insert sizes and clones are randomly selected from the library for sequencing from both ends (shotgun sequencing). The insert size of the clones being sequenced is typically 2 kb and the average sequence read length is about 650 bases. The randomly collected sequence reads are then assembled into the original molecule using assembly software, which will find overlaps in the sequence reads to accomplish the assembly process.

The features of the vectors and library construction protocols for shotgun DNA sequencing are very rigorous due to the high cost of sequencing, the technical challenge of the assembly of shotgun sequence reads, and the cost of closing the remaining gaps after the assembly. The libraries need to have a very low incidence of clones that contain no inserts, or the no-insert-containing clones need to be readily identifiable and excluded from the sequencing pipeline. The libraries should have a narrow insert size range. This allows the assembly software to use the distances between the sequences obtained from each end of the clone. The libraries cannot contain chimeric DNA inserts as these will confound the assembly process. The libraries need to be as truly random as is technically achievable to minimize the number of gaps in the sequence after the assembly of the sequence reads. This requires that the insert DNA be sheared instead of the more traditional technique of partial restriction digestion to reduce the size of the DNA fragments to that required for cloning.

Vectors for Genomic DNA Libraries

Vectors for genomic DNA libraries are selected based on the projected application for the library as discussed above. **Table I** summarizes the properties of the vectors covered in this review. A brief introduction to each type of vector and the procedure for preparing the vector for library construction follows below.

Plasmid Vectors

Plasmids are small extrachromosomal circular double-stranded DNA molecules that replicate independently of the chromosome or chromosomes of a microorganism. Their copy number in the cell is maintained by control systems built into the plasmid's gene content but varies depending on the replication system of the plasmid. For example, pUC-based plasmid vectors are maintained at a copy number of 500–700 per cell. Plasmid vectors derived from the *E. coli* F plasmid are rigorously maintained at a copy number of one.

Naturally occurring bacterial plasmids have been engineered to serve as vectors for the propagation of exogenous DNA fragments. To serve as a vector the plasmid must have in addition to a replication system, a selectable marker (typically an antibiotic resistance gene) and a cloning site, a unique restriction site in a nonessential region of the plasmid for the insertion of the exogenous insert DNA.

One of the historical limitations on the use of plasmids for genomic libraries was the inefficient procedure of chemical transformation for transferring the recombinant plasmid DNA constructs into *E. coli*. For library construction, that technique has been replaced by the use of very high efficiency electroporation. A high-voltage electric field is applied briefly to cells, producing transient holes in the cell membranes through which plasmid DNA enters. Electroporation allows for the efficient transfer of plasmid DNA as large as 200 kb into cells.

Preparation of Plasmid Vectors for Library Construction

Vector preparation in general includes procedures to remove unwanted DNA fragments and to generate the desired ends at the cloning site. The process of making a plasmid genomic library is illustrated in **Figure I**. Additional steps may be incorporated to reduce or eliminate the ability of the vector to be replicated in the absence of an insert fragment.

For plasmid vectors, preparation for use in library construction typically involves digestion with the

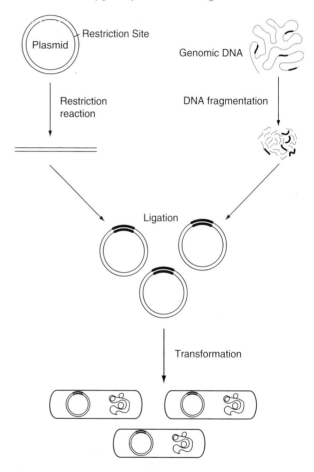

Figure I Construction of a genomic library in a plasmid vector.

Table I Properties of the vectors used in construction of genomic libraries

Vector	Cloning capacity (kb)	Applications
Plasmids	0.1–12	Single gene cloning; shotgun sequencing libraries
Bacteriophage lambda	10–20	Single gene cloning
Cosmids	35–45	Single gene cloning; genome mapping and sequencing
Bacteriophage P1	30–90	Genome mapping and sequencing
BACs	30–300	Genome mapping and sequencing
YACs	100–1000	Genome mapping and sequencing

appropriate restriction enzyme to give the appropriate sticky (single-stranded tails of typically four bases) or blunt ends followed by the removal of the 5′ phosphates using a phosphatase enzyme. The absence of the 5′ phosphates eliminates the ability of the plasmid to be ligated back into a circular molecule. This phosphatase step is required to minimize the number of clones in the library that contain no inserts.

Bacteriophage lambda Vectors

Bacteriophage lambda is a virus that infects *E. coli*. The typical infection cycle results in the lysis of the *E. coli* cell and the release of about 100 progeny phage particles, each capable of infecting another cell. When lambda is plated at low density on a lawn of *E. coli* cells on agar medium, the resulting pattern of clearings (plaques) in the lawn caused by the lysed cells identify the location of individual lambda clones. Harvesting the phage particles from a plaque (picking a plaque) provides a stock of the phage clones for subsequent rounds of propagation.

Like the plasmid vectors, wild-type lambda has been extensively engineered for use as a vector. Genes not essential for the lambda life cycle described above have been removed to make room for carrying exogenous insert DNA. The early popularity of lambda as a cloning vector for genomic library construction is a consequence of the very efficient pathway for getting lambda DNA into *E. coli* cells. This was in contrast to the inefficient chemical transformation used for plasmids, particularly for larger constructs. The bacteriophage lambda DNA or recombinant lambda DNA-containing inserts is packaged into infectious phage particles using an efficient *in vitro* packaging reaction. Once the particles are formed, each one can inject its DNA into an *E. coli* cell. The limit on how much exogenous DNA can be propagated in a lambda vector results from the packaging capacity of the phage particle, approximately 35–50 kb of DNA. Because of the requirement for lambda genes for a productive infection, the amount of insert DNA is restricted to 10–20 kb depending on the specific vector. **Figure 2** illustrates the process of constructing a genomic library in a lambda vector.

Preparation of lambda Vectors for Library Construction

Lambda vectors come in two different forms because of the restriction on DNA size that is a feature of the packaging system. Lambda insertion vectors are those that simply require that the lambda DNA be cut at a unique restriction site for the insertion of exogenous DNA. This is why they are termed insertion vectors. After the restriction digestion, a phosphatasing step to remove the 5′ phosphates is frequently also included

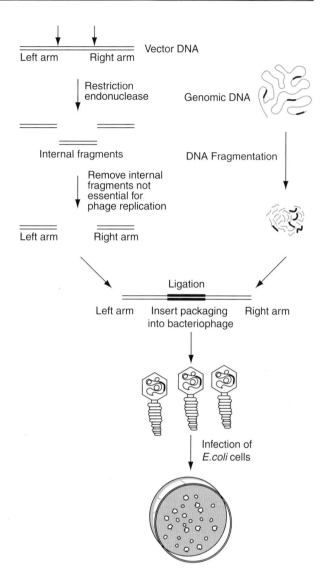

Figure 2 Construction of a genomic library in a bacteriophage lambda vector.

to minimize the number of clones in the library without inserts. Since the packaging limitations are from about 35–50 kb, insertion vectors can handle exogenous DNA fragments up to only about 15 kb because the vector itself must be at least 35 kb to be packageable and therefore viable.

The second form of lambda vector is termed a replacement vector. These vectors have a stuffer fragment that is removable by restriction digestion and DNA size fractionation. In the course of removing the stuffer fragment, the cloning site sticky ends are also generated. The removal of a stuffer fragment allows for the insertion of a larger DNA fragment. Typical replacement vectors can propagate inserts with sizes up to 25 kb. The phosphatase step is also incorporated into the vector preparation process with lambda replacement vectors. Removal of the 5′ phosphatase is used to prevent vector to vector ligation that will reduce the efficiency of the *in vitro* packaging reaction. After the removal of the stuffer fragment, the remaining arms are sufficiently small that if ligated together without inserts will not be packaged. This will prevent the propagation of vectors without inserts.

Cosmid Vectors

Cosmid vectors are plasmids that can be packaged into infectious bacteriophage lambda particles via the lambda *in vitro* packaging system. All that is required for this to occur is for the plasmid to contain the lambda cos DNA sequence. Since the plasmid does not require the lambda genes necessary to form progeny phage particles on infection, there is more capacity in the cosmid for containing insert DNA than with a bacteriophage lambda vector. As a result, cosmids can generally accept 30–40 kb insert fragments. Once the phage particle containing the cosmid clone injects its DNA into an *E. coli* cell, the cosmid is replicated through its plasmid replication system and cells containing the cosmid clone are selected by the antibiotic resistance marker in the vector. So a cosmid clone is simply a 35–50 kb plasmid which can be efficiently packaged into bacteriophage lambda particles and injected into *E. coli* cells. One of the advantages of cosmids for constructing genomic libraries of organisms with large genomes is that they have a cloning capacity about twice that of lambda vectors, i.e., they can accept inserts of up to about 40 kb whereas lambdas are restricted to about 20 kb. A disadvantage is that some cosmid clones are unstable on propagation in *E. coli* due to the high copy number plasmid replication system.

Preparation of Cosmid Vectors for Library Construction

Cosmid vectors are prepared in much the same manner as plasmids. The cloning site sticky ends are generated by digestion with a restriction enzyme and a phosphatase is used to remove the 5′ phosphates from the vector to prevent vector to vector ligation. The insert fragments must be size selected so that the *in vitro* packaging of the DNA into bacteriophage lambda particles occurs efficiently.

P1 Vectors and PACs

The P1 cloning system was developed by Nat Sternberg for use in large genome mapping and sequencing projects. P1 vectors are much like cosmids in that they are plasmids that can be packaged in a phage particle for efficient injection into *E. coli*. The bacteriophage P1 phage head can hold about 110 kb of DNA. The vectors designed for use with the P1 packaging system are up to about 30 kb in size so that the cloning capacity

of P1 systems is 70–100 kb. The P1 cloning vectors feature two replication systems. The P1 replicon functions after the DNA is injected into the cell. This replicon maintains the P1 plasmid at a copy number of one that minimizes the possibility that the insert will rearrange. An inducible replicon is available to increase the copy number by 20–30 fold immediately before DNA purification. The P1 cloning system does not exhibit the instability of the cosmid system due to the low copy number propagation of the clone in *E. coli*.

P1 artificial chromosomes, PACs, use the same P1 cloning vectors but do not go through the *in vitro* packing step. Instead, the inserts ligated to the vector molecules are electroporated directly into *E. coli* cells. Without the need to be packaged into bacteriophage particles, the size of the inserts can be increased to greater than 100 kb. The properties of PAC libraries and the procedures for making and manipulating them are similar to the BAC libraries discussed below.

Preparation of P1 and PAC Vectors for Library Construction

P1 vectors are prepared in essentially the same way as bacteriophage lambda vectors (see above). PAC vectors are prepared in essentially the same way as BAC vectors (see below).

Bacterial Artificial Chromosome Vectors

Bacterial artificial chromosome vectors (BACs) were developed to permit the cloning and stable maintenance of large (100–200 kb) pieces of DNA in *E. coli*. Their stability and ease of handling have made these vectors increasingly popular for whole genome mapping and sequencing projects from microbes, plants, and animals. The copy number of these cloning vectors is rigorously maintained at one by the BAC replication system derived from the *E. coli* F plasmid. The use of these vectors with a recombination-deficient host allows DNA that is unstable in higher copy number cloning systems to be propagated without incurring deletions or rearrangements. Large insert libraries constructed in BAC vectors have served as the starting point for the sequencing of several organisms with large genomes including the human, mouse, the model plant *Arabidopsis thaliana*, and rice.

Preparation of BAC or PAC Vectors for Library Construction

As a result of the reduced electroporation efficiency of 100 kb circles (BAC clones) relative to 8 kb circles (BAC vectors) even a minor amount of recircularized vector present after ligation with inserts will yield a major fraction of the colonies containing only vector molecules. To avoid this, considerable effort must be expended in preventing the formation of BACs without inserts in the library construction process.

The vector molecules are digested with the appropriate restriction enzyme and the 5′ phosphates removed by treatment with a phosphatase. Linear vector molecules of the correct size are obtained by recovery from an agarose gel after sizing by electrophoresis. The recovered linear vector molecules are then self-ligated in bulk to form multimers from those molecules still retaining the 5′ phosphates. These ligation products are removed by another round of agarose gel electrophoresis and size selection for the linear BAC monomer.

At this point aliquots of the vector are self-ligated and ligated with a test insert in separate reactions. The reaction products are electroporated into *E. coli* and the colony count compared. The vector only electroporation should yield few to no colonies and a large number of colonies should be obtained with the test insert. If the vector only electroporation does give a consequential number of colonies, then another round of self-ligation and size purification is required. Once the desired result is obtained, the vector is ready to receive inserts.

Yeast Artificial Chromosomes

Yeast artificial chromosomes (YACs) provide the largest insert capacity of any cloning system. This system, developed by Burke and Olson in 1987, supports the propagation of exogenous DNA segments hundreds of kilobases in length. YACs representing contiguous stretches of genomic DNA (YAC contigs) have provided a physical map framework for the human, mouse, and even *Arabidopsis* genomes.

The YAC vector itself provides the essential elements for propagation of DNA as a chromosome in the yeast *Saccharomyces cerevisiae*. These elements include a yeast centromere, two functional telomeres, and auxotrophic markers for selection of the YAC in an appropriate yeast host. A problem encountered in constructing and using YAC libraries is that they typically contain clones that are chimeric, i.e., contain DNA in a single clone from different locations in the genome.

Preparation of YAC Vectors for Library Construction

YAC vectors are like P1s and lambda replacement vectors in that they require the isolation of two vector arms and removal of the 5′ phosphates from the arms to prevent vector to vector ligation and recirculation. Since recircularized vector DNA can transform yeast with a high frequency, the presence of even a small quantity of such molecules can produce a high background of vector only transformants.

Average Insert Size and Representation of the Genome

When the purpose of a genomic library is to screen for a single gene, if the gene is there the library representation of the genome is sufficient. If the library is intended for genome-wide studies, the usefulness of the library depends on maximizing the fraction of the entire genome present in the library. Genomic libraries are usually characterized by the size of the library, i.e., the number of clones in the library and the average insert size of the clones.

The typical way of presenting library size is to determine the ratio of the amount of genomic DNA in the library to the amount of DNA in the genome. For example, if a human BAC library has an insert size of 100 kb and contains 300 000 clones, the library contains 30 Gb of human DNA. Since the haploid human genome contains 3 Gb of DNA, the library is 10 times larger than the genome. This is sometimes referred to as a 10 × library or a 10-hit library. This ratio is the library coverage value.

The coverage value indicates on the average how many times a particular sequence is present in the library. This does not mean that libraries larger than 1× in coverage contain the entire genome. Since the probability of finding a sequence in a library follows a Poisson distribution, assuming random cloning, some sequences will be present less often, or even absent, and others more often than the number indicated by the coverage value. For a 1× library, the probability of finding one or more clones containing a particular sequence is 0.632. For a 10× library this probability increases to 0.99995. A 5× to 7× library with probabilities ranging from 0.99 to 0.999 reflects what is considered to be the typically useful library size.

Preparation of the Inserts

The ideal genomic library contains all sequences present in the genome of the subject organism. In addition to the considerations of library size discussed above, sequences are absent from the library as a result of either of two additional circumstances. The first circumstance is that a particular sequence in a cloning vector results in the killing of the host or the DNA sequence itself is unstable in the host. If all of the cells containing a sequence are killed, then that sequence will not be present in the library. If the sequence is deleted or rearranged in the host, it will not be found or recognized in the library. These kinds of sequences are called unclonable sequences. The problem of unclonable sequences will be treated below. The other circumstance occurs when a nonrandom method such as restriction digestion is used for fragmenting the genomic DNA.

This results in some specific fragments that are too big or too small to be cloned in a particular vector. There are two approaches to the fragmentation of genomic DNA for the construction of libraries. One is by partial restriction digestion and the other is by physical shearing. Both methods are reviewed below.

Genomic DNA Fragmentation by Partial Restriction Digestion

Ideally, the fragmentation of genomic DNA for library construction is accomplished by a process that breaks DNA randomly. Physical shearing is the only way to generate truly random fragments. The first genomic libraries and the majority of libraries made to date were constructed utilizing fragmentation of the genomic DNA by partial digestion with a restriction endonuclease. Partial digestion with restriction enzymes can be used to break the DNA in an approximately random manner. For enzymes with a 4-base recognition sequence, the restriction site will statistically be present about every 200 bases. For 6-base-recognizing enzymes this statistical frequency is about once every 4000 bases. The primary advantage of using restriction enzymes is that the sticky ends and blunt ends generated by the enzymes can be efficiently ligated to a vector. Frequently a 4-base-recognizing enzyme is selected for genomic DNA fragmentation which will give a sticky end compatible to the sticky end of a 6-base recognition site in the vector that is used as the cloning site.

Conditions for the partial restriction digestion are determined empirically on an analytical scale. Genomic high-molecular-weight DNA is incubated with limiting amounts of the selected restriction enzyme for variable lengths of time. Samples of the digested DNA are removed at different time intervals and analyzed by agarose gel electrophoresis to determine the size range of the digested DNA. The time point of the digestion containing the largest amount of DNA in the desired size range is used as the guide for the preparative digestion reaction.

DNA of the desired size is isolated from the preparative partial digestion reaction by size fractionating the DNA by low-melting-point agarose gel electrophoresis or sucrose gradient centrifugation. The agarose gel technique is the more versatile and can be used for obtaining insert fragments from the hundred base pair to the Mb size range in the appropriate gel system.

Digestion with a restriction endonuclease requires that additional considerations be reviewed for library construction. The presence of compatible stickly ends on both the ends of the vector molecule and the inserts allow for the vector to ligate to itself without an insert and also allows the insert fragments to ligate to each other. This can lead to vector only clones and chimeric clones in the library. Procedures to minimize these outcomes must be employed.

Genomic DNA Fragmentation by Physical Shearing

Physical shearing fragments DNA in a random fashion. Depending on the desired size of DNA inserts, different procedures are applied to accomplish the fragmentation. Sonication can be used to reduce the size of genomic DNA fragments into the hundreds of base pairs range. Nebulization using a disposable medical nebulizer can be used to achieve reproducible fragmentation to obtain products in the 1500 bp to 10 kb size range. The desired size ranges are obtained by using different gas pressure to achieve the nebulization. The minimum pressure that achieves slow nebulization (about 5–6 lb/sq. in. = 35–53 kPA) is used to obtain fragments in the 10 kb range. Higher pressures shear the DNA to smaller fragments. DNA fragments larger than 10 kb can frequently be obtained directly from the DNA purification procedure. Alternatively, a BAL31 exonuclease digestion can be used to reduce the size of genomic DNA obtained after extraction from the cells.

The use of a physical shearing process to obtain DNA fragments for library construction necessitates additional steps before the inserts can be cloned into the library vector. Sheared DNA has ragged ends in contrast to the defined sticky or blunt ends generated by restriction enzyme cutting. Before sheared fragments can be cloned into a vector, the ragged ends must be repaired. In doing this they are either repaired to blunt ends, which can then be blunt-end-cloned into the vector or repaired to blunt ends and modified by the addition of oligonucleotide adaptors to give sticky ends, which can then be sticky-end- cloned in to the vector. Since blunt-end- cloning is inherently inefficient, the preferred method is to modify the fragment ends with adaptors or linkers. Adaptors are small synthetic pieces of DNA that contain one blunt end and one sticky end compatible with a restriction enzyme-generated end. Linkers are small completely double-stranded blunt-ended pieces of DNA containing the recognition sequence for a restriction enzyme. An important feature of the adaptor strategy is that the adaptors must have sticky ends that are not self-complimentary so that the adaptors cannot ligate together through their sticky ends. The most commonly used adaptor for this strategy is the commercially available BstXI adaptor.

The steps in the adaptor strategy process are:

1. after completion of the genomic DNA preparation, a quantity of the DNA is sheared to the desired size. The size range of the sheared DNA is verified by analytical agarose gel electrophoresis.

2. The DNA is size-fractionated to obtain fragments of the desired size typically using preparative low-melting-point agarose gel electrophoresis.
3. After extraction from the gel, the ragged ends of the fragments are repaired to blunt ends using T4 DNA polymerase.
4. The now blunt ends on the fragments are ligated to oligonucleotide adaptors. The adaptors are selected to have sticky ends compatible with those on the library vector.
5. The excess adaptors and any chimeric fragments generated in the ligation reaction are removed by recovering the fragments of the desired size from low-melting-point agarose. At this point the insert fragments are ready for ligation to the library vector.

Future Genomic Libraries

There are more genomic libraries being made now than at any time in the past. These libraries are being made to support genome-wide mapping and sequencing projects. The scale and scope of these projects demand very high-quality libraries as discussed earlier. Most of these requirements result from the high cost of DNA sequencing and from the need to assemble the sequence reads from both ends of a clone into contiguous sequence. When the sequence is assembled in these projects, unclonable sequences remain as gaps in the assembly. These gaps are expensive and time-consuming to fill. At The Institute for Genomic Research, Rockville, MD, and elsewhere the issue of vector design to minimize the incidence of unclonable sequences is being investigated. Some sequences are unclonable because the DNA is unstable in *E. coli* or because the RNA or protein product of a sequence is toxic to *E. coli*. Using *E. coli* host strains that are recombination deficient, which is common practice, minimizes the unstable DNA problem. The deleterious consequences of unstable DNA and toxic products are ameliorated by use of a vector that is maintained at a lower copy number. Plasmid vectors with replication systems that maintain copy number from 500–700 (pUC) down to 1 (BAC), and at many copy number levels in between, can be explored for genomic library applications.

An additional issue of clone viability is transcription of the insert region or transcription originating within the insert. The first will express toxic products coded by the insert, the second may initiate transcription that may interfere with replication as transcription extends around the plasmid vector circle. An approach to dealing with this issue is to design a vector in which the entire cloning region is isolated from RNA transcription. Strong promoters oriented toward the cloning site, such as the *lac* promoter contained in the pUC series of vectors, should not be present. Such promoters can lead to expression of toxic peptides coded by the insert, and might contribute to transcription-stimulated recombination events in the insert region. In addition, it would be desirable to enclose the insert region within strong transcription terminators. The terminators serve a dual purpose. Firstly, they prevent strong promoters that might be present in the cloned insert from transcribing into the vector sequence and possibly interfering with plasmid replication. Secondly, they prevent transcription arising in the surrounding vector sequence from reading into the insert.

As the vectors and associated library construction strategies continue to develop in supporting genome sequencing projects, the quality of the libraries will continue to increase. The level of coverage of the genome will improve as more sequences in the genome are removed from the unclonable category by library vector design and by the use of physical shearing for fragmentation of the genomic DNA. Additionally, library construction strategies will be used that minimize the incidence of chimeric clones in libraries. The development of genomic library technology in these directions will result in better libraries being available for any application.

Further Reading

Birren B, Green ED, Kapholz S et al. (eds) (1999) *A Laboratory Manual: Cloning Systems, vol. 3, Genome Analysis.* Plainview, NY: Cold Spring Harbor Laboratory Press.
Sambuod J and Russell D (2001) *Molecular Cloning: A Laboratory Manual*, 3rd edn. Plainview, NY: Cold Spring Harbor Laboratory Press.

See also: **Genome; Human Genome Project; Phage λ Integration and Excision; Plasmids; Vectors**

Genomics

Copyright © 2001 Academic Press
doi: 10.1006/rwgn.2001.2131

Genomics is the term for the study of the genome, the DNA content of a cell.

See also: **Functional Genomics; Genome; Genome Organization; Genome Size**

Genotype

L Silver

Copyright © 2001 Academic Press
doi: 10.1006/rwgn.2001.0561

For any one organism, its genotype is the set of alleles present at one or more loci under investigation. At any one autosomal locus, a genotype will be either homozygous (with two identical alleles) or heterozygous (with two different alleles).

See also: **Heterozygote and Heterozygosis; Homozygosity**

Genotypic Frequency

A Clark

Copyright © 2001 Academic Press
doi: 10.1006/rwgn.2001.0562

Populations consist of assemblages of individuals each having its own genotype. Considering the entire genome, in all but exceptional cases, like identical twins or clonal organisms, each individual has a unique genotype. If we restrict our attention to one or a few genes at a time, then there will be many individuals having the same genotype. Considering just a single gene, there may be one, two, or more alleles segregating in the population. If there is only one allele, then all genotypes are the same and the genotypic frequency is 1. If there are two alleles, say A and a, then there may be as many as three genotypes, AA, Aa, and aa. The 'genotypic frequency' is defined as the count of a genotype divided by the total count of individuals in the sample.

Numerical Example

If a sample of genotypes from a population consists of 16 AA, 48 Aa, and 36 aa, then the frequency of genotype AA is $16/100 = 0.16$. Similarly the frequencies of Aa and aa are 0.48 and 0.36, respectively. Note that the sum of the frequencies of all genotypes is 1. Notice also that these are estimates of the genotypic frequencies. Out of the entire population, the true frequency of genotype AA may be slightly different from 0.16. We can estimate our statistical confidence in the genotypic frequency estimate by assuming that the sampling was done by randomly drawing individuals from the population. Under this kind of sampling, the variance of a genotype with frequency x is approximately $x(1 - x)/n$, where n is the sample size. It should be clear that the larger our sample is, the smaller will be this variance, and the better will be our estimate of genotypic frequency.

Hidden Variation

Whenever we determine what genotypes individuals have, we nearly always restrict attention to one or a few genes. Calculation of the genotypic frequencies of the genes we observe is done in the same way, whether we score only the one gene or many other genes. There will always be hidden or unobserved variation lying within each of the genotypic classes. Another kind of hidden variation that makes calculation of genotypic frequencies difficult is dominance. The simplest kind of dominance occurs when genotypes AA and Aa both have the same phenotype. In this case we cannot directly count up the genotypes, so only indirect estimation of genotype frequencies is possible. In this case, it would be necessary to make some additional assumptions about the population before it would be possible to estimate genotypic frequencies. In this example, if we were willing to assume that the population is in Hardy–Weinberg equilibrium, then we could estimate the frequencies of AA and Aa from the frequencies of the A and a alleles.

See also: **Allele Frequency; Hardy–Weinberg Law**

Germ Cell

T Schedl

Copyright © 2001 Academic Press
doi: 10.1006/rwgn.2001.0563

Germ cells are a central component of sexual reproduction in animals. They are the route by which the genome and cytoplasmic components are transferred to the next generation. This route utilizes meiosis and gametogenesis, processes that are unique to germ cells. Germ cells differentiate to produce male and female gametes, sperm and unfertilized eggs (oocytes or ova), and undergo meiosis to produce a haploid set of chromosomes. Haploid gametes then unite to form a diploid zygote that develops into a new individual. Germ-cell-mediated sexual reproduction thus creates genetic diversity, which is essential for evolution: meiosis and gamete fusion generates offspring that are genetically dissimilar from each other and distinct from either parent. In many animals, there is a germ-line lineage, composed of germ cells that will form gametes, and a somatic lineage, containing the majority of cells, which form the rest of the organism (tissues

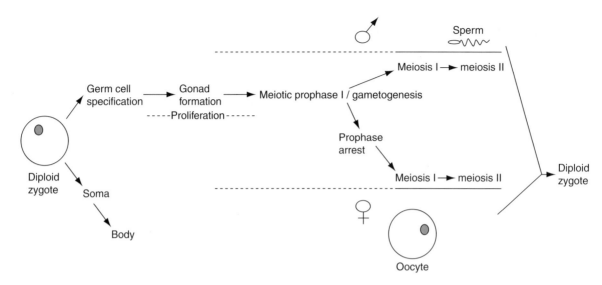

Figure 1 Cycle of the germline. Soon after formation of the diploid zygote, germ cells become specified as distinct from somatic cells that will give rise to the rest of the organism. These primordial germ cells migrate and then interact with specific somatic cells to form the gonad. Germ cells proliferate and then initiate meiotic development (enter meiotic prophase). The timing of proliferation and entry into meiotic prophase depends on the species and the sex. For example, in female mammals all germ cells have entered meiotic prophase prior to birth, while in male mammals, proliferation and entry into meiotic prophase is continuous in sexually mature animals. Reciprocal recombination between homologs occurs during meiotic prophase I. Many of the activities of gametogenesis occur contemporaneously with late stages of meiotic prophase I. For male germ cells, progression through meiotic prophase I and the divisions of meiosis I and meiosis II occur without pause. Female germ cells of most species arrest late in meiotic prophase. Following external signals, the oocyte matures and progresses through meiosis I. A number of species have a second arrest point (e.g., vertebrate oocytes arrest in meiosis II). Fertilization relieves the arrest, resulting in the completion of meiosis and the initiation of a new round of zygotic development.

such as gut, limbs, etc.). One can take the view that the *raison d'être* for an organism's somatic cells is to facilitate the function of the germline so that their genetic material is passed to the next generation. Features of germ cells and their development are described below in general terms and with specific organismal examples, which give them their unique character.

Germline Development

The development of germ cells is similar among animals (**Figure 1**), although the details often differ between species and between sexes of the same species. Germ cells are usually separated from somatic cells in early development. In a number of nonmammalian species, cytoplasmic 'germ plasm' in the unfertilized egg (also called pole plasm, germinal granules, or P-granules, depending on the organism) may specify the germ cell identity. In the fruit fly *Drosophila*, the nematode *Caenorhabditis elegans*, and various amphibians, embryonic cells that contain 'germ plasm' usually develop as germ cells. The 'germ plasm' is prelocalized in the *Drosophila* oocyte or becomes asymmetrically segregated to certain blastomeres during cleavage divisions in *C. elegans*. By contrast,

in the mouse (and likely other mammals), cell–cell interactions are important for germline specification and localized maternal 'germ plasm' appears not to be involved. Once specified, these primordial germ cells migrate and populate the forming gonad, i.e., the female ovary or the male testis. Germ cells in the gonad enter the meiotic pathway and undergo either oogenesis (development of the egg) or spermatogenesis. The mechanism by which the sex of germ cells is determined depends on both germ cell autonomous influences (sex chromosomes, X/A ratio, and/or maternal factors) and signals from the somatic gonad, and varies widely among different species.

Meiosis

The process by which diploid germ cells produce haploid gametes that contain only one of each homologous chromosome is called meiosis. Following the last mitotic division, germ cells initiate meiosis/gametogenesis, undergoing a period of DNA synthesis such that both the maternal and paternal homologous chromosomes are duplicated, resulting in each containing two sister chromatids. Prior to the first meiotic division, the 4n germ cells are in prophase (prophase I) for a prolonged period, which can last more than 40 years for mammalian oocytes. While meiotic prophase I resembles G_2 of the mitotic cell cycle, it is distinct in two important ways: (1) the chromosomes proceed through a series of stages (lepotene, zygotene, pachytene, diplotene, and diakinesis) that are associated with the process of reciprocal recombination (crossover) between homologs; and (2) there are many synthetic activities and morphological changes associated with gamete differentiation. Early in meiotic prophase I, the maternal and paternal homologous chromosomes pair and synapse and initiate recombination (lepotene and zygotene stages). The paired homologs are assembled into an elaborate structure called the synaptonemal complex that is maintained throughout the pachytene stage. The paired homologous chromosomes, each with two closely opposed sister chromatids, is called a bivalent or tetrad. The chromosomes desynapse and the synaptonemal complex dissolves during diplotene. At this time, chiasmata are formed that are attachment points between recombined homologs; chiasmata are considered to be the morphological consequence of prior crossovers between two nonsister chromatids of a bivalent. These processes prepare the chromosomes for the specialized two successive cell divisions of meiosis.

The meiosis I reductional division (MI) separates the two homologous chromosomes. There are at least three features of the MI reductional division that differ from mitosis. First, chiasmata and sister chromatid cohesion distal to the chiasmata serve an essential function, analogous to a mitotic centromere, of holding and aligning the maternal and paternal chromosomes until MI anaphase. Second, the kinetochores (the site of attachment of the spindle microtubules) of the two sister chromatids of a homolog behave as a single unit, insuring that both proceed to the same pole. Third, in the transition from metaphase to anaphase of MI, the previously closely opposed sister chromatids become unglued and the chiasmata dissolve leading to segregation of the homologs to opposite poles. Following MI, the meiosis II equational division (MII) occurs, without an intervening period of DNA synthesis, where the sister chromatids are separated to opposite poles in a similar way to a mitotic division. Meiosis thus generates genetic diversity in two ways: random assortment of chromosomes at the MI and MII divisions and the reshuffling of genetic material through recombination during prophase I. Disruption of any of the steps in meiosis can cause abnormal chromosome segregation (called nondisjunction) producing aneuploid gametes. The resulting progeny may have birth defects as a consequence of nondiploid chromosome number (e.g., Down syndrome, trisomy 21).

Gametogenesis

The sperm and egg are highly specialized for their different tasks. Sperm are small, highly motile and efficient in the process of fertilization. In many species, sperm also provide the centrioles necessary for zygotic development. The egg is very large, ranging from a thousand- to more than a millionfold the mass of a typical somatic cell, depending on the species. The egg supplies organelles (e.g., mitochondria), nutrients, precursors, RNAs, proteins, and a protective covering or shell. The stored RNAs and proteins provide the materials necessary to direct embryogenesis until expression from the zygotic genome is initiated. For many nonmammalian species, the unfertilized egg contains molecular determinants that are either prelocalized or become localized following fertilization, which provide polarity information for the developing embryo. In addition, for nonmammalian species, the stored materials allow embryogenesis to occur externally without further support from the mother.

In the production of gametes, the nuclear events of meiosis and the cellular differentiation of oogenesis and spermatogenesis are intimately intertwined. Much of the RNA and protein synthetic activity necessary for gametogenesis occurs in pachytene and diplotene. For oocytes, the massive growth usually occurs in diplotene. The large accumulation of material in the oocyte is also often assisted by somatic gonad cells (follicle cells) and, for many invertebrates, can also be aided by other germ cells called nurse cells. For many nonmammalian species, yolk is synthesized outside the ovary and is transported to growing oocytes. Spermatogenesis usually occurs continuously, without arrest, in reproductively mature males. The meiotic divisions produce four haploid spermatids, of equal size, which then undergo extensive postmeiotic differentiation to produce mature spermatozoa. Oogenesis has a number of features that are distinct from spermatogenesis. To generate the large size of the egg, the meiotic divisions are unequal; MI generates a large diploid oocyte (often called the secondary oocyte) and a small first polar body and MII produces a large haploid egg and a small second polar body. Oogenesis is often arrested in prophase to allow oocyte growth and to provide a means of regulating egg release. In most vertebrates, the prophase arrest is in diplotene. The release from prophase arrest (called meiotic maturation) is regulated by external cues (e.g., hormonal signals from the menstrual or estrus cycle). In many vertebrate species, there is a second arrest at metaphase of MII. Following ovulation, where the egg is discharged from the ovary, fertilization releases the arrest resulting in the completion of meiosis and the initiation of zygotic development. The point in oocyte/egg development at which fertilization occurs varies from late prophase to after the MII division, depending on the species.

Immortality and Totipotency

The life history of the germline thus marches from fertilization to fertilization, proceeding through the stages of germ cell specification, migration and gonad formation, proliferation, and entry into and progression through meiosis and gametogenesis (**Figure 1**). This cycle of the germline, from generation to generation, is a central feature of the continuum of multicellular life. Because the germline is essentially continuous from generation to generation, the germline lineage can be thought of as being 'immortal,' although individual germ cells are not.

The fertilized egg is totipotent as it will give rise to all the cell types and cell assemblies that constitute the organism. Since germ cells form the zygote, they can be considered as carrying the property of totipotency. In certain cases (e.g., mouse), cells from cell lines derived from primordial germ cells (embryonic germ (EG) cells), as well as cell lines from early embryos (embryonic stem (ES) cells), have been experimentally demonstrated to be totipotent. These cell lines have been very useful for genetic manipulations in the mouse, allowing targeted mutations to be generated and studied in the whole organism.

Further Reading

Alberts B, Bray D, Lewis J, Raff M, Roberts K and Watson JD (1994) Germ cells and fertilization. In: *Molecular Biology of the Cell*. New York: Garland Publishing.

Gilbert SF (1997) Saga of the germ line. In: *Developmental Biology*. Sunderland, MA: Sinauer Associates.

Handel MA (1998) Meiosis and gametogenesis. In: *Current Topics in Developmental Biology*, vol. 37. San Diego, CA: Academic Press.

Roeder GS (1997) Meiotic chromosomes: it takes two to tango. *Genes and Development* 11: 2600–2621.

Wylie C (1999) Germ cells. *Cell* 96: 165–174.

See also: Gametogenesis; Meiosis

Giemsa Banding, Mouse Chromosomes

M T Davisson

doi: 10.1006/rwgn.2001.0564

Identification of individual chromosomes of the laboratory mouse (genus *Mus*) was virtually impossible until the development of methods for staining metaphase chromosomes to reveal their differential banding patterns. A method for banding mouse chromosomes was first developed using quinacrine mustard fluorescence by Lore Zech and Torbjörn Caspersson in 1969–1970. During the early 1970s, several laboratories developed methods using Giemsa stain and various combinations of heat and trypsin treatment, called the ASG (acetic acid–saline–Giemsa) or ASG/trypsin methods. Edward P. Evans was one of the key scientists involved in developing high quality Giemsa banding (G banding) of mouse chromosomes. The Giemsa stain used in these methods is the same as that traditionally used for staining blood smears. In the mid 1990s, fluorescence banding of chromosomes returned with the use of DAPI and related stains to identify mouse chromosomes with fluorescent *in situ* hybridization (FISH) gene mapping methods. G banding, however, remains the best method for high resolution identification of banding patterns in mouse chromosomes and chromosomal aberrations.

The basis of all these banding methods appears to be the frequency of A-T versus C-G base pairs in a stretch of chromosomal DNA. An extensive literature was published during the mid 1970s on 'chromosomal banding.' It should be noted that even G-banded mouse chromosomes can be difficult for the novice to identify and classify. Although banding patterns of individual chromosomes are nonvariant (except for pericentromeric heterochromatin C bands), they may appear different at different stages of chromosomal contraction. In 1984, Cowell produced a good guide to classification with photographs of mouse chromosomes at different stages of contraction (Cowell, 1984).

A standard method for preparing G-banded metaphase chromosomes from living mice is outlined below; details on technique and sources of reagents may be found in Davisson and Akeson (1987). The same method can be used to prepare G-banded chromosomes from any mitotic tissue in the mouse. For example, suspensions of bone marrow cells can be washed out of femurs with a 23 to 25 gauge needle or solid tissues such as the spleen can be minced and pipetted to obtain cell suspensions. To prepare metaphase chromosomes from live mice, approximately 70 µl of blood is drawn by retroorbital or tail vein bleeding and mixed immediately with 0.1 ml sterile sodium heparin (500 USP units ml^{-1}). Blood is cultured in 16 × 125 mm disposable culture tubes. 0.2 ml of whole blood/heparin mixture is inoculated into 0.95 ml of RPMI 1640 culture medium containing glutamine, Hepes buffer, and gentamicin solution (final concentration, 0.1 mg ml^{-1}), and supplemented with 0.15 ml of fetal bovine serum, 0.1 ml of 750 µg ml^{-1} lipopolysaccharide (LPS) and 0.1 ml of 60–90 µg ml^{-1} purified PHA (phytohemagglutinin; concentration determined by a dose–response curve for each batch of PHA). The cultures are incubated at an approximately 45° angle for 43 h at 37 °C in a shaking water bath. Colchicine (0.15 ml of a 50 μg ml^{-1} solution) is added to each culture for the last 15–20 min. Cells are harvested by centrifugation, resuspension in hypotonic 0.56% (0.75 mol) potassium chloride for 15 min, centrifugation, and fixation in methanol:glacial acetic acid (3:1). After 30 min cells are centrifuged and resuspended in three sequential washes of the methanol:glacial acetic fixative.

The method of slide preparation is important because well-spread metaphases are critical for high quality G-banded preparations. Precleaned slides are soaked in fixative at least 15 min prior to use. Air-dried metaphases are prepared by dropping a few small drops of cell suspension onto a precleaned slide, allowing it to spread, and then rapidly blowing dry when the drop begins to contract and rainbow colors appear at the edges. Some cytogeneticists believe spreading is improved by dropping a very small drop of clean fixative onto the preparation just as it starts to dry and allowing the slide to dry in a horizontal position. G bands appear sharper if slides are aged at room temperature for 7–10 days.

To prepare G-band chromosomes slides are incubated in Coplin jars (no more than five to six per jar) in 2 × SSC at 60–65 °C for 1.5 h, transferred to 0.9 % NaCl at room temperature, then each slide is rinsed individually in fresh 0.9 % NaCl and drained. Thorough rinsing is critical. Slides are stained for 5–7 min in a trypsin–Giemsa solution (1.0 ml Gurr improved Giemsa R66, 45 ml Gurr pH 6.8 phosphate buffer, 4 drops 0.0125% trypsin), then transferred to Gurr phosphate buffer diluted 1:1 with distilled water, then slides are rinsed individually in two changes of buffer–distilled water solution and blown dry. Factors that influence chromosomal response to trypsin treatment and, therefore, G-band quality, include chromosome length (contracted chromosomes are more sensitive than elongated ones), chromosome dryness (recently made preparations are more sensitive than aged ones), and chromosome fixation time (sensitivity is inversely proportional to fixation time or chromosome hardness).

Further Reading

Akeson EC and Davisson MT (2000) Analyzing mouse chromosomal rearrangements with G-banded chromosomes. In: Jackson I and Abbott C (eds) *Mouse Genetics and Transgenics: A Practical Approach Series,* 2nd edn, pp. 144–153. Oxford: Oxford University Press.

Committee on Standardized Genetic Nomenclature for Mice (1972) Standard karyotype of the mouse, *Mus musculus. Journal of Hereditary* 63: 69–71.

Lyon MF, Rastan S and Brown SDM (eds) (1996) *Genetic Variants and Strains of the Laboratory Mouse,* 3rd edn. Oxford: Oxford University Press.

References

Cowell JK (1984) A photographic representation of the variability in the G-banded structure of the chromosomes in the mouse karyotype. *Chromosoma* 89: 294–320.

Davisson MT and Akeson EC (1987) An improved method for preparing G-banded chromosomes from mouse peripheral blood. *Cytogenetics and Cell Genetics* 45: 70–74.

See also: Chromosome Banding

Gilbert, Walter

W C Summers

Copyright © 2001 Academic Press
doi: 10.1006/rwgn.2001.0565

Walter Gilbert (1932–), an American molecular biologist, was born 21 March 1932 in Boston, Massachusetts. He was educated at Harvard University and the University of Cambridge, receiving the PhD with a thesis on particle physics in 1957. He did postdoctoral work in physics at Harvard, and in 1959 joined the Physics faculty at Harvard. In the summer of 1960 he joined James Watson and François Gros in Watson's laboratory in research on messenger RNA. This initial exposure to molecular biological research redirected his career from theoretical physics to molecular biology, where he has made his major scientific contributions, and he subsequently transferred to the faculty in Biochemistry and Molecular Biology at Harvard. In 1982 he left Harvard to head the Swiss biotechnology company, Biogen, but returned to Harvard in 1984. Among his many honors, Gilbert received the Nobel Prize in Chemistry in 1980, sharing it with Frederick Sanger and Paul Berg.

His early research focused on the utilization of mRNA and the mechanisms of protein synthesis, especially the relationships between the messenger RNA, the ribosome, and the transfer RNA. In the mid-1960s, Gilbert and Benno Müller-Hill isolated the protein that functions as the repressor of the lactose operon in *Escherichia coli*, the first example of a genetic control element. This work led to his investigation of the physical basis of gene regulation by study of the interaction of the lac repressor with RNA polymerase and fragments of DNA. In 1968 Gilbert and David Dressler proposed the 'rolling-circle model' for DNA replication which gave the first clear indication as to how certain small phages might replicate their DNA. This model was quickly extended to many other systems and subjected to experimental tests. In the mid-1970s, Allan Maxam and Gilbert developed an ingenious method to determine the sequence of nucleotides in DNA by base-specific chemical cleavages of end-labeled DNA fragments followed by size fractionation by gel electrophoresis. This method, often called the chemical method or the 'Maxam–Gilbert' method, was widely used in the early stages of DNA sequence analysis until it became supplanted by the simpler enzymatic methods developed by Fred Sanger.

As an outgrowth of nucleic acid sequencing, Gilbert was an early proponent of genomics, the use of sequence databases to study genome structures, sequences, organization, and evolution. He has written extensively on the evolutionary origins and significance of the intron/exon structure of eukaryotic genes as well as the possible relationship of splicing, exon shuffling, and gene rearrangements to modular protein evolution.

See also: Genome Organization; Repressor; Rolling Circle Replication; Sanger, Frederick

Glioma

V P Collins

Copyright © 2001 Academic Press
doi: 10.1006/rwgn.2001.1575

Gliomas are neoplasms composed of tumor cells that on histopathological examination show varying degrees of phenotypical similarity to adult or developing macroglia. The macroglia form the main subgroup of the neuroglia and include astrocytes, oligodendrocytes, and ependymal cells. More than 20 types of glioma are recognized and the histological criteria for their diagnosis defined in the World Health Organization (WHO) classification of tumors of the central nervous system. The tumors may in addition be malignancy graded in grades I–IV on the basis of histological attributes defined by WHO. The malignancy grade is an estimation of the degree of malignancy usually encountered in each type of tumor, where grade I is the least and grade IV the most malignant. Response to contemporary therapy is individual to each tumor type and malignancy grade. The cells of origin for these phenotypically diverse tumors are unknown. The various tumor types have different genetic abnormalities. The commonest form of glioma in adults is the highly malignant glioblastoma, the tumor cells of which show phenotypical similarities to astrocytes. In children, the commonest glioma is the relatively benign pilocytic astrocytoma. Gliomas are more common in males than in females.

See also: Genetic Diseases

Globin Genes, Human

DJ Weatherall

Copyright © 2001 Academic Press
doi: 10.1006/rwgn.2001.0567

The globin genes determine the structure and synthesis of the globin chains that constitute the different hemoglobins that are produced in the human embryo, fetus, and adult. Human beings make different hemoglobins as they develop as an adaptive response to the variation in oxygen requirements between embryonic, fetal, and adult life.

All the normal human hemoglobins have the same basic structure. They consist of two different pairs of globin chains, that is, long strings of amino acid which fold into a complex three-dimensional structure. Each of the four globin subunits that makes up a hemoglobin molecule has a heme group, the oxygen-carrying moiety, embedded in its surface. The different globin chains are named after letters of the Greek alphabet. Adult and fetal hemoglobins have α chains associated with β (hemoglobin A, $\alpha_2\beta_2$), δ (hemoglobin A_2, $\alpha_2\delta_2$) or γ chains (hemoglobin F, $\alpha_2\gamma_2$), whereas in the embryo, embryonic α-like chains called ζ chains combine with γ (hemoglobin Portland, $\zeta_2\gamma_2$) or ε chains (hemoglobin Gower 1, $\zeta_2\varepsilon_2$), and α and ε chains combine to form hemoglobin Gower 2 ($\alpha_2\varepsilon_2$). The embryonic hemoglobins are so-called because they were first characterized at University College Hospital in Gower Street, London, and in Portland, Oregon.

Since each globin peptide chain is the product of a gene locus, it follows that there must be α, β, γ, δ, ε, and ζ globin genes.

Hemoglobin Genes Organized in Clusters

The globin genes are organized into two clusters which are situated on different chromosomes (**Figure 1**). The α-like genes, which are encoded on chromosome 16, are found in the order $5'$-ζ-$\varphi\zeta$-$\varphi\alpha2$-$\varphi\alpha1$-$\alpha2$-$\alpha1$-$\theta1$-$3'$. The β-like globin genes, on chromosome 11, occur in the order $5'$-ε-$^G\gamma$-$^A\gamma$-$\varphi\beta$-δ-β-$3'$. The $5'$ to $3'$ nomenclature indicates the order of the genes, from left to right.

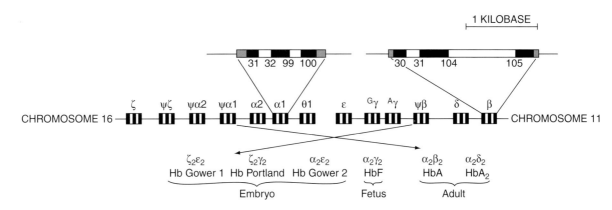

Figure 1 The human globin gene clusters on chromosomes 11 and 16.

Both clusters contain two genes which are duplicated; there are two α chain genes, α2 and α1, and two γ chain genes, $^Gγ$ and $^Aγ$. The G and A refer to the amino acids glycine and alanine; the products of the two γ genes are identical except at amino acid residue 136, at which one contains glycine and the other alanine. The product of the pairs of α genes are identical. The other feature of these clusters is the presence of pseudogenes which are given the prefix φ. They are thought to be evolutionary remnants of once-active globin genes.

Structure of Globin Genes and their Clusters

The structure of the globin genes has been highly conserved throughout evolution. Their transcribed regions, that is the parts of the gene which form the template for messenger RNA production, contain three coding regions, or exons, separated by two introns, or intervening sequences (IVS), of variable length. From the CAP site, the start of transcription, the first exon encompasses approximately 50 bp of untranslated sequence (UTR) and the codons for amino acids 1–31 in the α and 1–30 in the β globin genes. Exon 2 encodes amino acids 32–99 and 31–104 respectively, those portions of the globin chains that are involved in heme binding and in contacts between the α and β chains that are critical for the normal function of hemoglobin as an oxygen carrier. The third exon encodes the remaining amino acids, 101–141 for the α, and 105–146 for the β chains, together with a 3′ untranslated region of about 100 bp. The sizes of the introns vary between different genes. In the α globin genes they are both small, 117–149 bp, while in the ζ gene IVS-1 is ~886 and IVS-II is ~239 bp. IVS-1 in the β genes is also small, 122–130 bp, while IVS-II is much larger, 850–904 bp.

As well as the exons there are other sequences in the globin genes which are highly conserved. Removal of the intervening sequences from the initial messenger RNA transcript, and joining the exon sequences to form the definitive messenger RNA, is dependent on the specific sequences of the borders between exons and introns. At the 5′ end of each intron there is always the dinucleotide GT, and at the 3′ end AG. Adjacent nucleotides are also conserved to form a consensus sequence. Mutations that involve these regions in certain inherited disorders of hemoglobin interfere with the normal processing of messenger RNA to such a degree that no gene product is produced. Processing also involves the addition of a track of adenylic acid (A) residues at the 3′ end of the messenger RNA. The signal in each globin gene for this process is AATAAA, which is conserved in the 3′ untranslated region, approximately 10–30 nucleotides upstream of where the initial transcript is to be cut and polyadenylated.

How Globin Genes Are Regulated

A complete account of the regulation of the globin genes would explain why they are only active in appropriate tissues, that is, in the red cell precursors in the bone marrow, how their expression is controlled such that they synthesize relatively large amounts of globin in a way which ensures that the output of the α and β chains is almost synchronous, and how the different globin genes are activated and repressed at different stages of development. Currently, it is impossible to answer these questions fully, although some progress has been made.

Transcription of genes is dependent on the attachment of a transcription complex including the enzyme RNA polymerase, at their 5′ ends. Appropriate positioning of the transcription machinery is brought about by recognition of specific DNA sequences in the region upstream of the transcriptional start site, known as the promoter. Like many genes the globin genes have boxes of DNA homology, TATA and CCAAT, found 30 and 70 bp upstream of the CAP site. In addition to these regions, many erythroid-specific genes, including the globin genes, have a CACCC homology box in the promoter, upstream of the CCAAT box. This particular region is found in most of the β-globin-like promoters and is duplicated in the β globin gene but not in the α globin gene promoters. This sequence is also missing from the promoter of the δ gene.

In addition to the promoter sequences, more distal sequences are found in the β globin gene clusters which increase the levels of gene transcription. Five regions with this property, called enhancers, have been identified in the α and β globin gene complexes. In addition, both complexes have major regulatory elements which, if deleted, completely inactivate all the genes in the complex. The β globin locus control region (LCR) lies upstream from the ε globin gene and is marked by five DNase hypersensitive sites. Similarly, there is a region 40 kb upstream from the α globin genes which is also marked by a site of this kind, and hence which is called HS-40. Again, if this is lost by deletion the entire α globin gene cluster is inactivated. These regulatory regions, and a variety of other regions throughout the globin gene clusters, are marked by DNA binding motifs for a variety of transcription factors, some of which, including GATA-1 and NF-E2, are erythroid-specific, while others are for ubiquitous factors, transcription factors which are active in many different tissues.

Currently it is believed that the β LCR together with other enhancers, a variety of transcription factors, and other regulatory proteins becomes opposed sequentially to the different genes of the β globin gene cluster, resulting in their activation.

The mechanisms for turning on and off the ε and γ globin genes, and for activating the β and δ globin genes at different stages of fetal development are not understood. It seems likely that there may be developmental-stage-specific transcription factors although these have not been identified in the case of the human hemoglobin genes.

How Human Hemoglobin Genes Evolved

Globin genes arose early in evolution and are found in fungi, plants, and invertebrates, as well as in all vertebrate species. It seems likely that gene duplication, followed by selection of adaptive sequence changes, resulted in the production of diverse globin chains with specialized functions. This process presumably allowed what were originally monomeric forms of hemoglobin to evolve into the tetrameric proteins that are now found in all higher animals. Different α and β globin chains are found in all vertebrates, suggesting that they originated before ∼4–5 million years ago. In fish and amphibians the genes for the two

types of chains are linked together in a single cluster. In other species chromosomal rearrangements must have resulted in the separation of the α and β gene clusters, certainly by the time that birds evolved.

In the α globin gene cluster, duplication leading to a specialized embryonic (ζ) globin chain occurred ∼400 million years ago, while the α gene underwent a further duplication in many species. Duplication of the primitive β chain gene occurred independently in birds and mammals ∼180–200 million years ago to give rise to the embryonic ε gene. Before the divergence of the mammals (∼85 million years ago) further duplication events of both genes gave rise to the ε and γ proto-gene in one case and the adult proto-δ and proto-β genes in the other. Other duplications must have given rise to the various pseudogenes that are seen in the α and β gene clusters. Interestingly, in most mammals the proto-γ gene has remained as an embryonically expressed gene and was only recruited to the fetal stage of development after the emergence of primates (55–60 million years ago). Its duplication occurred about 35–55 million years ago and has been maintained in the lineages leading to the apes.

Normal Variation of Structure of Globin Genes

The globin gene clusters show a considerable amount of variability in their base composition. This can easily be identified when a single nucleotide change produces or removes a cutting site for a restriction enzyme; these harmless changes are called restriction fragment length polymorphisms (RFLPs). These do not occur at random but form a series of patterns, or haplotypes, which occur at varying frequencies among different populations of the world. In the β globin genes there are two separate haplotype regions separated by an area where there is frequent recombination. In this gene cluster there are only single nucleotide RFLPs. However, although the α globin gene cluster contains no 'hotspots' for recombination it is even more highly polymorphic, containing a number of single nucleotide RFLPs and several highly variable regions of DNA, that is repeat sequences which vary considerably in length and hence provide valuable genetic markers. The RFLP haplotypes of the globin gene clusters are of considerable value for population genetics and for evolutionary studies. They are also useful markers for studying the distribution and evolution of different mutations of the β globin genes.

Mutations of Globin Genes

The mutations of the globin gene clusters result in the commonest genetic diseases in man. They cause either structural hemoglobin variants, or thalassemias, disorders that are due to a reduced rate of production of either the α or β chains of hemoglobin. The particularly common disorders of the globin genes, sickle cell anemia and the different thalassemias, have reached their high frequency in the world population because of heterozygote advantage against malaria.

Further Reading

Bunn HF and Forget BG (1986) *Hemoglobin: Molecular, Genetic and Clinical Aspects.* Philadelphia, PA: WB Saunders.

Fraser P, Gribnau J and Trimborn T (1998) Mechanisms of developmental regulation in globin loci. *Current Opinion in Hematology* 5: 139–144.

Grosveld F, Dillon N and Higgs D (1993) The regulation of human globin gene expression. *Clinical Haematology* 6: 31–55.

Stamatoyannopoulos G, Perlmutter RM, Marjerus PW and Varmus H (eds) (2000) *Molecular Basis of Blood Diseases*, 3rd edn. Philadelphia, PA: WB Saunders.

Weatherall DJ, Clegg JB, Higgs DR and Wood WG (2001) The hemoglobinopathies. In: Scriver CR, Beaudet AL, Sly WS *et al.* (eds) *The Metabolic and Molecular Bases of Inherited Disease*, 8th edn, pp. 4571–4636. New York: McGraw-Hill.

See also: Sickle Cell Anemia; Thalassemias

Glucose 6-Phosphate Dehydrogenase (G6PD) Deficiency

L Luzzatto

Copyright © 2001 Academic Press
doi: 10.1006/rwgn.2001.1520

The glucose 6-phosphate dehydrogenase (G6PD) gene is a prototype housekeeping gene, as it is ubiquitously expressed in most organisms and cell types, and its product performs a general, important function in cell metabolism. Specifically, G6PD is an enzyme that catalyzes the oxidization of glucose 6-phosphate (G6P) to 6-phosphoglucono lactone (6PG), coupled with the reduction of the coenzyme NADP to NADPH. Because 6PG can then be decarboxylated to a pentose sugar, the G6PD reaction is often referred to as the first reaction in the pentose phosphate pathway; at the same time, NADPH is essential as an electron donor in numerous biosynthetic pathways and in the defense of cells against oxidative stress. There is evidence from evolutionary data and from genetic inactivation of the G6PD gene in microorganisms and in mammalian cells that G6PD is indeed indispensable for these functions, but not for pentose synthesis.

Formal and Molecular Genetics

The G6PD gene is highly conserved in evolution. The alignment of all available sequences from a wide range of organisms highlights regions with the highest degree of conservation, for instance, the active center and the NADP-binding domain. In mammals the G6PD gene is X-linked, and in humans it maps to the tip of the long arm of the X chromosome (cytogenetic band Xq28). The human gene spans some 13 kb, and it consists of 13 exons, encoding a polypeptide chain of 515 amino acids; the active enzyme is a dimer of this polypeptide chain. Each subunit is folded into a globular structure including 9 α-helices and 9 β-sheets; there is no covalent bond between the two subunits, and the subunit interface in the dimer consists of β-sheets and α-helices, which form a kind of barrel. Like in many housekeeping genes, the promoter region is highly GC-rich, with several Sp1 and Ap2 binding sites, the functional role of which has been characterized by deletion analysis and mutagenesis. Within this region, a 630-bp promoter has been shown to retain housekeeping gene expression in transgenic mice.

Since the G6PD gene is X-linked, women heterozygous for G6PD deficiency are genetic mosaics in their somatic cells after X chromosome inactivation. For instance, about half of their red cells will be G6PD normal and the other half will be G6PD deficient. However, in some cases, owing to drift or to somatic cell selection, there may be an excess of one or the other cell types, giving a completely normal or a completely deficient phenotype. Thus, the extent of clinical consequences of G6PD deficiency (see below) will be a function of the proportion of G6PD deficient cells. For this reason G6PD deficiency should formally be regarded not as recessive but as codominant.

Evolutionary Genetics

G6PD is very ancient in an evolutionary context: it is found in all organisms except in some of the Archaea that live in anaerobic environments and some intracellular microorganisms that seem to be able to exploit the G6PD activity of their respective host cells. The G6PD sequence shows evidence of conservation throughout all living phyla, with some regions being identical in disparate organisms, for example, the active center (which includes the G6P binding site) and the NADP binding site.

G6PD Deficiency

Investigations of patients who developed acute hemolytic anemia upon exposure to certain antimalarial drugs revealed, in 1956, that their red blood cells had a markedly reduced G6PD activity, and that this trait was inherited. Thus, G6PD deficiency emerged as the first example of a blood cell disease caused by a specific enzyme abnormality. It quickly became apparent that this inherited abnormality predisposes to hemolysis in response to several other factors. The wide range of factors that can trigger hemolysis in G6PD-deficient subjects is related to the fact that all of them impose an oxidative stress on red cells. The response to this type of stress involves, in particular, glutathione (GSH). Since G6PD activity is rate-limiting for regeneration of GSH, normal red cells can withstand such stress, but G6PD-deficient red cells succumb. G6PD deficiency is due to mutations in the G6PD gene. There are some 130 mutations known to date: all of them are in the coding region, and almost all of them are point mutations causing single amino acid replacements. In most cases of G6PD deficiency the activity of the enzyme in red cells is reduced to about 10–20% of normal activity; in some cases it may be as low as 1–2%. However, there is always some residual activity. In a few instances these amino acid replacements may affect the catalytic function of the enzyme, but in the majority of cases they cause G6PD deficiency because they cause the protein to become unstable. The absence of large deletions, frameshifts, or nonsense mutations supports the notion that complete G6PD deficiency would be lethal. This notion has been confirmed recently by targeted homologous recombination in mouse embryonic stem (ES) cells: when 'G6PD knock-out' ES cells are injected into blastocysts heterozygous female mice can be obtained, but hemizygous male mutants die *in utero* at about 10 days of gestation.

Population Genetics

In many human genes pathogenic mutations are often regarded as being in a different category from 'polymorphisms.' In the case of G6PD it is quite remarkable that many mutations, which are potentially pathogenic because they cause G6PD deficiency, are also polymorphic. Indeed, these mutant genes have frequencies of up to 10–20% and even greater in many human populations. Since the G6PD gene is X-linked, in any population in which G6PD deficiency is common, the frequency of G6PD-deficient hemizygous males will be higher than that of G6PD-deficient homozygous females but lower than that of females heterozygous for G6PD deficiency. Interestingly, different

allelic mutants account for the overall prevalence of G6PD deficiency in different parts of the world, and in many populations several polymorphic alleles coexist (see **Figure 1**). All of these populations are in malaria-endemic areas, or in areas that have been malaria-endemic until recently, suggesting that each one of these alleles represents an example of balanced polymorphism. In fact, there is evidence from clinical studies that subjects with G6PD deficiency have a relative resistance to *Plasmodium falciparum* malaria, decreasing significantly the risk of death from this condition. *In vitro* studies have shown that G6PD-deficient red cells parasitized by *P. falciparum* are phagocytosed by autologous macrophages more effectively than G6PD normal red cells. The fact that so many independently arisen G6PD deficiency mutations have become prevalent wherever malaria has existed for a long time virtually eliminates the possibility that G6PD deficiency has become common merely by genetic drift. Indeed, the multitude of these G6PD-deficient alleles is in itself a strong argument for the notion of balanced polymorphism in the sense of convergent evolution.

Clinical Genetics

As stated above, it was the clinical manifestation of acute hemolytic anemia (AHA) that led to the discovery of G6PD deficiency; AHA can be triggered by a variety of drugs, including antimalarials, aspirin, some sulfate drugs, and some antibiotics such as nalidixic acid. G6PD-deficient subjects can also develop AHA in concomitance with a variety of infections, or after ingestion of fava beans (see Favism, a well-characterized syndrome which in children is life-threatening). The most important approach to these clinical problems is prevention, by helping people at risk to avoid the offending agents. In cases of severe AHA blood transfusion may be imperative. In addition, G6PD deficiency can cause a predisposition to severe neonatal jaundice, which can result in long-term neurological damage. Phototherapy is sufficient in preventing such damage in most cases, but exchange transfusion may be required in severe cases.

A small proportion of patients with G6PD deficiency present with a more severe disease, namely chronic nonspherocytic hemolytic anemia (CNSHA), even in the absence of any triggering agent. These patients have anemia and jaundice, and may require regular blood transfusion, which can bring about iron overload and the need for iron chelation. The association of G6PD deficiency with CNSHA and with AHA is an excellent example of genotype–phenotype correlation. Indeed, not surprisingly, the mutations

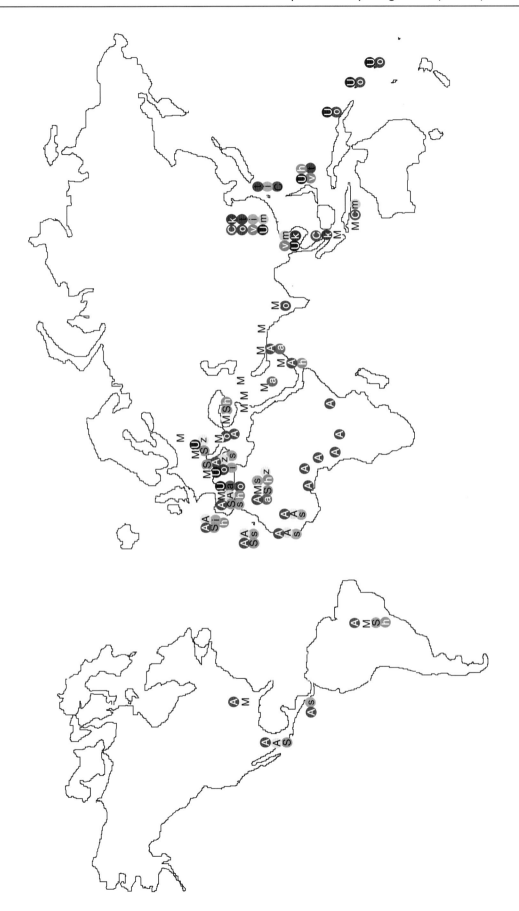

Figure 1 Worldwide distribution of polymorphic variants of G6PD. The variants in each country are shown in order of prevalence according to these symbols: U = Union; C = Canton; M = Mediterranean; A = A – (202A); k = Kaiping; t = Taipei; v = Viangchan; m = Mahidol; h = Chatham; l = Coimbra; p = Local variant; S = Seattle; s = Santamaria; a = Aures; z = Cosenza; A = A – (968C). See Color Plate 9.

that cause AHA are different from those that cause CNSHA: the latter mutations are invariably those that cause amino acid replacements that compromise the stability of the enzyme most drastically. A large proportion of the mutations map to the region of the molecule involved in the dimer interface, because they make the dimer structure unstable.

***See also:* Balanced Polymorphism; Embryonic Stem Cells; Favism**

Glutamic Acid

doi: 10.1006/rwgn.2001.2081

Glutamic acid (Glu or E) is one of the 20 amino acids commonly found in proteins. It has a negatively charged side chain and exists as glutamate. Its chemical structure is shown in **Figure 1**.

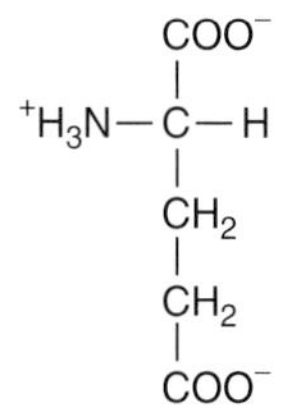
COO⁻
⁺H3N—C—H
CH2
CH2
COO⁻

Figure 1 Glutamate.

***See also:* Amino Acids; Proteins and Protein Structure**

Glutamine

J Read and S Brenner

doi: 10.1006/rwgn.2001.2080

Glutamine (Gln or Q) is one of the 20 amino acids commonly found in proteins. Its side-chain contains a polar amide group, which can interact strongly with water by forming hydrogen bonds. Its chemical structure is shown in **Figure 1**.

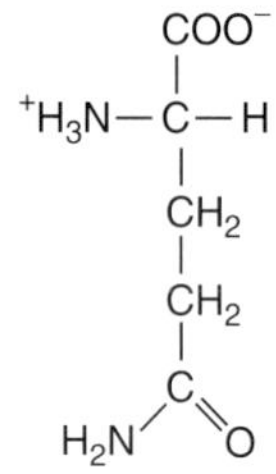
COO⁻
⁺H3N—C—H
CH2
CH2
C
H2N
O

Figure 1 Glutamine.

***See also:* Amino Acids; Proteins and Protein Structure**

Glycine

J Read and S Brenner

doi: 10.1006/rwgn.2001.2074

Glycine (Gly or G) is the smallest of the 20 amino acids commonly found in proteins and has no special hydrophobic or hydrophilic character. Its chemical structure is shown in **Figure 1**.

Glycine

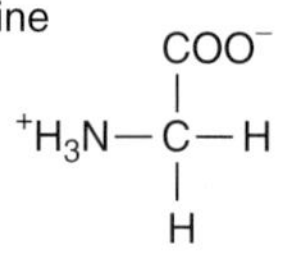
COO⁻
⁺H3N—C—H
H

Figure 1 Glycine.

***See also:* Amino Acids; Proteins and Protein Structure**

Glycine max (Soybean)

P Gresshoff

doi: 10.1006/rwgn.2001.1664

Soybean is the common name for *Glycine max* (Merrill), an amphidiploid grain legume ($2n = 2x = 40$), part of the genus *Glycine* Willdenow, family Leguminosae, subfamily Papilionoideae, tribe Phaseolae. *Glycine* genus has its origins in Asia and Australia, first named by Linnaeus (*Genera Plantarum*, 1737) based on the Greek *glykys* = "sweet" (from the sweet tubers of *Glycine apios* L. which now correctly is classified as *Apios americana*). *Glycine max* is congenic with the wild soybean *Glycine soja*, with which fertile hybrids can be obtained. Soybean is self-fertile but outcrossing at about 1–3% is possible. Biparental inheritance of some mitochondrial DNA markers suggests the possibility of mixed cytoplasms. Soybean is a major crop, being used for animal feed, vegetable oil, lubricants, industrial paints, ink, mayonnaise, soaps, and pharmaceuticals such as isoflavone phytoestrogens (genistein and daidzein) and anticancer treatment (naranginin which stimulates cytochrome P-450 mono-oxygenase). Flowering is controlled by maturity and daylength; genetic variation produced different maturity groups ranging from 000 (high latitudes) to X (= 10) in tropical regions. Average yield is about 1.5 tonnes per hectare. World production (1999) was 156 million tonnes selling as a commodity on the Chicago Board of Trade at a cyclically low price of about US$ 190 per tonne. Soybean seeds contain about 20% oil and 40% (range 35–45%) protein. Average seed size is 15 g per 100 seeds. The average soybean plant grows to 1 m in height, and develops determinate (non-meristematic, spherical) nitrogen-fixing nodules in symbiosis with bacterial cells of *Bradyrhizobium japonicum* and *Sinorhizobium fredii*. At present about 100 genes from the bacterial microsymbiont being involved in nitrogen fixation or nodule initiation have been cloned and characterized. Despite this component of genetic information in the prokaryotic partner, most of the key regulatory functions of the soybean nodule symbiosis are encoded in the plant genome. The haploid genome size of soybean is about 1050–1100 Mb, consisting of about 35% highly repeated DNA, 30% moderately repeated DNA, and 35% unique or near-single-copy DNA. The karyotype reveals two large, 14 intermediate, and four small chromosomes, with extensive centromeric heterochromatin, allowing pachytene discrimination. Trisomics for each chromosome are available. Telomere-associated sequences have been sequenced and contain the canonical TTTAGGG sequence. One nucleolus is visible matching molecular data for one rRNA locus. Two major satellite DNA types of 92 bp and 132 bp have been cloned. The 92 bp satellite is clustered in four regions with about 70 000–100 000 copies per haploid genome. The 132 bp satellite is dispersed. The genome of soybean has been found to contain several transposable elements, although phenotypic evidence for their action is scarce. Numerous retrotransposons have been discovered. Isoenzymes and biochemical mutants (e.g., nitrate reductase, lipoxygenase) are available as markers and tools of molecular physiology.

Several genetic maps are available comprising phenotypic markers such as seed coat, hilum, flower and pubescence color, root fluorescence, viral, cyst nematode and fungal resistance, male sterility, pubescence density, dwarfism, leaf shape, and nodulation. Recessive EMS and fast neutron mutations leading to non-nodulation and supernodulation demonstrate that the plant genome controls major components of the nodulation and nitrogen fixation process. The classical genetic maps have been improved through the integration of molecular markers such as random RFLP clones, EST clones, AFLP, RAPD, and DAF polymorphisms, and microsatellites (simple sequence repeats, SSRs) allowing marker-assisted breeding as well as map-based cloning. The total genome size is about 3300 cM. Physical mapping in one region (pA36 marker on linkage group H) suggests that 1 cM represents about 400 kb. BAC libraries arrayed on nylon filters are available as are expressed sequence tagged (EST) libraries from different tissues and developmental stages. EST collections have been arrayed on microarrays for molecular expression studies.

Soybean was first transformed by *Agrobacterium tumefaciens* and by biolistic particle bombardment in 1988, leading to the development of one of the first GMO products in agriculture, the Round-Up Ready soybean. This transgenic plant is resistant to lethal doses of the herbicide Round-Up (phosphono-methyl-glycine) and has led to considerable public debate and antagosism towards its inventors, the Monsanto Company. Other transgenic products with altered insect resistance and oil composition are being developed.

Further Reading

http://www.unitedsoybean.org/soystats
http://www.ag.uiuc.edu/~stratsoy/new/

See also: Nodulation Genes; Symbionts, Genetics of; Transfer of Genetic Information from *Agrobacterium tumefaciens* to Plants

Glycolysis

F K Zimmermann

Copyright © 2001 Academic Press
doi: 10.1006/rwgn.2001.0570

Glycolysis, a centrally important metabolic pathway in almost all organisms, degrades hexoses to pyruvate with the concurrent production of adenosine triphosphate (ATP) and reduced nicotinamide adenine dinucleotide (NADH) (**Figure 1**).

Glycolysis in *Saccharomyces cerevisiae*

The genetics of the glycolytic enzymes has been fully explored in the yeast *Saccharomyces cerevisiae*, where many genes coding for regulatory factors have been identified. There is a large set of genes coding for hexose uptake facilitators with different regulation and kinetic parameters. Two hexokinases (genes *HXK1* and *HXK2*), with a 76% amino acid identity, are not only catalysts but also sensors for internal glucose and fructose and thus trigger carbon catabolite repression. Their activity is modulated by an essential feedback inhibition by trehalose-6-phosphate, as shown by the drastic effects of mutants deficient in trehalose synthesis. A specific glucokinase (gene *GLK1*) accounts for about 20% of the total glucose phosphorylating activity. It is not involved in carbon catabolite repression or sensitive to trehalose-6-phosphate. Phosphoglucose isomerase (gene *PGI1*) is required for growth not only on glucose but also on fructose, because the formation of the essential regulator trehalose-6-phosphate starts from glucose-6-phosphate.

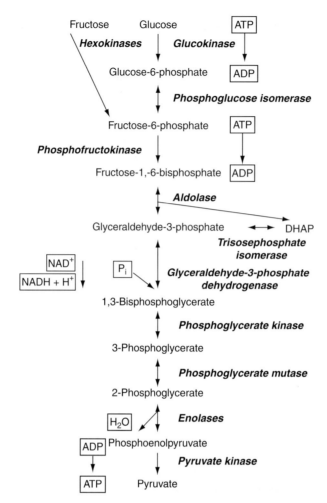

Figure I Glycolysis: metabolites, enzymes, and products. Glycolysis can be represented by:

$$Glucose + 2ADP + 2P_i + 2NAD^+ = 2Pyruvate + 2ATP + 2NADH + 2H^+$$

ATP, adenosine triphosphate; ADP, adenosine diphosphate; NADH, reduced nicotinamide adenine dinucleotide; P_i, inorganic phosphate; DHAP dihydroxyacetone phosphate.

Heterooctameric phosphofructokinase consists of two different subunits, with about 50% amino acid identity (genes *PFK1* and *PFK2*). Its activity is subject to numerous effectors, most prominently by activating fructose-2,6-bisphosphate generated by two 6-phosphofructo-2-kinases (genes *PFK26* and *PFK27*). A block in glycolysis requires deletion of *PFK1* and *PFK2*. A single deletion of *PFK2*, but not of *PFK1*, slightly reduces growth on glucose. Double mutants without *PFK26* and *PFK27* cannot form fructose-2,6-bisphosphate but grow normally on hexoses. However, all these mutants have altered levels of glycolytic metabolites.

Yeast aldolase (gene *FBA1*) belongs to the prokaryotic type II aldolases. Mutants deleted for *FBA1* are inhibited by glucose and grow very poorly on a mixture of acetate and low amounts of galactose. The deduced amino acid sequence of triosephosphate isomerase (gene *TPI1*) shows about 50% identity to vertebrate forms. There are three genes coding for glyceraldehyde-3-phosphate dehydrogenases, *TDH1*, *TDH2*, and *TDH3*, with over 90% amino acid sequence identities and amounting to 10–15%, 25–30%, and 50–60% of the total activity, respectively. Mutant strains with all three genes deleted cannot be obtained, suggesting that this type of protein is essential for growth. Gene *PGK1*, coding for phosphoglycerate kinase, with 65% amino acid sequence identity to the human enzyme, has been used to construct heterologous expression cassettes in yeast, and the regulatory components of the promoter have been studied in great detail. Phosphoglycerate mutase, gene *PGM1*, shares about 50% identical amino acids with the human erythrocyte bisphosphoglycerate mutase.

Two enolases, which differ only in 20 out of 436 amino acids, are encoded by constitutively expressed

Table I Human enzyme deficiencies and genetic disease

Glycolytic enzyme	Mutation-associated demonstrated or possible defects
Hexokinase I	Nonspherocytic hemolytic anemia
Hexokinase II	Nonspherocytic hemolytic anemia; insulin resistance; possible cause of increased glycolysis in cancer cells
Glucokinase	Gestational diabetes; hyperinsulinism of the newborn; maturity-onset diabetes of the young
Phosphoglucose isomerase	Nonspherocytic hemolytic anemia
Phosphofructokinase	Exercise intolerance and compensated hemolysis (Tarui disease)
Aldolase B	Hereditary fructose intolerance
Triosephosphate isomerase	Multisystem disease, lethality in early childhood
Glyceraldehyde-3-phosphate dehydrogenase	Diverse nonglycolytic functions, could be involved in, e.g., prostate cancer, age-related neurodegenerative disease
Phosphoglycerokinase	Chronic hemolytic anemia
Phosphoglycerate mutase	Exercise intolerance
Enolase I	Deregulation of c-*myc* oncogene
Pyruvate kinase	α-Hereditary hemolytic anemia

ENO1 and *ENO2*, which is strongly induced when glucose-6-phosphate levels increase. There are also two pyruvate kinase genes in yeast that share about 40% of the amino acids with the mammalian isoenzymes. *PYK1* codes for the major enzyme that is induced by increased levels of both glucose-6-phosphate and fructose-6-phosphate. This enzyme requires fructose-1,6-bisphosphate for activation. Lack of this enzyme blocks growth on glucose. *PYK2* codes for a pyruvate kinase with about 70% amino acid identity to the *PYK1*-encoded protein. However, it is fully active without fructose-1, 6-bisphosphate, and transcription is repressed by glucose. Pyruvate kinase converts phosphoenolpyruvate to pyruvate under glycolytic conditions, whereas, under conditions of gluconeogenesis, phosphoenolpyruvate is formed from oxaloacetate by phosphenolpyruvate carboxykinase. A simultaneous activity of both enzymes would create a futile ATP-wasting cycle under gluconeogenic conditions. Strains producing the fructose-1,6-bisphosphate-independent enzyme at the level of the glycolytic pyruvate kinase grew at the normal rate under gluconeogenic conditions, suggesting the existence of an additional control mechanism preventing such metabolic waste (Boles *et al.*, 1997). The rate of glycolysis as determined by the rate of ethanol production could not be increased by the overproduction of individual enzymes or several combinations of glycolytic enzymes.

Glycolysis in Humans

The genetics of glycolysis in humans is complicated (1) by the presence of tissue and cell type-specific isoenzymes and (2) because several glycolytic enzymes and their genes have additional functions beyond a strictly catalytic role. The expression of the glycolytic enzymes is stimulated by glucose in several cell types via glucose-6-phosphate and a hypoxia-inducible helix–loop–helix transcription factor. Numerous genetic diseases are caused by enzyme deficiencies in the glycolytic pathway (**Table I**). Deficiency in hexokinase type I causes hemolytic anemia. Hexokinase II is a leading enzyme and glucose 'sensor' in insulin-sensitive tissues, and a defect causes type 2 diabetes. Many tumor cells have increased rates of glucose catabolism, which can promote cell proliferation. Certain tumor-associated p53 mutant proteins cause a significant activation of the type II hexokinase promoter. Glucokinase is the glucose sensor, and low-activity and low-stability mutants can explain in part the maturity-onset diabetes of the young (MODY), because glucose metabolism of the β-cells controls insulin secretion, and amino acid substitutions have been associated with this syndrome. Different amino acid substitutions of the muscle phosphofructokinase cause an exertional myopathy and hemolytic syndrome (Tarui disease). A stop codon in position 145 of the triosephosphate isomerase locus has been associated with neurological disorders. Glyceraldehyde-3-phosphate dehydrogenase has a subunit that participates in RNA export and DNA replication and repair. Mutant forms of this enzyme could be involved in several disease syndromes. Phosphoglycerate kinase deficiency has been found in patients with myoglobinuria. The gene coding for the α-enolase isoenzyme is transcribed into a single mRNA species which, when translated from the first initiation codon, yields enolase. Another AUG codon 400 bp downstream starts the translation of a protein, MBP-1, binding and thus downregulating the promoter of the c-*myc* gene which, when overexpressed, causes cancer. Thus the human *eno1* gene could be a tumor suppressor gene. Many well-defined mutations affecting erythrocyte pyruvate kinase enzymic parameters cause severe hemolytic anemia.

Recent findings support the view that nuclear genes for the enzymes of glycolysis in eukaryotes were acquired from mitochondrial genomes (Liaud *et al.*, 2000).

References

Andradeab MA and Borkab P (2000) Automated extraction of information in molecular biology. *FEBS Letters* 476(30): 12–17.

Boles E, Schulte F, Miosga T et al. (1997) Characterization of a glucose-repressed pyruvate kinase (Pyk2p) in *Saccharomyces cerevisiae* that is catalytically insensitive to fructose-1, 6-bisphosphate. *Journal of Bacteriology* 179: 2987–2993.

Liaud MF, Lichtle C, Apt K, Martin W and Cerff R (2000) Compartment-specific isoforms of TPI and GAPDH are imported into diatom mitochondria as a fusion protein: evidence in favor of a mitochondrial origin of the eukaryotic glycolytic pathway. *Molecular Biology and Evolution* 17: 213–223.

Online Mendelian Inheritance in Man (OMIM), http://www3.ncbi.nlm.nih.gov/Omim/

Zimmerman FK and Entian K-D (eds) (1997) *Yeast Sugar Metabolism*. Lancaster, PA: Technomic Publishing.

See also: Enzymes; Mitochondrial Genome; Tumor Suppressor Genes

Glycosylase Repair

J Laval

Copyright © 2001 Academic Press
doi: 10.1006/rwgn.2001.0571

A large number of intrinsic and extrinsic mutagens induce structural damages to cellular DNA, as well as errors occuring during DNA replication. These DNA damages are cytotoxic, miscoding, or both, and are believed to be the origin of cell lethality, tissue degeneration, aging, and cancer. In order to counteract immediately the deleterious effects of such lesions, leading to genomic instability, cells have evolved a number of DNA repair mechanisms including the direct reversal of the lesion, sanitization of the dNTPs pools, and three different DNA excision pathways: mismatch repair, nucleotide excision repair, and base excision repair (BER). In the BER pathway, the process is initiated by a DNA glycosylase excising the modified or mismatched base by hydrolysis of the glycosidic bond between the base and the deoxyribose of the DNA, generating a free base and an abasic site (AP site) which is cytotoxic and mutagenic. In turn an AP-endonuclease or an AP-lyase incises the phosphodiester bond next to the AP site that is further processed by the sequential action of either dRPase or 5′ termini removing activity, DNA polymerase and DNA ligase and other accessory proteins, in order to restore the integrity of the information contained in DNA. The BER pathway is highly critical for cells since it is conserved from *Escherichia coli* to humans. The pioneering investigations were performed using bacteria and led to the concept of a new pathway for the repair of uracil residues, the deaminated product of cytosine, then to the demonstration that the initial steps for the repair of alkylated bases was mediated by the sequential action of two repair proteins, then to the identification of the various DNA glycosylases, the cloning of the genes coding for the respective proteins, and the identification or the construction of mutant strains deficient in these activities. These investigations greatly facilitated subsequent work in human cells.

DNA glycosylases remove lesions generated by deamination of bases, alkylating agents, oxidative stress, ionizing radiation, or replication errors. All these lesions cause little perturbation of DNA structure. Most DNA glycosylases excise a wide variety of modified bases, while few of them have, so far, a very narrow substrate specificity. The fact that BER enzymes perform more than one step in the BER pathway is another piece of evidence of their versatility. There are two types of DNA glycosylases, the monofunctional devoid of any other associated activity and the bifunctional with an associated AP-lyase activity (β or β-δ-lyase activity) incising the phosphodiester bond 3′ to the AP site and leaving a 5′ phosphate termini or a 3′ phosphate–5′phosphate gap. The biological role of this latter activity is still unknown. As a general rule, the free modified base excised is an extremely poor inhibitor of its respective DNA glycosylase. The best inhibitors known of the activity of DNA glycosylases are transition-state analogs of the reaction catalyzed by these proteins. The goal of DNA glycosylases is to locate fast and efficiently the aberrant base amongst a huge excess of normal ones. Very little is known how these proteins achieve this goal. Based upon the known structures of DNA glycosylases bound to their substrates or inhibitors, it appears that different types of distortions occur in DNA leading to the insertion of the aberrant nucleotide of the DNA substrate into a pocket of the active site by a process termed base flipping or nucleotide flipping and first described in the case of a cytosine 5-DNA methyltransferase acting on DNA. The comparison of the crystal structures of a number of DNA glycosylases revealed structural homologies leading to the concept of a superfamily of BER glycosylases, the helix–hairpin–helix (HhH) superfamily, having similar HhH fold and a Gly/Pro-rich stretch with nearby Asp (GPD) motifs, although very little sequence similarity. This HhH motif plays an important role in the flipping out of the modified base.

The number of known DNA glycosylases remained constant for a long time; however, by identifying the active core region of some of these enzymes then searching for homologs to this core, new DNA glycosylases have been identified. By improving functional predictions for uncharacterized genes by evolutionary analysis, one could expect to identify new DNA glycosylases.

The BER pathway has been reconstituted *in vitro* with cell-free extracts of *E. coli*, or human cells, or using proteins purified at homogeneity. The major proteins performing this process are well defined but the accessory proteins required to obtain an optimal repair are not yet completely identified. Since the damage-specific initial step is carried out by either a monofunctional or a bifunctional DNA glycosylase, it yields abasic sites with different structures. The processing of the resulting AP site, a mutagenic repair intermediate, presumably by the major mammalian AP-endonuclease, HAP1/APEX, occurs via two alternative pathways: the short-patch (filling a one-nucleotide gap) and the long-patch (resynthesis of two to six nucleotides) BER. These two pathways involve some common proteins but also some specific ones. For example, in the short-patch pathway, Pol β is involved in the resynthesis step, whereas PCNA and Pol β/δ/ε are implicated in the long-patch pathway. The results obtained so far suggest that lesions recognized by monofunctional DNA glycosylases are processed by both the short- and the long-patch pathways, whereas those recognized by bifunctional DNA glycosylases are processed via the short-patch pathway. Moreover the selection of the BER pathway could be cell-cycle dependent, the long-patch one might be postreplicative. However the rates of repair measured are not yet optimal and should be improved by the identification and the use of accessory proteins. Although some proteins such as poly(ADP-ribose) polymerase are involved in the repair of lesions induced by simple alkylating agents, the precise role of this protein in the resistance of the cells to alkylating agents remains unclear. The recent identification of new DNA polymerases able to replicate efficiently and accurately miscoding and mutagenic modified bases have to be taken into account in the understanding of BER.

In the case of oxidative damages generated by hydroxyl radicals caused by a track of ionizing radiations, clustered multiple damaged sites have been observed, most of them being modified bases rather than DNA strand breaks. These modified bases are within half a turn of the double helix, i.e., five nucleotides, some of them on the two strands, and they therefore present a challenge to the cell for their repair. The precise mechanisms are so far very poorly understood.

Since, so far, no human diseases have been linked to defects of protein involved in BER, DNA repair genes functionally expressed in mammalian cells and now transgenic mice having a null mutation in the gene coding for BER proteins are very important tools to ascertain the biological role of these proteins in mammalian cells. It has been surprising to notice that, apart from a few examples of targeted deletion of genes encoding some BER proteins in mice leading to embryonic lethality (for example the AP-endonuclease), the genotype of the other knockout mice (such as a number of DNA glycosylases) does not show any striking particularity in term of predisposition to cancer or aging for example, raising the possibility of back-up pathway(s) that have yet to be identified. One could expect important breakthroughs from crosses between different strains to produce double knockouts to identify the possible back-up systems, the processes involved in regulation, and the interactions of the different pathways.

Detailed understanding of the mechanisms leading to the coordination of various proteins involved in the molecular reaction of BER is of paramount importance for gaining insights into the efficiency and fidelity of this key pathway for genome stability, prevention of cancer, resistance to chemotherapeutic agents, degenerative diseases, and more recently in some aspects of teratogenicity.

See also: **Excision Repair**

Grasses, Synteny, Evolution, and Molecular Systematics

E A Kellogg

doi: 10.1006/rwgn.2001.1728

The grass family (Gramineae or Poaceae) is descended from a single common ancestor, thought to have lived sometime between 70 and 55 million years ago (mya) in tropical forest margin habitats. The major radiation of the grasses was much later, probably around 35 mya and correlates with an acquired ability to tolerate drought. Today there are about 10 000 species of grasses, occurring on all continents and covering about 20% of the earth's land surface. Members of the family provide food for most humans, and include rice, maize, wheat, oats, barley, rye, sugarcane, sorghum, and the various species known as millet. Other grasses are the main source of feed for livestock. Because of their economic importance, the grasses have been studied extensively by biologists and have become important model systems on which our knowledge of plant biology is based. This is particularly true for maize, which has an excellent genetic map and an enormous collection of mutants, and rice, whose genome is now almost entirely sequenced.

The evolutionary history is now well known thanks to numerous investigations by molecular systematists. From these studies, a classification has been derived that follows the evolutionary history. Because the family is so large, it is divided into twelve subfamilies for convenience. The most important of these are the Panicoideae, which includes about 3200 species, the Pooideae, which includes about 3300 species, the Chloridoideae, which includes about 1350 species, and the Bambusoideae, with about 1000 species. The Panicoideae and Chloridoideae include many species that exhibit the C4 photosynthetic pathway, which appears to be an adaptation to hot, dry environments.

The nuclear genomes of the grasses are approximately colinear, with large blocks of genes in the same order in all species investigated. The blocks of genes are then arranged in different ways, so that the number of chromosomes varies. For example, the genes on chromosome 10 of rice are all found in the same order in maize and other grasses in the subfamily Panicoideae. In the panicoids, however, rice 10 is not a separate chromosome, but is inserted into the middle of rice 3. Combination of some chromosomes gives the panicoids a smaller number of chromosomes (9 or 10) than rice, which has 12.

Gene order is conserved in spite of large changes in genome size. The amount of DNA in the nucleus varies among grasses by a factor of 20, with rice and foxtail millet having among the smallest genomes and wheat and barley among the largest. The greatest differences in size are caused by the amount of noncoding DNA between the genes. This noncoding DNA appears to be largely an accumulation of retrotransposons.

The forces that maintain colinearity are unknown. Although most grasses have relatively few rearrangements, a few have extensive changes in gene order. The amount of rearrangement does not correlate with evolutionary relationship. For example, although rye is more closely related to wheat than it is to barley, wheat and barley have nearly identical gene orders, whereas rye has multiple differences.

Colinearity of the genomes is potentially useful in positional cloning of genes. The enomorous size of the wheat genome makes chromosome walking virtually impossible, even with a very precisely mapped gene. If a gene can be localized well enough, however, it is possible to find the corresponding region in the rice genome and locate the gene in the rice genomic sequence. The orthologous wheat gene can then be identified by sequence similarity to the rice gene. This approach could in principle be used to investigate variation in any grass, not just well-studied crop species.

See also: **Genome Relationships: Maize and the Grass Model;** *Hordeum* **Species;** *Oryza sativa* **(Rice);** *Triticum* **Species (Wheat)**

Gravitropism in *Arabidopsis thaliana*

P H Masson

Copyright © 2001 Academic Press
doi: 10.1006/rwgn.2001.1681

Life on earth has evolved in the presence of gravity. Hence, it is not surprising that many organisms have acquired ways to use that inherent vectorial information to guide specific processes. Plants are no exception: they have acquired the ability to use gravity to orient the growth of their organs. This response, named gravitropism, is of primary importance to these sessile organisms. Indeed, it allows the shoots to grow upward, above the soil, where they can photosynthesize, and the roots to grow downward into the soil, where they can take up the water and mineral ions required for plant growth and development.

Gravitropism is also important in agriculture and horticulture. It promotes upward growth of crop shoots prostrated by the action of wind and rain, thereby keeping seeds away from soil moisture and pathogens and amenable to mechanical harvest. On the other hand, gravitropism is responsible for some unwanted shoot bending that occurs during transport and/or storage of cut flowers.

Plant organs grow using a combination of cell division in their apical meristems, and cell expansion in their subapical regions. Cells that are laid down by the division of initials in the apical meristem undergo an expansion process before full differentiation. Cell expansion is a highly controlled process, and is the primary target for environmental signals that guide organ growth. Thus, when a plant organ is reoriented within the gravity field, it responds with differential cellular elongation (expansion along the longitudinal axis) on opposite flanks of the elongation zone. The differential growth results in the development of a curvature that brings the organ tip back to an acceptable orientation (gravitational set point angle).

The existence of a gravitropic response implies that plant organs can sense a change in their orientation within the gravity field, and transduce this physical information into a physiological signal. The physiological signal is then transmitted from a site of sensing to the site of response (elongation zone), where it promotes a differential cellular elongation on opposite flanks, responsible for the curvature. A great deal of information on the gravitropic response of plant organs has recently been obtained through the molecular genetic analysis of gravitropism in the model plant *Arabidopsis thaliana*.

Arabidopsis thaliana as a Model for the Study of Gravitropism

Arabidopsis thaliana is a powerful model for the study of growth and development processes in plants. It is a small plant that has a short generation time (~6 weeks), and grows well under laboratory conditions, on shelves at room temperature, with limited amounts of light. It reproduces by self-pollination, although cross-pollination can be easily accomplished. It generates approximately 10 000–30 000 seeds. Its nuclear genome is small (125 Mb) and has been completely sequenced. The plant can be transformed very easily by *Agrobacterium tumefaciens*, and large collections of T-DNA-insertion and transposon-mobilized lines have been generated and are available for forward and reverse genetic studies.

Importantly for the field of gravitropism, *Arabidopsis thaliana* is a small plant that generates tiny seeds. Upon germination, these seeds give rise to small seedlings that can be grown under sterile conditions in petri dishes, under controlled environmental conditions. Hence, it is possible to subject individual seedlings to changing levels of a specific environmental parameter, while maintaining other growth conditions constant.

This ability to grow a large number of *Arabidopsis* seedlings under highly controlled environmental conditions has allowed the development of large-scale screens to examine many mutagenized plants for identification of gravitropic mutants. These screens have typically involved growing seedlings on or in vertical agar-containing media for a few days. Then, young seedlings were gravistimulated by rotating the plates

by 90°. Under these conditions, wild-type seedlings reoriented the growth of their primary organs within 12 h, resuming vertical upward and downward growth for hypocotyls and roots, respectively. Gravitropic mutant seedlings were not able to reorient well in response to gravistimulation. Rather, their roots and hypocotyl grew more randomly along the gravity vector than the wild-type, even before plate rotation.

Similar procedures have been developed to identify mutants affected in inflorescence stem gravitropism. In this case, plants are germinated and grown in soil until bolting. When inflorescence bolts reach a few centimeters, they are cut, inserted in a block of solidified medium, and placed horizontally. Here again, wild-type shoots reorient upward, while mutant shoots do not. It is interesting to note at the outset that mutations were identified that affect the gravitropic response of all three organs (roots, hypocotyls, and inflorescence stems), while others were specific to one or two of these organs. This reflects both the redundancy that exists at some steps of the gravity signal transduction pathway, and the fact that some of the steps in gravity signal transduction are common between all three organs, while others are specific to one or two of them.

Gravity Sensing and Signal Transduction

Gravity sensing appears to occur in a few specialized cells of each plant organ, named statocytes. In roots, statocytes are located in the center of the cap, an organ that covers the root apical meristem. In shoots, the statocytes appear to be located in the starch sheath, an endodermal cell layer that surrounds the vasculature. The statocytes are highly polarized cells that contain sedimentable amyloplasts, named statoliths, which are starch-filled plastids whose density is 1.5 times higher than that of the surrounding cytoplasm. Hence, upon reorientation within the gravity field, amyloplasts sediment to the new physical bottom of the statocytes. The starch-statolith hypothesis proposes that the statocytes are capable of sensing amyloplast sedimentation, or the pressure exerted by these plastids on unknown gravity receptors.

Amyloplast sedimentation appears to be the primary gravity-sensing mechanism in higher plants, although alternative models have been proposed that may account for some aspects of the response. Magnetophoretic studies involving a lateral mobilization of the diamagnetic amyloplasts within the statocytes by high-gradient magnetic fields have demonstrated that amyloplast sedimentation is sufficient for the promotion of shoot and root tip curvature.

Consistent with a primary role of amyloplast sedimentation in gravity sensing, starch-deficient mutants show strong defects in gravitropism. For instance, mutations in the *phosphoglucomutase (PGM)* gene of *Arabidopsis* affect both shoot and hypocotyl gravitropism. Phosphoglucomutase is an enzyme involved in starch biosynthesis, and some of the *pgm* mutants are unable to accumulate starch in their statocytes. Interestingly, magnetophoresis does not promote statolith displacement or organ-tip curvature in these mutants.

The *scr (SCARECROW)* and *shr (SHORT-ROOT)* mutations affect the formation of ground tissue (cortex and endodermis) in *A. thaliana* roots and shoots. Mutant organs lack one cell layer at the position normally occupied by the ground tissue. The remaining layer in this position has characteristics of both tissue types in *scr*, while they lack any endodermal specification in *shr*. In both mutants, ground-tissue cells lack statoliths, while endodermal cells in wild-type shoots and hypocotyls do contain them. Interestingly, shoots and hypocotyls of *scr* and *shr* mutant seedlings did not respond to gravistimulation, while their roots did. As the root statocytes are located in the cap, not in the endodermis, the results provide good correlative evidence for the starch-statolith hypothesis described above.

Even though amyloplast sedimentation appears sufficient to promote the development of a curvature at the tip of a plant organ, it is not clear how the corresponding physical information is transduced into a physiological signal within the statocytes. Physiological evidence points to Ca^{2+}, IP_3, and pH as possible second messengers in this pathway. However, genetic evidence for this conclusion has yet to come.

So far, mutations in only three genes, *ARG1*, *ARL2*, and *RHG*, have been shown to affect the signal transduction phase of gravitropism. Mutant seedlings develop an altered gravitropic response in hypocotyls and roots, without affecting their phototropic competency (ability to curve toward or away from a light source, respectively). Because gravitropism and phototropism appear to involve similar differential cellular elongation responses promoted by the redistribution of a specific plant growth regulator (auxin: see below), this result strongly suggests that these genes are involved in early phases of gravity signal transduction. The *ARG1* and *ARL2* genes encode similar dnaJ-like proteins that carry a coiled coil domain at their C-terminus. In ARG1, this domain is similar to coiled coils found in a number of cytoskeleton-binding proteins. Hence, it was postulated that ARG1 might regulate gravity signal transduction either by promoting the formation of a signal transduction complex in the vicinity of the cytoskeleton, or by altering the general organization of the cytoskeleton. It is interesting to note that dnaJ-like proteins have been implicated as

molecular chaperones in the facilitation of a number of signal transduction pathways, as well as in general protein folding, translocation, or degradation. Although the molecular function of ARG1 has not been fully elucidated yet, it is important to note that this protein is probably not important for general protein folding, translocation, or degradation, considering the specificity of the Arg1 phenotype.

Hence, ARG1 and ARL2 could act in gravitropism by serving as chaperones in the folding of specific components of the gravity signal transduction pathway, or their targeting to specific cellular subcompartments. Interestingly, genetic modifiers of *arg1* have been identified. Modified seedlings appear to develop a more dramatic phenotype, displaying an almost random orientation of their organs, with some tendency to an opposite orientation compared to wild-type. The molecular analysis of these genetic modifiers promises to unravel important clues on the molecular function of *ARG1* in gravitropism.

Signal Transmission to the Responding Zone

The composition of the physiological signal that is generated upon perception of a gravistimulus within the statocytes and informs the elongation zone of a need to respond to the stimulus has not yet been fully elucidated. However, this signal appears to include a plant growth regulator, named auxin. Early physiological studies showed that auxin may be redistributed across the gravistimulated organ in response to the activated gravity signal transduction pathway. The corresponding cross-organ gradient is then transmitted to the elongation zone where it promotes a differential growth response.

In plants, auxin is transported through cell files in a polar fashion. It enters successive cells in the file through an influx carrier or by passive diffusion through the plasma membrane, and exits them through a complex auxin efflux carrier. The efflux carrier is made of a transmembrane protein, a regulatory protein that may bind the cytoskeleton and appears to be the target for a number of transport blockers, and a putative linker protein. Polarity of transport appears to be mediated by the polar distribution of this auxin efflux carrier complex within the transporting cells.

Interestingly, several mutations that affect gravitropism in *A. thaliana* were recently shown to affect the transport of auxin. Mutations in *AUX1* result in altered root gravitropism and increased root growth resistance to auxin. The gravitropism phenotype of *aux1* seedlings can be rescued by adding a low concentration of 1-NAA, a synthetic auxin that appears to diffuse through the cellular membranes quite efficiently, but not by adding 2, 4D or IAA to the medium. Because the latter two auxins are believed to require a transporter to penetrate the cells, it was hypothesized that *AUX1* encodes an influx carrier of auxin. The *AUX1* gene encodes a transmembrane protein that shares homologies with tryptophane (TRP) transporters. Because the molecular structure of auxin is quite similar to that of TRP, it has been postulated that *AUX1* encodes an auxin influx carrier involved in the local transport of auxin at the root tip. Auxin-transport studies have since confirmed this conclusion.

Other gravitropism mutations of *A. thaliana* have been shown to affect a transmembrane component of the auxin efflux carrier complex. Indeed, *agr1* mutant seedlings are more sensitive to high concentrations of 1-NAA, more resistant to ethylene, and more resistant to blockers of the auxin efflux carrier (NPA, TIBA) than wild-type plants. Mutant roots are also defective in their ability to transport radioactively labeled auxin in a basipetal fashion, supporting a role for the corresponding gene in auxin transport in roots. The *AGR1* gene (also named *EIR1*, *PIN2*, or *WAV6*) encodes a transmembrane protein that is localized on the basal membrane of root elongation-zone cells. When expressed in yeast, this protein allows for better auxin export activity. Taken together, these results support a direct role for AGR1 in cellular auxin efflux.

Auxin is a growth regulator that has multiple roles in plant growth and development, including embryo axis formation, vasculature development, lateral root formation and development, apical dominance, and tropisms. However, *aux1* and *agr1* show very specific defects in gravitropism. In fact, *AUX1* and *AGR1* belong to large gene families, and one can speculate that specific members of each family have different functions in a subset of these growth and developmental processes. For instance, the *PIN1* gene appears to mediate the polar transport of auxin in inflorescence stems. Hence, a better insight into the function of each member of these important gene families will enhance our understanding of the role(s) played by auxin in multiple phases of plant growth and development.

Although auxin appears to be an important component of the physiological signal that dictates organ tip curvature in response to gravistimulation, it is not the only player. Indeed, auxin transport and auxin response mutants (see below) still appear to develop some remnants of a gravitropic response. Also, a robust gravitropic response is still observed even when corn or *Arabidopsis* roots are exposed to high auxin levels, otherwise sufficient to completely inhibit root growth. Furthermore, the differential cellular elongation that occurs on opposite flanks of the root elongation zone in response to gravistimulation is very complex, and cannot be explained by a simple redistribution of auxin across the root. Hence, it appears that gravitropism also involves an auxin-gradient-independent process. Although there is no clear understanding of this auxin-gradient-independent phase of gravitropism, physiological experiments suggest that it might involve electrical signals. The availability of ion channel mutants in *A. thaliana*, and of efficient reverse-genetic procedures to disrupt the expression of other channel genes identified by the completed genome-sequencing project, should allow experimental testing of this model.

The Curvature Response

A number of auxin-response mutants have been isolated in *A. thaliana*. Most of these mutants were also shown to be defective in their ability to respond to gravistimulation. Molecular analysis of the corresponding genes revealed interesting features of the auxin-response pathway.

Auxin appears to regulate cellular elongation by altering the activity of the plasma membrane proton pump, by affecting cell wall extensibility and by regulating the expression of a number of genes important for these processes. Auxin has been shown to bind to a number of proteins within plant cells. However, only the auxin-binding protein ABP1 has been postulated to act as an auxin receptor in the control of cell expansion. Upon auxin binding, this predominantly ER-localized protein would somehow regulate the activity of the proton pump, and promote cell expansion. The details of its mode of action are yet to be elucidated.

Some aspects of auxin signal transduction leading to differential gene expression have recently been elucidated. The *AXR1* gene of *A. thaliana* is important for gravitropism and other aspects of auxin response. It encodes a nuclear protein that interacts with ECR1 to activate members of a the RUB/NEDD8 family of ubiquitin-related proteins. Interestingly, the AXR1/ECR1 complex appears to mediate the rubination of another protein, named cullin. Cullin belongs to a protein complex that also includes ASK and the F-box containing TIR1 protein, which is also essential for gravitropism and auxin response. The ASK/cullin/TRI1 complex is similar to the yeast SKp1-Cok 53-F-box-protein (SCF) complex which has been implicated in ubiquitin-mediated protein degradation. The targets of this TIR1-containing SCF-like complex appear to be repressors of early auxin-response genes that may be targeted to destruction by the proteasome in an ubiquitin-dependent manner. The AXR2, AXR3, and SHY2 proteins may constitute such targets. These short-lived proteins interact with auxin-response transcription factors, and may negatively regulate the expression of other auxin-response genes. These three genes are also important for gravitropism and auxin response. Hence, a gene-regulation cascade appears to be activated by this complex auxin-dependent pathway, even though the site of auxin action in the pathway remains elusive.

Future Prospects

Our understanding of the molecular mechanisms that drive gravitropism in plant organs has improved through the analysis of gravitropic-response mutants in *A. thaliana*. This analysis has contributed to substantiate the starch-statolith hypothesis, even though the data remain purely correlative at this time. A role for auxin as a component of the gravitropic signal transmitted from the site of sensing to the site of response has been confirmed. Also, some of the proteins involved in polar auxin transport have been identified and are being characterized, thus opening the door to an elucidation of the multiple roles played by auxin transport in plant growth and development. Finally, a clear involvement of ubiquitin-mediated proteolysis in the auxin signal transduction pathway has been elucidated, and a number of target regulatory genes for that pathway have been uncovered.

Many things remain to be done, however, before one can fully understand the multiple mechanisms involved in gravitropism in higher plant organs. The gravitropic receptor that is activated by amyloplast sedimentation or pressure in the statocytes has to be identified and characterized. The molecules involved in transducing the corresponding signal within the statocytes have yet to be characterized. Physiological and physicochemical evidence suggest the existence of an alternative mode of gravity sensing in higher plants, possibly involving perception of the pressure exerted by whole protoplasts on their cell walls and intracellular cytoskeleton networks. The relative contribution of each gravity-sensing mechanism remains to be elucidated. A better understanding of the mechanisms involved in auxin redistribution is needed, as well as the identification of additional components of the signal transmitted to the responding zone. Finally, a complete elucidation of the mechanisms involved in the cellular responses to these signals is needed.

Fortunately, an unprecedented number of tools derived from genetics, reverse genetics, genomics, proteomics, and biochemistry in *Arabidopsis*, rice, corn, and other plant species have recently been added to an already impressive arsenal of physiological, cytological, and physicochemical techniques. A multidisciplinary approach is now possible, and should improve our ability to answer these important questions of plant biology. Thus, we can anticipate some important breakthroughs in our understanding of the molecular mechanisms that allow plant organs to use gravity and other environmental stimuli to control their growth patterns and generate some truly amazing growth behaviors.

Further Reading

Baluska F and Hasenstein KH (1997) Root cytoskeleton: its role in perception of and response to gravity. *Planta* 203: S69–S78.

Blancaflor EB, Fasano JM and Gilroy S (1998) Mapping the functional roles of cap cells in the response of *Arabidopsis* primary roots to gravity. *Plant Physiology* 116: 213–222.

Chen R, Rosen E and Masson P (1999) Gravitropism in higher plants. *Plant Physiology* 120: 343–350.

Evans ML and Ishikawa H (1997) Cellular specificity of the gravitropic motor response in roots. *Planta* 203: S115–S122.

Gray WM and Estelle M (2000) Function of the ubiquitin–proteasome pathway in auxin response. *Trends in Biochemical Science* 25: 133–138.

Palme K and Galweiler L (1999) PIN-pointing the molecular basis of auxin transport. *Current Opinion in Plant Biology* 2: 375–378.

Rosen E, Chen R and Masson P (1999) Root gravitropism: a complex response to a simple stimulus? *Trends in Plant Science* 4: 407–412.

Sack F (1997) Plastids and gravitropic sensing. *Planta* 203: S63–S68.

Swarup R, Marchant A and Bennett MJ (2000) Auxin transport: providing a sense of direction during plant development. *Biochemical Society Transactions* 28: 481–485.

Tasaka M, Kato T and Fukaki H (1999) The endodermis and shoot gravitropism. *Trends in Plant Science* 4: 103–107.

Weise SE and Kiss JZ (1999) Gravitropism of inflorescence stems in starch-deficient mutants of *Arabidopsis*. *International Journal of Plant Science* 160: 521–527.

See also: *Arabidopsis thaliana*: The Premier Model Plant; Photomorphogenesis in Plants, Genetics of; Root Development, Genetics of

Group Selection

M J Wade

Copyright © 2001 Academic Press
doi: 10.1006/rwgn.2001.0573

Basic Concepts

Natural selection occurs in any system whose members have the properties of replication, variation, and heredity (Lewontin, 1970; Maynard Smith, 1976). When the system consists of cells, the variation among cell lineages in replication and death rates, and the similarity of daughter to mother cells, gives rise to among-cell selection, which determines tissue shape. When such a process operates among cells within the germline, it can result in gametic selection or 'meiotic drive,' one of the strongest evolutionary forces known. When selection occurs among individuals, among groups, or among species, it is called individual selection (sometimes mass selection), group selection, or species selection, respectively.

Group selection has been a controversial topic in evolutionary biology for several reasons (Williams, 1966; Wade, 1978; Wilson, 1980). First, it is difficult to establish that groups of individuals have the necessary properties of replication, variation, and heredity. Groups can be formed in so many different ways and the processes of group formation determine, in large part, whether biologically significant variation among groups can exist and, if it exists, whether or not it is heritable (Wade, 1996). Secondly, if groups do have the requisite properties, it is not clear what category of adaptations or patterns in nature can be better explained as a unique result of group selection than by the more familiar individual selection. It is for this reason that much of the group selection controversy has been focused on adaptations that are good for the group but harmful for the individual or on adaptations such as sex which might favor group 'evolvability' (Williams, 1975; Maynard Smith, 1976). Such adaptations would be the distinctive signature of group selection (Wilson, 1992). Thirdly, whenever individual and group selection operate simultaneously, the number of episodes of individual selection is likely to be greater than that for group selection, because individual birth and death rates are higher than group colonization and extinction rates. (This criticism does not apply to D.S. Wilson's trait group selection – what (Wade, 1978) has called 'intrademic group selection'.) Fourthly, the common wisdom subscribes to a naive form of group selection when it incorrectly describes adaptations of all sorts as being "for the good of the species." This attribution is a serious misunderstanding of the Darwinian logic and evolutionary dynamic. Countering this misconception and misuse of naive group selection as a causal explanation has instilled a profound bias against the entire concept of group selection in some biologists (e.g., Williams, 1966; Dawkins, 1976).

Illustration of Group Selection

Geographic and physical barriers often constrain the movements of individuals and thereby impose a degree of genetic subdivision or population genetic structure on most species. In addition, individuals tend to aggregate or cluster together whenever resources are patchily distributed. It is this spatial aggregation of individuals and the expression of social behaviors within aggregations that results in novel ecological and evolutionary processes involving group selection. Whenever an individual's behavior affects its own fitness and the fitness of conspecifics, group selection will affect the evolution of that behavior in a genetically subdivided population (Wade, 1978; Wilson, 1980).

Consider a hypothetical species with two kinds of individuals, benefactors and recipients (**Figure 1**). The benefactors provide a fitness benefit to other members of the group and do so at a cost to their own fitness. Recipients do not engage in provisioning behaviors but benefit from the behavior of benefactors and experience increased fitness whenever they are around benefactors. This difference in behavior and its fitness effects makes the benefactor–recipient interaction an example of the frequently discussed altruism–cheater interactions. Darwin believed that the existence of such benefactor adaptations could be "fatal to my whole theory" of evolution by natural selection (Darwin, 1859, p. 236) because, by definition, the benefactor lowers its fitness while increasing the fitness of the recipient. Natural selection should operate to eliminate such behaviors, yet they appear prevalent in some of the major taxonomic groups of insects and mammals, e.g., the sterile castes in colonies of bees, ants, or wasps, 'helpers at the nest' in some birds, or group feeding in the social spiders.

Darwin solved this problem by postulating that group selection, among colonies or families, operated in opposition to individual selection within colonies or families (Darwin, 1859 p. 237). We can illustrate how selection operates in different directions at different levels using the benefactor–recipient illustration. First, consider two groups of birds (**Figure 2**). Each group consists of five birds, but the groups differ from one another in the frequency of benefactors. Group 1 is rich in benefactors, with a frequency of 0.80, while group 2 is relatively poor in benefactors, with a frequency of only 0.20. Thus, the groups meet the first criterion for the existence of group selection, variability, specifically, variability in the frequency of

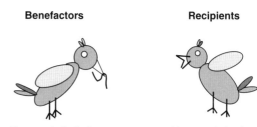

Benefactors	Recipients
Bear cost of altruism and bestow benefit on others	No cost of altruism but reap benefits from others

Figure 1 Individual variation in social behavior.

benefactors. Indeed, there are two components of variation in the frequency of benefactors: (1) among birds within groups, and (2) among groups.

Each component of variation has a selective effect or consequence for both individual and group replication. Within each group, benefactors experience reduced fitness (**Figure 3**). In group 1, the frequency of benefactors declines from 0.80 to 0.78. Similarly, in group 2, the frequency of benefactors declines from 0.20 to 0.17. Individual selection operating within groups selects against the benefactors and their frequency declines as a consequence. The magnitudes of the decline in benefactor frequency are −0.02 and −0.03, in groups 1 and 2, respectively. The total decline in the frequency of benefactors by individual selection is −0.027. This is the weighted average decline, where the weights are determined by the size of the group relative to the total after individual selection.

The among-group component of variation also has a selective effect (**Figure 4**). A group with a high frequency of benefactors has a higher growth rate than a group with a lower frequency of benefactors. This positive effect of benefactor frequency is the opposite of the negative fitness effect of being a benefactor within a group. Group 1, with an initial frequency of 0.80 benefactors, increased in size from five

to nine birds, while group 2, with a lower initial frequency of benefactors (0.20), increased only from five to six birds. The relative fitness of group 1 is 1.2, which is calculated as a per-head growth rate of 1.8 (i.e., 9/5) relative to the mean growth rate of 1.5 (i.e., 15/10). This is much higher than the relative growth rate of group 2, which is 0.80, i.e., a per-head growth rate of 1.2 (6/5) relative to the mean of 1.5. This difference in growth rate of groups also causes a change in the frequency of benefactors. Hence, group selection favors benefactors and results in a positive change in their frequency equal to +0.06.

The total change in the frequency of benefactors equals the sum of the changes caused by the two opposing levels of selection (**Figure 5**): individual selection against benefactors and group selection favoring benefactors. The total change in the frequency of benefactors is positive despite the opposition of individual selection against benefactors within every group. In this example, group selection is stronger than opposing individual selection. This kind of interesting interaction between individual and group selection and behavioral evolution has been experimentally demonstrated in laboratory populations of flour beetles (Wade, 1980a), in farm populations of chickens (Muir, 1996), in field populations of willow leaf beetles (Breden and Wade, 1989; Wade, 1994), jewelweed (Stevens *et al.*, 1995), and social spiders (Aviles, 2000). (See Goodnight and Stevens, 1997, for a recent review of experimental studies of group selection.)

Group Genetic Structure

Group genetic structure is often characterized in hierarchical terms associated with the components of genetic variation among individuals within groups

Group 1: $p_{benefactors} = 0.80$ Group 2: $p_{benefactors} = 0.20$

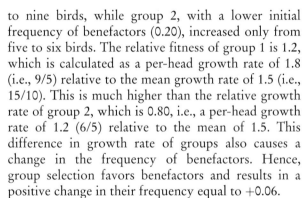

Figure 2 Variability: groups differ in frequency of benefactors.

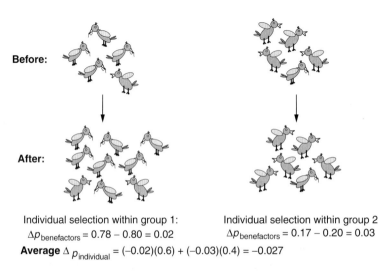

Before:

After:

Individual selection within group 1:   Individual selection within group 2
$\Delta p_{benefactors} = 0.78 - 0.80 = 0.02$ $\Delta p_{benefactors} = 0.17 - 0.20 = 0.03$
Average $\Delta p_{individual} = (-0.02)(0.6) + (-0.03)(0.4) = -0.027$

Figure 3 Individual selection within groups opposes benefactors.

$$\Delta p_{group} = \{(0.8)^*(1.2) + (0.2)^*(0.8)\}/2 - 0.5 = +0.06$$

Figure 4   Group selection favors benefactors.

$$\Delta p_{group} = (0.8)^*(1.2) + (0.2)^*(0.8) - 0.5 = +0.06$$

$$\text{Average } \Delta p_{individual} = (-0.02)(0.6) + (-0.03)(0.4) = -0.027$$

$$\Delta p_{total} = \{\text{🐦/Total}\} - \{\text{🐦/Total}\}$$

After   Before

$$\Delta p_{total} = \{8/15\} - \{5/10\} = +0.033$$

$$\Delta p_{total} = \Delta p_{individual} + \Delta p_{group} = -0.027 + 0.060$$

Figure 5   Total selection favors benefactors.

and among groups. When quantified using Wright's F statistics (Wright, 1969, 1978), group genetic structure describes the fraction of the total genetic variance accounted for at a given level of metapopulation subdivision. For our example, the total variance in the frequency of benefactors is $(0.5)^2$. This total variance can be partitioned into two components: (1) the mean variance within groups, which is 0.16 $\{[(0.8)(0.2) + (0.2)(0.8)]/2\}$; and (2) the variance among groups, which is 0.09 $\{[(0.8 - 0.5)^2 + (0.2 - 0.5)^2]/2\}$. Note that, in this example, the variance among groups is approximately only 36% of the total variance so that F, the fraction of the variance among groups, equals 0.36, which is half of that within groups. In fact, the among-group variance is only 56% as large as the mean variance within groups. Note also that the genetic variance among groups is also the genetic correlation among individuals within groups (Cockerham, 1954). Thus, whenever individuals live in groups of genetic relatives, there will necessarily be genetic variation among groups (Wade, 1980b).

The value of F is influenced by a large number of factors, including the numbers of breeding adults per group ('effective' group size, N_e), the rate and pattern of gene flow among groups (m), the extinction and

	aa	*Aa*	*AA*
bb	0.50	0.75	1.00
Bb	0.75	0.75	0.75
BB	1.00	0.75	0.50

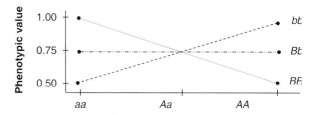

Figure 6   Two-locus, 'additive-by-additive' epistasis.

colonization of local groups (Whitlock and McCauley, 1990), group fission and fusion (Breden and Wade, 1989; Whitlock, 1992), and group density regulation (i.e., hard versus soft selection: Wade, 1985; Kelly, 1992, 1994). The effects of these factors on the among-group genetic variance have been reviewed elsewhere (Wade, 1996). Wright noted that variation in offspring numbers, variation in breeding sex ratio, and fluctuations in the size of breeding groups (Wright, 1931, 1941, 1952) all tend to reduce N_e. It is important to emphasize that natural selection itself reduces N_e and, thus, increases F: Whenever natural selection occurs, the variance in fitness exceeds random, by definition, and consequently N_e is reduced to less than N. This inevitable reduction in N_e that accompanies natural selection is called the Hill–Robertson effect (Hill and Robertson, 1966; Barton, 1995).

Genetic Structure of Adaptations in Relation to Group Selection

Arguments in the controversy over individual versus group selection tend to overlook the genetic architecture of adaptations. Most adaptations are not determined by alternative alleles at single genes, but rather by epistasis, the integrated action of many genes. Whenever multiple loci determine a trait, individual selection becomes significantly less efficient and group selection more efficient, an important feature unique to interaction systems and not captured by single-gene models. Epistasis both enhances the evolutionary potential of group selection and simultaneously diminishes that of individual selection. To see this, consider a simple two-gene interaction (**Figure 6**). In **Figure 6**, there is a simple additive-by-additive genetic interaction between the A and B loci

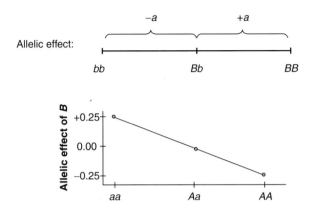

Figure 7 Variation in the sign of the allelic effect of B with genetic background.

in affecting an individual's phenotypic value. When the genetic background at the A locus is homozygous *aa*, the effect of the B allele is to additively increment phenotypic value. By 'additively increment,' I mean that a homozygous *BB* individual with two B alleles has twice the phenotypic value of a heterozygous individual *Bb* with only one B allele. However, when the genetic background at the A locus is changed to be homozygous *AA*, the effect of the B allele is the opposite: it additively decrements phenotypic value.

The effect of the B allele is not a property of the allele itself, but rather a property of the interaction system (**Figure 7**). This has profound effects on the evolution of the B allele, especially in genetically subdivided populations where the frequency of the A allele changes from group to group. This kind of epistasis represents a 'genetic constraint' on individual selection. In groups with a high frequency of the A allele, B will increase by virtue of its positive effect on phenotypic value. However, in groups, with a low frequency of the A allele, B will decrease in frequency by virtue of its negative effect on phenotypic value. Because positive and negative values of Δp_B are combined and averaged to determine the effect of individual selection (see section "Illustration of group selection"), the change in the frequency of the B allele by individual selection is reduced. The greater the value of F, the greater the variation in genetic background among groups. In contrast, group selection will favor those groups in which gene combinations result in high group fitness (see section "Illustration of group selection").

Summary and Conclusion

Although group selection remains a controversial topic in evolutionary biology, experimental studies, in both laboratory and field, have shown that groups have

the necessary properties of replication, variation, and heredity. Indeed, given the known processes of group formation, these essential properties must be common in nature. It is also clear that group selection can affect the evolution of many traits, especially those with a complex genetic basis, and not only adaptations, which are good for the group but harmful for the individual. Even in those circumstances where the number of episodes of individual selection exceeds that for group selection, epistasis for fitness can severely limit the efficiency of individual selection at the same time that it opens unique opportunities for group selection.

It remains important, however, to avoid naive group selection when attempting to explain the origin of adaptations. Recognizing the hierarchy of biological levels, to which the Darwinian logic and evolutionary dynamic apply, does not constitute an endorsement of causal explanations based on "the good of the species."

References

Aviles L (2000) Nomadic behaviour and colony fission in a cooperative spider: life history evolution at the level of the colony? *Biological Journal of the Linnean Society* 70: 325–339.

Barton NH (1995) Linkage and the limits to natural selection. *Genetics* 140: 821–841.

Breden FJ and Wade MJ (1989) Selection within and between kin groups in the imported willow leaf beetle. *American Naturalist* 134: 35–50.

Cockerham CC (1954) An extension of the concept of partitioning hereditary variance for analysis of covariances among relatives when epistasis is present. *Genetics* 39: 859–882.

Darwin C (1859) *On the Origin of Species.* (A Facsimile of the 1st Edition reprinted 1964). Cambridge, MA: Harvard University Press.

Dawkins R (1976) *The Selfish Gene.* Oxford: Oxford University Press.

Goodnight CJ and Stevens L (1997) Experimental studies of group selection: what do they tell us about group selection in nature? *American Naturalist* 150: S59–S79.

Hill WG and Robertson A (1966) The effect of linkage on the limits to artificial selection. *Genetical Research* 8: 269–294.

Kelly JK (1992) Restricted migration and the evolution of altruism. *Evolution* 46: 1492–1492.

Kelly JK (1994) The effect of scale dependent processes on kin selection: mating and density regulation. *Theoretical Population Biology* 46: 32–57.

Lewontin RC (1970) The units of selection. *Annual Review of Ecology and Systematics* 1: 1–18.

Maynard Smith J (1976) *The Evolution of Sex.* New York: Cambridge University Press.

Muir WM (1996) Group selection for adaptation to multiple-hen cages: selection program and direct responses. *Poultry Science* 75: 447–458.

Stevens L, Goodnight CJ and Kalisz S (1995) Multi-level selection in natural populations of *Impatiens capensis*. *American Naturalist* 150: S59–S79.

Wade MJ (1978) A critical review of the models of group selection. *Quarterly Review of Biology* 53: 101–114.

Wade MJ (1980a) An experimental study of kin selection. *Evolution* 34: 844–855.

Wade MJ (1980b) Kin selection: its components. *Science* 210: 665–667.

Wade MJ (1985) Hard selection, soft selection, kin selection, and group selection. *American Naturalist* 125: 61–73.

Wade MJ (1994) The biology of the imported willow leaf beetle, *Plagiodera versicolora* (Laicharting). In: Jolivet PH, Cox ML and Petitpierre E (eds) *Novel Aspects of the Biology of the Chrysomelidae*, pp. 541–547. Amsterdam: Kluwer.

Wade MJ (1996) Adaptation in subdivided populations: kin selection and interdemic selection. In: Rose MR and Lauder G (eds) *Evolutionary Biology and Adaptation*, pp. 381–405. Sunderland, MA: Sinauer Associates.

Wade MJ and McCauley DE (1988) Extinction and recolonization: their effects on the genetic differentiation of local populations. *Evolution* 42: 995–1005.

Whitlock MC (1992) Nonequilibrium population structure in forked fungus beetles: extinction, colonization, and the genetic variance among populations. *American Naturalist* 139: 952–970.

Whitlock MC and McCauley DE (1990) Some population genetic consequences of colony formation and extinction: genetic correlations within founding groups. *Evolution* 44: 1717–1724.

Williams GC (1966) *Adaptation and Natural Selection*. Princeton, NJ: Princeton University Press.

Williams GC (1975) *Sex and Evolution*. Princeton, NJ: Princeton University Press.

Wilson DS (1980) *The Natural Selection of Populations and Communities*. Menlo Park, CA: Benjamin/Cummings.

Wilson DS (1992) Group selection. In: Keller EF and Lloyd EA (eds) *Keywords in Evolutionary Biology*, pp. 145–148. Cambridge, MA: Harvard University Press.

Wright S (1931) Evolution in Mendelian populations. *Genetics* 16: 97–159.

Wright S (1943) Isolation by distance. *Genetics* 28: 114–138.

Wright S (1951) The genetical structure of populations. *Annual Review of Eugenics* 15: 323–354.

Wright S (1969) *Evolution and the Genetics of Populations*, vol. 2. Chicago, IL: University of Chicago Press.

Wright S (1978) *Evolution and the Genetics of Populations*, vol. 4. Chicago, IL: University of Chicago Press.

See *also*: Behavioral Genetics; Fitness; Natural Selection; Population Genetics; Population Substructure

Growth Factors

J K Heath

Copyright © 2001 Academic Press
doi: 10.1006/rwgn.2001.0574

'Growth factors' is a generic term applied to define a specific set of polypeptides that act via association with high-affinity transmembrane receptors to induce intracellular signals, which mediate cell proliferation, differentiation, and survival. Growth factors are the principal means of intercellular communication in the development and regeneration of metazoan organisms. Mutation of either growth factors or their cognate receptors can have profound effects on organismic development and physiological function.

Although there is considerable structural and functional diversity amongst growth factors, some common themes regarding their mechanism of action can be defined. Growth factors, unlike classical endocrine hormones, generally act locally within tissues rather than between organs. Many growth factors exhibit biochemical features which constrain their activity to cells in close proximity to the source of synthesis. These may include association with nonsignaling components such as specific binding proteins or extracellular matrix components, anchorage to the plasma membrane, or a requirement for proteolytic cleavage to elicit biological activity.

Growth factors and their receptors can be grouped into 'families,' based upon shared features of amino acid sequence, and into 'superfamilies,' based upon shared structural folds. Many growth factor families display significant evolutionary conservation in sequence; for example, homologs of the fibroblast growth factor (FGF), epidermal growth factor (EGF), and transforming growth factor beta (TGF-beta) families can be found in nematodes, echinoderms, and *Drosophila*, as well as higher vertebrates such as mouse and humans. A common finding is that higher vertebrates have larger growth factor families than invertebrates. For example, there are currently 22 members of the FGF gene family in the human genome, but only one in *Drosophila* and *Caenorhabditis elegans*. Some growth factor superfamiles such as chemokines, whose primary action is in infection and immunity, are found in gene clusters and exhibit significant divergence in sequence and gene number between closely related mammalian species.

A key feature of the divergence and elaboration of growth factor and receptor gene number in higher vertebrates is that it results in diversification of receptor recognition specificity; it is frequently observed that each member of a growth factor family has a unique repertoire of receptors with which it can interact. In addition, individual members of growth factor families can exhibit widely divergent patterns of gene expression *in vivo*. Collectively this means that individual members of growth factor families can display characteristic physiological defects upon mutation; for example, homozygous null mutants of FGF-4 result in a peri-implantation lethal defect in the mouse, whereas homozygous null mutants of FGF-5 are viable but exhibit an 'angora' hair phenotype.

As might be expected from their biochemical functions, mutations in growth factors and their receptors have important consequences for human disease. Somatic mutations in particular growth factors or receptors have been associated with carcinogenesis and exhibit the properties of oncogenes. These mutations are generally dominant in character and result, by a variety of different means, in activation of intracellular signaling pathways. For example, mutations in receptors which result in receptor oligomerization (such as fusion to a dimeric partner protein) are associated with particular human malignancies. Ectopic activation of growth factor expression by retroviruses has been been associated with retroviral-induced carcinogenesis in experimental systems. Some inherited dominant mutations are associated with developmental dysplasia; for example, Crouzon is a congenital craniofacial syndrome which results from mutations in FGF receptor-2, leading to receptor activation in the absence of FGF ligand. Achondroplasia is a dominant-acting congenital dwarfism syndrome which results from specific mutations in FGF receptor-3. Recessive, homozygous loss-of-function mutations in growth factors and receptors are much rarer in natural populations and frequently arise from forced selective breeding for desirable physiological traits; for example, the 'double muscle' phenotype of Belgian Blue cattle results from homozygous recessive mutation of the gene encoding the TGF-beta family member myostatin.

Finally, certain growth factors have significant practical utility in genetics research. The ability to cultivate embryonic stem (ES) cells in culture is dependent upon a specific growth factor, leukemia inhibitory factor (LIF). In the presence of LIF, ES cells can be selected for specific mutations, which can be introduced back into the germline by transplantation of the genetically modified ES cells into the host embryo.

See also: Achondroplasia; Embryonic Stem Cells; Oncogenes

GSD (Gerstmann–Straussler Disease)

J Hodgkin

Copyright © 2001 Academic Press
doi: 10.1006/rwgn.2001.0575

Gerstmann–Straussler disease (GSD), also known as Gerstmann–Straussler–Scheinker disease (GSSD), is a human neurodegenerative disease characterized by cerebellar ataxia and progressive dementia. Like the related diseases Creutzfeldt–Jakob disease (CJD) and familial fatal insomnia (FFI) it is associated with alterations in the prion protein. Most cases of GSD are familial, in contrast to CJD, and are caused by certain missense mutations in the prion gene.

See also: Creutzfeldt-Jacob Disease (CJD); Familial Fatal Insomnia (FFI); Spongiform Encephalopathies (Transmissible), Genetic Aspects of

GT Repeats

See: Microsatellite, CA Repeats

GT–AG Rule

Copyright © 2001 Academic Press
doi: 10.1006/rwgn.2001.1852

The GT–AG rule describes the presence of these invariable dinucleotides at the first two and last two positions of introns in nuclear DNA.

See also: Introns and Exons

GTP (Guanosine Triphosphate)

E J Murgola

Copyright © 2001 Academic Press
doi: 10.1006/rwgn.2001.0576

Guanosine-5′-triphosphate (GTP) is synthesized in the cell by phosphorylation of guanosine diphosphate (GDP), catalyzed by a nucleoside diphosphate kinase, with ATP as the phosphate donor:

$$GDP + ATP \rightleftharpoons GTP + ADP$$

For the synthesis of deoxyguanosine triphosphate (dGTP), a precursor of DNA, the 2′ hydroxyl group of the ribose moiety of GTP is replaced by a hydrogen atom. The final step in this conversion is catalyzed by ribonucleotide reductase.

GTP is an energy-rich, activated precursor for RNA synthesis that also plays important roles in several other cellular processes such as protein synthesis, protein localization, signal transduction, visual excitation, and hormone action. The free energy of hydrolysis of GTP can be used to drive reactions that otherwise are energetically unfavorable. For example, for translocation of a protein through a membrane of the endoplasmic reticulum, GTP hydrolysis is probably needed to insert the signal sequence into the channel and is required to release the signal recognition particle from its receptor. GTP may act as an allosteric effector, causing a protein to change shape slightly. Its hydrolysis can then lead to a cyclic variation in macromolecular shape and functioning of the protein. This is seen in the GTP-dependent release of photoexcited rhodopsin from transducin.

In mRNA-programmed, ribosome-dependent protein synthesis, GTP plays a role at all three stages: initiation, elongation, and termination. For initiation, the binding of GTP to a protein initiation factor leads to formation of the small subunit initiation complex. Subsequent hydrolysis of GTP results in the association of the large subunit with the complex. In elongation, GTP has more than one role. It binds to elongation factor (EF) Tu (EF-1 in eucaryal cells) to facilitate the delivery to the ribosome of each successive aminoacyl-tRNA as dictated by the mRNA sequence. The aminoacyl-tRNA is delivered as part of a ternary complex composed of itself, EF-Tu, and GTP. After GTP hydrolysis, EF-Tu is released, GTP is regenerated, and the cycle continues for the next designated aminoacyl-tRNA. To ensure the accuracy of the match between the incoming aminoacyl-tRNA and the mRNA codon in the A-site of the ribosome, the binding of GTP and its EF-Tu-dependent hydrolysis play a role in the process of proofreading. After peptide bond formation, translocation of the peptidyl-tRNA from the A-site to the P-site requires the binding of GTP to EF-G (EF-2 in eucaryal cells) and its subsequent EF-G-dependent hydrolysis.

Finally, after the presence of a termination codon (UGA, UAA, or UAG) in the A-site is recognized by the combined action of a protein release factor (RF) and specific regions of ribosomal RNA, a signal is transmitted to the hydrolytic center of the ribosome (in the large subunit) to hydrolyze the peptidyl-tRNA in the P-site. The binding of GTP to another RF and

its subsequent RF-dependent hydrolysis functions to promote the release of the codon-dependent RF from the ribosome, preparing the way for dissociation of the subunits, release of mRNA, and utilization of the ribosomal subunits in another round of polypeptide synthesis.

See also: **Protein Synthesis; Ribosomal RNA (rRNA)**

Guanine

R L Somerville

Copyright © 2001 Academic Press
doi: 10.1006/rwgn.2001.1724

Guanine is a purine (molecular formula $C_5H_5N_5O$) found within RNA in the form of a ribonucleotidyl residue and in DNA in the form of a deoxynucleotidyl residue. In DNA, guanine is usually base-paired via three hydrogen bonds with the pyrimidine cytosine. A number of low-molecular-weight, guanine-containing nucleotide coenzymes are also found within cells, where they serve as substrates for RNA and DNA biosynthesis, energy sources in protein biosynthesis, and donors of sugar residues in the synthesis of polysaccharides. Guanine residues in DNA are uniquely susceptible to alteration by reactive oxygen species. When guanine in DNA undergoes oxidation to 8-oxoguanine, its base-pairing properties change and it acquires the ability to pair with adenine. Most cells have an active base-excision repair system that removes 8-oxoguanine residues from DNA, thereby avoiding the potentially hazardous creation of a transversion mutation.

See also: **Purine**

Guide RNA

See: **RNA Editing in Trypanosomes**

Gynogenone

W Reik

Copyright © 2001 Academic Press
doi: 10.1006/rwgn.2001.0578

Gynogenetic embryos have maternal genomes only (haploid or diploid) but arise from oocytes that have been fertilized by sperm. In natural gynogenesis (occurring, for example, in some fish species), the paternal, sperm-derived genome is inactivated or lost. Experimentally, gynogenetic embryos can be made using irradiated sperm. The haploid gynogenetic embryos can then be diploidized. In the mouse, gynogenetic embryos are made by pronuclear transplantation. Following fertilization and formation of pronuclei, the male pronucleus is removed by microsurgery and replaced by a second female pronucleus.

Gynogenetic embryos are useful for genetic mapping or for the rapid recovery of mutations. In mice gynogenetic development only progresses midway through gestation, to early postimplantation stages. Such embryos have particularly deficient development of extraembryonic membranes such as the trophoblast and yolk sac, which may explain the failure of further development. This developmental failure is explained by the phenomenon of genomic imprinting, whereby certain genes in eutherian mammals are expressed from only one of the parental chromosomes. Gynogenetic embryos thus lack gene products that are only made by the paternal genome, and have overexpression of gene products made by the maternal genome.

Parthenogenetic embryos are those that have been activated to develop without sperm. Diploid parthenogenetic embryos have the same developmental potential as gynogenetic ones, showing that imprinting is a purely nuclear phenomenon. The existence in nature of gynogenetically or parthenogenetically reproducing species, or normal development following experimental production of gynogenones, indicates that genomic imprinting is largely absent in these species.

See also: **Imprinting, Genomic; Parthenogenesis, Mammalian**

Gyrase

Copyright © 2001 Academic Press
doi: 10.1006/rwgn.2001.1853

Gyrase is a type II topoisomerase of *Escherichia coli* that is able to generate negative supercoils in DNA.

See also: **DNA Supercoiling; Topoisomerases**

H19

K N Gracy and S Brenner

doi: 10.1006/rwgn.2001.0620

H19 is an imprinted gene in which one of the parental copies of the gene is silenced. It encodes for a nonprotein-coding RNA and is closely linked to the reciprocally imprinted gene Igf2, which encodes a fetal growth factor. On the parental chromosome, H19 is not transcribed and Igf2 is active, while on the maternal chromosome H19 is transcriptionally active and Igf2 is not. Differences in methylation distinguish the parental origin of the gene and methylation of nearby silencer and enhancer elements play an important role in the gene's regulation. The differentially methylated domain (DMD) located upstream of H19 is essential for the imprinting of both H19 and Igf2. H19 is located on mouse distal chromosome 7 and on the Beckwith–Wiedemann region on human chromosome 11p15.5.

See also: **Gene Silencing;** *Igf2* **Locus; Imprinting, Genomic**

H2 Locus

See: **Major Histocompatibility Complex (MHC)**

Hadulins

W Broughton

doi: 10.1006/rwgn.2001.1643

Hadulins are proteins that are synthesized during root-hair development, but especially during the deformation and curling that accompanies invasion by symbiotic bacteria (rhizobia). An example is a nonspecific lipid transfer protein (LTP), the expression of which is upregulated in root hairs by rhizobia and their Nod-factors (Krause *et al.*, 1994).

Reference

Krause A, Christian JA, Christian CJA, Dehning I *et al.* (1994) Accumulation of transcripts encoding a lipid transfer-like protein during deformation of nodulation-competent *Vigna unguiculata* root hairs. *Molecular Plant–Microbe Interactions* 7: 411–418.

See also: **Nod Factors;** *Rhizobium*

Hairpin

doi: 10.1006/rwgn.2001.1854

A hairpin is a double helical region formed by base-pairing between adjacent (antiparallel) complementary sequences in a single strand of RNA or DNA. It comprises a stem and loop structure.

See also: **Antiparallel**

Hairy Cell Leukemia (HCL)

D Catovsky

doi: 10.1006/rwgn.2001.1576

Hairy cell leukemia (HCL) is a malignancy of mature B lymphocytes with cytoplasmic 'hairy' projections involving the peripheral blood, bone marrow, and red pulp of the spleen. HCL comprises about 2% of adult leukemias and affects predominantly males (male: female ratio 5:1) with a median age of 52 years. Clinically the main features are splenomegaly, anemia, thrombocytopenia, and leukopenia. The low leukocyte count, chiefly neutrophils and monocytes, is responsible for opportunistic infections in untreated patients. Large abdominal nodes are a feature in a minority. Hairy cell express surface Ig (IgM+/− D, G, or A) with a single light chain, and B-cell antigens CD19, 20, 22, 79a, and express strongly CD11c, CD25, and CD103. There is no consistent cytogenetic abnormality but there is overexpression of cyclin D1 in about

50% in the absence of t (11;14) or BCL1 rearrangement. Prolonged remission can be obtained with the nucleoside analogs pentostatin and cladribine. Median survival is greater than 10 years. A rare variant form of the disease has the same histological features as typical HCL but the leukocyte count is high (50 × $10^9 1^{-1}$), the hairy cells have a prominent nucleolus and the response to therapy and overall prognosis are poor.

See also: **Leukemia**

Haldane, J.B.S.

K R Dronamraju

Copyright © 2001 Academic Press
doi: 10.1006/rwgn.2001.0579

John Burdon Sanderson Haldane ("J.B.S.") (1892–1964) (**Figure 1**) was widely acknowledged as the last of the polymaths, a renaissance man, and a scholar of ancient classics, who contributed significantly to physiology, genetics, biochemistry, and biometry while possessing no academic qualification in science. He was a highly skilled and versatile popularizer of science, who regularly contributed to numerous magazines and newspapers.

J.B.S. Haldane was born in Oxford, England, on 5 November 1892. He was the son of John Scott Haldane, a distinguished Oxford physiologist, and Louisa Kathleen Trotter, who came from a comfortable south Scottish family whose ancestors served with distinction in India. Haldane's childhood was marked by episodes of precocious intellectual feats which occasionally, though not always, portend a future genius. Haldane was educated at Eton and Oxford, graduating with distinction in the classics in 1914, but received no formal training in any branch of science. From an early age, his father encouraged him to assist in physiological experiments and taught him the fundamentals of science. The rest was self-taught.

Scientific Work

Haldane's contributions to genetics were largely theoretical and mathematical. Yet few scientists have had more influence on the steady growth of genetics than Haldane during his long career. Haldane's first contribution to genetics, which dealt with the measurement of linkage in mice, was published in 1914. His research was interrupted by World War I, but in 1919 Haldane worked out a more accurate method of

Figure 1   J.B.S. Haldane, arriving in India, 1957.

detecting linkage and a way of relating the map distance to the frequency of recombination ('mapping function'). He suggested the use of 'centimorgan' (cM) as a unit of chromosome length. In collaboration with others, he undertook a series of linkage studies in the following years, extending the linkage theory to polyploids, demonstrating the effect of age on the frequency of recombination in the fowl, and demonstrating partial sex-linkage in the mosquito *Culex molestus*. In his book *New Paths in Genetics* (1941) Haldane introduced '*cis*' and '*trans*' to replace the terms 'coupling' and 'repulsion' that were in vogue at that time.

Perhaps the most famous aspect of Haldane's genetical work is his generalization concerning the offspring in interspecific crosses, which he formulated in 1922, called 'Haldane's rule':

When in the first generation between hybrids between two species one sex is absent, rare or sterile, that sex is always the heterogametic sex.

This rule has stood the test of time since Haldane first proposed it in 1922, having shown to be valid in different species across several taxa in the animal kingdom.

As early as 1920, Haldane was already referring to the gene as a nucleoprotein molecule, emphasizing that enzymes are products of gene action, and introducing the concept of one gene–one enzyme. Although experimental evidence was produced in 1941, using *Neurospora*, Haldane's early emphasis on the biochemical interpretation of gene action prepared the ground for the ready acceptance in later years of the experimental results of Beadle, Ephrussi, Tatum, and Lederberg.

Population Genetics

Haldane is best remembered as a founder of population genetics, an honor he shared with R.A. Fisher and S. Wright. Population genetics is best described as the offspring of the union between Mendelian genetics and the Darwinian theory of evolution. In a series of papers, entitled "A mathematical theory of natural selection," which were published between 1924 and 1934, Haldane investigated the conditions required to maintain a balance between selection intensity and mutation pressure, under varying intensities of selection, inbreeding, size of the population, frequency of a character, reproductive isolation, type of inheritance, and environmental interaction. Haldane later commented that adequate quantitative data are rarely available to test the mathematical models that he and others had developed.

Haldane's contributions to population genetics were quite extensive. He showed that the probability that a single mutation will ultimately become established in a population of finite size is proportional to its selective advantage, but for dominant mutations is independent of the population size. He showed further that mutant genes which are harmful singly, but become advantageous in combination, could accumulate in small, isolated populations, leading eventually to speciation.

Haldane further showed, mathematically, that the impact of a mutation on a population depends merely on the rate of recurrence of the mutation and not on the degree of severity of selection against it. This principle was later applied to measure the impact of genetic damage resulting from high-energy radiation by the US National Academy of Sciences. From an evolutionary point of view, Haldane's paper on "The cost of natural selection" broke new ground in its approach to measuring one of the major factors determining the rate of evolution. His calculations showed that, during the course of evolution, the substitution of one gene by another involves a number of deaths that is equal to 30 times the number in a generation (on average) and that the mean time taken for each gene substitution is about 300 generations. He concluded: "This accords with the observed slowness of evolution." Subsequently, Haldane's work became the basis for Motoo Kimura's neutral theory of evolution.

Human Genetics

Haldane's contributions to human genetics were of particular importance. He was a pioneer who laid its foundations, and shaped and nursed its growth from its infancy to a mature discipline. Furthermore, through numerous popular writings, Haldane prepared the ground for the acceptance of human genetics and an appreciation of its importance in the public domain. He developed statistical methods for the study of genetic traits in families and populations and the analysis of gene–environment interaction. He estimated the first mutation rate of a human gene (hemophilia) and prepared the first human gene map, involving the traits on the X chromosome, hemophilia and color blindness.

Of special importance was Haldane's suggestion that resistance to malaria and other infectious diseases played a significant role in recent human evolution, resulting in greater genetic diversity and greater prevalence of certain diseases such as sickle cell anemia. This has stimulated a great deal of epidemiological research of considerable importance in recent years. Haldane's books include: *Daedalus or Science and the Future* (1923), *Possible Worlds and Other Essays* (1928), *The Causes of Evolution* (1932), *Science and the Supernatural* (1935), *Heredity and Politics* (1938), *New Paths in Genetics* (1941), and *The Biochemistry of Genetics* (1954). He was also the author of a popular children's storybook, *My Friend, Mr. Leaky* (1937). For several years during the 1940s, Haldane embraced Marxism, but there is no evidence to indicate that it had a significant influence on his scientific work.

Further Reading

Clark RW (1968) *JBS: The Life and Work of J.B.S. Haldane.* London: Hodder & Stoughton.

Dronamraju KR (ed.) (1968) *Haldane and Modern Biology.* Baltimore, MD: Johns Hopkins University Press.

Dronamraju KR (ed.) (1995) *Haldane's Daedalus Revisited.* Oxford: Oxford University Press.

Haldane JBS (1924) A mathematical theory of natural and artificial selection. Pt. I. *Transactions of the Cambridge Philosophical Society* 23: 19–41.

Haldane JBS (1932) *The Causes of Evolution.* London: Longman, Green & Co.

Haldane JBS (1938) *Heredity and Politics.* London: Allen & Unwin.

Haldane JBS (1954) *The Biochemistry of Genetics.* London: Allen & Unwin.

Haldane JBS (1957) The cost of natural selection. *Journal of Genetics* 55: 511–524.

References

Haldane JBS (1941) *New Paths in Genetics.* London: Allen & Unwin.

See also: Fisher, R.A.; Haldane–Muller Principle

Haldane–Muller Principle

J F Crow

Copyright © 2001 Academic Press
doi: 10.1006/rwgn.2001.1425

The Haldane–Muller principle, as the name suggests, was discovered independently by J.B.S. Haldane and H.J. Muller. It relates the mean fitness of a population to the mutation rate. The impact of recurrent harmful mutation on fitness is a function, not of the deleterious effect per mutation, but of the mutation rate itself.

This is perhaps counterintuitive. It can be explained simply as follows. If a mutation has an effect so drastic that it kills the individual carrying it, the mutation causes one death. If, in contrast, it causes a 1% probability of death, it will persist in the population for an average of 100 generations before being eliminated and will therefore affect 100 individuals. If, in a system of mutation cost-accounting, 100 individuals, each with a 1% probability of death, are equated to one individual with 100% probability, then each causes one death. The effect may be reduced fertility rather than survival, but the principle is similar. Muller called each premature death or failure to reproduce a genetic death.

Algebraically, in a population of size N, $2N\mu$ dominant mutations occur per generation, where μ is the mutation rate per locus per generation. Each generation NQs mutations are eliminated by selection, where Q is the number of mutations per individual and s is the individual probability of elimination. At equilibrium these two processes must balance, hence $NQs = 2N\mu$, and $Q = 2\mu/s$. Now, if each mutant causes a fitness reduction equal to s, the mutation load is $Qs = 2\mu$. Summing over all relevant loci, the mutation load is $2\Sigma\mu$ or twice the total mutation rate per gamete. If the mutations are recessive, then two are eliminated by each genetic death, and the load is only half as large.

The Haldane–Muller principle may then be stated, as Haldane did, that the total effect of mutation on fitness is the total haploid mutation rate per generation, multiplied by a factor of 1 or 2 depending on whether the mutation is recessive or dominant. If n mutations are eliminated with each genetic death, as might be true with extreme epistasis, then the load is $1/n$ as large as if they were eliminated independently.

J.L. King made this more precise by saying that the mutation load is twice the mutation rate divided by the difference between the frequency of mutations in individuals eliminated by selection and that before selection. This principle can be written in more general form. The mutation load is

$$L = 2U/(z - x + 2U)$$

in which $U = \Sigma\mu(1 - q)$, q is the mutant allele frequency, x is the mean number of mutations per individual before selection and z is the mean number of individuals eliminated by selection per generation (Kondrashov and Crow, 1988).

With epistasis the mutation load can be considerably decreased by permitting several mutations to be eliminated with each genetic death. Such epistasis is generated by truncation selection, which may be the way in which many organisms survive a high mutation rate (see Mutation Load).

References

Kondrashov AS and Crow JF (1988) King's formula for the mutation load with epistasis. *Genetics* 120: 855–856.

See also: Genetic Load; Mutation Load

Haldane's Mapping Function

See: Mapping Function

Hamilton's Theory

B Brembs

Copyright © 2001 Academic Press
doi: 10.1006/rwgn.2001.0581

Selfish Genes and Cooperation

Paradoxically, inheritance is the basis of evolutionary change. Without safe transmission of genetic information from one generation to the next, there would be random arrangement of the genetic building blocks. Constant randomization of information carriers obviously cannot lead to meaningful information. Thus, the cornerstone of evolution is genetics. Only after conserving well-tried genes can there be competition (selection) between new, yet untested ones (i.e., mutations). Charles Darwin (1809–1882) was the first to formulate a theory of gradual evolutionary change caused by adaptive mutations that are selected out of a number of other random variants. In a relentless "struggle for existence" many slightly different variants are competing with each other and only few survive. Darwin's notion of "survival of the fittest" seems to convey the picture of a war in which everyone fights everyone. Nature is "red in tooth and claw," a merciless killing in which only the strongest and meanest can prevail. Victory (i.e., evolutionary success or 'fitness') is granted according to the reproductive success of the survivor. Again, only if the trait that led to successful reproduction is safely transmitted to the offspring, will this trait spread and eventually be represented as a feature of the species. Of course, if the trait in addition leads to procreation at a competitor's expense, the animal not only gains fitness itself, but also reduces the fitness of those animals it is exploiting, increasing its odds even further. It is no wonder that parasitism and exploitation are widespread phenomena and virtually universal across the living world. Darwin himself emphasized:

No instinct has been produced for the exclusive good of other animals, but each animal takes advantage of the instincts of others.

Indeed this is one of the few truly falsifiable test statements in the Darwinian theory. And it seems so easily falsifiable: is there not ample evidence of cooperation in the animal kingdom? Parental care, shoaling fish, cooperatively hunting wolfs or lions, the mycorhiza symbiosis between the fungus and the plant, the subterranean colonies of the naked mole rat, coalition forming in primates or the social insects are but some of the most well known examples. Darwin was well aware of the problem and described it as:

One special difficulty, which at first appeared to me insuperable, and actually fatal to the whole theory.

Group Selection

When describing the above problem, Darwin was referring to the social insects in particular. At that time, it was already common knowledge that hymenopteran colonies (honeybees, wasps, bumble bees, and ants) usually consist of one reproducing queen and a multitude of sterile workers. This particular case of sociality is termed 'eusociality.' In addition to sterile individuals cooperatively helping the fertile

animals to raise their offspring, eusociality is characterized by another trait: At least two generations overlap in life stages in which they are capable of contributing to colony labor, so that the offspring can assist their parents during part of their life cycle.

The abandonment of reproduction by the worker caste was the huge dilemma to which Darwin devoted an entire chapter in his book *On the Origin of Species*. While the omnipresence of exploitation is well in accord with the rule "reproduce at the cost of your competitors," the equally obvious existence of all degrees of altruism up to the complete sacrifice of reproductive success in favor of another organism seemed an insurmountable obstacle. How can individuals without their own offspring exist if reproduction and inheritance are the foundation of the whole theory?

Darwin's own solution was to assume that the colonies formed some sort of superorganism that competes against other colonies in a very similar way as individuals do. To perceive animal colonies as superorganisms with their members as rough analogs of cells has long been known and is a very useful concept, even today for certain studies. The idea of family or 'group selection' placated Darwin's contemporaries and was still widely accepted well into the twentieth century. According to this idea, the unit of selection for altruistic alleles of an originally selfish gene would be the colony or deme, not the individual. The altruistic, cooperative allele spreads in the species, as colonies without a high occurrence (gene frequency) of this allele become extinct. However, in order for interdemic selection to be effective, one has to assume that there is no migration between the groups and that there is sufficient selection pressure, i.e., the rate of colony extinction is very high. Furthermore, individual selection will always be faster than group selection, as the number of individual organisms is much larger than that of populations and the turnover rate of individuals is much higher. Thus, group selection can never counteract individual selection. Because of these considerations, group selection was eventually abandoned as the prime explanation for the evolution of cooperation. Then, in 1964, William Donald Hamilton's principle of 'kin selection' was published in the *Journal of Theoretical Biology*. At the time, it was so innovative that it almost failed to be published and was largely ignored for a decade. When finally noticed, its influence spread exponentially until it became one of the most cited papers in the field of biology. It is the key to understanding the evolution of altruistic cooperation among related organisms, such as the social insects. Cooperation among unrelated individuals is beyond the scope of this article and is treated elsewhere (see Further Reading).

Kin Selection

Why should there be a distinction between cooperation among unrelated individuals and that among related individuals? We have learnt that genetics is the basis upon which evolutionary change is taking place. Fitness was defined above in terms of successful reproduction, i.e., the number of offspring carrying the selected allele. The more offspring, the 'fitter' the parent. Darwin's "struggle for existence" is a struggle for reproduction. With sexual reproduction, however, only one half of an organism's genome is transferred to one of his offspring at a time. Therefore, any particular trait – depending on its mode of inheritance – is often transmitted from a parent to its offspring with a probability of less than one. Thus, in order to transmit as many of one's genes into the next generation as possible (and hence be evolutionarily successful), an organism has to produce as many surviving offspring as possible in order to maximize the probability of transmitting all its genetic information. This might constitute a difficult task, however, since all its competitors try to do the same. But there are other sources of one's own genes available: relatives.

An ordinary diploid, sexually produced organism shares 50% of its genes with either of its parents. Accordingly, it shares about 50% of its genes with its siblings, 25% with its uncles, aunts, grandparents, grandchildren, etc. (coefficient of relatedness, see **Figure 1**). Hamilton's stroke of genius was to reformulate the definition of fitness as the number of an individual's alleles in the next generation. Or, more precisely, *inclusive fitness* is defined as an individual's relative genetic representation in the gene pool in the next generation:

$$inclusive\ fitness = \frac{\begin{array}{c}(own\ contribution \\ + \ contribution\ of\ relatives)\end{array}}{\begin{array}{c}average\ contribution \\ of\ the\ population\end{array}} \quad (1)$$

Thus, fitness denotes the capability of an allele to spread in a population: if the fitness value for a given allele is larger than one it will increase in frequency and if it is smaller it will decrease in frequency. It is evident that such *genic* (as opposed to group) *selection* will favor an allele that not only enhances reproductive success of its carrier, but also of all other individuals sufficiently related to it. But could an allele that reduces the fitness of its carrier while enhancing the fitness of its relatives be adaptive? Would it spread in a natural population? This is not a trivial question and it takes some computational effort to solve it. Let us try to formulate the inclusive fitness w of an individual i. As noted in equation (1), w should be composed of the fitness a of the focal individual and the contribution x of its relatives:

$$w_i = a_i + x \quad (2)$$

The contribution x to individual i's inclusive fitness w is then the sum of all alleles in the gene pool that are shared by i and its relatives j:

$$x = \sum_j r_{ij} b_{ij} \quad (3)$$

where r is the coefficient of relatedness between individual i and its relative j, and b is the fitness of j. Note that r is always ≤ 1 and therefore j's contribution to w_i depends critically on its relatedness to i. We can thus reformulate equation (2) to:

$$w_i = a_i + \sum_j r_{ij} b_{ij} \quad (4)$$

Obviously, if the allele in question infers a fitness cost (i.e., $a_i < 1$), w_i will only be greater than one if r is sufficiently high (given that the higher fitness b of the relative also means higher cost). Reformulating equation (4) into a cost (C)/benefit (B) ratio describing the necessity of w_i being greater than one if the allele of interest is to spread, yields

$$1 - C + rB > 1 \quad (5)$$

which can be easily rearranged to produce Hamilton's rule:

$$rB - C > 0 \text{ or } \frac{C}{B} < r \quad (6)$$

Put into words, the relatedness of the individual that profits from the altruistic act of the focal individual must be higher than the cost/benefit ratio this act imposes. Thus, the question as to whether 'cooperative' genes may spread even if the cooperation infers fitness costs, can be solved both by simulation to find out the critical ranges of the parameters in question and experimentally by measuring the relevant parameters and comparing them with the simulated results.

A very simple example will explain the concept. Consider a pair of brothers ($r = 0.5$, see **Figure 1**), one of whom sacrifices all of his fitness ($C = 1$) by not reproducing, but helping his brother to successfully rear offspring. In order for C/B to become smaller than $r = 0.5$, the altruist's act must at least double the receiver's fitness in order for the altruist to gain representation in the next generation. Evidently, a

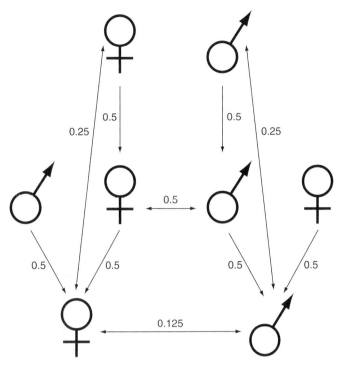

Figure 1   The coefficient of relatedness. In diploid organisms, every parent (top row) transmits 50% of its genetic information to each offspring (middle row). On average, therefore, siblings share half of each parent's contribution to their genome, adding to a coefficient of relatedness $r = 0.5$. Consequently, cousins share an $r = 0.125$ or $r = 1/8$ (bottom row). Likewise, these cousins are related to their common grandparents by 1/4 or $r = 0.25$. It might also be said that r is a measure for the probability that any given allele is shared by two individuals.

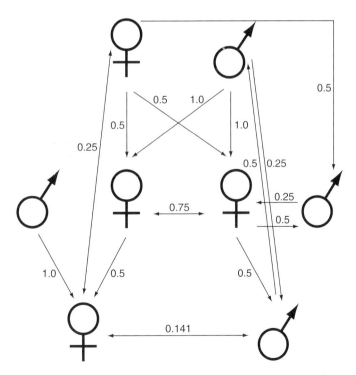

Figure 2   The coefficient of relatedness with haplo-diploid sex determination. Note how the coefficients are skewed with respect to the diploid system depicted in **Figure 1**. For example, sisters (middle row) are more related to each other ($r = 0.75$) than they are to their mother (top row; $r = 0.5$).

very high coefficient of relatedness is needed to overcome high fitness costs due to sterility or a decrease in life expectancy, or both. The benefit of altruism decreases rapidly with declining relatedness. It becomes clear that the distinction between cooperation among related and among unrelated individuals is vital for understanding the evolution of cooperation.

Hamilton's Rule and Social Insects

Why did Hamilton's theory have such an impact on modern evolutionary biology? The main reason for this was because it explained the evolution of a significant part of all the cooperation occurring in nature, without having to resort to group selection and its very restrictive assumptions. But there is another piece of evidence that adds embellishment to a beautiful theory: the haplo-diploid sex determination of the social hymenopterans, i.e., the bees, ants, and wasps – the very insects that posed such a severe puzzle to Darwin.

While most animal genera have a hetero- and a homogametic sex (i.e., a different set of sex chromosomes for the different sexes), hymenopterans universally produce males from unfertilized (i.e., haploid) eggs and females from fertilized (i.e., diploid) eggs. This system skews relatedness in an almost perfect way for eusociality to evolve (see **Figure 2**). Consider a female worker. Half of her genome comes from the father (haploid) and half from the mother (diploid). That means she carries all of her father's genes and half of her mother's genes, and so does her sister, implying that they share the entire genome of their common father (i.e., already 50% of their genome), plus, on average, a quarter of their mother's genome, yielding a coefficient of relatedness of 0.75. Thus, altruistically helping their mother (the queen) and her offspring (new founding queens and workers) need only yield a small benefit (compared to a 'normal' diploid organism) in order to spread through the population. Accordingly, the hymenopterans are the order with the highest occurrence of eusociality in the animal kingdom: eusociality has arisen at least eleven times independently during the evolution of the hymenopterans. Only a few species within the Arthropoda are known to be eusocial, such as the termites (*Isoptera*) and some aphids (*Hemiptera*). Outside the Arthropoda, the only species known to form eusocial colonies is the naked mole rat (*Heterocephalus glaber*). This prevalence of eusociality within the hymenopterans is very suggestive of Hamilton's rule having a deep impact on their evolutionary path. (Note that haplodiploidy is not sufficient, however, to create sociality because most hymenopteran species are solitary.) In the light of the theory of kin selection, even Darwin's notion of family or group selection can be seen in a different light: the otherwise weak interdemic selection can act together with genic selection, and against individual selection, to spread cooperative genes in a population.

Further Reading

Brembs B (1996) Chaos cheating and cooperation: potential solutions to the Prisoner's Dilemma. *Oikos* 76: 14–24. http://brembs.net/ipd

Futuyma DJ (1986) *Evolutionary Biology*. Sunderland, MA: Sinauer Associates.

Hamilton WD (1996) *Narrow Roads of Gene Land*. Oxford: Oxford University Press.

Hölldobler B and Wilson EO (1990) *The Ants*. Cambridge, MA: Belknap Press.

See also: Fitness; Frequency-Dependent Selection; Population Genetics

Handedness, Left/Right

J Hodgkin

Copyright © 2001 Academic Press
doi: 10.1006/rwgn.2001.1128

Handedness, or left–right asymmetry, can refer to asymmetry at three or more levels of organization.

Molecular Handedness

Molecular handedness refers to the chirality of molecules resulting from asymmetrically substituted carbon atoms, or to differently handed arrangements of larger assemblages of atoms, for example in right-handed and left-handed DNA structures.

Developmental Handedness

Developmental handedness or laterality usually refers to differences between the left and right sides of bilaterally organized animals. Some animals, such as fruit flies, are bilaterally symmetric in their anatomy, but most animals exhibit anatomical differences between left and right sides, to a greater or lesser extent. How these differences are created is a special case of pattern formation in development, and some progress has been made in understanding the genetic and molecular basis of laterality in vertebrates and nematodes, although the mechanisms do not seem to be conserved. Some animals, or organs within them, also exhibit helical anatomy which can take one of two hands. The two testes of male fruit flies are bilaterally symmetric in placement, but both develop into left-handed helical tubes, spiraling counterclockwise. Snail shells can occur in either left-handed or right-handed spirals. In this case the direction of the spiral is genetically determined by the maternal genotype, which affects the handedness of the first spiral cleavage in developing eggs. Many plants exhibit helical growth patterns, with one hand or the other preferred.

Behavioral Handedness

Behavioral handedness in animals refers to the preferential use of one limb or organ as compared to its contralateral homolog. The fact that most, but not all, humans are right-handed suggests that this behavioral asymmetry is adopted partly at random, but is biased towards right-handedness by developmental asymmetry. There is no convincing evidence for a separate genetic influence on behavioral handedness in humans.

See also: **Maternal Effect; Pattern Formation; Right/Left Handed DNA**

Haploid Number

M A Ferguson-Smith

doi: 10.1006/rwgn.2001.0583

The term haploid number refers to the number of chromosomes contained within each gamete. During gametogenesis the chromosome number is reduced to half the number present in somatic cells. This is achieved in the first meiotic (reduction) division, at which the chromosomes in the pregametic cell pair with their homologs to form bivalents, which process allows each member of the pair to separate from one another during first anaphase into different daughter cells prior to the second meiotic division (see Meiosis).

See also: **Diploidy; Meiosis**

Haploinsufficiency

M A Cleary

doi: 10.1006/rwgn.2001.0582

Haploinsufficiency is the requirement for two wild-type copies of a gene for a normal phenotype. For haploinsufficient genes, when one copy of a gene is deleted or contains a loss-of-function mutation, the dosage of normal product generated by the single wild-type gene is not sufficient for complete function. Diseases resulting from haploinsufficiency are usually caused by mutations in genes encoding proteins required in large amounts, or in genes encoding regulatory molecules whose concentrations are closely titrated within the organism. Human diseases associated with haploinsufficiency include Greig syndrome, which results from loss of the transcriptional regulatory protein GLI-3, and Williams syndrome, which results from a deletion of the gene encoding the extracellular matrix protein elastin.

Haplotype

D E Bergstrom

doi: 10.1006/rwgn.2001.0584

The term 'haplotype' refers to a particular set of alleles at linked loci that are present on one of two homologous chromosomes. During the course of a gene-mapping experiment involving backcross or intercross mating schemes, geneticists use a process known as 'haplotype analysis' to place genetic markers in a precise order. For purposes of this discussion, we will use a backcross as an example. To initiate a backcross mapping experiment, two inbred parental strains, A and B, are mated to produce an F_1 hybrid. By definition, strain A is inbred and can be considered to be homozygous for A alleles (A/A) at all autosomal loci. Likewise, strain B can be considered to be homozygous for B alleles (B/B) at all autosomal loci. F_1 hybrids derived from these parental strains must be heterozygous (A/B) at all autosomal loci. Meiotic events within the germline of the F_1 hybrid generate recombinant chromosomes in which A and B alleles are placed in new combinations along the length of the chromosome. By backcrossing the F_1 hybrid back to an inbred parental strain (of strain B in this case), one can determine the haplotype of these recombinant chromosomes by genotyping the resulting progeny. If, for example, five closely linked markers, 1–5, were genotyped in a single offspring and had the haplotype 1^A (A at locus 1), 2^B, 3^A, 4^A, and 5^B, one can conclude that loci 1, 3, and 4 are on one side of a point of recombination, and that loci 2 and 5 are on the opposite side. Similarly, by determining the haplotypes of additional progeny in which recombination has occurred between different sets of markers, one can begin to subdivide these groups and refine the order of markers even further.

By using mapping panels containing DNA from hundreds of progeny that have been previously genotyped with thousands of markers, one can quickly establish the map location of any new genetic marker that is polymorphic between the two parental strains. Two such well-characterized mouse community mapping crosses include The Jackson Laboratory Backcross Mapping Panels and the European Collaborative Interspecific Mouse Backcross Mapping Panel. By utilizing distantly related parental strains, these mapping panels provide useful community resources that exploit polymorphism at a large number of loci.

Specialized mapping panels are also frequently established that are segregating for an investigator's phenotype of interest. Using these mapping panels, a phenotype (for which no molecular basis has been elucidated) can also be mapped with respect to nearby genetic markers. This provides the basis for positionally cloning the gene underlying the mutant phenotype.

The term 'haplotype' can also be used to describe particular sets of alleles present at linked loci within naturally occurring populations; for example, t haplotypes occurring within a specialized region of mouse chromosome 17 known as the t complex.

See also: Gene Mapping; Linkage Map; t Haplotype

Hardy–Weinberg Law

K E Holsinger

Copyright © 2001 Academic Press
doi: 10.1006/rwgn.2001.0585

When Mendel's laws were rediscovered, many biologists had difficulty understanding why recessive traits were not lost from populations. "If recessive traits are not expressed in the presence of a dominant allele," they reasoned, "they should eventually disappear from populations." Observations showed that this does not happen. W. E. Castle explained why this was so in 1903, although his explanation was ignored until G.H. Hardy, a British mathematician, and W. Weinberg, a German biologist, published papers independently in 1908 that provided mathematical justification for Castle's intuitive argument.

Hardy's and Weinberg's papers pointed out another important consequence of Mendel's rules as applied to populations: if individuals choose mates at random and several other important assumptions apply, there is a simple relationship between allele frequencies and genotype frequencies. If x_{ij} is the frequency of

a genotype carrying alleles A_i and A_j, and if p_k is the frequency of allele A_k, then:

$$x_{ij} = 2p_ip_j \quad \text{if} \quad i \neq j \quad \text{and} \quad x_{ii} = p_i^2$$

A population in which this relationship between genotype and allele frequencies holds is said to have its genotypes in Hardy–Weinberg proportions.

Deriving Hardy–Weinberg Proportions

In many types of population genetic problems it is useful to construct a mating table. From this table we can calculate the frequency of genotypes and alleles among offspring produced according to any specified mating pattern. **Table I** shows the mating table for the six conditions sufficient to guarantee that genotypes in a population segregating for two alleles at one locus will be found in Hardy–Weinberg proportions.

Meiosis is Fair

The first of the six conditions required for derivation of the Hardy–Weinberg proportions is that segregation in heterozygotes produces equal proportions of the two types of gametes. While this assumption is usually met, there are exceptions. At the t-allele locus in house mice or the segregation distorter locus in *Drosophilia*, for example, some alleles may be found in more than 90% of gametes produced by heterozygotes. If segregation distortion occurs, then the genotype proportions among progeny of any matings involving heterozygotes will be quite different from those shown in **Table I**. If A_1 is found in 90% of the gametes produced by A_{12} heterozygotes, for example, then 90% of the progeny of a mating between A_1A_1 and A_1A_2 will be A_1A_1 and only 10% will be A_1A_2.

Table I Mating table for Hardy–Weinberg proportions

Mating	Frequency	Offspring genotypes		
		A_1A_1	A_1A_2	A_2A_2
$A_1A_1 \times A_1A_1$	x_{11}^2	1	0	0
$A_1A_1 \times A_1A_2$	$x_{11}x_{12}$	1/2	1/2	0
$A_1A_1 \times A_2A_2$	$x_{11}x_{22}$	0	1	0
$A_1A_2 \times A_1A_1$	$x_{12}x_{11}$	1/2	1/2	0
$A_1A_2 \times A_1A_2$	x_{12}^2	1/4	1/2	1/4
$A_1A_2 \times A_2A_2$	$x_{12}x_{22}$	0	1/2	1/2
$A_2A_2 \times A_1A_1$	$x_{22}x_{11}$	0	1	0
$A_2A_2 \times A_1A_2$	$x_{22}x_{12}$	0	1/2	1/2
$A_2A_2 \times A_2A_2$	x_{22}^2	0	0	1

No Input of New Genetic Material

If mutations occur while gametes are being produced, alleles will be passed from parents to progeny with probabilities different from those in the absence of mutation. If A_1 can mutate to A_2, for example, there is a chance that some progeny of a mating between two A_1A_1 individuals will be A_1A_2 or even A_2A_2. In fact, if the A_1A_1 mutates to A_1A_2 with a frequency μ, A_1A_2 progeny will occur in this cross with a frequency $2\mu(1 - \mu)$ and A_2A_2 progeny will occur with a frequency μ^2. Similarly, if new individuals become part of the population through migration, then the frequency with which different types of matings occur will depend on the genotype frequency of migrants, not just on the genotype frequency of the resident population. Thus, either the genotype proportions or the mating frequencies in **Table I** would have to be changed if this assumption were violated.

Individuals Mate at Random

The assumption of random mating is the one most commonly identified with the Hardy–Weinberg law. In conjunction with the assumption that there is no migration into the population (see section "No input of new genetic material"), this assumption allows us to calculate the frequency with which each type of mating occurs. The probability that a particular pair of genotypes mates at random is equal to the probability that two individuals we select at random from the population have those genotypes. For example, the probability that we select an A_1A_2 individual at random is x_{12}. The probability that we select an A_1A_1 individual at random is x_{11}. Thus, the probability of an $A_1A_2 \times A_1A_1$ mating is $x_{12}x_{11}$, and the probability of an $A_1A_1 \times A_1A_2$ mating is $x_{11}x_{12}$. (Recall that it is conventional to describe matings with the genotype of the maternal parent first, so these are two different types of matings.) Similarly, the frequency of an $A_2A_2 \times A_2A_2$ mating is x_{22}^2.

The Population is Effectively Infinite

In a small population the 'actual' frequency of offspring genotypes observed in matings involving a heterozygote may be different from the 'expected' frequency listed in the **Table I** for the same reason that a fair coin tossed four times will not always give two heads and two tails. If meiosis is fair, the gametes that participate in fertilization are a random sample of all gametes produced, and in a small sample the observed and expected frequencies may be different from one another. Similarly, the 'actual' frequency of a mating in a small population may differ from the 'expected' frequency. As with sampling of gametes to form zygotes, the matings that actually occur are a sample of all those that could have occurred.

The sampling of gametes and matings are two sources of the phenomenon of genetic drift. In a very large population the actual and expected frequencies will almost always be very close to one another, so we can neglect the difference between them. If there are 50 matings between A_1A_1 and A_1A_2, each of which produces one offspring, for example, there is a 5% chance that the frequency of heterozygotes from these matings will either be less than 36% or more than 64%. If there were 5000 such matings, however, there is a 95% chance that the frequency of heterozygotes will be between 48% and 52%.

All Mated Pairs Produce the Same Number of Offspring

The preceding assumptions allow us to calculate the frequency with which different types of matings occur and the frequency of different genotypes among those matings. If we also assume that all mated pairs produce the same number of offspring regardless of their genotype, we can calculate the frequency of the different genotypes among newly formed progeny. Specifically:

$$
\begin{aligned}
x'_{11} &= x_{11}^2 + \tfrac{1}{2}x_{11}x_{12} + \tfrac{1}{2}x_{12}x_{11} + \tfrac{1}{4}x_{12}^2 \\
&= x_{11}^2 + x_{11}x_{12} + \tfrac{1}{4}x_{12}^2 \\
x'_{12} &= \tfrac{1}{2}x_{11}x_{12} + x_{11}x_{22} + \tfrac{1}{2}x_{12}x_{11} + \tfrac{1}{2}x_{12}^2 \\
&\quad + \tfrac{1}{2}x_{12}x_{22} + x_{22}x_{11} + \tfrac{1}{2}x_{22}x_{12} \\
&= x_{11}x_{12} + 2x_{11}x_{22} + x_{12}x_{22} + \tfrac{1}{2}x_{12}^2 \\
x'_{22} &= \tfrac{1}{4}x_{12}^2 + \tfrac{1}{2}x_{12}x_{22} + \tfrac{1}{2}x_{22}x_{12} + x_{22}^2 \\
&= \tfrac{1}{4}x_{12}^2 + x_{12}x_{22} + x_{22}^2
\end{aligned}
$$

where the ′ is used to distinguish genotype frequencies among offspring from those in their parents.

All Genotypes Survive with the Same Probability

If all genotypes survive with the same probability, then the frequency of each genotype in the offspring generation is equal to its frequency in newly formed zygotes. The frequency of A_1A_1 among adults, for example, will be:

$$
\begin{aligned}
x'_{11} &= x_{11}^2 + x_{11}x_{12} + \tfrac{1}{4}x_{12}^2 \\
&= (x_{11} + \tfrac{1}{2}x_{12})^2 \\
&= p^2
\end{aligned}
$$

Similarly, the frequency of A_1A_2 among adults will be $2pq$ and the frequency of A_2A_2 among adults will be q^2. Thus, p^2, $2pq$, and q^2 are the Hardy–Weinberg proportions for one locus with two alleles. Notice that the allele frequency in offspring is equal to the allele frequency in parents.

Importance of Hardy–Weinberg Proportions

Given the many assumptions needed to derive the Hardy–Weinberg Law, it may come as a surprise to learn that it plays a central role in the theory of population genetics. It does so for two reasons. First, it provides a way to estimate allele frequencies for a trait in which heterozygotes are indistinguishable from one of the homozygotes, provided we are willing to assume that all of the assumptions apply to the population in which we are interested. Second, it tells us what will happen in a population in the absence of any evolutionary forces. As the philosopher Elliott Sober has pointed out, it plays a role in population genetic theory similar to the role that the first and second laws of motion play in Newtonian mechanics.

The first and second laws of motion tell us that an object at rest will tend to remain at rest and an object in motion will tend to remain in motion (in a straight line at a constant speed) unless acted on by outside forces. They are 'zero-force laws' that tell us what to expect when no forces are operating on an object. Moreover, they allow us to judge the magnitude and direction of any forces operating on an object by the acceleration to which it is subject.

The Hardy–Weinberg law is population genetics' zero-force law. It tells us what a population will look like if neither genetic drift nor any evolutionary forces affect it. If all of the assumptions of Hardy–Weinberg apply, then the population **must** have genotypes in Hardy–Weinberg proportions. Moreover, a single generation in which those assumptions apply is sufficient to put genotypes into those proportions, and neither the allele frequency nor the genotype frequencies will change so long as they continue to apply. If genotypes are not in Hardy–Weinberg proportions, then one or more of the assumptions **must** have been violated in this population, and the direction in which genotypes depart from Hardy–Weinberg proportions is often a clue to the cause of the departure. If, for example, fewer heterozygotes are observed than expected, some form of inbreeding is a likely cause.

It is important to remember, however, which inferences can be made with the Hardy–Weinberg law and which cannot:

1. If the assumptions apply, genotypes will be in Hardy–Weinberg proportions.
2. If genotypes are not in Hardy–Weinberg proportions, one or more of the assumptions has been violated.

It is tempting to conclude that if genotypes are in Hardy–Weinberg proportions, all the assumptions apply. But this conclusion is not justified. Suppose, for example, genotypes differ in their ability to survive, but all the other assumptions apply. Then genotypes will be found in Hardy–Weinberg proportions among newly formed zygotes, but they will not be found in Hardy–Weinberg proportions in adults.

Further Reading

Castle WE (1903) The laws of heredity of Galton and Mendel, and some laws governing race improvement by selection. *Proceedings of the American Academy of Arts and Sciences* 39: 233–242.

Hardy GH (1908) Mendelian proportions in a mixed population. *Science* 28: 49–50.

Hartl DL and Clark AG (1997) *Principles of Population Genetics*, 3rd edn. Sunderland, MA: Sinauer Associates.

Provine WB (1971) *The Origins of Theoretical Population Genetics*. Chicago, IL: University of Chicago Press.

Sober E (1984) *The Nature of Selection: Evolutionary Theory in Philosophical Focus*. Cambridge, MA: The MIT Press.

Weinberg W (1908) On the laws of heredity in man. I. General part. *Zeitschrift für induktive Abstammungs- und Vererbungslehre* 2: 276–330.

See also: **Equilibrium Population; Genetic Drift; Inbreeding; Natural Selection; Panmixis; Segregation Distortion, Mouse**

Harlequin Chromosomes

O J Miller

Copyright © 2001 Academic Press
doi: 10.1006/rwgn.2001.0586

Harlequin chromosomes are metaphase chromosomes whose two sister chromatids show reciprocal patterns of lightly and darkly stained segments along their length. These patterns are the result of multiple sister chromatid exchanges that have been made visible by incorporating bromodeoxyuridine (BrdU) into one strand of the DNA of one chromatid during a previous S-phase and preferentially destroying the BrdU-containing strand before staining with a DNA-binding dye.

See also: **Bloom's Syndrome; DNA Repair**

Heat Shock Proteins

E P M Candido

Copyright © 2001 Academic Press
doi: 10.1006/rwgn.2001.0588

Heat shock proteins (Hsps) are specific proteins that are made when cells are briefly exposed to temperatures above their normal growth temperature. Because Hsps may also be produced by cells exposed to harmful chemicals or to other conditions that cause cellular stress, they are sometimes called stress proteins. The synthesis of Hsps results from a turning on or induction of the genes encoding these proteins, following the temperature increase.

It was first observed in the fruit fly *Drosophila melanogaster* that when either isolated tissues or whole flies were subjected to a heat shock, new proteins, not detectable in unshocked cells, were made. Furthermore, other specific proteins that were present in unshocked cells were made in much greater amounts following a heat shock. Both of these categories of proteins were defined as heat shock proteins (Hsps). The synthesis of Hsps is a universal phenomenon, occurring in all plant and animal species studied, including humans. Hsps are also made by prokaryotic cells, namely the bacteria and archaea. The temperature at which heat shock proteins are induced varies depending upon the normal growth temperature of the species. For instance, fruit flies, normally grown at 25 °C in the laboratory, are heat shocked at 35–37 °C, whereas human or mouse cells are induced to make Hsps when the temperature is raised to several degrees above their normal body temperature of 37 °C, for instance 41–42 °C. Chemicals that can induce Hsps in many cell types include heavy metal ions and arsenite.

Heat shock proteins can be distinguished on the basis of their molecular masses, and are thus conveniently named according to their sizes. Major Hsps in animal cells have molecular masses of approximately 90 000 daltons (Hsp90), 70 000 daltons (Hsp70), 60 000 daltons (Hsp60), and 25 000–30 000 daltons (Hsp25 or Hsp30). These four groups also make up distinct families of Hsps which have characteristic amino acid sequences, three-dimensional structures, and mechanisms of action.

One of the properties of Hsps is the ability to prevent partially unfolded proteins from aggregating to form insoluble complexes. Since Hsps are able to prevent such undesirable interactions, they are also referred to as molecular chaperones. In cells, unfolded or partially unfolded proteins may include those in the process of being made on ribosomes, and therefore not yet folded to their mature state, pre-existing proteins that have become unfolded due to physical or chemical stresses, and proteins that are partially unfolded in the process of their transport across a cell membrane. Thus most Hsps have roles in interacting with unfolded proteins in normal, unstressed cells, and are also of particular importance during exposure to heat or other stressors.

With respect to mechanism of action, the best understood heat shock protein is Hsp60. Hsp60 occurs in mitochondria and chloroplasts of eukaryotic cells, and in the cytoplasm of bacteria. The bacterial Hsp60 is also known as GroEL. GroEL/Hsp60 forms a barrel-shaped complex made up of two stacked seven-membered rings, and acts as a catalyst of protein folding. Partially unfolded protein substrates are bound inside the barrel, where repeated cycles of binding and release lead to their refolding. GroEL/Hsp60 utilizes adenosine triphosphate (ATP) as a source of energy to drive the changes in shape that cause the binding and release of the protein substrate. A large number of proteins made in bacteria rely on GroEL/Hsp60 to attain their correct folded shape, and this chaperone system is essential for the life of the bacterial cell under all temperature conditions.

One of the most prominent Hsps in most cells is Hsp70, and most eukaryotic cells contain several types of Hsp70 with specialized functions. Hsp70 can bind to exposed hydrophobic regions of unfolded protein chains, recognizing lengths of seven to eight amino acids. Like Hsp60, Hsp70 binds and utilizes ATP as a source of energy to power its changes in shape associated with binding and release of substrate proteins. Many proteins that are transported from the cytoplasm of the cell into the mitochondrion are bound by Hsp70, which keeps them in an unfolded state so that they may be threaded through channels in the mitochondrial membrane before they become refolded and functional inside the mitochondrion.

The wide range of functions carried out by heat shock proteins in both normal and stressed cells has made them objects of intense research. They are of interest in a variety of medical studies, including investigations of their roles in stress tolerance, immunity, aging, and neurodegenerative diseases.

Heavy/Light Chains

See: **Globin Genes, Human**

Helicases

D M J Lilley

Copyright © 2001 Academic Press
doi: 10.1006/rwgn.2001.0590

Helicases are ubiquitous enzymes that actively unwind the helical structure of nucleic acids, using the free energy of hydrolysis of nucleoside triphosphates (generally ATP). Both DNA and RNA helicases exist and are important in virtually every transaction undergone by nucleic acids. Unpairing of DNA by DNA helicases is essential in replication, recombination, repair, and chromatin remodeling, while RNA helicases are required in translation, transcription, splicing, RNA processing, editing, mRNA export, and degradation. Common eubacterial DNA helicases include: the PriA protein, which is involved in the assembly of the primasome; DnaB, which acts at the replication fork; Rho, in transcriptional termination; UvrAB, in DNA repair; and RecBCD, in homologous recombination; while SV40 T-antigen is a eukaryotic example. An example of a prominent eukaryotic RNA helicase is the eIF4a protein required for the initiation of translation, one of the common set of RNA helicases containing the amino acid motif DEAD.

Helicases exist in superfamilies, containing common sequence elements. Some of these are the Walker A and B boxes that form nucleotide binding pockets. The free energy of hydrolysis of NTP is used to unwind the nucleic acid and to translocate along it at a high rate. Thus, in general, helicases are DNA- or RNA-dependent ATPases that act as molecular motors. Translocation of DNA helicases is normally unidirectional, but can be either 5′ to 3′ (defined relative to the enzyme-bound strand) or the reverse. The translocation may be more or less processive.

Most helicases act in multimeric form, and a number, exemplified by the Rho protein and T7 Gp4, form hexameric ring structures. Others act in dimeric form, while some such as PcrA and UvrD act as monomers. The structures of some DNA helicases have recently been solved by X-ray crystallography.

See also: **ATP (Adenosine Triphosphate); Nucleic Acid; Rho Factor**

Helicobacter pylori

F Carneiro and C Caldas

Copyright © 2001 Academic Press
doi: 10.1006/rwgn.2001.1578

Helicobacter pylori organisms are spiral, microaerophilic, gram-negative bacteria that colonize the human stomach. *H. pylori* bacteria were identified for the first time in 1982 by Warren and Marshall in Perth, Australia. However, there is evidence to suggest that these organisms colonized the stomach well before we became humans. *H. pylori* infection is one of the most common chronic infections in man. It is believed that until the twentieth century nearly all humans carried *H. pylori* or closely related bacteria

in their stomach. Presently, it is calculated that the infection chronically affects up to 50% of the world's human population. However, there are wide geographical differences in the distribution of *H. pylori*. In most developing countries the infection affects 90% or more of the population, while in developing countries its prevalence ranges from 20 to 50%. The decline of the infection in some parts of the world is most probably related to the improvement of socioeconomic conditions, sanitation, and nutrition. *H. pylori* is thus becoming a 'submerging' rather than an 'emerging' pathogen. The infection is acquired in childhood and, if not treated, persists for the lifetime of the host. *H. pylori* causes acute and chronic inflammation in the stomach (gastritis). The magnitude of the inflammation varies from host to host. Most infected individuals remain asymptomatic throughout their lives; however, in 20–30% of people, organic diseases will develop in the stomach or duodenum such as duodenal ulcer, gastric ulcer, gastric cancer (adenocarcinoma), or mucosa-associated lymphoid tissue (MALT) lymphoma. Diversity of clinical outcomes of *H. pylori* has been attributed to different factors, such as environmental factors (mainly diet), host factors (characteristics of the mucus layer covering the gastric mucosa, immune response, etc.) and virulence factors of *H. pylori* strains.

Genetic studies indicate that *H. pylori* strains are enormously diverse. The complete genomic sequences of two distinct *H. pylori* strains were published in 1997 and 1999. About 1500 genes exist in *H. pylori* strains; a large majority of these have been functionally characterized and a good proportion seems to be *H. pylori* specific. A few genes have been shown to be associated with virulence of the strains, namely *vac*A, *cag*A, and *ice*A genes. The product of the *vac*A gene is a protein with cytotoxic activity that induces vacuolization of human cells. Two distinct regions exist in *vac*A gene, the s (signal) region and the m (middle) region. Within each of these regions several variants can be identified (s1a, s1b, s1c, and s2 in the s region; m1, m2a, and m2b in the m region). Strains typed as s1/m1 have the highest cytotoxic activity. The gene *cag*A encodes a high molecular weight protein whose function is not fully elucidated. This gene is one member of a genomic region that exists in only about 60% of *H. pylori* strains. This region is designated as the pathogenicity island (PAI) and encompasses several virulence-associated genes: the *cag*A gene is considered as a 'marker' of this island; the *ice*A gene is induced by contact with epithelium and exists as two variants, *ice*A1 and *ice*A2. The genetic constitution of the strains has clinical relevance. In Western countries it was shown that a person colonized by a *cag*A+, *vac*A s1 (and *ice*A1?) strain is more likely to develop gastric or duodenal ulcer or gastric cancer. In contrast, people infected with *cag*A⁻, *vac*A s2, and *ice*A2 strains will most probably remain asymptomatic despite developing gastritis. There is a wide variation in the *H. pylori* genotypes colonizing different parts of the world. The similarities between several populations (for instance in the Iberian Peninsula and South America) with respect to the prevalence of specific *H. pylori* genotypes suggest comigration and coevolution of *H. pylori* and humans. These similarities may reflect historical, cultural, and socioeconomic relationships between different areas of the world.

See also: **Adenocarcinomas; Bacterial Genetics**

Helix-Loop-Helix Proteins

See: **DNA-Binding Proteins**

Helix–Turn–Helix Motif

J Read and S Brenner

Copyright © 2001 Academic Press
doi: 10.1006/rwgn.2001.2170

A helix–turn–helix motif is a protein motif that is able to recognize and bind to specific DNA sequences. The motif comprises two α-helix separated by a short β-sheet. One helix interacts with the major groove of the DNA while the other inserts into the DNA and interacts with the bases. Such motifs are commonly found in transcription factors.

See also: **DNA Structure**

Helper Phage

E Kutter

Copyright © 2001 Academic Press
doi: 10.1006/rwgn.2001.0594

Bacteriophage replication normally requires a variety of functions that are routinely supplied by their bacterial hosts. However, some bacteriophages also lack certain additional essential components for their own replication. A second homologous or heterologous phage that can supply such missing components and permit replication and packaging of the other phage is termed a helper phage. The best-studied natural heterologous system requiring such a relationship is the bacteriophage P2–P4 system. Bacteriophage P4

is missing all the genes for structural phage proteins. It can only complete its replication cycle and make phage particles when the cell is simultaneously infected with another temperate phage like P2. This is a particularly interesting case, since the two phages have no sequence homology and the P4 head is only about one-third as large as that for P2, reflecting the relative sizes of their genomes. Somehow, a P4 protein (*sid*) is able to tell the P2 capsid protein, gpN, to assemble into a very different structure than it would normally make. The same five P2 genes are also required to make the P2 and P4 capsids. The switch is total; no P2 phage are made under these circumstances when P2 is acting as a helper phage for P4 assembly. The term 'helper phage' can also be applied to a second phage in the same family that permits the replication of a phage damaged by UV, chemicals, or X-rays or one mutated in genes that are essential to survival. Molecular biologists have also designed a number of clever systems for packaging foreign DNA into phage particles; most of these involve especially designed helper phages that provide the necessary components and packaging machinery without themselves being replicated under the conditions used for packaging the foreign DNA.

See also: **Bacteriophages; Temperate Phage**

Hemizygote

Copyright © 2001 Academic Press
doi: 10.1006/rwgn.2001.1855

The term 'hemizygote' refers to a nucleus, cell, or organism that possesses only one of a normally diploid set of genes.

See also: **Heterogenote; Homozygosity**

Hemoglobin

See: **Globin Genes, Human**

Hemophilia

F Giannelli

Copyright © 2001 Academic Press
doi: 10.1006/rwgn.2001.0597

Hemophilia is the name shared by two X-linked recessively inherited defects of blood coagulation. These manifest as spontaneous or excessive bleeding following minor surgery or trauma. The bleeding episodes may occasionally threaten life and in the long run may cause serious disability, especially by damaging joints. Hemophilia is due to defects in either the gene for coagulation factor VIII or that for factor IX. Mutations of the factor VIII gene cause hemophilia A, or classic hemophilia, while those of the factor IX gene cause hemophilia B, or Christmas disease.

Population Genetics

Both hemophilias have been maintained in the population by an equilibrium between mutation and selection against the affected males. The latter causes a loss of hemophilia genes at each generation equal to $[(1 - f)\, I/3]$, where f is the chance that a patient will produce offspring relative to that of a normal male, and I is the incidence of the disease. The value of f was about 0.5 prior to the introduction of modern treatment for both hemophilias, so that existing hemophilia genes were lost from the population and were replaced by new mutations at a rate of 1/6 per generation. As a result both diseases show a high degree of mutational heterogeneity. In the second half of the last century the value of f is thought to have increased in developed countries because of treatment and better patients' health. In this situation, since the mutation rates are expected to remain unchanged, the incidence of the disease rises, eventually to reach a new equilibrium between mutation and selection. Currently, in the UK, the incidence of hemophilia A and B is respectively 1 per 5000 and 1 per 30 000 males.

Factor VIII Gene

This gene is in band Xq28, 1.5 Mb from the telomere. It spans 186 kb, contains 26 exons, and is oriented so that the promoter lies telomeric to the rest of the gene. A CpG island in intron 22 of the factor VIII gene is the origin of two nested genes: F8A and F8B. The first is a 1.8 kb intronless gene entirely contained within intron 22 and transcribed in opposite orientation to the factor VIII gene. The second is transcribed in the orientation of the factor VIII gene, and its message contains a specific first exon followed by exons 23 to 26 of the factor VIII gene. The CpG island at the origin of the F8A and F8B genes is part of a 9503 bp segment of intron 22 of the factor VIII gene called *int22h* that is found repeated in opposite orientation 350 and 450 kb telomeric to the factor VIII gene. These three repeats are designated *int22h-1, −2,* and *−3* according to their increasing distance from the centromere. The factor VIII gene produces a mRNA of 9028 nucleotides.

Factor IX Gene

The factor IX gene is near the boundary between Xq26 and Xq27. It spans 33.5 kb, contains 8 exons, and produces a message of 2802 nt. This gene appears to derive from an ancestral gene that gives rise to three more genes encoding proteins of blood coagulation: factors VII and X, and protein C.

Factor VIII and Factor IX

Factor IX is a serine protease that cleaves and activates factor X in the proteolytic cascade that results in the conversion of fibrinogen into fibrin, and hence in blood coagulation. Factor VIII is the cofactor that associates with factor IX to ensure physiologic levels of factor X activation. The complex of factors VIII and IX is, more generally, responsible for maintaining the coagulation cascade after its initiation by factor VII and tissue factor.

Factor IX is synthesized with a signal peptide consisting of a prepeptide that is cleaved upon transport to the endoplasmic reticulum and a propeptide that is cleaved prior to secretion. The latter is important for interaction with the enzyme that γ-carboxylates the first 12 glutamates of circulating factor IX. This circulating protein consists of 415 amino acids organized in the following domains:

1. The gla domain, containing the γ-carboxylated glutamates important for Ca^{2+} binding and affinity for phospholipidic membranes.
2. Two epidermal growth-factor-like domains important for protein–protein interactions.
3. An activation domain that is cleaved to release residues 146–180 and activate factor IX.
4. The catalytic or serine protease domain homologous to trypsin and other members of this family of proteases. Factor IX posttranslational modification includes γ-carboxylation, N- and O-glycosylation of different residues, and partial β-hydroxylation of aspartate 64.

Factor VIII is synthesized with a prepeptide of 19 residues that is cleaved off prior to secretion. The remaining 2332 residues of factor VIII are organized in the domain structure $A_1a_1A_2Ba_2A_3C_1C_2$, where A_{1-3} are homologous to the domains of ceruloplasmin (a copper ion-binding protein), a_1 and a_2 are small acidic peptides, B is a unique domain encoded by an exon (number 14) of 3106 bp, and C_1 and C_2 are homologous to milk-fat globule-binding protein. Prior to secretion the protein is extensively modified by N- and O-glycosylation of several residues and sulfation of six tyrosines. In addition, it is cleaved at the B/a_2 boundary and at variable positions within the

B domain. The heterodimer ($A_1a_1A_2B + a_2A_3C_1C_2$) is the inactive circulating form of factor VIII and is carried by a large multimeric protein: von Willebrand factor. This protects factor VIII and slows down its clearance. Mutations of von Willebrand factor that only affect its factor VIII binding property may therefore mimic hemophilia A.

Factor VIII is activated by cleavage at the A_1a_1/A_2 and a_2/A_3 boundary while any residual B domain is eliminated by cleavage at the A_2/B boundary. The heterotrimeric ($A_1a_1 + A_2 + A_3C_1C_2$) active form of factor VIII is unstable and may become inactive either by spontaneous dissociation of the A_2 chain or by enzymatic cleavage at the A_1/a_1 boundary or within the A_2 chain. Activated protein C operates these cleavages as part of a negative feedback control on blood coagulation.

Factor VIII is homologous to coagulation factor V. This, however, has a clearly distinct B domain and lacks the a_1 and a_2 acidic peptides.

Mutations Causing Hemophilia A

The severity of hemophilia is a function of the gene mutation and is directly related to the deficit of coagulant factor activity. A quarter of all hemophilia A cases is due to gross gene rearrangements and, in particular, 5% is due to gross gene deletions and 20% to inversions of 500 or 600 kb breaking intron 22 of the factor IX gene. These inversions are due to intrachromosome or intrachromatid homologous recombination between the *int22h-1* sequence of the factor VIII gene and either *int22h-2* or *int22h-3*. However, *int22h-3* is involved five times more frequently than *int22-2*. The inversions appear to occur at the rate of $4–7 \times 10^{-6}$ per gamete per generation and account for nearly half the patients with severe hemophilia A.

Approximately 75% of hemophilia A mutations are base substitutions or small deletions/insertions. These may act by (1) leading to abnormal RNA splicing through damage to normal or creation of abnormal splicing signals; (2) causing premature termination of translation (frameshifts, nonsense codons); or (3) producing subtle protein changes such as amino acid deletions or amino acid substitutions. So far, 228 different missense mutations (i.e., mutations causing amino acid substitutions) have been found in the factor VIII gene of hemophilic patients but more than 500 different missense mutations are expected to be capable of causing hemophilia A. Promoter mutations causing hemophilia A have not been reported so far.

Most hemophilia A mutations arise in the male germline and this appears especially true of the inversions breaking intron 22.

Mutations Causing Hemophilia B

Less than 2% of hemophilia B cases is due to gross rearrangements, represented generally by gross deletions that may even remove the entire gene. The other mutations are base substitutions and small deletions/insertions.

Three per cent of the mutations affect the promoter of the factor IX gene and usually cause a disease that markedly improves after puberty and may become asymptomatic. This is called Leyden-type hemophilia B. The exceptions, so far, are two different substitutions at nucleotide 26 that cause nonimproving hemophilia. Since residue 26 is part of an androgen receptor binding site, it can be argued that binding of the ligand saturated androgen receptor at this site may restore the promoter activity impaired by the mutations causing Leyden-type hemophilia B, while serious damage to the same site irretrievably damages promoter activity.

Other mutations damage or create RNA splicing signals ($\sim 12\%$); cause frameshifts ($\sim 4\%$); generate nonsense codons ($\sim 12\%$); or result in amino acid deletion ($\sim 2\%$) or substition ($\sim 63\%$). So far, 425 different missense mutations have been found in the factor IX gene of hemophilia B patients.

Factor IX mutations occur eight to nine times more frequently in the male than in the female germline.

Genotype–Phenotype Correlations

In both hemophilia A and B frameshifts and nonsense mutations tend to cause severe disease with absence of coagulant protein in circulation; in hemophilia B this seems to be true irrespective of the position of the premature translation stop signal. Splicing and missense mutations may cause mild, moderate, or severe disease. Missense mutations may simply impair the function of the coagulant factor; cause gross reduction or virtual absence of the coagulation factor in the blood; or decrease the amount of protein in circulation as well as reducing its specific activity.

Mutations expected to prevent the synthesis of 'near-normal' coagulant proteins such as gross or complete gene deletions, frameshifts, nonsense mutations, and inversions breaking intron 22 of the factor VIII gene, predispose to the inhibitor complication. This entails the development of antibodies against the coagulant factor used in replacement therapy, so that the patient becomes refractory to standard treatment. Predisposition to manufacture such antibodies is probably due to failure to develop tolerance to the relevant coagulation factor because of inadequate exposure to the factor during maturation of the immune system.

Mutational Heterogeneity and its Relevance to Genetic Counseling

Most hemophilia A and B mutations are unique, but frequent repeats of some mutations may occur. In some instances this is due to founder effects and tends to be restricted to the populations the founders belonged to, while in others it is due to mutational hotspots such as the *int22h* regions for the common inversions causing hemophilia A, or CpG sites. The latter undergo transition mutations at 10 times the rate of other sites.

In general, hemophilia A and B mutations are of recent origin, and a significant proportion is less than three generations old. This, together with the small size of modern families, allows as many as half the hemophilia families to appear sporadic. These families are unsuited to methods for carrier and prenatal diagnosis based on the analysis of the intra-familial segregation of polymorphic markers, and instead direct characterization of the gene defect is needed.

A strategy that allows optimal genetic counseling and rapid progress in the understanding of the molecular biology of the relevant disease is based on the construction of national confidential databases of mutations and pedigrees. The databases are assembled by characterizing the mutation of an index patient from each family and collecting the family's pedigree. Carrier and prenatal diagnoses can then be based on detection of the defect specific to each family, and can be made for all the at-risk blood relatives of the index patient, for generation after generation. In the UK such a database has been constructed for hemophilia B, and that for hemophilia A is being assembled.

The high mutational heterogeneity of the hemophilias makes the analysis of natural mutants a very efficient way of investigating the features that are important to the function of factors VIII and IX and their genes.

Treatment of Hemophilias A and B

Replacement therapy is available for hemophilia A and B and is based on intravenous administration of concentrates of factor VIII and factor IX, respectively. These factors are either purified from blood donations or from cultures of cells expressing the recombinant factors.

Work for the development of gene therapy is ongoing but is still at the animal-experimentation phase. Human application requires safe and efficient methods of gene delivery capable of ensuring satisfactory and stable gene expression.

Further Reading

Tuddenham EGD and Cooper DN (1994) *The Molecular Genetics of Haemostasis and its Inherited Disorders*, Oxford Monographs on Medical Genetics no. 25. Oxford: Oxford University Press.

Bloom AL, Forbes CD, Thomas DP and Tuddenham EGD (eds) (1994) *Haemostasis and Thrombosis*. Edinburgh: Churchill Livingstone.

See also: **Genetic Counseling; Genetic Diseases; Sex Linkage**

Hereditary Diseases

D E Wilcox

Copyright © 2001 Academic Press
doi: 10.1006/rwgn.2001.0599

Hereditary diseases are those whose causation has a genetic component. This component is caused by transmissible change(s) in the genetic material. The heritability of a disease is a measure of the relative proportions of genetic and environmental factors. A single gene disorder with little environmental influence, e.g., Duchenne muscular dystrophy in humans, will have a high heritability. A multifactorial disorder, e.g., congenital heart disease, will have a lower heritability. The hereditary component of a disease may be caused by a single gene, multiple genes (polygenic), or various chromosome abnormalities such as deletions or translocations.

See also: **Clinical Genetics; Congenital Disorders; Genetic Diseases**

Hereditary Neoplasia

L M Mulligan

Copyright © 2001 Academic Press
doi: 10.1006/rwgn.2001.1579

Cancers or groups of related cancers which occur with an increased frequency in families, as compared to the general population, due to genetic risk factors may be termed hereditary neoplasia. These diseases result from the inheritance of a mutation in a tumor suppressor gene or (rarely) an oncogene which makes the individual susceptible to developing the specific tumor type(s). Although risks of developing an hereditary neoplasm may be very high for individuals within such families, overall, only 10–15% of cancer falls into this category. Characteristic features of hereditary neoplasia include an early age of disease onset and the occurrence of multiple primary tumors. It should be stressed that hereditary neoplasia refers to inheritance of a susceptibility allele or alleles but not to inheritance of a cancer phenotype *per se*.

See also: **Cancer Susceptibility; Tumor Suppressor Genes**

Heritability
W G Hill

Copyright © 2001 Academic Press
doi: 10.1006/rwgn.2001.0600

Heritability is a commonly used and important term to describe properties of the inheritance of quantitative traits, such as stature in man or milk yield of cows. Informally, heritability (h^2) is the proportion of the variation in the trait due to genetic differences between individuals, but a more precise definition of heritability is important because the term is both widely used and widely misused. Correlations among relatives and response to directional selection are proportional to the heritability. Although a property of a specific trait in a specific population, it is found that heritabilities of similar traits take similar values in different species and populations. The first use of the word 'heritability' is uncertain, but it is most often associated with Jay L. Lush, who applied the theory of quantitative genetics of Sewall Wright and R.A. Fisher to animal breeding.

Definition

The observed performance or phenotypic value, P, of an individual for a quantitative trait can be partitioned in a simple additive model into two components, genotypic value (G) and environmental deviation (E) as:

$$P = G + E$$

A genotype × environment interaction term, GE, can also be included in the model, but cannot usually be distinguished from the environmental deviation, E, as each environmental deviation is unique. Because individuals transmit only one gene at each locus to their offspring, the other copy coming from the second parent, in describing correlations among most relatives and in predicting responses to selection it is necessary to consider the average performance of individuals who receive one copy of a specified gene and

the other at random from the population. Fisher described this in 1918 as the average effect of the gene. The breeding value, A, of an individual is the sum of the average effects of the genes it carries. More simply and practically, the breeding value of an individual is defined as twice the expected deviation of the mean of its progeny, if randomly mated, from the population mean; but these definitions are the same unless there is epistasis. The dominance deviation, D, defines differences between genotypic value and breeding value which are due to interactions between genes at individual loci, and the epistatic deviation, I, defines differences due to interactions between different loci. (I can be further partitioned into additive × additive, additive × dominant and other terms.) Hence a fuller model is:

$$P = A + D + I + E$$

Variation among individuals in phenotypic value, V_P, can now be partitioned into components:

$$V_P = V_G + V_E = V_A + V_D + V_I + V_E$$

assuming that correlations between or interactions of genotype and environment can be ignored or catered for in other ways.

There are two different definitions of heritability:

- Heritability in the broad sense: $H^2 = V_G/V_P$
- Heritability in the narrow sense, or simply 'heritability': $h^2 = V_A/V_P$

In most situations, particularly when describing correlations among relatives or in predicting response to selection, heritability in the narrow sense is the more useful quantity and is implied. Heritability appears as a squared term, because h was first defined by Sewall Wright in 1918 as the path coefficient from genotype (or breeding value, since he included no dominance term) to phenotype. As h also equals the correlation between breeding value (A) and phenotype (P), h is the accuracy of selection on phenotype. It also follows that h^2 is the regression of breeding value on phenotype. Further, since $h = \mathrm{corr}(A, P)$, the variance in A which is not explained by P is $V(A|P) = (1 - h^2)V_A$.

Magnitude

Because the amount of genetic variance depends on the frequencies and effects of genes at many loci, and the environmental variation depends on the environment in which individuals are kept, heritability differs among traits, species, populations within species, and over time. In practice, however, it turns out that each

Table 1   Heritability values for different species and traits

Species and traits	h^2 (%)
Drosophila[1]	
Life history traits (longevity, fecundity, development time)	12
Behavioral traits (locomotion, mating activity, geo- and phototaxis)	18
Morphological traits (bristle number, wing and thorax size)	32
Pig[2]	
Reproductive rate (litter size)	10
Growth rate (daily gain, feed intake, and conversion efficiency)	30
Morphology (backfat, carcass lean %)	45
Humans	
IQ (meta-analysis)[3]	34
	48
	(broad sense)
Stature[4]	65
Finger ridge count[5]	>95

[1]Roff and Mousseau (1987).
[2]Rothschild and Ruvinsky (1998).
[3]Devlin *et al.* (1997).
[4]Roberts *et al.* (1978).
[5]Holt (1955).

kind of trait has a typical heritability value, which is often similar among very different species. Some general values are given in **Table 1**.

Estimation

Relatives resemble each other because they have genes in common, and the closer the relationship the more likely they are to share genes and the more highly correlated are their phenotypes for quantitative traits. Similarly, the higher the heritability, the more highly correlated are the phenotypes of relatives. Heritability is therefore estimated from the resemblance between relatives, scaled to take account of the relationship.

There are two major problems in estimating heritability. The first is to avoid confounding of the correlations among relatives by nongenetic causes such as shared environment. It is therefore much easier to get good estimates in well-designed experiments in laboratory animals than in man. The second problem is to get sufficient data to provide accurate estimates; and while estimates from distant relatives may be less confounded by common environment, they estimate only a fraction of h^2 and so have to be scaled up and have a high sampling error.

Heritability in the broad sense can be estimated from the correlation of phenotypes of individuals which have the same genotype, i.e., clones or identical twins. In plants this can be feasible, whereas in humans identical twins share the pre- and postnatal environment. In cattle, identical twins formed by embryo splitting can be reared in different foster mothers; in humans adoptive twins do not share postnatal environment.

Parent and Offspring

The covariance of parent and offspring, which have precisely one (autosomal) gene in common at each locus, and therefore half their genotype, equals $V_A/2$. Hence if individuals are sampled at random, the correlation of phenotype of offspring and individual parent is $h^2/2$, and similarly the regression of phenotype of offspring on one of its parents is $h^2/2$. (The word regression was coined by Galton to describe the fact that extreme parents had less extreme offspring.) Hence if a set of data on parent and offspring are collected and the regression ($\pm$SE) of progeny on parent phenotype is 0.2 ($\pm$ 0.1), then the estimate of heritability is 0.4 ($\pm$ 0.2). Because it is easier to deal with large numbers of offspring and because the estimate is not biased by selection on the trait (providing it is only on that trait), it is usual to use the regression rather than correlation as the estimator. If the phenotype of offspring is regressed on parental average for a trait measured on both parents, the regression coefficient estimates h^2. Maternal effects can bias estimates from offspring–parent regression or correlation, for example in body weight due to family environment in man or the association between dam's milk production and weight in cattle. If there is nonrandom mating among parents, as in humans for stature, the regression or correlation of offspring on individual parent is biased (upward with positive assortative mating), but the regression on mid-parent is not.

Full and Half Sibs

Full sibs share 0, 1, or 2 parental genes at each locus, with respective probabilities 1/4, 1/2, and 1/4, and half sibs 0 or 1, each with probability 1/2. It follows that the genetic covariance of full sibs equals $V_A/2 + V_D/4$ and of half sibs $V_A/4$ (plus some epistatic terms). Typically, however, full sibs also share a common environment, for example both pre- and postnatal in mammals, which contributes a variance, V_C, to the variance among families or covariance between family members. The environmental correlation is often called the c^2 term, where $c^2 = V_C/V_P$. Of course, there are designs in which V_C is eliminated among full sibs (e.g., embryo transfer) and others where it is present among half sibs (e.g., in plants where maternal half sibs are the norm, and in animals where half sibs are raised together). Data from experiments or field trials are typically subjected to analysis of variance, and heritability is estimated from the intraclass correlation. Assuming there is no confounding, this correlation is an estimate of $h^2/2$ for full sibs and $h^2/4$ for half sibs. In mammals in which each male has several mates, both the full and half-sib correlations can be estimated for the same experiment. The half-sib estimate is usually taken because it is less likely to be confounded by common environment and dominance, although it has a higher sampling error. As for the offspring–parent correlation, positive assortative mating can increase the correlation among sibs.

Twins

There are considerable problems in eliminating common environment effects for heritability estimation in man. The use of twins provides a route, specifically by comparing the correlations of identical (monozygous, MZ) and nonidentical (dizygous, DZ) twins. If the MZ correlation is assumed to equal $h^2 + c^2$ and the DZ correlation $h^2/2 + c^2$, then an estimate of heritability is $2[\mathrm{corr(MZ)} - \mathrm{corr(DZ)}]$. This is, however, biased upward by all nonadditive genetic effects (for dominance the MZ covariance includes V_D and the DZ includes $V_D/4$) and epistasis, and by any extra similarity of environment that MZ share over DZ through their treatment or behavior.

Combination of Information

So as to make best use of information on all relatives, particularly from field data, sophisticated models and computer-intensive statistical methods using (restricted) maximum likelihood or Bayes' theorem (via Gibbs sampling) are adopted. These incorporate correlations among all relatives, suitably weighted for relationship and numbers of records, and account for identifiable environmental differences such as farms or years of birth. Such methods are replacing simple regression or correlation analyses in many applications because they are efficient, enable the precision of an estimate of heritability to be computed accurately, and enable successively more complicated models to be fitted and tested. Thus the shape of the likelihood curve describes the degree of support for a particular value of the heritability. In a Bayesian context, the posterior distribution of heritability fulfills a similar role. Also, for example, a likelihood ratio test can be used to check whether a nonadditive genetic component is important.

Selection Response

As the regression of offspring on mid-parent phenotype equals heritability and is linear or close to linear under polygenic inheritance (exactly linear under multivariate normality), the regression of the offspring of a group of individuals on the mean of their parents' phenotype also equals h^2. Hence, if a group of individuals are selected which differ in phenotype by an amount S, the selection differential, their offspring will be expected to deviate in performance from those of unselected parents by an amount $h^2 S$. This is the selection response, given by $R = h^2 S$, the classical prediction equation of quantitative genetics. Therefore, providing environmental change over generations can be eliminated, or corrected for by maintaining an unselected control population alongside the selected population, and heritability can be estimated from the response to selection as Falconer's 'realized heritability,' $h^2 = R/S$. If selection is practiced over several generations the heritability may not change much, in which case the (realized) heritability can be estimated from the regression of cumulative response over generations on the cumulative section differential.

Uses

Heritability tells us no more than the additive genetic variance and phenotypic variance do separately, but it is a useful summary and descriptive parameter. Just as the correlation among relatives can be used to estimate heritability, so the heritability can be used to predict the correlation of relatives. Prediction of the expected phenotype of offspring of selected individuals (equal to the breeding values of these individuals) and thus of selection response is probably the most important practical use of the heritability estimate. A comparison between the heritability predicted from collateral relatives such as half sibs and the realized heritability or selection response provides a check on quantitative genetics theory (whereas comparison of realized heritability and regression of offspring on parent does not, for they are based on the same principles).

Discrete Traits

Although primarily used for traits with continuous expression such as stature, heritability can also be applied to traits with discrete phenotypes. Traits with many categories such as litter size in pigs can be treated as continuous. There are alternative methods for traits which have only two or so classes, but no simple Mendelian expression, such as survival to weaning, incidence of twinning in man or cattle, or incidence of a congenital defect such as club foot, can also be analyzed. One way is simply to regard the traits as having two values, say 1 (affected) and 0 (unaffected), and ignore any nonlinearity or heterogeneity of variance. More naturally within the quantitative genetics framework, the discrete (all-or-none) trait can be considered as the expression of some underlying continuous variable liability, such as level of circulating hormone or strength of immune reaction, with a threshold value above which affected individuals lie. Heritability on the all-or-none and on the underlying liability scale are functions of each other: the former is always lower, the difference widening the further the incidence of the trait departs from one-half. Methods were developed by Falconer to estimate heritability on the liability scale directly from the frequencies of the trait in the population as a whole and in the relatives of affected individuals, by analogy with a selection experiment in which the latter play the role of offspring in the next generation of selected (affected) individuals.

Some Misinterpretations

The magnitude of the heritability does not tell us a lot of things. For example, as it applies to individuals within populations, it cannot be used to predict genetic differences between races or other populations from phenotypic differences, whether or not they share the same environment.

The prediction formula $R = h^2 S$ applies only (other than in very special circumstances) if selection is practiced on the trait on which response is measured. If selection is practiced on some trait or combination of traits other than the one of interest, the regression of response on selection differential is not therefore an unbiased estimate of heritability, but depends *inter alia* on the genetic and phenotypic correlations among the traits. This is a serious problem in inferences about selection in nature, where the actual selection applied is not known. Methods exist to overcome this problem, but require that records be available on **all** traits on which selection is practiced or to which fitness is related.

As the heritability is a summary parameter over loci, it does not tell us about either the numbers of genes that affect a quantitative trait or the magnitude of their effects. It is not therefore a constant as a population changes. But heritability is nevertheless a useful concept when properly used.

Further Reading

Falconer DS and Mackay TFC (1996) *Introduction to Quantitative Genetics*, 4th edn. Harlow, UK: Longman.

Hartl DL and Clark AG (1997) *The Genetics of Populations*, 3rd edn. Sunderland, MA: Sinauer Associates.

Kearsey MJ and Pooni HS (1996) *The Genetical Analysis of Quantitative Traits*. London: Chapman & Hall.

Lynch M and Walsh B (1998) *Genetics and Analysis of Quantitative Traits*. Sunderland, MA: Sinauer Associates.

Roff DA (1997) *Evolutionary Quantitative Genetics*. New York: Chapman & Hall.

References

Devlin B, Daniels M and Roeder K (1997) The heritability of IQ. *Nature* 388: 468–471.

Holt SB (1955) Genetics of dermal ridges: frequency distribution of total finger ridge count. *Annals of Human Genetics* 20: 270–281.

Roberts DF, Billewicz WZ and McGregor IA (1978) Heritability of stature in a West Indian population. *Annals of Human Genetics* 42: 15–24.

Roff DA and Mousseau TA (1987) Quantative genetics and fitness: lessons from *Drosophila*. *Heredity* 58: 103–118.

Rothschild MF and Ruvinsky A (eds) (1998) *The Genetics of the Pig*. Wallingford, UK: CAB International.

See also: **Additive Genetic Variance; Artificial Selection; Genetic Variation**

Hermaphrodite

M A Ferguson-Smith

Copyright © 2001 Academic Press
doi: 10.1006/rwgn.2001.0601

A hermaphrodite is an individual that possesses both male and female gonads, theoretically capable of producing both sperm and ova. The situation is normal in some species of plants, can occur uncommonly in some amphibia, birds, and fish but only rarely in mammals, where it is usually associated with infertility. Humans with both testicular and ovarian tissue are usually described in the scientific literature as *true* hermaphrodites, to distinguish them from male and female pseudohermaphrodites who may show sex reversal in the presence of testes and ovaries respectively.

Various anatomical varieties of true hermaphroditism are described. Lateral hermaphrodites have a testis on one side and an ovary on the other. Spermatogonia may be observed in the testis and oogonia in the ovary in lateral hermaphroditism. More commonly a compound gonad, or ovotestis, is present either unilaterally or bilaterally. Ooogonia and developing oocytes may be present in the ovarian part of the ovotestis but the testicular structure is usually devoid of spermatogonia after puberty; in fact, degenerating oocytes are occasionally seen within testicular tubules. Differentiation of the internal genital ducts depends on the nature of the ipsilateral gonad. An ovary is always associated with a normal fallopian tube and at least partial development of the uterus and absence of the Wolffian ducts on the same side. In lateral hermaphroditism this results in a unicornuate uterus and tube associated with the ovary, and a vas deferens, seminal vesicle, and regression of the uterus and tube on the side of the testis. An ovotestis is usually associated with development of the Mullerian ducts and regression of the Wolffian ducts. In all types of true hermaphroditism, the presence of testicular tissue leads to ambiguity of the external genitalia with posterior fusion of the labial folds and clitoral enlargement. At puberty, there is breast development with the formation of both glandular and ductal components and menstruation may occur.

In most patients with true hermaphroditism no cause can be found and the chromosome constitution is indistinguishable from that of a normal female. A small number of cases are described with mosaicism for XXY and XX cells. In rare cases there is true chimerism in which both normal 46, XY (male) cells and 46, XX (female) cells coexist in the same individual. A double contribution of alleles from each parent at a number of genetic loci confirms an origin from the fusion of two fertilized eggs, or the double fertilization of a diploid egg. In equally rare cases, the condition is due to abnormal recombination between the X and the Y during paternal meiosis whereby the sex-determining region of the Y is transferred to the end of the short arm of the X. It is presumed that random X inactivation leads to the development of testis-inducing and ovary-inducing populations of cells in the early embryo, a situation analogous to XX/XY chimerism. It is noteworthy that experimental XX/XY chimerism in mice, produced either by blastocyst fusion or by injection of donor embryonic stem cells into the recipient blastocyst, may lead to hermaphroditic phenotypes identical to those found in true hermaphroditism in humans. It is also of interest that most examples of XX/XY chimerism in mice are associated with an unambiguous male phenotype. X–Y interchange in humans also most often leads to a male phenotype in infertile, so-called XX males with features of Klinefelter syndrome (see Klinefelter Syndrome). Very rarely, XX males and XX true hermaphrodites have been identified in the same pedigree; the cause is so far unexplained.

See also: Chimera; Intersex; Klinefelter Syndrome; Sex Reversal

Hershey, Alfred

W C Summers

Copyright © 2001 Academic Press
doi: 10.1006/rwgn.2001.0603

Alfred Day Hershey (1908–97), an American geneticist, was born 4 December 1908 in Owosso, Michigan, and received his BS (1930) and PhD (1934) from Michigan State College (East Lansing). He was a faculty member in the Department of Bacteriology at Washington University (St Louis) from 1934 to 1950, when he joined the Department of Genetics of the Carnegie Institution of Washington at Cold Spring Harbor, New York. His research focused on the genetics of bacteria and bacteriophages and he made important contributions to the understanding of the nature of genes, their replication and recombination. Among many honors, he received the Nobel Prize in Physiology or Medicine in 1969, sharing it with Max Delbrück and Salvador Luria. He died 22 May 1997.

Hershey's early research at Washington University was carried out in collaboration with Jacques Bronfenbrenner, a well-known immunologist and early bacteriophage worker. They studied the metabolism of bacteria before and after phage infection. In 1943 Hershey, Delbrück, and Luria initiated a series of periodic meetings to discuss their mutual interests in bacteriophage biology, an event which is often viewed as the start of the research school now known as the "American Phage Group."

In his work during the 1940s, Hershey developed the bacteriophage T2 as a genetic organism. He found both host-range and plaque-morphology mutants and showed that coinfection with two different parental phage allowed detection of genetic recombination in bacteriophage. Through this work, he showed that T2 phage was an ideal organism to study basic genetic mechanisms. One class of his plaque-morphology mutants turned out to be an unusual type of host-range mutant as well, the rapid-lysis (r) mutants. Analysis of the rII locus in T-even phage provided deep insight into the nature of the gene and the genetic code.

Study of the process of phage infection and multiplication led Hershey to devise methods to interrupt phage infection by hydrodynamic shearing of the bacterium–bacteriophage complex. With this technique, Hershey and his collaborator Martha Chase carried out their most famous work, an experiment that came to be known as the 'Hershey–Chase Experiment.' (Because they used a common food blender to shear the bacterial culture, the experiment is also called the 'Blender Experiment.') Using newly available radioactive tracers for metabolic labeling of the protein (^{35}S) and nucleic acid (^{32}P) components of phage T2, they sheared the phage-infected complexes after a time when shearing would not prevent intracellular phage production. They found that the protein and nucleic acid components of the phage dissociated upon infection, with most of the protein remaining susceptible to removal by shearing while most of the nucleic acid had entered the bacterial cell and was thus protected from the external shear forces.

The interpretation they cautiously presented was that the proteinaceous phage coat remained outside the cell, while the DNA was injected into the cell. This result was immediately taken as confirmation that the DNA was the substance which was associated with the genetic continuity of the phage and that the protein coat was merely a transport vehicle. This experiment is usually described in idealized terms, although the actual data presented by Hershey and Chase certainly allowed for some possible protein to accompany the DNA into the cell.

In the 1960s Hershey turned his attention to the lysogenic phage lambda and devised simple yet elegant approaches to study the physical states of the lambda DNA. He pioneered methods for dealing with large DNA molecules, which are highly sensitive to breakage by shear forces in solutions. His methods for DNA extraction (phenol) and zone sedimentation (in sucrose gradients) allowed him to show that lambda DNA existed in both linear and circular forms, and that it has unpaired (presumably complementary) cohesive termini. This work was seminal in developing our current understanding of lysogeny as well as in the applications of lambda bacteriophage in recombinant DNA technologies.

See also: Bacteriophages; Delbrück, Max; Luria, Salvador

Heteroallele

F W Stahl

Copyright © 2001 Academic Press
doi: 10.1006/rwgn.2001.0604

Heteroalleles are alternative mutant forms of a given gene resident at the same locus.

Heteroallelic Complementation

A heteroallelic diploid is characteristically mutant in phenotype. A heteroallelic diploid that has wild-type or quasi wild-type phenotype is said to manifest interallelic (or intragenic) complementation. Such complementation often reflects either a multimeric state of the functional protein product of that gene or two or more domains within the protein manifesting more or less independent functions.

Heteroallelic Recombination

When the altered nucleotide sequences defining the two heteroalleles are not overlapping, interallelic (intragenic) recombination can generate the wild-type as well as the doubly mutant allele. When genes are small (intron-free), recombination between heteroalleles usually occurs by gene conversion.

History

In the 1940s and 1950s, demonstrations of interallelic complementation and recombination strained the classical definition of a gene. Complementation between mutations is a classical demonstration that two mutations are in separate genes, defined as units of function. However, understanding of quarternary protein structure soon rationalized the exceptional cases of heteroallelic complementation. Recombination between mutants is a classical demonstration that the two mutations are in separate genes, defined as units of recombination. However, analysis of the *r*ll gene of bacteriophage T4 combined with the Watson–Crick hypothesis for DNA structure established the modern view that a gene is a segment of a continuous DNA duplex with recombination possible between any pair of adjacent nucleotides (Benzer, 1955).

Reference

Benzer S (1955) Fine structure of a genetic region in bacteriophage. *Proceedings of the National Academy of Sciences*, USA 41: 344–354.

See also: Complementation Test; Gene Conversion

Heterochromatin

A T Sumner

Copyright © 2001 Academic Press
doi: 10.1006/rwgn.2001.0605

Heterochromatin was originally defined by Heitz in 1928 as chromosome segments that failed to decondense at the end of telophase, but which remained condensed throughout interphase, and which appeared as condensed segments at the following prophase, that is, it showed positive heteropyknosis. Subsequently, it was realized that there is more than one class of heterochromatin. 'Constitutive heterochromatin' is found at virtually all stages of an organism's life cycle, in the same place on both of a pair of homologs, can be stained by specific methods, and generally contains distinctive types of DNA. 'Facultative heterochromatin,' on the other hand, only occurs in one of a pair of homologs, cannot generally be stained distinctively, and necessarily contains the same type of DNA as that found in the nonheterochromatic homolog. The best-known example of the latter is the inactive X chromosome of female mammals.

Constitutive Heterochromatin

Constitutive heterochromatin is most easily demonstrated using C-banding; a variety of other chromosome banding methods produce specific staining of certain heterochromatic regions of chromosomes in certain species. Characteristically, constitutive heterochromatin consists largely of highly repetitive ('satellite') DNA, although blocks of heterochromatin may not necessarily consist exclusively of such DNA, and in some species moderately repetitive rather than highly repetitive DNA seems to be present. The DNA of constitutive heterochromatin is late-replicating, and in mammals, its cytosines are often methylated. A number of proteins have been described that are either specific to, or concentrated in, constitutive heterochromatin; such proteins may well be involved in the condensed state of heterochromatin. Heterochromatin has generally been regarded as genetically inert. The quantity in the genome can vary extensively without any apparent phenotypic effects. In *Drosophila* it is not replicated during polytenization of chromosomes, and in certain other organisms heterochromatin is eliminated in somatic cells, and retained only in the germline. The highly repetitive DNA sequences found in most heterochromatin could not be translated into proteins. Nevertheless, constitutive heterochromatin is not without effects. It can have profound effects on the position and number of chiasmata at meiosis; induce the inactivation of genes close to it (position-effect variegation); and in *Drosophila* can contain Y-chromosome fertility factors, factors involved in pairing and disjunction of achiasmate chromosomes, and certain other unconventional genetic factors such as *Responder* and *ABO*. The genetics of few organisms have been studied as intensively as that of *Drosophila*, and it may yet turn out that constitutive heterochromatin in many species contains nonconventional factors.

Facultative Heterochromatin

The best-known example of facultative heterochromatin is the inactive X chromosome of female mammals, in which one of the X chromosomes is permanently inactivated early in development, apparently as a means of dosage compensation, so that the amount of X-chromosome gene products produced is similar in males (with only one X) and in females (with two X chromosomes). (It should be noted that in birds, with an independently evolved ZW/ZZ sex chromosome system, there appears to be no dosage compensation, and no facultative heterochromatin, while in *Drosophila* dosage compensation is achieved by increased transcription from the single X chromosome in males.) Like constitutive heterochromatin, the facultative heterochromatin of the mammalian inactive X is late-replicating, and its DNA is more methylated than that of its euchromatic homolog; however, the inactive X cannot be stained distinctively by chromosome banding techniques.

The other reasonably well-known system of facultative heterochromatin occurs in the mealybugs. In the males of this insect, the entire paternal set of chromosomes becomes heterochromatinized, although this does not appear to be related to sex determination. In somatic cells, the heterochromatin replicates less than the euchromatin, while in male meiosis, two wholly heterochromatic and two wholly euchromatic nuclei form, of which only the two latter develop into spermatozoa.

Heterochromatin: Substance or State?

In the past, it was argued whether heterochromatin was a substance or a state. We can now answer that question. Constitutive heterochromatin is evidently a substance, since it consists of specific DNA fractions combined with specific proteins. Conversely, facultative heterochromatin is evidently a state, as its DNA sequence is identical to that of its euchromatic homolog, and in rare cases its heterochromatinization is reversible. Euchromatin inactivated as a result of position-effect variegation, when the inactivation spreads from an adjacent region of constitutive heterochromatin, is clearly also a state of chromatin. Nevertheless, there are occasional systems in which typical constitutive heterochromatin becomes decondensed, for example in the early stages of development in *Drosophila*, when the rate of division is very high, and there may perhaps be no time to condense the heterochromatin. In spite of these exceptions, it is still useful to make the distinction between constitutive and facultative heterochromatin.

See also: **Chromosome Banding; Heteropyknosis; Position Effects; X-Chromosome Inactivation**

Heterochronic Mutation

A E Rougvie

Copyright © 2001 Academic Press
doi: 10.1006/rwgn.2001.0606

The term heterochronic is derived from the Greek *heteros*, meaning other or different, and *khronos*, meaning time. Thus, a heterochronic mutation is a mutation that alters the relative timing of events as an organism develops. Heterochronic mutations have been identified in many organisms; among the best studied are certain cell lineage mutants of the nematode *Caenorhabditis elegans*.

Developmental Timing in *Caenorhabditis elegans*

Genetic analysis has been used to study the temporal progression of pattern formation during postembryonic development in *C. elegans*. The heterochronic mutations identified in these studies alter the timing of certain stage-specific postembryonic developmental events relative to other unaffected events. One of the events studied is the terminal differentiation of lateral epidermal cells (called hypodermal cells in *C. elegans*), which is illustrated in **Figure 1A** and **B**. The nematode hatches from an egg and develops through four larval stages (L1 to L4) in the process of reaching adulthood (A). During the first three larval molts in wild-type animals, these lateral hypodermal cells divide and synthesize a larval-type cuticle (**Figure 1B**). During the final molt they terminally differentiate; they do not divide and they synthesize an adult-type cuticle containing a set of ridges, termed adult alae, that extend along the lateral length of the animal. Heterochronic mutations have been identified that cause hypodermal cell terminal differentiation to occur too early or too late relative to the properly timed gonadal development when compared with wild-type animals. The genes defined by these mutations, the heterochronic genes, have been analyzed in considerable detail.

Inactivation of the heterochronic gene *lin-14* (*lin-14(0)*) results in the precocious execution of hypodermal cell terminal differentiation during the L3 molt (**Figure 1B**). Conversely, a gain-of-function mutation in *lin-14* (*lin-14(gf)*), which causes inappropriately high levels of *lin-14* activity at late developmental times, results in a 'retarded' phenotype, i.e., the indefinite delay of hypodermal cell terminal differentiation. These animals execute a larval-type developmental program during the fourth molt, and this program is repeated during extra molting cycles not observed in wild-type animals.

The biological basis of the altered time of hypodermal cell differentiation in *lin-14* mutants has been traced to cell lineage defects. Mutations in *lin-14* alter the time at which certain stage-specific cell division patterns occur. The wild-type hypodermal cell division pattern of each stage is denoted as S1–S4, with A representing the terminally differentiated adult state (**Figure 1B**). During the L1 stage, the hypodermal cell V6 divides once – the S1 pattern. During the L2 stage, a double division is executed, the S2 pattern,

and so on until terminal differentiation occurs in adults (A). In *lin-14(0)* animals, the S1 pattern is deleted and the remaining patterns are each executed one stage early: S2→S3→S4→A. The net result of this temporal transformation in cell fate is that terminal differentiation occurs during the third, rather than the fourth, molt. In *lin-14(gf)* mutants, the S1 pattern is reiterated indefinitely. This interpretation of the *lin-14(gf)* defect is best illustrated by examining the lineage of a tail hypodermal cell (T, **Figure 1B**). The S1 T cell division pattern is characterized by seven cell divisions and one programmed cell death (x). The S2 pattern is much simpler and consists of a single cell executing a double division. Loss of *lin-14* activity results in this double division during the L1 stage, while in the presence of extra *lin-14* activity, the cell that normally divides in the L2 stage still divides, but instead of undergoing the simple S2 division pattern it behaves like its grandparent and executes the complex S1 pattern.

Other identified heterochronic genes in *C. elegans* include *lin-4*, *lin-28*, *lin-29*, *lin-42*, and *daf-12*. These genes are each also required for the correct temporal patterning of the lateral hypodermis and mutations in these genes cause cells to express developmental programs that are normally reserved for a different stage. Lateral hypodermal cell lineage patterns for *lin-4*, *lin-28*, and *lin-29* mutants are summarized in **Figure 1C**. As for *lin-14*, loss of *lin-28* activity results in precocious execution of hypodermal terminal differentiation; however, S1 patterns are executed normally and the S2 pattern is omitted. In contrast, loss of *lin-4* or *lin-29* activity results in a retarded phenotype, although the cell lineage defects caused by these mutations differ. Genetic analysis has demonstrated that *lin-4* is a negative regulator of *lin-14* and *lin-28* and that these genes in turn negatively regulate *lin-29*. *lin-29* activity triggers the switch to the adult program; in its absence, larval cell division patterns are observed during the fourth and subsequent molts.

Molecular Analysis of Heterochronic Genes

The opposite phenotypes exhibited by gain-of-function and loss-of-function *lin-14* alleles reflect the key role that *lin-14* plays in the heterochronic gene pathway. Molecular analysis of *lin-14* has revealed that it encodes a nuclear protein (LIN-14) that accumulates in hypodermal cells of newly hatched L1 larvae and decreases to an undetectable level by the early L2. This disappearance of LIN-14 is required for the switch from the S1 to the S2 cell division pattern. In *lin-14(gf)* mutants, LIN-14 remains present in the hypodermis throughout development and the S1 pattern is reiterated. The normal disappearance of

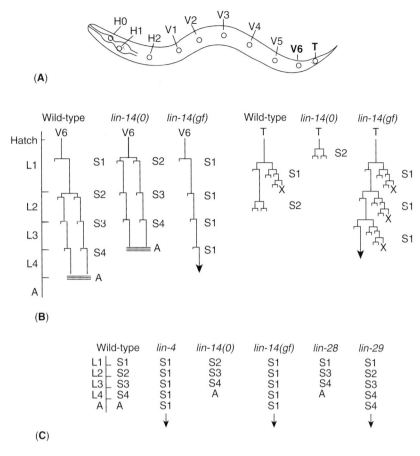

Figure I   Illustration of phenotypes resulting from heterochronic gene mutations in *C. elegans*. (A) A schematic LI stage larva is shown indicating the positions of the left lateral hypodermal blast cells. This pattern is repeated on the right lateral side of the animal. (B) The cell lineage of the V6 and T cells are shown for wild-type and *lin-14* null (0) and gain-of-function (gf) mutants. The vertical axis indicates developmental time, showing the four larval stages and the adult stage. The marks on the Hatch verticle axis indicate the molts. In the lineage diagrams, vertical lines indicate cells and horizontal lines indicate cell divisions. The triple horizontal bars indicate terminal differentiation and synthesis of the adult cuticular ridges termed alae. VI–V4 lineage patterns resemble the V6 lineage and the remaining hypodermal blast cell lineage patterns contain slight variations. Arrows indicate that the division pattern is repeated through additional molting cycles not observed in wild-type animals. Cells that undergo a programmed cell death are indicated with an 'X.' SI–S4 and A are used to denote the stage-specific division patterns in wild-type animals. (C) The cell division patterns defined in (B) are used to summarize the phenotypes of heterochronic mutants *lin-4*, *lin-14*, *lin-28*, and *lin-29*.

LIN-14 protein in young L2 larvae requires wild-type *lin-4* activity. In *lin-4* mutants, LIN-14 remains inappropriately high, again resulting in reiteration of S1 patterns. The functional *lin-4* product is not a protein, but rather a small RNA molecule with antisense complementarity to sequences present in the 3′ untranslated region (UTR) of the *lin-14* mRNA. These complementary sequences are deleted in *lin-14(gf)* mutants, rendering the mutant *lin-14* mRNAs insensitive to *lin-4* activity and preventing down-regulation of LIN-14 levels.

lin-28 encodes a cytoplasmic protein with RNA binding motifs and is also downregulated through a *lin-4*-complementary site within its 3′ UTR. The disappearance of the *lin-14* and *lin-28* gene products during early larval stages ultimately allows accumulation of LIN-29 in hypodermal cells during the L4 larval stage. *lin-29* encodes a transcription factor with five Cys2-His2 type zinc finger motifs and triggers the switch to the adult program by regulating the expression of other genes, including stage-specifically expressed cuticle collagen genes.

Coordination of Developmental Time Throughout the Organism

Cell division defects in *lin-29* mutants are limited to the hypodermis. Thus *lin-29* is a downstream effector of the timing genes in a specific cell type. In contrast, the upstream genes in the heterochronic pathway, *lin-4*, *lin-14*, and *lin-28*, are more global temporal regulators. In addition to controlling stage-specific division patterns in the hypodermis, they also regulate temporal patterning in several other cell types including muscle, neurons, and intestine. These genes act in the temporal coordination of developmental events throughout the organism, presumably by controlling genes with effector functions analogous to that of *lin-29*.

Developmental Timing Mutants in Other Organisms

The molecular mechanisms that control the timing of developmental events in other organisms are also being elucidated. Mutations have been identified in several organisms that cause alterations in the time of onset of certain developmental events and define genes with roles in the temporal progression of patterning. Analogous to the heterochronic gene mutations in *C. elegans*, mutations in these genes either advance or retard the expression of specific developmental programs. For example, in *Dictyostelium*, mutations in *rde* cause premature terminal differentiation of stalk and spore cells, and in maize, mutations in the *Teopod1*, *Teopod2*, and *Teopod3* genes retard the transition between the expression of juvenile and adult characteristics in shoot development, while mutations in *glossy15* cause premature expression of adult characteristics. In *Drosophila*, mutation of the *ana* gene causes certain neuroblasts to proliferate too early.

Finally, one example of a developmental timing abnormality described in humans is altered time of onset of puberty. Puberty, or sexual maturation, is a developmental event that is normally timed to occur in the early teenage years, triggered by the synthesis of hormones which must be produced and function at the correct developmental time. Individuals have been described in which puberty is triggered at the wrong time, resulting in premature or delayed puberty. A variety of molecular defects can cause these condition. In males, precocious puberty can be caused by a dominant gain-of-function mutation in the luteinizing hormone receptor. Luteinizing hormone (LH) binds this receptor causing specific cells in the testes to synthesize testosterone, thus triggering sexual maturation. The receptor mutation causes the receptor to behave as if LH is present when it is not and testosterone is produced abnormally early, leading to precocious sexual maturity. Conversely, individuals with an inactive LH receptor fail to undergo sexual maturation at puberty, an abnormality that may be interpreted as retarded expression of the juvenile program.

Relationship to Heterochrony

The term heterochrony is usually applied in an evolutionary context, referring to a change in the timing of a developmental event in an organism relative to when that event occurred in its ancestors. Naturally occurring heterochronic mutations analogous to those described here could, if stably incorporated into a population, result in heterochrony and provide a mechanism for evolutionary variation between species.

Further Reading

Ambros V (1997) Heterochronic genes. In: C. elegans II, pp. 501–518. Plainview, NY: Cold Spring Harbor Laboratory Press.

Slack F and Ruvkun G (1997) Temporal pattern formation by heterochronic genes. *Annual Review of Genetics* 31: 611–634.

See also: Caenorhabditis elegans; Cell Division in Caenorhabditis elegans

Heterochrony
See: **Neoteny**

Heteroduplexes
P J Hastings

Copyright © 2001 Academic Press
doi: 10.1006/rwgn.2001.0607

Hybrid DNA is formed from complementary single DNA strands from two different parental molecules. The parental molecules must be homologous with each other, that is, they have the same sequence of base pairs overall. This does not exclude the possibility that there are allelic differences between the parental molecules, in which case, there will be mismatched base pairs within the hybrid molecule. Hybrid DNA with such mismatches is called heteroduplex DNA. The term heteroduplex is also sometimes used to mean hybrid DNA, whether or not it contains a mismatch. A mismatched base pair is a pair of bases in complementary nucleotide chains that are unable to form the correct hydrogen bonds between them, despite being chemically correct. The mismatches will cause distortion of the DNA molecule, often with the bases swinging into a position outside the double helix (extrahelical bases). Mismatches also occur as single nucleotides or short deletions and insertions, forming loops of unpaired single strands. Substantial heterologies

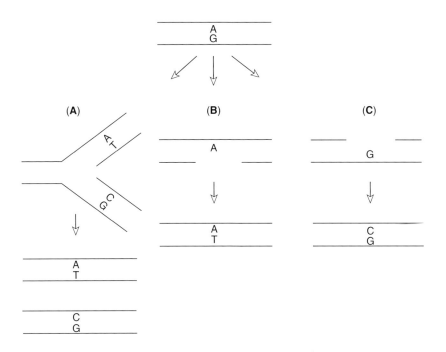

Figure 1 Resolution of mismatched base pairs in heteroduplex DNA. (A) Resolution by replication. When the replication fork passes a mismatch, the two chains are separated and replicated faithfully, so that each daughter molecule is now of one genotype or the other. (B and C) Resolution by mismatch repair. The mismatch is recognized and excised on one strand or the other. Copying the remaining strand restores homoduplex.

(nonhomologous sequences) can be incorporated into heteroduplex, in which case there will be a large unpaired loop.

Evidence for heteroduplex was described in 1952 based on the occurrence of bacteriophage bursts, derived from a single phage particle, that were found to contain the genotype of both parents. This is interpreted as each DNA strand having carried one genotype. When the heteroduplex is replicated, each strand is copied faithfully and the first two progeny each have one of the two genotypes (see **Figure 1A**). Not long after heteroduplex was first described, it was detected in meiotic tetrads of spores in several different fungi. These fungi have eight spores derived mitotically from the four meiotic products. Mitotic spore pairs were seen that differed in genotype from each other. Whereas other pairs of alleles had segregated from each other during meiosis, these mixed spore pairs were evidence that segregation could also occur during the following mitosis. This phenomenon is therefore known as postmeiotic segregation. These observations gave rise to the idea that recombination proceeded by the formation of hybrid molecules joined by complementary base pairing. Those mismatches that do not show postmeiotic segregation have been resolved to homozygosity by a mismatch repair system (see **Figure 1B** and **C**).

Natural Occurrence of Heteroduplex

Ideas on how heteroduplex DNA is formed during the process of recombination are discussed in detail elsewhere (Recombination, Models of). Although single strands of DNA can anneal spontaneously and quite rapidly, *in vivo* the process is catalyzed by a class of proteins of which RecA from *Escherichia coli* is the best-known example. Eukaryotic homologs of RecA are known as Rad51, after a RecA homolog found in *Saccharomyces cerevisiae*. These proteins can also catalyze the invasion of a duplex by a single strand and, once the reaction has begun, the reciprocal exchange of strands between two duplex molecules. This generates hybrid DNA reciprocally on two DNA molecules.

Making Heteroduplex in the Laboratory

Heteroduplex is also generated in the laboratory for use in experiments on mismatch repair mechanisms. This is readily done by use of certain bacteriophage DNA that occurs both as duplex DNA while growing in the infected cell, and as single strands in the mature viral particles. Separation of the strands of the duplex and reannealing with an excess of single strand DNA from phage of a different genotype yields heteroduplex. Another method is available for use with bacteriophage lambda, which has two strands of different density. The separated linear single strands can be isolated individually by density gradient centrifugation and then annealed with complementary single strands of a different genotype producing heteroduplex molecules.

Mismatch Repair

The best known mismatch repair system is the Mut system of *E. coli*. Homologous systems are found in eukaryotes. It is called Mut because mutations in this system cause cells to have a mutator phenotype. This is because the mismatch repair system acts on mismatches generated by replication errors, as well as those occurring in heteroduplex DNA. The mismatch repair system acts on mismatches in heteroduplex at two different levels. It prevents the formation of heteroduplex between molecules that have substantial divergence in their sequence, as would be encountered in interspecific crosses. In intraspecific heteroduplex, where mismatches are few, the mismatch repair system recognizes the mismatch and excises one strand over a distance of a few hundred base pairs. The resulting gap is then filled by DNA synthesis that copies the remaining strand. This results in homoduplex of one genotype or the other (see **Figure 1B** and **C**). Mismatch repair of heteroduplex is the major mechanism of gene conversion.

Different mismatches are recognized by the mismatch repair system with different efficiency. The frequency of DNA-mediated transformation in pneumococcus varies with the efficiency of mismatch repair. Mismatches that are readily recognized are excised from the donor strand so that incorporation into the genome is rare, while those that escape recognition are incorporated very frequently. This observation was interpreted as showing the effects of mismatch correction as early as 1966. The C–C base pair is poorly recognized in several organisms. These differences underlie many marker effects, that is, situations in which the nature of the heterozygosity present in a cross has an effect on the outcome of the experiment.

Further Reading

Ephussi-Taylor H and Gray TC (1966) Genetic studies of recombining DNA in pneumococcal transformation. *Journal of General Physiology* 49 (suppl.): 211–231.

Hershey AD and Chase M (1952) Genetic recombination and heterozygosis in bacteriophage. *Cold Spring Harbor Symposia on Quantitative Biology* 16: 471–479.

See also: Marker Effect; Mismatch Repair (Long/Short Patch); Recombination, Models of

Heterogenote

J H Miller

Copyright © 2001 Academic Press
doi: 10.1006/rwgn.2001.0608

'Heterogenote' is a term meaning the same as heterozygote, viz., a diploid organism having different alleles for one or more genes that therefore produces different gametes.

See also: Heterozygote and Heterozygosis

Heterokaryon

F Ruddle

Copyright © 2001 Academic Press
doi: 10.1006/rwgn.2001.0609

A heterokaryon is a cell containing two or more nuclei of different origin or in different states in a common cytoplasm. Examples include: (1) a mouse and a human nucleus as separate and distinct organelles within a single cell; or (2) two nuclei in different epigenetic states, one from a liver cell, the other from a pancreatic cell within a common cytoplasm; or (3) nuclei at different positions within the cell cycle bounded by a cell membrane. Heterokaryons are produced by bringing two different cells into contact and then inducing membrane fusion to produce a single cell with a common cytoplasm and containing multiple donor nuclei. Heterokaryon analysis has been useful in determining nuclear cytoplasmic interactions and particularly the influence of cytoplasmic factors on nuclear gene expression.

See also: Nuclear Transfer

Heteropyknosis

A T Sumner

Copyright © 2001 Academic Press
doi: 10.1006/rwgn.2001.0610

Heteropyknosis is the attribute of chromatin that shows condensation behavior different from that of 'normal' chromatin (generally equivalent to euchromatin). Heterochromatin typically shows 'positive heteropyknosis' by remaining condensed in interphase. Chromosomal regions that show less condensation than the rest of the chromosome during prophase or metaphase are said to show 'negative heteropyknosis.'

***See also:* Chromatin; Heterochromatin**

Heterosis

J F Crow

Copyright © 2001 Academic Press
doi: 10.1006/rwgn.2001.0611

Heterosis is a synonym for hybrid vigor: the increased size, performance, resistance, and strength of hybrids. Heterosis is particularly pronounced in crosses between inbred strains. Early in the twentieth century, after the rediscovery of Mendelian inheritance, it became obvious that hybrids had greater heterozygosity than their parents. The word 'heterosis' was coined by G.H. Shull as a descriptive term to avoid such cumbersome expressions as 'the stimulus of heterozygosis;' it is not intended to favor any genetic hypothesis.

The weakening effect of inbreeding and the vigor of hybrids has been known since classical antiquity. The hardiness and strength of mules were recognized and made use of by the Greeks and especially the Romans. In the nineteenth century many botanists noticed that species hybrids regularly exceeded their parents in size. The most thorough analysis was done by Charles Darwin, whose book *The Effects of Cross- and Self-Fertilization in the Vegetable Kingdom* (Darwin, 1876) can still be read with profit. In this he says:

> The first and most important conclusion which may be drawn from the observations given in this volume, is that cross-fertilization is generally beneficial and self-fertilization injurious.

An understanding of heterosis in genetic terms had to await the rediscovery of Mendel's laws in 1900. It was immediately apparent that hybrids are more heterozygous than their parents. A decrease in the number of heterozygotes implied an increase in the number of homozygotes. This immediately gave rise to two explanations. The 'dominance' hypothesis notes that most recessive mutants are deleterious, so inbred lines are weakened by having an increase in the number of homozygous recessive genes. Hybrids, in contrast, are stronger because the recessives from each parent are usually concealed by dominants from the other. The 'overdominance' hypothesis assumes that there are some loci at which the heterozygote is superior to either homozygote. Although the two ideas are not mutually exclusive, the dominance hypothesis is now generally favored. This explanation also applies to variety and species hybrids, because the hybrids are always more heterozygous than their parents, the more so as the parents diverge. The contrast is greatest, however, when the parents are highly homozygous inbred lines.

The greatest practical impact of heterosis has been from hybrid corn. Inbred lines have been developed and crossed to produce hybrids that are grown by the farmer. The inbred lines are selected not only for their own performance, but for producing superior hybrids. Since the introduction of hybrid maize in the 1930s, the yield of corn has increased about five-fold. It represents a high point in modern agriculture. About 70% of the improvement is the result of superior hybrids, while the remainder is due to improved agronomic practices.

Although less widely applied than in maize, other horticultural and cereal crops also show heterosis. In many cases the corn model of crossing inbred lines has been productive. In others the heterosis is not so great and greater practical results are obtained by more conventional breeding methods.

Reference

Darwin C (1876) *The Effects of Cross- and Self-Fertilization in the Vegetable Kingdom*. London: John Murray.

***See also:* Overdominance**

Heterotrimeric G Proteins

H C Korswagen and R H A Plasterk

Copyright © 2001 Academic Press
doi: 10.1006/rwgn.2001.0572

Heterotrimeric guanine nucleotide-binding proteins (G proteins) form an ancient family of signaling molecules that connect seven-helical transmembrane receptors (7-TM receptors) to a limited set of intracellular effectors. 7-TM receptors are one of the largest receptor families in vertebrates and function in a variety of cellular processes. Thus, 7-TM receptors are required for the response to hormones and neurotransmitters, but are also required for light detection in the visual system and odorant sensation in olfactory cells. Downstream effectors of 7-TM receptors and G proteins can be enzymes such as adenylyl cyclases, phosphodiesterases and phospholipases, ion channels or other intracellular proteins. G protein activation can stimulate or inhibit such effectors, resulting in the generation or breakdown of second messengers. An important property of G-protein-coupled signal transduction is that at each step in the pathway there is a considerable amplification of the signal.

Heterotrimeric G proteins consist of a guanine nucleotide-binding Gα subunit and a closely associated Gβγ subunit, both of which are linked to the plasma membrane through lipid modifications. Based on sequence similarity and shared intracellular effectors, mammalian Gα subunits can be divided into four subfamilies: G_s, G_i, G_q, and G_{12}. Both the Gα and the Gβγ subunit have signaling capabilities and can interact with specific targets in the cell.

Heterotrimeric G proteins act as molecular switches in signal transduction. In the inactive state, the Gα subunit is associated with a molecule of GDP and is complexed with the Gβγ subunit. Ligand binding by an appropriate 7-TM receptor will induce the Gα subunit to exchange GDP for GTP, which results in the dissociation of the two subunits, enabling them to interact with their specific targets in the cell. The intrinsic GTPase activity of the Gα subunit hydrolyzes the bound GTP back to GDP, allowing reassociation with the Gβγ subunit to restore the inactive heterotrimeric complex. The relatively slow GTPase activity of the Gα subunit cannot completely account for the fast GTP hydrolysis observed *in vivo*. A family of RGS domain (Regulator of G protein)-containing proteins is responsible for enhancing the slow GTPase activity of specific Gα subunits.

Structural analysis of heterotrimeric G proteins has resulted in considerable insight into the molecular mechanism of GTPase activity and the molecular interaction of the Gα subunit with its effectors. The crystal structure of Gα shows that the Gα subunit consists of a guanine nucleotide-binding domain that is structurally similar to small G proteins such as Ras and elongation factor Tu, and a helical domain that is unique to heterotrimeric G proteins. Thus, the helical domain has functions that are performed by separate proteins in small G proteins. Thus, the helical domain prevents dissociation of GDP from the guanine-nucleotide-binding core and functions in GTP hydrolysis. The catalytic mechanism of the Gα GTPase activity and the conformational changes necessary for Gβγ dissociation and effector interactions were determined from the structures of Gα·GDP, Gα·GTP, and the complete heterotrimeric complex. The general picture that emerges from these studies is that differential binding of guanine nucleotides induces specific conformational changes in the Gα subunit that allow it to release or bind the Gβγ subunit and enable it to interact with its effectors.

Multiple G-protein-coupled signal transduction pathways may function in a single cell. Consequently, G proteins form complex signal transduction networks *in vivo*. Insight into the complexity of

G-protein-coupled signal transduction pathways can be gained from genetic studies. Model organisms such as the yeast *Saccharomyces cerevisiae*, the slime mold *Dictyostelium discoideum*, the nematode *Caenorhabditis elegans*, the fruit fly *Drosophila melanogaster*, and the mouse have been used to study G protein signaling *in vivo*. In yeast and *Dictyostelium*, G proteins transmit developmental signals such as a pheromone and aggregation signal. In the metazoan organisms *C. elegans* and *Drosophila*, G proteins have been adapted to transduce a more complex set of developmental, endocrine, and sensory signals. Clear homologs of the four mammalian subfamilies of Gα subunits are present in these organisms and they serve as an important model for conserved G-protein-coupled signal transduction. The powerful genetic tools available for *C. elegans* and *Drosophila* allow detailed genetic dissection of G protein signaling. Using genetics, novel players of G-protein-coupled signal transduction pathways have been discovered. An example is the family of RGS proteins, which was first identified as a negative regulator of G-protein signaling in yeast and *C. elegans*.

See also: Signal Transduction

Heterozygote and Heterozygosis

D E Wilcox

Copyright © 2001 Academic Press
doi: 10/1006/rwgn.2001.0612

A heterozygote is an individual whose DNA molecules in a homologous pair of chromosomes differ in sequence at a particular genetic locus. (A homozygote is an individual whose DNA sequences at a locus are identical.) Usually this locus will be a gene and the different forms of the gene are called alleles. A locus is said to be in heterozygosis when two alternate alleles are present. If the phenotype of the heterozygote is normal, the effects of the alternate allele are said to be recessive to the normal allele. Conversely, if the phenotype of the heterozygote is abnormal then the effects of the alternate allele are said to be dominant. A major task in human medical genetics is identifying whether a patient with a normal phenotype is a heterozygote (carrier of a disease allele).

Alleles and Heterozygotes in Populations

The alternate DNA sequence, or allele, may occur rarely or commonly in the whole population. When the alternate allele is very rare, it is often called a mutant and the common allele is called the normal or wild-type. Shorthand for the wild-type homozygote is +/+, for the wild-type/mutant heterozygote is +/m, and for the mutant homozygote is m/m. If the mutant is dominant to normal its symbol is capitalized, i.e., +/M. At some loci the alternate alleles have more equal proportions, such as the three common alleles of the human ABO blood group. When the frequency of a variant allele in a population is too high to be explained by recurrent mutation, it is called a polymorphism. Even though there may be three or more alleles of a locus in the population, an individual with normal chromosomes can only have a maximum of two alleles at that locus, one for each chromosome. (Sex-linked loci will either have one or two alleles depending on the individual's sex and thus number of each sex chromosome.)

Allelic Origins in a Heterozygote

A heterozygote carrying a common variant such as a blood group antigen will have inherited it from a parent who also has that variant either as a heterozygote or as a homozygote. If the alternate sequence is unique to that individual, with neither of the parents carrying the variant, then it will have arisen as the result of mutation. In population terms, an allele that is rare will more commonly be present in heterozygotes than in homozygotes. The exact proportions can be calculated using the Hardy–Weinberg Law; as an example, approximately 1 in 20 people in Scotland are heterozygous carriers of one cystic fibrosis mutation (+/cf), while only 1 in 1600 people are affected and are homozygous for two cystic fibrosis mutations (cf/cf).

Single Nucleotide Polymorphisms

Not all alleles will have an effect on the phenotype of the individual. Some DNA sequence changes will have no effect on the final structure and function of the protein coded by the gene. Nonetheless, when identified, they can be used to track rarer, unidentified disease causing mutations to which they are linked by being situated nearby on the same DNA molecule. Some silent variants in DNA sequence effect a change at a single nucleotide only (single nucleotide polymorphism, SNP) and occur at regular intervals throughout the genome. The study of genetic components of common diseases such as hypertension will be revolutionized by comparing SNPs in healthy and affected members of the population.

Phenotypic Effects of Alleles in Heterozygotes

Alleles that are recessive have no effect on the phenotype of a heterozygote (+/m). Recessive alleles usually involve changes to the coded protein which result in loss of normal function. In heterozygotes, the wild-type allele on the other chromosome produces sufficient normal protein to maintain healthy function and phenotype and the disease phenotype is only seen when an individual is homozygous with two mutant recessive alleles (m/m). Dominant mutant alleles will affect the phenotype of the heterozygote (+/M). In this situation, the mutant protein may have gained a new function that affects the phenotype even in the presence of the normal protein. The mutant protein may not be processed or broken down at the same rate as the normal protein. Another possibility is that the protein may function normally by forming polymers or chains. In this case, a heterozygote will form polymers that are a mixture of normal and mutant proteins. The resulting compound polymer will have a different structure and function to the normal polymer.

Reproductive Fitness of Heterozygotes

The reproductive fitness of a heterozygote is only affected if the phenotype is altered. Thus, genetic selection can act on heterozygotes for a dominant mutation. If the heterozygotes for a disease mutation have a low reproductive fitness, the mutant allele will only be maintained in the population by the process of new mutation. In recessive disorders the heterozygotes have a normal phenotype and genetic selection can only act on the affected homozygotes. Since the majority of mutant alleles in a population are present in healthy heterozygotes, the frequency of the two alleles will change very little from generation to generation, even if none of the mutant homozygotes reproduce and their alleles are lost to the population each generation.

Heterozygote Advantage

In some circumstances, the effects of a recessive mutation can affect the phenotype and thus reproductive fitness of heterozygotes. This is not always a negative effect as can be seen in the condition human sickle-cell anemia. Sickle-cell carriers have a heterozygote advantage over the reproductive fitness of normal homozygotes in some environments. In most populations, sickle-cell anemia is a rare mutation, but in malarial regions of Africa as many as one in three of the population are carriers of the mutation in the hemoglobin gene. The presence of the mutant hemoglobin in heterozygotes interferes with the malarial parasite's life cycle. Heterozygotes are therefore more resistant to the debilitating effects of malaria than the normal homozygotes. This heterozygote advantage in many sickle-cell carriers outweighs the severe reproductive disadvantage of the rarer sickle-cell homozygotes. This maintains the mutation in this population at a high frequency as a polymorphism.

See also: **Balanced Polymorphism; Heterosis; Sickle Cell Anemia**

Hfr

S M Rosenberg and P J Hastings

Copyright © 2001 Academic Press
doi: 10.1006/rwgn.2001.0613

Hfr strains of bacteria such as *Escherichia coli* are strains carrying an integrated conjugative plasmid such as the ∼100 kb F (for fertility) factor (see: F Factor). This enables them to transfer their chromosomal DNA to other bacteria into which the DNA can recombine. The existence of Hfr strains of *E. coli* was observed by their high frequency of recombination with other bacteria. This was possible because some of the strains mixed by Joshua Lederberg and Edward Tatum in early mating experiments (Lederberg and Tatum, 1946b) or contained the F conjugative plasmid and others did not. In cultures of cells carrying an F, some of the cells are Hfrs (have an integrated F). The non-F-carrying strains are called female or recipient bacteria and the Hfr or F-carrying strains are male or donors. Transfer of conjugative plasmid DNA, or chromosomal DNA in Hfr cells, is unidirectional, i.e., male to female (Hayes, 1952). Males can be recipients only at much lower efficiency, or under special environmental conditions.

Hfr strains form because the F carries transposable elements that are also carried by the *E. coli* chromosome: two copies of the insertion sequence IS3, one IS2, and one copy of transposon Tn*1000* (also called

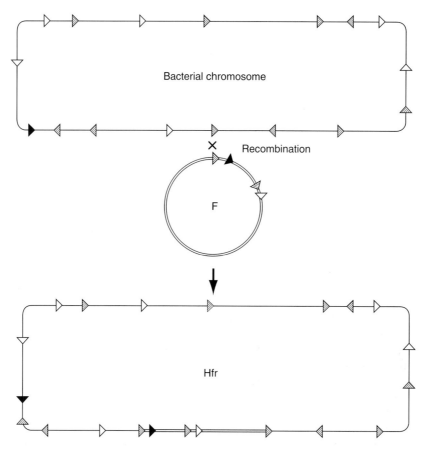

Figure 1   Formation of an Hfr by recombination of the F plasmid with the *Escherichia coli* chromosome. Single lines represent duplex DNA. Triangles represent transposable genetic elements IS3 (▷), IS2 (▷), and Tn*1000* (▶) that are present in the F (double lines) and also in the *E. coli* chromosome (single lines). These elements provide regions of DNA sequence identity between which homologous recombination can occur (represented by an X), incorporating the F into the chromosome and producing an Hfr.

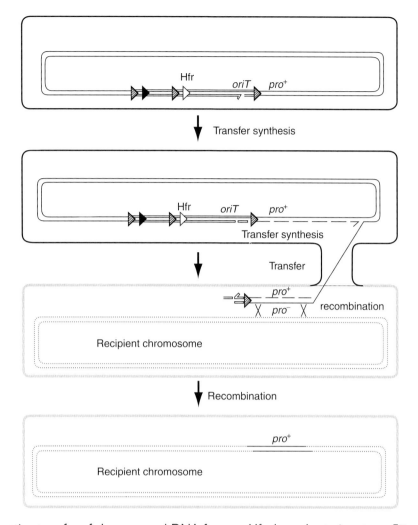

Figure 2 Conjugative transfer of chromosomal DNA from an Hfr donor bacterium to an F⁻ recipient bacterium. Single lines represent single strands of DNA; dashed lines represent newly synthesized DNA and arrowheads represent 3′ ends. Transfer begins with single-strand cleavage of F DNA (------) at the transfer origin, *oriT*, followed by DNA synthesis, which displaces a single strand that is transferred into the recipient. Synthesis of the complementary strand occurs in the recipient, and the duplex fragment (——) can be incorporated into the recipient chromosome (══) by recombination.

γδ) (**Figure 1**). These elements are regions of DNA sequence homology between the chromosome and the F that allow the F to recombine with the chromosome and become incorporated into the chromosome (**Figure 1**). The F can integrate at many different sites in the chromosome, making many different Hfr strains. Each different Hfr is capable of high frequency transfer of the chromosomal DNA next to itself, or if given enough time to mate without interruption, the whole 4.7 megabase *E. coli* chromosome.

Transfer of chromosomal DNA from an Hfr to a female cell is depicted in **Figure 2**. Transfer begins by action of an F-encoded single-strand endonuclease and helicase, TraI, on the F origin of transfer, *oriT*. Leading strand synthesis is primed from the 3′ end at the nick and displaces the 5′ DNA strand. Continued

synthesis displaces that strand extending into the contiguous bacterial DNA, and the single DNA strand displaced is transferred into a female bacterium that has become attached to the male in a mating pair. Transfer stops at random locations when the synthesis tract encounters a DNA break in the donor template, or due to breakage of the transferred strand. The occurrence of such random disruptions produces a gradient of transfer, with DNA near *oriT* being transferred most efficiently, and decreasing transfer efficiency with increasing distance from *oriT*. Once inside the female, lagging strand synthesis of the complementary strand takes place, creating a double-strand linear DNA fragment. The transferred DNA will be lost unless it recombines into the recipient chromosome, which it can do (Xs in **Figure 2**) using the cell's RecBCD system of homologous recombination of linear DNA and double-strand break repair. This results in homologous replacement of a segment of recipient DNA with sequences derived from the donor chromosome. If that segment contains different genetic information (prototrophic *pro*⁺ information is depicted in the transferred piece entering an auxotrophic *pro*⁻ recipient in **Figure 2**), the recipient can become genetically recombinant. Recombinant strains made by Hfr conjugation do not usually become male (Hfr) upon acquisition of donor DNA, because the F transfer genes are the last to be transferred and are not homologous with the recipient DNA.

Hfr crosses provided the first demonstration of genetic recombination in bacteria and in so doing encouraged the idea that bacteria, like other organisms, possess genes. Hfr crosses were also the first tools used for exploration of the proteins and enzymes that catalyze DNA recombination, leading to the discovery of, for example, RecA (Clark and Margulies, 1965), a universal recombination and DNA repair protein of which there are orthologs in all eubacterial, eukaryotic, and archaeal species examined to date. For descriptions of the *E. coli rec* genes discovered using Hfr crosses, the recombination systems and pathways, and double-strand break repair machinery of *E. coli*.

Further Reading

Brock TD (1990) *The Emergence of Bacterial Genetics*. Plainview, NY: Cold Spring Harbor Laboratory Press.

Clark AJ and Sandler SJ (1994) Homologous genetic recombination: the pieces begin to fall into place. *Critical Reviews in Microbiology* 20: 125–142.

Low KB (1996) Hfr strains of *Escherichia coli* K12. In: Neidhardt FC, Curtiss III R, Ingraham JL *et al.* (eds) Escherichia coli *and* Salmonella: *Cellular and Molecular Biology*, 2nd edn, vol. 2, pp. 2402–2405. Washington, DC: ASM Press.

References

Clark AJ and Margulies AD (1965) Isolation and characterization of recombination deficient mutants of *Escherichia coli* K12. *Proceedings of the National Academy of Sciences, USA* 53: 451.

Hayes W (1952) Recombination in *Bact. coli* K12: unidirectional transfer of genetic material. *Nature* 169: 118–119.

Lederberg J and Tatum EL (1946a) Gene recombination in bacteria. *Nature* 158: 558.

Lederberg J and Tatum EL (1946b) Novel genotypes in mixed cultures of biochemical mutants of bacteria. *Cold Spring Harbor Symposia on Quantitative Biology* 11: 113–114.

See also: Bacterial Genetics; Conjugation, Bacterial; F Factor; Genetic Recombination; Rec Genes; RecA Protein and Homology; RecBCD Enzyme, Pathway; Recombination, Models of; Recombination Pathways; RuvAB Enzyme; RuvC Enzyme; Transposable Elements

Hin/Gin-Mediated Site-Specific DNA Inversion

S K Merickel and R C Johnson

Copyright © 2001 Academic Press
doi: 10.1006/rwgn.2001.0566

Inversion Systems

The DNA invertases catalyze a recombination reaction that inverts a segment of DNA between two specific recombination sites. The best-characterized invertases, Hin from *Salmonella typhimurium* and Gin from bacteriophage Mu, catalyze site-specific inversion reactions that result in alternate gene expression. The Hin invertase regulates flagellar phase variation in *Salmonella*, allowing the bacterium to evade a host immune response (**Figure 1A**). In one orientation, a promoter located within the invertible segment of DNA directs the expression of the H2 flagellin gene (*fljB*), as well as a repressor of the H1 flagellin gene (*fljC*). After Hin catalyzes a site-specific inversion event, the promoter becomes inverted and can no longer drive the expression of these genes. Consequently, the H1 flagellin gene is expressed from its unlinked site. The Gin invertase of bacteriophage Mu controls the alternate expression of tail fiber genes (**Figure 1B**). Each orientation of the invertible segment in bacteriophage Mu encodes a different C-terminal portion of the tail fiber protein S. Site-specific inversion catalyzed by Gin switches the expression of the C-terminal part of the protein, which determines the host specificity range for the phage. The Cin-mediated reaction of phage P1 performs a similar function. Due to the homology of these proteins and the similarity of their recombination substrates, the characterized invertases are functionally interchangeable. The invertases belong to the resolvase/invertase (also known as the serine) family of recombinases which currently has over 50 members. Site-specific DNA inversions can also be catalyzed by recombinases belonging to the phage integrase (also known as tyrosine recombinase) family.

Site-Specific Inversion Reaction

Site-specific inversion by Hin and Gin has been studied extensively both *in vivo* and *in vitro*. The invertases require a supercoiled DNA substrate that contains two inversely oriented recombination sites. The 26-bp recombination sites have partial dyad symmetry with the central two base pairs being the site of DNA strand exchange (**Figure 1C**). For efficient recombination the invertases also require

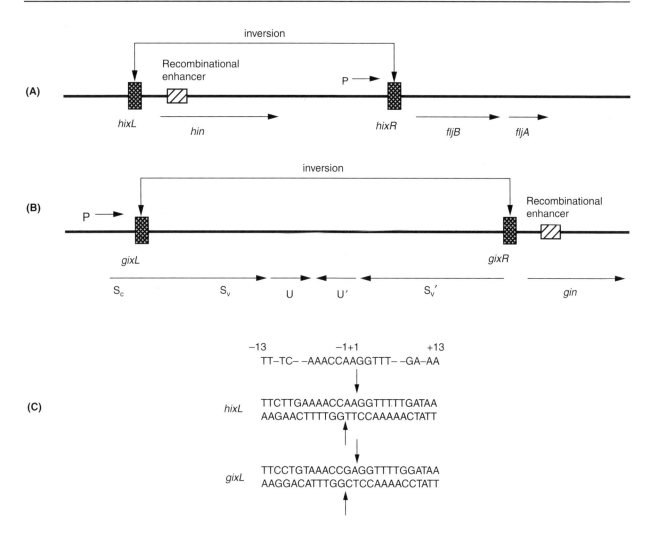

Figure 1 Regulation of gene expression by site-specific DNA inversion. (A) *Salmonella* invertible DNA segment. The *hixL* and *hixR* recombination sites are shown as dark rectangles. The recombinational enhancer is depicted as a striped rectangle. The 1 kb invertible segment, located between the two recombination sites, contains the *hin* gene and a promoter (P) that directs the expression of the flagellar genes, *fljB* and *fljA*. Hin-catalyzed inversion switches the orientation of the invertible segment such that the promoter can no longer direct the expression of *fljB* and *fljA*. (B) Phage Mu invertible DNA segment. The *gixL* and *gixR* recombination sites are depicted as dark rectangles. The recombinational enhancer, illustrated as a striped rectangle, and the *gin* gene are located outside of the ~3 kb invertible segment. A promoter (P) located outside of the invertible segment controls the expression of the S and U tail fiber genes. The constant N-terminal portion of the S tail fiber gene (Sc) is also located outside of the invertible segment, while the variable C-terminal portion of the S tail fiber gene (Sv) and the U gene are located within the invertible segment. Gin-catalyzed inversion of the invertible segment alternates the expression of the Sv and U genes with the Sv′ and U′ genes. (C) Sequence of the invertase recombination sites. The recombination site consensus sequence for the invertase family of recombinases is shown at the top. The *hixL* and *gixL* recombination site sequences are shown below the consensus sequence. The arrows mark the sites of 2 bp staggered double-strand DNA cleavage. The relative orientation of the recombination sites are determined by these two core nucleotides.

another *cis*-acting DNA element called the recombinational enhancer. Each recombination site is bound by a dimer of the Hin or Gin recombinase, and the enhancer contains two binding sites for the dimeric protein Fis. Once bound to their respective DNA sites, Hin/Gin and Fis dimers are able to assemble into a higher order nucleoprotein complex called

an invertasome (**Figure 2iii**). The DNA bending protein HU also aids in the formation of the invertasome complex in the Hin system by facilitating the bending of a small loop of DNA between one recombination site and the enhancer. Once assembled in the invertasome structure, Fis stimulates Hin/Gin to catalyze recombination. The inversion reaction can be broken down into two basic catalytic steps: DNA cleavage and strand exchange. The recombination sites are concertedly cleaved, producing 2 bp staggered double-strand DNA breaks (**Figure 1C**). In this reaction, a serine nucleophile in each invertase subunit bound to the recombination sites attacks the phosphate backbone, resulting in a phosphoserine bond with the 5′ recessed end of the DNA. After DNA cleavage, the DNA ends are exchanged and the recombination sites are religated in a recombinant configuration through a reversal of the phosphoserine linkage.

Invertasome

The three DNA sites must synapse in a highly specific fashion to form an invertasome complex. The Fis-bound enhancer interacts with the invertase-bound recombination sites at a branch in plectonemically supercoiled DNA (**Figure 2iii**). The recombination sites pass on either side of the enhancer such that two negative DNA nodes are trapped within the complex. Immunoelectron microscopy of crosslinked invertasome complexes has provided direct evidence for the three-looped DNA structures containing Hin

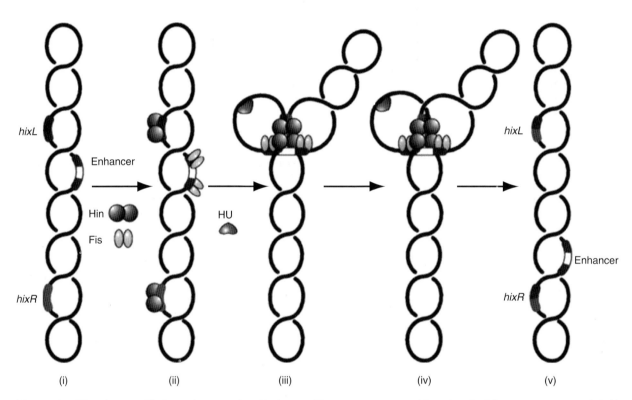

Figure 2 The site-specific inversion reaction. Pathway of invertasome assembly using the Hin system as a model. (i) Supercoiled DNA substrates contain two inversely oriented recombination sites, *hixL* and *hixR*, and a recombinational enhancer. (ii) A Hin recombinase dimer binds to each recombination site and two Fis dimers bind to the recombinational enhancer. (iii) Hin and Fis assemble into an invertasome complex with the aid of the DNA-bending protein HU. In the invertasome complex the recombination sites associate with the enhancer at a branch in the supercoiled DNA. (iv) Once assembled in the invertasome complex, Fis activates Hin to catalyze DNA cleavage and strand exchange. (v) Recombination results in an inversion of the segment of DNA located between the recombination sites.

and Fis. The sizes and positions of the DNA loops in the invertasome complex were consistent with the enhancer associating with both recombination sites at a branch in supercoiled DNA. The formation of these structures was absolutely dependent on DNA supercoiling.

The specific topology of the DNA strands in the invertasome complex has been determined through several experimental approaches. The change in linking number observed in the DNA molecules after inversion by Gin suggested that two negative DNA nodes were trapped in the invertasome complex. In addition, the stereostructure of knotted DNA products generated from iterative rounds of Hin/Gin recombination provided strong evidence for this specific configuration of DNA strands at synapsis. The knotted DNA products also indicated that each round of recombination results in a 180° right-handed rotation of the DNA ends. Since the invertase is covalently associated with the DNA ends during strand exchange, this observation implies that the recombinase subunits must also undergo a rotation. Direct experimental evidence for exchange of subunits between dimers accompanying strand exchange, however, is lacking thus far.

Regulation of Inversion Reaction by Fis and the Enhancer

The DNA invertases catalyze recombination very weakly on their own even when two recombination sites have formed a complex. When Hin/Gin and Fis assemble the topologically correct invertasome complex, however, Fis activates each of the invertase subunits to initiate the chemical steps of recombination. The enhancer is located 90–500 bp from the closest recombination site in the characterized inversion systems. However, it can be artificially positioned many kilobases from the recombination sites and still function effectively to activate the reaction. Although the position of the enhancer relative to the recombination sites is very flexible, the relative positions of the Fis binding sites within the enhancer (48 bp between their centers) is critical for efficient activity. The precise positioning of the Fis dimers on the enhancer enables both Fis dimers to contact the DNA invertases to assemble the invertasome.

Effective regulation of invertase activity is essential to avoid unwanted chromosomal rearrangements at secondary recombinase binding sites found throughout the genome. Fis and the recombinational enhancer perform this control function by (1) limiting the location of recombination to the vicinity of the enhancer and (2) strongly biasing the type of recombination to DNA inversion rather than a deletion or intermolecular fusion. The weak association between the DNA invertase and Fis is overcome by DNA supercoiling. DNA supercoiling directs the appropriate three-site collision at the base of a plectonemic branch to form the invertasome structure where alignment of the recombination sites specifies inversion.

Mutational and structural studies have shown that the N-terminal region of Fis is responsible for activating the invertases to catalyze recombination. This region contains two mobile β-hairpin arms that extend over 20 Å from the Fis dimer core, although only one of these arms is required to activate the DNA invertase (**Figure 3A**). A triad of amino acids near the tip of one of these β-arms is believed to form the critical contact region with the invertase. The opposite end of the Fis dimer structure contains helix–turn–helix DNA binding motifs. The two DNA recognition helices within the Fis dimer are separated by only 25 Å rather than the usual 32–34 Å, requiring the DNA to bend significantly when bound by Fis.

Structure and Mechanism of Activation of DNA Invertases

The 180–190 amino acid DNA invertases are organized in a two-domain structure similar to the resolvases (**Figure 3B**). The crystal structure of the C-terminal 52 amino acid DNA-binding domain of Hin revealed a 3 α-helix fold that displays aspects of both a bacterial helix–turn–helix motif and a eukaryotic homeodomain. The N-terminal catalytic and dimerization domain, which is located on the opposite side of the DNA from the DNA-binding domain, is believed to closely resemble the structure of the catalytic domain of γδ resolvase. In the resolvase–DNA crystal, the active site serines that form an ester linkage with the DNA upon cleavage are not located close to their sites of attack. Thus, it is likely that a conformational change must occur within the recombinase structure in order to initiate catalysis. Fis–invertase interactions may induce a conformational change upon invertasome assembly that repositions the active sites within each invertase dimer to promote DNA cleavage. Several lines of evidence suggest that this Fis-induced repositioning of the active sites may involve a quaternary change in the invertase dimer interface. A dimer containing a disulfide bond that covalently links the subunits is able to form synaptic complexes but is catalytically inactive. Certain detergents that partially destabilize the Hin dimer increase the rate of DNA cleavage by Hin over 30-fold. Additionally, a subset of amino acid substitutions within the dimer interface result in hyperactive mutants that are able to catalyze recombination without the presence of a recombinational enhancer or Fis. Reactions performed without

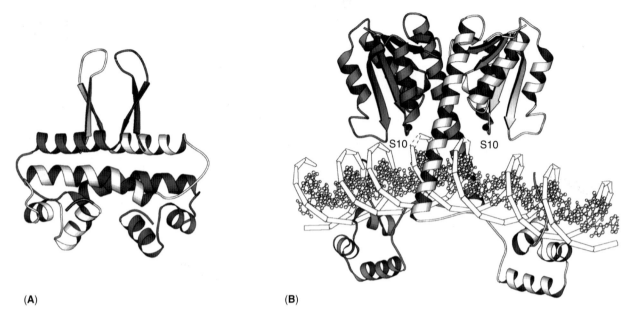

Figure 3 Model of Fis and Hin dimers. (A) Structure of a Fis dimer. The N-terminal β-hairpin arms (protruding from the top of the structure in the figure) are responsible for stimulating invertase activity. The helix–turn–helix DNA-binding domains are located in the C-terminal end of the protein. (B) Model of a Hin dimer bound to a recombination site. The structure of the Hin C-terminal DNA-binding domain bound to a recombination half site was determined by X-ray crystallography. The N-terminal catalytic domain is modeled after the structure of the homologous recombinase γδ resolvase. The location of the active-site nucleophile serine 10 is marked with a black ball. In this figure, the catalytic domains are located above the DNA and the DNA-binding domains are located below the DNA.

Fis using the Fis-independent mutant DNA invertases, efficiently catalyze deletions as well as inversions since random collision of recombination sites yield catalytically active synaptic complexes.

Current and Future Research

Although the first steps are well established in this relatively simple recombination system, there are many questions yet to be answered. Researchers in the field are investigating the precise molecular arrangement of the proteins and DNA sites in the invertasome complex, the conformational changes that accompany catalytic activation, and the mechanics of DNA strand exchange.

Further Reading

Johnson RC (2001) Site-specific DNA inversion in bacteria. In: Craig NL, Craigie R, Lambowitz AM and Gellert M (eds) *Mobile DNA II*. Washington, DC: American Society for Microbiology.

Johnson RC (1991) Mechanism of site-specific DNA inversion in bacteria. *Current Opinion in Genetics and Development* 1: 404–411.

See also: Gene Expression

Hirschsprung's Disease

S Malcolm

Copyright © 2001 Academic Press
doi: 10.1006/rwgn.2001.1580

In Hirschsprung's disease there is an obstruction of the intestine due to aganglionosis of the gut. Germline mutations of a receptor tyrosine kinase and proto-oncogene, *RET*, have been found in approximately 50% of familial cases and 30% of isolated cases but the disorder is a model for a complex disorder. Mutations have been found in a few instances in four other genes, all of which are within functional pathways involving *RET*. Glial cell line derived neurotrophic factor is a soluble ligand of *RET* in which mutations have been found. Two components of a further signaling pathway involving *RET*, endothelin B receptor (EDNRB), and its ligand endothelin 3, are mutated in about 5% of cases as is SOX10 which regulates EDNRB expression. There is preliminary evidence for interactions between variants of these genes affecting the penetrance and severity of the disorder.

See also: RET Proto-Oncogene

Histidine

J Read and S Brenner

Copyright © 2001 Academic Press
doi: 10.1006/rwgn.2001.2075

Histidine (His or H) is one of the 20 amino acids commonly found in proteins. Although it contains a positive charge it is only a weak base at neutral pH. Its chemical structure is:

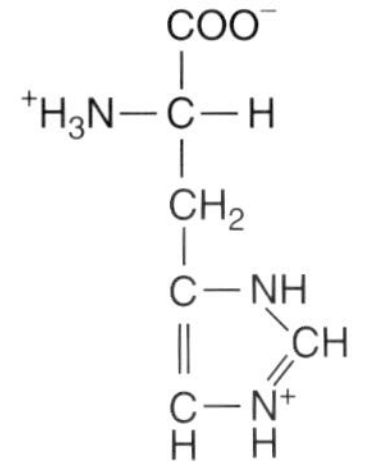

Figure 1 Histidine.

See also: **Amino Acids; Proteins and Protein Structure**

Histidine Operon

P Alifano

Copyright © 2001 Academic Press
doi: 10.1006/rwgn.2001.0615

Histidine Operon as Model System

Studies of the biosynthetic pathway leading to the synthesis of the amino acid histidine in prokaryotes and lower eukaryotes began more than 40 years ago. This effort resulted not only in an elucidation of the chemical intermediates in the pathway, but also in the unravelling of many fundamental mechanisms of biology. The histidine system was of the utmost importance in the definition and refinement of the operon theory and of the one operon–one messenger theory of transcription. Together with *lac* and *trp*, the *his* operon was used as a model system to study the phenomenon of polarity. Another area in which the *his* operon system played a fundamental role was the study of regulatory mutants and of the mechanisms governing operon expression in general. Together with early studies on the *trp* operon these studies were the basis for the characterization of a novel mechanism of gene regulation, termed attenuation. Studies of the mechanisms by which the first enzyme in the pathway was inhibited by feedback inhibition provided important insights into the allosteric regulation of biochemical reactions.

Histidine Biosynthetic Pathway

The biosynthesis of histidine has been studied extensively in *Salmonella typhimurium* and *Escherichia coli*. In these microorganisms, a single operon composed of eight adjacent genes encodes the complete set of enzymes required for the biosynthesis of histidine. Three (*hisD*, *hisB*, and *hisI*) of the eight genes of the operon encode bifunctional enzymes, while two (*hisH* and *hisF*) encode enzymes that catalyze single steps, for a total of 10 enzymatic reactions.

The first step in histidine biosynthesis (**Figure 1**) is the condensation of ATP and 5-phosphoribosyl 1-pyrophosphate (PRPP) to form N'-5'-phosphoribosyl-ATP (PRATP). This reaction is catalyzed by N'-5'-phosphoribosyl-ATP transferase, the product of the *hisG* gene. This reaction is the one involved in feedback inhibition by the end product of the pathway, histidine. The inhibitory effect of histidine requires the presence of the product of the reaction, PRATP, and is further increased by AMP. Synergistic inhibition by the product of the first reaction and the end product of the pathway is a sophisticated variation of the general principle of feedback control, that has been found to also regulate the activity of glutamine synthetase. The inhibitory effect of AMP supports the energy charge theory of D.E. Atkinson and is logical in view of the high energy input required for histidine biosynthesis.

The product of the transferase reaction, PRATP, is hydrolyzed to N'-5'-phosphoribosyl-AMP (PRAMP). This irreversible hydrolysis is catalyzed by an activity associated with the C-terminal domain of the enzyme encoded by the *hisI* gene. The other activity, localized within the N-terminal domain of the bifunctional enzyme, is a cyclohydrolase, that opens the purine ring of PRAMP. This leads to the production of an imidazole intermediate, the N'-[(5'-phosphoribosyl) formimino]-5-aminoimidazole-4 carboxamide ribonucleotide (abbreviated to 5'-ProFAR). The fourth step of the pathway of histidine biosynthesis is an internal redox reaction, also known as an Amadori rearrangement, involving the isomerization of the aminoaldose 5'-ProFAR to the aminoketose N'-[(5'-phosphoribulosyl) formimino]-5-aminoimidazole-4 carboxamide ribonucleotide (abbreviated to 5'-PRFAR).

Although the pathway of histidine biosynthesis was almost completely characterized by 1965, the biochemical event leading to the synthesis of imidazole-glycerol phosphate (IGP) and 5-aminoimidazole 4-carboxamide ribonucleotide (abbreviated to AICAR or ZMP) from 5'-PRFAR remained unsolved for a long time. The protein products of the *hisH* and *hisF* genes were known to be involved in the overall

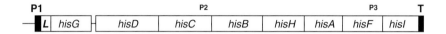

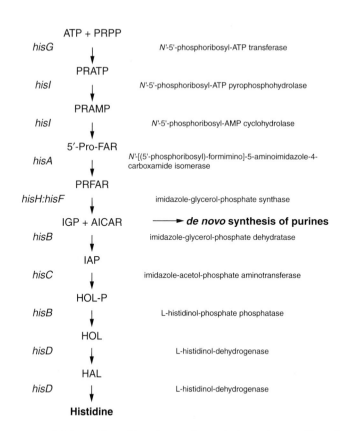

Figure 1 Structure of the *his* operon of *Salmonella typhimurium* and metabolic pathway of histidine biosynthesis. Top: the relative positions of **P1** (primary promoter), **P2** and **P3** (internal promoters) and **T** (rho-independent bifunctional transcription terminator) are indicated below the genetic map. **L** represents the leader regions preceding the structural genes. Bottom: biosynthetic steps from ATP and PRPP to histidine. Abbreviations are specified in the text.

process in eubacteria, but the catalytic events were elusive. The last blind spot of histidine biosynthesis has recently been clarified. The protein encoded by the *hisF* gene has an ammonia-dependent activity that converts PRFAR to AICAR and IGP, while the product of the *hisH* gene has no detectable catalytic properties. However, in combination, the two proteins are able to carry out the above reaction with glutamine as a nitrogen donor, without releasing any free metabolic intermediate. The *hisH* and *hisF* gene products form a stable 1:1 complex that constitutes the IGP synthase holoenzyme. AICAR, which is produced in the reaction catalyzed by IGP synthase, is recycled into the *de novo* purine biosynthetic pathway. The other product, IGP, is dehydrated by an activity of a bifunctional enzyme encoded by *hisB*. The resulting enol is ketonized nonenzymatically to imidazole-acetol phosphate (IAP). The seventh step of the pathway consists of a reversible transamination between IAP and glutamate. The reaction, catalyzed by a pyridoxal-P-dependent aminotransferase encoded by the *hisC* gene, generates α-ketoglutarate and L-histidinol phosphate (HOL-P). The HOL-P is converted to L-histidinol (HOL) by a phosphatase activity situated in the N-terminal domain of a bifunctional enzyme encoded by the *hisB* gene. In the final steps of histidine biosynthesis, HOL is oxidized to the corresponding amino acid L-histidine (His). This irreversible four-electron oxidation proceeds via the unstable amino aldehyde L-histidinal (HAL), which is not released as a free intermediate. A single enzyme, L-histidinal dehydrogenase, encoded by *hisD* catalyzes both oxidation steps. This prevents the decomposition of the unstable aldehyde intermediate. This enzyme is one of the first examples of a bifunctional NAD^+-linked dehydrogenase.

Mutants bearing nonfunctional enzymatic activities that are required for histidine biosynthesis grow normally in minimal medium when supplied with exogenous histidine. On the basis of this evidence, the histidine pathway was presumed to lack any branch point leading to other metabolites required for growth. Nevertheless, the two initial substrates of histidine biosynthesis, PRPP and ATP, play key roles in intermediary and energy metabolism and link this pathway to the biosynthesis of purines, pyrimidines, pyridine nucleotides, folates, and tryptophan. Moreover, the purine and histidine biosynthetic pathways are connected through the AICAR cycle. AICAR, a by-product of histidine biosynthesis, is also a purine precursor. The conversion of AICAR to purines involves a folic acid-mediated transfer of a one-carbon unit. Following treatment thought to lower the folic acid pool, the unusual nucleotide 5-aminoimidazole-4-carboxamide riboside-5′-triphosphate (ZTP) accumulates in *S. typhimurium*. On the basis of this and additional evidence, the rare nucleotide ZTP was proposed to be an alarmone signaling C-1 folate deficiency and to mediate a physiologically beneficial response to folate stress.

Organization of Histidine Genes

In many of the species where *his* genes were identified and characterized, they were not dispersed throughout the genome but were clustered with other genes in complete or partial operons. The same is partly true for operonless fungi, in which some of the *his* genes resulted from the fusion of different segments bearing homology to different bacterial genes. The organization of genes into *his* operons or clusters varies among different species, indicating that during evolution, genes were separated or linked, apparently without severe constraints. In other bacterial operons that have been characterized in several species, such as the *trp* operon, gene order was largely invariant. The recently determined organization of *his* gene clusters in several microorganisms is presented in **Figure 2**.

Regulation of Histidine Biosynthesis

It has been calculated that 41 ATP molecules are consumed for each histidine molecule made. The considerable metabolic cost required for histidine biosynthesis accounts for the evolution in different organisms of multiple and complex strategies to fine tune the rate of synthesis of this amino acid in response to environmental changes. In *S. typhimurium* and in *E. coli*, the biosynthetic pathway is under the control of distinct regulatory mechanisms that operate at different levels. Feedback inhibition by histidine of the activity of the first enzyme of the pathway almost instantaneously adjusts the flow of intermediates along the pathway in response to the availability of

exogenous histidine. Transcriptional attenuation at a regulatory element, located upstream of the first structural gene of the cluster, allows coordinate regulation of the levels of the histidine biosynthetic enzymes in response to the changing of histidyl tRNA. Two prominent features of the leader region of the *his* operon account for *his*-specific translational control of transcription termination, which is the essence of attenuation control: (1) a short coding region that includes numerous tandem histidine codons (7 histidine codons in a row of 16); and (2) overlapping regions of dyad symmetry that can fold into alternative secondary structures, one of which includes a rho-independent terminator. In the termination configuration, base pairing involves regions A and B, C and D, and E and F (**Figure 3**). The stable stem–loop structure E:F followed by a run of uridylate residues constitutes a strong intrinsic terminator. In the antitermination configuration, base pairing between B and C and between D and E prevents formation of the terminator, thus allowing transcriptional readthrough. The equilibrium between these alternative configurations is determined by the ribosome occupancy of the leader RNA, which in turn depends on the availability of charged histidyl tRNA. Low levels of the specific charged tRNA will cause ribosomes to stall on the leader region at the histidine codons, thereby disrupting A:B pairing by masking region A. Under these circumstances, the antitermination configuration will be favored. Conversely, in the presence of high levels of charged histidyl tRNA, ribosomes will rapidly move through the histidine regulatory codons, thereby occupying both the A and B regions. Pairing between C and D and between E and F leads to premature transcription termination.

In addition to histidine, the system is also regulated by other molecules whose levels reflect the energetic and metabolic state of the cell. It has been previously mentioned that PRPP and ATP stimulate the activity of the first enzyme of the pathway, whereas AMP enhances the inhibitory effect of histidine on this enzyme. Moreover, the alarmone guanosine 5′-diphosphate 3′-diphosphate (ppGpp), the effector of the stringent response, positively regulates *his* operon expression by stimulating transcription initiation at the level of the primary *hisP1* promoter. Stimulation occurs under conditions of moderate amino acid starvation and in cells growing in minimal medium. In addition to *hisP1*, two weak internal promoters, designated *hisP2* and *hisP3*, have been localized proximally to *hisB* and *hisI*, respectively. Although such internal promoters are quite common in large bacterial operons, the physiological significance of these genetic elements is controversial. Although these promoters may be physiologically

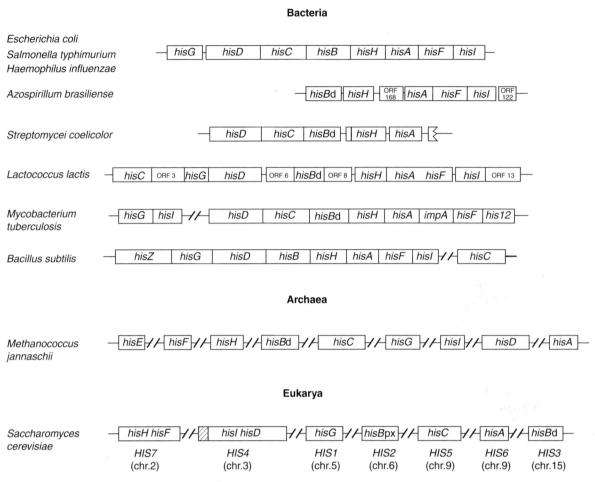

Figure 2 Organization of the histidine genes in different organisms. The single gene encoding a bifunctional enzyme, formerly known as *hisIE*, has been renamed *hisI* in *E. coli*. I have therefore used *hisI* for organisms with a single gene and *hisI* and *hisE* for organisms with two independent genes. Another gene encoding a bifunctional enzyme, *hisB*, is often split into two separate genes in different organisms. They are referred to as *hisB* proximal (*hisBpx*) encoding the HOL-P phosphatase, and *hisB* distal (*hisBd*) encoding the IGP dehydratase. In *Mycobacterium tuberculosis*, a gene encoding inositol monophosphatase, *impA*, is located between *hisA* and *hisF*. In *Bacillus subtilis* the structural gene encoding the histidyl tRNA synthetase is located proximally to the biosynthetic *his* cluster.

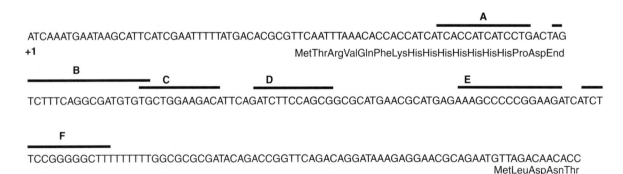

Figure 3 Features of the leader region of the *his* operon of *Salmonella typhimurium*. The nucleotide sequence of the leader region from the transcription initiation site (+1) to the first structural gene (*hisG*) is reported. The amino acid sequence of the leader peptide and of the amino-proximal region of the *hisG* gene product are indicated below the nucleotide sequence. Solid lines above the nucleotide sequence correspond to regions (**A** to **F**) capable of forming mutually exclusive secondary structures.

unimportant and their presence merely fortuitous, their presence in homologous genomic regions of related microorganisms supports their physiological relevance. They could reinforce the expression of distal cistrons of large operons, thereby alleviating the effects of natural polarity. Alternatively, they could allow regulation of an operon in a noncoordinate fashion and cause differential expression of certain genes under specific growth conditions. Based on several features of the nucleotide sequence, the internal promoters, as well as the primary *hisP1* promoter, belong to the $E\sigma^{70}$ class of promoters.

Transcription of the *his* operon is also modulated at the level of intracistronic rho-dependent terminators by a nonspecific mechanism operating during the elongation step. Terminators account for the polarity exhibited by several nonsense and frameshift mutations. Polarity is a phenomenon observed in polycistronic operons, by which certain mutations that prematurely arrest translation not only affect the gene in which they occur, but also reduce the expression of downstream genes. Although polarity was first described in the lactose system, the coordinate effect of polar mutations on downstream gene expression and the existence of polarity gradients were defined with precision in the *his* system by using a large collection of polar mutations. The phenomenon of polarity has been explained by postulating the existence of cryptic intracistronic rho-dependent terminators. According to a general model of transcriptional polarity, premature arrest of translation would favor the binding of rho to the nascent transcript via cytosine-rich and guanosine-poor regions. Using the energy of ATP hydrolysis, rho moves along the nascent transcript, overtakes elongating RNA polymerase, and precipitates release of the transcript. The physiological significance of rho-dependent intracistronic termination should be to prevent further elongation of nontranslated or infrequently translated transcripts.

Finally, it has recently been documented that post transcriptional events contribute substantially to *his* operon expression. In *S. typhimurium* and *E. coli*, the unstable native 7300 nucleotide-long polycistronic *his* message is degraded with a net 5′ to 3′ directionality, generating products that decay at different rates. The decay process generates three major processed species, 6300, 5000, and 3900 nucleotides in length (Pr1, Pr2, and Pr3), that encompass the last seven, six, and five cistrons, respectively, and have increasing half-lives (5, 6, and 15 min, respectively). RNase E controls the decay of the native transcript. Active translation of the 5′-end-proximal cistrons of the processed Pr1 and Pr2 species is required to temporarily stabilize these species. The overall process of decay may have

functional relevance in balancing the expression of the promoter-proximal and the promoter-distal genes. The most distal 3900 nucleotide-long processed species has a half-life of about 15 min. The specific processing event leading to production of this species is mechanistically complex. It requires sequential cleavage by two endoribonucleases, RNase E and RNase P.

As discussed above, the regulation of *his* operon expression in *E. coli* and *S. typhimurium* has been the subject of intensive studies and the general mechanisms and molecular details of the process are fairly well established. On the other hand, very few studies in this area have been performed with other prokaryotic cells. In general, it seems that while the biochemical reactions leading to histidine biosynthesis are the same in all organisms, the overall genomic organization, the structure of the *his* genes, and the regulatory mechanisms by which the pathway is regulated differ widely among taxonomically unrelated groups. For these topics, the interested readers are referred to specialized reviews that cover this subjects.

Further Reading

Alifano P, Fani R, Liò P et al. (1996) Histidine biosynthetic pathway and genes: structure, regulation, and evolution. *Microbiological Reviews* 60: 44–69.

Winkler ME (1996) Biosynthesis of histidine. In: Neidhart FC et al. (eds) Escherichia coli *and* Salmonella *Cellular and Molecular Biology*, 2nd edn, vol. 1, pp. 485–505. Washington, DC: American Society for Microbiology Press.

See also: Operon

Histocompatibility

J Read and B J Smith

Copyright © 2001 Academic Press
doi: 10.1006/rwgn.2001.0616

Histocompatibility is required for one individual to accept tissue grafts from another individual. It has long been recognized that successful blood transfusion/tissue transplantation are dependent on matching donor and recipient red blood cells. This led Gorer to the identification of a group of antigens in mice which, when matched between donor and recipient animals, greatly improved the success of a tissue graft. These antigens are known as histocompatibility antigens.

Different antigens are recognized by different T cell types. For example, in man, cytotoxic T cells involved in the recognition of noncompatible tissue grafts and/or virally infected cells recognize HLA-A and HLA-B antigens on the surface of foreign cells and, in conjunction with T cells, will destroy the foreign cells.

See also: **Antigen; Major Histocompatibility Complex (MHC)**

Histocompatibility Complex Genes

See: **Major Histocompatibility Complex (MHC)**

Histone Genes

A P Wolffe[†]

Copyright © 2001 Academic Press
doi: 10.1006/rwgn.2001.0618

Background

Histone genes were among the first eukaryotic genes to be characterized. Their cloning and isolation in the 1980s was facilitated by their repetition in metazoans, their small size, the abundance of their mRNAs, and the early sequence characterization of the histone proteins. Interest in histone genes derives from their regulated transcription, the control of histone mRNA stability, and the regulation of histone mRNA 3' processing. The histone genes provide a paradigm for the study of DNA replication (S-phase)-dependent transcription. They have also been exceptionally useful in the investigation of the determinants of tissue-specific and embryo stage-specific transcriptional control. Pioneering studies on the assembly of specialized architectures within chromatin made effective use of histone gene sequences. Research in the 1990s has led to recognition of the importance of histone protein sequence in the packaging of DNA for transcription, replication, recombination, and repair, together with the maintenance of chromosome stability and chromosome segregation. The histone genes have been subjected to an extensive mutational analysis, with the consequences for DNA metabolism of histone gene ablation, deletion, and point mutation investigated by many research scientists.

Histone Gene Organization

Some simple eukaryotes such as the budding yeast *Saccharomyces cerevisiae* have only two copies of each gene encoding the core histones H2A, H2B, H3, and H4. This lack of diversity and low copy

number greatly expedite mutational analysis (see below). In metazoans, the genes for all four core histones are normally clustered together and tandemly repeated 5 to 20 times. For example, *Xenopus laevis*, the clawed frog, has two predominant types of tandemly repeated clusters that differ in the precise gene arrangement and in the presence of genes for particular linker histone H1 genes. The regulatory DNA and coding sequence for each core histone gene within each cluster occupy less than 1 kb. Each cluster generally occupies less than 10 kb and appears to possess the capacity to assemble a unique regulatory nucleoprotein complex within chromatin. The vast majority of core histone genes are found clustered together. This organizational strategy is likely to facilitate coordinate expression. The clustered majority of core histone genes are almost invariably expressed as a cohort during S-phase. These replication-dependent genes lack introns and utilize a specialized processing mechanism for generating their 3' ends that is distinct from polyadenylation. A smaller group of core histone genes are not primarily regulated in response to cell-cycle signals, but are either constitutively expressed at low levels in somatic cells, or they can be expressed in differentiation specific patterns during metazoan development. These core histone variants, whose mRNAs are encoded by these replication-independent histones, can accumulate to very high levels only in cells that have ceased to divide. Nondividing cells have also stopped synthesizing the replication-dependent histones. This facilitates the replacement of replication-dependent histones by replication-independent variants, especially on DNA sequences at which regulated chromatin disruption might occur during transcription and repair. The replication-independent histone genes differ in the *cis*-acting elements controlling promoter activity from the replication-dependent genes. Replication-independent genes can also have introns and can be polyadenylated. Thus, the replication-independent histone genes look much more like normal genes transcribed by RNA polymerase II. They are also normally not present in the large clusters. These differences can also be extended to the linker histone genes. The normal histone H1 somatic gene in *Xenopus* is found in a cluster with the core histone genes, lacks introns, and is transcribed in S phase. In contrast, the specialized maternal linker histone B4 gene is transcribed throughout oogenesis in the absence of replication and contains introns. The contrast between replication-independent and replication-dependent histone genes serves to emphasize the many unusual features of specialized organization and control utilized by the replication-dependent genes to ensure very high expression at a single time in the cell cycle.

[†]deceased

Transcriptional Control of Histone Genes

Replication-dependent core histone gene transcription is generally regulated through a three- to tenfold range during the cell cycle. Control is mediated by *cis*-acting elements that are within 200 bp of the start site of transcription. In *S. cerevisiae* the histone H2A and H2B genes share common regulatory elements with other genes controlled by the cell cycle, including the HO endonuclease (see below). Negative and positive regulatory elements have been identified. In humans, the H2B gene is regulated by three elements: the TATA box, an octamer motif (ATTTGCAT), and a distal activating domain including the CCAAT box. The TATA box is recognized by the basal transcriptional machinery including TFIID, the octamer motif is recognized by the ubiquitous octamer-binding transcription factor (OTF-1), and the proteins binding the distal activating domain have not yet been fully characterized. However, the constitutive activator NF-Y is an excellent candidate for interaction with the CCAAT box in vertebrate cells. In the replication-dependent H4 promoters, the octamer motif is replaced by other regulatory elements shared with replication-dependent linker histone gene promoters. The molecular definition of the transcription factors binding to these sites is still at a rudimentary stage of development. It appears that the core histone genes utilize a diverse group of constitutively expressed transcription factors to control transcription. It is probable that their expression is coordinated through the recruitment of common transcriptional coactivators such as the p300/CBP protein. Consistent with this hypothesis is the observation that the *Drosophila* and *Xenopus* core histone genes are assembled into specific regulatory nucleoprotein architectures independent of cell-cycle-regulated transcription. This result demonstrates that the regulatory DNA for the histone genes is always occupied by the DNA-binding transcription factors, and that it is the efficiency with which these preassembled complexes recruit RNA polymerase II that is regulated.

The regulation of the replication-independent and differentiation-specific core and linker histone genes is more complex. The promoters of these genes, such as the oocyte-specific histone B4 gene in *Xenopus* or the erythroid-specific histone H5 gene in the chicken, depend upon specific regulatory factors for transcriptional activation in particular tissues. For example, the accumulation of histone H5 protein in avian erythrocytes occurs during the differentiation of the erythroid cell, correlating with the shut down of replication and a decrease in transcriptional activity. The accumulation of histone H5 mRNA is predominantly controlled at the transcription level. Erythroid-specific and ubiquitous elements control expression of this gene in erythroid lineages. The activity of the gene is low in early erythroid precursors and rises as differentiation proceeds. Activation during erythropoiesis is essential due to the action of three enhancers, two of which lie upstream and one downstream of the transcription start site. The tissue specificity of these enhancers is related to the presence of several sites for an erythroid-specific transcription factor GATA-1. However, the activity of GATA factors alone cannot account for the activation of H5 gene expression and ubiquitous transcription factors seem also to play a central role in this process. The proximal promoter region of H5 contains a segment showing extensive similarity with a region of the H4 gene proximal promoter. A positive transcriptional regulatory element has been identified in this region which binds specifically the histone gene-specific factor, H4TF2, in proliferative cells. However, it does not seem to be essential for the activity of the gene in differentiated cells. In contrast, a neighboring GC-rich sequence element is required for gene activity in both the proliferative precursors as well as in the early stages of cell differentiation. Finally, the basal transcription of this promoter seems to involve sequences located downstream of the initiation site.

Posttranscriptional Control of Histone mRNA

At the end of S-phase when DNA replication stops, the half-life of the replication-dependent histone mRNAs decreases from 30–60 min to 10–15 min. This destabilization of histone mRNA depends on the regulated association of proteins with the 3' terminus of histone mRNA. A stem–loop structure in the 3' terminus controls the processing nucleocytoplasmic export of histone mRNA, translational efficiency, and mRNA stability. Exactly how this is accomplished is unknown. There is also a possible autoregulatory contribution to the regulation of mRNA abundance, since individual core histones and linker histones have been reported to induce the destabilization of histone mRNA *in vitro*.

Importance of Histone Gene Sequences for Transcriptional Control in Eukaryotic Nucleus

The core histones, H2A, H2B, H3, and H4, are among the most evolutionarily conserved of all eukaryotic proteins. They consist of two domains: a basic N-terminal domain and a histone-fold

C-terminal domain. The histone-fold domain has two defined functions: it heterodimerizes with a second histone – H3 with H4, H2A with H2B – and, once heterodimerized, it wraps DNA in the nucleosome. The basic N-terminal 'tail' domains lie outside the nucleosome and do not have any defined structure. Although extensive protein–protein and protein–DNA interactions can potentially explain the sequence conservation of the histone-fold domains, the N-terminal tails of histones H3 and H4 show comparable conservation from yeast to man. The reasons for this conservation have been enigmatic, but two nonexclusive explanations have been proposed.

The first suggested explanation is that the H3 and H4 N-terminal tails represent the sites at which signal transduction pathways impact on chromatin structure. The N-terminal tails are known to be sites of histone phosphorylation, acetylation, and methylation, and these modifications are closely correlated with changes in the functional properties of chromatin. Sequence conservation at the N-terminus might be required to transduce the activities of various targeted and ubiquitous histone modification enzymes involved in chromatin assembly and transcription. The second suggested explanation is that the N-terminal tails might represent the sites of interactions between histones and regulatory proteins that have direct structural and functional roles in the transcription process. Such specific interactions have now been shown to occur. Histone modifications are predicted not only to alter chromatin structure, but also the interactions between the N-terminal tails and histone-binding regulatory proteins. The first genetic experiments suggesting that the histone tails play a part in the regulation of specific eukaryotic genes concerned the establishment of silent mating-type loci in S. cerevisiae. Subsequent work has firmly established that the N-terminal tail domains of histones H3 and H4 are essential for repression of the silent mating type loci, as well as of genes placed close to the telomeres in yeast. Transcriptional repression at these chromosomal sites also depends on the silent-information regulatory proteins SIR2, SIR3, and SIR4. SIR3 and SIR4 interact with each other and with the DNA-binding protein RAP1. Together, they direct the compartmentalization of yeast chromosomal telomeres to the vicinity of the nuclear envelope.

Mutations in the N-terminal tail of histone H4 that alleviated silencing can be suppressed by single amino acid substitutions in SIR3, suggesting that the two proteins directly interact. Biochemical experiments have confirmed that SIR3 binds directly to the N-terminal tail of H4, and also to the N-terminal tail of H3. The data suggests that SIR4 interacts in a similar way with these two histones. The specificity of these interactions was demonstrated by the failure of either SIR3 or SIR4 to interact with the N-terminal domains of H3 and H4 are also required for the assembly of SIR3 into telomeric chromatin, and consequently for the association of the telomere with the nuclear envelope.

A model for transcriptional silencing at yeast telomeres predicts that RAP1 interacts with the telomeric repeats and recruits SIR3 and SIR4, which polymerize along nucleosomal arrays through interactions with the N-terminal tails of H3 and H4. At the silent mating type loci, a distinct repressive mechanism (yet to be definitively characterized) also leads to the recruitment of SIR3 and SIR4. This model proposes that transcriptional silencing is dependent on the assembly of an extended domain of repressive chromatin structure, where transcription factors and RNA polymerase are excluded both by SIR3 and SIR4, and by the entrapment of this chromatin domain in a perinuclear compartment.

This second set of experiments that link the histones to the transcriptional regulation of specific genes concerns the C-terminal histone-fold domain and the SWI/SNF general activator complex. A substantial component of transcriptional regulation is increasingly perceived to depend upon the interplay of transcription factors and histones at specific sites within the enhancers and promoters of eukaryotic genes. In the yeast S. cerevisiae, the outcome of this interaction is influenced by the products of the SWI1/ADR6, SWI2/SNF2, SWI3, SNF5, and SNF6 genes. All five of these proteins are found within a single 'general activator' complex, required for the transcriptional induction of many yeast genes. Genetic and biochemical studies of the yeast proteins and their larger eukaryotic homologs suggest that the general activator complex serves as a molecular machine that functions to help transcription factors overcome the specific repressive effects of nucleosome assembly on transcription.

In the early 1980s, Herskowitz and colleagues discovered that mutations in a set of 'SWItch' genes – SWI1, SWI2, and SWI3 – reduce expression of the HO gene, which encodes a endonuclease involved in yeast mating-type switching. Simultaneous experiments by Carlson and colleagues defined sucrose nonfermentation mutations of the genes SNF2, SNF5, and SNF6, which reduced expression of the SUC2 invertase gene. Both sets of mutations reduced target gene induction by two orders of magnitude; moreover, SWI2 was found to be identical to SNF2, suggesting that both the SWI and SNF gene products functioned through a common mechanism.

Over the subsequent decade, a dozen other inducible genes were found to be dependent on SWI or

SNF gene activities for transcriptional stimulation. More recent experiments have shown that the *Drosophila fushi tarazu* and *bicoid* gene products, mammalian steroid receptors, and yeast transcription factor GAL4 all stimulate transcription through mechanisms dependent on SWI/SNF activities. *Drosophila*, mouse, and human homologs of the SWI2/SNF2 subunit exist and have similar roles in facilitating transcriptional activation of a variety of genes. Taken together, these results clearly indicate that the general activator complex has a central role in the regulation of eukaryotic transcription, but how is this transcriptional activation function exerted?

A major clue to the molecular mechanism by which the general activator complex exerts its function came from a genetic screen for mutations of genes that would allow transcription of HO in the absence of SWI1. Two genes, SIN1 and SIN2, were identified that, when mutated, led to SWI-independent transcription. Both of the SIN genes isolated in this way encode components of chromatin. SIN1 is a highly charged nuclear protein, somewhat similar to mammalian HMG1/2 proteins. The HMG1/2 proteins have been found to be associated with nucleosomes, most probably interacting with linker DNA. Every nucleosome contains 165–220 bp of DNA, of which 146 bp are wrapped in 1.75 turns around the octamer of core histones in the nucleosome core. The additional DNA that lies between nucleosome cores is the linker DNA. Linker histones (such as H1, H5, and H1°) normally bind to linker DNA, however, in certain circumstances, they may be replaced by HMG1/2.

A more direct association with nucleosomal structure is found for SIN2, which encodes histone H3. Kruger, Peterson, Herskowitz, and colleagues also identified SIN alleles of the H4 gene, after reintroduction *in vivo* of the *in vitro* mutagenized gene. The location of the amino acid changes in histone H3 and H4 that lead to the SIN phenotype offer additional insight into potential roles for the general activator complex. However, in order to appreciate the structural significance of the mutations, it is important to know their position within the nucleosome. The carboxy-terminal histone-fold domains of each core histone are predominantly α-helical, with a long central helix bordered on each side by a loop segment and a shorter helix. Each of the loop segments has some β-strand character. Histone dimerization leads to the loop segments from each half of the dimer being paired to form eight, parallel β-bridge segments, two of which are found within each of the histone heterodimers – H3, H4 and H2A, H2B. Each β-bridge segment is associated with a least two positively charged amino acids, which are available to make contact with DNA on the surface of the histone octamer.

The second repeating motif within the nucleome is assembled from the pairing of the amino-terminal end of the first helical domain of each of the histones in the heterodimers. These four 'paired-ends-of-helices' motifs also appear to contact DNA. Thus, each of the four heterodimers within the core can make at least three, pseudosymmetrical, contiguous contacts with three inward-facing minor grooves of DNA. The parallel β bridges and four paired-ends-of-helices provide 12 potential DNA-contact sites that are regularly arranged along the ramp on which the double helix is wound. The SIN mutants in histones H3 and H4 cluster in one β-bridge motif within the heterodimer. Because of the juxtaposition of two (H3, H4) heterodimers at the dyad axis of the nucleosome, the SIN mutations have the potential to disrupt histone–DNA interactions involving the central turn of DNA at the dyad axis. This could have a major impact on the integrity of both the nucleosome and higher order chromatin structures.

These two examples of transcriptional regulation have in common the highly selective recognition of individual core histones by a variety of regulatory proteins. These interactions can be targeted by sequence-specific DNA-binding proteins, and provide an explanation for the highly selective activation or repression of particular genes following mutation of individual histones. The inclusion of histones as architectural components within regulatory nucleoprotein complexes further strengthens the evidence for their essential role in eukaryotic transcription. The reasons for the conservation of the primary sequence of the core histones and their genes thus go beyond merely conserving the internal architecture of the nucleosome, and include the functional requirement of conserving interactions with the regulatory proteins that modulate chromatin function. These results also suggest that novel families of proteins remain to be defined that will contain conserved regions capable of specifically recognizing histone domains both outside and inside the nucleosome. Defining the nature of these proteins that truly 'hang on' to the histones will offer much insight into how regulatory events occur within chromosomes.

Further Reading

Gargiulo G, Razvi F, Ruberti I, Mohr I and Worcel A (1985) Chromatin specific hypersensitive sites are assembled on a *Xenopus* histone gene injected into *Xenopus* oocytes. *Journal of Molecular Biology* 181: 33–349.

Heindl LM, Weil TS and Perry M (1988) Promoter sequences required for transcription of *Xenopus laevis* histone genes in injected frog oocyte nuclei. *Molecular and Cellular Biology* 8: 676–682.

Hinkley C and Perry M (1991) A variant octamer motif in a *Xenopus* H2B histone gene promoter is not required for transcription in frog oocytes. *Molecular and Cellular Biology* 11: 641–654.

Khochbin S and Wolffe AP (1994) Developmentally regulated expression of linker-histone variants in vertebrates. *European Journal of Biochemistry* 225: 501–510.

Schumperli D (1988) Multilevel regulation of replication-dependent histone genes. *Trends in Genetics* 4: 187–191.

Wolffe AP (1998) *Chromatin: Structure and Function.* San Diego, CA: Academic Press.

See also: Eukaryotic Genes

Histones

Histones are conserved proteins found in the nuclei of all eukaryotic cells where they are complexed to DNA forming the nucleosome, the basic subunit of chromatin. Histones are of relatively low molecular weight and are basic, owing to their high arginine/lysine content.

See also: Chromatin

Hitchhiking Effect

M Kreitman

DNA, the genetic material, is packaged into chromosomes ranging in length from thousands to many tens of millions of base pairs. Now consider the fate of two independent mutations that have occurred by chance on one specific copy of a chromosome in a population. Imagine one of these mutations is a selectively favorable mutation that natural selection will increase in frequency in the population each generation (see Selective Sweep). The other mutation on this same chromosome is a selectively neutral mutation (see Neutral Mutation), one whose fate will be governed under normal circumstances by genetic drift (see Genetic Drift). Its association with the favorable mutation on the same chromosome guarantees that as the adaptive mutation increases in frequency by natural selection, so the 'linked' neutral mutation will also deterministically increase in frequency. The 'hitchhiking effect' is the associated change in frequency of a nonselected mutation resulting from its physical linkage to a different mutation under selection on the same chromosome.

The magnitude of genetic hitchhiking is related directly to the recombination rate between the mutations under consideration. The animal mitochondrial genome, for example, a maternally inherited circular genome, is expected to be particularly susceptible to hitchhiking events because it is a nonrecombining genome. In one species of the fruit fly, *Drosophila simulans*, a maternally inherited microorganism, called *Wolbachia*, has a mechanism by which it provides a strong selective advantage to females carrying the infection when they are introduced into a population without the infection. This strong selective advantage and maternal inheritance of both the advantageous bacteria and the mitochondrial genome has been shown to cause the mitochondrial variant in the infected female to increase in frequency as it hitchhikes up along with the frequency of *Wolbachia* infection.

A curious feature of genetic hitchhiking accompanying the fixation of a selectively favored mutation (see Selective Sweep) is that other mutations within a tightly linked interval spanning the site under positive selection, if they do not undergo a recombination event during the course of fixation of the favored mutation, will also all go to fixation (or extinction). Therefore, one telltale sign of a selective sweep of a favorable mutation is a region of the genome that has lower than expected levels of nucleotide polymorphism in a population sample. Several such signatures of a selective sweep have been reported in this manner, especially in the *Drosophila*.

A second type of hitchhiking is also possible, and it involves the hitchhiking to extinction (rather than to fixation) of mutations that are linked to a selectively deleterious mutation, i.e., one that is doomed to be eliminated from the population by natural selection. When a deleterious mutation arises in a population, it is generally eliminated, but often this elimination requires tens or hundreds of generations to complete. During this time, any other mutation that also arises on this doomed chromosome, unless it is strongly advantageous or it is sufficiently loosely linked and can recombine away, will also be eliminated in due course. Under this scenario, called 'background selection,' only those chromosomes in the population without any deleterious mutation will contribute to the future ancestry of the population. Theory shows that this fraction of unmutated chromosomes is approximately $f(0) = e^{\mu/s}$ where μ is the deleterious mutation rate and s is the selective disadvantage of the mutation. In *Drosophila*, certain regions of a chromosome have very much lower recombination rates, as measured by the recombination rate per kilobase of DNA, than other regions of the same chromosome. In these regions, background selection is predicted to reduce the standing crop of neutral mutations by the fraction $f(0)$. In fact, strong reductions in variation have consistently been found in these low-recombining regions of the genome, providing a modicum of support for the prevalence of background selection (but see Selective Sweep for an alternative explanation for this observation).

Further Reading

Charlesworth B (1996) Background selection and patterns of genetic diversity in *Drosophila melanogaster*. *Genetic Research* 68(2): 131–149.

Kaplan NL, Hudson RR and Langley CH (1989) The "hitchhiking effect" revisited. *Genetics* 123: 887–899.

Kim Y and Stephan W (2000) Joint effects of genetic hitchhiking and background selection on neutral variation. *Genetics* 155(3): 1415–1427.

See also: Background Selection; Genetic Drift; Neutral Mutation; Selective Sweep

HIV

See: **Virus**

Hodgkin's Disease

M J S Dyer

Copyright © 2001 Academic Press
doi: 10.1006/rwgn.2001.1581

Hodgkin's disease (HD) is a collection of disparate lymphoid disease, defined histologically by the presence of multinucleated Hodgkin or Reed–Sternberg (H/RS) cells. The first eponym derives from the *postmortem* description of six cases with lymphadenopathy and splenomegaly by Thomas Hodgkin at Guy's Hospital, London in 1832. The H/RS cells were described by Dorothy Reed in 1902, and Sternberg in 1898. There are four distinct histological subtypes: nodular sclerosing (NS), mixed cellularity (MC), lymphocyte depleted (LD), and lymphocyte predominant (LP). NS is the most common and is found mainly in young adults. The LP subtype is distinct, lacking H/RS cells and having instead populations of large 'lymphocyte and histiocytic' or L&H cells, which derive from mature B cells. In contrast, the H/RS cells of the other histological subtypes express molecules associated with a number of hemopoietic lineages including T-cell antigens such as CD2 and CD4, myeloid antigens such as CD15 as well as B-cell antigens.

The etiology of HD remains unknown and given the wide differences in the histological appearances, it is likely that the etiology of each subtype will be distinct. Familial clustering of HD has been reported. Whether this represents a common genetic predisposition and/or exposure to some common environmental agent is not clear. Epstein–Barr virus (EBV) may have a role to play in some cases as some H/RS cells contain EBV genomes and express LMP1 which is known to have oncogenic potential in B cells. Cytogenetic analysis of primary material and derived cell lines has shown no recurrent abnormalities; HD-derived cell lines are notable for their cytogenetic complexity. Recently, comparitive genomic hybridization (CGH) studies have shown gains of chromosome 2p13 and high-level amplification of chromosomes 4p16, 4q23–q24, and 9p23–p24. The last region, which is also amplified in mediastinal B-cell lymphomas, contained *JAK2*.

The H/RS and L&H cells represent the malignant cells of HD. The major problem with the study of HD is that these malignant cells comprise only a small subset, often less than 1% of the tumor, with the remainder composed of infiltrating reactive T cells, B cells, neutrophils, and fibrotic tissue. For a long time the cell of origin of both H/RS and L&H cells remained unknown. To overcome this problem, microdissection and amplification of RNA and DNA has been undertaken. Using these techniques, along with high-throughput sequencing of HD cDNA libraries (http://www.hodgkins.georgetown.edu./) and gene profiling methods, the origins and pathophysiology of HD are being revealed. Analysis of the microdissected H/RS cells from patients with NS HD has shown that these cells not only exhibit rearranged immunoglobulin heavy (*IGH*) chain gene segments but also have mutations consistent with their exposure to antigen in the germinal center of the lymph node. Furthermore, analysis of rare patients with concurrent HD and B-cell lymphoma showed the same clonal *IGH* rearrangements with an overlapping pattern of somatic mutations within the variable region (*VH*) gene segments. Together, these data indicate a B-cell origin for at least some if not all, H/RS cells.

Concerning the pathophysiology, study of HD cell lines has revealed constitutive activation of NF-κB and secondly, autocrine stimulation via IL 13. Nuclear NF-κB promotes cell survival through the transcriptional upregulation of a number of antiapoptotic genes. However, in most normal cells, NF-κB is retained in the cytoplasm due to the presence of inhibitory (IκB) proteins. Concurrent deletion and mutation of IκBα alleles, resulting in protein truncation and loss of inhibitory activity, has been reported in a number of cell lines and primary cases. Secondly, constitutive IL 13 secretion has been detected by gene profiling of HD lines and shown in primary material by *in situ* hybridization. Moreover, in one HD cell line, neutralizing antibodies to IL 13 blocked proliferation, suggesting that this might be a new therapeutic target in some cases of HD.

Further Reading

Jarrett RF and MacKenzie J (1999) Epstein–Barr virus and other candidate viruses in the pathogenesis of Hodgkin's disease. *Seminars in Hematology* 36: 260–269.

Joos S, Kupper M, Ohl S *et al.* (2000) Genomic imbalances including amplification of the tyrosine kinase gene JAK2 in CD30+ Hodgkin cells. *Cancer Research* 60: 549–552.

Rose M (1981) *Curator of the Dead: Thomas Hodgkin (1798–1866)*. London: Peter Owen.

Staudt LM (2000) The molecular and cellular origins of Hodgkin's disease. *Journal of Experimental Medicine* 191: 207–212.

Reference

Hodgkin's disease cDNA libraries: http://www.hodgkins.georgetown.edu/

See also: Epstein–Barr Virus (EBV); Reed–Sternberg Cells

Hogness Box

A J Berk

Copyright © 2001 Academic Press
doi: 10.1006/rwgn.2001.0621

Also known as the 'TATA box,' the Hogness box is an 8-bp AT-rich promoter sequence in eukaryotes and Archaea that is the binding site for the TATA-box binding protein (TBP), a subunit of the TFIID initiation factor in metazoans. TBP functions as an initiation factor without additional TBP-associated factors in Archaea and at many promoters in *Saccharomyces cerevisiae*. The first base of the sense strand consensus sequence T-A-T-A-T/A-A-T/A-N is approximately 30 bp upstream of RNA polymerase II transcription start sites in metazoans, Archaea, and some fungi. In *S. cerevisiae* the Hogness (TATA) box occurs ~90 bp upstream of the transcription start site. The match to the consensus sequence (determining the affinity for TBP) is an important determinant of promoter strength.

See also: Consensus Sequence; Promoters

Holliday Junction

P J Hastings

Copyright © 2001 Academic Press
doi: 10.1006/rwgn.2001.1429

A Holliday junction is the structure formed by the exchange of single DNA strands between two

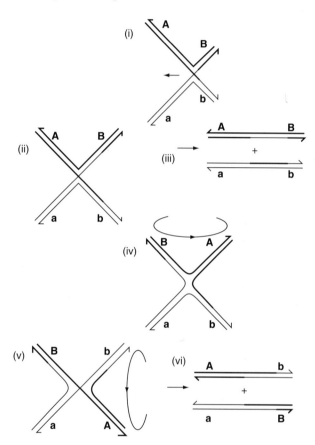

Figure 1 A Holliday junction with hybrid DNA on both molecules. Single strands of one parent are distinguished from the other parent by the thickness of the line. Hybrid DNA can be seen as thick and thin strands within the same molecule. (i) Migration of the Holliday junction toward the left (small arrow) has extended the two lengths of heteroduplex. (iii) Cleavage of the junction by resolvase from the structure (ii), cutting the crossing strands, yields two molecules with parental combinations of markers A and B or a and b, although each includes a length of hybrid DNA. (iv) Rotation of the upper arms, shown by the circle, shows the same structure from a different point of view. (v) A further rotation, this time of the two arms on the right (shown by a circle), reveals the alternative isomer. (vi) Cleavage of the isomer in (v) by cutting the crossing strands gives two molecules with the recombinant combination of the markers, A and b or a and B. Each has a length of hybrid DNA.

homologous DNA molecules. The structure is named after Robin Holliday who first proposed the structure in 1964 (Holliday, 1964). **Figure 1** shows how the two molecules are held together by the presence of hybrid DNA, that is, DNA formed with one strand from one parental molecule and the other strand from the other parent. Physical models of DNA show that it can adopt this structure without strain and with all bases remaining paired.

The Holliday junction is central to recombination theory because it has three interesting properties. First, it can isomerize, i.e., take on an alternative structure (see Isomerization (of Holliday Junctions)). Second, it can migrate, leading to extension or shortening of the lengths of hybrid DNA (see Branch Migration). Third, it can be resolved by a special class of enzymes (resolvases) that cut the structure symmetrically to give two separate molecules (see Resolvase).

Isomerization occurs spontaneously. Resolving the structure while in one isomer is expected to lead to crossing-over. In the other isomer, resolution restores the parental combination of flanking regions, but lengths of single strands have been exchanged. Migration of a Holliday junction may be able to occur by random drift, but it is an enzyme-mediated process in *Escherichia coli*, where the RuvABC proteins acting together are able to catalyze both branch migration and resolution.

Reference

Holliday R (1964) A mechanism for gene conversion in fungi. *Genetic Research* 5: 282–304.

See also: Branch Migration; Isomerization (of Holliday Junctions); Resolvase

Holliday's Model

P J Hastings

Copyright © 2001 Academic Press
doi: 10.1006/rwgn.2001.0622

In 1964, Robin Holliday proposed the basic model of recombination by the formation of hybrid DNA coupled with correction of mismatched base pairs. In this model, initiation of recombination occurs by cutting a single DNA strand (nicking) at identical positions on the like strands of two homologous DNA molecules, as shown in part (1) of **Figure 1**. Both of these strands become unwound from the nick (2) and anneal with the homolog so that the two displaced strands have changed places thereby forming hybrid DNA (3). The structure so formed is called a Holliday

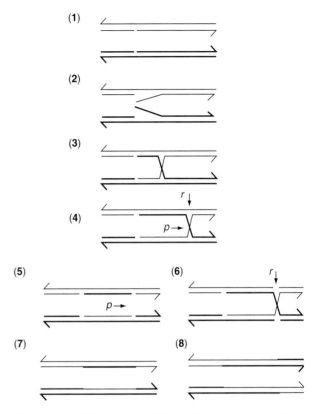

Figure 1 Holliday's model. Each line represents a single DNA strand. Thick and thin lines distinguish the DNA of two homologous molecules. Arrows on the strands indicate polarity. The figure is described in the text.

junction. The Holliday junction can migrate in either direction. If it migrates away from the site of initiation, hybrid DNA is extended on both DNA molecules (4). If the Holliday junction migrates towards the initiation site, the lengths of hybrid DNA will diminish symmetrically.

Holliday proposed that the interacting DNA molecules could be separated by strand breakage at the Holliday junction. If the breaks occur on the inner strands at the positions marked *p* in (4), the two molecules could separate (5) without recombination of markers flanking the event, and ligation would yield two noncrossover products in which lengths of single strands have been exchanged locally (7). If the outer strands labeled *r* in (4) are broken, as seen in (6), ligation of the ends would result in a crossover (8).

If there is an allelic difference between the interacting molecules, the hybrid DNA will contain one or more mismatched base pairs or unmatched nucleotides (single-strand loops). Such a hybrid molecule is called a heteroduplex. Holliday proposed that a mismatch repair system will operate on the mismatch in the heteroduplex to excise one genotype or the other, and replace it by copying the remaining single strand. This correction process may then either convert a

DNA molecule to the genotype of the homolog or restore the parental genotype. Uncorrected heteroduplex DNA would persist until the next replication of the chromosomes, when the two daughter chromosomes would be of different genotypes. This explains the phenomenon of postmeiotic segregation, where a single meiotic product is seen to have both parental genotypes even though it has only one copy of any one DNA molecule.

By this simple form of the model, lengths of heteroduplex DNA are necessarily symmetrical, that is, they have the same length on the two participating DNA molecules. However, it was known that the distribution of conversion may be asymmetrical. Holliday overcame this problem by proposing that the Holliday junction might migrate back toward the initiation site after mismatch correction has occurred on only one chromatid. This could have the effect of leaving an asymmetrical length of conversion.

Further Reading

Holliday R (1964) A mechanism for gene conversion in fungi. *Genetic Research* 5: 283–304.

See also: Heteroduplexes; Holliday Junction

Holocentric Chromosomes

D G Albertson

Copyright © 2001 Academic Press
doi: 10.1006/rwgn.2001.0623

Holocentric chromosomes are distinguished by the structure of the kinetochore, which extends along the poleward face of the metaphase chromosome. Microtubule attachment is distributed along holocentric chromosomes, in contrast to monocentric chromosomes where the kinetochore and hence microtubule attachment is localized to one region. In meiosis, the nonlocalized kinetochore is absent and the ends of the chromosomes are said to adopt 'kinetic activity,' referring to the observation that in the meiotic divisions the chromosomes move end on toward the spindle poles. Holocentric chromosome organization has been described for certain plants, protozoa, nematodes, and insects. A review of the earlier literature describing the cytological observations on holocentric chromosome behavior in various groups is available (White, 1973). In recent years, the nematode *Caenorhabditis elegans* has been the subject of extensive cytological, molecular, and genetic studies, which have contributed to the understanding of various aspects of holocentric chromosome behavior in this organism. Research on mitotic and meiotic segregation in *C. elegans* indicates that these holocentric chromosomes have features and behaviors in common with the more familiar monocentric chromosomes (Albertson *et al.*, 1997).

Mitotic Behavior

The nonlocalized kinetochore becomes visible at the ultrastructural level in prophase. By metaphase, it is typically a well-differentiated trilaminar structure resembling the kinetochore of monocentric chromosomes and probably extends the entire length of the chromosome. Holocentric chromosomes appear as stiff rods under the light microscope and lack the primary constriction that demarcates the centromere of monocentric chromosomes. At metaphase, the chromosomes align parallel to the equator of the metaphase spindle and lie entirely within the spindle. Microtubule attachments are distributed along the kinetochore, so that at anaphase the chromosomes move broadside on to the spindle poles. Studies in *C. elegans* have also demonstrated that these holocentric chromosomes terminate in telomere sequences similar to those of mammalian telomeres.

Chromosome Rearrangements

Holocentric chromosome organization allows the stable propagation of chromosome rearrangements that are not mitotically and meiotically stable in organisms with monocentric chromosomes. Translocation chromosomes involving two entire holocentric chromosomes align and segregate to a single spindle pole, whereas in organisms with monocentric chromosomes, the linkage of two chromosomes results in the formation of dicentric chromosomes that fail to segregate properly. Fragments of holocentric chromosomes may also be propagated, because they retain the capability to attach to the spindle apparatus. In contrast, fragmentation of monocentric chromosomes results in the generation of mostly acentric fragments that are lost. Indeed, before visualization of the holocentric kinetochore by electron microscopy was possible, this differential behavior of holocentric and monocentric chromosome fragments formed the basis of a test for holocentric organization.

Meiotic Behavior

Holocentric chromosomes typically behave differently in meiosis and mitosis. In meiosis, in most organisms that have been examined at the ultrastructural level, no kinetochore structure is seen. Instead,

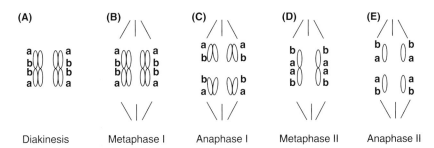

Figure 1 Orientation and segregation of axially oriented holocentric chromosomes in meiosis. (A) Two holocentric meiotic bivalents are shown in diakinesis. The homologs are associated at the ends labeled (**b**). (B–E) Segregation of meiotic chromosomes on meiotic spindles drawn with the long axis vertical. The spindle microtubules are indicated by lines and are shown converging toward the poles. (B) Alignment of the bivalents on the metaphase I spindle with the ends labeled **a** proximal to the spindle poles and the ends labeled **b** on the spindle equator. (C) At anaphase I, homologs separate to opposite spindle poles with the ends labeled **a** leading the way to the spindle poles. (D) Axial orientation of sister chromatids with ends labeled **b** proximal to the spindle poles and the ends labeled **a** on the equator of the spindle. (E) Anaphase II segregation with ends **b** leading the way to the spindle poles.

microtubules appear to project directly into the chromatin. At diakinesis of meiotic prophase, the bivalents of holocentric chromosomes are composed of homologous chromosomes, which appear to be held together in an end-to-end association. In earlier literature, this association was attributed to terminalization of chiasmata, but whether there is terminalization in organisms for which meiosis I is reductional is now being questioned. It seems more likely that the extreme condensation of the chromatin obscures cytological manifestations of distributed crossovers and gives rise to the apparent end-to-end association of the homologs. Furthermore, proper disjunction of the homologs requires a crossover event, and it appears that the location of the crossover determines which of the two ends of the homologs are associated in the bivalent.

The orientation of the bivalents on the metaphase I spindle varies from species to species. The bivalents may adopt the equatorial orientation and align parallel to the equator of the spindle, or they may align parallel to the spindle pole axis, adopting the axial orientation. If the bivalent aligns axially, then the sister chromatids segregate to the same pole at anaphase I, so that the first meiotic division is reductional, as occurs in meiosis in species with monocentric chromosomes. For equatorially oriented bivalents, the order is reversed and the first meiotic division is equational. In *C. elegans* and in some heteropteran species, it has been possible to use cytological markers to study the segregation of axially oriented homologs. As shown in **Figure 1**, the chromosomes align axially at metaphase I and move end on toward the spindle pole at anaphase I. On completion of meiosis I, the sister chromatids remain in association at the ends that were poleward

in metaphase I. They align axially with these ends on the equator of the metaphase II spindle, and then at anaphase II, the opposite ends of the chromosomes lead the way toward the spindle poles. Thus, in these organisms, it has been established that both ends of the chromatids adopt 'kinetic activity' in meiosis, with first one end performing this function at meiosis I and the other at meiosis II.

Our understanding of the behavior of holocentric chromosomes in mitosis and meiosis is based largely on cytological observations in a variety of species. These observations raise a number of questions regarding the structure and function of the holocentric 'centromere.' For example, how are kinetochores assembled on the metaphase chromosomes, how does a holocentric metaphase chromosome become oriented toward only one spindle pole, are there underlying centromeric DNA sequences distributed throughout the genome, and how is kinetic activity restricted first to one end and then the other of meiotic chromosomes? Future application of molecular and genetic approaches should help to provide answers to these questions as they relate specifically to holocentric chromosomes and to the behavior of chromosomes in general.

References

Albertson DG, Rose AM and Villeneuve AM (1997) Chromosome organization, mitosis and meiosis. In: Riddle DL, Blumenthal T, Meyer BJ and Priess JR (eds) C. elegans II, p. 47. Plainview, NY: Cold Spring Harbor Laboratory Press.

White MJD (1973) *Animal Cytology and Evolution*. Cambridge: Cambridge University Press.

See also:* Cell Division in *Caenorhabditis elegans

Holophyly

E Mayr

This is the process by which all the descendants of a stem species, no matter how divergent, are combined into a single holophyletic lineage (cladon). Such a cladon was erroneously called monophyletic but Haeckelian monophyly is a very different concept. Hennig's monophyly was therefore renamed holophyly by Ashlock (1971). Holophyly is a property of a branch of the phyletic tree (cladogram), while monophyly is a property of a taxon in a Darwinian classification.

Further Reading

Haeckel E (1866) *Generelle Morphologie der Organismen.* Berlin: Georg Reiner.

References

Ashlock PD (1971) Monophyly and associated terms. *Systematic Zoology* 20: 63–69.

See also: Cladograms; Monophyly

Homeobox

T Bürglin

The homeobox was identified independently by Bill McGinnis and Mike Levine in the laboratory of Walter Gehring in Switzerland, and Matthew Scott and Amy Weiner working with Thomas Kaufman and Barry Polisky at Indiana University in 1983–1984. When the sequences of several cloned homeotic genes were compared, it was found that they shared a common, conserved stretch of approximately 180 bp. This sequence element has been termed 'the homeobox' (in previous literature, also 'homeo box,' 'homoeobox,' etc.), because it was discovered in homeotic genes. The homeobox encodes a protein domain, the homeodomain, that has now been found in many developmental control genes. In essence, homeobox genes code for transcription factors and most of them play important roles during the development of multicellular organisms. They have been found in plants, fungi, and animals, as well as slime molds.

Structure of Homeodomain

The typical homeodomain is 60 amino acids long. The structure of several highly divergent homeodomains from yeast, flies, and vertebrates has been determined using X-ray and nuclear magnetic resonance (NMR) analysis. The different homeodomains are essentially very similar, even though the primary sequence similarity can be very small. The core of the homeodomain consists of three α-helixes (**Figure 1**). Helix 2 and helix 3 are linked via a short turn and form a structural motif called a 'helix–turn–helix motif' that is shared with many bacterial DNA-binding transcription factors and repressors. Helix 1 crosses over helix 3 so that the three helixes form a hydrophobic core that stabilizes the structure (**Figure 1**). A substantial part of the DNA-binding activity is located in helix 3, which lies in the major groove of the DNA and provides most of the sequence-specific contacts. In particular, residue 9 of helix 3 is a key residue that provides DNA-specific contacts; most homeodomains have a glutamine at that position. The flexible N-terminal arm of the homeodomain can reach into the minor groove of the DNA and provide additional contacts, although the mode of contact of this arm is subject to more variation between different types of homeodomains.

In several different classes of homeobox genes, insertion events have expanded or contracted the size of the homeodomain. The insertion points for extra residues are either in the loop between helix 1 and helix 2, such as in the TALE class of homeobox genes, or in the turn between helix 2 and helix 3.

Classes of Homeobox Genes

The homeobox genes can be divided into different classes depending on their sequence and gene structure. Many homeobox genes encode not only a conserved homeodomain but also additional conserved domains that are located N- or C-terminally of the homeodomain. The largest diversity of homeobox genes is found in animals, thus, unless noted, the described classes and families are found only in animals.

Hox Cluster Genes

The perhaps best-known homeobox genes are those located in the Hox cluster (**Figure 2**). Some of the first homeobox genes cloned were genes such as *Antp*, *Ubx*, and *ftz*. With the exception of the *Abd-B* genes, all Hox cluster genes have a recognizable, small, five- to six-amino acid motif upstream of the homeodomain, called the 'hexapeptide'. While the genes in the center of the cluster are very similar to each other, the outermost genes can be very different from each

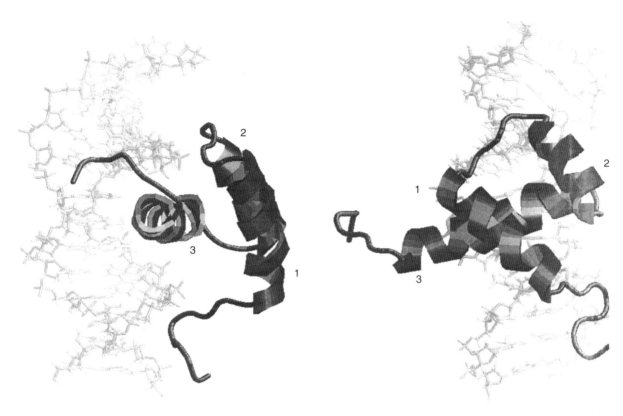

Figure 1 Structure of the homeodomain. Two schematic views of the Antennapedia homeodomain bound to DNA as determined by M. Billeter, G. Otting, and colleagues in the laboratory of K. Wüthrich. The NMR data were modeled in RasMol V2.6, the DNA is shown as a stick model in light gray, while the protein backbone (side chains not shown) is displayed as a dark ribbon. The numbers indicate the three α-helixes. Helix 3 sits in the major groove of the DNA.

other, often sharing less than 50% identity in the homeodomain. The vertebrate *Evx* genes, though not Hox genes, are part of the vertebrate Hox cluster; in other species such as flies, this gene family has separated from the cluster.

Dispersed Hox-like Genes and Other Clusters

A number of homeobox gene families share similarities with the Hox cluster genes. For example, the *empty spiracles* (*ems*) and *caudal* (*cad*) genes play a role in anterior–posterior patterning and have a hexapeptide upstream of the homeodomain. A second group of genes of the NK-2, NK-1, Tlx (which also has a hexapeptide), and ladybird (lbx) families reside in another gene cluster, the NK cluster, in *Drosophila*. Several NK and NK-related goes are also linked in vertebrates: analysis of human genome data suggests that an NK cluster was duplicated and subsequently broken in vertebrate evolution. A further small cluster, termed the 'ParaHox cluster' has been found in amphioxus, with the gene families cad, Xlox, and Gsx. Some other gene families that are dispersed through the genome, but are more similar to the Hox, NK, and

ParaHox cluster genes than to other classes, are msh, Mox, Dll, Hlx, en, NEC, ceh19, Bar, Xnot, and Hex; some of these may have originally been part of the Hox or NK clusters.

POU Class

These homeobox genes encode a POU-specific domain upstream of a distinct type of homeodomain, the POU homeodomain. The POU-specific domain is a DNA-binding domain of about 80 amino acids that contains a helix–turn–helix motif like the homeodomain. The POU domain was first found in the mammalian transcription factors Pit-1, Oct-1, and Oct-2, and the *Caenorhabditis elegans* gene *unc-86*. A special feature of the POU homeodomain is the cysteine residue at position 9 of helix 3. Six families, POU-I to POU-VI, have been defined.

prd Class

The paired class of homeobox genes is named after its first member, the *Drosophila* gene *paired*. This class is characterized by having a prd domain upstream of the homeodomain. The prd domain is about 130 amino acids in length and binds DNA. The Paired domain is

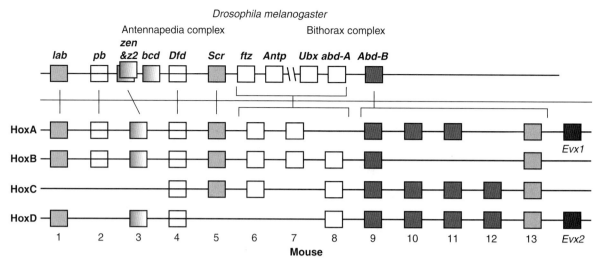

Figure 2 Hox clusters in *Drosophila melanogaster* and mouse. The Hox cluster of *D. melanogaster* contains 12 homeobox genes: *labial* (*lab*), *proboscopedia* (*pb*), *zerknüllt-related* (*zen2*), *zerknüllt* (*zen*), *bicoid* (*bcd*), *Deformed* (*Dfd*), *Sex combs reduced* (*Scr*), *fushi tarazu* (*ftz*), *Antennapedia* (*Antp*), *Ultrabithorax* (*Ubx*), *abdominalA* (*abd-A*), *Abdominal B* (*Abd-B*). The cluster is in fact split into two parts, one part is the Antennapedia complex and the other, the Bithorax complex. Although located in homeotic complexes, several of the homeobox genes in the cluster are not homeotic genes: *zen*, *zen2*, *bcd*, and *ftz*. In mouse and other vertebrates, there are four paralogous Hox clusters (apart from fish, which have more). The duplications of the cluster from a single ancestral cluster probably happened at the beginning of vertebrate evolution. In the course of evolution, some of the paralogous Hox genes were lost, so that the present-day mammalian cluster contains 39 Hox genes, as well as two genes of the even-skipped family (*Evx1* and *Evx2*), which probably formed part of the ancestral cluster, too. The genes can be grouped into 13 paralog groups. The lines between the mouse and the fly cluster show the evolutionary relationships between the homeobox genes. Thus, the Hox genes of paralog group 1 are orthologous to *lab* in flies, paralog groups 9–13 are homologous to *Adb-B*. Not all genes have 1:1 paralogs: for example, in the central part of the cluster the fly genes *Antp*, *Ubx*, and *adb-A*) and the mouse paralogs *Hox6*, *Hox7*, and *Hox8* may have arisen through independent duplication events from a single ancestral gene. Likewise, *zen2* is a relatively recent duplication from *zen*, and *bcd* also is derived from an ancestral *zen*/*Hox3* gene.

actually comprised of two similar domains, each containing a helix–turn–helix motif. The homeodomain distinguishes itself from other homeodomains by having a serine residue at position 9 of helix 3.

prd-Like Class

This group of homeobox genes is related to the paired class of homeobox genes through their homeodomain. Some prd-like homeodomains are more than 70% identical to prd class homeodomains. However, they do not contain a prd domain, and residue 9 of helix 3 in the homeodomain is not a serine residue. More than 15 families have been described.

LIM Class

The LIM class of homeobox genes contain two LIM domains upstream of the homeodomain. The LIM domain is composed of two so called zinc fingers, which contain conserved cysteine, histidine, and aspartate residues that bind zinc. The LIM-domain zinc fingers are distinct from other zinc fingers, and, unlike many of the other zinc-finger families that are involved in DNA-binding, the LIM domains of LIM homeobox genes are involved in protein–protein interactions with other factors. At least six conserved families are found.

ZF Class

The ZF (zinc-finger) class of homeobox genes are an unusual group of genes. They contain classic zinc-finger domains such as have been found in zinc-finger transcription factors that bind DNA plus one or more homeodomains. The combination of these domains can take quite bizarre proportions, as in the mammalian gene ATBF1, which contains 17 zinc-finger domains and 4 homeodomains.

cut Class

The cut class genes are characterized by a variable number of cut domains upstream of the homeodomain. Three separate families exist, having either 1, 2, or 3 cut domains. The cut domain is also a DNA-binding domain.

SO/SIX Class

The sine oculis/Six class of homeobox genes contain a large conserved domain of presently unknown function N-terminally adjacent to the homeodomain. Several families exist.

HD-ZIP Class

Immediately C-terminal of the domain is a so-called Leucine-Zipper family of homeo box genes found in plants. Within the homeodomain is a so-called leucine-zipper, a region that forms coiled-coil structures involved in dimerization. Four different families have been defined.

TALE Class

The TALE group of homeobox genes is very ancient; their homeodomain is 63 amino acids long. TALE stands for "three amino acid loop extension," because of the three extra residues in the loop between helix 1 and helix 2. TALE homeobox genes are found in plants (two classes: KNOX and BEL), in fungi (two classes: CUP and M-ATYP), and animals (four classes: PBC, MEIS, TGIF, and IRO). The KNOX, PBC, and MEIS classes each contain large conserved domains upstream of the homeodomain. Sequence comparison has shown that the KNOX, PBC, and MEIS domains share weak sequence similarity, suggesting a common ancestry.

pros Class

The prospero class is a highly divergent class of homeodomain proteins. The homeodomain has three extra residues between helix 2 and helix 3, and a pros domain of about 100 amino acids follows immediately after the homeodomain.

Evolution

Homeobox genes are found in plants, fungi, and animals, and even in slime molds (*Dictyostelium*). Although now several prokaryotic genomes have been sequenced, no true homeobox gene has been found in these organisms. Thus, it appears likely that the first homeobox appeared sometime in eukaryote evolution, probably derived from a helix–turn–helix factor. In the ancestral organism from which eventually plants, fungi, and animals were derived, at least two different homeobox genes must have existed already: one a typical 60-amino acid homeobox gene, and one TALE homeobox gene. This ancestral TALE homeobox gene had a conserved upstream domain from which the KNOX, MEIS, and PBC domains are derived. While in plants and fungi some proliferation of different types of homeobox genes has taken place, by far the largest expansion has happened

in animals, where there are now dozens of different classes and families of homeobox genes (see Classes of Homeobox Genes). The emergence of the different classes of homeobox genes seems to have happened early in metazoan evolution, since in sponges and cnidaria many different types of homeobox genes are found, and in the Bilaterialia phyla essentially all classes and many families are present.

Function

Given the widespread nature of homeobox genes in higher eukaryotes, it is not surprising that their function is very diverse. Nevertheless, most of them play important roles in the development of their respective organisms. Homeobox genes have been found to function at the earliest points in development as well as in the very latest cell differentiation events. Some examples follow.

The Hox cluster genes of animals are involved in patterning and specification of identity in regions along the anterior–posterior body axis. A striking aspects is that the order of the genes in the cluster is collinear with their function along the anterior–posterior axis. Thus, the *Drosophila* gene *labial* (*lab*) functions in the very anterior of the animal, while the gene *Abdominal-B* (*Abd-B*) functions in the 5th to 8th abdominal segments. In vertebrates, the Hox cluster genes are likewise involved in patterning along the body axis. Since the binding site of Hox cluster genes is rather short, additional cofactors are necessary to provide DNA-binding specificity. Two TALE class homeobox genes have been identified as cofactors for Hox cluster genes: in flies the two genes *extradenticle* (*exd*), a PBC class gene, and *homothorax* (*hth*), a MEIS class gene, have been shown to form complexes with Hox proteins such as *Ubx* or *lab*.

One of the earliest developmental homeobox genes is found in *Drosophila*. The gene *bicoid* plays a key role in setting up the anterior–posterior axis in the embryo. Mutant embryos lack head and thorax and develop posterior structures at the head. The *bcd* RNA is provided maternally, and the RNA as well as the protein are localized at the anterior pole of the embryo. Despite the crucial role in early development for *Drosophila*, the *bcd* gene is a relatively new gene in evolutionary terms; it is most likely derived from an ancestral gene of the Hox3 group. Furthermore, while most homeobox genes bind DNA and function as transcription factors, *bcd* also plays a regulatory role at the level of messenger RNA. It can bind RNA and regulate expression of other genes at the translational level.

The *C. elegans* prd-like gene *unc-4* is involved in the specification of motor neurons. Mutations in this gene lead to abnormal synaptic connectivities so that the VA neurons receive synaptic input that is normally appropriate only for VB neurons (which are the sister cells of the VA neurons). The consequence of these wiring defects is that the animals cannot move backwards anymore. *unc-4* is expressed in the VA motoneurons; it confers VA identity to these neurons, and is thus one of the final steps in differentiating a subset of motoneurons.

The POU genes Oct-1 and Oct-2 were first identified as transcription factors because of their biochemical properties of binding the octamer sites in the promoter region of the immunoglobulin enhancers. This provided compelling evidence that homeobox genes are transcription factors.

In yeast, the mating-type locus contains two homeobox genes, Mat1 and Mat2, the latter a TALE homeobox gene. These two genes are involved in mating-type switching, i.e., they regulate and switch between the two cell fates that yeast can adopt.

In plants, the gene *shootmeristemless* (*STM*) of *Arabidopsis thaliana* encodes a KNOX class homeobox gene. Mutations in *STM* fail to develop a shoot apical meristem. Converse phenotypes have been found when the closely related gene *Knotted 1* from maize is overexpressed in tobacco. Thus, also in plants, homeobox genes are involved in developmental processes.

Further Reading

Bürglin TR (1998) The PBC domain contains a MEINOX domain: coevolution of Hox and TALE homeobox genes? *Development Genes and Evolution* 208: 113–116.

Gehring WJ, Affolter M and Bürglin TR (1994) Homeodomain proteins. *Annual Review of Biochemistry* 63: 487–526.

de Rosa R, Grenier JK, Andreeva T *et al.* (1999) Hox genes in brachiopods and priapulids and protostome evolution. *Nature* 399: 772–776.

Pollard SL and Holland PWH (2000) Evidence for 14 homeobox gene clusters in human genome ancestry. *Current Biology* 10: 1059–1062.

See also: Homeotic Mutation

Homeotic Genes

doi: 10.1006/rwgn.2001.1857

A homeotic gene is one that contains a homeobox, whose level of expression is set during embryogenesis in response to positional cues, and which

subsequently directs the later formation of tissues and limbs appropriate to that part of the organism.

See also: Homeobox; Homeotic Mutation

Homeotic Mutation

T Bürglin

doi: 10.1006/rwgn.2001.0626

The term 'homeosis' was coined by William Bateson in 1894 to describe particular types of biological variation whereby "something has been changed into the likeness of something else." More than 20 years later, the first mutation that causes homeosis – a homeotic mutation – was described by C.B. Bridges in *Drosophila*, and many more were subsequently discovered. These homeotic mutations lead to partial or complete transformations of particular body regions in the fly. For example, a segment can be transformed such that it resembles its anterior neighbor, as in the case of particular mutations in the ultrabithorax (*Ubx*) gene, which cause partial transformations of the third thoracic segment into the second thoracic segment. In the most extreme case, when several *Ubx* alleles are combined, a fly can have four wings instead of two wings and two halteres, because the halteres of the third thoracic segment are converted into wings.

Another well-known gene is Antennapedia, dominant mutations in which can cause transformations of antennae into legs. The first homeotic genes cloned were found to contain a conserved sequence element that was termed the 'homebox.' However, subsequent research showed that not all homeotic genes are homeobox genes, and not all homeobox genes are homeotic genes. For example, the *Drosophila* homeotic gene spalt encodes a zinc-finger protein, and the homeotic gene fork head was the founding member of another family of transcription factors that contain a fork head domain.

Several of the homeotic genes in flies are located in two gene clusters: the bithorax complex (BX-C) and the antennapedia complex (ANT-C). Collectively, these two complexes are often referred to as the homeotic complex (HOM-C). Genes in these two complexes control the development of the *Drosophila* body along the anterior–posterior body axis. Intriguingly, the gene order on the chromosome is collinear with the respective gene function along the body axis. In vertebrates, the corresponding clusters of genes are called HOX clusters. The genes in the vertebrate HOX clusters are highly conserved with their fly counterparts. Functional analysis of these genes using knock-out techniques revealed that in vertebrates, too, they function in patterning along the anterior–posterior body axis and cause homeotic transformations. The HOM-C and HOX clusters harbor the perhaps most well-known developmental control genes; Ed Lewis was awarded the Nobel Prize for his ground-breaking studies of BX-C.

While the term 'homeotic mutation' is mainly known from mutations in segmentation genes in *Drosophila*, the original definition of homeosis is very broad. Thus, other mutations that cause transformations have also been termed homeotic. For example, in the nematode *Caenorhabditis elegans* the gene *lin-12*, which encodes a transmembrane receptor, has been termed a homeotic gene, because many cell lineages (patterns of cell divisions) are transformed into other cell lineages. In plants, many homeotic mutations are known that cause transformations of leaves and flowers.

See also: Homeobox; Lewis, Edward

Homogeneously Staining Regions

G Levan

Copyright © 2001 Academic Press
doi: 10.1006/rwgn.2001.1582

This is one of the cytogenetically visible signs of gene amplification, the other being 'double-minute chromosomes.' It is known that the homogeneously staining regions (*hsr*) just as the *dmin*, will contain copies of an amplified DNA segment (the amplicon), leading to cellular overexpression of the genes contained in the segment. In a single *hsr* there are usually many amplicon copies arranged in tandem array. Characteristically, *hsr* can be detected after chromosome banding in metaphase preparations as a large chunk of diffusely staining chromatin somewhere inside an ordinary chromosome. The mechanism for generating the *hsr* is not known exactly, but it is generally assumed that the amplification can take place during an episomal phase, in which a circular DNA molecule is replicating autonomously relative to the bulk of chromosomal DNA. The episomes may be transferred into chromatin bodies visible in the light microscope (*dmin*) and the *dmin* will subsequently be integrated at a (random) chromosomal site to generate the *hsr*. However, various other schemes have been proposed for the origin of *hsr* and it is quite likely that several

different molecular mechanisms may be functional leading to the same end results (Schwab, 1999).

Reference

Schwab M (1999) Oncogene amplification in solid tumors. *Seminars in Cancer Biology* 9: 319–325.

See also: Amplicons; Double-Minute Chromosomes; Gene Amplification

Homologous Chromosomes

J R S Fincham

Copyright © 2001 Academic Press
doi: 10.1006/rwgn.2001.0627

Homologous means having a common origin by descent. Chromosomes that are homologous in this broad sense may have diverged to a very considerable extent. In the long evolutionary term, chromosomes may undergo structural rearrangements, so that homology between different species and genera is often a property of chromosome segments rather than whole chromosomes. Between the chromosomes of mice and humans, for example, there is quite a high degree of patchwork homology, with blocks of similar genes in locally similar sequences (syntenic genes) in very different larger-scale arrangements.

However, in the context of experimental genetics, homology most usually refers to the close similarity of the pairs of chromosomes in the diploid organism. The classical criterion for assessing homology in this stricter sense is ability to pair. A fully homologous chromosome pair will be closely associated all along their lengths at the pachytene stage of meiosis. The pairing can be seen under the light microscope in organisms with reasonably large chromosomes (e.g., very clearly in *Zea mays* and not at all in yeasts, except by fluorescent *in situ* hybridization, also known as FISH), but the synaptonemal complex, which is formed between the paired homologues, can usually be clearly visualized with the electron microscope, even in yeasts, by virtue of its staining with silver ions. In *Drosophila* species (as well as in other flies), homologous pairing can be seen in unrivalled detail in the giant nuclei of salivary gland cells, where maximally extended chromosomes are amplified over 100-fold in thickness by repeated replication without separation (polytene chromosomes), and homologs are closely paired. The close pairing of chromosomes, either at pachytene of meiosis or in *Drosophila* giant nuclei (where much more detail can be seen), reveals structural differences between homologs due to inversions, interchanges, or deletions of chromosome segments.

Whereas homology, as judged by pairing, is virtually complete between the two chromosome sets within a diploid species, it is usually much less evident between species, even when the species are closely related taxonomically. Interspecific hybrids seldom show regular chromosome pairing and bivalent formation at meiosis; that is the usual reason for hybrid sterility. Nevertheless, the chromosomes of related species are often similar in number, in relative sizes and, so far as it can be determined, in function, and are obviously homologous in the sense of related by descent. A good example is provided by wheat, *Triticum aestivum*, which is a 42-chromosome hexaploid, with three different diploid sets of 14 chromosomes, derived from three different species. At meiosis, wheat regularly forms 21 bivalents, with pairing restricted to chromosomes from the same ancestral diploid. However, this stringent specificity of pairing is under genetic control, and when a certain chromosome (5B, the fifth chromosome of the B genome) is removed by selective breeding, pairing also occurs between corresponding chromosomes from different ancestral diploids. The lower degree of homology so revealed is sometimes called homeology, a term used mainly by cereal breeders, though it could have a wider application. For example, it is used to describe recombination in *Saccharomyces cerevisiae* with a chromosome from a closely related species or between closely related but diverged, duplicated genes.

Still lower degrees of homology exist between chromosomes, or segments of chromosome, of non-hybridizable species, but have to be demonstrated by methods based on DNA technology such as *in situ* hybridization of DNA probes to chromosomes ('chromosome painting,' FISH).

See also: Meiosis; Polytene Chromosomes; Segmental Interchange; Synapsis, Chromosomes; Synaptonemal Complex; Synteny (Syntenic Genes)

Homologs

Copyright © 2001 Academic Press
doi: 10.1006/rwgn.2001.1858

Homologs are chromosomes that carry the same genetic loci. A diploid cell has two copies of each homolog, one derived from each parent.

See also: Chromosome; Homologous Chromosomes

Homology

E O Wiley

Copyright © 2001 Academic Press
doi: 10.1006/rwgn.2001.0628

Although (Owen, 1843) is generally credited with coining the word 'homolog,' the idea that parts of organisms are comparable in some fundamental sense can be traced back at least to Aristotle. Owen characterized the term 'homolog' to denote the comparative similarity in structure between parts of two different organisms "under every variety of form and function." For example, the right forelimb of a bird would be considered homologous with the right forelimb of a human in spite of differences in function and considerable differences of form. This was contrasted with the term analogy which denoted similar function without necessary underlying similarity (wings of birds and butterflies). Although many consider these words as having a complementary meaning, this was not their original intent (Pachen, 1994). Homologous parts can have analogous functions (wings of birds and wings of bats), just as nonhomologous parts can have analogous functions (wings of bumble bees and wings of birds). After the general acceptance of the general theory of evolution (descent with modification), most biologists used the term homolog to denote comparable (similar or identical) characters shared through common descent. This generated a whole new set of terms to denote similarity of form gained independently (e.g., convergences, parallelisms, paralogs, etc.).

Characters, Character States, and Homology

It is common to distinguish between characters as a general description of the part of an organism or taxon, and character state, the specific feature of a particular organism or taxon. Thus, one might term the character 'base pair position 148 in cytochrome *b*' and the character state 'guanine nucleotide present.' However, this is simply restating two characters: one that exists at a higher level (presence of that base position in the gene) and another that exists at a lower level (guanine nucleotide). Because homologous characters have a history that is tied directly to a hierarchy of descent, the distinction between character and character state is not necessary (Wiley, 1981; Ax, 1987; Patterson, 1988). Essentially, homologous characters are simply recorded features of two different organisms that are thought to have a particular relationship, and some level of similarity between structure and position of the characters of two different organisms seems necessary to be able to do so. The practice of distinguishing characters and character states grew from the use of data matrices where columns of data were given a general name and the characters of organisms a specific name. But, columns really represent initial hypotheses of homology. Characters placed in a single column of data are initially thought to be good candidates for having a homologous relationship. Whether this is true in the end is another matter.

Homology at the Taxon Level

Just as with species concepts, concepts of taxic homology are numerous and what constitutes homology between parts of two organisms is hotly debated. (Wagner, 1994 and earlier papers) has distinguished three concepts of homology: historical, morphological, and biological. The question is, should there be three (or more) kinds of taxic homology, or are some kinds simply a manifestation of a larger concept? With the rise of the evolutionary paradigm, what we take as the fundamental nature (or ontology) of homology became associated with descent with modification. Homologous parts are comparable, not because they are derivations from an archetype *per se*, but because they are inherited from a common ancestor in modified or unmodified form. Wiley (1975), Patterson (1982), and others have taken this to a logical conclusion: homologs at the level of taxa are apomorphies (derived characters, evolutionary novelties) at some point in their history. Perhaps a thought experiment is in order. Imagine that we have the entire tree of genealogical descent mapped out at our feet. If we place all the similarities and differences observed among organisms on this tree at the point where they arose and followed their fates, we would see the coalesced homologies as apomorphies that diagnose species (autapomorphies) and monophyletic groups (synapomorphies). We would see the homoplasies (nonhomologous similarities) and analogies (functionally similar but structurally dissimilar) scattered thoughout the tree in different groups. Interestingly, what we would not see are symplesiomorphies, shared primitive homologies. This is because every symplesiomorphy is actually a synapomorphy higher in the phylogeny and the reason we have the term 'symplesiomorphy' is because we do not consider the entire tree at any one time. Symplesiomorphies are simply homologies that arose in ancestors more ancient than those that are logically included in the restricted tree. Under this concept, there is a single concept of taxic homology of which other concepts of homology are special (and perhaps perfectly valid) cases.

Origin of Homologous Characters

If homology is a concept that extends below the level of taxa, then it should be obvious that homology at lower levels cannot simply be apomorphy. Haszprunar (1991) suggests four levels of homology: (1) iterative homology is the correspondence serial homologs in the same individual at the same time; (2) ontogenetic homology is the correspondence of parts at different times in the same individual; (3) polymorphic homology is the correspondence of parts between individuals of the same species lineage; and (4) supraspecific homology is the correspondence of parts between taxa (taxic homology). Ontogenetic and polymorphic homology are directly related to the origin and eventual fixation of apomorphies, while iterative homology is related to serial homology and homonomy (mass homology). Iterative homology may or may not be translatable into taxic homology (see below, "Conjunction test"). The origin of taxic homology, suggests Haszprunar, lies with the origin of apomorphies within species where they coexist, for a time, with their plesiomorphic homologs. Further, their origin on the molecular level may not be unique but recurrent. It is possible for gene alleles to be identical by descent and yet remain polymorphic over speciation events, creating homoplasy at the taxic level, while being homologous at the gene level. Such phenomena and others create differences between gene trees and species trees.

Nature of Homologs

Exactly what constitutes homology from the ontological perspective is also debated. Although we understand that homologs gain their 'comparability' through descent, and we understand that homologies appear as apomorphies on phylogenetic trees, we also understand that the homologies being compared do not actually have descent relationships. That is, right hands do not actually give rise to other right hands, nor does guanine at position 158 in a cytochrome *b* gene sequence give rise to a descendant guanine at that same position in a descendant mitochondrion. Rather, the relationship between homologs is always indirect, being mediated by ontogeny at the morphological level and semiconservative replication at the DNA level (and other processes at intermediate levels). This has led authors such as Van Valen (1982), Hausperger (1991), and Roth (1994) to characterize homology as a manifestation of the flow of biological information between generations and over phylogeny. This concept of information should not be confused with sequence information; it includes epigenetic information as well. Under this concept, homologous structures are the observable manifestation of information flow over time and through descent. If so, then this general concept of homology can be easily extended to behavioral and functional characters (see Greene, 1994 and Lauder, 1994, for examples). Further, it solves certain conundrums such as how to homologize Meckel's cartilage in vertebrates where the structure is induced by different tissues (see review in Wagner, 1994). In such cases, epigenetic constraints on the developing phenotype may allow for considerable variation in the actual way that a particular structure is built during ontogeny.

Homology and Homoplasy at Different Levels of Organization

Not all comparable features of organisms are apomorphies at some level in a phylogeny. Even identical characters such as the same base residue at the same position can evolve independent of each other (and thus be apomorphies at different levels or in different places in the phylogeny). Given homologies, what of similar but nonhomologous characters? The general concept of homoplasy can apply to characters that show some level of structural, behavioral, ontogenetic, or genetic similarity, but that do not qualify as homologies because they have independent evolutionary origins. The complexities of homology and homoplasy can be seen in molecular systems where there are three levels of homology, two at the taxic level and one at the gene level. At the first level, orthologous genes (Fitch, 1970) are strictly comparable between organisms, so their sequence variation can contain homologs. At the second level, the level of the organism, orthologs are candidates for taxic homology. That is, the presence of orthologous genes in the taxa of which the organisms are part can be a synapomorphy of a monophyletic group containing these organisms.

Paralogous genes (Fitch, 1970) are related among themselves in gene trees, but because of gene duplication, two or more paralogous genes exist in the same organism. At the level of the organism, paralogs may contain similar bases at the same base position, but these similarities are nonhomologous in terms of taxic homology. That is why we call the genes by different names. Using a mix of sequence data from α- and β-hemoglobin (α from one species, β from another, etc.) would lead to spurious results since positional homology between the genes does not exist relative to the organisms of which the paralogous genes are a part. But as parts of organisms themselves, the distribution of paralogous genes can be used to test relationships because the presence of various gene copies can act as synapomorphies. So, among vertebrates, the presence of β-hemoglobin is a synapomorphy of jawed vertebrates, Gnathostomata (Goodman et al., 1987), while other paralogs are synapomorphies at higher and lower levels in the phylogeny. Thus, among gene families there are two levels of taxic homology relative to organisms. The first level is the level of sequence variation among orthologs. At this level, analysis of homologous base positions leads to an hypothesis of the relationships among taxa in the same manner as the analysis of homologous morphological characters. The second level is the distribution of orthologs and paralogs among the organisms in a phylogeny. The distribution of members of a gene family leads to an hypothesis of relationship among organisms in the same manner as sequence variation among orthologs or the distribution of morphological homologs. The third level obtains among paralogs and their gene descent. While sequence positional homology might not obtain between paralogs relative to taxa or relative to organisms within taxa, it does obtain between paralogs relative to their own descent in gene trees. This level does not pertain to the organisms *per se*, but to the descent of the genes from their own gene ancestors. **Figure 1** illustrates these levels and concepts.

Independence of Different Homologs

Atomization refers to the ability of an investigator to gather characters of organisms into suites of supposedly homologous characters. This activity is best seen in the construction of a data matrix for purposes of analyzing phylogenetic relationships. Columns of data are hypothesized to represent different and independent suites of homologous characters. (Indeed, all phylogenetic algorithms treat different data columns as independent.) At the level of gene sequences, this may be an easy task because base position of orthologous gene sequences provides a rationale for recognizing data columns that contain homologous nucleotides. In morphology, behavior, and function, the issue of how to atomize characters can be more complex, but in general some judgement is made that divides the features observed into the smallest comparable units that the investigator can justify.

Given this atomization, there remains the issue of character independence among different suites of homologs. In systematic analysis, this issue can be framed rather crisply: how many independent columns of data actually exist as compared to the total number of data columns. For example: if the investigator is analyzing sequences from a ribosomal gene, are the data columns that record base pair complements really independent of each other? If we examine the distribution of synapomorphies over a phylogenetic tree, we can partly address this question.

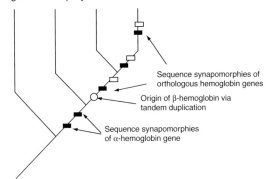

Figure 1 A phylogeny of selected vertebrate groups illustrating two levels of homology for orthologous and paralogous genes (upper), and a gene tree of the globin family of genes illustrating sequence homology between paralogous genes at the level of gene trees.

Synapomorphies from different suites of homologous characters that appear at different points on the phylogeny are independent in the evolutionary sense. They may come to be dependent where they occur together in the same group, but their origins are not coupled. The same cannot be said when synapomorphies co-occur on the same branch. In such cases, other studies (ontogenetic, for example) would have to be applied to demonstrate that they are independent characters. In some cases, such as synapomorphies from different genes, different gene regions, or different functional complexes, the case for independence may seem to be evident. In other cases, such as complementary base pairs in stem regions of ribosomal genes, the case for independence may be suspect. In evolutionary studies, especially those concerned with phylogeny reconstruction, the issue of independence is closely tied to the issue of support for a tree. Four synapomorphies that are functionally or ontogenetically linked may only be one synapomorphy (one evolutionary event with four manifestations) rather than four synapomorphies. If an alternative monophyletic group is diagnosed with one, two, or three different

synapomorphies, then what appears to be the most parsimonious tree (four dependent synapomorphies) may actually have less support than the alternative tree.

Recognition and Testing of Homologs

Patterson (1982, 1988) outlined and discussed the tests that can be applied to parts of organisms hypothesized to be homologs and has used these distinctions to characterize many types of nonhomologous similarities. His analysis was made under the assumption that homologies are apomorphies.

Similarity Test

Parts that are dissimilar are not likely candidates for hypotheses of homology. Testing may take the form of (Remane, 1956) tests of similarity, topological position, and special correspondence. Base position forms the major criterion of similarity in DNA sequence data (Remane's criterion of topological position within the gene). Base similarity forms the major criterion among bases that occupy the same base position (similar bases are presumed homologous as an initial hypothesis). As Hennig (1966) stressed, characters that pass similarity tests must always be assumed homologs in the absence of contrary evidence (such as that provided by the two additional tests detailed below). This assumption of homology is necessary to avoid *ad hoc* dismissal of evidence.

Conjunction Test

"If two supposed homologues are found together in one organism, they cannot be homologous" (Patterson, 1988, p. 605). In morphological characters, similarities that are found in two to many 'copies' are termed homonomies. Homonomies (iterative homologies at lower levels) may take the form of 'serial homologs' in the case of metameristic repeats of body segments or 'mass' or 'general' homologies in the case of hair in mammals. Homology statements mixing parts of homonomous body segments would result in spurious taxic homology statements (as in paralogous genes). However, the evolutionary novelty that produced the serial homology may act as a synapomorphy at a higher level in the phylogeny. In the case of such characters as mammalian hair, general presence of the mass homology may act as a synapomorphy in spite of the fact that it may be difficult to impossible to provide a one-to-one homology statement about individual hairs.

In genetic systems, similar characters that fail the conjunction test may take the form of paralogous genes or xenologous genes (paralogy is discussed above). Some paralogous gene families show concerted evolution and in some circumstances their copies (plerologs) may be treated as a single gene for the purposes of phylogenetic analysis (but see Hillis (1994), for further discussion). Xenology obtains when genes of the same gene family are spread by lateral gene transfer rather than common descent. On a phylogeny of organisms, paralogous genes are expected to form nested sets of characters that reflect descent of the gene family (**Figure 1**). There is no expectation of such a pattern in xenologs whose spread is not historically constrained.

Congruence Test

Similarities that pass both the similarity and conjunction tests whose distribution on a phylogenetic tree are congruent with many other similarities are deduced to be candidates for the status of uncontested homologies. Under the assumption that homology is apomorphy, the congruence test provides the final arbitrator for accepting or rejecting parts that pass the first two tests. Similarities that fail the congruence test are frequently termed parallelisms if they are very similar, or convergences if they are dissimilar upon reinspection. (The distinction between parallel and convergent characters is debatable, see Homoplasy.)

Homology Issues in Molecular Genetics

One basic issue is the extent to which molecular homology differs from morphological, behavioral, and other kinds of homology. Patterson (1988) suggested that there was a difference because similarity was used to establish homology and the basis of this similarity was statistical (the probability that sequence similarity is due to chance is rejected.) However, if we treat this issue as one of identity or an issue of relationships among comparable entities, then there is no need to conclude that homology is fundamentally different on the molecular and morphological levels. Statistical similarity may lead to the conclusion that the genes of two organisms belong to the same class of gene (e.g., they are both α-hemoglobin), or to the hypothesis that they are members of the same gene family (e.g., they are members of the globin gene family). However, testing the hypothesis that two apparent paralogs are members of a gene family would seem to be a matter of establishing their gene tree relationships and that requires synapomorphies at the gene level.

Hillis (1994) provides a detailed discussion of issues of homology particular to molecular biology: an incomplete summary of some of the major issues is given below:

1. Positional homology. Positional homology refers to the position of a single nucleotide site within a gene, a ribosome, or an amino acid site within a protein. Since nucleotides and amino acids have the same structure regardless of their evolutionary origins, the similarity criterion applied to the sequence and amino acid levels of analysis refers only to positional homology. An adenine and a thymine at a well-established homologous position are regarded as homologous in spite of their obvious nonsimilarity on the structural level of the nucleotide. For orthologous genes, accuracy of positional homology is dependent on correct alignment of the sequences or amino acids. Alignment of sequences of orthologous genes of the same length is relatively easy. Difficulties arise when genes contain introns or diverge such that they are of different lengths. Such cases require an understanding of gene architecture (in the case of exon–intron relationships; loop and stem architecture of the functional ribosomal sequence, etc.) and explicit rules for aligning the obtained sequences that include costs for introducing gaps.

2. DNA hybridization. DNA hybridization provides a measure of overall similarity of cross-hybridized sequences but does not distinguish between orthology, paralogy, and positional homoplasy. DNA hybridization is not useful for explicit hypotheses of homology.

3. Restriction enzyme analysis. Restriction site mapping can yield homologous characters because restriction sites are composed of specific base recognition sites along orthologous genes. Restriction fragment homology determination is more problematic, especially between species because of various sources of error.

4. Random amplified polymorphic DNA (RAPD). Fragments produced in RAPD studies have the same sources of error as other fragment data. In addition, studies have suggested that amplification of paralogs and nonhomologous loci may yield fragments of the same size and that RAPD-based phylogenetic inferences are incongruent with well-established phylogenies.

5. Allozyme electrophoresis. Allozyme electrophoresis is particularly valuable for studying the distribution of paralogs among taxa and studies of the differential expression of paralogous genes in different tissues of the same organism. Criteria for determining the orthologous or paralogous nature of expressed products is well established. For electromorphs of an orthologous gene, homology is determined by electrophoretic mobility coupled with the congruence test and works best for closely related species.

References

Ax P (1987) *The Phylogenetic System*. New York: John Wiley.

Fitch WM (1970) Distinguishing homologous from analogous proteins. *Systematic Zoology* 19: 99–113.

Goodman M, Miyamoto M and Czelusniak J (1987) Pattern and process in vertebrate phylogeny revealed by the coevolution of molecules and morphologies. In: Patterson C (ed.) *Molecules and Morphology in Evolution*, pp. 141–176. Cambridge: Cambridge University Press.

Greene HW (1994) Homology and behavioral repertoires. In: Ha BK (ed.) *Homology: The Hierarchial Basis of Comparative Anatomy*, pp. 369–391. San Diego, CA: Academic Press.

Haszprunar G (1991) The types of homology and their significance for evolutionary biology and phylogenetics. *Journal of Evolutionary Biology* 5: 13–24.

Hennig W (1966) *Phylogenetic Systematics*. Urbana, IL: University of Illinois Press.

Hillis DM (1994) Homology in molecular biology. In: Ha BK (ed.) *Homology: The Hierarchial Basis of Comparative Anatomy*, pp. 339–368. San Diego, CA: Academic Press.

Lauder GV (1994) Homology, form, and function. In: Ha BK (ed.) *Homology: The Hierarchial Basis of Comparative Anatomy*, pp. 151–196. San Diego, CA: Academic Press.

Owen R (1843) *Lectures on Comparative Anatomy and Physiology of the Invertebrate Animals* (delivered at the Royal College of Surgeons). London: Longman, Brown, Green & Longman.

Pachen AL (1994) Richard Owen and the concept of homology. In: Ha BK (ed.) *Homology: The Hierarchial Basis of Comparative Anatomy*, pp. 21–62. San Diego, CA: Academic Press.

Patterson C (1982) Morphological characters and homology. In: Joysey KA and Friday AE (eds) *Problems in Phylogenetic Reconstruction*, pp. 21–74. London: Academic Press.

Patterson C (1988) Homology in classical and molecular biology. *Molecular Biology and Evolution* 5: 603–625.

Remane A (1956) *Die Grundlagen des naturlichen Systems der vergleichenden Anatomie und Phylogenetik 2*. Leipzig: Geest und Portig K.G.

Roth L (1994) Within and between organisms: replicators, lineages, and homologs. In: Ha BK (ed.) *Homology: The Hierarchial Basis of Comparative Anatomy*, pp. 301–337. San Diego, CA: Academic Press.

Van Valen L (1982) Homology and causes. *Journal of Morphology* 173: 305–312.

Wagner G (1994) Homology and the mechanisms of development. In: Ha BK (ed.) *Homology: The Hierarachial Basis of Comparative Anatomy*, pp. 273–299. San Diego, CA: Academic Press.

Wiley EO (1975) Karl R. Popper, systematics and classification: a reply to Walter Bock and other evolutionary taxonomists. *Systematic Zoology* 24: 233–243.

Wiley EO (1981) *Phylogenetics: The Theory and Practice of Phylogenetic Systematics*. New York: John Wiley.

See also: **Gene Trees; Homoplasy; Orthology; Paralogy; Phylogeny**

Homoplasy

E O Wiley

Copyright © 2001 Academic Press
doi: 10.1006/rwgn.2001.0629

Lankester (1870) introduced the term 'homoplasy' to describe all resemblances that were not homologous. Lankester included such resemblances of serial and general homologs within the concept, but most modern biologists restrict the concept to analogous, convergent, and parallel similarities shared among species or other taxa. In general, taxic homoplasies are similarities in either form or function that fail one or more of the three tests: similarity, conjunction, and congruence (see Homology).

Analogous Similarities

There is considerable confusion concerning the concept of analogy (see Analogy). Analogous similarities, as the term is usually applied in systematics, are similarities in function and frequently do not appear as similarities in underlying structure. As such, they do not usually appear as homoplasies because they are screened before analysis and would appear in different data columns in a matrix of characters. That is, the investigator would not enter the analysis with an underlying hypothesis that the wings of bats and the wings of insects were homologous. However, analogous similarities can appear as homoplasies if the underlying structures are homologous but modified to perform a similar function. For example, one could imagine that a matrix of all vertebrates would contain a column containing the character 'wings present' versus 'forelimbs present,' and that the resulting analysis would show 'wings present' as homoplastic, appearing as a synapomorphy of birds and another synapomorphy of bats independently. However, even a cursory examination of the character 'having wings' would reveal that the structure of the wings of bats and birds are different relative to the details of wing architecture. Analogies fail the similarity test (see Analogy for further discussion).

Convergence and Parallelism

Patterson (1988) reviews the history of the distinction between parallel and convergent similarities. Some authors find the distinction to be arbitrary and use only the term homoplasy (e.g., Wiley, 1981; Ax, 1987). Others suggest that while the concepts are not easily separated, convergences are similarities exhibited by groups that are not closely related (e.g., enlarged canines of marsupial cats and saber-toothed cats), while parallelisms are exhibited by groups that are closely related. Parallelisms and convergences are similar in that both are identified through the congruence test. Patterson (1988) suggested that convergences fail the similarity test at some level (like analogies) while parallelisms pass the similarity test. Of course, since all characters of organisms, including the homologous ones, are dissimilar at some level, a certain amount of arbitrariness might be involved in the assessment. Hennig (1966) used a special term for some parallelisms, homoiology, to denote those parallelisms that arise repeatly from a common genetic base. Some authors, such as Wagner (1989) consider parallelisms to be homologous under some concepts of homology. (This reasoning finds its analog in the idea that paraphyletic groups are a kind of monophyletic group.) Parallelisms are not considered homologous under the concept that taxic homologs are apomorphies at some level in the phylogeny.

Parallelisms and convergences can be found at several levels of organization. The most basic level is the level of sequence variation in DNA, amino acid variation in protein sequences, electromorph variation at the allele level, and part descriptions at the morphological level. In each case, similar characters (base residues, amino acids, electronmorphs, flower color, bone shape, etc.) that pass the criteria of similarity and conjunction fail the test of congruence. That is, they appear in a phylogenetic tree more than once, indicating that they originated in two or more lineages. At the level of genes, lateral gene transfer may result in the presence of genes in quite distantly related organisms. Such genes are termed xenologous genes. Patterson (1988) suggested the term 'paraxenolog' for the case in which more than one copy of a xenologous gene family was present in the same organisms. He suggested that this was somewhat analogous to homeosis at the morphological level. At morphological levels of organization, parallelism or convergence may take the form of presence or absence of an entire structure that has been lost or gained independently in several lineages.

Homonomy

At the level of taxa, homonoms are similarities that fail the conjunction test because two or more similar structures are found in the same taxon. Some homonomies are termed 'general homologies' or 'mass homologies' and their presence versus absence is treated as a synapomorphy. For example, although it is difficult to homologize any two mammalian hair

follicles, the presence of hair is treated as a synapomorphy of Mammalia. In other cases, body parts are duplicated through metamerism in development (legs and antennae of insects). The classic example of homoplasy in this example is the mouth parts of arthropods where function of feeding is allocated to appendages that belong to different segments in different groups. Homologous structures can be found among organisms within a homologous segment, but comparison of similar appendages that belong to different segments would lead to a mistake in taxic homology determination. Paralogous genes are examples of homonomous parts at the organism level of organization. Just as with 'mass homology,' the presence of a particular gene family may be treated as a synapomorphy if it passes the congruence test (see Homology).

References

Ax P (1987) *The Phylogenetic System*. New York: John Wiley.

Hennig W (1966) *Phylogenetic Systematics*. Urbana, IL: University of Illinois Press.

Lankester ER (1870) On the use of the term homology in modern zoology, and the distinction between homogenetic and homoplastic agreements. *Annals and Magazine of Natural History* 6 (4): 34–43.

Patterson C (1988) Homology in classical and molecular biology. *Molecular Biology and Evolution* 5: 603–625.

Wagner G (1989) The biological concept of homology. *Annual Review of Ecology and Systematics* 20: 51–69.

Wiley EO (1981) *Phylogenetics: The Theory and Practice of Phylogenetic Systematics*. New York: John Wiley.

See also: Analogy; Homology; Monophyly; Paraphyly; Phylogeny; Synapomorphy

Homozygosity

L Silver

Copyright © 2001 Academic Press
doi: 10.1006/rwgn.2001.0630

Homozygosity is the genetic state in which a diploid organism carries two identical alleles at a locus of interest. In this situation, the organism is considered to be homozygous at this locus. The contrasted state is heterozygosity.

See also: Heterozygote and Heterozygosis

Hordeum Species

J W Snape and W Powell

Copyright © 2001 Academic Press
doi: 10.1006/rwgn.2001.1673

The genus *Hordeum*, the barleys, comprises a group of grass species, the most economically and socially important of which is the cultivated form, *Hordeum vulgare*, which is the fourth most widely grown cereal after wheat, rice, and maize. *Hordeum*s belong to the tribe Triticeae which also includes the wheats, rye, and oats, in the grass family Poaceae, with a basic chromosome number of $2n = 14$. However, they exist in diploid, tetraploid, and hexaploid forms. The morphology of the *Hordeum*s is rather specialized and they are characterized taxonomically by having spikelets with single flowers borne together in triplets on the main axis of the spike (the rachis). The central spikelet is generally sessile and male and female fertile, whereas the two lateral flowers, which are stalked in most species, may be fertile (as in six-row cultivated barely) or sterile (as in two-row cultivated barley). In *Hordeum*s the glumes are reduced compared to most Triticeae and situated on the dorsal side of each spikelet with long awns both on the lemmas as well as the the glumes. Species differ in reproductive behavior and life cycle with *H. vulgare* being annual and self-pollinating, and species such as *H. bulbosum* being perennial and obligatory cross pollinating by virtue of having a self-incompatibility mechanism.

Origins and Phylogeny

The genus comprises about 30 species distributed through the temperate regions of most of Eurasia, North and South America, Africa, and Australia. With respect to cultivated barley, it is generally recognized that there are three gene pools that can be exploited for barley improvement. The primary gene pool consists of cultivated barley, *H. vulgare* subsp. *vulgare*, and its wild progenitor which grows predominantly in the Middle East, *H. vulgare* subsp. *spontaneum* (known generally as *H. spontaneum*). Crosses between *H. vulgare* and *H. spontaneum* are easily obtained and hybrids are self-fertile. The secondary gene pool comprises only *H. bulbosum*, which exits in two forms, a diploid form ($2n = 14$) and an autotetraploid form ($2n = 28$). *H. vulgare* can be hybridized with both forms and hybrids are easily obtained with the use of embryo rescue techniques. Although the hybrids are generally sterile, seed can be obtained by backcrossing the hybrid as female to *H. vulgare*, and genetic recombination between the

genomes has recently been obtained. One peculiarity of *H. vulgare* × diploid *H. bulbosum* crosses is that generally, after a hybrid zygote forms, the *H. bulbosum* chromosomes are eliminated at cell divisions giving rise to the embryo, so that the end product is a haploid *H. vulgare* plant. This process is now used as a breeding tool to produce barley doubled haploid populations.

The tertiary gene pool comprises all the other *Hordeum* species. Cultivated barley can be hybridized to many of these to give sterile hybrids, but few authenticated reports of gene transfer from these hybrids into cultivated barley have been reported. However barleys can also be hybridized to many other Triticeae species, for example, wheat and rye, and this is very useful for genetical and cytogenetical analysis.

Cultivated barley is an annual species, but varieties have been bred that are suitable for sowing either in the autumn (winter barley) or in the late winter, early spring (spring barley). This difference is genetically determined and particular varieties are adapted to each of these different life cycles. Winter varieties have a requirement for a period of low temperatures treatment (vernalization) before floral initiation can commence, whilst spring varieties do not. Winter varieties tend to be more frost tolerant and are generally adapted to resist or tolerate a different disease spectrum to spring barleys.

Uses of Barley

The grain of cultivated barley has two major uses, first for malting to produce beer and spirits, and second for animal feed. Plant breeding has produced varieties that are specialized for malting, and all others not suitable for this are used for animal feed. These latter varieties tend to be the highest yielding. Some grain is used directly for human food products, for example, in certain countries such as Ethiopia and Nepal, but overall, this is a minor use. Malting varieties are bred for a particular grain composition which includes low protein and β-glucan content, and high enzyme activity, although the final product is also affected by the environmental conditions under which the variety is grown. To produce malt for the brewing industry, grains of barley are germinated so that enzymes are released for digestion of the cell walls and endosperm. The digested grains are then heat treated and dried. This produces malt which is a mixture of enzymes and substrates, mainly starch, proteins, and β-glucans. The malting process thus involves degradation of the cell wall material by β-glucanases, digestion of starch by α-amylases, and hydrolysis of the protein matrix. The malt is then used as a substrate for fermentation by yeasts in the brewing process. Different malts and brewing additives result in different types of beers.

Breeding of feed varieties concentrates on maximizing the yield through improved agronomic characteristics. Little research has been done on selecting for improved nutritional aspects of barley, although some research has tried (unsuccessfully) to increase the lysine content of barley. Lysine is an essential amino acid needed for animal growth and is limiting when barley is fed in isolation. Generally, however, barley is used as the energy component of the animal diet with protein coming from other sources such as legumes.

Cytogenetics of Barley

Barleys have a basic chromosome number of seven. The chromosomes are large enough to be identified individually by light microscopy, particularly if they are differentially stained using C-banding or N-banding where blocks of heterochromatin reveal distinctive patterns for each chromosome. This also reveals species relationships, such as the close relationship between the genomes of North and South American species. The H symbol (with or without a species superscript) is conventionally used to designate chromosomes of the genus so as to indicate homoeology with chromosomes of other species of Triticeae. Cultivated barley chromosomes are thus designated 1H to 7H, and *H. bulbosum* chromosomes or *H. chilense* chromosomes H^b and H^{ch}, respectively. Barley geneticists originally designated the chromosomes of cultivated barley from 1 to 7 and the relationship between the old and new (H) nomenclature is 1 = 7H, 2 = 2H, 3 = 3H, 4 = 4H, 5 = 1H, 6 = 6H, and 7 = 5H. Cytogenetical and genetical analysis in barley has been greatly assisted by the availability of a range of aneuploid stocks including a complete barley trisomic series and 11 telotrisomic lines. Amongst the most useful aneuploid stocks for genetical analysis in barley are those obtained by interspecific hybridization. In particular, the chromosomes of cultivated barley and *H. chilense* added to bread wheat, and substitution lines derived from these. In addition, cultivated barley has a whole range of other cytogenetically defined stocks including over 1000 reciprocal translocations, inversions, deletions, and duplications. Deletion breakpoints have been used to make comparisons between physical and genetic maps of barley.

Genetic Markers and Genetic Maps of Barley

Various molecular assays have been developed to detect polymorphism at the DNA level in barley. Restriction fragment length polymorphism (RFLP),

relying on the use of restriction enzymes, has been complemented by assays arising from development of the polymerase chain reaction (PCR). These include: random amplified polymorphic DNA (RAPD), amplified fragment length polymorphism (AFLP), and simple sequence repeats (SSRs) or microsatellites. Both RAPD and AFLP allelic polymorphisms are inherited in a dominant manner, whereas SSR polymorphisms are transmitted in a codominant manner. The convenience and high information content of SSRs have resulted in this class of molecular marker being very popular with barley researchers. Currently there are over 560 functional barley SSRs. The detection and quantification of single nucleotide polymorphisms (SNPs) is in its infancy in barley. However, it is anticipated that this form of biallelic marker has great potential to improve the efficiency of marker-assisted selection and provide a means of relating sequence diversity to phenotype.

Using standard segregation analysis, more than 80 loci for morphological and disease resistance characters were assigned to the seven barley chromosomes by 1962. Genetic maps have been created by monitoring the segregation of alleles from F$_2$, backcross, recombinant inbred, and doubled haploid families. Developments in molecular biology, coupled with access to computer software and mapping algorithms, have resulted in a recent explosion of information. Extensive genetic maps, incorporating morphological, biochemical, and molecular marker data are now being created. In addition, composite maps represented by data from multiple mapping populations have been generated.

Barley genetic maps are now viewed as an important resource to localize qualitative and quantitative traits for marker-assisted breeding. They also and provide a platform for the map-based cloning of genes for simple and complex phenotypes.

Breeding Barley

Barley is a natural inbreeder and most breeding schemes follow a pedigree selection scheme with minor variations in detail. All schemes are based on the principle of identifying the desirable recombinant whilst progressing to homozygosity. Conventional breeding schemes tend to be lengthy (up to 10 years) but have been successful in contributing an average 1% annual increase in grain yield. Both single seed descent and doubled haploid methods are being used to augment conventional breeding methods. These approaches reduce the time scale and improve the efficiency of selection by creating homozygous material for evaluation. Molecular marker technology is being used to enhance the effectiveness of barley breeding by identifying new sources of allelic variability and for targeted backcross conversion programmes. Genotype by environment interaction is one of the factors that has limited barley breeding for low input environments. Decentralization of the breeding process together with farmers' participation is being deployed in developing countries.

Pests and Diseases

With respect to fungal pathogens of barley, powdery mildew (*Erysiphe graminis*) is of major significance and interest. Major gene resistance loci have been located on the seven barley chromosomes, and two genes (*Mlo* and *Mla*) responsible for resistance have been isolated and characterized. Cereal rusts (*Puccinia graminis*, *P. hordei*, *P. striformis*) are a second class of obligate biotrophic pathogens of economic significance. In addition *Rhynchosporium secalis* (scald), *Pyrenophora teres* (net blotch), *P. graminea* (leaf stripe), and *Cochliobolus sativa* (spot blotch) are important pathogens. For many pathogens, major gene resistance genes have been recognized and localized to chromosomes. Resistance to barley yellow dwarf virus (BYDV) conferred by *Yd2* is located at the centrometric region of chromosome 3L. The barley yellow mosaic virus complex comprises two different strains: barley mild mosaic virus (BaMMV) and barley yellow mosaic virus (BaYMV). Cereal cyst nematode *Heterodea avenae* is an important pest of barley with resistance genes being identified on chromosome 2L.

Genetic Engineering of Barley

The genetic transformation of barley is now possible using a variety of techniques. This has opened up the possibility of genetically engineering barley using cloned genes from any biological source, be it other plants, microorganisms such as bacteria and viruses, and even animals. The predominant technique of transforming barley is to use 'biolistics,' that is, shooting isolated pieces of DNA coated onto gold particles into target tissue. Target tissues are generally isolated microspores (immature pollen grains) or immature embryos excised from developing grains. After shooting, the target tissue is placed on a medium which allows the development of callus tissue, and transformed callus selected by the presence of an introduced selectable marker gene in addition to the target gene. Usually the selectable marker is the *Bar* gene, conferring resistance to the herbicide Bialophos, so that when the callus is cultured on media containing the herbicide only transformed tissue grows. Thus, most barley varieties transformed for a particular desired trait, are also herbicide-resistant. Biolistic methods of transformation are random with respect to where the target genes are introduced into the genome, and usually several copies can be introduced in one or a few loci. Recently, barley has also been successfully transformed using *Agrobacterium*, and this may have the advantage of allowing more control of the gene integration process.

Present commercial targets for the genetic engineering of barley include modifications for improved malting quality, better pest and disease resistance, and greater nutritional quality of the grain, although no transgenic barley has been released commercially in the world up to the beginning of the new millennium.

See also: **Grasses, Synteny, Evolution, and Molecular Systematics; Polyploidy; Transfer of Genetic Information from *Agrobacterium tumefaciens* to Plants; *Triticum* Species (Wheat)**

Horizontal Transfer

M G Kidwell

Copyright © 2001 Academic Press
doi: 10.1006/rwgn.2001.0632

Horizontal gene transfer is generally defined as the lateral transfer of a gene, or other DNA sequence, from one genome to another. Transfer between contemporary individuals of different species is usually implied. However, a special case involves the horizontal transfer of DNA between chloroplast, or mitochondrial, and nuclear genomes. Horizontal transfer is distinct from the normal mode of vertical transfer by which genetic information is passed from parent to offspring. In addition to entire genes, parts of genes, such as exons or introns, may also be transferred in this way. Sometimes horizontal transfer is also used to denote the transfer of a parasite, or endosymbiont, from its association with one host species to that of another. Although horizontal transfer is more likely to be successful between closely related than distantly related species, it does occur between species as divergent as those found in different kingdoms. This review focuses on horizontal transfers involving eukaryotic organisms.

Frequency of Horizontal Transfer

Until quite recently, it was widely believed that horizontal transfer was mostly restricted to bacteria, and that this process, if it occurred at all, had little importance for the understanding of evolution in eukaryotes.

With the advent of large-scale DNA sequencing it has become apparent that both the frequency and significance of this phenomenon have been considerably underestimated. Not only is there evidence that horizontal transfer is rampant among contemporary bacteria, but it also seems to have dominated the evolution of early life before modern cells came into being. Horizontal transfer appears to have become increasingly less frequent with the evolution of increasingly more complex eukaryotic cells and the erection of barriers to the promiscuous exchange of DNA between divergent lineages. It is important to note that only those transfers affecting the germ cells that produce the next generation are of any significance from an evolutionary perspective. When the germline is sequestered in specialized organs, as it is in humans, its reduced accessibility provides an additional barrier to horizontal gene transfer. Although the frequency of horizontal transfer involving eukaryotes appears to be extremely low compared to that among prokaryotes, it can, in some instances, have important evolutionary consequences, as described below.

Mechanisms of Horizontal Transfer

Horizontal transfer is an endproduct of a process, rather than a specific mechanism. Apart from instances of horizontal transfer following rare matings between closely related species that are usually reproductively isolated, mechanisms of horizontal transfer are by their very nature mating-independent. Transfers from eukaryotes to prokaryotes, by means of transformation, commonly occur in contemporary molecular biology laboratories. Transfers from prokaryotes to eukaryotes may occur by transformation or conjugation in nature. A large variety of bacterial plasmids can stimulate conjugal transfer of DNA from bacteria to a broad range of organisms, including other bacteria, yeast, fungi and plants. Conjugal plasmids can survive in host species during normal vertical evolution and they have the ability to adapt to their new host following horizontal transfer. Other possible mechanisms of non plasmid-mediated transfer into eukaryotic cells include endocytosis, mediated by mammalian cell transfer and fungus-to-fungus endoparasitism. In theory, viruses have many of the properties necessary to enable them to carry DNA sequences between species. However, many viruses have limited host ranges and well-documented examples of viral transfer in nature have been difficult to find. Although parasitic wasps and mites may also serve as transfer vectors, the identity of the specific vector in most cases of horizontal transfer remains enigmatic.

Detection of Horizontal Transfer

The discovery of an outstanding discontinuity in the phylogenetic distribution of a gene or other DNA sequence, or the incongruence between gene trees and species trees, often provide reasons to suspect that horizontal transfer may have occurred. However, there are several pitfalls in making quick conclusions from such observations alone because a number of other mechanisms can also lead to incongruent phylogenetic trees. These include unequal rates of nucleotide substitution, ancestral polymorphisms, convergent evolution and inappropriate comparisons between paralogous, rather than orthologous, members of multigene families.

Transkingdom Horizontal Transfer

A number of possible instances of horizontal transfer between different kingdoms have been proposed, but the supporting evidence is much stronger for some claims than others. The horizontal transfer of glucose-6-phosphate isomerase between a eukaryote (plant) and a prokaryote (ancestor of the bacterium *Escherichia coli*) provides one well-supported example. Another such example is the transfer of Fe-superoxide dismutase between a prokaryote and the eukaryotic protist *Entamoeba histolytica*. Endosymbiotic gene transfer is a special case of transkingdom horizontal transfer that was initiated by the import of certain bacteria into the bodies of early eukaryotes. These imports later evolved into the organellar genomes of mitochondria and chloroplasts that are now a universal component of plant cells and the mitochondria found in all animal cells. Although these organelles have kept the majority of proteins that are integral to the eubacterial nature of their metabolisms, mitochondria and chloroplasts have subsequently relinquished the majority of their remaining genes to the nucleus by horizontal transfer. Some genes of eubacterial origin have replaced their nuclear homologs subsequent to transfer. In other instances, the products of other transferred genes were rerouted during evolution to compartments other than those from which the genes were donated.

Horizontal Transfer between Eukaryotes

In contrast to prokaryotes in which horizontal transfer is rampant, only relatively few cases involving eukaryotes have been well documented. These may conveniently be divided into two groups, depending on whether, or not, transposable genetic elements are involved. Some transposable elements naturally possess the molecular machinery for inserting their DNA into different locations of a host genome – an important prerequisite for successful horizontal transfer. Transposable elements routinely use this machinery for transposition to different sites within a single host genome, but occasionally it is used for jumping between genomes of different host species. However, unlike some viruses, transposable elements do not have the ability to survive outside of the environment of a host cell. Therefore, they are dependent on other organisms for transfer between species. In most instances the identity of these transfer vectors is not known.

Prominent among well-documented examples of horizontally transferred transposable elements are the *P* and *mariner* elements that were first described in *Drosophila* species. Both these elements transpose by means of a DNA–DNA intermediate. The *mariner* element is capable of spectacular interkingdom jumps because it does not depend on host factors to integrate into the genome of a new species. In contrast, the *P* element does require host factors for integration and has a host transfer range that is apparently restricted to a few insect orders. Recent evidence indicates that *copia* and some other retroelements that use reverse transcriptase for transposition also have the ability for horizontal transfer between species.

Significance

The existence of horizontal gene transfer in nature has important implications for both basic and applied science. However, because of the infancy of studies in this area, the full significance of this process is not yet known. The strictly bifurcating tree of life as envisaged by Darwin assumes no exceptions to the vertical transmission inherent in normal parent-to-offspring inheritance of genetic material. In contrast, horizontal transfer introduces crosslinks into the phylogenetic trees of those genes that are transferred and incongruities between the phylogenies of different gene sequences. Thus if horizontal transfer is frequent, our picture of the tree of life is changed significantly and serious practical difficulties can arise when attempts are made to infer phylogenies from horizontally transferred sequences. The existence of natural horizontal transfer also has important implications for artificial gene transfer in medicine and agriculture.

See also: **Conjugation; Symbiosis Islands; Transfer of Genetic Information from *Agrobacterium tumefaciens* to Plants**

Host-Lethal Gene

E Kutter

Large virulent bacteriophages like coliphage T4 and subtilis phage SPO1 generally have many weapons in their arsenal, each of which by itself is capable of killing or seriously damaging the host cell. The products of these genes are involved in shutting off host transcription, translation, DNA replication and/or cell division. The phage may also encode nucleases that selectively degrade the host DNA. If cloned into a host cell, each of these individual host-lethal genes can kill the host or drastically slow its growth if any expression of the gene occurs during growth of the host cell, even if the gene is not intentionally being expressed. Regions encoding such genes are generally missing from cloning libraries and/or contain many mutations, since only cells where the lethal functions have been lost can survive. Understanding the mechanisms involved in the virulence of such viral genes can provide new insights into key aspects of host physiology. The genes they target are presumably important to bacterial survival; thus, identifying those host genes can suggest potential targets and approaches for developing new classes of chemical antibiotics involving molecules that can mimic the effects of these phage proteins

Cloning of such host-lethal genes is challenging, since readthrough of terminator sites generally permits a basal level of transcription of the entire plasmid even if the cloned gene is missing its own promoter and is put under the control of a promoter that can be carefully controlled. Special vectors have been developed to aid in cloning such genes in bacterial systems and to permit very tightly controlled overexpression of the cloned protein. For example, many of the pET vectors carry the *lac* operator region adjacent to the cloning site along with the gene for the *lac* repressor in the opposite orientation following the cloning site. This blocks readthrough into the cloned gene both through binding of the *lac* repressor and through the synthesis of antisense messenger from the *lac* promoter. The pET vectors also have a bacteriophage T7 late promoter, recognized only by the efficient T7-encoded RNA polymerase, in front of the cloning site. Thus, expression can be obtained by transferring the plasmid into a host with an inducible T7 polymerase gene under tight control or, in the case of very host-lethal proteins, by growing the cells into mid-log phase and then infecting them with a special lambda phage into which the T7 polymerase gene has been cloned. This method works even for very host-lethal proteins like the T4 gp*alc*, which shuts off the elongation of transcription on all templates containing cytosine in their DNA – a good strategy for T4, since it uses hydroxymethylcytosine rather than cytosine in its DNA. The *alc* protein provides a valuable tool for looking at the process of transcription elongation, since it is the only factor known that can produce termination only when the RNA polymerase is actively elongating, not when it is pausing or moving slowly.

See also: **Bacteriophages; Elongation Factors; Translation**

Host-Range Mutant

J H Miller

Copyright © 2001 Academic Press
doi: 10.1006/rwgn.2001.0634

The host range of a phage is the spectrum of cells that they can infect and lyse. For instance, the bacteriophage T4 may infect a series of *Escherichia coli* strains, its host range. Host-range mutants of phage such as T4 can be found that change the spectra of strains the phage can infect, now allowing infection of certain strains that could not be infected before. Often, the mutations that cause the altered host-range phenotype are in the phage tail fiber protein that adsorbs to specific receptor sites on the cellular exterior.

See also: **Bacteriophages**

Hot Spot of Recombination

F W Stahl

Copyright © 2001 Academic Press
doi: 10.1006/rwgn.2001.0637

Enzyme systems for generalized recombination can effect recombination anywhere along a pair of homologous chromosomes. However, the rate of such recombination per internucleotide bond is not uniform. A short segment of chromosome with a conspicuously higher than average rate of recombination is a hot spot.

Basic Properties of Meiotic Hot Spots

Early-described hot spots for meiotic recombination, *cog* in *Neurospora crassa* and *M26* in *Schizosaccharomyces pombe*, manifest features that have characterized most subsequently discovered hot spots: they can

mutate to an inactive state; they can function when the hot spot is present on only one of the two homologs; they can increase recombination up to several kilobases away; and they promote meiotic gene conversion unidirectionally – genetic markers near and in *cis* to an active hot spot tend to be lost.

Molecular Basis of Meiotic Hot spots

Extrapolating from studies in *Saccharomyces cerevisiae*, hot spots of meiotic recombination are sites at which chromatids are cut on both strands by a meiosis-specific endonuclease. Repair of these cuts is carried out with the help of an intact chromatid, usually from the paired homolog. The homologous chromatid serves as a jig to align the two segments of the broken chromatid and as a template for the replacement of nucleotides lost subsequent to the cutting. The resulting intermediate, which contains two Holliday junctions, is resolved in a manner that recombines the segments of DNA flanking the intermediate approximately half the time. Whether the resolution effects such crossing-over or not, genetic markers between the junctions are subject to recombination by gene conversion, a local violation of the 2:2 rule of Mendelian segregation resulting from the loss and replacement of DNA segments during or after formation of the intermediate.

In *S. cerevisiae*, meiotic hot spots are manifested physically by the detection of meiosis-specific double-strand breaks and genetically by the high rates of conversion they impose on markers within a few kilobases and by the high rates of crossing-over they impose on markers flanking the region of conversion. One known hot spot confers a conversion rate on adjacent markers approaching 50%. More commonly, rates of 5%–10% are reported. These values are higher than those reported in other fungi, and they are atypically high for *S. cerevisiae*. The rate of conversion falls with distance from a hot spot, resulting in a conversion gradient.

Meiotic hot spots correspond to regions of the chromosome that are highly susceptible to cutting *in vitro* by endonucleases whose rate of cutting is limited by chromatin structure. These nuclease-sensitive regions tend to correspond to promoters of transcription. Large regions of some chromosomes have higher rates of recombination than other large regions, which may sometimes reflect the relative concentration of transcription promoters.

Hot Spots in Prokaryotes

In prokaryotes also, hot spots correspond to DNA double-strand cut sites. In phage lambda, whose circular replicating form is linearized at *cos* prior to packaging of the chromosome into a phage head, *cos* is a hot spot for recombination. The role of double-strand breaks as hot spots is illustrated also by lambda crosses in which the chromosome of one parent has a site for cutting by a restriction system carried by the host cell. Recombination in such a cross is focused close to the restriction site.

In bacteria, the primary recombination pathway is dependent on proteins homologous to the RecA and RecBCD proteins of *Escherichia coli*. The pathway is activated by double-strand breaks, which serve as entry points for the RecBCD enzyme, which then unwinds the duplex processively from the double-strand break, cutting the resulting single strands as it does so. This destruction stops, with low probability per base pair, when the enzyme undergoes a transition that diminishes the nuclease but not the helicase activity of the enzyme. This transition occurs with high (about 50%) probability at species-specific nucleotide sequences, called Chi. In *E. coli*, the fully active Chi sequence is 5' GCTGGTGG 3'. The intact single strands resulting from the unwinding of DNA distal to Chi are recombinagenic after becoming coated with RecA protein. Thus, Chi is a hot spot of recombination because it limits the extent of DNA degradation occurring at a double-strand break. This recombination system helps maintain normal rates of DNA replication by promoting recombinational repair of accidentally broken replication forks.

Some recombination systems are specialized to cut and rejoin DNA at specific nucleotide sequences. *att* of phage lambda is a specific site for recombination effected at high rate by lambda's Int system. Int can also effect homologous recombination at low levels nearby the *att* site. Int-mediated recombination requires that both participants have an *att* site.

A hot spot in the gene *34–35* region of phage T4 is absent in the closely related phage T2. DNA glucosylation, which differs between the two phages, is required for the hot spot activity in T4.

See also: Chi Sequences; Gene Conversion; RecBCD Enzyme, Pathway; Recombination, Models of

Hot Spots

J H Miller

Copyright © 2001 Academic Press
doi: 10.1006/rwgn.2001.0635

A hot spot is a site in the DNA that is significantly more mutable then normal. Seymour Benzer first established the concept of hot spots in his classic studies of the *rII* locus in bacteriophage T4 in the late 1950s and early 1960s. Benzer mapped a very large series of mutations in the *rII* locus, assigning each mutation to a specific site. Two sites had an enormous number of recurrences of mutations and were clearly extraordinary hot spots. Statistical methods could show that other sites were also more mutable than normal. Subsequent work from different laboratories has revealed that hot spots are a general phenomenon. The molecular basis for some spontaneous hot spots are now understood. Benzer's hot spots, as well as others in different genes, result from repeat-tract sequences i.e., tandemly repeated mono-, di-, or even tetranucleotides. For instance, in the *lacL* gene of *Escherichia coli*, the sequence 5'-CTGGCTGGCT-GG-3' appears in the wild-type. More than 70% of the spontaneous mutations in *lacL* are the addition or deletion of one of the tandemly repeated units, CTGG. In mismatch repair deficient backgrounds, repeat-tract sequences respresent very powerful hot spots. In the *E. coli xylB* gene, 90% of the spontaneous mutations in a mismatch repair deficient background are deletions or additions of a -G- at a run of eight Gs (-GGGGGGGG-) in the wild-type *xylB* gene.

5-methylcytosine residues also result in hot spots in many cases, since the deaminations at the 5-methylcytosine result in thymine across from guanine, that can lead to mutations at the next round of replication if not repaired. Mutagen-induced mutations are rarely randomly distributed, resulting in hot spots at certain points. Neighboring pyrimidines are favored sites of UV-induced mutations, since several photoproducts occur at pyrimidine–pyrimidine sequences. Even among these sequences, however, hot spots still occur, for reasons that are not presently understood.

In certain cases, hot spots for mutations are programmed into natural DNA sequences, to allow for more frequent variation and sometimes to avoid host immune responses. For instance, in *Haemophilus influenzae*, the intergenic region between the fimbriae protein encoding *hifA* and *hifB* genes has 10 repeats of the -TA- sequence in the promoter. When the sequence mutates to 11 or 9 repeats, transcription is lowered, or abolished, respectively. The number of tandem repeats is so high that the resulting hot spot allows 0.1–1% variation in a typical population.

See also: Mutation, Spontaneous; Tandem Repeats

Housekeeping Gene

M Goldman

Copyright © 2001 Academic Press
doi: 10.1006/rwgn.2001.0639

While different tissues in higher organisms are distinct phenotypically, they generally have the same set of genes. The phenotypic differences are brought about by differential regulation of gene expression. The genes that are expressed differentially are called 'tissue-specific genes' (or sometimes 'luxury genes'). Housekeeping genes, on the other hand, are expressed in all tissues, and are generally assumed to be involved in key steps in cellular metabolism such as DNA synthesis, protein synthesis, transcription, or energy metabolism.

As there is a vast difference in how tissue-specific in contrast to housekeeping genes must be regulated, it is not surprising that the promoters of these genes differ as well. Most housekeeping genes utilize a promoter lacking the common TATA and CAAT boxes, and having instead a series of GC boxes (consensus sequence GGGCGG). GC boxes provide binding sites for the transcription factor Sp1 and, like the TATA box, direct the start of transcription. Since there are several GC boxes in the promoters of many housekeeping genes, the transcription start site is ambiguous. Indeed, many housekeeping gene transcripts have heterogeneous 5′ start sites. The coding function of these genes is not impaired, however, because all of the alternative start sites are within the 5′ untranslated region of the mRNA. As an example, the human c-*Ha-ras* oncogene promoter has about 80% G+C content, 10 GC boxes, and at least four transcription start sites.

The products of housekeeping genes may be needed in all cells, but in limited quantities. Therefore the housekeeping gene promoters are often weak, representing a baseline level of transcription.

While some genes such as *Hprt* and *Pgk* fall squarely into the housekeeping category, as they are involved in nucleotide and energy metabolism, others are not so easily categorized. The metallothionein gene, for instance, is relatively quiescent, but is stimulated in the presence of heavy metals. This gene, however, is available for transcription in all cell types, even though it may not actually be transcribed at a particular point in time.

See also: **Gene Regulation; TATA Box; Transcription**

Hox Genes

A Gavalas and R Krumlauf

Copyright © 2001 Academic Press
doi: 10.1006/rwgn.2001.0640

Hox genes are the homologs of the homeotic genes of the fruit fly *Drosophila*. The *Drosophila* homeotic genes were first identified through mutations that caused the transformation of a particular segment of the fly body into the likeness of another, hence the term homeotic from the Greek word *homeo*, which means similar. With the advent of molecular biology these genes were isolated and found to encode proteins that play fundamental roles in controlling regulation of many other genes. The *Hox* genes share a 60-amino-acid DNA-binding motif, the homeodomain, and in association with other homeodomain-containing proteins act as transcription factors to regulate gene expression. Today we know that these homeobox (*Hox*) genes have been widely conserved during metazoan evolution and they are present in organisms ranging from primitive chordates to humans. They are generally linked in chromosomal clusters. In simple ancestral organisms there is a single cluster. In association with genome-wide duplications in higher animals, this gave rise to the four *Hox* clusters that encompass a total of 39 *Hox* genes present in nearly all vertebrates, including mice and humans.

A distinguishing hallmark of *Hox* clusters is the correlation between the physical arrangement of these genes along the chromosome and their temporal and spatial order of expression in the developing embryo. Genes located closer to the 3′ end of the chromosomal clusters will be expressed earlier and in more anterior domains than genes located closer to their 5′ ends. This property is known by the term temporal and spatial colinearity and is thought to reflect the mechanism that regulates the expression of these genes. *Hox* genes encode key developmental regulators, which specify the regional character of cells along the antero-posterior body axis of all three germ layers in both vertebrate and invertebrate embryos.

Studies using the mouse and other vertebrates as model systems have shown that genetic mutations in some of the *Hox* genes or changes in their expression patterns result in abnormalities in a large number of tissues. This can cause defects in the nervous system, limbs, skeleton, and many organs. In some cases the defects are much milder than expected, but genetic studies have shown that some of these *Hox* genes work together and can compensate for each other. Hence a defect in one gene is corrected by the similar activities of other Hox proteins. In humans, specific *Hox* genes have been implicated in genetic disorders affecting development of the limbs and the genitourinary tract. Several studies have suggested that *Hox* genes are also required for proper function of adult tissues. Specific *Hox* genes function together to control development of the mammary gland in response to pregnancy, whereas others may be involved in human endometrial development and implantation. Recent studies have also shown direct involvement of deregulated *Hox* genes in the development of human leukemias.

Since the description of the first homeotic mutations by Bateson in 1894 and the discovery of the homeodomain in 1984 there has been tremendous progress in understanding the function of these important genes. These genes represent important control points in the processes that regulate morphogenesis or how tissues are formed and patterned. To build a picture of how this entire process occurs we still need to determine the immediate gene and cellular targets of their action in order to understand how they regulate cell growth and differentiation.

See also: **Homeotic Genes**

Hsp

See: **Heat Shock Proteins**

HTLV-1

M J S Dyer

Copyright © 2001 Academic Press
doi: 10.1006/rwgn.2001.1583

Human T-cell lymphotrophic virus 1 (HTLV-1) is a 9032 bp human C-type retrovirus that was isolated in 1979 from T-cell lymphoma cell lines maintained *in vitro* with IL2. It was the first human pathogenic retrovirus to be described. HTLV-1 is the causative agent for at least two diseases, firstly a malignancy of mature CD4 T cells (adult T-cell lymphoma/leukemia or ATLL) and secondly, a neurological disorder known as either tropical spastic paraparesis (TSP) or HTLV-1-associated myelopathy (HAM); only the former is discussed here.

Like other retroviruses, HTLV-1 contains *Env* (encoding receptor binding protein), *Gag* (core protein), and *Pol* (RNA-dependent DNA polymerase) genes, but also *Tax* and *Rex*, genes involved in the regulation and splicing of viral RNA. HTLV-1 lacks an obvious transforming oncogene. HTLV-1 may infect several different cell types *in vitro* but only replicates efficiently in CD4+ T cells. The virus is endemic in the tropics and the prevalence may reach over 20% in some areas. Transmission may be vertical, from mother to infant by breastfeeding, or horizontally, via intravenous drug abuse, sexual contact, or transfusion of contaminated blood. The percentage of infected individuals developing either disease is very low and the factors necessary for the development remain unknown.

ATLL

In 1977, a rapidly progressive and uniformly fatal T-cell lymphoproliferative disorder in patients from the south west of Kyushu, Japan was described by Takatsuki and Uchiyama. An identical disease in patients from the Caribbean was described by Catovsky and colleagues in 1982 in London. Subsequent investigations showed the presence of antibodies to HTLV-1 and monoclonal proviral integration in tumor cells. Other cases have now been reported from a number of other geographical sources including southeastern USA, South America (Chile and Brazil), and West Africa. In southwest Japan, ATLL constitutes a major health problem. Various clinical types of ATLL have been described, but ultimately, all forms progress and are fatal; treatment is often associated with opportunistic infections. Patients present with enlarged lymph nodes and skin rash often accompanied by hypercalcemia. Diagnosis is usually made on the presence of cells with a characteristic convoluted nuclear morphology ('flower cells') in the peripheral blood. These cells are characteristically CD4+ and CD25+, the latter being a component of the IL2 receptor. There are no consistent cytogenetic abnormalities in ATLL patients and the mechanisms that promote transformation of infected CD4+ T cells are not known. Proviral integration appears to be random. Comparative genomic hybridization studies have shown amplification of 14q32 and 2p13 in some patients, although the nature of the target genes is not known. Recent work indicates that viral *Tax* may result in constitutive NF-κB activation and therefore prolonged cell survival through its interaction with the IKKβ/IKKγ complex of controlling kinases. Expression of antiapoptotic proteins such as BCL-$_{XL}$ may also be upregulated.

See also: **Retroviruses**

Human Chromosomes

M A Ferguson-Smith

Human chromosomes were probably first observed in cancer cells by Arnold in 1879. Hansemann in 1881 and Flemming in 1898 attempted to count the number in serial sections of mitotic cells producing crude estimates of approximately 24. Quite different results were produced in 1912 by de Winiwarter. He was probably the first to study gonadal material and found 47 chromosomes in testis and 48 in ovary. He concluded that humans, like the locust, had an XX female/X male sex-determining mechanism. Painter in 1923 repeated this work on sections of testis material, in which he detected the small Y chromosome which de Winiwarter had apparently missed. He concluded that 48 and not 47 was the correct number for humans of both sexes, but mentioned in his publication that in the clearest mitotic figures he could only count 46.

There matters stood until 1956 when Tjio and Levan, working on colchicinized cell cultures treated with hypotonic fluid before fixation, regularly counted only 46 chromosomes, in samples from different cultures. This number was confirmed as the correct number by Ford and Hamerton using testis material later the same year.

More widespread interest in human chromosomes immediately followed the discovery by Lejeune and colleagues of an additional small chromosome in cells cultured from five children with Down syndrome. The observation that such a gross genetic abnormality could occur in a live, albeit handicapped, individual led to a search for similar chromosome abnormalities in other clinical syndromes. However, it was the paradoxical sex chromatin findings in the Turner and Klinefelter syndromes (see Klinefelter Syndrome; Turner Syndrome) which led to the next discovery of sex chromosome aneuploidy in these disorders later in 1959.

These early results on human chromosome aberrations were made on fibroblast cultures from skin biopsies or from bone marrow samples obtained by sternal aspiration. A major technical advance was made in 1960 when Moorhead and colleagues developed the short-term culture of lymphocytes from peripheral blood samples. Chromosome analysis thus became more widely applicable for the investigation of human chromosome aberrations. Cytogenetic laboratories have flourished ever since, using increasingly sophisticated methods for the identification of

even smaller defects. The latest methodology now exploits multicolor fluorescent *in situ* hybridization and a wide range of other molecular genetic techniques.

See also: **Klinefelter Syndrome; Sex Determination, Human; Turner Syndrome**

Human Genetics

M A Ferguson-Smith

Human genetics is the study of genetics and biological variation in *Homo sapiens*. Its various branches are population genetics, cytogenetics, biochemical genetics, and genome studies including biodiversity and human evolution. Clinical genetics is that part of human genetics that studies genetic variation associated with the pathogenesis of disease (see Clinical Genetics).

See also: **Biochemical Genetics; Clinical Genetics; Cytogenetics; Ethics and Genetics; Genetic Diseases; Human Chromosomes; Human Genome Project; Population Genetics**

Human Genome Project

D Seemungal and G Newton

The Human Genome Project (HGP) is an international 13-year effort to sequence and discover all human genes (the human genome) and make them accessible for further biological study. The collaborative project began formally in October 1990, and involves 20 groups from the USA, UK, Japan, France, Germany, and China. Originally, the project was expected to last 15 years, but technological advances have brought forward the completion date to 2003.

The total size of the human genome is estimated to be about 3 billion base pairs, arrayed in 24 distinct chromosomes (autosomes 1–22 plus X and Y). The chromosomes range in size from 50–250 million bases (megabases) long, too large to be sequenced directly, so each chromosome is first broken into relatively large fragments about 150 000 bp long. The large fragments are inserted into bacterial artificial chromosomes (BACs), and genome mapping techniques are used to determine the position of each of these fragments in the genome. The next stage involves 'shotgunning' – cutting each of the fragments into smaller, overlapping pieces for sequencing (about 500 bp each). Shotgunning at random, but repeatedly, ensures that some of the fragments will contain overlapping regions. Finally, the small DNA pieces are sequenced, the sequences assembled into the full sequence of the original BAC fragment, and the sequences of the BACs assembled to give the full chromosomal sequence.

An alternative strategy for sequencing a genome is termed the 'whole-genome shotgun' method. This method does not involve mapped bacterial clones; instead, the whole genome is broken up into small pieces at random, the pieces are sequenced and the sequence reassembled. This method can produce sequence more rapidly, but reassembly of the information is more difficult, especially since about half of the human genome is composed of highly repetitive sequences.

Chromosome 22 – the first human chromosome to be sequenced – was completed in December 1999. An initial 'working draft,' which covers more than 90% of the euchromatic part of the genome (which contains most of the genes) at an accuracy of about 99.9%, was completed in June 2000. The final 'gold standard' standard genome sequence – produced by sequencing each piece of the genome about 10 times – to an accuracy of 99.99% – is due for completion in 2003.

All the DNA sequence produced by the Human Genome Project is released freely onto the Internet. The sequence is then analyzed to find the estimated 30–40 000 human genes encoded within it – but which comprise only about 5% of the entire genome. Other studies are examining variations in human genome sequences, in particular the single-nucleotide polymorphisms (or SNPs) which occur about once every 1000 bases and account for most of the variation between individuals. Using the genome sequence, functional genomics studies are examining how and when genes are expressed (transcriptomics) and the structure and function of the proteins encoded by the genes (proteomics). As these studies advance, the human genome sequence will undoubtedly have a significant impact on our understanding of biological processes and advance the treatment of disease.

References:

Dunham I et al. (1999) The DNA sequence of human chromosome 22. Nature 402: 489–499.

International Human Genome Sequencing Consortium Announces "Working Draft" of Human Genome (Press Release), The Sanger Centre, Monday 26 June 2000.

http://www.ncbi.nlm.nih.gov/genome/seq/page.cgi?F=HsHome.shtml&ORG=Hs

International Human Genome Sequencing Consortium (2001) Initial Sequencing of the Human Genome. Nature 409: 860–921.

See also: Artificial Chromosomes, Yeast; BAC (Bacterial Artificial Chromosome); DNA Sequencing; Gene Mapping; Nucleotides and Nucleosides; Shotgun Cloning

Hunter Syndrome

K M Beckingham

doi: 10.1006/rwgn.2001.0645

Hunter syndrome (type II mucopolysaccharidosis) is a rare, recessive X-linked genetic disorder almost exclusively limited to Caucasian males. The symptoms arise from a loss of iduronate sulfatase activity, an enzyme required for degradation of the mucopolysaccharide components of connective tissues. Partially degraded mucopolysaccharides accumulate in the bones and connective tissues producing characteristic developmental defects such as facial distortions, dwarfism, and a hunched posture with flexed limbs. In the mild form of the disease, average life expectancy is about 20 years. Intellectual impairment is minimal and death is typically due to cardiac complications. For the more severe form, life expectancy is about 12 years. Progressive neurological deterioration, seizures, and emaciation characterize the later stages and death usually results from pulmonary failure. Bone marrow transplantation has been attempted as a corrective measure, but enzyme replacement, via protein or gene therapy, would appear to be the most hopeful future possibility.

See also: Gene Therapy, Human; Sex Linkage

Huntington's Disease

D C Rubinsztein

doi: 10.1006/rwgn. 2001.0646

Huntington's disease (HD) is an autosomal dominant neurodegenerative condition associated with abnormal movements, cognitive decline, and psychiatric disturbances. Symptoms most commonly appear between the ages of 35 to 50 years, but the disease can present at any age. Death occurs about 15–20 years after the initial symptoms. HD neuropathology is characterized by neuronal loss in the caudate nucleus, the putamen, and the cerebral cortex. HD is caused by abnormal expansions of a $(CAG)_n$ trinucleotide repeat tract in the coding portion of a gene of currently unknown function, which maps to 4p16. The $(CAG)_n$ repeats are translated into a polyglutamine tract. This mutation confers a deleterious new function on the mutant protein. The formation of abnormal ubiquitinated protein aggregates, containing the polyglutamine-containing protein of the HD protein, are a characteristic of the pathology.

Epidemiology and Clinical Features

HD varies in prevalence in different populations. It is particularly common in the Zulia region of Venezuela, near the shores of Lake Maracaibo, where there is a cluster of cases derived from a single ancestor. This extensive pedigree of about 7000 individuals contains over 100 living affected cases. HD is rare in Japan (<0.5 per 100 000) and among Black South Africans (1 per 100 000). Its prevalence in the UK and USA ranges from about 5 to 10 per 100 000.

HD generally presents insidiously. In adults, the motor features include chorea, abnormal eye movements, dysphagia, dysarthria, rigidity, and gait disturbances. Swallowing difficulties often lead to death, either from suffocation or from starvation. Juvenile-onset HD often presents with a different picture, where bradykinesia, rigidity, and dystonia are dominant features and chorea may be absent.

The overt cognitive features of HD generally start to manifest around the same time as the motor features present, although this is not universal. The patients develop a form a subcortical dementia which is progressive and becomes more global in the late stages of the disease. Subtle neuropsychological abnormalities have been detected in HD patients before any overt clinical features have manifested.

HD patients can develop a range of psychiatric disturbances. Depression is the most frequent problem and may be found in up to 40% of patients. The depression seen in HD is often a primary feature of the disease process, rather than a secondary reaction to the diagnosis or other symptoms. Irritability and apathy are also common features, while HD patients can develop obsessive–compulsive disorder and, rarely, schizophrenia-like features.

Genetics

HD is associated with abnormal expansions of a $(CAG)_n$ repeat in the 5′ end of the coding region of a large gene called IT15. Normal chromosomes are polymorphic with respect to repeat number and have 35 or fewer perfect repeats, while disease chromosomes are associated with 36 or more repeats. The mutant allele is expressed at the protein level and the $(CAG)_n$ repeats are translated into a polyglutamine tract.

HD shows the clinical feature of anticipation, where the age at onset of symptoms tends to decrease in successive generations. This phenomenon can be explained by a combination of the following two observations. First, while normal chromosomes have low mutation rates, the number of repeats on disease chromosomes frequently changes in successive generations. Increases in repeat number tend to be more common than decreases when the mutation is passed through the male line, although this mutational bias is not an obvious feature of female transmissions. Second, age-at-onset of symptoms correlates inversely with repeat number, with juvenile-onset cases having particularly long alleles. The CAG repeats on disease chromosomes account for about 70% of the variance in the age at onset of symptoms.

The penetrance of HD is not always complete, as some individuals with 36–39 repeats have lived into their ninth and tenth decades without clinical or neuropathological features of the disease. It has been suggested that genotype variation at the GluR6 kainate receptor locus may modify the age at onset of the primary mutation in the HD gene.

HD is one of the rare diseases where homozygotes do not appear to have a more severe phenotype than heterozygotes in the same family.

Pathology

Neuronal loss in HD is particularly severe in the caudate nucleus, putamen, and cerebral cortex. However, in advanced cases, there is overall atrophy and brain weight can be reduced by up to 25%. The cell loss in the caudate nucleus and the putamen (which together comprise the corpus striatum) is selective. The earliest loss is in the dorsal and medial regions and this progresses laterally and ventrally as the disease takes its course. Within the striatum, the medium spiny neurons, particularly those synthesizing enkephalin and γ-aminobutyric acid (GABA), show particular sensitivity to the HD mutation. In the cortex, the large neurons appear to be most severely affected, with greatest loss in layers VI, V, and III.

Pathological Mechanisms

The HD mutation confers a deleterious gain-of-function on the mutant protein. This model was suggested before the gene was cloned by observations in patients with Wolf–Hirschorn syndrome. These individuals have hemizygous deletions of the tip of 4p and are hemizygous for the HD gene but do not show the clinical features of HD. Subsequent to the HD mutation being identified, the gain-of-function mechanism has been confirmed. A woman with a balanced translocation disrupting the HD gene has been identified and shows no abnormalities. Transgenic mice expressing only one HD allele have no features of the disease, while HD 'null' mice have embryonic lethality. This lethality is rescued by transgenes with the HD mutation. Furthermore, a knock-in of the HD mutation into the endogenous mouse HD homolog is not associated with embryonic lethality, even in the homozygous form.

On the other hand, transgenic mice expressing exon 1 of the human HD gene with expanded repeats do show an abnormal neurological phenotype. These transgenic mice develop abnormal aggregates containing the expanded polyglutamine repeats in the nuclei of neurons. Subsequently, such neuronal intranuclear inclusions (NII) were found in brains of HD patients. It is not clear how these NIIs arise or how they relate to the neurodegeneration in HD. Two possible explanations for the mode of pathogenesis of the NIIs have come from work on the related disease spinocerebellar ataxia type 1, which is also caused by a $(CAG)_n$/polyglutamine expansion mutation. First, the inclusions appear to be alter matrix-associated structures, suggesting that this disease may result from disruption of nuclear function. Second, these inclusions are ubiquitinated and appear to sequester some of the cellular machinery responsible for the degradation of short-lived proteins. Since the levels of short-lived proteins have important regulatory consequences, it is possible that perturbation of these proteins levels may result in cell death.

It is not clear how these inclusions arise. Polyglutamine stretches in proteins may predispose to aggregate formation, as such sequences can form polar zippers. The formation of the NIIs may be partly mediated by transglutaminase, since inhibition of this enzyme partially reduces NII formation *in vitro*. The rate of aggregate formation may also be greatest in fragments of the mutant HD protein containing the expanded polyglutamines and slower in the full-length mutant protein. The formation of such fragments appears to be partly mediated by caspases; thus these enzymes may play an important part in the pathogenic pathway.

Relationships with Other Trinucleotide Repeat Diseases

HD is one of a class of diseases caused by abnormal expansions of $(CAG)_n$/polyglutamine repeats, including the spinocerebellar ataxias (SCA) types 1, 2, 3, 6, 7, spinobulbar muscular atrophy, and dentatorubral–pallidoluysian atrophy. In general, these disease are associated with repeat expansions above 36–40 glutamines, except for SCA6, which is associated with expansions of <30 repeats and may operate via a distinct mechanism. The other polyglutamine diseases appear to also be caused by gain-of-function mutations and intracellular aggregates have found in patients with SCA3, SCA1, SCA7, and DRPLA and in *in vitro* models of spinobulbar muscular atrophy. Thus, these disease are likely to share common pathophysiologies. However, it is not clear why the pattern of neurodegeneration in these diseases differs, particularly since the disease proteins are often widely expressed.

Further Reading

Harper PS (1996). *Huntington's Disease*, 2nd edn. London: WB Saunders.

Ross CA and Hayden MR (1998) Huntington disease. In: Rubinsztein DC and Hayden MR (eds) *Analysis of Triplet Repeat Disorders*, pp. 169–208. Oxford: Bios Scientific Publishers.

See also: **Genetic Counseling; Genetic Diseases; Microsatellite; Trinucleotide Repeats: Dynamic DNA and Human Disease**

Hurler Syndrome

T M Picknett and S Brenner

Copyright © 2001 Academic Press
doi: 10.1006/rwgn.2001.0647

Hurler syndrome is a genetic disorder resulting in a metabolic defect, and named after Gertrude Hurler, Austrian physician. Also known as gargoylism or mucopolysaccharidosis 1, Hurler syndrome is one of several rare genetic disorders involving a defect in the metabolism of mucopolysaccharides. Specifically, an autosomal mucopolysaccharidosis recessive storage disease in which α-iduronidase is absent, resulting in an accumulation of heparan and dermatan sulfates. Extensive deposits of mucopolysaccharide are found in gargoyle cells and in neurons.

Onset of the syndrome is in infancy or early childhood and affected individuals rarely live beyond adolescence. The disorder is characterized by severe mental retardation, large skull with wide-set eyes, heavy brow ridge and depressed nose bridge, hypertrichosis, short neck, large tongue and lips, poorly formed teeth, and clouding of the cornea. Individuals exhibit dwarfism with hunched back, short limbs and clawed hands, hirsutism, and deafness. Enlarged liver and spleen are common and coronary valves, vessels and heart muscles are often affected, leading to death from heart failure.

See also: **Genetic Diseases; Inborn Errors of Metabolism; Metabolic Disorders, Mutants**

Huxley, Thomas Henry

K Handyside, E Keeling, and S Brenner

Copyright © 2001 Academic Press
doi: 10.1006/rwgn.2001.0648

Thomas Henry Huxley (1825–95) was better known for his defence of Darwin's theory of evolution by natural selection than his own scientific research. He did more than even Darwin himself to gain acceptance for the theory among scientists and the public. His passion for the theory gained him the title of "Darwin's Bulldog."

His family was not wealthy and his only childhood education was two years at Ealing School. However, he schooled himself in science, history, philosophy, and German. Huxley began a medical apprenticeship at the age of 15 and a scholarship at Charing Cross Hospital meant that he could continue his studies. However, he did not pursue a career in medicine and instead joined the British Navy as an assistant surgeon on the frigate HMS *Rattlesnake* which was sent to chart waters in the South Pacific.

Returning to England in 1850 he found the research he had sent home on marine organisms had gained him entrance into the ranks of the English scientific establishment. He left the Navy in 1854 to go to the School of Mines in London and took up a lecturing position. For the next 40 years he was an active teacher, writer, and lecturer.

At first, Huxley was not an outspoken defender of Darwin's theory, disagreeing with certain ideas. But later, he began to accept evolutionary views and defended the cause in many debates. The most famous occasion, in June 1860, saw Huxley face the Bishop of Oxford, Samuel Wilberforce, at the British Association meeting in Oxford. All accounts describe it as an extremely heated debate, with Huxley declaring he would rather be descended from an ape than a bishop.

By profession he was a biologist but in fact covered the whole field of exact sciences. His most famous book was published in 1863, five years after Darwin's *On The Origin of Species*. Huxley's *Evidence on Man's Place in Nature* described what was known about

primate and human paleontology and ethology, linking evolution to homo sapiens.

Having had to fight his way into and to the top of the scientific profession he also helped set in place procedures for scientists to be awarded salaries. This gave all people, rich and poor, a chance to enter the scientific ranks.

See also: **Darwin, Charles**

Hybrid

L Silver

Copyright © 2001 Academic Press
doi: 10.1006/rwgn.2001.0649

Hybrid is the term for the offspring from two genetically distinct parents. When the two parents have no recent common ancestry, the offspring are referred to as F_1 hybrids.

See also: **F1 Hybrid**

Hybrid-Arrested Translation

Copyright © 2001 Academic Press
doi: 10.1006/rwgn.2001.1859

Hybrid-arrested translation is a technique used to identify cDNA representing an mRNA molecule, by virtue of its ability to base-pair with the RNA *in vitro* and thus to inhibit translation.

See also: **cDNA; Messenger RNA (mRNA)**

Hybrid Dysgenesis

M G Kidwell

Copyright © 2001 Academic Press
doi: 10.1006/rwgn.2001.0651

Hybrid dysgenesis is a term used to describe a suite of phenotypic abnormalities, referred to as dysgenic traits, which are simultaneously induced by intraspecific hybridization. These traits were first described in *Drosophila melanogaster*. They include increased rates of mutation and recombination, chromosomal rearrangements (such as inversions and translocations), and reduced fertility and viability. The genetic abnormalities result from the mobilization of certain families of transposable genetic elements (transposons) by intraspecific hybridization.

In many instances, hybrid dysgenic traits are observed to occur nonreciprocally. For example, given two interacting strains, A (carrying a particular transposon family) and B (lacking the relevant transposon family), only crosses between males of strain A and females of strain B will produce dysgenic hybrids; the reciprocal cross, between males of strain B and females of strain A, will produce normal offspring. Usually, but not always, the mobility of the transposons is restricted to the germline of the host; the somatic, or body, cells are not affected. This is thought to be an evolved trait that reduces the likelihood that unbridled activity of the transposon will reduce the fitness of its host, and thus decrease its own chances of survival.

Hybrid dysgenesis in nature appears to be associated with the arrival of an active transposon family in a new species by horizontal transfer, or introgression. Examples are the *P*, *I*, and *hobo* elements in *D. melanogaster* and the *Penelope* element in *D. virilis*. All four of these transposon families have invaded their new host species within the last century, possibly aided by increased human mobility and trade. Activation of the *P*, *I*, and *hobo* families of transposons is responsible for the P–M, I–R, and H–E systems of hybrid dysgenesis, respectively. There is no evidence for cross-mobilization of elements among any of these three systems. However, in a fourth system, found in *D. virilis*, hybrid dysgenesis results in the simultaneous activation of multiple families of transposons, including the *Penelope, Ulysses, Paris, Helena*, and *Telemac* families.

Partial or complete sterility is a signal trait commonly associated with hybrid dysgenesis. However, this hybrid sterility occurs in two distinctly different ways, referred to as GD sterility and SF sterility. GD sterility, or gonadal dysgenesis, describes the sterility associated with the *P*, *hobo*, and *Penelope* elements. In this case, one, or both, gonads of F_1 dysgenic hybrids are arrested at an early stage of development. If the arrested development is unilateral, then the individual will be fertile; individuals that are bilaterally affected are completely sterile. High temperatures applied at an early stage of development increase the frequency of gonadal dysgenesis. In contrast, the sterility caused by the I–R system of hybrid dysgenesis (SF sterility) is caused by partial or complete inviability of eggs laid by F_1 dysgenic hybrids. In this instance, low temperatures, applied early in development, increase the frequency of sterility.

In addition to *Drosophila*, hybrid dysgenesis-like phenomena are observed in other insects, such as the Mediterranean fruit fly, *Ceratis capitata*, and midges of the genus *Chironymus*. As reports of hybrid dysgenesis have so far been largely restricted to well-studied insect species, it is not clear whether this phenomenon is really phylogenetically limited, or whether, with additional study, its occurrence will be found to be more widespread.

Hybrid dysgenesis has evolutionary implications for the generation of new genetic variability with both negative and positive effects on the fitness of affected individuals. The discovery of the P–M system of hybrid dysgenesis led to the development of a new generation of tools for the genetic engineering of *Drosophila*. For example, the *P* element was developed as a transformation vector that allowed the production of transgenic flies through the manipulation of germline DNA.

See also: **Horizontal Transfer; Transposable Elements**

Hybrid Sterility, Mouse

S H Pilder

Copyright © 2001 Academic Press
doi: 10.1006/rwgn.2001.0655

Hybrid sterility, the phenomenon in which the hybrid offspring of parents from different populations fail to produce functional gametes, is a postzygotic reproductive isolating mechanism (RIM) that impedes gene exchange between diverging populations. This trait, generally thought to arise as an incidental by-product of genetic differentiation, is considered a causal hallmark of incipient speciation. Most instances of hybrid sterility follow Haldane's Rule, a generalization proffered by J.B.S. Haldane. He observed that when parents from divergent populations produce hybrid progeny, the absent, rare, or sterile sex among the offspring is always the heterogametic sex. In keeping with this 'rule,' hybrid sterility in the genus *Mus* (mouse) is male specific.

In *Mus*, hybrid sterility maps to seven genetic loci named *Hybrid Sterility 1–7 (Hst1–7)* are numbered by order of discovery. The *Hst1* phenotype appears to be governed by a single gene located in the third inversion from the centromere, *In(17)3*, of the region of proximal chromosome (Chr) 17 known as the *t* complex. This infertility trait is exhibited by male progeny of crosses between particular laboratory inbred strains of the species *Mus musculus domesticus (domesticus* and some wild mice from the closely related species, *Mus musculus musculus*. These two incompletely isolated species diverged from a common ancestor nearly one million years ago, and presently form a narrow hybrid zone across Europe through which introgression of some genes continues to occur.

Hst1 affected males suffer from spermatogenic arrest at pachytene I of meiosis, a defect which is germ cell autonomous. While the gene responsible for the *Hst1* phenotype has not yet been cloned, it is physically contained on a single 580 kb yeast artificial chromosome (YAC), and several testis-expressed candidate genes mapping to this YAC have been isolated. Because alleles of *Hst1* may interact epistatically with other hybrid sterility genes, the efficacy of these candidate genes to affect the *Hst1* phenotype may be difficult to determine.

The *Hst2* and *Hst3* phenotypes were originally identified on the basis of backcross analyses between *Mus spretus* and *M. domesticus*. These species diverged approximately three million years ago, and do not interact in the wild. However, they will occasionally interbreed in the laboratory if caged together. The hybrid male progeny of these matings are always sterile. The existence of *Hst2*, originally mapped to chromosome 9, has since been questioned, and its assignment to chromosome 9 has been retracted. *Hst3* has been mapped close to the pseudoautosomal region (PAR) of the X chromosome, tightly linked to the *Sxa* locus, thought to control X–Y chromosome association during meiosis. While *Sxa* could be *Hst3*, it is possible that the *Hst3* phenotype is caused by chromosomal rather than genic incompatibility between the PARs of different species. As yet, there is no definitive evidence in support of one possibility versus the other.

A unique phenotype of male-specific hybrid sterility was discovered when chromosome 17 from *Mus spretus* (S) was introgressed into the *domesticus* genetic background. In this case, the affected male offspring carried S and a *domesticus* homolog known as a *t* haplotype (*t*), a peculiar variant of the *t* complex region. This aberrant chromosome 17 polymorphism has been shown to house genetically interacting factors which perturb spermatogenesis in *domesticus*, so that +/*t* heterozygous males express a meiotic drive phenotype in which the *t* homolog is passed to the progeny of affected males at an abnormally high ratio. Interestingly, the same set of interacting genes that causes meiotic drive in the +/*t* heterozygote appears to be the basis of *t*/*t* homozygous male sterility, a phenotype that is absolute. In retrospect, the singular S/*t* hybrid sterility trait appears to result from an interaction of alleles on the S chromosome 17

homolog with mutant alleles on the *t* homolog, rather than wild-type, *domesticus* alleles carried on the *t* homolog.

The gross S/*t* hybrid sterility phenotype derives from the expression and/or epistatic interaction of four discrete *t* complex loci, *Hst4, 5, 6*, and 7. Three of these loci (*Hst4, 5*, and 6) are tightly linked to each other as well as to the strongest *t* haplotype meiotic drive locus within the confines of the largest and most distal of the *t* complex inversions, *In(17)4*. The fourth locus, *Hst7*, maps to the smallest and most proximal *t* complex inversion, *In(17)1*, to which another powerful enhancer of *t*-specific meiotic drive has been localized. While *Hst4, 5, 6*, and 7 also map in close proximity to *Hst1*, significant differences exist in the way in which *Hst1* and these other chromosome 17 genes manifest their effects on spermatogenesis. Unlike *Hst1*, which appears to be a meiotically expressed defect resulting in almost complete spermatogenic arrest, *Hst4, 5, 6*, and 7 are all expressed postmeiotically, affecting spermatid differentiation (axonemal assembly and/or mitochondrial sheath maturation) and sperm function (sperm motility, flagellar curvature, and/or sperm–egg penetration).

The most studied of these four loci is *Hst6*, mapping to a region of less than 1 centimorgan. Three genes map within this locus, two of which influence sperm flagellar curvature, while the third, sandwiched between the other two, plays a role in sperm–oolemma interaction. Additionally, in the *domesticus* background, homozygosity for the *spretus* allele of the proximal-most flagellar curvature gene causes a breakdown in the assembly of the sperm axoneme, the functional backbone of the sperm tail. Moreover, because both *t*/*t* homozygous males and *Hst6⁵*/*t* males express an indistinguishable abnormality in sperm flagellar curvature, it is feasible that *Hst6* is identical to the strong, distal *t* haplotype factor causing male meiotic drive and sterility in the *domesticus* species. Thus, an intensive effort to isolate the *Hst6* genes is currently underway.

Considerable work remains to be done in terms of understanding the process of speciation and the evolution of genetic diversity. In particular, a thorough molecular analysis of hybrid sterility in the mouse would be of benefit in elucidating the biological mechanism underlying Haldane's Rule, the roles of natural selection and genetic drift in generating hybrid sterility phenotypes in *Mus* as well as other genera, and the relationship between meiotic drive and hybrid sterility.

See *also*: Meiosis; Reproductive Isolation; Speciation

Hybrid Vigor

J A Fossella

Copyright © 2001 Academic Press
doi: 10.1006/rwgn.2001.0656

Hybrid vigor is the unusual health, stature, or fitness of offspring produced from the mating of unrelated inbred strains or between closely related species. 'Hybrid vigor' that occurs as a result of mating between unrelated inbred strains, also known as 'heterosis,' should be distinguished from hybrid vigor that can occur in matings between closely related species.

In the case of inbred parental strains, stature, health, and reproductive performance are commonly superior in F_1 hybrids. In these cases, the increase in fitness may arise from the complementation of deleterious recessive alleles fixed during inbreeding. This is also known as 'associative overdominance.' Vigor among F_1 hybrids may also arise from the synergistic interaction of alternate alleles at the same locus. This is referred to as 'true overdominance.' Both associative and true overdominance are the consequence of a phenomenon known as 'inbreeding depression,' the commonly observed decrease in fertility, health, and viability that occurs during the process of inbreeding.

In the case of closely related species, an increase in growth or stature in hybrids is frequently accompanied by defects in fertility such as 'hybrid sterility.' In interspecific crosses, where parents are taken from wild, noninbred populations, the complementation of deleterious recessive alleles is not responsible for hybrid vigor. The causes of hybrid vigor for interspecific hybrids is not well understood. One consistent trend is that reciprocal crosses between closely related species produce hybrid vigor in one direction but not in the reciprocal cross. One example of this common phenomenon are crosses between closely related species of *Peromyscus*, or common North American field mice. Crosses between *P. maniculatus* and *P. polionotus* yield large, vigorous F_1 pups when the father is *P. maniculatus*, but produce small, less fit F_1 offspring when the father is *P. polionotus*. The basis for this phenomenon may be any of a number of factors that underlie parent-of-origin effects such as sex chromosomes, maternal nourishment, maternal care, maternally transmitted episomes, and genomic imprinting.

See also: Heterosis; Hybrid Sterility, Mouse; Inbreeding Depression; Overdominance

Hybrid Zone, Mouse

L Silver

Copyright © 2001 Academic Press
doi: 10.1006/rwgn.2001.0657

Although mouse systematicists have reached a consensus on the structure of the *Mus musculus* group – with the existence of only four well-defined subgroups – there is still a question as to whether each of these subgroups represents a separate species, or whether each is simply a subspecies, or race, within a single all-encompassing house mouse species. The very fact that this question is not simply answered attests to the clash that exists between (1) those who would define two populations as separate species only if they could not produce fully viable and fertile hybrid offspring, whether in a laboratory or natural setting, and (2) those who believe that species should be defined strictly in geographical and population terms, based on the existence of a natural barrier (of any kind) to gene flow between the two populations.

The first question to be asked is whether this is simply a semantical argument between investigators without any bearing on biology. At what point in the divergence of two populations from each other is the magic line crossed when they become distinct species? Obviously, the line must be fuzzy. Perhaps, the house mouse groups are simply in this fuzzy area at this moment in evolutionary time, so why argue about their classification? The answer is that an understanding of the evolution of the *Mus* group in particular, and the entire definition of species in general, is best served by pushing this debate as far as it will go, which is the purpose of what follows.

Each of the four primary house mouse groups occupies a distinct geographical range. Together, these ranges have expanded out to cover nearly the entire land mass on the globe. In theory, it might be possible to solve the species versus subspecies debate by examining the interactions that occur between different house mouse groups whose ranges have bumped up against each other. If all house mice were members of the same species, barriers to interbreeding might not exist, and as such, one might expect boundaries between ranges to be extremely diffuse with broad gradients of mixed genotypes. This would be the prediction of laboratory observations, where members of both sexes from each house mouse group can interbreed readily with individuals from all other groups to produce viable and fertile offspring of both sexes that *appear* to be just as fit in all respects as offspring derived from matings within a group.

However, just because productive interbreeding occurs in the laboratory does not mean that it will occur in the wild where selective processes act in full force. It could be argued that two populations should be defined as separate species if the offspring that result from interbreeding are less fit in the real world than offspring obtained through matings within either group. It is known that subtle effects on fitness can have dramatic effects in nature and yet go totally unrecognized in captivity. If this were the case with hybrids formed between different house mouse groups, the dynamics of interactions between different populations would be quite different from the melting-pot prediction described above. In particular, since interspecific crosses would be 'nonproductive,' genotypes from the two populations would remain distinct. Nevertheless, if the two populations favored different ecological niches, their ranges could actually overlap even as each group (species) maintained its genetic identity – such species are considered to be 'sympatric.'

Species that have just recently become distinct from each other would be more likely to demand the same ecological niches. In this case, ranges would not overlap since all of the niches in each range would already be occupied by the species members that got there first. Instead, the barrier to gene flow would result in the formation of a distinct boundary between the two ranges. Boundary regions of this type are called hybrid zones because along these narrow geographical lines, members of each population can interact and mate to form viable hybrids, even though gene flow across the entire width of the hybrid zone is generally blocked.

The best-characterized house mouse hybrid zone runs through the center of Europe and separates the *domesticus* group to the West from the *musculus* group to the East. If, as the one-species protagonists claim, *musculus* and *domesticus* mice simply arrived in Europe and spread toward the center by different routes – *domesticus* from the southwest and *musculus* from the east – then upon meeting in the middle, the expectation would be that they would readily mix together. This should lead to a hybrid zone which broadens with time until eventually it disappears. In its place initially, one would expect a continuous gradient of the characteristics present in the original two groups.

In contrast to this expectation, the European hybrid zone does not appear to be widening. Rather, it appears to be stably maintained at a width of less than 20 km. Since hybridization between the two groups of mice does occur in this zone, what prevents the spreading of most genes beyond it? The answer seems to be that hybrid animals in this zone are less fit than those with pure genotypes on either side. One

manner in which this reduced fitness is expressed is through the inability of the hybrids to protect themselves against intestinal parasites. It has shown through direct studies of captured animals that hybrid zone mice with mixed genotypes carry a much larger parasitic load, in the form of intestinal worms. This finding has been independently confirmed. Superficially, these 'wormy mice' do not appear to be less healthy than normal; however, one can easily imagine a negative effect on reproductive fitness through a reduced life span and other changes in overall vitality.

Nevertheless, for a subset of genes and gene complexes, the hybrid zone does not act as a barrier to transmission across group lines. In particular, there is evidence for the flow of mitochondrial genes from *domesticus* animals in Germany to *musculus* animals in Scandinavia with the reverse flow observed in Bulgaria and Greece. An even more dramatic example of gene flow can be seen with a variant form of chromosome 17 – called a *t* haplotype – that has passed freely across the complete ranges of all four groups.

In contrast to the stable hybrid zone in Europe, other boundaries between different house mouse ranges are likely to be much more diffuse. The extreme form of this situation is the complete mixing of two house mouse groups – *castaneus* and *musculus* – that has taken place on the Japanese islands. So thorough has this mixing been that the hybrid group obtained was considered to be a separate group unto itself – with the name *Mus molossinus* – until DNA analysis showed otherwise.

In the end, there is no clear solution to the one-species versus multiple-species debate and it comes down to a matter of taste. However, the consensus has been aptly summarized by Bonhomme:

None of the four main units is completely genetically isolated from the other three, none is able to live sympatrically with any other. In those locations where they meet, there is evidence of exchange ranging from differential introgression...to a complete blending. It is therefore necessary to keep all these taxonomical units, whose evolutionary fate is unpredictable, within a species framework

Thus, in line with this consensus, the four house mouse groups are described by their subspecies names *M. m. musculus*, *M. m. domesticus*, *M. m. castaneus*, and *M. m. bactrianus*. *M. musculus* is used as a generic term in general discussions of house mice, where the specific subspecies is unimportant or unknown.

See also: Mus musculus; Mus musculus castaneus; Speciation; Sympatric

Hybridization

T M Picknett and S Brenner

Copyright © 2001 Academic Press
doi: 10.1006/rwgn.2001.1866

Hybridization (of nucleic acids) is a technique in which single-stranded nucleic acids are allowed to interact to form complexes, or hybrids with sufficiently similar complementary sequences. This technique allows the detection of specific sequences or may be used to assess the degree of sequence identity. Hybridization may be carried out in solution or more commonly on a solid-phase support, e.g., nitrocellulose paper. The hybrid of interest is often identified with a radioactively, or alternatively labeled nucleic acid probe or by digestion with an enzyme that specifically attacks single-stranded nucleic acids. Hybridization can be performed with combinations of DNA–DNA (heat-denatured to produce single strands), DNA–RNA, or RNA–RNA molecules. *In situ* hybridization of labeled nucleic acids with prepared cells or tissue sections is used to identify specific transcription or to locate genes on specific chromosomes (e.g., fluorescence *in situ* hybridization, FISH).

See also: **DNA Hybridization; FISH (Fluorescent *in situ* Hybridization); Probe**

Hydatidiform Moles

D K Kalousek

Copyright © 2001 Academic Press
doi: 10.1006/rwgn.2001.0658

There are two, genetically related, types of abnormal placental morphogenesis known as complete and partial hydatidiform moles. Their basic etiology is diagrammatically illustrated in **Figure 1**. Complete hydatidiform mole represents a proliferation of cells containing 46 chromosomes of paternal origin only, while partial hydatidiform mole is usually associated with triploidy (69 chromosomes) where two paternal and one maternal haploid complements are present. The dominance of the paternal sets is a common feature of both moles while the presence of a maternal set in partial mole and its complete absence in complete mole represents the main difference. It has been shown that both parental genomes are required for normal embryogenesis and that the paternal genetic contribution is essential for the development of placental (extraembryonic) tissues, whereas the maternal genetic contribution is more important in the development of the early embryo. This differential expression of genetic messages, depending on their maternal or paternal origin, is known as genomic imprinting (Hall, 1990).

Complete Hydatidiform Mole (CHM)

CHM is typically detected between the 11th and 25th week of pregnancy with an average gestational age of about 16 weeks. Excessive uterine enlargement occurs and may be accompanied by severe vomiting and pregnancy-induced hypertension. Ultrasonography often discloses a classic 'snowstorm' appearance. CHM is characterized by gross generalized villous edema with enlarged placental villi forming 'grape-like,' transparent vesicles, measuring up to 2 cm, absence of amnion, umbilical cord, and embryo/fetus. In all instances, when CHM is associated with an embryo or fetus, this finding represents a twin gestation (Lage *et al.,* 1992). For microscopic features, see **Table 1**.

The majority of complete moles have a 46,XX karyotype, resulting either from dispermy or duplication of haploid sperm in an anuclear ovum. This process is known as diploid androgenesis (Kajii *et al.,* 1984). The undisputable result of dispermy, XY moles, representing only some 4% of complete moles, originate from the fertilization of an anuclear ovum by two spermatozoa. No significant difference has been noted in the gross and microscopic findings between the XY and XX complete moles. Studies of invasive moles and choriocarcinomas have led to the suggestion that heterozygous complete moles (caused by dispermy) may have a more malignant potential than their homozygous counterparts arising through diploid androgenesis.

Partial Hydatidiform Mole (PHM)

PHM is more common than CHM. Morphologically, partial moles differ from that of a complete mole in three principle respects:

1. An embryo/fetus is usually present.
2. Microcystic pattern may be diffuse or focal and is not as prominent as in a complete mole and trophoblastic hyperplasia is both less prominent and strikingly focal.
3. Genetically partial hydatidiform moles are usually triploid with two paternal and one maternal haploid complements (Hall, 1990). They result from fertilization of a normal ovum either by a diploid sperm or by two different haploid sperm. Occasionally, tetraploidy, arising as a result of abnormal fertilization of a haploid ovum by sperm representing three

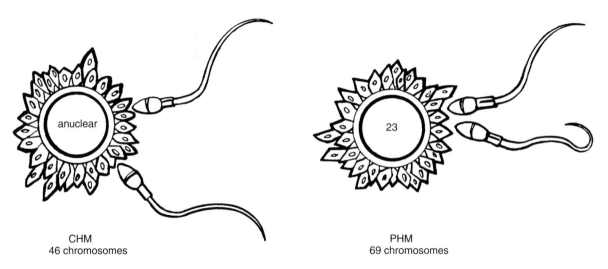

CHM
46 chromosomes

PHM
69 chromosomes

Figure 1 Origin of complete (CHM) and partial (PHM) hydatidiform moles.

Table 1 Differential features of complete and partial moles

Feature	Complete mole	Partial mole
Clinical presentation	Spontaneous abortion	Missed or spontaneous abortion
Gestational age	16–18 weeks	12–20 weeks
Uterine size	Often large for dates	Often small for dates
Serum hCG	++++	+
Cytogenetics	XX (over 90%) or XY (> 10%)	Triploid XXY (58%), XXX (40%), XYY (2%)
	Two paternal sets	Two paternal sets and one maternal set
Persistent gestational trophoblastic disease	10–30%	4–11% Same rate as in nonmolar pregnancies
Embryo/fetus	Absent	Present
Microscopic features		
Villous outline	Round	Scalloped
Hydropic swelling	Marked	Less pronounced
Trophoblastic proliferation	Circumferential	Focal, minimal
Trophoblastic atypia	Often present	Absent
Immunocytochemistry[a]		
βhCG	++++	+
αhCG	+	++++
PLAP	+	++++
PL	++	++++

[a]hCG, human chorionic gonadotropin; PLAP, placental alkaline phosphatase; PL, placental lactogen. (Modified from Silverberg SG and Hurman RJ (1992) *Atlas of Tumor Pathology: Tumors of the Uterine Corpus and Gestational Trophoblastic Disease*. Washington DC: Armed Forces Institute of Pathology.)

paternal chromosome sets, is detected. A few tri-somic conceptuses with partial mole-like morphology have been described.

The gross specimen in PHM shows hydropic villi like those seen in CHM mixed with nonmolar placental tissue. Evidence of an embryo or an amnion is usually present; stromal vasculature and vessels may contain fetal nucleated erythrocytes. Microscopic and differential features between CHM and PHM are

summarized in **Table 1**. However, the only conclusive means for the differential diagnosis is by cytogenetics or more practically flow cytometry (Lage *et al.*, 1992). It is important to distinguish between partial and complete moles, as the malignant transformation rate in partial hydatidiform mole is the same as in any nonmolar pregnancy.

The parental origin of the extra haploid set in triploidy has been shown to have a detectable effect on fetal phenotype in the second and third trimester. Two fetal phenotypes have been delineated: type I fetus with paternal sets dominance, associated with a large cystic placenta, has relatively normal fetal growth and microcephaly; type II fetus with maternal sets dominance, associated with a small noncystic placenta, is markedly growth retarded, and has a disproportionately large head (McFadden and Kalousek, 1991).

References

Hall JG (1990) Genomic imprinting: review and relevance to human disease. *American Journal of Human Genetics* 46(5): 857–873.

Kajii T, Kurashige M, Ohama K and Uchino F (1984) XY and XX complete moles: clinical and morphological correlation. *American Journal of Obstetrics and Gynecology* 150: 57–64.

Lage JM, Mark SD, Roberts DH *et al.* (1992) A flow cytometric study of 137 fresh hydropic placentas: correlation between types of hydatidiform moles and nuclear DNA ploidy. *Obstetrics and Gynecology* 79: 403.

McFadden DE and Kalousek DK (1991) Two different phenotypes of fetuses with chromosomal triploidy: correlation with parental origin of the extra haploid set. *American Journal of Medical Genetics* 38: 535–538.

See also: **Triploidy**

Hyperchromicity

Copyright © 2001 Academic Press
doi: 10.1006/rwgn.2001.2105

Hyperchromicity is the increase in optical density (OD) that occurs when DNA is denatured.

See also: **DNA Denaturation**

Hypervariable Region

Copyright © 2001 Academic Press
doi: 10.1006/rwgn.2001.1868

A hypervariable region is a region of either heavy or light chains of immunoglobulin molecules displaying great sequence diversity. This region specifies the antigen affinity of an antibody.

See also: **Constant Regions; Immunoglobulin Gene Superfamily**

I

Ichthyosis

F M Pope

Copyright © 2001 Academic Press
doi: 10.1006/rwgn.2001.0661

The ichthyoses (literally fish-scale dermatoses) are not only extremely heterogeneous, but are also a spectacular example of the application of modern molecular biology and protein chemistry to the wider biology of the epidermis. The latter has proven to be extraordinarily diverse and very much more complex and subtle than would have thought to be the case. Thus the molecular pathology of the ichthyoses afflicts basic structural components such as keratin intermediate filaments, cell envelope proteins, sulfating enzymes, desmogleins, and desmocollins.

Clinical classification includes autosomal dominant ichthyosis vulgaris, X-linked recessive ichthyoses, a variety of autosomal recessive erythrokeratodermas, and various other localized striate variants, often grouped under the term ichthyosis congenita or the collodion fetus. In other cases, there is overlap with proven keratin disorders such as epidermolysis bullosa simplex, Weber, Cockayne, and Dowling Meara. Some classifications are more complex than others and Mallory includes the ichthyoses under the term disorders of cornification (DOC) in which she also includes Darier disease as DOC 22 (Mallory and Leal-Khouri, 1994). However her groups DOC 1–7 correspond to ichthyosis vulgaris, steroid sulfatase deficiency, bullous epidermolytic hyperkeratosis, collodion baby, congenital erythrodermic, autosomal dominant lamellar ichthyosis, and the harlequin fetus respectively. Types 8 and 9 are ichthyosis hystrix and Netherton syndrome, respectively, whilst types 10 and 11 are Sjogren–Larssen and Refsum disease, respectively. Types 12–24 are either extremely rare or regarded as disorders of cornification but not strictly ichthyoses.

Ichthyosis Vulgaris Simplex

This is the commonest form of ichthyosis, with onset within the first 3 months of life. There is fine scaling of the extensor surfaces, sparing the trunk and flexures. There is criss-crossing of the palms and soles and histologically the granular cell layer is deficient, with epidermal hyperkeratosis. There is frequently associated atopy. Profillagrin deficiency has been identified (Sybert et al., 1985).

X-Linked Ichthyosis

This is also has a very early onset, at birth or within the first 3 months of age. The distribution of scaling differs substantially from ichthyosis vulgaris. Thus it involves the scalp, ears, neck, and flexures and affects the abdomen and the anterior trunk. Unlike ichthyosis vulgaris, the epidermis is hypertrophic, with a normal granular layer.

Both 3β-steroid sulfatase and aryl sulfatase are deficient causing estriol deficiency, delayed labor, and increased fetal loss. Postnatally, affected boys develop ichthyosis. In some families there is hypogonadism. The STS gene has been cloned and in most cases is completely deleted, but if not has 5′ misfunctional deletions (Basler et al., 1992). In other cases there are point mutations. The steroid sulfatase enzyme assay is also very reliable, and a simple staining assay for hexanol dehydrogenase provides rapid confirmation of diagnosis (Lake et al., 1991).

Epidermolytic Hyperkeratosis and Ichthyosis Bullosa of Siemens

In epidermolytic hyperkeratosis (EH) there are generally blisters or erosions at birth, followed later by generalized infantile or childhood scaling (**Figure 1A,B**), closely resembling hyperkeratosis. Histologically, the upper spinous layer is vacuolated with clumping of keratin filaments visible with electron microscopy (Haenke and Anton-Lamprecht, 1982). As such it is closely etiologically related to epidermolysis bullosa simplex (EBS). Like EBS there are mutations of keratins 1 and 10, usually in the highly conserved rod domains (Rothnagel et al., 1992).

Ichthyosis bullosa of Siemens is similar to EH, but has general erythema at birth, followed by erythema and blistering. Later large grey hyperkeratoses develop with lichenification. Siemens skin is more

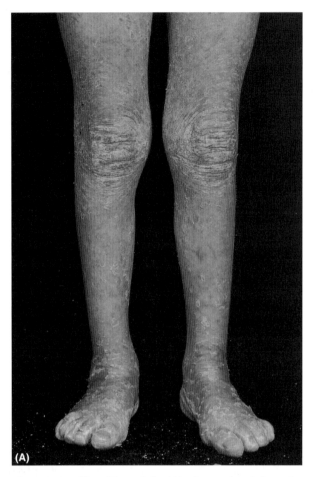
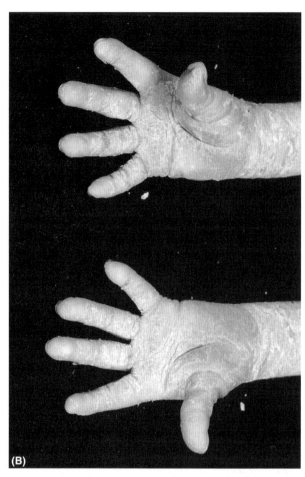

Figure 1 (See Plate 21) (A) Generalized hyperkeratosis with background erythema of the lower limbs; (B) palmoplantar hyperkeratosis extending proximally, typical of epidermolytic hyperkeratosis.

delicate than EH skin. However, like EH, there are keratin mutations, in this case in the rod domain of the keratin 2e gene on chromosome 12.

Lamellar Ichthyosis

Usually affected infants are collodion babies at birth. There are ectodermal dysplastic features, with poor sweating, dystrophic nails, alopecia, and ectropion. However, the very large branny scales are diagnostic and very typical and, furthermore, the teeth are normal. Inheritance is usually autosomal recessive.

Another variant has much more severe erythroderma and more severe collodion changes. Like the former type, there are very large adherent scales. The two differ histologically, with severe orthokeratosis and hyperkeratosis in the milder phenotype. The outcome is variable, some affected infants dying of dehydration, sepsis, or hypoproteinemia, whilst others heal and survive. There are two nonallelic gene loci, one of which at 14q11 is close to the transglutaminase gene (Russell *et al.*, 1995), and mutations have been detected (Parmentier *et al.*, 1995). There is a second

locus at 2q33–35 and a third locus, on chromosome 19 p12–q12. A fourth locus occurs at 3p21 and there is even further heterogeneity. The transglutaminases catalyze ε–γ-glutamyl lysine isopeptide bonds and are very important for keratin cross linking.

Harlequin Fetus

It is unclear whether this is allelic to the lamellar ichthyoses or a separate entity, (**Figure 2**). In any event, there is spectacular hyperkeratosis, with very severe facial edema and distortion. It is unclear whether or not this is allelic to any of the other lamellar ichthyoses.

Other Ichthyoses

These include Netherton syndrome (linear ichthyosis, with pili torti, or trichorrhexis). Refsum disease (ichthyosis vulgaris-like scaling with retinitis pigmentosa, peripheral neuropathy, and cerebellar ataxia), trichothiodystrophy, ichthyosiform erythroderma with

Figure 2 (See Plate 22) Generalized cutaneous features of a harlequin fetus.

sulfate deficient hair and photosensitivity, and numerous others.

References

Basler E, Grompe M and Parenti G et al. (1992) Identification of point mutations in three patients with X-linked ichthyosis. *American Journal of Human Genetics* 50: 483–491.

Haenke E and Anton-Lamprecht I (1982) Ultrastructure of blister formation in epidermolysis bullosa hereditaria. *Journal of Investigative Dermatology* 78: 219–223.

Lake BD, Smith VV and Judge MR et al. (1991) Hexanol, dehydrogenase activity shown by enzyme histochemistry on skin biopsies allows differentiation of Sjogren-Larsson syndrome from other ichthyoses. *Journal of Inherited Metabolic Diseases* 14: 338–340.

Mallory SB and Leal-Khouri S (1994) *An Illustrated Dictionary of Dermatologic Syndromes*, pp. 54–57. New York: Parthenon Publishing Group.

Parmentier L, Blanchet-Bardon C, Nguyen S et al. (1995) Autosomal recessive lamellar ichthyosis: evidence of a new mutation in transglutaminase I and evidence for genetic heterogeneity. *Human Molecular Genetics* 5: 555–559.

Rothnagel JA, Dominey AM, Dempsey LD et al. (1992) Mutations in the rod domains of keratins I and 10 in epidermolytic hyperkeratosis. *Science* 257: 1128–1130.

Russell LJ, Di Giovanna LJJ, Rogers GR et al. (1995) Mutations for the gene for transglutaminase I in autosomal recessive lamellar ichthyosis. *Nature Genetics* 9: 279–283.

Sybert VP, Dale BA and Holbrook KA (1985) Ichthyosis vulgaris: identification of a defect in synthesis of filaggrin correlated with an absence of keratohyaline granules. *Journal of Investigative Dermatology* 84: 191–194.

See also: **Clinical Genetics**

Identity by Descent

D L Hartl

Copyright © 2001 Academic Press
doi: 10.1006/rwgn.2001.0662

One of the most influential concepts in the theory of population genetics is 'identity by descent.' Two alleles of a gene are said to be identical by descent if, within the span of some specified number of generations, they originated by replication of a single allele in a common ancestor. In studies of pedigreed populations, the specified span of generations is usually short and the beginning often coincides with the most remote ancestors in the pedigree. For population studies, the span of generations is typically the time since the founding of any subpopulation in question. A number of important concepts in population genetics are based on the probability that two alleles are identical by descent. For example, the inbreeding coefficient equals the probability that the two alleles at a locus in an individual are identical by descent, and the coefficient of kinship (coefficient of consanguinity) equals the probability that a pair of homologous alleles, drawn at random, one from each of two individuals, are identical by descent. Conceived independently by Charles Cotterman (1940), and Gustave Malécot (1944), use of the concept of identity by descent and calculation of its probability soon reproduced all of the key results obtained previously by Sewall Wright using his method of path coefficients, which is related to partial regression coefficients. Because of its intuitive simplicity and ease of calculation, the concept of identity by descent soon replaced path coefficients in most applications in population genetics, especially in the theories of inbreeding and hierarchical population structure.

References

Cotterman CW (1940) *A Calculus for Statistico-Genetics.* PhD thesis, Ohio State University, Columbus, OH

Malécot GM (1944) Sur un problème de probabilités en chaîne que pose la génetique. *Comptes Rendus de l'Académie des Sciences* 219: 379–381.

See also: Population Genetics; Wright, Sewall

Idiogram

M A Ferguson-Smith

Copyright © 2001 Academic Press
doi: 10.1006/rwgn.2001.0663

An idiogram (or ideogram) is the diagrammatic representation of the karyotype of a cell, individual, or species. It is based on measurements of chromosome length and centromere position, and on the characteristic banding appearance revealed by staining techniques such as Giemsa banding. These bands provide landmarks for the identification of individual chromosomes and regions of chromosomes and act as an aid in the analysis of chromosome rearrangements.

See also: Chromosome Aberrations; Giemsa Banding, Mouse Chromosomes; Karyotype

Igf2 Locus

K N Gracy and B J Smith

Copyright © 2001 Academic Press
doi: 10.1006/rwgn.2001.0664

Igf2 is an imprinted gene in which one of the two parental alleles is inactivated, or silenced. The gene encodes for a potent fetal growth factor and is closely linked to the reciprocally imprinted *H19* gene. On the paternal chromosome, *Igf2* is transcriptionally active and *H19* is not transcribed, while on the maternal chromosome, *Igf2* is not transcribed and *H19* is active. This phenomenon and other expression regulation is mediated by control elements such as the differentially methylated domain (DMD), the mesodermal tissue silencer element DMR1, and a muscle-specific silencer element that is as yet unnamed. Loss of imprinting of *Igf2* is associated with Beckwith–Wiedemann syndrome, which is characterized by fetal overgrowth and childhood tumors. *Igf2* is located on mouse distal chromosome 7 and on the Beckwith–Wiedemann region on human chromosome 11p15.5.

Further Reading

Peters J (2000) Imprinting: silently crossing the boundary. *Genome Biology* 1: Reviews 1028.1–1028.4.

See also: Beckwith–Wiedemann Syndrome; Imprinting, Genomic

Igf2r Locus

L Silver

Copyright © 2001 Academic Press
doi: 10.1006/rwgn.2001.0665

The *Igf2r* locus encodes the insulin growth factor 2 receptor (IGF2R) polypeptide. This polypeptide sequesters – and thus modulates – the level of active insulin-like growth factor in the developing mammalian fetus. This modulation, in turn, adjusts the growth of the fetus. *Igf2r* is one of a small subset of mammalian genes that are subjected to a process known as genomic imprinting, where a gene is active or inactive depending on its parental origin. In the case of *Igf2r*, the maternal copy of the gene is active, while the paternal copy is suppressed. Genomic imprinting at *Igf2r* appears to be the result of an ancient battle between male and female parents attempting to maximize the survival and success of their offspring.

See also: Imprinting, Genomic

Illegitimate Recombination

D Carroll

Copyright © 2001 Academic Press
doi: 10.1006/rwgn.2001.0666

The distinction between legitimate and illegitimate recombination is based on the extent of homology between the DNA sequences undergoing recombination. Legitimate processes involve extensive homology, like that manifested by paired chromosomes in the act of meiotic recombination. Illegitimate recombination relies on very short homologies, and sometimes none at all. A special type of illegitimate process is site-specific recombination, in which particular sequences are recognized by proteins that catalyze breakage and rejoining events at those sites.

The definition of illegitimate recombination is imprecise, since there is no strict threshold in the amount of homology that defines legitimate events. When the junctions resulting from illegitimate recombination are examined – for example, in experiments with cultured mammalian cells – they often show matches of a few base pairs between the parental sequences. These microhomologies are not absolutely required, since some joints show no such matches. Typically the number of matched base pairs at the junction is 1–5, but occasionally longer matches are seen.

Illegitimate recombination is observed in most organisms. The origin of spontaneous illegitimate events cannot be traced, but such joints are clearly formed in response to double-strand breaks in chromosomal DNA. In cultured cells from multicellular eukaryotes, illegitimate end joining is the most common fate of linear DNAs introduced artificially into the cells. The yeast *Saccharomyces cerevisiae* and some other fungi have very efficient mechanisms of homologous recombination that predominate in the processing of chromosomal breaks and of DNA introduced during transformation; but illegitimate events can be detected if no homology is present, or if the capability of performing homologous recombination has been disabled by mutation.

The mechanism by which illegitimate recombination occurs in cells is not known. Two simple and attractive hypotheses describe mechanisms that very likely both contribute to the observed junctions. The first hypothesis is that DNA ends are simply joined by a DNA ligase (**Figure 1**). When there are complementary nucleotides appropriately situated in single-stranded regions, they stabilize an association between the ends and help set the register for the ligase. Joints of this type have been produced in crude extracts from eukaryotic cells and with some purified DNA ligases. There are also ligases – e.g., that encoded by bacteriophage T4 – that can join blunt DNA ends that have no single-stranded overlaps.

The second hypothesis is that rather long single-stranded tails are formed – presumably by the action of exonucleases – at broken ends (**Figure 2**). If these tails have free 3′ ends, microhomologies can support transient associations that can be stabilized by DNA synthesis through use of the transient joint as a primer-template complex by a cellular DNA polymerase.

While the details of the illegitimate recombination mechanism remain obscure, some information is available on proteins that participate in the process. In mammalian cells, a protein complex called Ku and its associated DNA-dependent protein kinase (DNA-PK) are required for efficient end joining. Yeast share the requirement for Ku and for a DNA ligase that is

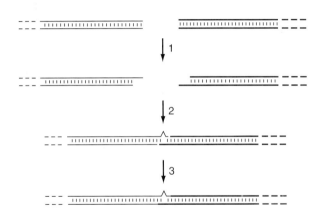

Figure 1 Illegitimate recombination by microhomology-directed ligation. The broken ends of two DNA molecules are indicated in the top diagram. Horizontal lines show the phosphodiester backbones and short vertical lines the Watson–Crick base pairs. In step 1, each end is partly degraded by a strand-specific exonuclease. In step 2, the single-stranded tails of the two DNAs come together, directed by the formation of two base pairs surrounding a mismatch. In step 3, the strands of the two DNAs are joined by a cellular DNA ligase.

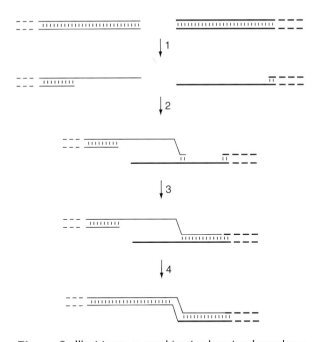

Figure 2 Illegitimate recombination by microhomology-directed DNA synthesis. The starting point is the same as in **Figure 1**. In step 1, each DNA is degraded more extensively by a 5′→3′ exonuclease. In step 2, the 3′ single-stranded tails come together through the formation of two base pairs. In step 3, the 3′ end of the thinner strand is extended by DNA polymerase, forming an extended base-paired region. In step 4, the thicker single strand is degraded, its 3′ end is used to prime synthesis in the remaining gap, and nicks at both ends of the new joint are sealed by DNA ligase.

different from the one utilized in DNA replication – DNA ligase IV and its associated XRCC4 protein. These same factors participate in site-specific recombination during immunoglobulin gene rearrangements (see below).

Site-specific recombination is seen with some viruses and transposable elements and in a few chromosomal situations that are probably derived from transposons. The hallmark of these processes is the involvement of at least one element-encoded protein that recognizes the DNA sequences that will be joined, helps hold them together, and catalyzes the breakage and rejoining reactions before releasing the products. In each case, the recombination event is a directed part of the life-style of the element.

An example of site-specific recombination is the integration of the bacteriophage lambda genome into the host *Escherichia coli* chromosome. The lambda-encoded Int protein recognizes the specific attachment sites of both DNAs and, in collaboration with host proteins, brings them together into a pre-integration complex. Recombination proceeds by a topoisomerase-like mechanism, in which hydroxyl groups on active site tyrosines in the Int protein attack specific phosphodiester bonds in the target DNA, producing covalent joints between Int and DNA as intermediates in integration. A subsequent transesterification reaction generates the new DNA joints and releases the protein.

An example of site-specific recombination in mammalian chromosomes is the generation of functional genes for antibodies, or immunoglobulins. Specific DNA sequences, called recombination signal sequences, are recognized by the RAG1 and RAG2 proteins that are expressed specifically in lymphoid cells. Recombination at these sites brings coding sequences for variable regions into proximity with the constant region coding sequences, allowing the production of a functional messenger RNA for the complete protein. The biochemical mechanism of this recombination has striking similarities to the mechanism of transposition by mobile DNA elements. Thus, it is hypothesized that immunoglobulin gene rearrangement is derived from an ancient transposable element.

Further Reading

Allgood ND and Silhavy TJ (1988) Illegitimate recombination in bacteria. In: Kucherlapati R and Smith GR (eds) *Genetic Recombination*, pp. 309–330. Washington, DC: American Society for Microbiology Press.

Berg DE and Howe MM (eds) (1989) *Mobile DNA*. Washington, DC: American Society for Microbiology Press.

Meuth M (1989) Illegitimate recombination in mammalian cells. In: Berg DE and Howe MM (eds) *Mobile DNA*, pp. 833–860. Washington, DC: American Society for Microbiology Press.

Roth D and Wilson J (1988) Illegitimate recombination in mammalian cells. In: Kucherlapati R and Smith GR (eds) *Genetic Recombination*, pp. 621–653. Washington, DC: American Society for Microbiology Press.

See also: **DNA Ligases; Genetic Recombination; Site-Specific Recombination**

Immunity

Copyright © 2001 Academic Press
doi: 10.1006/rwgn.2001.1870

Immunity in phages, plasmids, or transposons refers to the ability of a prophage, plasmid, or transposon to prevent another molecule of the same type from infecting the same cell (or for transposons, transposing to the same DNA molecule). Phage immunity (lysogenic immunity) is due to the synthesis of phage repressor by the phage genome. The ability of plasmids to confer immunity usually results from interference with the ability to replicate: transposon immunity results from a variety of mechanisms.

See also: **Plasmids; Prophage; Transposable Elements**

Immunoglobulin Gene Superfamily

L Silver

Copyright © 2001 Academic Press
doi: 10.1006/rwgn.2001.0669

The immunoglobulin gene superfamily is a very large family of genes present in all vertebrates. This gene superfamily consists of a series of gene families that each play a distinct role in the immune response. The superfamily is named after its most well-known and well-characterized gene family, the immunoglobulin gene family, which codes for polypeptides that form circulating antibodies (or immunoglobulins) in the bloodstream. Antibodies are one component of a two-pronged immune response that an animal mounts against invading bacteria and viruses. The antibody component of the immune response has been referred to as humoral immunity. The other component of the immune response is cellular immunity, carried out by cells called T cells and B cells.

Each gene member of the immunoglobulin gene super-superfamily contains immunoglobulin-like (Ig) domains and functions as a cell surface or soluble receptor involved in immune function or other aspects of cell–cell interaction. This superfamily includes the immunoglobulin gene families themselves, the major histocompatibility genes (called *H2* in mice), the T cell receptor genes, and many more. There are dispersed genes and gene families, small clusters, large clusters, and clusters within clusters, tandem and interspersed. Dispersion has occurred with the transposition of single genes that later formed clusters and with the dispersion of whole clusters *en masse*. Furthermore, the original Ig domain can occur as a single unit in some genes, but it has also been duplicated intragenically to produce gene products that contain two, three, or four domains linked together in a single polypeptide. The Ig superfamily, which contains hundreds (perhaps thousands) of genes, illustrates the manner in which the initial emergence of a versatile genetic element can be exploited by the forces of genomic evolution with a consequential enormous growth in genomic and organismal complexity.

See also: **Evolution of Gene Families**

Imprinting, Genomic

A C Ferguson-Smith

Copyright © 2001 Academic Press
doi: 10.1006/rwgn.2001.0672

A fertilized egg inherits a haploid set of chromosomes from both the egg and the sperm; however, in mammals these maternal and paternal gametes do not contribute equal genetic functions to the developing diploid embryo. This functional difference between the two sets of parental chromosomes is due to a process called genomic imprinting. Genomic imprinting is a mechanism that differentially 'marks' the maternally and paternally inherited chromosome homologs and results in particular genes being expressed or repressed in response to this parent-specific modification. Because the imprint affects gene activity, some imprinted genes are expressed only from the maternally inherited chromosome and others are expressed only from the paternally inherited chromosome (**Figure 1**). It is not known why such a process evolved and the precise mechanisms involved in the regulation of imprinted genes is not yet fully understood. However, it follows that the dosage of an imprinted gene can be doubled or lost completely if there is a uniparental duplication or deficiency involving the gene or chromosomal region

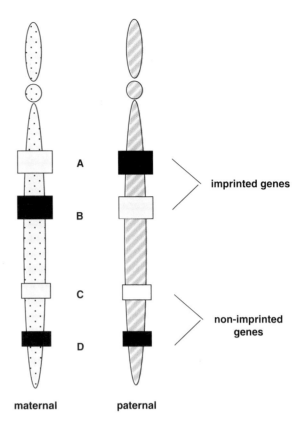

Figure 1 Schematic representation of a homologous chromosome pair with both imprinted (A, B) and nonimprinted genes (C, D). White boxes represent active alleles and black boxes inactive alleles. Imprinted genes show activity from one parental allele and repression at the other. The two neighboring imprinted genes, A and B, are said to be reciprocally imprinted: A is active on the maternal homolog and B is active on the paternal homolog. The nonimprinted genes, C and D, do not show differences in expression on the two parental alleles and are representative of the majority of genes in the genome.

in which it resides. Expression of an imprinted gene can also be affected if there is mutation in the chromosomal modifications responsible for its regulation. These effects on the dosage of an imprinted gene can have profound effects on mammalian embryonic development and in humans can result in recognized imprinting disorders.

Developmental Consequences of Imprinting

Genomic imprinting ensures the requirement for both a mother and a father to produce normal mammalian offspring as shown by the failure of bimaternal and bipaternal conceptuses to complete embryogenesis. Parthenogenesis, the development of an egg without fertilization by a sperm, is successful in some lower organisms. However, it is clear that parthenogenetic eutherian mammals cannot survive to term. In the mouse parthenogenesis, to create a diploid maternal egg, can be induced experimentally. Parthenogenetic embryos will survive to midgestation and appear morphologically relatively normal though growth retarded. The extraembryonic tissues, however, are underdeveloped and do not proliferate properly. Gynogenetic embryos, also containing a diploid maternal contribution though from two different mothers, exhibit the same properties as parthenogen-ones. Diploid paternal androgenetic conceptuses are made by replacing the female pronucleus in a newly fertilized egg with a second male pronucleus from another egg. These embryos fare worse than parthenogenones, with very poor development of the embryo which rarely develops beyond the 4-somite stage. In contrast to the parthenogenones, the extraembryonic tissues are well developed though not completely normal. In this respect the androgenone is reminiscent of the complete hydatidiform mole in humans. These conceptuses contain a genome derived solely from paternal chromosomes. The mole resembles a mass of cytotrophoblast without any embryonic components. Thus it appears that the parental genomes have reciprocal functions in embryogenesis, with the presence of a paternal genome generally being important for the development of the extraembryonic lineages and the maternal genome being required for the development of the embryonal components at these early stages. This reflects the properties of imprinted genes whose activity is either doubled or lost in the uniparental conceptuses.

Genetic Studies of Imprinting in the Mouse

It has been shown that the requirement for both parental genomes is limited to a subset of mammalian chromosomes. This has become evident using mouse translocation breeding experiments which result in embryos carrying uniparental duplications and corresponding deficiencies of whole chromosomes (uniparental disomy, UPD) or particular chromosomal regions. These duplications represent a subset of the whole genome duplications seen in the parthenogenetic and androgenetic embryos. Normal development of a UPD conceptus suggests that the duplicated region is not imprinted. These studies have shown that regions on mouse chromosomes 2, 6, 7, 11, 12, 17, and 18 are imprinted and hence the biparental requirement applies to a subset of the genome. On perturbation of the parental origin of these chromosomes, quite severe phenotypes are observed,

including lethality, growth defects, and behavioral anomalies. This indicates that developmentally important imprinted genes reside within these regions; however, it does not rule out the presence of imprinted genes elsewhere, which cause more subtle effects when their dosage is perturbed by uniparental duplication. Around 90% of the imprinted genes identified to date map to the regions identified in the genetic studies.

Imprinting in Disease

It became evident that imprinting had clinical implications through the study of patients with disorders that exhibit parental-origin effects in their patterns of inheritance. There are now several syndromes which are recognized as imprinting disorders. Imprinting mutations have also been implicated in the genesis of some tumours, notably Wilms's tumor and familial glomus tumors. These imprinting disorders show a normal autosomal dominant pattern of inheritance but from a parent of one sex – offspring of an affected individual of the opposite sex are completely unaffected. The disorder remanifests itself in a subsequent generation after inheritance through a phenotypically normal carrier individual of the appropriate sex. Males and females are equally affected which clearly distinguishes an imprinting pedigree from that of a sex-linked disorder. For example, benign familial glomus tumors show autosomal dominant inheritance but are only manifest in individuals inheriting the mutant gene from their fathers. Inheritance of the mutation from the mother results in normal offspring; however, her sons (if carriers) will have affected offspring at a frequency of 50%. Other imprinting disorders have been associated with a significant level of UPD. To date, all human chromosomes involved in these syndromes show evolutionary conservation with those in the mouse identified as imprinted chromosomes (see above). In addition to hereditary glomus tumors, imprinted disorders identified to date include Beckwith–Wiedemann syndrome and Silver–Russell syndrome which are growth defects, two neurological disorders – Angelmann syndrome and Prader–Willi syndrome, transient neonatal diabetes, and maternal UPD14 syndrome. The latter is a rare disorder associated with growth defects and premature puberty.

Mechanism of Imprinting

The mechanism causing parental-origin specific gene expression must allow the transcriptional machinery of the cell to distinguish between two chromosome homologs and differentially act on one or the other. The imprint is believed to be initiated late in the development of the egg and sperm and then acted upon in the zygote and developing conceptus to affect developmental gene activity. It is therefore likely that the imprint is a modification to the DNA and/or chromatin which must have the following properties:

1. It must be able to affect the transcription of the gene.
2. It must be heritable in somatic cells over many cell divisions and not lost during chromosome replication. This renders the imprint stable and allows it to have parental-origin specific memory. This step is known as maintenance.
3. Importantly, the imprints must be erased in the male and female germlines during gametogenesis to allow new imprints to be set down which are specific to the parental origin of the newly formed gametes.

There are only a few recognized mammalian genome modifications that might fulfil the above criteria. By far the best studied is DNA methylation. DNA methylation of CpG dinucleotides is known to affect gene activity. Indeed, methylation of CpG-rich regulatory portions of genes, for example on the inactive X chromosome in females, has long been associated with gene inactivity. More recently, it has been shown that imprinted genes contain regions that are differentially methylated on the two parental chromosomes; however, sometimes methylation is associated with the inactive allele and sometimes with the active allele. In the absence of the DNA methyltransferase gene, which encodes the methylating enzyme, the methylation imprint is lost from somatic cells and imprinted gene activity is perturbed. Thus, methylation is involved, at least, in the maintenance of imprinting. Whether methylation is the germline imprinting initiator remains to be proven; however, several differences in methylation have been found in the DNA of eggs and sperm in imprinted regions, which suggest that CpG methylation may have a role to play in the earliest imprinting events.

Other modifications may also be involved in the imprinting process. It is apparent that many imprinted genes show differences in their chromatin structure between the active and inactive alleles. However, in the case of imprinting, the relationship, if any, between a region's chromatin conformation and its methylation status is not understood. It is now well documented that modifications to chromatin-associated proteins, notably acetylation of core histones, have key roles to play in the regulation of gene expression and it is possible that these may be involved in the imprinting mechanism. Nonetheless, it seems that the imprints are acting both at short and long range, perhaps to provide a particular chromatin context within which individual genes can be further modified. It is likely that this context must differ between the two parental homologs.

Function and Evolution of Imprinting

Imprinting renders an autosomal gene functionally hemizygous and the potential benefit to the organism of this costly process remains unclear. Many of the imprinted genes identified to date are involved in the regulation of fetal and embryonic growth and are clustered in the genome. To date the most widely discussed theory to explain the evolution of imprinting is the 'parent–offspring conflict' theory. In promiscuous animals, the father seeks to promote the growth of his offspring at the expense of the resources of the mother who is likely to procreate with other males. The mother, in contrast must conserve her resources in order that she can maximize the chances of future pregnancies and many litters. The model predicts that in this parental 'tug-of-war,' paternally expressed genes will promote growth and maternally expressed genes will repress growth. This model is consistent with many of the growth defects observed in imprinted disorders in mouse and man and also with the function of many of the imprinted genes identified to date. Some disorders and imprinted genes do not fit this model and other theories have been proposed. These include the idea that imprinting arose to prevent parthenogenesis in mammals or ovarian teratomas in females; while this fits with the silencing of maternal genes it cannot explain the silencing of paternal genes. Others have suggested that imprinting is an extension of the bacterial host defense mechanism that guards against the invasion of foreign DNA via DNA methylation. However, while some imprinted genes are intronless retrotransposons of X linked genes, most are not and furthermore have important functions in mammalian development. It is likely that, as more imprinted genes are discovered and analyzed, these and other theories will be further scrutinized and the biological significance of this remarkable phenomenon will be better understood.

Further Reading

Bartolomei MS and Tilghman SM (1997) Genomic imprinting in mammals. *Annual Review of Genetics* 31: 493–525.

See also: Androgenone; Chromatin; CpG Islands; DNA Modification; Epigenetics; Hydatidiform Moles; *Igf2* Locus; *Igf2r* Locus; Parthenogenesis, Mammalian; Uniparental Inheritance; X-Chromosome Inactivation

In situ Hybridization

M A Ferguson-Smith

Copyright © 2001 Academic Press
doi: 10.1006/rwgn.2001.0697

In situ hybridization (ISH) is used to map and order genes and other DNA and RNA sequences to their location on chromosomes and within nuclei. The technique is based on the principle that double-stranded DNA denatures on heating to single-stranded DNA. On cooling, the single-stranded DNA reanneals with its complementary sequence into double-stranded DNA. If an appropriately labeled fragment of a DNA sequence (a DNA probe) is denatured and added to denatured nuclei or chromosomes on a routine, air-dried interphase preparation during the process of reannealing, some of the labeled DNA will hybridize to its complementary sequence in the chromosomal DNA. Detection of the labeled DNA probe under the microscope will identify the site of hybridization and thus the region of chromosomal DNA complementary to the DNA sequence in the labeled probe. If, for example, the DNA probe represents a sequence of more than 1 kb from a cloned gene, ISH has the capability of assigning that gene to its chromosomal location.

When ISH was introduced in 1970, DNA probes made from highly repetitive DNA fragments (satellite DNA) were labeled with tritium (^{3}H) or radioactive ^{125}I and detected by autoradiography using photographic emulsion applied directly to the microscope slide. The technique had poor resolution and was difficult to use with single-copy probes, even when they were cloned in phage or plasmid vectors.

Radioisotopic methods for ISH were replaced in the 1980s by nonisotopic alternatives such as biotin and digoxigenin, which are coupled to nucleotides and incorporated into the DNA probes by techniques such as nick translation using DNA polymerase. These probes are detected by fluorescence microscopy using fluorochromes coupled to avidin, streptavidin, or antibiotin antibodies in the case of probes labeled with biotin. The same fluorochromes coupled to anti-digoxigenin antibodies are used for probes labeled with digoxigenin. The fluorochromes most commonly used are fluorescein isothiocyanate (FITC), tetramethyl rhodamine isothiocyanate (TRITC), and aminomethyl coamarin acetic acid (AMCA). More recently the indirect systems using avidin and antibodies have been replaced by direct labeling methods in which fluorochromes such as FITC, Cy3, and Cy5 are coupled directly to the nucleotides (e.g., FITC-11-dUTP) that are used in labeling the DNA probes.

When exposed to a UV light source, each fluorochrome is excited by a different wavelength and each emits a distinctive fluorescence. In order to distinguish the various emissions produced by each fluorochrome, a series of exitation and emission filters are used that are specific for each fluorochrome. Combinations of filters allow the observation simultaneously of several fluorochromes excited by different wavelengths, and this, together with the development of digital fluorescence microscopy and image analysis, has led to the introduction of multicolor fluorescence ISH (M-FISH). M-FISH systems depend on the use of combinations of up to five different fluorochromes to label individual DNA probes so that a large number of probes can be distinguished in each preparation. This requires a sensitive, monochromatic, cooled charged-coupled device (CCD) camera and computerized image analysis. A gray-scale image of the fluorescence of each fluorochrome is acquired sequentially and merged to provide a false color on the computer screen, which is chosen on the basis of the relative intensities of the constituent fluorochromes.

DNA Probes Used in FISH

Total genomic probes are prepared by labeling DNA extracted from blood samples, cell cultures, or solid tissues. Chromosomes hybridized with these probes show an evenly distributed signal along their length, referred to as 'chromosome painting.' The main application of total genomic probes has been in the identification of human chromosome material in human-to-rodent interspecific somatic cell hybrids, including radiation-reduced cell hybrids.

'Chromosome-specific paint probes' are genomic probes that were prepared initially from chromosome-specific genomic libraries cloned in plasmid vectors. They can also be made from single-chromosome interspecific somatic cell hybrids. Most are now prepared from flow-sorted chromosomes and these tend to have the highest resolution. Each chromosome-specific paint is made from sorting 300–500 chromosomes and amplifying chromosomal DNA fragments by the random-primed polymerase chain reaction (DOP-PCR). Flow-sorted chromosomes can be obtained in high purity, and the PCR procedure amplifies over 90% of the chromosomal DNA. Chromosome-specific hybridization, free of background signal, is assured by prehybridization of the probe with itself before application to the test material. This ensures that highly repetitive signals are largely eliminated, and unique, conserved DNA sequences are available to paint all but the heterochromatic regions of the chromosomes. Chromosome-specific paint probes have wide application in the analysis of complex chromosome aberrations and are commercially available from several distributors either as single chromosome-specific paint probes or as complete probe sets in which each chromosome is labeled differently for M-FISH analysis. This allows the analysis of a complete cell in one hybridization.

The main disadvantage of chromosome-specific paint probes is that they are unable to identify intrachromosomal aberrations such as inversions, duplications, and insertions, and that areas containing repetitive sequences, especially telomeres and centromeres, are not painted. In these cases, region-specific paint probes prepared from amplified chromosome segments obtained by chromosome microdissection have found some application.

Chromosome-specific centromeric probes are prepared from cloned alphoid repeat sequences which are located adjacent to centromeres. Almost all human chromosomes have chromosome-specific sequences of this type. The exceptions are chromosomes 13, 14, 21, and 22. Chromosomes 13 and 21 have the same centromeric sequences, different from 14 and 22, which also share the same sequences. These probes are used to determine chromosome copy number in interphase nuclei. More than 80% of normal diploid nuclei will show two distinct signals when hybridized with a chromosome-specific centromeric probe. Centromeric probes are therefore used for aneuploidy detection in uncultured amniotic fluid cells, for preimplantation diagnosis in cells from the blastocyst, for the detection of residual disease in the management of certain hematological malignancies, and for the analysis of nondisjunctional abnormalities in sperm. Chromosome-specific sequences cloned in yeast artificial chromosome (YAC), bacterial artificial chromosome (BAC), or cosmid vectors replace the lack of specific centromeric probes for aneuploidy detection involving chromosomes 13, 14, 21, and 22.

The project to map and sequence the human genome has, as one of its by-products, a complete series of overlapping DNA clones from which reference probes can be produced which can be used as FISH markers to delineate any point on any chromosome. Cloned in a variety of cosmid and other vectors, they can be used to characterize specific breakpoints and to detect specific microdeletions (such as the DiGeorge syndrome on chromosome 22). These single-copy DNA sequence probes have wide application in clinical cytogenetics and in the mapping and cloning of disease genes.

Telomere-specific probes are now available for the ends of all human chromosomes. They have proved to be particularly valuable in the detection of reciprocal translocations which are beyond the resolution of conventional diagnostic cytogenetics.

Other Applications of FISH

Due to the condensation of the DNA fiber within metaphase chromosomes, the fluorescent signals from two cosmid clones can be resolved only if they are more than 2–3 Mb apart. At interphase the chromosomes are 10 times more extended than at metaphase, and so two cosmids more than 50 kb can usually be distinguished from one another. The order of several closely linked cosmids may be determined at interphase provided they are more than 50 kb and less than 1 Mb apart. The latter restriction is due to the tendency of a chromosome to coil back on itself.

The elemental DNA fiber may be further decondensed by techniques which release it from its associated histones and other proteins (see Chromosome Scaffold). Such preparations of DNA fibers on microscope slides can be used for hybridization with standard DNA probes. The technique permits the ordering of very closely linked single-copy DNA sequences and the analysis of the intrachromosomal relationships of various repetitive elements. It has also been used to identify small duplications and deletions within known genes (such as the Duchenne muscular dystrophy gene) and distances as short as 1 kb have been resolved.

While the genetic basis of cancers are well established and complex chromosome rearrangements are a common feature of malignancy, cytogenetic analysis of cancer cells has proved technically difficult. In part this is due to the difficulty in finding suitable metaphases in tumor material, and in part due to the complexity of the chromosomal rearrangements observed when suitable metaphases are found. One of the aims of cancer cytogenetics is to map regions of the chromosome complement which have been deleted and regions which have duplicated. Consistent patterns of abnormality may lead to the identification of key oncogenes or tumor suppressor genes important in the clonal evolution of the cancer. M-FISH techniques are now contributing to the detailed cytogenetic analysis of tumors. Comparative genome hybridization (CGH) has been a particularly informative method, because it has permitted the mapping of DNA amplifications of over 5–10 Mb and the deletion of chromosome segments over 10–20 Mb. In brief, the method involves the mixing of equal amounts of total genomic DNA from the tumor tissue labeled with FITC (green), with TRITC (red)-labeled total genomic reference DNA, and the hybridization of the mixture to normal metaphases. The relative amounts of tumor and normal DNA that anneal to a particular chromosome region depend on the number of copies of DNA complementary to that region in the test sample. If the tumor sample contains relatively more of a particular DNA sequence than the reference sample, this will be revealed by an increased green-to-red fluorescence ratio in the complementary region; similarly, chromosomal deletion in the tumor sample is revealed by a decreased green-to-red ratio. The method requires digital fluorescence microscopy in which the relative amounts of green and red fluorescence are measured along the length of the chromosome.

Mention should be made of the use of chromosome-specific paint probes in the study of comparative genomic and karyotype evolution. The conservation of genes between mammalian species is widely appreciated, and even widely divergent species such as the human, the fruit-fly, the nematode worm *Caenorhabditis elegans*, and yeast share a number of genes. Comparative mapping studies reveal that the X chromosome carries the same transcribed genes in all mammals, and also that large blocks of linked autosomal genes show similar conservation between species. In closely related species, these genetic linkage groups tend to be more extensive than in more distantly related species, sometimes representing whole chromosomes that are shared between the species. Cross-species chromosome painting has been used to demonstrate the extent of chromosome homology between species. Chromosome-specific paints from one species are hybridized to the chromosomes of a second species. The precise origin of a particular block of homology revealed by a paint probe from the first species can be determined by hybridizing chromosome-specific paint from the second species back to the chromosomes of the first species. In this way simple comparative maps can be constructed between species. If one of the species is a well-mapped species, such as human or mouse, a preliminary genetic map can be constructed for the unmapped species. This homology map can assist in more detailed mapping using genetic linkage and radiation hybrid techniques. Phylogenetic relationships between species can be studied, based on chromosome rearrangements revealed by chromosome painting and shared by species diverged from a common ancestor.

See also: **Chromosome; Chromosome Painting; Chromosome Scaffold; FISH (Fluorescent *in situ* Hybridization); Gene Mapping; Genome Organization**

In vitro Evolution

R L Dorit

Copyright © 2001 Academic Press
doi: 10.1006/rwgn.2001.0712

For evolution to occur, three conditions must be met. First, variation must be present among the evolving entities. Second, that variation must, to some extent, translate into differential survival and reproduction (fitness) among the evolving entities. Third, the variation responsible for differential fitness must be heritable: transmitted from parents to offspring. If these three conditions are met, the stage is set for a population to evolve.

In his seminal work, *On the Origin of Species*, Charles Darwin put forth this revolutionary understanding of the mechanisms of evolution. This Darwinian insight provides a materialistic account of the evolution and diversity of organisms on earth, an explanation that has, since its inception, endured. Evolution is an ongoing process and its consequences are constantly on display. The emergence of antibiotic resistance in bacteria, of insecticide resistance in agricultural pests and of herbicide resistance in weeds are but a few obvious examples of evolution at work. More desirable instances of the power of selection to shape organisms also surround us in the form of crops, domesticated animals, and livestock. Darwin discerned in nature a parallel to the practice of selective breeding, or artificial selection, which humans have been practicing for the past 10 000 years. Whether selecting for faster horses, higher milk yields from cows, or showier pigeons, humans have shown that the selective breeding of individuals exhibiting the desired traits will usually lead to changes in the population – and to an accentuation of the selected trait over generations.

Over the past 30 years, selective breeding has been brought to bear on an increasing variety of biological entities. Bacteria, viruses, nucleic acids, and proteins are now routinely evolved in the laboratory. These *in vitro* experiments in evolution occur in the beakers and test tubes of laboratories around the world. The motivations behind *in vitro* evolution experiments range from an interest in the mechanisms of adaptation to the determined pursuit of molecules exhibiting a desired feature. All of these varied investigations seek to harness the immense creative power of the evolutionary process. Such work has taught us much about evolutionary responses, about limits to adaptation, and about the genetic basis of novel features. Ultimately, this work may also help us to understand the mechanisms responsible for the emergence of life on this planet.

The History of *In Vitro* Evolution

The work carried out by Sol Spiegelman in the 1960s serves as an early landmark in the effort to examine the evolutionary process *in vitro*. In an elegant set of experiments, Spiegelman explored the evolution not of organisms, but of a particular molecule: the short RNA template molecules that can be copied by the enzyme Qβ replicase. This enzyme, the RNA-directed RNA polymerase of phage Qβ, uses an RNA template to synthesize new RNA molecules. This replicase can be made to operate in a cell-free system that contains only rNTPs (ATP, CTP, UTP, and GTP), salts, and a population of diverse short RNA molecules capable of acting as templates for the replicase. When Spiegelman's system evolved over several generations of replication, he realized that the character of the template population had changed significantly and now consisted almost entirely of a small subset of similar sequences. Those template sequences best suited to copying by the Qβ replicase had increased in frequency throughout the experiment, eventually coming to dominate the system (**Figure 1**). The conditions for evolution laid out in the introduction had all been met: 1) variation existed in the population (and was constantly resupplied by the errors committed by the Qβ replicase); 2) that variation led to differential reproduction – in this case copying by the Qβ replicase; and 3) those sequence features were passed on to the subsequent generation by the Qβ replicase through the template-directed synthesis of the complementary strand. The result was a succession of template RNA strands particularly well suited, in sequence and three-dimensional structure, to serve as templates for the Qβ replicase.

The field of *in vitro* evolution has expanded, exploded, really, since those early experiments. *In vitro* evolution is now an important aspect of both basic and applied research in the life sciences.

In Vitro Evolution: Basic Insights

The study of evolution is, for the most part, a retrospective endeavor. Until recently, evolutionary biologists dealt with products of evolution shaped over timespans far exceeding the lifetime of the investigator. The task of the investigator, then, was to reconstruct the evolutionary process based on its contemporary outcomes.

The idea of controlling and observing evolution directly, rather than reconstructing it *post hoc*, however, holds immense appeal. The power of retrospective approaches can now be supplemented by results obtained from experimental evolution. Furthermore, the validity of our methods of evolutionary reconstruction can now be tested directly by comparing reconstructions to observed events. Increasingly, since the 1970s, evolutionary experiments are being carried out using bacteria, phages, viruses, and even cell-free systems.

Over the past two decades, a number of investigators have used *in vitro* evolution to explore molecular function directly. Much of this pioneering work again focuses on RNA, which had been shown to be both an information-conveying molecule (as most nucleic acids are), and a molecule capable of carrying out precise biochemical function. The discovery of catalytic RNA immediately prompted questions about the catalytic range of RNA and about the possibility that an entire rudimentary metabolism could be based on RNA alone. Central to this conjecture of an 'RNA world' was the assumption that RNA could catalyze a variety of reactions, possibly even

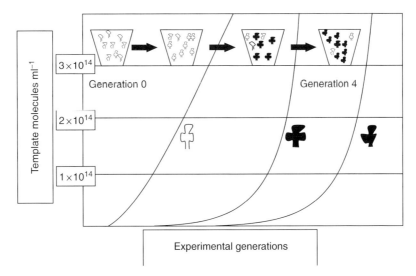

Figure 1   A diagram of the results obtained by Sol Spiegelman in his early *in vitro* evolution experiments on Qβ replicase. As can be seen, the particular templates best replicated by the Qβ replicase rise in frequency; as new variation is constantly introduced, new, even better templates emerge and come to dominate the system. The composition of the template population changes and evolves at every generation.

including the template-directed synthesis of new RNA molecules.

In vitro evolution methods provide a powerful tool with which to explore this assumption, and two strands of research quickly emerged. The first of these searched for RNAs capable of binding with high affinity to particular molecules or molecular features. Such RNA sequences (often referred to as 'aptamers') would confirm the ability of RNA to adopt the precise three-dimensional configuration required to bind substrates and cofactors in enzymatic reactions. Such aptamers would also confirm the ability of RNA to stabilize transition states, the critical intermediate molecular configuration adopted by reactants in a chemical reaction. Unless RNA could be shown to be capable of a high-affinity interaction with defined chemical species, it was impossible to argue for the plausibility of RNA-based metabolisms. A series of experiments was quickly undertaken to demonstrate the existence of high-affinity aptamers capable of binding synthetic and naturally occurring molecules. The overall design of these SELEX (systematic evolution of ligands by exponential amplification) experiments consists of the generation of starting populations containing an immense variety (10^{10}–10^{15}) of RNA sequences. Such starting pools are created by synthesizing either fully randomized sequences flanked by conserved regions (used in subsequent amplification of selected molecules) or by partially randomizing a pre-existing functional molecule. This population then is passed through a column composed of inert material covered in the target molecular species (the 'ligand'). Those RNA molecules in the population comparatively best able to bind the ligand would then be slowed in their passage through the column; conversely, other RNA molecules would flow through freely. After the entire RNA population has been passed through the column, bound RNA molecules are stripped from the ligand and used as the progenitors of the next round of *in vitro* selection. This simple cycle, successfully completed multiple times, leads to an increase in the mean affinity of the evolving pool for the target ligand and to the eventual isolation of RNA molecules showing enhanced ligand affinity. Note that *in vitro* 'selection,' the eventual isolation of desired molecules from a large starting population, should be contrasted with *in vitro* 'evolution,' where, in addition to a selection step, new variation is constantly reintroduced into the population (see **Figure 2**).

One characteristic SELEX experiment began with a pool of 10^{13} versions of a RNA 100-mer, in pursuit of molecules capable of binding a synthetic dye (Cibachron Blue). This pool was estimated to contain 1 in 10^{10} molecules capable of binding the dye with noticeable affinity; after six rounds of selection, more than

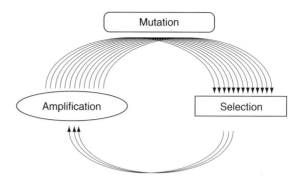

Figure 2 The basic elements of a simple *in vitro* evolution system. Mutation introduces variation into the system; a selection step sorts among the available variants, allowing only those best suited for the particular function to emerge and be amplified ('reproduce') in the subsequent step. A mutation step then restores variation. This basic cycle, iterated multiple times, results in the evolution of a population of molecules. *In vitro* selection experiments follow the same basic design, but mutation is only introduced at the outset, and subsequent cycles involve an alternation of the amplification and selection steps.

60% of the pool exhibited high-affinity binding. Similar *in vitro* evolution strategies have resulted in the isolation of aptamers capable of binding to specific nucleic acid and protein sequences with micromolar or submicromolar affinities ($K_d < 1$ μM). Aptamer selections have now been directed to a broad variety of compounds including amino acids, nucleotides, cofactors, and antibiotics.

In a second strand of research, scientists have successfully used *in vitro* evolution to explore the versatility and limitations of RNA's catalytic ability. To do this, studies begin with a pool of variants based on an existing, catalytically active ribozyme, or, in some instances, with a fully randomized pool of longer (>60 nucleotides) RNA molecules. These studies have different objectives. They may seek to modify an existing RNA catalytic activity (e.g., by changing the ion dependency of the GpI intron from Mg^{2+} to Ca^{2+}) or to expand the catalytic activity of a ribozyme to a new substrate or reaction (e.g., evolving DNA-cleaving derivatives of the Gp I and RNaseP RNA ribozymes) (**Figure 2**). Such studies also aim to isolate RNA molecules capable of performing a particular function, such as self-cleavage, ligation, aminoacylation, and peptide bond synthesis. Ongoing experiments also explore the dynamics and interactions in molecular ecosystems involving multiple molecular species.

More recently, pools of DNA variants have been subjected to a similar battery of *in vitro* evolution regimes, resulting in the identification of DNA aptamers and of DNA enzymes (DNAzymes, or

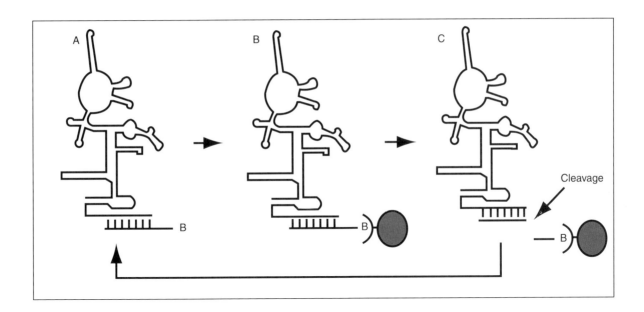

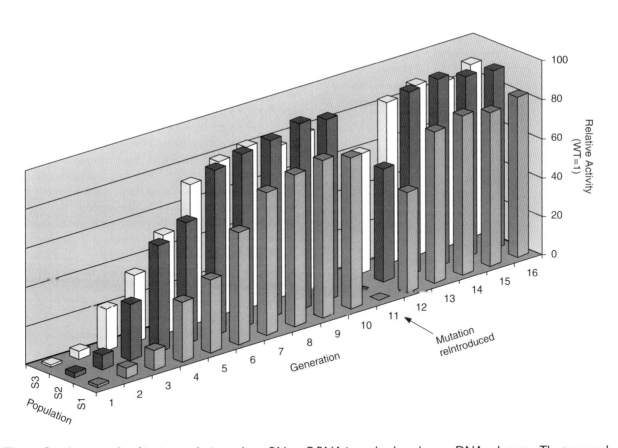

Figure 3   An example of *in vitro* evolution, where RNase P RNA is evolved to cleave a DNA substrate. The top panel shows a diagram of the selection scheme, where variant RNA molecules anneal to a DNA substrate, which in turn attaches to a column (via a biotin 'B' molecule). Those variants that can cleave the DNA substrate are eluted from the column, and are amplified for the next generation of selection. The bottom panel shows the response of three parallel RNase P RNA populations under *in vitro* evolution. Note the increase in the overall activity of the population, as well as the dramatic, but transient drop in activity that accompanies the reintroduction of mutations into the evolving population.

deoxyribozymes) capable of enhancing the rate of biochemically important reactions. The success of these DNA selections expand the perspective of workers in the field, and have led to the realization that any biopolymer that can be copied with some degree of fidelity can, in principle, serve as the raw material for *in vitro* evolution.

Taken together, *in vitro* results underscore the tremendous functional versatility of nucleic acid polymers. The fact that *in vitro* evolution experiments frequently succeed attests to the density of functional solutions scattered throughout sequence space. Phrased differently, a fully randomized pool of RNA 100-mers could theoretically contain 4^{100} or $\sim 10^{60}$ variants. A typical experiment will thus sample only $10^{13}/10^{60}$ or 1 of 10^{47} possible sequences. Even with this extremely sparse sampling, functional sequences are almost always retrieved. While there may be certain functions for which a viable solution is so rare that it cannot be captured by *in vitro* selection, theoretical and empirical results paint a different picture of functional space. Indeed, multiple solutions appear to exist for any given catalytic challenge, and these solutions seem both to lie in close proximity and to be accessible from practically any given starting point. The presence of so many peaks on the RNA functional landscape may well account for the rapid emergence of organization early in the history of life.

In Vitro Evolution: Applied Research

Early in the history of the field, the isolation of aptamers directed to particularly visible targets, such as the Rev protein and reverse transcriptase of the HIV virus, hinted at the applied potential of *in vitro* evolution methods. Although naturally occurring, high-affinity interactions between proteins and nucleic acids are integral to all known metabolisms, it soon became apparent that any protein could, in principle, be targeted using *in vitro* approaches. In fact, high-specificity aptamers capable of binding with disease-causing molecules or pathways in some cases show stronger binding affinities than those typically associated with antibodies. This binding has obvious implications for the diagnosis of disease conditions. In those cases where binding interferes with the operation of molecules involved in disease pathways, aptamers show significant therapeutic promise. Over the past decade, a number of aptamers of potential diagnostic or therapeutic importance have been developed using *in vitro* evolution. These aptamers target a diverse collection of disease-related proteins (e.g., thrombin, antibodies involved in autoimmune conditions, proteases). More recently, scientists have isolated aptamers that can be directed not only at specific proteins, but at particular diseased tissues (e.g., sclerotic arterial deposits).

Future Directions

The uses of *in vitro* evolution continue to expand, limited only by the ability to identify targets of interest and to design effective selection strategies. The raw materials for *in vitro* evolution are now more diverse. For example, synthetic nucleotides and nucleotide analogs have been incorporated into the sequences constituting the initial pool for *in vitro* evolution. This increase in the complexity of the nucleotide sequences increases the number of potential three-dimensional interactions, and, by extension, the number of shapes that can be assumed by the sampled pool. Similarly, recent methods have succeeded in coupling peptides to their coding sequences. This significant advance allows for the *in vitro* isolation of proteins with desirable properties, followed by the replication of their cognate coding sequences. The substantially wider repertoire of side groups provided by a 20 amino-acid alphabet may well allow for the *in vitro* evolution of catalysts capable of a broader range of chemical reactions (**Figure 3**).

The power of the *in vitro* approach is now being directed toward the more subtle issues of emergence and complexity. Studies now underway seek not just to evolve novel molecules or to expand the catalytic repertoire of single molecular species, but instead to evolve metabolic networks. Such enclosed networks, composed of multiple interacting molecular species, serve as a model for the earliest protocells and their rapidly evolving metabolic potential. The field of *in vitro* evolution is still in its early phase, and its potential still enormous. In effect, *in vitro* evolution allows us to explore sequence, structure, and function space beyond the solutions already present in living systems. This ability to compare existing functional solutions with possible (but unrealized) solutions, to probe not just the actual but the possible, adds a radically new tool to the arsenal of comparative biology.

***See also:* Bacterial Genetics; Biochemical Genetics; Evolution; RNA World; Selection Techniques**

In vitro Fertilization

R Edwards

Copyright © 2001 Academic Press
doi: 10.1006/rwgn.2001.0713

In vitro fertilization (IVF) opened up the prospects of many genetic studies on human conception. Some of them were of greater clinical interest, such as embryo transfer for the alleviation of various forms of male and female infertility. Others were more genetic and academic, yet have now been developed clinically. For example, controlling the growth of the human embryo *in vitro* enabled its various forms of growth to be classified and related to differing chromosomal, nuclear, and cytoplasmic anomalies typical of early human development. Astonishingly, high numbers of human embryos grown *in vitro* carry several such anomalies, and limited evidence suggests the same is true for those developing *in vivo* after natural conception. Some anomalies do not seem to be correlated with identified disorders in the human oocyte and preimplantation embryo. Others are recognizably abnormal and involve chromosomal disorders such as aneuploidy, haploidy, polyploidy, and mosaicism, which effectively terminate development before implantation or soon afterward. A few continue to later stages of gestation. A very significant feature is the suprisingly low implantation rate per embryo, i.e., 20% or even less after growth *in vitro* or *in vivo*. Coupled with this evidence of serious weaknesses in oogenesis and embryogenesis, enormous numbers of human spermatozoa are weakly immotile or misshapen, so that as few as 14% of normal forms is considered to indicate a highly fertile man! Human gametogenesis and embryogenesis thus seem to be highly flawed, yet women ovulate only one egg per month. Humans seem to have serious flaws in the control of reproductive systems, unlike other mammals where strong selective pressure apparently maintains highly effective systems in reproduction with implantation rates of 80–90% and few disorders in preimplantation growth. Perhaps a highly effective mother–child bonding or some similar highly adaptive and protective system has relaxed the human need for tight controls over the close cell cycle and over meiosis, fertilization, and cleavage.

IVF has helped to gain a deeper understanding of other genetic aspects of conception. Very severe infertility in men has an unusual genetic basis, owing to large deletions in three distinct regions of the Y chromosome. Characterizing these regions has provided a superb understanding of Y chromosome genetics and the exact sequences undergoing deletion. For such men, the intracytoplasmic injection of a single spermatozoon into an egg (ICSI) enables their extreme oligozoospermia to be overcome by using the very rare spermatozoa in ejaculates, epididymis, or testis, or even spermatids. Fewer spermatozoa are collected from some of these men than the number of oocytes collected from their wives. This finding has necessitated great care in searching for mutants among the children, although at present more attention is paid to disordered chromosome constitutions. Treating other extreme forms of male infertility has also uncovered genetic defects rarely found in a normal-conceiving human population, such as cystic fibrosis variants that distort the formation of the vas deferens. Applying ICSI in these cases can risk the health of the child when wives are carriers for cystic fibrosis. Separating human X and Y spermatozoa is now possible and reliable and also more easily achieved by applying ICSI for the limited spermatozoa available until techniques improve to produce sufficient spermatozoa for artificial insemination.

Preimplantation genetic diagnosis for inherited disease is also becoming more widespread in IVF programs. A single cell excised from an 8-cell embryo or half a dozen cells from the trophectoderm of a blastocyst can be used to type genetic disorders in human embryos. Many single-gene disorders can be identified in the embryos, and also highly complex translocations, chromosome errors, and complex variants, such as those involved in Duchenne muscular dystrophy. Improvements in array technology promise to permit hundreds or even thousands of genes to be identified in preimplantation embryos. Such knowledge might provide a genetic blueprint of the growth of the embryo, with considerable social and ethical implications. Other genetic-related advances stemming from IVF include the potential cloning of human embryos for spare-parts surgery. It is notable that cloned embryos and offspring have enormous anomalies and very high death rates, and the effects of such epigenetic changes will presumably be present in embryo stem cells. Cloning was not attempted in hundreds of IVF laboratories practicing ICSI, which could enable cloning to be introduced. The UK government's decision to permit cloning of human embryos to make tolerant embryo stem cells for organ repair has just been announced.

Knowledge about the genetic regulation of the human oocyte and embryo is now accumulating rapidly. Polarities have been identified in oocytes, cleaving embryos, and blastocysts, and genes affecting early growth have been identified. This information, together with that gained from the mouse and human genome projects, indicates that hundreds or thousands of genes are expressed in preimplantation mammalian embryos, with blocks of closely linked genes acting in concert to regulate successive cleavage stages.

See *also*: Ethics and Genetics; Fertilization

In vitro Mutagenesis

M Arkin

Copyright © 2001 Academic Press
doi: 10.1006/rwgn.2001.0714

In vitro mutagenesis methods, especially site-directed mutagenesis, have revolutionized our understanding of protein function and gene regulation. *In vitro* mutagenesis describes the process by which a researcher alters one or more base pairs in a cloned gene; expression of the gene yields a protein with one or more altered amino acids. These mutant proteins may show a change in function, such as lost or altered activity. The ability to manipulate precisely the chemical nature of a gene – and therefore the protein encoded by this DNA – has enabled biologists to identify protein function, characterize protein structure, and manipulate the activity of a protein *in vivo*. Furthermore, 'protein engineers' have used site-directed and random mutagenesis procedures to create new proteins designed to have unique or improved function.

In vitro mutagenesis has been enabled by a number of breakthroughs in biotechnology. Other articles in this encyclopedia describe discovery and uses of recombinant DNA, DNA polymerases, the polymerase chain reaction (PCR), and restriction endonucleases. This article will describe the application of these technologies to the mutagenesis of recombinant genes.

Nonselective Mutagenesis

Deletions

Nested deletion mutagenesis has been used to identify functional domains of proteins and RNA. By this method, the plasmid containing the gene of interest is linearized at a restriction site near the gene. The gene is then cleaved for discrete amounts of time by the enzyme exonuclease III, which removes bases from duplex DNA containing a 5′ overhang. The result is a 'nested set' of plasmids in which the gene fragments vary in length from one side of the gene and contain a common end. These partially digested genes are then recloned into a plasmid vector and transformed into *Escherichia coli*. In one early example of this method, researchers studying 5S ribosomal RNA used exonuclease III to delete bases from the 5′ end and identified regions within the 5S rRNA gene which control its transcription initiation.

Chemical Damage and Enzymatic Misincorporation

Chemical mutagenesis and enzymatic misincorporation techniques cause a small number of mutations throughout a piece of DNA. Both methods yield a library of mutations which are cloned into a plasmid and then screened or selected for function. Commonly used chemicals include sodium bisulfite, formic acid, and hydrazine. Sodium bisulfite causes the deamination of cytosine to uracil; during DNA synthesis, the altered base is paired with adenosine instead of guanine. Hydrazine and formic acid remove bases from the DNA strand, creating abasic sites that can pair with any one of the four bases during enzymatic synthesis. Nucleotides can also be altered at random sites through misincorporation of deoxyribonucleotide triphosphates (dNTPs) during DNA synthesis. For example, DNA polymerase runs with impaired fidelity in the presence of manganese ions, and occasionally adds an incorrect base. Alternatively, when one of the dNTPs is added in very low concentrations, the enzyme will sometimes misincorporate one of the other three bases. Certain dNTPs, such at N^6-hydroxydeoxycytidine, are also mutagenic, and can cause mispairing mutations. In all cases, the frequency of mutation is increased by using DNA polymerase without a proofreading function (such as Klenow fragment from *E. coli*). The modern version of enzymatic misincorporation, error-prone PCR, is regularly used to make mutant DNA libraries.

Site-Directed Mutagenesis

Site-directed mutagenesis involves the specific substitution of one DNA base for another. Unlike the nonspecific mutations described above, site-directed mutagenesis allows precise control of the number, placement, and base substitution of mutants. The two classes of site-directed mutagenesis include methods that use double-stranded DNA cassettes and those that use single-stranded oligonucleotide primers. All of the techniques described here can give high yields of the desired mutations; the choice of mutagenesis method is largely a matter of convenience and personal preference.

Site-directed mutagenesis is possible because of the invention of automated chemical synthesis of DNA and the overexpression of DNA-processing enzymes. Through chemical DNA synthesis, defined oligonucleotides up to ~100 bases can be prepared reproducibly and inexpensively. Synthetic oligonucleotides are used extensively for site-directed mutagenesis, as primers for DNA polymerase and as oligonucleotide cassettes. Equally important has been the identification and overexpression of DNA-modifying enzymes, including restriction endonucleases for cleaving DNA at specific recognition sites and DNA polymerases for generating double-stranded DNA from a single-stranded template. Furthermore, the discovery of thermophilic DNA polymerases has enabled PCR-based methods for site-directed mutagenesis.

Cassette Mutagenesis

In cassette mutagenesis, a synthetic double-stranded oligonucleotide 'cassette' containing the desired mutations is docked between two restriction enzyme sites on a plasmid vector. In the simplest procedure, the restriction sites are separated by no more than 100 base pairs; the ends of the oligonucleotide duplex are complementary to the restriction cleavage sites so that the cassette can be readily ligated into the plasmid (**Figure 1A**). Since dozens of restriction enzymes are commercially available, it is often possible to identify restriction sites near the sequence of interest.

One clever cassette design takes advantage of restriction endonucleases such as BspMI and BcgI which cleave DNA several base pairs away from their recognition sequences. BcgI, for instance, cleaves DNA at any sequence 10 bases away from each side of the enzyme's specific binding site, while BspMI cleaves on one side of an asymmetric recognition sequence. The recognition sequence and product of BcgI cleavage are shown below, where N is any nucleotide:

$$N_{10}\text{-CGA-}N_6\text{-TGC-}N_{12}$$
$$N_{12}\text{-GCT-}N_6\text{- ACG-}N_{10}$$

To prepare a BcgI cassette, these recognition sequences are added into a cloned gene by PCR such that the restriction site replaces the region to be mutated (**Figure 1A**). A cassette is synthesized to contain (1) the gene sequence which was removed from the vector, (2) the desired mutations, and (3) ends complementary to the products of BcgI cleavage. The site-directed mutant is then made by cutting the plasmid with BcgI and ligating in the cassette. The advantage of these vectors is that the restriction enzyme sites are cut out of the gene when the cassette is added. Thus, the recombinant gene does

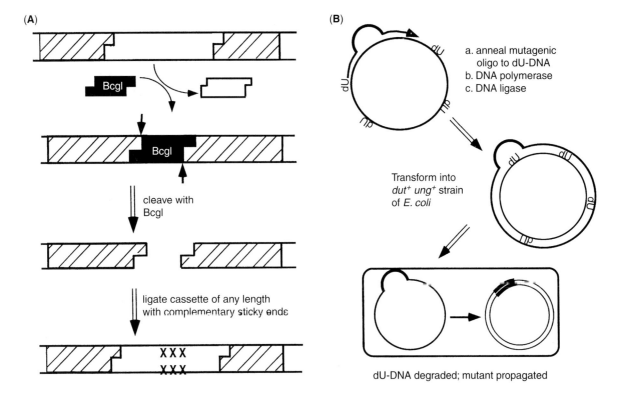

Figure 1 Methods of oligonucleotide-directed mutagenesis. (A) Cassette mutagenesis with BcgI-containing plasmid. The BcgI-containing plasmid is constructed by removing the region to be mutagenized and replacing it with a BcgI recognition sequence. Short arrows show sites of BcgI cleavage. Following cleavage by the restriction enzyme, the mutagenic cassette is ligated into the gene. Note that the restriction sites are removed during mutagenesis. (B) dU method for primer-based mutagenesis. A dU-containing single-stranded plasmid is prepared from an M13 vector in a *dut⁻ ung⁻* strain of *Escherichia coli*. The mutagenic oligonucleotide is hybridized to the dU template (the mutagenic primer shown will create an insertion in the gene of interest). The rest of the second strand is filled in by DNA polymerase and ligated by DNA ligase. Transformation into a *dut⁺ung⁺* strain of *E. coli* results in degradation of the dU strand and propagation of the mutant.

not need to contain unique restriction sites, and the wild-type vector can be readily distinguished from the mutant by restriction digest. Furthermore, since linear DNA is not readily transformed and replicated in cells, precutting with the restriction enzyme before transformation will increase the yield of mutant clones. This type of cassette has been used in the mutational analysis of HIV reverse transcriptase.

Primer-Directed Mutagenesis

General methods

Site-directed mutagenesis can also be accomplished using an oligonucleotide containing the desired mutation, called a mutagenic oligonucleotide, as a primer for DNA synthesis. By this technique, the single-stranded oligonucleotide is hybridized to a single-stranded plasmid, using bases complementary to the wild-type gene. The mutagenic region of the oligonucleotide can contain several single base mismatches, or it can be much longer or shorter than the wild-type sequence (yielding insertions or deletions in the mutated gene). DNA polymerase initiates synthesis of the DNA at the oligonucleotide and fills in the second strand; addition of DNA ligase seals the nick in the newly synthesized strand. Transformation of this heteroduplex plasmid produces both wild-type and mutant plasmids in *E. coli*, but several methods (see below) have been devised to increase the proportion of mutants.

DNA templates for oligonucleotide-based mutagenesis are readily prepared using the single-stranded DNA bacteriophage M13. Commercially available plasmids contain M13 replication initiation sites as well as cloning sites with regulated promoters. Thus, a single plasmid can be used for cloning, M13 mutagenesis, and protein expression.

dU Method

Variations of the primer-based method increase the yield of mutant by preferentially degrading the template strand. A commonly used technique, first described by Kunkel, takes advantage of *dut⁻ ung⁻* strains of *E. coli* (**Figure 1B**). Whereas most bacteria will degrade DNA containing uracil (dU-DNA), *dut⁻ ung⁻* strains are are deficient in the degradadation of both dUTP (*dut⁻*) and dU-DNA (*ung⁻*). Thus, M13 templates isolated from *dut⁻ ung⁻* bacteria will contain some dU in place of dT. After hybridization of the mutagenic oligonucleotide, DNA synthesis and ligation, the heteroduplex plasmid is transformed into a *dut⁺ ung⁺* strain of *E. coli* which degrades the wild-type, dU-containing template strand but not the newly synthesized mutant strand. Thus, mostly mutagenic plasmid is propogated. Other methods use similar approaches by adding methyl-dC or thiophosphate-dC

during *in vitro* DNA synthesis; these modifications make the mutagenic strand resistant to degradation by certain restriction enzymes.

Polymerase Chain Reaction

PCR-mediated mutagenesis is similar to the oligonucleotide methods described above, in that a mutagenic oligonucleotide is used as a primer for DNA synthesis. An advantage of the PCR method lies in the inherent amplification of the mutagenic DNA, which requires only a small amount of the wild-type DNA as template. PCR mutagenesis can be performed on linear pieces of DNA, such as restriction fragments, as well as on circular plasmids. **Figure 2** pictures some of the methods discussed below.

PCR Mutagenesis of Linear DNA

If the desired mutation is found near a restriction enzyme site, the mutation can be incorporated by preparing one PCR primer containing the mutation and the restriction site and a second primer containing a downstream restriction site. The PCR product is then treated with the restriction enzymes and ligated into the plasmid as a DNA cassette. If there are no restriction sites near the mutagenic sequence, 'overlap-extension' PCR and 'megaprimer' PCR can be used to introduce the mutations. Overlap-extension PCR requires four primers and three PCR steps (**Figure 2A**). The first two PCR steps produce two overlapping DNA fragments, both containing the desired mutation. The final PCR step uses the outside primers to stitch together the two fragments into the full-length cassette. Megaprimer PCR, a variant of the overlap-extension method, uses three primers and two PCR steps. The first step yields a DNA fragment containing one restriction site and the mutations. This long DNA fragment is used as a megaprimer in the second PCR step along with a primer containing the second restriction site.

PCR Mutagenesis of Circular DNA

PCR mutagenesis can also be used to amplify the entire plasmid containing the gene of interest. One straightforward method, termed 'inverted' or 'counter' PCR, uses back-to-back primers (**Figure 2B**); one PCR primer serves as the mutagenic oligonucleotide and the other oligonucleotide primes from the opposite strand, adjacent to the mutagenic primer. The PCR product is a full-length, linear plasmid which is then phosphorylated and ligated before transformation. This method can readily be used to make deletion mutants by creating a gap between the primers. Variants of this method include 'recombinant circle' PCR (**Figure 2C**) and 'recombination' PCR, both of which rely on recombination of linear plasmids. In these

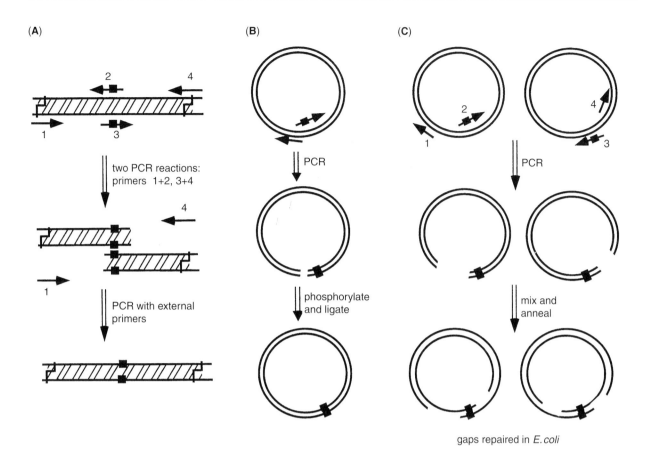

Figure 2 Methods of PCR mutagenesis. (A) Extension-overlap PCR generates mutations between two restriction enzyme sites. Four primers are prepared; two containing the restriction sites (primers 1 and 4) and two containing the mutagenic sequence (primers 2 and 3). After two PCRs, fragments 1–2 and 3–4 are combined and stitched together by PCR using primers 1 and 4. The long product 1–4 is restricted and ligated into the vector. (B) Inverted PCR uses two back-to-back primers, one containing the mutations of interest. PCR yields the full-length, linear plasmid which is made into closed circular DNA by DNA ligase. (C) Recombinant circle PCR uses two sets of primers. Primers 2 and 3 contain the mutagenic sites and prime opposite strands; primers 1 and 4 prime from different positions on the plasmid. The PCR products 1–2 and 3–4 are truncated, linear versions of the plasmid; recombination *in vitro* gives gapped plasmids which are repaired in *E. coli*.

techniques, two inverse PCRs are performed with gapped primers at different sites. These two mutant plasmids are then recombined *in vitro* by mixing and annealing (recombinant circle PCR) or *in vivo* (recombination PCR). The gaps are then repaired by the bacterial DNA repair machinery.

Libraries of Mutations

Combinatorial and random mutagenesis methods create libraries of DNA which are subsequently screened or selected for function. By analyzing large numbers of clones simultaneously, a small number of active mutants can be separated from a pool of millions of variants. The preparation of libraries, selection of 'winners' and amplification of these selectants is often called '*in vitro* evolution.'

Doped versus Saturation Mutagenesis

Libraries of mutant DNA molecules can be designed such that a small number of random mutations are introduced throughout the gene – analogous to the nonselective mutagenesis described above – or large number of mutations are focused on a small region of a gene. When all possible DNA mutations can be found at a given site with equal frequency, the site is described as 'saturated.' When the number of mutations at a given site is small, the site is said to be 'doped' with the mutation. Saturation mutagenesis is readily accomplished through automated DNA synthesis. During synthesis, discrete bases are added in sequence to the growing DNA chain; to saturate a position on this chain, equal amounts of all four bases are added simultaneously. Doping can similarly be accomplished by mixing a measured fraction of mutagenic base at a given site. After these doped or saturated mutagenic oligonucleotides have been synthesized, *in vitro* mutagenesis proceeds as usual via cassette mutagenesis or primer-based mutagenesis. Error-prone PCR offers an alternate method for doping mutants throughout a gene; the rate of mutagenesis is approximately 0.7% for *Taq* DNA polymerase. Both saturation and doping strategies have been used to identify critical protein residues and to create novel binding or catalytic functions. Examples include peptides that antagonize or agonize cell-surface receptors, and enzymes that are active in nonaqueous environments.

Mutagenesis and Recombination

An increasingly popular method for generating libraries of a gene utilizes an *in vitro* recombination technique called 'DNA shuffling.' In DNA shuffling, one or more genes are randomly chopped into smaller pieces of DNA by a nuclease and reconnected with a DNA polymerase. During this reconstruction phase, homologous fragments of DNA can anneal and prime each other, creating a recombined gene. Mutations are incorporated into the genes via errors during DNA polymerization. DNA shuffling has been used to optimize the function of proteins as well the activity of whole operons and viruses.

Prospects

In vitro mutagenesis has become an integral part of genetic analysis. Controlled mutagenesis has identified the function of new genes, a process termed 'reverse genetics,' and allowed dissection of the mechanism of known proteins. Additionally, site-directed mutagenesis has become an important tool in biotechnology. For example, the design of non-immunogenic antibodies for human therapeutics underscores the practical benefits of mutagenesis and protein engineering. The availability of DNA-modifying enzymes, cloning vectors, and synthetic DNA make site-directed mutagenesis straightforward in most laboratories; its applications are limited only by the imagination.

Further Reading

Boyer PL and Huges SH (1996) Site-directed mutagenic analysis of viral polymerases and related proteins. In: Kuo LC, Olsen DB and Carroll SS (eds) *Methods in Enzymology*, pp. 538–555. San Diego, CA: Academic Press.

Chen K and Arnold FH (1993) Tuning the activity of an enzyme for unusual environments: sequential random mutagenesis of subtilisin E for catalysis in dimethylformamide. *Proceedings of the National Academy of Sciences, USA* 90: 5618–5622.

Kunkel TA (1985) Rapid and efficient site-specific mutagenesis without phenotypic selection. *Proceedings of the National Academy of Sciences, USA* 82: 477–492.

Riechmann L, Clark M, Waldmann H and Winter G (1988) Reshaping human antibodies for therapy. *Nature* 332: 323–327.

Smith M (1985) *In vitro* mutagenesis. *Annual Review of Genetics* 19: 423–462.

Stemmer WPC (1994) DNA shuffling by random fragmentation and reassembly: *in vitro* recombination for molecular evolution. *Proceedings of the National Academy of Sciences, USA* 91: 107–147.

Tao BY and Lee KCP (1994) Mutagenesis by PCR. In Griffin HG and Griffin AM (eds) *PCR Technology: Current Innovations*, pp. 69–83. Boca Raton, FL: CRC Press.

Watson JD, Gilman M, Witkowski J and Zoller M (1992) *Recombinant DNA*, 2nd edn. San Francisco, CA: WH Freeman.

See also: DNA Sequencing; *In vitro* Evolution; Mutant Allele; Mutational Analysis; Polymerase Chain Reaction (PCR); Recombinant DNA; Restriction Endonuclease; Screening

In vitro Packaging

I Schildkraut

Copyright © 2001 Academic Press
doi: 10.1006/rwgn.2001.0715

In vitro packaging is the method of reconstituting a virus *in vitro* by mixing the protein components of the virus with nucleic acid. The protein components of the virus are prepared from extracts of infected cells by eliminating the nucleic acids in the extract. The nucleic acid component to be packaged is usually an *in vitro* recombinant DNA construct. *In vitro* packaging is useful as a means to efficiently introduce a DNA fragment recombined with a viral vector into a cell by using infective properties of viral particles to pass through the cell wall/membrane.

See also: Vectors

Inborn Errors of Metabolism

T M Picknett and S Brenner

Copyright © 2001 Academic Press
doi: 10.1006/rwgn.2001.1805

An inborn error of metabolism is a biochemical or genetic lesion that gives rise to an inherited metabolic block. Many are due to the inability to synthesize an individual protein or the production of a biologically inefficient form of a protein.

See also: **Genetic Diseases; Metabolic Disorders, Mutants**

Inbred Strain

L Silver

Copyright © 2001 Academic Press
doi: 10.1006/rwgn.2001.0674

An inbred strain is a population of animals that result from a process of at least 20 sequential generations of brother–sister matings. The resultant animals are essentially clones of each other at the genetic level. When two animals have the same strain name – such as BALB/c or C57BL/6 – it means that they can both trace their lineage back through a series of brother–sister matings to the *very* same mating pair of inbred animals. With the use of the same standard inbred strain, it is possible to eliminate genetic variability as a complicating factor in comparing results obtained from experiments performed in any laboratory in the world.

The Generation of Inbred Strains

The offspring that result from a mating between two F_1 siblings are referred to as members of the 'second filial generation' or F_2 animals, and a mating between two F_2 siblings will produce F_3 animals, and so on. An important point to remember is that the filial (F) generation designation is only valid in those cases where a protocol of brother–sister matings has been strictly adhered to at each generation subsequent to the initial outcross. Although all F_1 offspring generated from an outcross between the same pair of inbred strains will be identical to each other, this does not hold true in the F_2 generation which results from an intercross where three different genotypes are possible at every locus. However, at each subsequent filial generation, genetic homogeneity among siblings is slowly recovered in a process referred to as 'inbreeding.' Eventually, this process will lead to the production of inbred animals that are genetically homogeneous and homozygous at all loci.

The process of inbreeding becomes understandable when one realizes that at each generation beyond F_1, there is a finite probability that the two siblings chosen to produce the subsequent generation will be homozygous for the same allele at any particular locus in the genome. If, for example, the original outcross was set up between animals with genotypes AA and aa

at the A locus, then at the F_2 generation, there would be animals with three genotypes AA, Aa, and aa present at a ratio of 0.25:0.50:0.25. When two F_2 siblings are chosen randomly to become the parents for the next generation, there is a defined probability that these two animals will be identically homozygous at this locus. Since the genotypes of the two randomly chosen animals are independent events, one can derive the probability of both events occurring simultaneously by multiplying the individual probabilities together according to the 'law of the product.' Since the probability that one animal will be AA is 0.25, the probability that both animals will be AA is $0.25 \times 0.25 = 0.0625$. Similarly, the probability that both animals will be aa is also 0.0625. The probability that either of these two mutually exclusive events will occur is derived by simply adding the individual probabilities together according to the 'law of the sum' to obtain $0.0625 + 0.0625 = 0.125$.

If there is a 12.5% chance that both F_2 progenitors are identically homozygous at any one locus, then approximately 12.5% of all loci in the genome will fall into this state at random. The consequence for these loci is dramatic: all offspring in the following F_3 generation, and all offspring in all subsequent filial generations will also be homozygous for the same alleles at these particular loci. Another way of looking at this process is to consider the fact that once a starting allele at any locus has been lost from a strain of animals, it can never come back, so long as only brother–sister matings are performed to maintain the strain.

At each filial generation subsequent to F_3, the class of loci *fixed* for one parental allele will continue to expand beyond 12.5%. This is because all fixed loci will remain unchanged through the process of incrossing, while all unfixed loci will have a certain chance of reaching fixation at each generation.

After 20 generations of inbreeding, 98.7% of the loci in the genome of each animal should be homozygous. This is the operational definition of 'inbred.' At each subsequent generation, the level of heterozygosity will fall off by 19.1%, so that at 30 generations, 99.8% of the genome will be homozygous and at 40 generations, 99.98% will be homozygous.

These calculations are based on the simplifying assumption of a genome that is infinitely divisible with all loci assorting independently. In reality, the size of the genome is finite and, more importantly, linked loci do not assort independently. Instead, large chromosomal chunks are inherited as units, although the boundaries of each chunk will vary in a random fashion from one generation to the next. As a consequence, there is an ever-increasing chance of complete homozygosity as animals pass from the 30th to 60th generation of inbreeding. In fact, by 60 generations, one would be virtually assured of a homogeneous homozygous genome if it were not for the continual appearance of new spontaneous mutations (most of which will have no visible effect on phenotype). However, every new mutation that occurs will soon be fixed or eliminated from the strain through further rounds of inbreeding. Thus, for all practical purposes, animals at the F_{60} generation or higher can be considered 100% homozygous and genetically indistinguishable from all siblings and close relatives.

See also: Homozygosity; Mutation, Spontaneous

129 Inbred Strain

L Silver

doi: 10.1006/rwgn.2001.0929

129 is the name given to a group of related inbred strains of mice that are commonly used in germline genetic manipulation experiments. The various 129 strains have been used as the source of a series of embryonic stem cell lines that can be readily manipulated in tissue culture and then directed back into the mouse germline through a process of chimera formation.

See also: Chimera; Embryonic Stem Cells; Inbred Strain

Inbreeding

See: Inbred Strain

Inbreeding Depression

L Silver

doi: 10.1006/rwgn.2001.0677

The major hurdle that must be overcome in the development of new inbred strains from wild populations is inbreeding depression which occurs most strongly between the F_2 and F_8 generations (second through eighth generation of sequential brother–sister mating). The cause of this depression is the load of deleterious recessive alleles that are present in the genomes of wild animals as well as all other animal species. These deleterious alleles are constantly generated at a low rate by spontaneous mutation but their number is normally held in check by the force of negative selection acting upon homozygotes. With constant replenishment and constant elimination, the load of deleterious alleles present in any individual mammal reaches an equilibrium level of approximately ten. Different unrelated individuals are unlikely to carry the same mutations, and as a consequence, the effects of these mutations are almost never observed in large randomly mating populations.

However, it not surprising that during the early stages of inbreeding, many of the animals will be sickly or infertile, because deleterious recessive mutations present singly in one parent are likely to be homozygous in future inbred generations. At the F_2 to F_8 generations, the proportion of sterile animals is often so great that the earliest mouse geneticists thought that inbreeding was a theoretical impossibility. Obviously they were wrong. But, to succeed, one must begin the production of a new strain with a very large number of independent $F_1 \times F_1$ lines followed by multiple branches at each following generation. Most of these lines will fail to breed in a productive manner. But, an investigator can continue to breed the few most productive lines at each generation – these are likely to have segregated away most of the deleterious alleles. The depression in breeding will begin to fade away by the F_8 generation with the elimination of all of the deleterious alleles. Inbreeding depression will not occur when a new inbred strain is begun with two parents who are themselves already inbred because no deleterious genes are present at the outset in this special case.

See also: Breeding of Animals; Inbred Strain

Incompatibility

doi: 10.1006/rwgn.2001.1871

Incompatibility is the inability of certain plasmids to coexist in the same cell and is a cause of plasmid immunity.

See also: Immunity; Plasmids

Incomplete Dominance

J A Fossella

doi: 10.1006/rwgn.2001.0678

A mutant allele is said to show 'incomplete dominance' or 'semidominance' when its phenotypic effects as a heterozygote are distinctly dominant but less severe than when homozygous. For example, for a hypothetical locus *b* affecting hair growth, *bb* homozygotes have normal hair, *Bb* heterozygotes show partial baldness, while all *BB* homozygotes are completely bald; the *B* allele shows incomplete dominance since the heterozygous phenotype is less severe than that of *BB* homozygotes. In most cases, the phenotype of heterozygotes is intermediate relative to the wild-type and the homozygous states.

The term incomplete dominance is similar, but distinct in meaning to the term codominant. The distinction between codominance and incomplete dominance is that codominance refers to pairs of alleles, while semidominance refers to a single allele. Codominance is observed when individuals that are heterozygous for alternative alleles at the same locus express *both* phenotypes observed in the corresponding homozygotes, or when all three classes (both classes of homozygotes and one class of heterozygotes) are all distinguishable from each other.

Incomplete dominance may also be used with respect to fitness, rather than with respect to the visible effects of a gene, as described above. A novel dominant allele may show no visible phenotypic differences in homozygotes versus heterozygotes, but may have an effect on the overall fitness of an organism such that hetrozygotes may gain only a partial increase in fitness that is less than the benefits afforded by homozygosity.

See also: Codominance; Heterozygote and Heterozygosis

Incomplete Penetrance

See: Penetrance

Incross

L Silver

Copyright © 2001 Academic Press
doi: 10.1006/rwgn.2001.0681

A cross between two organisms that have the same homozygous genotype at designated loci, for example, between members of the same inbred strain.

See also: Backcross; Inbred Strain; Intercross; Outcross

Indel

W Fitch

Copyright © 2001 Academic Press
doi: 10.1006/rwgn.2001.0682

Two homologous molecular sequences are often of unequal length indicating that either one gene has suffered an insertion or the other a deletion. In the absence of further information, it is hard to tell which of the two possibilities is correct. It is thus easier to indicate such differences as indels.

See also: Deletion; Insertion Sequence

Independent Assortment

J Merriam

Copyright © 2001 Academic Press
doi: 10.1006/rwgn.2001.0683

Independent assortment is one of the two great principles annunciated by Mendel that underlie our awareness of genes as units of heredity. Mendel proposed that a hybrid individual produces two gamete types in equal frequency for each heterozygous character and that the choice of trait for each character is independent of the other character. We recognize this as producing four gamete types in equal frequency from a dihybrid, resulting in progeny in the ratios 9:3:3:1 from a self-cross or 1:1:1:1 from a test cross. In modern terminology the observation of independent assortment means the segregation of alternative alleles at one locus or gene is not influenced by the segregation of the alternative alleles at a second locus.

Although generations of students have struggled to keep the laws of segregation and independent assortment clear in their minds, independent assortment is the less important in understanding the biological mechanisms of heredity. What is important is that genes are part of chromosomes. The reductional (first) division of meiosis separates homologous parental chromosomes, providing a mechanism for the Mendelian segregation of alternative alleles into different gametes. This was confirmed by nondisjunction that results in the wrong inheritance of both parental homologous chromosomes with both parental alleles at a gene or neither chromosome nor parental allele. Genes located on different (nonhomologous) chromosomes show patterns of segregation that are independent of each other. Thus, one meiotic origin of independent assortment is that genes are located on nonhomologous chromosomes. By observing meiosis in species where chromosome size and shape differ sufficiently, individuals heterozygous for two distinct pairs of nonhomologous chromosomes can be seen to produce four gamete types equivalently from two different patterns of reductional divisions. Genes located on the same pair of homologous chromosomes may not show independent assortment. Linkage, recognized as the exception to independent assortment, locates genes on the same chromosome. The fact that genes can be traced to specific chromosomes and can be located within chromosomes is a key to identifying individual genes that is a major goal of genetics. Genes are commonly identified through their neighbors or their position on a chromosome. In that sense linkage, rather than independent assortment, is the more useful concept for the modern study of genes.

When comparing segregation patterns of alleles at two genes, such as with linkage studies, independent assortment is the default or null hypothesis. This is because it is constant whereas linkage results cannot be predicted in advance. There are two approaches to testing the observed results by means of the χ^2 test to see if the results fit the predictions of independent assortment. The usual way simply compares each observed number of progeny class against an expected value derived from the total progeny number $\times$ 25% (for a dihybrid test cross). The χ^2 test compares each difference between observed and expected numbers to arrive at a total value based on the differences. With four classes this test has three degrees of freedom. The χ^2 test leads to a conclusion that accepts the null hypothesis, when the expected and observed numbers are similar, or rejects the null hypothesis when the observed numbers are too different from expected. In that case it is the expected numbers that are rejected. In some situations, such as reduced viability of some of the progeny classes, the χ^2 value may be large and lead to rejecting the null hypothesis but may not mean linkage. A further test for linkage or independent assortment is to use a 2-by-2 contingency table to generate the predicted numbers based on the observed subtotals for each row and column. Based on independence the number expected for one cell in the table is one row subtotal $\times$ one column subtotal with the product divided by the total number of progeny. Comparing the actual and expected numbers usually gives a better fit by this approach, but the number of degrees of freedom is reduced to one.

There are two models to understand the meiotic origin of independent assortment. One is that the two marked genes are located on nonhomologous chromosomes. The other is that the two genes are located on the same pair of chromosomes but are far enough apart that recombination in the interval between them mimics independent assortment. On either model the observation that half the gametes are parental and half recombinant equals a map distance of 50 units between the genes, indicating independent assortment. The reason that recombination distances cannot exceed 50 units for a dihybrid cross is that a crossover involves just two of the four chromatids present in a meiotic prophase bivalent. Every cell with one crossover in an interval potentially yields two crossover-bearing gametes and two non-crossover-bearing products of meiosis. With multiple crossovers within an interval but no preferential distribution of which chromatids are involved with each, half the gametes will contain either zero or an even number of crossovers and half the gametes will contain one or an odd number of crossovers. The former will be scored as non-crossovers and the latter will be scored as crossover-bearing gametes.

In the end most pairs of genes assort independently of each other. That is why linkage is a powerful statement for genetic investigations. The chance that two genes will not assort independently can be assessed for a species by taking into account the number of chromosomes and the level of recombination. In species like the fruit fly *Drosophila*, with few chromosomes and moderate levels of recombination, the chance of independent assortment is perhaps about 80% for two genes chosen at random. In species like humans, with 23 chromosome pairs and low to moderate recombination levels the chance is closer to 99%. In species like the yeast *Saccharomyces cerevisiae*, with 16 chromosome pairs and high levels of recombination, independent assortment is almost always expected. The importance of independent assortment rests with shuffling the genome at each meiosis to create new combinations of alleles from those making up the parental generation. This permits the population to more rapidly change genotypes in response to environmental changes. This feature was probably necessary for the development of biological complexity such as multicellularity. And it underlies most theories of the origin of sexual reproduction.

See also: Linkage; Mendel's Laws; Mendelian Ratio

Independent Segregation

J R S Fincham

Copyright © 2001 Academic Press
doi: 10.1006/rwgn.2001.0684

When a diploid undergoing meiosis is heterozygous at two or more loci (*Aa*, *Bb*, *Cc*, etc.), the haploid meiotic products will each carry one or other of each of the pairs of alleles. With n heterozygous loci there will be 2^n different kinds of haploid product. If all the allelic differences are segregated independently, all combinations will be equally frequent, apart from sampling error and any differences in viability. For example, with three loci, there will be eight equally frequent meiotic products: *ABC, abc, Abc, aBC, ABc, abC, AbC,* and *aBc.*

Independent segregation, also called independent assortment, occurs when the allelic differences are associated with different chromosome pairs and hence different linkage groups and is explained by the fact that different bivalent chromosomes at the first metaphase of meiosis are oriented at random with respect to the spindle poles, as are the dyads at second division metaphase. Since nearly all eukaryotic organisms have several or many chromosome pairs, independent rather than linked segregation is the most common outcome of meiosis in double or multiple heterozygotes. It should be noted that allelic differences on the same chromosome can also segregate independently if their loci are sufficiently far apart. For a discussion of linked segregation see Three-Point Cross (Test-Cross).

When, exceptionally, different chromosome pairs fail to show independent segregation it may be because they have undergone a reciprocal exchange of segments (see Segmental Interchange).

Independent assortment also occurs when a diploid becomes haploid through random loss of chromosomes during mitotic growth, as can happen in such normally haploid fungi as *Aspergillus nidulans.*

See also: Aspergillus nidulans; **First and Second Division Segregation; Heterozygote and Heterozygosis; Linkage Group; Meiosis; Segmental Interchange; Three-Point Cross (Test-Cross); Translocation**

Inducer

An inducer is a small molecule that triggers gene transcription on binding to a regulator protein.

See also: **Induction of Transcription**

Inducible Enzyme, Inducible System

See: Induction of Transcription

Induction of Prophage

Induction of prophage is the excision of phage DNA from the host genome and entry into the lytic (infective) cycle. It occurs as a result of destruction of the lysogenic repressor.

See also: **Prophage**

Induction of Transcription
B Müller-Hill

All organisms have developed and continue to develop mechanisms to adapt to an ever-changing environment. For microorganisms like bacteria or yeast, the carbon sources they can use for growth may change fast and drastically. A suitable carbon source that was present in great amounts may disappear and may suddenly be replaced by another carbon source. Sensors may sense the presence or absence of carbon sources. To adapt and optimize the transcription frequency of the genes which code for the relevant permeases and enzymes involved in the metabolism of these carbon sources is the simplest way of responding to such changes. Such adaptation may happen by mutation or induction. The term induction was introduced into bacterial genetics in 1953 by Melvin Cohn, Jacques Monod, Martin Pollock, Sol Spiegelman and Roger Stanier at a time when its mechanism was not known (Cohn *et al.*, 1953). Then, it was believed that an inactive precursor of the enzyme would interact with the inducer. By folding in the presence of the inducer, the inactive precursor would be transformed into an active enzyme. This was called instruction theory.

Induction has been studied extensively in *Escherichia coli,* other bacteria, and yeast. We now know induction implies that a particular compound acts as an inducer and turns on the transcription of one or several genes. A particular enzyme, or at the extreme, a whole system of enzymes and proteins may thus be inducible. An inducer may work either by counteracting repression or by stimulating activation of transcription. The compound may also act as a corepressor, such as tryptophan with Trp repressor. It may also act indirectly and use signal transduction. The fact that the inducer or corepressor is often metabolized may obscure the analysis. The detailed description of an example is illuminating. The *lactose* system in *E. coli* may serve as a paradigm of enzyme induction. *E. coli* grown on lactose produce about 3000 molecules of tetrameric β-galactosidase per cell. In contrast, *E. coli* grown on glycerol produce about three molecules of β-galactosidase per cell. The steps which lead to induction will now be listed using the example of this case.

1. Inducer has to enter the cell in order to induce. *E. coli* is not freely accessible to chemicals from the outside. Every molecule on the outside has to be transported by a specific transporter or permease to the inside. At the start of induction, lactose is transported by one of the one or two Lac permease molecules which are produced by the *lac* operon in the absence of any inducer.

2. Lactose (1-4-galactosido-β-D-glucose) itself is not an inducer. It has to be metabolized to allolactose (1-6-galactosido-β-D-glucose) which then acts as an inducer. Lactose which has entered the *E. coli* cell meets there the very few molecules of β-galactosidase which are produced in the absence of inducer. They isomerize lactose into allolactose before hydrolyzing it into glucose and galactose. That lactose is not the inducer can be demonstrated in Z^- (β-galactosidase negative) cells. Lac permease, which belongs to the same operon as β-galactosidase, is not induced by lactose in such cells. However, it is induced by allolactose.

3. If one wants to study the process of induction in detail, one has to use an inducer which is not metabolized, a gratuitous inducer. Such inducers have been synthesized in large numbers for the *lac* system. In contrast to lactose or ordinary β-D-galactosides they are all 1-thio-β-D-galactosides which are not hydrolyzed by the amounts of β-galactosidase present. The structures and tests of such synthetic thiogalactosides indicate that the steric demands for an optimal inducer are very specific. Isopropyl-1-thio-β-D-galactoside (IPTG) is the best inducer of all 1-thio-β-D-galactosides. If a saturating amount of IPTG (10^{-3} mol l^{-1}) is added to *lac* wild-type ($I^+O^+Z^+Y^+$) cells growing on glycerol, newly synthesized β-galactosidase can be detected 3 min after the addition of the inducer: 3 min is the time it takes to synthesize the four subunits of β-galactosidase which form one molecule. From then on the rate of synthesis does not change any more. Such kinetic measurements were used by Jacques Monod to argue for *de novo* synthesis of β-galactosidase and against the instruction theory. Finally, compounds exist which counteract induction: they are called anti-inducers. *o*-Nitrophenyl-β-D-fucoside (i.e., *o*-nitrophenyl-β-D-6-deoxygalactoside) is the best-known example of an anti-inducer of the *lac* operon.

4. Inducers inactivate repressors or activate activators. Lac repressor occurs in two conformations. In the absence of inducer it binds tightly to *lac* operator DNA and thus represses transcription from the adjacent *lac* promoter. In the presence of inducer it changes its conformation and binds about 1000-fold less tightly to *lac* operator. In the wild-type situation induction of the *lac* operon depends on the second or third power of inducer (IPTG) concentration. A close analysis indicates that all four subunits participate in operator binding. Two subunits bind the main operator, the two other subunits bind an auxiliary operator. Thus only one monomer of tetrameric Lac repressor has to be occupied by inducer in order that repression decreases drastically. Indeed induction of Lac repressor does not follow the model of an allostery, where either all four or none of the subunits of Lac repressor would have to change their conformation. One subunit after the other binds to inducer as inducer concentration increases. Finally it should be pointed out that the exact mechanism of the detailed structural changes of Lac repressor during induction is unknown. *lac* mutants have been isolated, which still repress but in which inducer does not induce any more. This may happen either by destruction of the inducer binding site of Lac repressor or by destruction of the region where the structural changes caused by inducer binding occur. Such mutants are negative dominant. They are called I^S.

Induction was explained according to the well-analyzed paradigm of the *lac* operon. Inspection of other systems indicates that they act in principle in a similar manner but often differently in detail. Some examples will illustrate this. It was stated in the beginning that induction may work either through counteracting of repression or through stimulating activation. Like the *lac* system, the *gal* system of *E. coli* is induced by D-galactose which inactivates Gal repressor. In contrast to the *gal* system of *E. coli*, the *gal* system of yeast is indirectly controlled. It is induced by D-galactose which binds to GAL80 protein. GAL80 protein binds in the absence of galactose to the activator GAL4 and thus inactivates GAL4. In the presence of galactose, GAL80 no longer interacts with GAL4, thus allowing it to activate transcription. Finally the signal which leads to induction may not be a chemical. *E. coli* lysogenic for phage lambda may be induced by UV radiation. UV irradiation leads to the formation of thymine dimers in the DNA. The presence of thymine dimers triggers the turning on of the SOS pathway. This in turn leads to the proteolytic destruction of lambda repressor by RecA and so to the induction of phage lambda i.e., the liberation of phage lambda from repression.

Reference

Cohn M, Monod J, Pollock M, Spiegelman S and Stanier R (1953) Terminology of enzyme formation. *Nature* 172: 1096–1097.

See also: Beta (β)-Galactosidase; *lac* Operon

Infertility

P J Turek and R R Pera

Copyright © 2001 Academic Press
doi: 10.1006/rwgn.2001.0450

Introduction

Infertility is a common human health problem, in fact, almost as common as diabetes mellitus. Approximately 10–15% of couples of reproductive age are infertile. In women, common causes of infertility include tubal or pelvic disorders such as endometriosis, ovulatory dysfunction, or anatomical problems. In men, infertility can be caused by the presence of dilated blood vessels around the testes (varicoceles), blockage or absence of the spermatogenic tubules from infection or congenital absence of the vas deferens, and low or no sperm counts (oligospermia and azoospermia, respectively) from testicular failure. Genetic causes of infertility can lead to either defects in sperm or egg production or result in defects in anatomical development within the reproductive tract. This article will review our present understanding of these two kinds of genetic causes of infertility.

Genetic Infertility: Problems with Egg and Sperm Production

A frequent cause of infertility is the production of sex or germ cells (sperm or oocytes) in fewer than normal numbers or of poorer than normal quality. Germ cell production is complex and differs from that of any other cell type. Normal body (somatic) cells replicate by a process termed mitosis, in which identical daughter cells are created; no reduction in chromosome number occurs. However, when germ cells replicate, the process involves an extra cell division that reduces the number of chromosomes from 46 (diploid) to 23 (haploid). As a result of this extra step, a single diploid cell gives rise to four haploid progenitors. In males, all four cells derived from the diploid precursor cell

become sperm. In females, only one ovum is produced from this process; the remaining three cell products become nonfunctioning polar bodies. This sex cell replication pathway is termed meiosis. In well-studied organisms such as yeast or flies, meiosis involves hundreds of genes for its proper execution.

Genetic Infertility Associated with Egg Production Problems

Common conditions that directly affect the development of oocytes in the ovary are Turner syndrome, premature ovarian failure, and mutations in the follicle stimulating hormone (FSH) receptor. At present, little therapy exists to stimulate the production of oocytes in women with these conditions.

Turner syndrome

Turner syndrome is a well-studied disorder that is associated with structural abnormalities or absence of an X chromosome. Most women with Turner syndrome have a fairly characteristic appearance of short stature, webbed neck, shield chest, and an increased carrying angle of the elbow, associated with primary amennorhea (absence of menses) throughout life. The ovaries of women with Turner syndrome are described as 'streak' ovaries in that they lack oocytes and the normal associated follicular structures. In approximately 60% of Turner women, the karyotype is pure 45,X. In remaining individuals, the karyotype can show variable mosaicism in X or Y chromosome abnormalities (i.e., 45,X/46,XY).

Premature ovarian failure

Premature ovarian failure is also termed premature menopause. It is defined by secondary amennorhea (absence of menses) before age 40. It is believed that women enter menopause when oocyte reserves decrease from an initial population of approximately 500 000 at birth to approximately 1000. Premature ovarian failure, especially at a young age, can be caused by deletions on the long arm of the X chromosome. There are likely three or four different regions of the X chromosome required for oocyte production and deletions in any of these regions may cause premature ovarian failure. The genes that map to these regions have not yet been identified.

Normal development of oocytes is critically dependent upon the pituitary hormones, luteinizing hormone (LH) and follicle-stimulating hormone (FSH). The failure of oocyte development in women with a normal XX karyotype was considered to be unrelated to these pituitary hormones until a connection was made in studies of ovarian failure in Finnish women. In studies that used classical human genetic strategies to map a locus called ODG1 (*Ovarian*

DysGenesis 1) to a region of chromosome 2, close examination of this region revealed that it contained the gene that encodes for the FSH receptor. The open reading frame of the FSH receptor gene was sequenced from Finnish women with ovarian failure and revealed a number of mutations that alter the binding of FSH to its receptor. This work implies that the FSH receptor gene is required for normal oocyte development. However, the incidence of mutations in this gene in other ethnic groups is not yet known.

Genetic Infertility Associated with Sperm Production Problems

Genetic conditions that impair the development of sperm in the testicle tend to result from structural or numerical chromosomal abnormalities. Despite this, in even the most severe cases of low sperm production, biological paternity is possible with assisted reproductive technologies such as intracytoplasmic sperm injection (ICSI); illustrated in **Figure 1**. The use of ICSI in such cases virtually ensures that the genetic cause of infertility will be transmitted to offspring.

Klinefelter syndrome

Klinefelter syndrome is the most common genetic reason for azoospermia in men, accounting for 14% of cases. In this abnormality of chromosomal number, 90% of men carry an extra X chromosome (47,XXY) and 10% of men are mosaic with a combination of

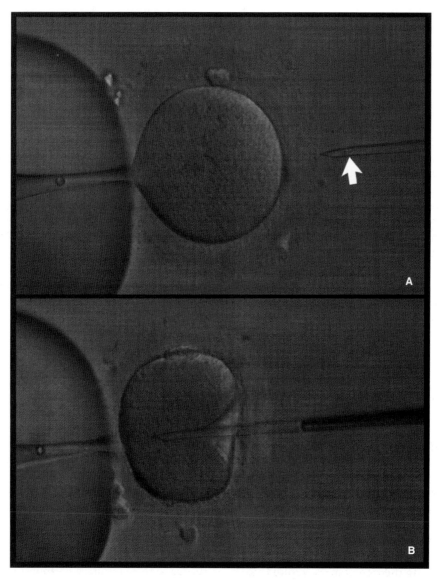

Figure 1 The intracytoplasmic sperm injection (ICSI) procedure. (A) A mature oocyte (left) is readied for injection with a sperm (arrow) in a micropipette under high-power microscopy. (B) The micropipette is placed directly into the oocyte and the sperm deposited in the cytoplasm.

XXY/XY chromosomes. This syndrome may present with increased height, decreased intelligence, varicosities, obesity, diabetes, leukemia, an increased likelihood of extragonadal germ cell tumors, and breast cancer (20 times higher than normal males). Paternity with this syndrome is rare, and more likely in the mosaic or milder form of the disease. Recently, paternity has been reported in several cases of pure XXY men with the use of ICSI.

XYY syndrome

XYY syndrome is based on another abnormality of chromosomal number and can result in infertility. Typically, men with 47,XYY have normal internal and external genitalia, but are taller than average. Semen analyses show either severe oligospermia or azoospermia. Testis biopsies may often demonstrate arrested germ cell development or complete absence of germ cells (Sertoli cell-only syndrome).

XX male syndrome

XX male syndrome is a structural and numerical chromosomal condition that presents as a male with azoospermia. Typically, there is normal male external and internal genitalia. Testis biopsy usually reveals an absence of spermatogenesis. The most obvious explanation for the disease is that the sex-determining region (SRY) or testis determining region is translocated from the Y to another chromosome. Thus, testis differentiation occurs, but other Y chromosome genes required for sperm production (see below) are not similarly translocated, with resultant sterility.

Noonan syndrome

Noonan syndrome presents phenotypically as a male Turner syndrome (45,X). However, the karyotype in these men is normal 46,XY and the chromosomal abnormality has not yet been identified. Typically, these men have dysmorphic features such as webbed neck, short stature, low-set ears and wide-set eyes. At birth, 75% will have cryptorchidism (undescended testes) that may limit fertility in adulthood.

Immotile cilia syndromes

Immotile cilia syndromes are a heterogeneous group of disorders in which sperm motility is reduced or absent. The sperm defects are based on abnormalities in the motor apparatus or axoneme of sperm and other ciliated cells. Normally, 10 pairs of microtubules within the sperm tail are connected by dynein arms (ATPase) that regulate microtubule and, therefore, sperm tail motion. In these conditions, various defects in the dynein arms cause deficits in ciliary motion and sperm activity. Most immotile cilia cases are diagnosed in childhood due to respiratory and sinus difficulties. Cilia within the retina and ear may also be defective and lead to retinitis pigmentosa and deafness (Usher syndrome). Men with immotile cilia characteristically have completely nonmotile but viable sperm in normal numbers. Depending on the severity of the ciliary defect, some sperm motility can be present.

Azoospermia gene(s)

Approximately 10–15% of men with azoospermia have structural changes in the Y chromosome. The sex-determining region (SRY) of the Y chromosome that controls testis differentiation is intact, but deletions may exist on the long arm of the chromosome (Yq) that result in azoospermia or severe oligospermia (**Figure 2**).

A relationship between the Y chromosome and spermatogenesis was originally postulated based on the finding of structural changes in the chromosome detected by karyotype in a population of men with azoospermia. This led to the hypothesis that the

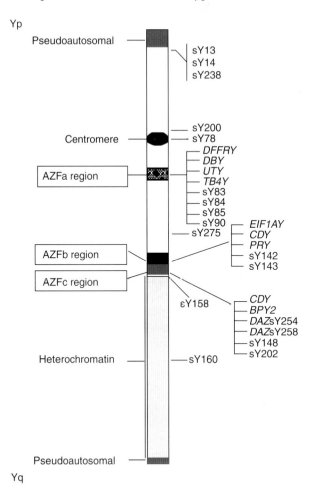

Figure 2 Three regions of the Y chromosome are required for fertility in men. They are termed the AZFa, AZFb, and AZFc regions.

Y chromosome held an 'azoospermia factor (AZF).' A mutation in, or absence of, AZF was thought to account for the azoospermia in men with observed deletions of Yq. Since then, more sophisticated analyses of the Y chromosome indicate that three gene sites may carry AZF genes. The exact function of these suspected genes in spermatogenesis has not yet been clearly delineated, as the gene products are only just beginning to be elucidated. Genes identified include *RBM* (*RNA-Binding Motif*), *DAZ* (*Deleted in AZoospermia*), and a number of others, as shown in **Figure 2**. It is likely that men who have these gene deletions will pass them to offspring if assisted reproductive technology is used to achieve paternity.

Genetic Infertility Associated with Reproductive Tract Abnormalities

Female Reproductive Tract Abnormalities

Infertility can often be traced to abnormal development of the female reproductive tract, including the ovaries, oviducts, uterus, and vagina. Although it is clear that genetic causes for abnormal development exist, they are likely to be polygenic or multifactorial in nature. Major female reproductive tract abnormalities include endometriosis, polycystic ovarian syndrome, and anomalies of uterine structure.

Endometriosis
Endometriosis is a complex disorder characterized by the presence of endometrial glands and stroma outside of the uterus. The most frequent sites of endometriosis are the ovaries, the uterosacral ligaments, the anterior and posterior cul-de-sac, and the posterior broad ligaments. It is estimated that 3–10% of women of reproductive age have endometriosis and that 25–35% of infertile women have endometriosis. No genes that cause endometriosis have been identified. Yet it is likely that genetic factors influence susceptibility to endometriosis. Numerous studies have found a 5- to 10-fold increase in the incidence of the disorder in first-degree relatives of patients with endometriosis when compared with control groups.

Polycystic ovarian syndrome
Polycystic ovarian syndrome is characterized by anovulation associated with the persistence of numerous cysts and a continuous secretion of gonadotropins and sex steroids. Similar to endometriosis, polycystic ovarian syndrome is common and may occur in 10–15% of women with normal reproductive function. Among infertile women with anovulation, polycystic ovaries are detected in 75% of cases. The genetics of polycystic ovarian syndrome are complex, yet studies suggest that, like endometriosis, there may be a 5- to

10-fold increase in the disorder in first-degree relatives of affected patients compared with controls. To date, no genes implicated in the disorder have been identified.

Uterine abnormalities
Uterine abnormalities due to defects in mullerian development (such as in Mayer–Rokitansky–Kuster–Hauser syndrome) are a relatively common cause of primary amennorhea. Abnormalities range from incomplete development of the vagina to the complete absence of all mullerian structures (fallopian tubes, uterus, and upper vagina). It is clear that the normal development of these structures requires proper function of the mullerian inhibiting substance (MIS) gene. Yet, mutations or structural alterations in the gene have not yet been identified in affected women. There are studies to suggest that in families affected by Mayer–Rokitansky–Kuster–Hauser syndrome, the uterine abnormality is likely to be caused by mutations in three or four different genes.

Male Reproductive Tract Abnormalities

In 10% of male infertility cases, there is abnormal development of the male reproductive tract. Abnormalities of wolffian duct development may affect the epididymis, vas deferens, seminal vesicles or associated ejaculatory apparatus, and generally result in obstruction to the flow of sperm from the testis. As with abnormal development of the female reproductive tract, such genetic conditions in men are predominantly polygenic or multifactorial in nature. This discussion excludes conditions that present at birth or childhood with ambiguous genitalia (intersex disorders).

Cystic fibrosis
Cystic fibrosis is the most common fatal autosomal recessive disorder in the United States. It is associated with more than 550 possible genomic mutations. The disease manifests with fluid and electrolyte abnormalities (abnormal chloride-sweat test) and presents with chronic lung obstruction and infections, pancreatic insufficiency, and infertility. Interestingly, 98% of men with cystic fibrosis (CF) also have wolffian duct abnormalities. The body and tail of the epididymis, vas deferens, seminal vesicles, and ejaculatory ducts are atrophic, fibrotic, or completely absent. Pituitary–gonadal hormones and spermatogenesis are usually normal. Fertility is possible with assisted reproductive technology such as ICSI.

Congenital absence of vas deferens
Congenital absence of the vas deferens (CAVD) accounts for 1–2% of all cases of infertility and up to 5% of azoospermic men. Men with this condition

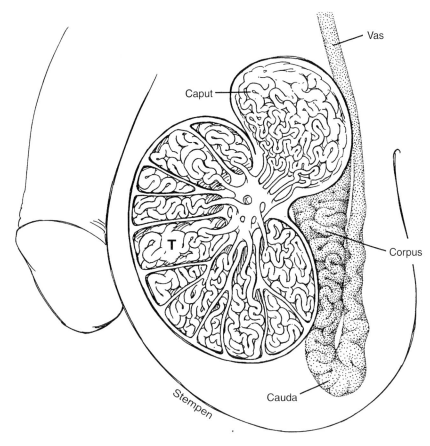

Figure 3 Illustration of scrotal anatomy. In congenital absence of the vas deferens (CAVD), there is a normal testis (T), but the epididymis and vas deferens (vas) are abnormal. The caput epididymis (caput) is present and attached to the testis, but the corpus and cauda epididymis and the vas deferens are absent (stipled areas).

have no palpable vas deferens (one or both sides) on physical examination (**Figure 3**). Similar to CF, the rest of the wolffian duct system may also be abnormal and is largely unreconstructable. Recently, this disease has been shown to be a genetic *form fruste* of CF, even though the vast majority of these men fail to demonstrate any symptoms of CF. In men with bilateral vasal absence, 65% will harbor a detectable CF mutation. In addition, 15% of these men will have renal malformations, most commonly unilateral renal agenesis. In patients with unilateral vasal absence, the incidence of detectable CF mutations is lower, and the incidence of renal agenesis approaches 40%. Pituitary–gonadal hormones are usually normal, as is spermatogenesis.

Young syndrome

Young syndrome presents with the clinical triad of chronic sinusitis, bronchiectasis, and obstructive azoospermia. The obstruction is located in the epididymis, usually near the junction of the head and body. Since obstruction may not occur until well after puberty, fertility is possible in some patients. The pathophysiology of the condition is unclear but may involve abnormal ciliary function or abnormal

mucus quality. Pituitary–gonadal hormones and spermatogenesis are normal in these men. Reconstructive microsurgery can be attempted in these men but usually meets with lower success rates than observed with other obstructive conditions.

Idiopathic epididymal obstruction

Idiopathic epididymal obstruction is a relatively uncommon, but well-recognized condition found in otherwise healthy azoospermic men in which the small ducts within the epididymis are obstructed. It can be successfully treated with microsurgical reconstruction. There is recent evidence linking this condition with CF: in one series, 47% of men so obstructed were seen to harbor a gene mutation associated with CF. This implies that up to one-half of patients with obstruction in the epididymis may in fact have a genetic predisposition for the problem.

Summary

Although our understanding of genetic causes of female and male infertility is still quite naive, it is already obvious that research in this field has the potential to decipher the origins of many cases of presently unexplained infertility. It is also important for patients to understand that genetic infertility may be passed to offspring, given the recent revolutionary developments in the field of assisted reproduction.

Further Reading

Desjardins C and Ewing LL (1993) *Cell and Molecular Biology of the Testis*. New York: Oxford University Press.

Mak V and Jarvi K (1996) The genetics of male infertility. *Journal of Urology* 156: 1245–1257.

Seibel MM (1997) *Infertility: A Comprehensive Text*, 2nd edn. Stamford, CT: Appleton & Lange.

Speroff L, Glass RH and Kase NG (1994) *Clinical Gynecologic Endocrinology and Infertility*. Baltimore, MD: Williams & Wilkins.

See also: Ethics and Genetics; Fertilization

Influenza Virus

K N Gracy and W Fitch

Copyright © 2001 Academic Press
doi: 10.1006/rwgn.2001.0690

Influenza, caused by the influenza virus, is a highly contagious infection of the nose, throat, bronchial tubes, and lungs. Its severity and recurrence is caused by the ability of the virus to mutate quickly and thus reinfect populations that have already built up antibodies to the virus through a previous infection.

The virus evolves in two ways. Mutations gradually build up through continued replication of the viral RNA. This antigenic 'drift' allows the virus to evade the immune system of the host, even if it has been previously infected with an older version of the virus. The virus can also mutate through abrupt replacements of the hemagglutinin and neuraminidase genes that make up part of its protein coat. This antigenic 'shift' results in a new subtype of virus that has no immunological relation to the previous subtype, thus accounting for the disease's virulence when a new form enters the population.

There are three types of influenza virus: type A causes the most severe infections and type C the most mild. Type A viruses undergo both antigenic drift and shift, while type B viruses change only through antigenic drift. Type C viruses cause mild illness and do not lead to epidemics. Type A viruses are further categorized by differences in their hemagglutinin and neuraminidase coat proteins. There are at least 15 varieties of hemagglutinin (H) and nine varieties of neuraminidase (N) that can combine to create different strains, so viruses are named according to the type of H and N proteins they produce. Because of the virus's ability to mutate quickly and constantly (1% change per year in hemagglutinin), inoculations against it are only temporarily effective. The World Health Organization and the Centers for Disease Control and Prevention oversee influenza surveillance and make recommendations for the next year's vaccines based on the virus's mutations the previous year. The most common subtypes of influenza A are designated A(H1N1) and A(H3N2). These, along with an influenza type B strain are included in the trivalent vaccines produced each year against the influenza virus.

Influenza infection can be severe or even lethal. Death, particularly in the young, elderly, or immunocompromised, is generally caused by cardiopulmonary or upper respiratory complications associated with the infection. Influenza infection peaks during the winter months and, when an antigenic shift occurs, can spread pandemically. The most notable of these, the Spanish Flu Pandemic of 1918, caused between 20 and 40 million deaths worldwide. One-fifth of the world population was infected. More localized epidemics occur frequently and have led to the recommendation of yearly flu vaccinations, particularly in susceptible populations.

Further Reading

Center for Disease Control and Prevention:www.cdc.gov/ncidod/disease/ flu/fluinfo/htm

Landon Pediatric Foundation: www.medmall.org/Profu/

See also: Virus

Inheritance

J Merriam

Copyright © 2001 Academic Press
doi: 10.1006/rwgn.2001.0691

The evidence for inheritance includes circumstantial anecdotes, such as the ancient statement that "like begets like," or the more contemporary statement that "traits run in families." It also includes sophisticated recognition that the experimental manipulation of traits through breeding, with consistent and predictable results, requires a mechanism of determinative factors or genes that are transmitted from parent to offspring. Inheritance refers to the mechanism that genes, and more specifically the permanent condition, or allele, that can be distinguished from other alleles of the same gene, is transmitted from parent to offspring. From examining the observation that related individuals share traits and picking out distinct, rare traits to follow, details of the inheritance mechanism that explain, for instance, how certain traits reappear after seeming to skip generations, have been worked out. This has progressed so well that inheritance can refer to the mechanism for either the transmission of specific traits or the transmission of all the biological information required for life without specifying genotypes.

See also: Mendelian Inheritance; Quantitative Inheritance

Inherited Rickets

J L H O'Riordan

Copyright © 2001 Academic Press
doi: 10.1006/rwgn.2001.1375

Rickets is a disorder in which there is a failure of mineralization of bone and an accompanying defect in remodeling of growing bone in children. The mineral deposit is primarily made up of calcium and phosphate, as hydroxyapatite, and so it is to be expected that disorders of both calcium and phosphate regulation could cause of rickets.

The calcium disorders that cause inherited rickets are primarily those related to the metabolism or action of vitamin D. Cholecalciferol (vitamin D_3) can be formed in the skin by UV irradiation of 7-dehydrocholesterol or it can be absorbed from the diet. It is metabolized in the liver to 25-hydroxy-vitamin D, which can be further hydroxylated in the kidney by a 1-hydroxylase enzyme to the active form, namely 1,25-dihydroxycholcalciferol. Ergocalciferol (vitamin D_2) is a product of plants that can be metabolized similarly. The term '1,25-dihydroxyvitamin D' includes both the vitamin D_2 and the vitamin D_3 forms. It can be regarded as a hormone in that it is produced by one organ, namely the kidney, and enters the circulation to act on another organ, particularly the intestine, where it increases calcium absorption. In the circulation it is bound to a transport protein, vitamin D-binding protein, so it is carried like other steroid hormones. Within the target cells is also behaves like a steroid hormone, being transported to the nucleus by a vitamin D receptor protein, which, when complexed to ligand, can bind to DNA through its zinc fingers to modulate transcription, for example, of calcium-binding protein in the intestine.

The phosphate disorders that cause inherited rickets are those involving defects in renal tubular phosphate reabsorption. This is an active process that involves sodium-dependent phosphate transporters, the genes for which have been cloned. However, mutations in these genes for the phosphate cotransporters have not been shown to occur and these genes are therefore not relevant to the conditions described here. When there is excessive renal tubular loss of phosphate, hypophosphatemia develops and that leads to defective mineralization of bone and so to rickets.

There are five situations in which rickets develops on the basis of a known gene mutations. Two of these are related to vitamin D, and one is a consequence of a failure of calcium reabsorption in the renal tubule. The other two forms are due to failure to reabsorb phosphate in the renal tubules.

Rickets due to Inherited Abnormalities in the Synthesis or Action of Vitamin D

Defect in Vitamin D_1 Hydroxylation

The clinical features of this disorder (OMIM 264 700) were first described as pseudovitamin D-deficiency rickets. It seems likely to be an autosomal recessive disease. The phenotype consists of severe rickets. There is hypocalcemia and the characteristic feature is finding a normal concentration of circulating 25-hydroxyvitamin D with a low concentration of circulating 1,25-dihydroxyvitamin D. The rickets in these patients heals completely after treatment with small doses of 1,25-dihydroxyvitamin D.

The 1-α hydroxylation of 25-hydroxyvitamin D occurs in the renal tubules, under the influence of 25-hydroxy-1-α hydroxylase (P450c1-α). The human gene maps to locus 12q14 and has been cloned; it consists of nine exons spanning a region of approximately 4.8 kb. A transcript of 2.5 kb has been detected in renal tissue. The 508-amino acid P450c1-α protein has a predicted topology that is similar to that of mitochondrial cytochrome P-450 enzymes, with a putative N-terminal mitochondrial signal sequence and conserved ferredoxin- and heme-binding sites. Mutations of this gene (also called *CYP27P1*) have been found in a number of families of varying ethnic origin. These mutations include single base-pair substitutions, causing alterations in single amino acid residues, as well as deletions, resulting in loss of function of the enzyme.

Vitamin D-Resistant Rickets with End Organ Unresponsiveness

Patients with this condition (OMIM 277 440) have resistance to treatment with vitamin D in any form. Some of the patients have associated alopecia. It has been suggested that with the presence of alopecia the rickets is more severe, but this is not the case. The condition is inherited as an autosomal recessive and occurs particularly in the Arab countries but also in Japan and the Philippines. Biochemically the hallmark of the condition is the presence of high circulating concentrations of 1,25-dihydroxyvitamin D. The disease can be treated effectively by infusions of calcium, overcoming the defect in calcium absorption, but the infusions need to continue for a long time, usually daily, for about a year. It is remarkable that with this treatment the healing can be complete and that relapse does not occur for several years after the treatment has stopped.

The condition is generally due to mutations in the gene for the vitamin D receptor (locus 12-q14). The gene consists of 11 exons, spanning approximately 75 kb. Exons 2 and 3 encode the two zinc fingers that are responsible for binding to DNA, while exons 7, 8, and 9 encode the ligand binding domain, which complexes 1,25-dihydroxyvitamin D. Mutations of either domain can cause rickets. However, in one patient with the typical phenotype, no mutation was found, despite sequencing the whole of the coding region and large parts of the noncoding regions. A knockout model in mice, with deletion of the vitamin D receptor gene, produces a phenotype that includes alopecia as well as rickets. The effects of mutations in the DNA binding domain can be analyzed at the crystallographic level, by comparing the known crystal structure of the DNA-binding part of the glucocorticoid receptor, which is presumed to have a similar structure to the corresponding part of the vitamin D receptor, for which only the amino acid sequence is known. All the mutations in the DNA-binding domain of the vitamin D receptor that cause rickets affect conserved residues that have a particular function (such as hydrogen bonding between the proteins and the DNA) in the crystal structure of the glucocorticoid receptor-DNA complex. The larger ligand-binding domain of the vitamin D receptor has itself been crystallized and its structure when complexed to ligand has been established. There is considerable homology in structure between this receptor and that for thyroid hormone and for the retinoid receptors. The mutations in its ligand-binding domain that cause rickets affect residues that are important for dimerization of the vitamin D receptor to the retinoid X receptor, dimerization which is necessary for action of the vitamin. Thus in this case it is possible to consider the effects of mutations causing rickets at the Angstrom level.

Hypercalciuric Rickets

Dent's disease (OMIM 300 009) was originally described as a combination of rickets and hypercalciuria. It later became apparent that in the same families a variety of other phenotypes could occur, including renal tubular proteinuria, nephrocalcinosis and renal calculi, and the development of renal failure. Within any one family, the phenotype is variable, and in some families rickets does not occur. As a result of this variable phenotype, the condition has had various names, including 'X-linked recessive nephrocalcinosis.'

The disease maps to Xp11.22. Mutations in this condition led to the discovery of the voltage-gated chloride channel gene *CLCN5*. The gene is organized into 12 exons spanning 25–30 kb of genomic DNA. Mutations of the same gene have been found in X-linked recessive Dent's disease, X-linked recessive nephrocalcinosis, and X-linked recessive hypophosphatemic rickets, implying that these are all variants of the same disease phenotype. The way in which this chloride channel affects the handling of calcium and protein in the renal tubule remains to be established.

Rickets due to Renal Tubular Phosphate Leak

X-Linked Dominant Hypophosphatemic Rickets

This condition (OMIM 307 800) is characterized by severe rickets, with a low serum phosphate concentration and inappropriately raised level of urinary phosphate excretion. Paradoxically bone density in this condition is raised, even though there is the defect of mineralization. In adults, the increased bone density may be associated with ossification of intraspinus ligaments, and there may occasionally be cord compression due to exostoses. The condition responds partially to treatment with oral phosphate supplements, which have to be accompanied by vitamin D treatment, since phosphate on its own produces hypocalcemia.

In this condition there are mutations of the *PEX* gene. This acronym refers to the gene being involved in phosphate handling, having homology with endo-peptidases, on the X chromosome (locus Xp22.2–p22.1). The gene is also known as the *Phex* gene. The predicted protein has a small intracellular region, a single transmembrane domain, and a large extracellular catalytic domain. The homology with metalloproteinases, particularly neutral endopeptidase, was unexpected, and the mechanism whereby this mutation causes rickets is not clear. It is possible, by analogy with a tumor-associated form of rickets, that the enzyme is acting on a putative phosphate-regulating hormone, 'phosphatonin.' The gene is expressed in bone cells and in the kidney but its role remains unclear. In X-linked hypophosphatemic rickets, about three-fourths of patients have mutations of this gene which can be detected. These mutations include deletions that may be large or small, or there may be point mutations leading to single amino acid changes or splice-site alterations. There are two mouse homologs; one of these is the Hyp mouse. This was the result of a spontaneous mutation, while the second model is the Gyr mouse, in which there is hypophosphatemic rickets plus a gyratory movement. In both mutations of the mouse homolog, *Pex* has been found; in the Gyr mouse this is a deletion which includes also an adjacent gene. In the Hyp mouse, there is evidence that a hormonal mechanism is involved, which provides some support for the possibility that the *PEX* gene product is acting upon the yet unidentified hormone.

Autosomal Dominant Hypophosphatemia

This condition (OMIM 193 100) is similar to X-linked dominant hypophosphatemic rickets, but increased bone density seems not to be a feature, and in fact osteoporosis may become a problem in later life. It is remarkable that the biochemical and clinical features of autosomal dominant hypophosphatemic rickets can disappear in late childhood, although they may subsequently recur in later life. This can make it difficult to establish the true phenotype in an adult, especially since the severity of the condition can vary within the same family.

The gene causing this condition has recently been identified and mutations have been established. The gene encodes a protein that is homologous to fibroblast growth factors and has been given the name *FGF23*. The mechanism whereby alterations in such a protein lead to defects in renal tubular phosphate reabsorption are not clear.

In conclusion, it should be pointed out that, in identifying five genes, mutation of which leads to rickets, the nature of the relevant gene product was quite unexpected in three of them.

Further Reading

Online Mendelian Inheritance in Man (OMIM), http://www.ncoi.nlm.gov.omim/.

See also: **Growth Factors; Sex Linkage; Vitamins**

Initiation Factors

A J Berk

Copyright © 2001 Academic Press
doi: 10.1006/rwgn.2001.0692

Initiation factors are proteins other than the RNA polymerase required for correct initiation of transcription (transcription initiation factors). They are also proteins which, in addition to ribosomal proteins, are required for initiation of translation (translation initiation factors). In eubacteria, transcription initiation factors are called σ factors. In eukaryotes, transcription initiation factors usually refer to the general transcription factors required for transcription initiation from most promoters. For RNA polymerase II these are TFIIA, TFIIB, TFIID, TFIIE, TFIIF, and TFIIH; for RNA polymerase III these are TFIIIB, as well as TFIIIC, for tRNA genes, and TFIIIA and TFIIIC for 5S rRNA genes; and for RNA polymerase I these are SL1 and UBF in humans, and TBP, Rrn3, core factor, and upstream activating factor in *Saccharomyces cerevisiae*. Bacterial translation initiation factors are called IF1, IF2, and IF3. Eukaryotic translation initiation factors include eIF1, eIF2, eIF3, eIF4, and eIF5. The eIF4E subunit of eIF4 binds to the 5′ cap structure on eukaryotic mRNAs.

See also: **Transcription**

Insertion Sequence

M Chandler

Copyright © 2001 Academic Press
doi: 10.1006/rwgn.2001.0696

Discovery

Insertion sequences (ISs) are small pieces of DNA which move within or between genomes using their own specialized recombination systems. They were discovered in the mid-1960s in studies of gene expression in *Escherichia coli* and its bacteriophages. Initially recognized by their ability to generate highly polar but unstable mutations in the *gal* and *lac* operons and in the 'early' genes of bacteriophage lambda, they were later identified by electron microscopy as short insertions of DNA. The repeated isolation of a limited number of identical DNA sequences associated with these unstable mutations led to their being named: insertion sequences. The similarity of ISs and the mobile genetic elements described by Barbara McClintock in *Zea mays* in the 1940s became clear when it was realized that ISs formed an integral part of the *E. coli* genome and that their mutagenic activity was a result of their movement to new genetic locations. At about this time, transmissible resistance to antibiotics was also observed. Genetic studies of this phenomenon implicated an analogous mechanism of gene mobility in the distribution of these drug resistance genes among the conjugal plasmids and phage involved in this transmission. Subsequently, insertion sequences were shown in many cases to play a key role in mobilizing these genes.

General Structure

ISs are genetically compact (**Figure 1**), typically less than 2.5 kb in length, and carry only the genes necessary for their transposition. They comprise a single, or sometimes two, open-reading frames covering almost the entire length of the element. The products are specialized recombinases called transposases (Tpases). ISs characteristically terminate in small flanking (10–40 bp) inverted repeat sequences (IRs) with imperfect homology. By convention, the terminal inverted repeat proximal to the Tpase promoter is defined as the left repeat, IRL, while the distal IR is defined as the right repeat, IRR. In the majority of cases, ISs are flanked by small directly repeated duplications in the target DNA, which they generate on insertion. The length of this duplication is specific for each element and ranges from 2 to 13 bp.

Occurrence and Variety

ISs form an integral part of the chromosomes of many bacterial species and their extrachromosomal elements such as plasmids and bacteriophages. They have also been found in the genomes of many eukaryotes. ISs can represent a significant fraction of genomic and plasmid DNA. Although individually each IS is mobile at a low frequency (of the order of 1×10^{-7} to 10^{-9}/cell per generation), such movements rarely become established on a population scale. The localization of many ISs is sufficiently stable within their host genomes to provide a specific and characteristic profile, which has made certain ISs useful markers in

epidemiological studies. ISs have been characterized from most bacterial species analyzed to date and over 600 have been described. They can be grouped into at least 17 families based on their genetic organization, similarities in their IRs and transposase sequences, the number of target base pairs duplicated on insertion, and their preference for given target DNA sequences (**Table 1**). As more of these elements are characterized, this classification will certainly continue to evolve. Such groupings have provided significant insights into important conserved features, which not only assist in understanding their phylogenetic relationships but also contribute to understanding different aspects of their function.

Role of ISs in Gene Transfer and Expression

In early studies of antibiotic resistance, resistance genes were often observed to be flanked by DNA sequences of between 1 and 2 kb in direct or inverted orientation. These segments of DNA proved to be ISs. By acting in concert, the flanking ISs are able to mobilize the intervening DNA segment. Such structures are known as compound or composite transposons. The mobilized genes are not limited to antibiotic resistance but can include virulence determinants and catabolic genes. ISs are thus important in the sequestration, assembly, and transmission of sets of accessory functions in bacteria. Moreover, many elements can control expression of neighboring genes either by initiating transcription from indigenous IS promoters or, more commonly, by formation of hybrid promoters as a result of insertion. Many ISs carry outwardly directed −35 hexamers in their IRs and can generate functional promoters when inserted at the correct position with respect to a −10 element upstream from a host gene.

Note that compound transposons differ fundamentally in organization from a second large class, transposons of the Tri3 family, where the genes specifying accessory functions form an integral part of the transposon. However, this family of transposons also includes elements resembling ISs in addition to more elaborate elements in which the typical accessory genes have been integrated.

Terminal Inverted Repeats

Transposition requires DNA cleavage at the ends of the element and transfer of these DNA ends into a target molecule. The signals for recognition and processing by the transposase reside in the terminal IRs. Analysis of several different IRs suggests the presence of at least two functional domains (**Figure 1**). One,

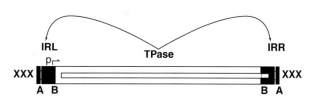

Figure 1 General organization of IS elements. The open box represents the IS element. Terminal inverted repeats (IRL and IRR) are shown as shaded boxes. A single open reading frame is shown within the IS. It stretches the entire length of the element and, although not always the case, is shown here to terminate within IRR. The indigenous Tpase promoter is shown located (by convention) in IRL. The arrows show that the protein acts on the ends of the element. The domain structure of the IRs is indicated by A (the region recognized by Tpase and which is involved in cleaveage) and B (the region to which Tpase binds in a sequence-specific way). XXX represents the short direct target repeat sequence which is duplicated during the insertion event.

Table I   The IS families

Family	Groups	Size range	Direct target repeats (bp)	Ends	IR	ORF	TPase
IS1	–	770	9 (8–11)	GGT	Y	2	lambda integrase ?
IS3	IS2	1200–1550	5	TGA	Y	2	DDE
	IS3		3 (4)			2	
	IS51		3 (4)			2	
	IS150		3–5			2	
	IS407		4			2	
IS4	–	1300–1950	9–12	C(A)	Y	1	DDE
IS5	IS5	800–1350	4	GG	Y	1	DDE
	IS427		2–3	Ga/g		(2)	
	IS903		9	GGC		1	
	IS1031		3	GAG		1	
	ISH1		8	–		1	
	ISL2		2–3	–		1	
IS6	–	750–900	8	GG	Y	1	DDE
IS21	–	1950–2500	4 (5, 8)	TG	Y	2	DDE
IS30	–	1000–1250	2–3	–	Y	1	DDE
IS66	–	2500–2700	8	GTA	Y	> 3	–
IS91	–	1500–1850	N	–	N	1	ssDNA Rep
IS110	–	1200–1550	N	–	N	1	Site-specific recombinase
IS200/IS605	–	700–2000	N	–	N	1(2)	Complex organization
IS256	–	1300–1500	8–9	Gg/a	Y	1	DDE eukaryote relatives
IS630	–	1100–1200	2	–	Y	1	DDE eukaryote relatives
IS982	–	1000	?	AC	Y	1	DDE
IS1380	–	1650	4	Cc/g	Y	1	–
ISAS1	–	1200–1350	8	C	Y	1	–
ISL3	–	1300–1550	8	GG	Y	1	–

Size range in base pairs (bp) represents the typical range of each group. N, no; less frequently observed lengths are included in parentheses; Ends, typical nucleotide sequences at the very ends of the element. Presence (Y) or absence (N) of terminal inverted repeats is indicated. DDE represents the common acidic triad presumed to be part of the active site of the transposase. ssDNA Rep indicates that the enzyme is a polymerase of the rolling circle type.

located within the IR (B), is involved in Tpase binding and probably assures correct sequence-specific Tpase positioning at the ends. The second (A) corresponds to 2–4 base pairs located at the tip of the IRs and is necessary for efficient cleavage and strand transfer. These bases, generally identical at both ends of the element, are presumably in intimate contact with the catalytic pocket of the Tpase and determine the specificity of the cleavage (and/or strand transfer) reactions. IRL and IRR are tacitly assumed to interact in a similar way with Tpase and their contribution to the reaction is thought to be identical. This may, however, prove to be an oversimplification and the subtle sequence differences found between the IRs of certain elements may prove to reflect differential activity of the ends.

Indigenous IS promoters are often partially located in IRL (Figure 1). This arrangement would facilitate autoregulation of transposase expression by transposase binding. In addition to carrying sites for Tpase

and RNA polymerase binding, binding sites for other host-specified proteins involved in regulation of Tpase expression or in modulating the transposition activity of the ends may be located within or proximal to the IRs.

Members of a small number of IS families (Table I) do not exhibit terminal IRs and are also the only families which do not generate direct target repeats on insertion. This is presumably because such elements have adopted fundamentally different transposition mechanisms.

Transposases: Domain Structure and Catalytic Site

Many Tpases encoded by ISs share a similar overall organization. A region involved in recognition of the ends is located in an N-terminal domain, while the catalytic core of the enzyme is located toward the C-terminal end. These enzymes also function as multimers and carry domains involved in multi-merization. Indeed, in several cases, multimerization appears to be essential for DNA binding (see below).

Sequence alignment of most bacterial Tpases and the functionally related retroviral integrases, IN (which catalyze integration of the double-stranded viral cDNA into the host genome), revealed a common triad of acidic amino acids with a characteristic spacing, the DDE motif (**Table 2A**). This similarity was subsequently shown to include additional conserved amino acids and has also been detected in many other major transposons (**Table 2B**) and IS

families (**Table 1**). Extensive mutagenesis both of IN and a limited number of Tpases has shown that the DDE motif is intimately involved in catalysis. Determination of the three-dimensional structure of IN and several Tpases confirmed the close juxtaposition of these residues and demonstrated that these enzymes share related topological folding. This structural similarity is not limited to IN and Tpases but is also seen in RNase H and in *RuvC*, the endonuclease which processes recombination intermediates. These observations have led to the definition of a 'superfamily' of phosphoryltransferases.

Table 2 The DDE motif showing representative transposases from various insertion sequence families (A) and transposases from other bacterial transposons (B)

A	N2	N3	C1
HIV-1 (IN)	64 wql **D** cth (51)	116 vht **D** ngsnf (35)	152 ynpqsQgvi **E** smNKel **K**
IS911 (IS3)	207 wcg **D** vty (59)	287 fhs **D** qgshy (35)	323 gncwNspm **E** rffRsl **K**
IS10 (IS4)	97 vlv **D** wsd (63)	161 ivs **D** agfkv (130)	292 niyskRmq **E** etfRdl **K**
IS50 (IS4)	119 siq **D** ksr (67)	188 avc **D** readi (136)	326 diythRwri **E** efHKaw**K**
IS903 (IS5)	121 lvi**D** stg (71)	193 asa **D** gaydt (65)	259 niyskRmqi **E** eftRdl**K**
IS26 (IS6)	78 whm **D** ety (59)	138 int **D** kapay (36)	173 qikylNNvi **E** cdHgkl**K**
IS30	237 weg **D** lvs (55)	293 ltw **D** rgmel (33)	327 qspwqRgtn **E** ntNgli **R**
IS21 (IstA)	122 lqh **D** wge (61)	184 vlv **D** nqkaa (46)	230 rrartKgkv **E** rmvKyl **K**
IS630	181 fye**D** evd (80)	261 liv **D** nyiih (35)	297 vyspwvNhv **E** rlwQal
IS982	112 sii **D** sfp (79)	192 vlg **D** mgylg (45)	237 nfskrRKvi **E** rvsfl
IS256	167 lmt **D** vly (65)	233 vis **D** ahkgl (107)	341 nrlkstNli **E** rlNQev **R**
Tc1	86 iws **D** esk (90)	177 fqq **D** ndpkh (108)	286 spspdlNpi **E** hmweele **R**

B	N2	N3	C1
Mu (MuA)	269 ing **D** gyl (66)	336 iti**D** ntrga (55)	392 kgwgqaKpv **E** rafgvg
Tn7 (TnsA)	28 (hgk **D** yip) (85)	114 mst**D** flvdc (34)	149 erleKlel **E** rrywqq**K**
Tn7 (TnsB)	273 yei **D** ati (87)	361 lla**D** rgelm (34)	396 rrfdaKgiv **E** stfRel
Tn552	166 wqa **D** htl (73)	240 fyt **D** hgsdf (35)	276 gvprgRgki **E** rffQtv
Tn3	689 asa **D** gmr (75)	765 imt**D** tagas (129)	895 riltqlNrg **E** srHava**R**

Large bold letters indicate highly conserved residues, smaller bold letters indicate partially conserved residues. Bold figures above each line indicate the coordinates in amino acid residues and figures in parentheses indicate the number of residues between the conserved DDE. Part A includes an example of the HIV integrase protein to show its similarity to Tpases and Tc1, a member of the eukaryote mariner/Tc insertion elements.

Not all ISs exhibit a well-defined DDE triad (**Table 1**). For example, the Tpases of members of the IS91 family show strong similarities with replicases involved in rolling circle plasmid and bacteriophage replication. Members of the IS110 family appear to encode a novel type of site-specific recombinase, while the IS1 transposase shows limited similarity to phage lambda integrase.

Transposition Strategies

Endonucleolytic cleavage of the phosphodiester bonds at the ends of the transposable element and their transfer into a target DNA molecule generally requires the assembly of a synaptic complex including the Tpase, the transposon ends, and target DNA. There are two principal modes of transposition, conservative and replicative, based on whether or not the element is copied in the course of its displacement. This is dictated by the nature and order of the cleavages at the ends (**Figure 2**): whether the transposon is liberated from its donor backbone by double strand cleavages or whether it remains attached following cleavage of only a single strand.

The DNA cleavage and strand joining reactions necessary for transposition of many transposable elements with Tpases of the DDE type are remarkably similar. These Tpases catalyse endonucleolytic cleavage at each 3′ transposon end to liberate 3′ OH groups, which are then used in a concerted nucleophilic attack on the target molecule. An important feature of the transposition reaction is therefore the way in which the 5′ end (second strand) is processed.

Replicative transposition entails cleavage of only one strand at each transposon end and transfer into a target site in such a way as to create a replication fork (**Figure 2**). Some IS elements do not appear to process the second strand and simply undergo replicative transposition, or more precisely, 'replicative integration.' These include members of the Tn3 and IS6 families and perhaps IS1. If transposition is intermolecular, replication from the nascent fork(s) generates cointegrates (replicon fusions), where donor and target replicons are separated by a directly repeated copy of the element at each junction. Resolution of these structures to regenerate the donor and target molecules, each carrying a single copy of the element, is accomplished by recombination between the two elements. This proceeds for some transposons by site-specific recombination promoted by a specialized transposon-specific enzyme distinct from the Tpase, the 'resolvase' (e.g., Tn3 family), or is taken in charge by the host homologous recombination system.

In conservative or 'cut-and-paste' transposition, the element is excised from the donor site and reinserted into a target site without replication. This implies cleavage of both DNA strands at the ends of the element and their rejoining to target DNA to generate a simple insertion. The original donor DNA molecule is either degraded or repaired by host-specified enzymes. Different IS elements have adopted various strategies to separate themselves from the donor DNA backbone. For the IS4 family members, IS10 and IS50 (**Figure 2**), the two breaks are not analogous. 3′ cleavage occurs before 5′ cleavage and the free 3′ OH generated by 3′ cleavage is itself used as the nucleophile in attacking the second strand. This generates a hairpin structure at the transposon ends; this is subsequently hydrolyzed to regenerate the final 3′ OH ends, which will undergo transfer to the target. The free ends are retained in a relatively stable complex with Tpase and generate a noncovalently closed excised transposon circle. This mechanism is reminiscent of V(D)J recombination used in generating the immunoglobin repertoire, although the V(D)J hairpin is generated on what might be considered as the donor backbone ends. This chain of controlled consecutive reactions allows the repeated use of a single Tpase molecule bound to each end of the element.

A second strategy is used by IS2, IS3, IS150, and IS911 and presumably by other members of this large IS3 family. Here, Tpase promotes single strand cleavage at one end of the transposon and its site-specific transfer to the same strand of the opposite end (**Figure 2**). This circularizes a single transposon strand leaving the complementary strand attached to the donor backbone. This second transposon strand is then resolved to generate a double-stranded covalently closed transposon circle, in which the transposon ends are abutted. The resolution mechanism but could involve simple cleavage and repair or replication promoted by host proteins. The covalently attached ends can then undergo simultaneous single strand cleavage and transfer to a target. This strategy of separating the transposon from its donor molecule may have also been adopted by members of the IS21 and IS30 families. While site-specific strand transfer from one end of the element to the other generates transposon circles, it can also occur between two elements carried by the same molecule. Transfer of ends between the two IS copies in a plasmid dimer, for example, would be expected to generate head-to-tail IS tandem dimers. This type of structure has been observed for IS21, IS2, IS30, and IS911, and is extremely active in transposition.

Of those ISs that do not carry a well-defined DDE triad, only IS91 has been analyzed in detail. As suggested by the similarity of its Tpase with rolling circle type replicases, IS91 appears to have adopted a polarized rolling circle transposition mode requiring a

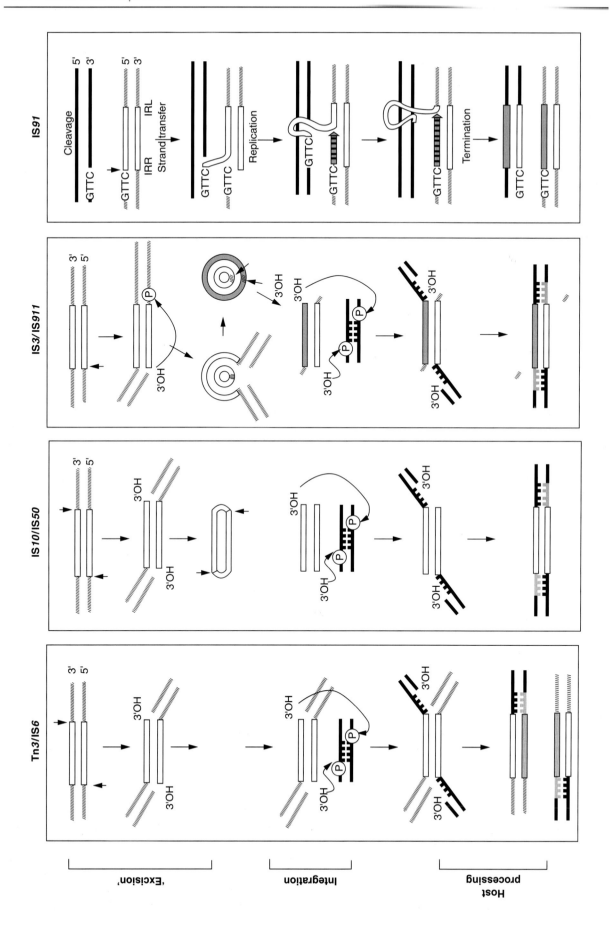

specific tetranucleotide target sequence which abuts IRR (**Figure 2**). 'One-ended' transposition products occur at high frequency in the absence of IRL. They carry a constant end defined by IRR and a variable end defined by a copy of the target consensus located in the donor plasmid. It is thought that donor strand cleavage results in a covalent complex between the 5′ IRR end and Tpase and is followed by single-strand transfer into the target DNA at a site containing a consensus tetranucleotide. The attached single strand of the IS is displaced by replication in the donor molecule. Termination is triggered when the complex reaches either the 3′ IRL end or a tetranucleotide consensus sequence in the donor (**Figure 2**). This scheme does not, however, address how the element is replicated into the target molecule.

Target Specificity

ISs show differing degrees of selectivity in their choice of target DNA sites. Sequence-specific insertion is exhibited to some degree by several elements and varies considerably in its stringency. It is strict in the case of IS91, which requires a GTTC/CTTG target sequence, but less strict for members of the IS630 (and the related eukaryotic mariner/Tc) family, which require a TA dinucleotide in the target, for IS10, which prefers (but is not restricted to) the symmetric 5′-NGCTNAGCN-3′ heptanucleotide, and for IS231, which shows a preference for 5′-GGG(N)$_5$CCC-3′. In the case of IS10, sequences immediately adjacent to the consensus have also been shown to influence target choice. A demonstration that IS10 Tpase directly influences target choice has been obtained by isolation of Tpase mutants which exhibit altered target preference. Other elements show regional preferences such as DNA segments rich in GC (IS186) or AT (IS1), which could reflect more global parameters such as local DNA structure. Indeed, the degree of supercoiling (IS50), bent DNA (IS231), replication (IS102), transcription (IS102, Tn5/Tn10), and possibly protein-mediated targeting to (or exclusion from) transcriptional control regions have all been evoked as parameters which influence target choice.

Another phenomenon which may reflect insertion site specificity is the interdigitation of various intact or partial IS elements noted repeatedly in the literature. These are presumably the scars of consecutive but isolated transposition events resulting from selection for acquisition (or loss) of accessory genes. Some indication of the statistical significance of this is expected to emerge from the many bacterial genome sequencing projects underway. On the other hand, several ISs show a demonstrable preference for insertion into other elements: IS231 inserts into the terminal 38 bp of the transposon Tn4430, which includes both the sequence-specific and conformational components described above, while IS21 has been reported to show a preference for insertion close to the end of a second copy. In the latter case, the site-specific DNA binding properties of the Tpase are presumably implicated. At the mechanistic level, this phenomenon might be related to the capacity of IS10 Tpase to form synaptic complexes with IS10 ends located on separate DNA molecules.

Control of Transposition Activity

High levels of transposition are likely to be disadvantageous to the host cell under normal growth conditions and ISs have adopted a variety of mechanisms to restrain this activity. The location of Tpase promoters partially in IRL would permit autoregulation by Tpase binding. Some ISs such as IS1 encode specific repressor proteins. Additionally, binding sites for a range of host encoded proteins are found within or close to IS ends. These proteins include IHF, FIS, and DnaA. Not only can their binding regulate transposition activity *per se*, but can also provide rather subtle changes in the type of transposition products obtained. In some elements, Tpase promoter activity is also regulated by the state of methylation of neighboring sites. One example is Dam methylation of a GATC sequence in the Tpase promoter of IS10. Transcription directed by the promoter is reduced when the site is fully methylated (on both strands) compared with hemimethylated DNA. Methylation has the double effect of lowering transposition activity of the end and of timing bursts of transposase synthesis with the passage of a replication fork which produces transiently hemimethylated DNA. This assures duplication of the element prior to transposition, an important consideration for elements which transpose in a conservative mode. An additional level of regulation at the level of transcription is by premature termination and mRNA processing. Transcription terminator

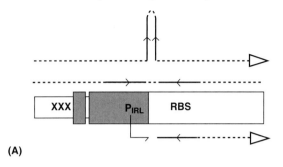

(A)

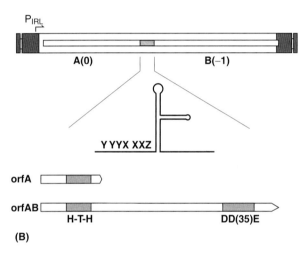

(B)

Figure 3 Control strategies. (A) Sequestration of ribosomal binding sites. The figure shows the left end of an IS with its terminal inverted repeat represented by two shaded boxes. The transcript impinging from outside the element is shown as a dotted line above and transcription driven by the indigenous promoter (P$_{IRL}$) is indicated below. Internal inverted repeat sequences are indicated by bold lines and their relative orientation is shown by arrows. (B) Programmed translational frameshifting. Two consecutive open-reading frames (A and B) together with their relative reading phases (0 and −1, respectively) and the region of overlap ( ) are shown within the IS element. Below (bold line) is shown the overall secondary structure of the corresponding mRNA. The group of codons which permit the ribosome to slide back one nucleotide is also indicated. The bottom of the figure shows how frameshifting can assemble two different functions into one protein. Here this is represented by a helix–turn–helix motif (H–T–H) in the N-terminal region, which permits sequence-specific binding of the Tpase to the ends of the IS and a DD(35)E motif in the C-terminal region which is essential for catalysis.

sequences have been uncovered within the Tpase genes of IS1 and IS30 and are undoubtedly widespread.

Integration into a highly active gene would also be expected to activate expression of IS genes. Many elements have adopted specific strategies to reduce such adventitious activation. One such strategy is to sequester translation initiation signals (**Figure 3A**). Here, an internal inverted repeat sequence carrying the Tpase ribosome binding site (rbs) is located close to IRL. Transcripts invading IRL from neighboring DNA will carry the inverted repeat and form a stem-loop structure trapping the rbs. Translation initiation signals in transcripts from the resident promoter, however, remain accessible since these carry only the proximal repeat (**Figure 3A**). This has been demonstrated for the IS4 family members, IS10, and IS50, but many other ISs carry appropriately placed potential hairpin structures. Another level of control operates at translation initiation and involves synthesis of an antisense RNA which sequesters translation initiation signals. This type of control has been well documented for IS10, where it is responsible for multicopy inhibition in which the presence of an IS10 copy on a high copy-number plasmid inhibits the activity of a copy located in the chromosome.

Additional regulation may occur at the level of translation elongation. Several ISs carry two partially overlapping open reading frames (ORFs). In one case, the IS21 family, this arrangement may give rise to translational coupling. In its simplest form, this may use an overlap of the last base of the termination codon of the upstream ORF (in phase 0) with the first of the initiation codon of the downstream ORF arranged in phase −1 (TGATG). A second mechanism which regulates transposase synthesis involves programmed translational frameshifting (**Figure 3B**). A −1 frameshift occurs by slippage of the translating ribosome one base upstream. Translation then continues in the alternative (−1) phase. This occurs at the position of 'slippery' codons in a heptanucleotide sequence generally of the type Y YYX XXZ in phase 0 (where the bases paired with the anticodon are underlined), which is read as YYY XXX Z in the shifted −1 phase. The sequence A AAA AAG is a common example of this type of heptanucleotide. Ribosomal shifting of this type is stimulated by structures in the mRNA that tend to impede the progression of the ribosome, such as potential ribosome binding sites upstream or secondary structures (stem-loop structures and potential pseudoknots) downstream of the slippery codons. Translational control of transposition by frameshifting has been demonstrated for IS1 and for members of the IS3 family, but may also occur in several other IS elements (e.g., one subgroup of the IS*5* family).

Other control mechanisms may occur at translation termination. In some cases, the translation termination codon of Tpase genes is located within their IRR sequences, while in others the transposase gene simply does not possess a termination codon. Among the latter cases, the IS is known to insert into a specific target sequence in which the target direct repeat produced on insertion itself generates the Tpase termination codon. This has been observed for certain members of the IS*630* family. The significance of these arrangements may be to couple translation termination, transposase binding, and transposition activity.

Early studies of IS*1* and IS*50* demonstrated that impinging transcription from outside reduces transposition activity. Transcription may disrupt the formation of the transposition complexes known as transpososomes in which transposase and the transposon ends are intimately bound.

Tpase stability can also contribute to control of transposition since it can limit activity both temporally and spatially. This may explain the observation that several Tpases function preferentially in *cis* (see below). Derivatives of the IS*903* Tpase that are more resistant to the *E. coli* Lon protease than the wild type protein are more active and exhibit an increased capacity to function in *trans* (see below).

Early studies indicated that transposition activity of some elements was more efficient if the transposase is provided by the element itself or by a transposase gene located close by on the same DNA molecule. This preferential activity in *cis* reduces the probability that transposase expression from a given element will activate transposition of related copies elsewhere in the genome. The effect can be of several orders of magnitude. It presumably reflects a facility of the cognate transposases to bind to transposon ends close to their point of synthesis and is likely to be the product of several phenomena such as expression levels and protein stability. Another contributing factor may derive from the domain structure of known transposases (see above) in which the DNA binding domain is located in the N-terminal end of the protein. This arrangement would permit preferential binding of nascent transposase polypeptides to neighboring binding sites. Indeed, the N-terminal portion of several Tpases exhibits a higher affinity for the ends than does the entire transposase molecule, suggesting that the C-terminal end may mask the DNA binding activity of the N-terminal portion.

Further Reading

Berg DE and Howe MM (eds) (1989) *Mobile DNA*. Washington, DC: American Society for Microbiology Press.

Chaconas G, Lavoie BD and Watson MA (1996) DNA transposition: jumping gene machine, some assembly required. *Current Biology* 6: 817–820.

Haren L, Ton-Hoang B and Chandler M (1999) Integrating DNA: transposases and retroviral integrases. *Annual Review of Microbiology* 53: 245–281.

Mahillon J and Chandler M (1998) Insertion sequences. *Microbiology and Molecular Biology Reviews* 62: 725–774.

Mizuuchi K (1992) Transpositional recombination: mechanistic insights from studies of Mu and other elements. *Annual Review of Biochemistry* 61: 1011–1051.

Mizuuchi K (1997) Polynucleotidyl transfer reactions in site-specific DNA recombination. *Genes to Cells* 2: 1–12.

Rice P, Craigie R and Davies DR (1996) Retroviral integrases and their cousins. *Current Opinion in Structural Biology* 6: 76–83.

Saedler H and Gierl A (eds) (1996) *Transposable Elements, Current Topics in Microbiology and Immunology*, Vol. 204. Berlin: Springer-Verlag.

See also: *Escherichia coli*; Transposable Elements

Insertion, Insertional Mutagenesis

See: Chromosome Aberrations; DNA Cloning; *In vitro* Mutagenesis; Mutation

Insulinoma

C S Grant

Copyright © 2001 Academic Press
doi: 10.1006/rwgn.2001.1584

Insulinoma occurs primarily in one of two principal forms, sporadic and familial, specifically as one component of the multiple endocrine neoplasia type 1 (MEN-1). MEN 1 is a clinical syndrome inherited in an autosomal dominant pattern, and includes primary hyperparathyroidism, multiple duodenopancreatic endocrine tumors (of which insulinoma is one type), and pituitary adenomas. No specific genetic abnormality has been consistently identified as the cause of sporadic insulinomas, whereas the recently cloned gene responsible for inheritance of MEN-1 has been mapped to chromosome 11q13. This gene contains 10 exons that encode a 610-amino acid protein product, menin. Research suggests that the MEN-1 gene is a tumor suppressor gene.

Further Reading

Chandrasekharappa SC, Guru SC, Manickam P et al. (1997) Positional cloning of the gene for multiple endocrine neoplasia-type 1. *Science* 276: 404.

Larsson C, Skogseid B, Oberg K, Nakamura Y and Nordenskjold M (1988) Multiple endocrine neoplasia type 1 gene maps to chromosome 11 and is lost in insulinoma. *Nature* 332: 85–87.

See also: **Adenoma; Multiple Endocrine Neoplasia**

Integrase

N Grindley

Copyright © 2001 Academic Press
doi: 10.1006/rwgn.2001.0698

The term integrase is used to describe the following two enzymes:

1. An enzyme (Int) responsible for catalyzing the breakage and rejoining of DNA during the insertion of a bacteriophage genome into (and its excision from) its chromosomal attachment site by the process of site-specific recombinases. Most Int proteins belong to the tyrosine recombinase family of site-specific recombinases but several examples of serine recombinases are also known.
2. An enzyme (IN) encoded by retroviruses that is responsible for the 3′ processing of retroviral DNA and insertion of the processed DNA into a genomic target. IN proteins are derived by proteolysis from the C-terminus of the gag-pol polyprotein, and belong to the DD(35)E family of transposases.

See also: **Integrase Family of Site-Specific Recombinases; Phage λ Integration and Excision; Retroviruses; Site-Specific Recombination; Transposable Elements**

Integrase Family of Site-Specific Recombinases

A Landy

Copyright © 2001 Academic Press
doi: 10.1006/rwgn.2001.1452

The Int family of recombinases belongs to the general class of proteins that act on specific DNA sequences to effect deletion, insertion, or inversion of large segments of genomic DNA. The approximately 100 known proteins in this family are found in archaebacteria, eubacteria, and eukaryotes. Sometimes referred to as the tyrosine integrase family, they are distinguished by their use of a tyrosine nucleophile and five highly conserved basic residues to catalyze DNA cleavage and ligation reactions in the absence of high-energy cofactors. Another hallmark for the recombinases in this family is a sequential strand exchange mechanism that generates a four-way DNA junction (Holliday junction) as a recombination intermediate.

The biological roles of various Int family members include copy number control and stable inheritance of circular replicons, the integration and excision of viral chromosomes into and out of the chromosomes of their respective hosts, the regulation of expression of cell surface proteins, conjugative transposition, the movement of antibiotic resistance genes into and out of transposable elements and plasmids, and the relaxation of positive and negative supercoils during eukaryotic DNA replication repair, recombination, and transcription.

The Reaction

The minimal Int family target on DNA consists of a single binding site for a topoisomerase monomer. DNA strand cleavage involves activation of the scissile phosphate by the highly conserved pentad of active site residues (Arg, Lys, His, Arg, His) and formation of a nick with a 5′ OH and 3′-phosphotyrosine linkage to the recombinase. This transient covalent intermediate releases one superhelical turn, via mechanics that are not completely understood, and the nick is resealed by a simple reversal of the cleavage step.

In the case of Int family-mediated recombination, the minimal DNA target consists of two recombinase binding sites that are positioned as inverted repeats separated by 6–8 bp (called the 'overlap region'). Synapsis and proper alignment of two such recombination partners generates a tetrameric complex in which each recombinase protomer carries out the cleavage and ligation of one DNA strand, executed as two sequential pairs of cleavage/ligation reactions. In the first pair of reactions one strand in each partner DNA helix is cleaved and the first three or four bases of the free 5′ hydroxyl-terminated strands of the overlap region are swapped and then ligated. This forms a Holliday junction with four continuous DNA strands (see **Figure 1**). After some rearrangements within the Holliday junction, the intermediate is 'resolved' by a reciprocal strand swapping of the second pair of strands so that all four DNA strands have new junctions and two recombinant DNA helices have been generated. The formation of Holliday juction intermediates distinguishes the Int family from the resolvase/invertase family of site-specific recombinases, which use a serine nucleophile to carry out a pair of concerted (rather than sequential) strand exchanges, and from the transposase family of

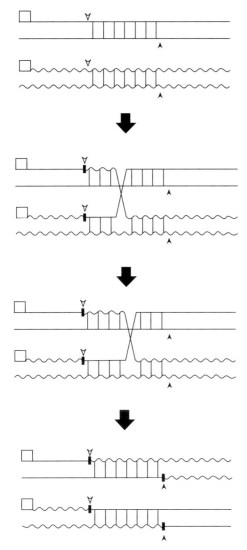

Figure 1 Holliday junction formation and resolution by Int family members. One strand of each recombining partner duplex is cleaved (open arrowhead) at the left boundary of the overlap region (in this example, 7 bp denoted by short vertical lines) forming a covalent 3′ phosphotyrosine intermediate (not shown). A short segment (three bases in this example) of single-stranded DNA from each partner is swapped between the duplexes and the phosphotyrosine linkages are disengaged by the formation of new phosphodiester linkages (heavy vertical bars) at the recombinant joints (second panel). Following this pair of ligation reactions the Holliday junction intermediate undergoes rearrangements that include movement of the crossover point and some conformational changes (third panel) which set the stage for the second pair of strand exchanges on the other side of the overlap region (filled arrow heads). Now it is the bottom strands that are cleaved, swapped, and religated to form the second pair of recombinant joints and the recombinant product duplexes.

reactions, which do not involve covalent protein–DNA intermediates and additionally require some DNA synthesis to complete recombination.

The Overlap Region

The sequential pair of reciprocal strand exchanges that generate and then resolve a Holliday junction are separated not only temporally but also spatially, by six to eight base pairs that are referred to as the 'overlap' region. The overlap region is precise and characteristic for each recombinase. Because of this stagger in cleavages the resulting overlap region in the recombinant DNA helices is 'heteroduplex,' i.e., it has one strand from each of the two parental helices. If the overlap region DNA sequence were not identical in the two parental helices the recombinant heteroduplex region would have base pair mismatches. In most (but not all) Int family pathways such mismatches are not tolerated and there is a strict requirement for sequence identity in the overlap regions of recombining partners. It is thought that the reciprocal (and reversible) strand swaps of 3–4 bp is where sequence identity is recognized and not at some earlier step such as the synapsis or alignment of parental helices.

Target Specificity

There are two sources for the target specificity in the Int family recombinations. One is the requirement for overlap region identity described above. This source of specificity serves primarily to match two targets to each other because there is a wide latitude of DNA sequences that can function in the overlap region for a given recombinase. The second source of specificity resides in the DNA-binding sites of the recombinase and any required accessory proteins. In the simplest Int family reactions this amounts to four protein binding sites, one for each of the required recombinase monomers. These 7–9 bp recognition sequences, which occur as inverted repeats flanking the overlap regions, are allowed and may be favored by some degeneracy. Thus, the overall target specificity is equivalent to a DNA sequence of approximately 15–20 bp. For some Int family members target specificity is further enhanced by the addition of 'extra' recombinase binding sites that are not essential for the minimized reaction but may play a role in nature. Even higher specificity is to be found among those Int Family members that contain a second specific DNA-binding domain and/

or depend upon several sequence-specific accessory proteins

Sub-Families

Monomeric Targets
Based upon the number of reaction components, the eukaryotic type IB topoisomerases exemplify the most basic Int family reaction: a single protomer executing one cleavage/ligation reaction on one strand of duplex DNA. The two best-studied examples of this subgroup, and for which X-ray crystal structures are available, are the human topo I and the vaccinia virus-encoded topoisomerase.

Dimeric Targets
After the topoisomerases, the most basic recombination pathway requires four identical protomers and is described for the most part by the reaction scheme outlined above. The two best-studied examples of this group are the Cre recombinase of the *Escherichia coli* bacteriophage P1 and the Flp recombinase of the 2 μm plasmid of the yeast *Saccharomyces cerevisiae* (see **Table 1**). Cre recombinase acts on two DNA target sites (*lox* sites) to reduce multimers of P1 plasmid to monomeric circles (each containing a single *lox* site) and thereby numerically favoring the passively dispersive inheritance of P1 to both daughter cells. FLP recombinase also has the biological function of enhancing plasmid inheritance but uses a different strategy. The 2 μm plasmid contains two Flp target sites (*frt* sites) oriented with respect to each other such that recombination between them results in inversion (rather than deletion) of the intervening DNA. The effect of this inversion is to convert two divergent DNA replication forks into two tandem forks with a rolling circle mode of replication that generates multiple copies of plasmid DNA. Neither Cre nor Flp exhibit topological or orientation selectivity. That is, they are capable of recombining sites on different molecules or on the same molecule. When the sites are on the same molecule they can be direct repeats, leading to excision of the intervening DNA (called resolution), or they can be inverted repeats, leading

to inversion of the intervening DNA, as is found in nature for Cre and Flp, respectively.

Accessory Proteins
The next step up in Int family complexity is best exemplified by the XerC/XerD pathway of *E. coli*, in which the first pair of strand exchanges is executed by XerC and the second pair by (the closely related) XerD. Additionally, two site-specific DNA-binding proteins, ArgR and PepA, which have other roles in *E. coli*, are incorporated as structural elements in the synaptic complex between two XerC/XerD recombination sites. They act at accessory sequences such that approximately 180 bp of DNA adjacent to each core recombination site are interwrapped approximately three times in a right-handed fashion. The topology of the interwrapped synapsed sites ensures that recombination occurs only between directly repeated sites on the same molecule, a constraint consistent with the role of this pathway in converting plasmid multimers into monomers. This pathway is also responsible for maintaining the *E. coli* chromosome as a monomer and does so at a site called *dif*. It is interesting to note that XerC/XerD core recombination sites with a 6 bp instead of an 8 bp overlap sequence do not require the accessory proteins and lose the orientation and topological selectivity. However, when supplied with accessory sequences and accessory proteins selectivity is restored.

Heterobivalent Recombinases
The third level of Int family complexity has been best studied in the pathways of lysogenic viruses that catalyze the integration and excision of viral chromosomes into and out of the chromosomes of their hosts. Ironically, the first Int family member to be identified genetically and characterized biochemically was the integrase (Int) of bacteriophage lambda, one of the well-studied exemplars of the most complex pathways in this family. The distinguishing feature of this subgroup is that they possess an additional DNA-binding domain that binds with high affinity to 'arm-type' sites that are different and distant from the core-type binding sites where strand exchange takes place. The apparent paradox raised by a heterobivalent recombinase was resolved by the finding that several essential accessory proteins are sequence-specific DNA-bending proteins with binding sites that fall between the two different types of Int binding sites. The introduction of 'U-turn' bends in the DNA delivers Int bound at the high affinity arm-type sites to the lower affinity core-type sites where catalysis takes place. In the lambda pathway example, two of the accessory bending proteins, IHF (integration host factor) and Fis (factor for inversion stimulation), are encoded by

Table 1 Levels of complexity in different Int family reactions

Recombinase	Accessory factors	Heterobivalent Int
Cre	None	–
Flp	None	–
Xer	ArgR, PepA	–
λInt	IHF, Xis, Fis	+

the *E. coli* host where they play important roles in the regulation of DNA transcription and replication, and the third accessory bending protein, Xis (excision factor), is encoded by the viral genome. This additional complexity affords mechanisms by which the viruses can both control the direction of recombination (the presence or absence of Xis is required for excisive versus integrative recombination, respectively) and modulate its efficiency (the levels of IHF and Fis have opposite effects on the efficiency of excisive recombination).

Structures

Crystal structures have been determined for four recombinases and two topoisomerases. Each of the structures captures a different view of the protein: a monomeric catalytic domain of lambda Int, a dimeric catalytic domain of HP1 Int, the full length protein of XerD, a Cre tetramer covalently bound to a Holliday junction recombination intermediate, the catalytic core of vaccinia virus topoisomerase, and fragments of human topoisomerase I complexed with DNA. The most dramatic and informative of the structures are those of the cocrystals with their respective target DNAs, where many of the biochemical insights into Int Family reaction mechanisms have been visualized and extended.

A number of informative generalizations also emerge from a comparison of the structures and especially from the structures involving protein–DNA cocrystals. Despite the great divergence in primary amino acid sequences there are extensive regions where the six structures possess a similar tertiary fold, but because of the differences in primary sequence these structural similarities are punctuated by insertions and deletions. A critical region, involving several of the conserved active site residues and the tyrosine nucleophile is surprisingly not part of the highly conserved tertiary fold. However, it is thought that these differences are likely due to the different multimerization states and the presence or absence of DNA in the crystals. It remains to be determined whether these differences might reflect structures that are relevant at different steps in the reaction or whether they are idiosyncrasies of the particular crystallization states. As expected, the crystal structures of the Int family site-specific recombinases comprise the foundation and impetus for further sharpening our understanding of this fascinating class of DNA transactions.

Further Reading

Grainge I and Jayaram M (1999) The integrase family of recombinase: organization and function of the active site. *Molecular Microbiology* 33: 449–456.

Sadowski PD (1993) Site-specific genetic recombination: hops, flips and flops. *FASEB Journal* 7: 760–767.

Shuman S (1998) Vaccinia virus DNA topoisomerase: a model eukaryotic type IB enzyme. *Biochimica et Biophysica Acta* 1400: 321–337.

Stark WM, Boocock MR and Sherratt DJ (1992) Catalysis by site-specific recombinases. *Trends in Genetics* 8: 432–439.

See also: **Chromosome Dimer Resolution by Site-Specific Recombination; Cre/lox – Transgenics; Flp Recombinase-Mediated DNA Inversion; Holliday Junction; Integrase Family of Site-Specific Recombinases; Phage λ Integration and Excision; Resolvase; Resolvase-Mediated Deletion; Site-Specific Recombination; Topoisomerases; Transposase**

Integration

N Grindley

Copyright © 2001 Academic Press
doi: 10.1006/rwgn.2001.0699

Integration is the insertion of one DNA molecule into another to form a single product. It is commonly used to describe the insertion of a viral genome or a plasmid into the chromosome of its host cell.

Integrons

R M Hall

Copyright © 2001 Academic Press
doi: 10.1006/rwgn.2001.0700

In its general sense, the term integron is used to describe genetic entities that are able to capture small mobile elements known as gene cassettes (see Gene Cassettes) and thus have the capacity to incorporate new genes at a specific internal location. Integrons include three characteristic features (**Figure 1**): an *intI* gene, an *attI* site, and a P$_c$ promoter. The *intI* gene encodes a site-specific recombinase (IntI) belonging to the tyrosine recombinase or integrase family. The adjacent *attI* site is a recombination site recognized by the integrase. The IntI integrase also recognizes the 59-be (59-base element) recombination sites found in gene cassettes and incorporates the cassette into the *attI* site. The third key feature of an integron is a promoter (P$_c$), facing toward the *attI* recombination site, that directs transcription of the cassette-associated genes. Thus, integrons are natural cloning vehicles that act

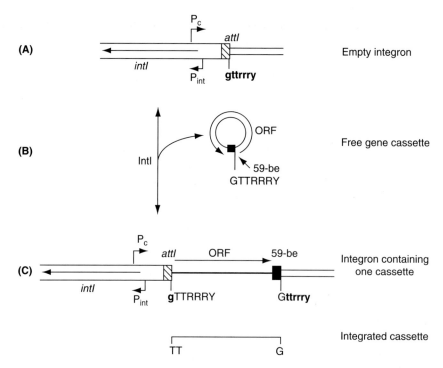

Figure 1 Integrons capture gene cassettes. (A) An empty integron, showing the key features of an integron, an *intI* gene that encodes the IntI integrase, an adjacent recombination site, *attI* (hatched box), and promoters P_c and P_int. (B) A circular gene cassette consisting of a gene or open reading frame (ORF) and a 59-be recombination site (filled box). (C) An integron containing one gene cassette, showing the boundaries of the integrated cassette below. Gene cassettes are inserted into the integron by IntI-catalyzed recombination between *attI* in the integron and the 59-be in the circular cassette. The ORF in the inserted gene cassette can now be transcribed from P_c. The 7-bp core sites surrounding the recombination crossover point in the *attI* site of the integron and in the 59-be of the circular cassette are represented by gttrrry and GTTRRRY, respectively, and the configuration of these bases after incorporation of the cassette is shown in (C). Further cassettes may be inserted at *attI* in like manner, leading to arrays of integrated cassettes.

both as agents of gene capture and as expression vectors for the captured genes.

There Are Many Different Classes of Integrons

Integrons were discovered relatively recently as a consequence of observations made in the 1980s. Heteroduplexes formed between DNA derived from different bacterial plasmids (or transposons) that contain one or more antibiotic resistance genes revealed that several quite different antibiotic resistance genes were flanked by identical, or very closely related, regions of DNA. As sequences became available, the identity of the flanking regions (the integron) was confirmed and the very precise nature of the boundaries between the conserved segments and the various regions containing the resistance genes (the gene cassettes) was revealed. A site-specific recombination mechanism was implied and this was subsequently demonstrated experimentally.

The term integron was originally coined by Stokes and Hall in 1989 to describe this specific group. However, as further different integrons have since been found, this group are now designated class 1 integrons. Class 1 integrons include the characteristic features of an integron, as now defined in its more general sense, but are also mobile elements. They are widely distributed in clinical and environmental gram-negative bacteria and are responsible for the dissemination of many different cassette-associated antibiotic resistance genes. Because the mobility of class 1 integrons is an important factor in spreading resistance genes, for this group the original definition of an integron continues to be used, i.e., they are defined as including the whole mobile element (transposon or defective transposon derivative), which includes *intI1*, *attI1*, and P_c. Hence class 1 integrons are both transposons and integrons and this dual nature allows them to move onto plasmids and hence to become widely distributed in the bacterial world.

Several different *intI*/*attI* units that are associated with gene cassettes have been found and it is likely that many more remain to be discovered. To distinguish them, integrons are classified using the sequence of the *intI* gene and IntI recombinase. Members of the same class have the same (>98% identity) integrase. The known IntI proteins all share significant levels of identity and form a distinct family within the integrons (or tyrosine recombinase) superfamily. Overall, integrons fall into two groups: those that are mobile and those that are an integral part of a bacterial chromosome. The three classes of integrons found in antibiotic-resistant clinical isolates are all mobile. The best-characterized example of the chromosomal integrons is situated in the small chromosome of *Vibrio cholerae*. Chromosomal integrons are also found in other Vibrionaceae and it appears that an integron found its way into the small chromosome of the common ancestor before speciation occurred. Other bacterial species also include an integron as part of their genome.

Integrons Usually Contain Arrays of Gene Cassettes

An integron does not necessarily include any gene cassettes and empty class 1 integrons (**Figure 1A**) have been found in the wild and created experimentally. However, it is most common for one or more cassettes to be found in any individual integron. When cassettes are present they are viewed as part of the integron though, strictly speaking, such integrons are composite structures made up of the integron backbone and an array of gene cassettes. Furthermore, any individual integron–cassette combination can be described by listing the cassettes in order. Arrays of one to five cassettes containing antibiotic resistance genes are most common in mobile integrons but, in the chromosomal integrons, the array can include over 150 gene cassettes as is the case for the one in the recently sequenced *Vibrio cholerae* small chromosome. The *Vibrio* cassette array is a highly variable region of the chromosome and differs from strain to strain. Though most of these cassettes contain an open reading frame (ORF) whose function is not known, genes encoding toxins, virulence factors, restriction and modification enzymes, and a lipoprotein, as well as a few potential antibiotic resistance genes, have all been identified among them. Thus, these long cassette arrays may act as storage depots for cassettes containing a wide range of genes. The cassettes in chromosomal integrons can presumably be picked up and moved out into other organisms by passing plasmids that carry a mobile integron. Indeed, chromosomal integrons are likely to be the source of the cassettes that carry antibiotic resistance genes. As there are a vast number of gene cassettes, each of which can be incorporated into the *attI* site of an integron, and as more than one gene cassette can be integrated at the *attI* site to create arrays containing multiple gene cassettes, a potentially infinite number of configurations are possible.

Integrons Capture Gene Cassettes

The main function of integrons is to capture gene cassettes. Integrons differ from most other *int*/*att* units in that they do not mobilize the entity in which they are contained, rather they act in *trans* to mobilize cassettes. Gene cassettes are the simplest of the known mobile elements and consist of a single gene, or occasionally two genes, and a downstream recombination site. These cassette-associated recombination sites are called 59-be (see Gene Cassettes) and they have a different architecture to that of the *attI* sites. Available information on the integron recombination system and on cassette uptake and loss is largely restricted to studies using the class 1 IntI1/*attI1* system. IntI1 has been shown to recognize both *attI1* and 59-be sites and can catalyze integrative site-specific recombination between any pair of primary sites, *attI1* × *attI1*, *attI1* × 59-be, and 59-be × 59-be, and excisive recombination between *attI1* and a 59-be or between two 59-be. Recombination between *attI1* and a 59-be is the preferred integrative reaction catalyzed by IntI1 and integration of free, circular gene cassettes (**Figure 1**) occurs via a single IntI1-mediated site-specific recombination reaction between *attI1* in the integron and the 59-be in the cassette. Though integrative recombination between two 59-be sites also occurs with high efficiency, it seems to play no part in the integration of gene cassettes. When one or more cassettes are already present in the integron, further cassettes are inserted preferentially at the *attI1* site. Excision of cassettes occurs via both *attI1* × 59-be and 59-be × 59-be reactions. No accessory factors have been identified to date and IntI1 appears to be sufficient for both integration and excision reactions.

Expression of Cassette-Associated Genes

Gene cassettes are compactly organized and the vast majority do not include a promoter. Expression of the cassette-associated genes is thus dependent on the presence of an upstream promoter (see Gene Cassettes). Cassettes are integrated in only one orientation and, in general, the relationship of the gene and 59-be is such that in this orientation the P_c promoter supplied by the integron lies upstream as shown in **Figure 1**. For class 1 integrons containing more than one cassette, it has been shown that all of the cassette genes are transcribed from P_c. Integrons thus create new operons containing a wide variety of genes and gene orders. The level of expression is highest for the gene in the P_c proximal cassette and falls progressively for genes in downstream cassettes. Consequently, a cassette needs to be located relatively close to P_c if its gene is to be expressed. The P_c promoters of other classes of integrons have not been located, but in some cases their presence is implied because antibiotic resistance genes in associated cassettes are expressed. Whether the genes and ORFs found in cassettes that are part of the very long cassette array in the *V. cholerae* chromosome are expressed remains to be established. However, it is possible that only the genes in cassettes located closest to the *attI* site or in a cassette that contains a promoter can be expressed, while downstream genes remain silent.

The *attI* sites

The structure of the *attI1* site has been examined experimentally and is shown in **Figure 2**. Cassettes are incorporated precisely between the G and TT in the right-hand core site (Gttrrry) of the *attI1* simple site. This position is indicated by an arrow in **Figure 2**. A region of 65 bp is required for the reaction between *attI1* and a 59-be. This region includes a simple site, made up of two inversely oriented IntI1 binding domains, and two further IntI1 binding domains that are located to the left and act as recombination enhancers. This enhancement effect is not seen when *attI1* recombines with a second *attI1* site. Differences in the architecture of 59-be and *attI1* sites presumably underlie these preferences.

The sequences of the other integron-associated *attI* sites (*attI2*, *attI3*, etc.) are not closely related either to one another or to *attI1*, but, like *attI1*, do not share the characteristic features of the cassette-associated 59-be sites. The identifiable features shared by the *attI* sites are currently limited to a pair of inversely oriented putative IntI binding domains equivalent to those that make up the simple site in *attI1*. Whether the other *attI* sites also include further IntI binding regions that enhance recombination remains to be established. Available evidence indicates that the various IntI recombinases recognize only their adjacent (cognate) site, though IntI1 is also able to recognize other *attI* sites with low efficiency. Thus, each *attI* site must include distinctive features that permit this selectivity.

Integrons of Different Classes Share Gene Cassettes

It is known that integrons of different classes share cassettes because identical gene cassettes have been found in integrons from more than one class. All of the cassettes that have been found in the cassette arrays of class 2 and class 3 integrons have also been found associated with class 1 integrons. Thus, it appears that the known IntI-type integrases can all recognize the same cassette-associated 59-be sites, though, to date, this has been demonstrated experimentally only for IntI1 and IntI3. This is in contrast to the strong preference of each integrase for its own *attI* site. However, many distinct groups of 59-be have been found and the 59-be in the cassette arrays of any individual chromosomal integron are generally from a single group. In contrast, mobile integrons contain cassettes with many different 59-be types. Hence, it is possible that the different IntI recombinases recognize one type of 59-be more efficiently than others.

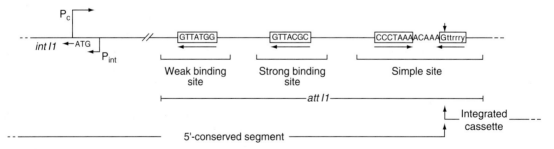

Figure 2 Structure of *attI1* and the promoter region of class I integrons. *attI1* contains four IntI1 binding domains that include a central 7-bp sequence related to the core site consensus sequence GTTRRRY. These core sites are boxed, and arrows indicate their relative orientations. The simple site of *attI1*, within which the recombination crossover (vertical arrow) occurs, is at the right-hand end. The binding sites found to the left of the simple site enhance recombination efficiency. Bases to the left of the crossover belong to the 5′-conserved segment that is found in all class I integrons. Bases to the right of the crossover point (lower case letters) are part of the first integrated cassette, if a cassette is present. The genes in cassettes are transcribed from the promoter P_c, located within the *intI1* gene, and *intI1* is transcribed leftward from P_{int}.

Structure of Class 1 Integrons

Class 1 integrons have a variety of structures resulting from the incorporation of other genes (e.g., the *sul1* sulfonamide resistance gene) and of insertion sequences (IS) that have caused subsequent deletion and rearrangement events leading to loss of some or all of the transposition genes. The presumed progenitor, exemplified by Tn*402*, is a transposon that includes both the integron functions (*intI1*, *attI1*, and P$_c$) and a set of transposition genes (*tniA, B, Q, R*) and is bounded by 25-bp inverted repeats IR$_i$ and IR$_t$. However, this structure is rare and most class 1 integrons are transposition-defective derivatives. Generally, they retain the transposon terminal inverted repeats and hence they can and do move using transposition proteins supplied in trans. Because these structures cannot legitimately be numbered as transposons, they are designated In and numbered to distinguish the many variations in the backbone structure.

Further Reading

Hall RM and Collis CM (1995) Mobile gene cassettes and integrons: capture and spread of genes by site-specific recombination. *Molecular Microbiology* 15: 593–600.

Hall RM and Collis CM (1998) Antibiotic resistance in gram-negative bacteria: the role of gene cassettes and integrons. *Drug Resistance Updates* 1: 109–119.

Partidge SR, Recchia GD, Scaramuzzi C et al. (2000) Definition of the *attI1* site of class 1 integrons. *Microbiology* 146: 2855–2864.

Recchia GD and Hall RM (1995) Gene cassettes: a new class of mobile element. *Microbiology* 141: 3015–3027.

See also: Gene Cassettes; Integrase Family of Site-Specific Recombinases; Site-Specific Recombination

Intelligence and the 'Intelligence Quotient'

T J Crow

Copyright © 2001 Academic Press
doi: 10.1006/rwgn.2001.0702

The concept of intelligence developed from the attempt to quantify human cognitive abilities in the early years of the twentieth century. There is no doubt that individuals can be ranked in terms of their ability to complete tests of verbal and nonverbal ability, that is to say, the ability to use words and visuospatial constructs, as well as more complex capacities such as the ability to read and to handle mathematical symbols. These abilities develop in the course of childhood. It was discovered early on through multivariate analysis that a general factor of ability can be extracted from the performances of individuals on batteries of tests constructed to assess the development of cognitive ability. From these analyses emerged the concept of the 'intelligence quotient' which attempts to assess the extent to which an individual differs from the mean of the population of his/her age group, a calculation that is based on a population mean of 100. Standard batteries of tests (e.g., the Stanford–Binet and the Wechsler adult intelligence scale, WAIS) have been constructed and widely used both for assessing learning disability and for the purposes of educational and occupational selection.

Once generated, the abstract concept of 'intelligence' acquired an autonomous life that left unanswered questions concerning its reality and origin. Controversy centered on whether intelligence is unitary or a composite of component abilities and, if the latter, which of these are fundamental. Equally importantly, the origins of the variation within populations and the extent to which it can be regarded as genetically determined have been widely and sometimes acrimoniously disputed, with claims being made for differences between populations that cannot be accounted for by environmental factors such as educational opportunity.

The possibility that some part of the variation is genetic raises the further interesting questions of what sort of genes might be responsible and what selective pressures these genes might be under. One can also ask whether the variation is specific to *Homo sapiens* or whether similar variation might be detected in other primates and other mammals.

These questions suggest a quite different approach to human cognitive abilities and that the whole concept of intelligence can be placed in an alternative context. This is the suggestion that what is characteristic of *Homo sapiens* is not intelligence (or a particular degree of intelligence) but the capacity for language, and that this arose as a result of discrete genetic changes in the course of hominid evolution. Language, according to the linguists N. Chomsky and D. Bickerton, for example, is a capacity that has no obvious precedents in the communicative abilities of other primates. It is the defining feature of modern *Homo sapiens*.

The salient candidate for the genetic change is that the brain lateralized, i.e., that the functions of the two hemispheres became differentiated, and that this occurred on the basis that development of the hemispheres became subtly asynchronous across the anteroposterior axis, i.e., from right frontal to left occipital lobes. One component of language, probably the phonological sequence, is localized in the 'dominant,' usually the left, hemisphere. Dominance for this component of language is reflected in directional handedness (85–90% of most populations is right-handed) and this also appears to be a characteristic that distinguishes humans from the chimpanzee.

Handedness, reflecting cerebral dominance, is a trait that is associated with quantitative variation. Whether this variation is a correlate of human cognitive ability, as would seem plausible if it underlies the specific characteristic of language, has been much debated, but it now appears that lesser degrees of lateralization ('hemispheric indecision') are associated with delay in the development of verbal, and also nonverbal, ability (Crow et al., 1998). Thus it appears that lateralization is associated with significant variation in the rate at which words acquire meaning, and that this variation reflects a dimension that is specific to *Homo sapiens*. The genetics of lateralization reflects the mechanism of transition from a precursor hominid to modern *Homo sapiens*. Of particular note is the fact that there are sex differences both in handedness (girls on average are more right-handed and less likely to be left-handed than boys) and verbal ability (girls acquire words faster). There is an obvious possibility that the relevant gene(s) is sex-linked and an X-Y homologous locus has been suggested.

These considerations cast the question of human 'intelligence' in a new and perhaps more biological perspective. In particular, they emphasize the species-bound nature of the variation and the survival value of the core characteristic of language. There remains the problem of the genetic nature of the variation and its persistence. Such questions touch on the evolutionary significance of species transitions and the maintenance of species boundaries.

Reference

Crow TJ, Crow LR, Done DJ and Leask SJ (1998) Relative hand skill predicts cognitive ability; global deficits at the point of hemispheric indecision. *Neuropsychologia* 36: 1275–1280.

See also: Heritability

Intercross

L Silver

doi: 10.1006/rwgn.2001.0703

An intercross is a cross between two organisms that have the same heterozygous genotype at designated loci. An example would be a cross between sibling F_1 hybrid organisms that were both derived from an outcross between two inbred strains.

See also: Backcross; Incross; Outcross

Interference, Genetic

L Silver

doi: 10.1006/rwgn.2001.0704

Multiple events of recombination on the same chromosome are not independent of each other. Instead, a recombination event at one position on a chromosome will act to interfere with the initiation of other recombination events in its vicinity. This phenomenon is known, appropriately, as 'interference.' Interference was first observed within the context of significantly lower numbers of double crossovers than expected in the data obtained from some of the earliest linkage studies conducted on *Drosophila*. Since that time, interference has been demonstrated in every higher eukaryotic organism for which sufficient genetic data have been generated.

Significant interference has been found to extend over very long distances in mammals. The most extensive quantitative analysis of interference has been conducted on human chromosome 9 markers that were typed in the products of 17 316 meiotic events. Within 10 cM intervals, only two double crossover events were found; this observed frequency of 0.0001 is 100-fold lower than expected in the absence of interference. Within 20 cM intervals, there were 10 double crossover events (including the two above); this observed frequency of 0.0005 is still 80-fold lower than predicted without interference. As map distances increase beyond 20 cM, the strength of interference declines, but even at distances of up to 50 cM, its effects can still be observed.

If one assumes that human chromosome 9 is not unique in its recombinational properties, the implication of this analysis is that for experiments in which fewer than 1000 human meiotic events are typed, multiple crossovers within 10 cM intervals will be extremely unlikely, and within 25 cM intervals, they will still be quite rare. Data evaluating double crossovers in the mouse are not as extensive, but they suggest a similar degree of interference. Thus, for all practical purposes, it is appropriate to convert recombination fractions of 0.25, or less, directly into centimorgan distances through a simple multiplication by 100.

See also: Linkage Map

Interphase

Interphase is the period between mitotic cell divisions, and is divided into three phases: G_1, S, and G_2.

See also: Cell Cycle

Intersex

M A Ferguson-Smith

The term intersex is used in clinical genetics to describe any individual with ambiguity of the internal and/or external genitalia. It is used more widely in animal genetics to indicate a phenotype in which the somatic sex is at variance with the genetic or chromosomal sex.

See also: Hermaphrodite; Sex Reversal

Interspecific, Intraspecific Cross

L Silver

A cross between organisms from two different, but closely related species (that can produce fertile offspring of at least one sex) for the purpose of taking advantage of the increased frequency of genetic differences to carry out linkage studies.

See also: Linkage; Linkage Map

Intervening Sequence

An 'intervening sequence' is another term for an intron.

See also: Introns and Exons

Intron Homing

M A Gilson and M Belfort

A mobile intron is defined as an intron that moves by an active mechanism to a new site on DNA, and upon establishment in the new site, continues to function as an intron. This active movement is mediated by an intron-encoded protein, usually an endonuclease. There are two types of intron mobility, homing and transposition. In the case of homing, an intron is copied from one site to the same position at a homologous but intronless site. Transposition occurs when an intron is copied into a heterologous site.

The DNA homing site is the segment of the cognate gene into which the intron inserts in the process of homing. The homing site consists of three parts: the endonuclease recognition sequence, the endonuclease cleavage site, and the intron insertion site. **Table 1** provides a listing of introns for which mobility has been demonstrated.

History

Mobile introns are widespread. They have been identified in bacteria and bacteriophage, archaebacteria, and eukaryotes. The RNA of most of these introns folds into a series of stems and loops. There are two different basic folding patterns, corresponding to the group I and group II introns. In addition to different RNA structures, introns in the two groups also have distinct autocatalytic splicing mechanisms. Mobility has been demonstrated for group I and group II introns and for a noncatalytic archaebacterial intron, but not for nuclear spliceosomal introns.

The first intron shown to be mobile, in the early 1970s, was the group I ribosomal large subunit (LSU) intron, formerly called the ω intron, of the yeast *Saccharomyces cerevisiae*. The DNA-based homing process was elucidated by experiments showing polarity of recombination in crosses between intron-plus and intron-minus alleles. The intron was mobilized so that more than 90% of the progeny were found to carry the intron-containing allele.

The first group II intron shown to exhibit homing was the aI1 intron, also of *S. cerevisiae*. The original papers refer to this as transposition, but it is in fact homing as defined above. Group II intron homing is distinguished from homing of group I introns by the involvement of the intron RNA in both templating and mediating the mobility event.

Table 1 Mobile introns. Based on the presence of endonuclease-encoding open reading frames and homology to known mobile introns, many more introns are likely to be mobile.

Intron Name	Organism	Reference
Group I		
LSU (ω)	*Saccharomyces cerevisiae*	Dujon B (1989) *Gene* 82: 91–114
coxI-3α	*Saccharomyces cerevisiae*	Szeczepanek T *et al.* (1994) *Gene* 139: 1–7
coxI-14a	*Saccharomyces cerevisiae*	Wenzlau JM *et al.* (1989) *Cell* 56: 421–430
coxI-15a	*Saccharomyces cerevisiae*	Moran JV *et al.* (1992) *Nucleic Acids Research* 20: 4069–4076
bi-2	*Saccharomyces capensis*	Lazowska J *et al.* (1992) *Comptes Rendus de l'Académie des Sciences* 315: 37–41
coxII	*Schizosaccharomyces pombe*	Shafer B *et al.* (1994) *Current Genetics* 25: 336–341
LSU-3	*Physarum polycephalum*	Muscarella DE and Vogt VM (1989) *Cell* 56: 443–454
LSU-5	*Chlamydomonas eugametos*	Turmel M *et al.* (1991) *Journal of Molecular Biology* 218: 293–311
LSU	*Chlamydomonas reinhardtii*	Durrenberger F and Rochaix JD (1991) *EMBO Journal* 10: 3495–3501
CobI-1	*Chlamydomonas smithii*	Colleaux L *et al.* (1990) *Molecular and General Genetics* 223: 288–296
td	T4 bacteriophage	Quirk SM *et al.* (1989) *Cell* 56: 455–465
sunY	T4 bacteriophage	ibid.
LSU	*Desulfurococcus mobilis*	Aagard C *et al.* (1995) *Proceedings of the National Academy of Sciences, USA* 92: 12285–12289
DiSSuI	*Didymium iridis*	Johansen S *et al.* (1997) *Molecular Microbiology* 24: 737–745
coxI	*Peperomia polybotrya*	Cho Y *et al.* (1998) *Proceedings of the National Academy of Sciences, USA* 95: 14244–14249
Group II		
al1	*Saccharomyces cerevisiae*	Meunier B *et al.* (1990) Group II introns transpose in yeast mitochondria. *In*: Quagliarello E *et al.* (eds) *Structure, Function and Biogenesis of Energy Transfer Systems*, pp. 169–174. Amsterdam: Elsevier.
al2	*Saccharomyces cerevisiae*	ibid.
L1.LtrB	*Lactococcus lactis*	Mills DA *et al.* (1997) *Journal of Bacteriology* 179: 6107–6111
RmInt1	*Sinorhizobium melliloti*	Martinez-Abarca F *et al.* (2000) *Molecular Microbiology* 35: 1405–1412
Xln6	*Pseudomonas alcaligenes*	Yeo CC *et al.* (1997) *Microbiology* 143: 2833–2840
P1DNA	*Podospora anserina*	Osiewacz HD *et al.* (1989) *Mutation Research* 219: 9–15
Cox 1.1	*Kluveromyces lactis*	Skelly PJ *et al.* (1991) *Current Genetics* 20: 115–120

Transposition has not been demonstrated for group I introns, but a bacterial group II intron is capable of transposition to ectopic sites, in addition to homing. Transposition also requires an RNA intermediate.

Homing Mechanism

Group I Mechanism

Intron homing requires homology of flanking exon sequences. Although extensive homology is favorable, homologous regions as small as 10 bp on either side of the intron are sufficient. Group I intron mobility is DNA-mediated. The intron-encoded endonuclease initiates the homing process by generating a double-strand break. This process is shown in **Figure 1**.

The DNA ends are then chewed back to form a gap by exonucleolytic activity. This gap is repaired by a gene-conversion event, with the intron-containing allele as a template. In addition to the insertion of the intron, genetic markers both upstream and downstream of the intron insertion site may be converted to those of the intron donor.

Group II Mechanism

Retrohoming

Group I introns with mutations that block RNA splicing remain capable of homing, but for group II introns splicing is a requirement for intron mobility, because the spliced intron RNA is active in the homing process. The prefix 'retro-' acknowledges the role of RNA in the group II homing mechanism. The intron-encoded proteins of group II introns are more complex than those of group I introns. They consist of a single multifunctional protein that generally encodes endonuclease, RNA maturase (for splicing enhancement), and reverse transcriptase functions. In all group II mobile introns, the open reading frame (ORF) is located in a large loop in the RNA secondary

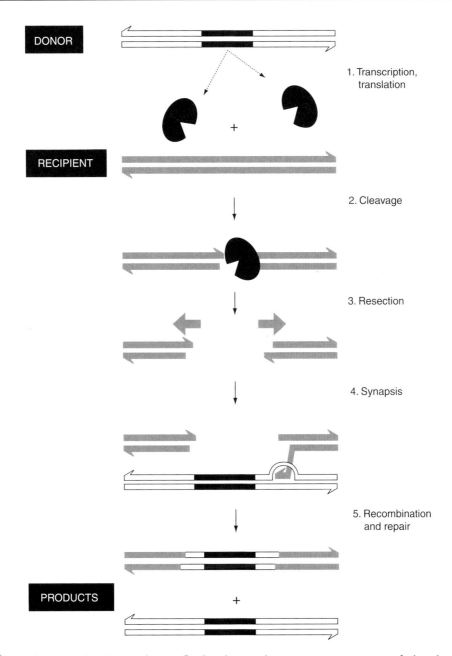

Figure 1 Group I intron homing pathway. Outlined strands represent sequence of the donor allele. Gray lines represent sequence of the recipient allele. Black lines symbolize the intron sequence, and the pac-man symbols represent the intron-encoded endonuclease. Arrowheads at the ends of the lines represent the 3′ end of the DNA.

structure. If the ORF is deleted from the intron, mobility is lost, but, when the intron-encoded protein is provided in *trans*, mobility is restored.

Group II intron homing is catalyzed by a ribonucleoprotein consisting of the intron-encoded protein and the spliced intron RNA. This is shown in **Figure 2**. The first step in homing is cleavage of the homing site of the intron-minus allele. The top strand is cleaved by the intron RNA, in a reverse-splicing reaction, while the bottom strand is cleaved by the

endonuclease function of the intron-encoded protein. Recognition of the target occurs primarily by base-pairing between intron RNA sequences and the DNA homing site. The inserted intron RNA is then copied into DNA by the reverse transcriptase moiety of the protein, using the 3′ end of the cleaved DNA as a primer. The mechanism by which this cDNA–RNA hybrid is resolved has not yet been elucidated, but the net result is the duplication of the intron in the intron-minus allele.

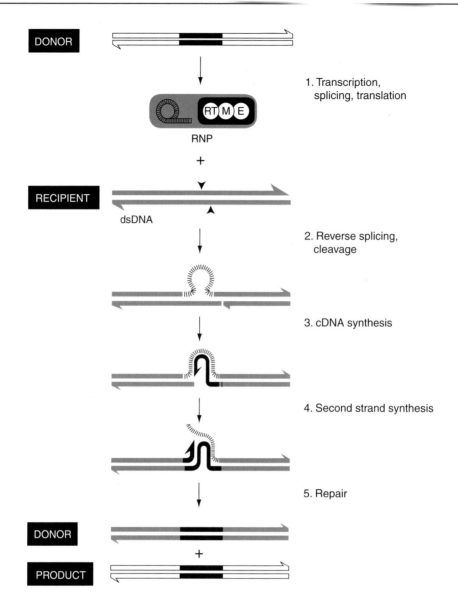

Figure 2 Group II intron retrohoming pathway. Outlined strands represent sequence of the donor allele. Gray lines represent sequence of the recipient allele. Black lines symbolize the intron sequence, with solid and dashed lines representing DNA and RNA, respectively. The staggered arrowheads mark the sites of intron insertion and endonuclease cleavage. The pathway shown is for a bacterial group II intron. A similar pathway, with some variations, occurs for the yeast introns. (dsDNA, double-stranded DNA; RNP, ribonucleoprotein; RT, reverse transcriptase; M, maturase; E, endonuclease).

Retrotransposition

Although transposition appears to occur by multiple pathways, the major transposition pathway is independent of the endonuclease function of the intron-encoded protein, as would be predicted for integration of the intron into single-stranded nucleic acid or the involvement of cellular nuclease(s). The new locations show some degree of homology to the intron homing target, specifically at the end of the first exon in a region that base-pairs with the intron RNA.

Intron-Encoded Proteins – Homing Endonucleases

The ORFs of mobile introns encode endonucleases that function to nick or cut the DNA of the insertion site to allow integration of the intron. Homing endonucleases are quite different from restriction endonucleases, which also cut DNA at a sequence-specific site. Restriction endonucleases generally recognize small (4–8 bases), often palindromic sites with strong sequence specificity at the cleavage site. In contrast, all the intron-encoded endonucleases characterized thus far have large (in the 12- to 40-bp range) recognition sites. The homing endonucleases exhibit a relaxed sequence specificity over these lengthy recognition sites and can tolerate many base changes. Yet the main requirement of homing endonucleases is simply to initiate cleavage of the DNA at the target site. An intron has been engineered to express the EcoRI restriction endonuclease, and this intron can home to an engineered EcoRI restriction site.

Homing endonucleases constitute a diverse group of proteins. While found in both mobile introns and inteins (mobile elements that splice at the protein level), they also exist in freestanding form, such as the HO endonuclease involved in mating-type switching in *S. cerevisiae*. Most homing endonucleases fall into four major classes, based on their conserved structural motifs: LAGLIDADG, GIY-YIG, H-N-H, and His-Cys. The name of the first three classes is the amino acid sequence of the motif in single-letter code. Enzymes in both the H-N-H and His-Cys classes share a common protein fold, and a metal ion involved in catalysis, so it has been proposed that they are diverse members of a single structural class.

Significance

Evolutionary Implications

Mobile introns have been found in all the kingdoms of life. They share with transposons, retrotransposons, and retroviruses the ability to integrate their DNA at new positions in the genome. Mobile introns may represent the ultimate 'selfish DNA,' since their mobility allows for efficient propagation, while the ability of the intron to splice prevents gene inactivation.

Although there is some controversy, it is generally believed that mobile introns arose through the invasion of introns by endonuclease genes. There are several lines of evidence to support this hypothesis. First, closely related mobile introns of similar sequence have been found to code for highly divergent endonucleases from different classes, suggesting independent invasion events of the intron by the endonuclease gene. Second, endonuclease ORFs are looped out of different secondary structure elements in different introns. Third, in a well-studied intron, the endonuclease ORF is flanked by sequence that closely resembles the intron homing site.

It is a provocative fact that the group II intron RNA is a ribozyme that acts catalytically to nick the DNA

at the homing site. The RNA world hypothesis postulates that the first enzymes were RNA-based. The group II ribonucleoproteins may represent ancient biochemistry and a transitional state between an RNA world and the DNA–protein world as we know it.

The splicing reactions of group II introns are mechanistically similar to those of the nuclear spliceosomal introns, which comprise about 15% of the human genome. It is widely hypothesized that group II introns evolved into spliceosomal introns, although direct evidence in terms of conservation of sequence and structure is lacking. It is also noteworthy that group II introns resemble retrotransposons that lack a long terminal repeat, both in mechanism of integration and in sequence of the reverse transcriptase moiety of the intron-encoded protein. These retrotransposons make up more than 17% of mammalian genomes. Group II introns and their close relatives have therefore played a major evolutionary role in shaping the human genome.

Potential Applications

Mobile introns offer a potentially valuable tool for gene manipulation. Because the homing sites are of such large size, the endonucleases are very useful as rare cutters. This enables specific digestion of DNA into large fragments. In group II introns, key features of the homing site are recognized by base-pairing between the DNA and the intron RNA. Thus it is theoretically possible to mutate group II introns to recognize and insert into any desired site in the genome. This may serve to inactivate a deleterious gene, or direct a beneficial gene to a benign location, because the intron-encoded protein can be provided in *trans*, and the intron can be engineered to carry a gene of interest. These introns are therefore useful both for gene targeting and as agents of gene delivery.

Further Reading

Belfort M, Derbyshire V, Parker MM, Cousineau B and Lambowitz AM (2001) Mobile introns: pathways and proteins. In: Craig N, Craigie R, Gellert M and Lambowitz AM (eds) *Mobile Elements*. Washington, DC: ASM Press.

Lambowitz AM and Belfort M (1993) Introns as mobile genetic elements. *Annual Review of Biochemistry* 62: 587–622.

Lambowitz AM, Caprara MD, Zimmerly S and Perlman PS (1999) Group I and group II ribozymes as RNPs: clues to the past and guides to the future. In: Gesteland RF, Cech TR and Atkins JF (eds) *The RNA World*, 2nd edn. Plain view, NY: Cold Spring Harbor Laboratory Press.

See also: Introns and Exons; Retrotransposons; Retroviruses; RNA World; Transposable Elements

Introns and Exons

A Stoltzfus

doi: 10.1006/rwgn.2001.0708

An intron (or 'intervening sequence') is a segment of RNA excised from a gene transcript, with concomitant ligation of flanking segments called 'exons.' This process of excision and ligation, known as 'splicing,' is one of several posttranscriptional processing steps that may occur prior to translation. Although 'intron,' in the strict sense, refers only to segments excised from RNA (and, by extension, the DNA segments that encode them), there exist developmental analogs of introns that are excised from DNA (the ciliate IES elements) or from protein (the printrons or inteins).

Diversity and Distribution

Introns of some type are found in every kingdom of cellular life, and also in viruses, bacteriophages, and plasmids. Different types of introns have different splicing mechanisms and distinctive patterns of distribution with respect to gene families, subcellular compartments, and taxonomic groups (e.g., protein-spliced tRNA introns are known only from tRNA genes in archaebacterial genomes or eukaryotic nuclear genomes). A single gene may have multiple introns and, rarely, introns of multiple types (e.g., some fungal mitochondrial genes have both group I and group II introns).

The most familiar introns are the 'spliceosomal' introns, which are excised by a ribonucleoprotein 'spliceosome,' and which typically have the sequence GU...AG. Spliceosomal introns are known only from genes in the eukaryotic nucleus (or nucleomorph) and in eukaryotic viruses. They range in length from less than 20 nt (nucleotides) to over 200 kilo-nt, while exons range in length from less than 10 nt to over 3 kilo-nt. The mean density of introns varies widely, from over 4 introns per kilo-nt of protein-coding sequence in the most intron-dense nuclear genomes (including those of vertebrates and vascular plants), to 0.04 in the yeast *Saccharomyces cerevisiae*.

Group I and group II introns are collectively known as 'self-splicing' introns, because the intron RNA plays a primary role in the biochemistry of splicing, in some cases being sufficient for splicing *in vitro*. Group I introns are the most broadly distributed mobile elements known, being found in the genomes of eubacteria and their phages, as well as in the nuclear, mitochondrial, and chloroplast genomes

of eukaryotes. Group II introns are known in eukaryotic organellar (but not nuclear) genomes, as well as in eubacterial chromosomes and plasmids. Though common in some organellar genomes, self-splicing introns are extremely rare elsewhere, and seem to be entirely absent from most prokaryotic genomes as well as many eukaryotic nuclear genomes.

Role in Gene Expression

In most cases, introns appear to be dispensable. Introns can be removed entirely from mitochondria of *S. cerevisiae* without obvious ill effect. Nevertheless, in a variety of cases, introns and splicing figure importantly in development. The delay caused by the transcription and splicing of a gene with many long introns can be important (e.g., the *knrl* gene of *Drosophila*). The intron may contain within itself some other feature: a DNA regulatory site (e.g., a promoter or enhancer), a structural RNA (e.g., intron-encoded snoRNAs in eukaryotic nuclear genomes), or a protein-coding region (e.g., intron-encoded maturases in organellar group I and II introns and homing endonucleases in bacteriophage introns). Splicing may join parts of two different RNA transcripts, a process known as 'trans-splicing' that is common in trypanosomes but rare or absent in most other organisms. Finally, the pattern of splicing of a single transcript may be variable, such that different mRNAs, and different protein products, are produced from the same pre-mRNA. Regulation of such 'alternative splicing' schemes plays a crucial role in sex determination in *Drosophila*. The frequency and importance of alternative splicing in most species is not well understood.

Mutation and Evolution

Introns are passively subject to the same mutational lesions that affect other genomic sequences; in some cases they contribute actively to the mutational process as mobile elements. Nucleotide substitutions that alter splicing have been implicated in many heritable diseases in humans. Such changes usually map to within a few nt of a splice junction. Over evolutionary time-scales, the internal sequences of spliceosomal introns diverge rapidly (by nucleotide substitutions as well as by short insertions and deletions), presumably because the demands of splicing impose no constraint on most internal sites. By contrast, group I and II introns evolve more slowly, and are densely packed with sequences that participate in splicing and mobility.

Rearrangement mutations involving introns also occur, sometimes based on recombination between repetitive elements within introns. In animal genomes, intron-mediated rearrangements have contributed importantly to the evolution of novel chimaeric genes by so-called 'exon shuffling.' On the scale of millions to hundreds of millions of years, homologous genes may diverge by loss and gain of introns. Loss of an intron may occur by way of reverse transcription and recombinational reincorporation of a spliced gene product. Insertion of introns by transposition has been observed experimentally for group I, group II, and spliceosomal introns. For group I and II introns, 'homing' to (intronless) allelic sites is also observed.

See also: **Eukaryotic Genes; Pre-mRNA Splicing**

Invariants, Phylogenetic

W Fitch

Copyright © 2001 Academic Press
doi: 10.1006/rwgn.2001.0710

Phylogenetic invariants is a method first proposed by Lake (1987). The 'invariants' derive from the fact that the addition and subtraction of the numbers of certain nucleotide distribution patterns are expected to remain constant (at zero) for all incorrect phylogenies. And thus can be used to distinguish among alternative phylogenetic trees. It is a property that is used on nucleotide sequences taken four at a time. For example, suppose that we had four such sequences that are homologously aligned from left to right, one under the other:

...AGA...
...AGT...
...CTT...
...CTA...

so that for any position in the alignment the four nucleotides produce a (vertical) pattern such as AACC. This might suggest that the first two sequences are sister sequences meaning that they are more closely related to each other than either of them is to the second two sequences (see **Figure 1A**). There are 256 possible patterns but some of them carry the same information. For example, the same relationship would be inferred if the pattern were GGTT. The method restricts itself to positions that have exactly two purines (A and/or G) and two pyrimidines (C and/or T) in their pattern as all the examples used here do. Their relationship is shown by the tree in **Figure 1A** where the arrowhead indicates that only a single transversion mutation is required to explain the observed nucleotides at the tips of this tree. (A transversion is the historical change from (or to) a purine to (or from) a pyrimidine; all other interchanges are called transitions.)

On the other hand, a pattern such as ACCA would suggest that sequences 1 and 4 were sisters rather than sequences 1 and 2 (see **Figure 1B**). The two relationships (trees) cannot both be true, but if sequences 1 and 2 really are the true sister sequences, then this third pattern can only have arisen by virtue of two transversions having occurred during the history of these sequences (see **Figure 1C**).

However, we can estimate how often the misleading case in **Figure 1C** arises. Note that in **Figure 1D** we have shown only three of the four nucleotides in the pattern. What could the fourth nucleotide be? As we only consider those patterns with two purines and two pyrimidines, there must be a pyrimidine. Which one? If we assume that there is no bias as to which nucleotide the mutation is to, then it can be either C (as in **Figure 1C**) or T (as in **Figure 1E**) with equal probability. But that means that, for the wrong tree, the number of occurrences of a pattern like that in **Figure 1C** should be the same as the number for the pattern like that in **Figure 1E**. Hence, subtracting those two numbers should give an number not statistically different from zero for the two tree structures that are wrong. (The third possible tree is for the pattern ACAC which suggests that sequences 1 and 3 are sisters.)

There are more details to the method but the preceding gives the spirit of the method. It is a method that is guaranteed to give the correct answer given sufficient lengths of the sequences being compared. This virtue, however, is more than offset by the answer to the question of how long the sequences must be to get that correct answer. It turns out that the sequences

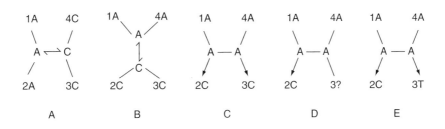

Figure 1

need to be incredibly long, sometimes greater than the size of the genome, as a consequence of which the method is not used.

Reference

Lake JA (1987) A rate-independent technique for the analysis of nucleic acid sequences: evolutionary parsimony. *Molecular Biology and Evolution* 4: 167–191.

See also: Phylogeny; Transition; Transversion Mutation

Inversion

N Grindley

Copyright © 2001 Academic Press
doi: 10.1006/rwgn.2001.0711

An inversion is a DNA rearrangement in which a segment of a chromosome is flipped (or reversed), so that the sequence reads in the opposite direction to the original. Genes contained within an inversion will map in the reverse order to normal and will be expressed in the opposite orientation.

See also: Hin/Gin-Mediated Site-Specific DNA Inversion; Site-Specific Recombination

Inverted Repeats

Copyright © 2001 Academic Press
doi: 10.1006/rwgn.2001.1881

Inverted repeats are two copies of the same DNA sequence repeated in opposite orientation in the same molecule.

See also: Repetitive (DNA) Sequence

Inverted Terminal Repeats

Copyright © 2001 Academic Press
doi: 10.1006/rwgn.2001.1882

Inverted terminal repeats are short related or identical sequences repeated in opposite orientation at the ends of some transposons.

See also: Transposable Elements

Isochromosome

M A Ferguson-Smith

Copyright © 2001 Academic Press
doi: 10.1006/rwgn.2001.0717

An isochromosome is an abnormal metacentric chromosome formed by the duplication of one arm of a normal chromosome with deletion of the other arm. Both arms of the metacentric chromosome are thus genetically identical. It may arise from transverse instead of longitudinal division of the centromere during cell division or, more often, by an isochromatid break and fusion of the daughter chromatids above the centromere. In the latter case the isochromosome is dicentric. One of the two centromeres of a dicentric isochromosome usually becomes nonfunctional, so that the chromosome segregates normally during cell division.

The commonest human isochromosome observed in livebirths is an isochromosome for the long arm of the X chromosome. This results in Turner syndrome (see Turner Syndrome), and it is found that the isochromosome is preferentially inactivated, forming larger than normal sex chromatin (Barr body; see Sex Chromatin). Isochromosomes of the Y chromosome are also found in livebirths, and can involve either the short or long arms. Short-arm Y isochromosomes cause male infertility as the testis-determining region is not lost despite the loss of spermatogenesis factors on the long arm. Long-arm isochromosomes of the Y are associated with female sex determination unless the isochromatid break lies distal to the sex-determining region of the Y.

Isochromosomes involving the human autosomes usually result in early spontaneous abortion; rare exceptions are isochromosomes for the short arms of chromosomes 9 and 12, and these are associated with severe mental and physical disability.

See also: Sex Chromatin; Turner Syndrome; X-Chromosome Inactivation

Isolation by Distance

N E Morton

Copyright © 2001 Academic Press
doi: 10.1006/rwgn.2001.1426

Sewall Wright pioneered the study of how genetic similarity declines with geographic distance. His work was based on a hierarchical model of local populations (demes) in successively larger regions, each with its own gene frequencies. This model is difficult to apply to real populations, and so has been superseded by the theory of Gustave Malecot for pairs of individuals born at a known distance d in a given region. Genetic similarity is measured by kinship ϕ_d, the probability that a gene drawn randomly from one individual be identical by descent with a random allele in the other individual. If the pair are spouses, kinship is the inbreeding F_d of their children. The Malecot equation is usually written as $\varphi_d = (1-L)a\ e^{-bd} + L$, where $0 < a < 1$ is kinship within a local population ($d = 0$) and $-1 < L \leq 0$ is kinship at large distance. If current gene frequencies are used in kinship bioassay on genotypes, phenotypes, or surnames, $L = -\varphi_R/(1 - \varphi_R)$, where φ_R is random kinship is the sampled region. If kinship in relation to founder gene frequencies is predicted from migration or genealogy, $L = 0$. The parameters a, b are functions of effective population size N and systematic pressure m largely due to migration. Validity of this equation depends on discreteness of local populations. More complicated expressions derived for continuous distributions and two or three dimensions are less accurate for real populations. Oceanic islanders and nomadic populations have small values of b compared with coastal islanders and agriculturists. Kinship increases rapidly in populations with preferential consanguineous marriage but then reaches a plateau that is not much greater than for isolates that avoid consanguineous marriage. The effect of migration is everywhere apparent. This evidence helped to resolve misunderstanding about the role of population structure in assessing forensic DNA identification.

In recent years the Malecot model has been useful for study of linkage disequilibrium. Distance between loci or nucleotide polymorphisms is measured along the physical or genetic map, usually in kilobases (kb) or centimorgans (cM), taking advantage of the fact that recombination acts on allelic association in the same way as migration acts on kinship. Isolation by distance has become a cornerstone of genetic epidemiology, as it has long been for population genetics and anthropology.

Further Reading

Wright S (1951) The genetical structure of populations. *Annals of Eugenics* 15: 323–354.

Malecot G (1969) *The Mathematics of Heredity.* San Francisco, CA: WH Freeman.

Lasker GW (1985) *Surnames and Genetic Structure.* Cambridge: Cambridge University Press.

Morton NE (1992) Genetic structure of forensic populations. *Proceedings of the National Academy of Sciences, USA* 89: 2556–2560.

See also: **Effective Population Number; Linkage Disequilibrium; Wright, Sewall**

Isoleucine

J Read and S Brenner

Copyright © 2001 Academic Press
doi: 10.1006/rwgn.2001.2076

Isoleucine (Ile or I) is one of the 20 amino acids commonly found in proteins. Its side-chain consists purely of hydrocarbons and it is only slightly soluble in water. Isoleucine belongs to the group of neutral-polar amino acids which includes glycine, alanine, valine, leucine, phenylalanine, proline, and methionine. These amino acids are usually found on the inside of protein molecules.

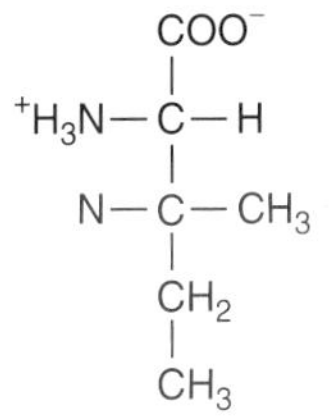

Figure 1 Isoleucine.

See also: **Amino Acids; Proteins and Protein Structure**

Isomerization (of Holliday Junctions)

P J Hastings

Copyright © 2001 Academic Press
doi: 10.1006/rwgn.2001.0718

A crossed-strand exchange or Holliday junction can be resolved endonucleolytically to restore the parental combination of flanking markers, or to give a reciprocal exchange – a crossover (see **Figure 1C**). This bifurcating decision need not require two different activities of the endonuclease making the cuts (the resolvase). Model building has shown that the Holliday junction itself can adopt an alternative form, such that the same enzyme activity gives the alternative results of crossover or noncrossover. Alternative forms of a

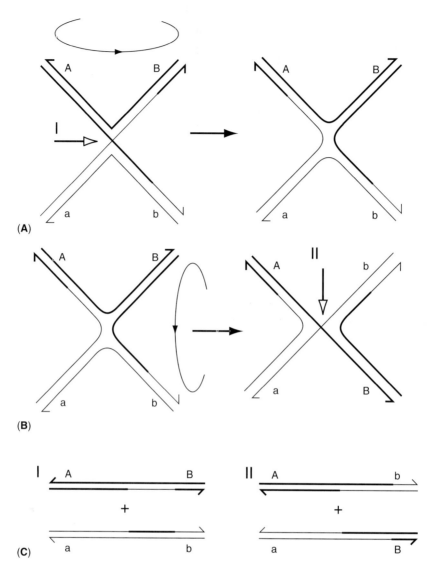

Figure I (A) With some base pairs unstacked, the Holliday junction takes on an X-form. The two DNA strands that cross in the middle of this X (labeled I) are those that exchanged places in the formation of the Holliday junction. Rotation of the upper arms, as shown by the circular arrow, reveals that the structure has a hole in its center. (B) A rotation of two side arms of this structure relative to the other two gives a configuration in which the other two strands cross in the center (shown at II). (C) If the crossing strands are cut at I in (A), the structure is resolved as a noncrossover. If the crossing strands are cut at II, a crossover results.

molecule are called isomers and the process by which a molecule adopts an alternative structure is called isomerization. These terms are applied to Holliday junctions.

Isomerization of a Holliday junction is conceived as beginning with some bases becoming unstacked. That is, the eight bases at the junction no longer interact with their neighbors in the same DNA strand. This allows the arms of the junction to open out so that the structure takes on the form of an X as shown in the first structure in **Figure IA**. The process of isomerization of a Holliday junction is described as two

rotations of pairs of arms of the structure, as shown in the **Figure IA** and **B**. The second rotation causes the two strands that cross each other to be a different pair than those that cross in the first structure. If the resolvase cuts only the crossing strands, the two isomers then give rise to the alternative outcomes. These rotations are constrained to occur in one direction only. DNA has sufficient flexibility for the rotating parts of the molecule to be local, rather than involving the whole length of DNA molecules. The process is reversible and the two isomers are expected to occur in a state of rapid equilibrium.

Further Reading

Meselson MS and Radding CM (1975) A general model for recombination. *Proceedings of the National Academy of Sciences, USA* 72: 358–361.

Sigal N and Alberts B (1972) Genetic recombination: the nature of a crossed strand-exchange between two DNA molecules. *Journal of Molecular Biology* 71: 789–793.

See also: **Cruciform DNA; Holliday Junction; Holliday's Model**

Isotype

Copyright © 2001 Academic Press
doi: 10.1006/rwgn.2001.1884

An isotype is a set of macromolecules sharing some common features, e.g., closely related immunoglobulin chains. The isotype describes the class, subclass, light chain type, and subtype of an immunoglobulin.

See also: **Immunoglobulin Gene Superfamily**

Isotype Switching

See: **Recombination in the Immune System**

J

J Gene

See: **Recombination in the Immune System**

Jackknifing

See: **Trees**

Jacob, François

K Handyside, E Keeling and S Brenner

The French biologist François Jacob (1920–) shared the 1965 Nobel Prize for Physiology or Medicine with André Lwoff and Jacques Monod for their discoveries concerning genetic regulatory mechanisms in bacteria.

After being severely wounded whilst in active combat in World War II, Jacob was forced to give up his studies for his chosen career as a surgeon. After gaining an MD degree in 1947 and a PhD in science in 1954 from the Faculty of Medicine and Faculty of Science in Paris respectively, he turned to biology.

While Jacob started as a research assistant at the Pasteur Institute in Paris in 1950, it was not long before he became the Laboratory Director. Within 10 years he had been promoted again to Head of the Department of Cellular Genetics. By 1965 he was also the Professor of Cellular Genetics at the Collège de France, and it was here that a position was created for him as the Professor of Cell Genetics.

With coworker Jacques Monod, Jacob studied the regulation of enzyme synthesis in bacteria. Later they made the significant discovery of 'regulator genes' and the mechanisms for controlling the expression of structural genes. The 'operon' theory of gene regulation (see Operon) is now central to today's understanding of genetic control. This discovery explained the mechanisms by which cells modulate the expression of genes in response to varying environmental conditions. Jacob, with Sydney Brenner and Matthew Meselson, also proved the existence of messenger RNA.

Jacob has been awarded a number of scientific awards and is an honorary member of numerous societies including the French Academy of Sciences, The National Academy of Sciences of the USA and the Royal Society of London. He has published many books on molecular biology.

Further Reading

http://www.nobel.se/medicine/laureathes/1965/jacob.bio.html.
http://www.rockefeller.edu/pubinfo/jacob.nt.html.

See also: **Brenner, Sydney; Gene Expression;** *lac* **Operon; Monod, Jacques; Operon**

Jukes–Cantor Correction

P Pamilo

The Jukes–Cantor equation provides an estimate of the actual number of nucleotide substitutions since the separation of two DNA sequences by correcting the observed differences for multiple substitutions at the same site. Two DNA sequences evolve from the same ancestral sequence by accumulating mutational differences. The number of nucleotide differences can be counted by comparing the aligned sequences. The observed number of nucleotide differences does not always show all the nucleotide substitutions that have occurred during the evolutionary past, because multiple substitutions at the same nucleotide position remain undetected. The Jukes–Cantor method corrects the estimate of sequence differentiation for such multiple hits. The method is based on a model which assumes that all four nucleotides are equally frequent, all types of nucleotide substitutions are equally common, and all nucleotide sites mutate with the same probability. Under these assumptions, it is easy to derive the relationship between the observed proportion (p) of nucleotide differences between two sequences and the frequency of nucleotide substitutions (d) that have occured. Let the proportion of nucleotide differences at time t since the common ancestor be

p and the probability of one nucleotide mutating per unit of time be α. The expected number of nucleotide substitutions per site in the two sequences, their evolutionary distance, is then $d = 2\alpha t$; two because mutations have occured in both lineages. In terms of the observed p, the evolutionary distance becomes

$$d = 2\alpha t = -(3/4)\log[1 - (4/3)p]$$

which is the Jukes–Cantor distance of the sequences. The estimate d is always larger than the observable differentiation measured by p. For small differences between the sequences (say $p < 0.2$), the observed nucleotide differences estimate well all the nucleotide substitutions as multiple mutations of the same nucleotide site are unlikely. With increasing differentiation, the estimate d starts to depart from p. When p becomes or exceeds the value of 0.75, the sequences are saturated by mutational differences and d becomes undefined. The limit of 75% difference (or 25% similarity) is what one gets by constructing two random sequences from four equally frequent nucleotides. The saturation makes any estimate of sequence differentiation unreliable. This is seen from the variance of the distance estimate d, which is

$$V(d) = p(1 - p)/[L(1 - 4p/3)^2]$$

where L is the length of the sequence (number of nucleotides). When p approaches the value of 0.75, the variance increases quickly.

The Jukes–Cantor method corrects for multiple substitutions of the same site but not for different mutational probabilities that depend on the type (A, G, T, or C) and position of the nucleotide. These can be taken into account, e.g., by Kimura's two-parameter model (allowing different probabilities for transitional and transversional mutations) or by distinguishing between synonymous and nonsynonymous substitutions in a coding sequence. Even though the Jukes–Cantor model oversimplifies the underlying evolutionary model, it has the advantage that it is robust and depends on the minimal number of model parameters (only the equal substitution rate is taken as a parameter). Estimators based on more complex substitution models are superior if the assumptions of the model are correct, but they can also become sensitive to departures between the assumptions of the model and the reality of molecular evolution.

If one has an estimate of the substitution rate α from known time of differentiation (e.g., based on fossil evidence) and the rate is close to constant over time in different evolutionary lineages, it becomes possible to estimate the time of separation of any lineages from $t = d/(2\alpha)$. The rate constancy can be tested by relative rate test. A matrix of pairwise Jukes–Cantor distances can be used for constructing phylogenies with distance-based methods, such as the neighbor-joining method.

Further Reading

Jukes TH and Cantor CR (1969) Evolution of protein molecules. In: Munro HN (ed.) *Mammalian Protein Metabolism*, vol. 3, pp. 21–132. New York: Academic Press.

Li W-H (1997) *Molecular Evolution*. Sunderland, MA: Sinauer Associates.

See also: **Kimura Correction; Molecular Clock**

Jumping Genes

See: **Horizontal Transfer; Transposon Excision; Transposons as Tools**

Karyotype

M A Ferguson-Smith

The karyotype is the chromosome complement of a cell, individual, or species, classified according to chromosome length, centromere position, and banding appearance produced by specific staining techniques.

The karyotype of a somatic cell is often arranged to show chromosome pairs in order of decreasing length and numbered accordingly. A diagram of the karyotype based on the analysis of a number of cells is referred to as an idiogram. The process of analyzing the chromosomes of a cell or individual and arranging them according to the species idiogram is known as karyotyping.

See also: **Idiogram**

kb (Kilobase)

kb (kilobase) is the abbreviation for 1000 base pairs.

See also: **Bases; DNA Structure**

Khorana, Har Gobind

R D Wells

Har Gobind Khorana (1992–) is one of the most outstanding geneticists in the world. Khorana may be best known for contributing to solving the genetic code in the 1960s. He has solved a number of important genetic problems from a chemical standpoint over the past five decades. His greatest contributions have been in the synthesis of oligonucleotides and small nucleic-acid-like molecules which culminated in the synthesis of a tRNA gene in the 1970s. His recent work has focused on the structure and function of rhodopsin and its role in signal transduction across membranes.

He was born in Raipur, India in 1922 and was educated in India, England, and Switzerland. He has served on the faculty of the University of British Columbia (1952–1960), the University of Wisconsin, Institute for Enzyme Research (1960–1970), and Massachusetts Institute of Technology, Departments of Chemistry and Biology (1970–present). He has received numerous awards and prizes including the Nobel Prize for Physiology or Medicine (shared with R.W. Holley and M.W. Nirenberg) in 1968. He has received at least 14 honorary doctorate degrees and has been elected to numerous honorary memberships to academic societies.

During his early work at the University of British Columbia, he pioneered the chemical synthesis of small ribo- and deoxyribonucleoside triphosphates using dicyclohexylcarbodiimide and was the foremost laboratory in the synthesis of dinucleotide and trinucleotide molecules of the deoxy- and ribo- types. In the 1960s at the University of Wisconsin, he developed methods for the synthesis of oligonucleotides as templates for DNA and RNA polymerases and/or substrates for kinases and ligases. This work culminated in solving the genetic code in 1966.

The total synthesis of a tyrosine suppressor tRNA gene with upstream and downstream control sequences was accomplished at MIT in 1970s. Over the past 25 years, Khorana and his colleagues have successfully investigated mutant bacteria rhodopsins to identify the amino acid residues involved in transport of protons across membranes.

See also: **Genetic Code; Nirenberg, Marshall Warren**

Kimura Correction

N Saitou

When we compare two homologous nucleotide sequences, we are often interested in estimating the number of nucleotide substitutions accumulated during the divergence of the two sequences. Let us assume that we obtained a reliable alignment for those two sequences. Then the simplest way is to count the number (m) of nucleotide differences between them. We often divide m by the number (n) of nucleotides compared. In this case, gap positions caused by insertions and deletions are not included. The proportion ($p = m/n$) is called the p distance. When the amount of divergence is small, it is intuitively clear that m or p reflects the actual number of nucleotide substitutions accumulated since the divergence of the two sequences. This is because parallel, backward, or successive substitutions at the same nucleotide site rarely occur under a low divergence. When the amount of divergence is relatively large, however, the probability of occurnce of those changes is expected to increase. Therefore, we need some kind of correction for m and p.

The simplest mathematical model for the correction is the one-parameter model. This model is also called the Jukes–Cantor model after the two researchers who first used this model. The four nucleotides are assumed to change with equal probability with each other under the one-parameter model. This simple situation clearly does not satisfy the real pattern of nucleotide substitution.

Kimura (1980) proposed two different rates of nucleotide substitutions, so this model is also called the two-parameter model. In practice, transitions usually outnumber transversions, and usually substitution rates for those two types are assumed to be different under the two-parameter model. Theoretically, however, any two substitution types can be considered in a two-parameter model.

The number (K) of nucleotide substitutions per site is estimated as:

$$K = -[1/2]\log\left[(1 - 2P - QW)\sqrt{(1 - 2)}\right]$$

where P and Q are proportions of transitional and transversional differences, respectively.

There is another Kimura correction for amino acid sequences (Kimura, 1983). Estimation of the number of amino acid replacements based on Dayhoff's PAM matrix is approximated by the following simple equation:

$$Kaa = -\log[1 - p - 1/(5p^2)]$$

where Kaa is the number of amino acid substitutions per site and p is the proportion of amino acid difference.

References

Kimura M (1980) *Journal of Molecular Evolution* 16: 111–120.
Kimura M (1983) *The Neutral Theory of Molecular Evolution*. Cambridge: Cambridge University Press.

See also: Jukes–Cantor Correction; Transition; Transversion Mutation

Kimura, Motoo

J F Crow

Motoo Kimura (1924–94) was a leading population geneticist, widely regarded as the successor to Wright, Fisher, and Haldane in developing the theory of population genetics and evolution. He is best known for his neutral theory of molecular evolution. Kimura was born in Okazaki, Japan on 13 November 1924. During his childhood he had a love of botany, but he also displayed a talent for mathematics. He attended Kyoto Imperial University during World War II and, although not in the military, suffered from wartime and postwar food shortages. On graduation he joined the staff of the National Institute of Genetics in Mishima and remained there for the rest of his life. After the war he was able to study in the United States and after one year at Iowa State College transferred to the University of Wisconsin, where he received his doctorate in 1956. In his later years he developed amyotrophic lateral sclerosis and died on his 70th birthday, 13 November 1994.

Kimura pioneered the use of the Kolmogorov diffusion equations. Although others had used the forward equation, he was one of the first to employ the backward equation and was particularly creative in its use. While still a graduate student he worked out the complete solution to the process of random genetic drift in a finite population from an arbitrary starting-point. He then proceeded to solve a number of important problems, including: the probability of fixation of a mutant gene, the time until fixation, conditions for a stable equilibrium with multiple alleles, and the evolution of closer linkage. Early in his career, he introduced the widely used stepping-stone model of population structure.

Kimura undertook a wide variety of problems, both deterministic and stochastic. He had a gift for formulating and solving problems, always with a particular genetic or evolutionary issue in mind. He was especially adept with partial differential equations, both in finding the appropriate boundary conditions and in finding solutions. His numerical solutions, often involving difficult approximations and worked out in the days before modern computers, have turned out to be remarkably accurate.

In 1968 Kimura became convinced that the rate of amino acid and nucleotide change in molecular evolution was too rapid to be accounted for by selection, and introduced his neutral theory – the idea that most molecular change is due to selectively neutral changes. Evolutionary change then becomes the result of mutation and random drift. For a strictly neutral gene, the rate of evolution, when viewed over a long time, is simply the mutation rate. This happy insight permitted a large number of tests of the neutral theory. At the same time he argued that molecular polymorphisms represent, for the most part, neutral sites in the process of fixation.

The neutral theory was greeted with great skepticism at the time it was introduced. Gradually it won acceptance, especially from molecular evolutionists. Over the years, partly as a result of Kimura's relentless advocacy, the theory has had a fairly wide acceptance. It is probably correct to say that the current consensus is that most nucleotide changes in higher animals and plants are due to random changes, but that the jury is still out on the relative number of random versus selected changes of amino acids.

Among biologists as a whole, Kimura is most widely known for his theory of molecular evolution. Among population geneticists, he is also greatly respected for his pioneering work in the mathematical theory of population genetics and evolution.

Further Reading

Kimura M (1983) *The Neutral Theory of Molecular Evolution.* Cambridge: Cambridge University Press.

Kimura M (1994) In: Takahata A (ed.) *Population Genetics, Molecular Evolution, and the Neutral Theory: Selected Papers.* Chicago, IL: University of Chicago Press.

See also: Kimura Correction; Neutral Theory

Kin Selection

See: **Hamilton's Theory**

Kinases (Protein Kinases)

J Hodgkin

Copyright © 2001 Academic Press
doi: 10.1006/rwgn.2001.0730

Kinases are enzymes that add phosphate groups to substrates. The most numerous and most extensively studied kinases are protein kinases, which phosphorylate specific target proteins and thereby modify their activities. Collectively, protein kinases represent the largest gene families in eukaryotes: about 2% of all genes in the yeast *Saccharomyces cerevisiae*, the nematode *Caenorhabditis elegans*, and the fruit fly *Drosophila melanogaster* are predicted to encode kinases – about 120, 400, and 300 genes, respectively. Extrapolation to vertebrate genomes suggests that these contain more than 1000 kinase genes.

Biochemically, protein kinases can be distinguished on the basis of the phosphorylated residue: histidine, serine, threonine, or tyrosine. Histidine kinases are primarily important in prokaryotes, in which they act as part of 'two-component' signaling systems. Few histidine kinases are known in eukaryotes, although they do occur in the slime mold *Dictyostelium*, in fungi, and in plants. The majority of eukaryotic kinases are serine/threonine kinases, which fall into dozens of different families. Tyrosine kinases seem to be absent from the yeast genome and therefore appear to have arisen during the evolution of multicellular organisms. They play major roles in development and oncogenesis: for example, of 21 characterized retroviral oncogenes, seven are tyrosine kinases (e.g., the *abl*, *src*, and *yes* oncogenes) and three are serine/threonine kinases (e.g., the *mos* and *raf* oncogenes).

Phosphorylation of target proteins by kinases can be reversed by protein phosphatases. Reversibility is a general advantage of phosphorylation as a regulatory strategy, in contrast to irreversible modifications such as proteolyis. However, although there are many specific protein phosphatases, some of which play significant roles in regulation, kinases are more numerous and usually more important.

Regulation of protein activities by kinases is ubiquitous in eukaryotic development, physiology, and metabolism. Control of glycolysis depends on phosphorylation, and the regulation of the cell cycle is centrally dependent on a variety of different kinases, most notably the cyclin-dependent kinases. The majority, and perhaps all, of signal transduction pathways involve kinases, sometimes in cascades of activity such as that discovered for mitogen-activated kinases (MAPK), which are regulated in turn by MAP kinase kinases (MAPKK) and MAP kinase kinase kinases (MAPKKK). These act to couple events outside the cell, or in the cytoplasm, to cytoplasmic or nuclear responses. The activity of many transcription factors is modulated, either postively or negatively, by the action of kinases. Similarly, ion channel properties can be altered by phosphorylation.

In animals, the initial responses of cells to external stimuli such as growth factors or other developmental signals are often mediated by receptor tyrosine kinases, which are membrane-spanning proteins with an extracellular ligand-binding domain, and a cytoplasmic tyrosine-kinase domain. These membranes act on specific cytoplasmic targets, affecting other kinases in turn. The multiple steps of phosphorylation in these signaling cascades, and the opportunity for crosstalk between different pathways, creates the opportunity for immensely elaborate modulation of cellular activity. Much of the complexity of cellular and neuronal function in higher eukaryotes appears to depend directly on their huge and versatile repertoires of protein kinases.

Reference

Hunter T, Plowman GD (1997) The protein kinases of budding yeast: six score and more. *Trends in Biochemical Sciences* 22: 18–22.

See also: **Cell Cycle; Enzymes**

Kinetochore

M A Hultén and C Tease

Copyright © 2001 Academic Press
doi: 10.1006/rwgn.2001.0731

The kinetochore is a proteinaceous region within the centromere to which spindle microtubules attach during mitosis and meiosis. The kinetochore is an active component of the cell checkpoint machinery that ensures the correct orientation and segregation of chromosomes at cell division.

Kinetochores behave in a contrasting manner at mitosis and meiosis. At mitosis, sister kinetochores of a chromosome attach to spindle microtubules and orient to opposite spindle poles. The sister chromatids separate at anaphase and pass to the spindle poles to ensure each daughter cell receives the full chromosomal complement. In the first meiotic division, however, the sister kinetochores of one chromosome orient to a single pole, while those of its homologous partner orient to the other. As a result, daughter cells receive half the original number of chromosomes. At meiosis II, the kinetochores orient in the same manner as mitosis resulting in chromatid segregation.

Structure

The somatic vertebrate kinetochore, when viewed by standard transmission electron microscopy, is a trilaminar structure on the surface of centromeric heterochromatin of each chromatid of a chromosome (**Figure 1**). A fourth, fibrillar layer can also be discerned adjacent to the trilaminar structure.

DNA Composition of Kinetochore-Associated Chromatin

Kinetochores generally form on chromatin with particular DNA sequences. However, there is no evidence that these sequences show evolutionary conservation. In the yeast *Saccharomyces cerevisiae*, for example, the minimal centromere contains 125 bp of DNA that falls into three distinct elements (CDE I, II, and III). All 17 chromosomes of *S. cerevisiae* carry this DNA at their centromeres. In *Drosophila melanogaster*, a 420-kb DNA sequence, composed of satellite arrays and various transposable elements, has been found at one centromere. Notably, these DNA sequences are also present in other chromosomal regions that do not form kinetochores. In humans, kinetochores are associated with alphoid satellite DNA (240 kb to several Mb in length); the kinetochore does not form along the whole array but within a restricted zone of this array.

The apparent absence of any consensus DNA sequence associated with kinetochores has led to the suggestion that formation of kinetochores may depend on particular, higher-order DNA–protein structures. Such chromatin might also be subject to some form of epigenetic modification that ensures formation of the kinetochore at this particular region in successive cell generations.

Centromere-Associated Proteins

Many kinetochore proteins have been identified although their functions have not been fully

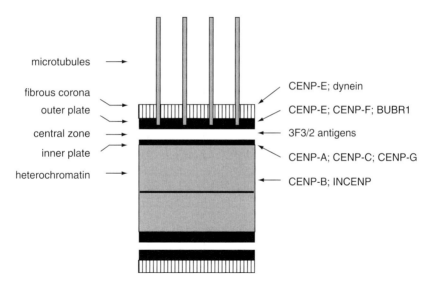

Figure 1 Schematic representation of somatic kinetochores as viewed by conventional electron microscopy. The various components of the kinetochore are identified and the locations of some of the centromere associated proteins are also indicated.

characterized. The locations of some of these proteins within the somatic vertebrate kinetochore are illustrated in **Figure 1**. In vertebrates, some centromere-associated proteins are present as constitutive elements of the kinetochore throughout the cell cycle, e.g., CENP-A, -B, -C. Others show a transient pattern of association, and are termed passenger proteins, being present usually from late G_2 to mitotic anaphase, e.g., CENP-E, -F, INCENP.

CENP-A and -C are essential for kinetochore function. In mice lacking these proteins, cell division is irregular and embryos die early in development. CENP-A is a histone H3-like protein; it may be involved in the epigenetic marking of kinetochore-associated chromatin and possibly also in the recruitment of CENP-C to the kinetochore. CENP-C is present at active centromeres, including neocentromeres (*de novo* sites of kinetochore activity outwith the centromere), but absent from inactive centromeres (e.g., in dicentric chromosomes). CENP-B binds to a specific 17bp DNA sequence that shows wide conservation in vertebrates. However, the functional role of this protein in kinetochore formation and activity is unclear. It is present on both active and inactive centromeres but is not present in neocentromeres.

Role in Mitotic Spindle Checkpoint

Kinetochores are important elements of a mitotic checkpoint. Failure of kinetochores to bind to spindle microtubules, or incorrect association such as when both sister kinetochores attach to microtubules from the same spindle pole, results in mitotic delay or arrest. Some proteins, for example mitotic-arrest-deficient

protein 2 (MAD2), may monitor microtubule binding to kinetochores. Others respond to the "tension" imposed on the kinetochore by the spindle microtubules by altering their phosphorylation state. These proteins are phosphorylated in misaligned kinetochores (and can be detected using an antibody, 3F3/2, that recognizes such epitopes) but dephosphorylated when kinetochores are correctly attached to the mitotic spindle. At present, we have an incomplete understanding of the pathway(s) through which kinetochores influence the spindle checkpoint.

Further Reading

Choo KHA (2000) Centromerization. *Trends in Cell Biology* 10: 182–188.

Craig JM, Earnshaw WC and Vagnarelli P (1999) Mammalian centromeres: DNA sequence, protein composition, and role in cell cycle progression. *Experimental Cell Research* 246: 249–262.

Pidoux AL and Allshire RC (2000) Centromeres: getting a grip of chromosomes. *Current Opinion in Cell Biology* 12: 308–319.

***See also:* Centromere; Meiosis; Mitosis**

Klinefelter Syndrome

M A Ferguson–Smith

Copyright © 2001 Academic Press
doi: 10.1006/rwgn.2001.0732

Klinefelter syndrome gets its name from a publication in 1942 by Klinefelter, Reifenstein, and Albright describing a series of patients with gynecomastia, small testes, aspermatogenesis, androgen deficiency, and increased levels of follicle stimulating hormone. Of particular interest at that time was the association of primary hypogonadism with high levels of gonadotrophins. The condition was soon found to be a common cause of male hypogonadism. Klinefelter syndrome attracted little attention until 1956 when Plunkett and Barr demonstrated the sex chromatin body in somatic cell nuclei, suggesting that those affected were sex-reversed females. However, in 1959 it was shown that this was incorrect and that the sex chromatin-positive cases had an XXY sex-chromosome constitution. Later variants of the syndrome were observed with XXXY and XXXXY sex chromosome complements, and others with sex chromosome mosaicism, such as XY/XXY and XX/XXY.

The paradoxical sex chromatin findings prompted nuclear-sexing surveys of various populations, using buccal mucosal cell smears as a readily obtained source of test material. Thus, Klinefelter syndrome was found to be one of the commonest causes of male infertility due to azoospermia and extreme oligozoospermia, accounting for over 10% of such cases. Also, approximately 1% of males with severe learning difficulties were found to be affected by Klinefelter syndrome. Overall, 1 in 1000 of all male births are affected with the disorder.

In adults with XXY Klinefelter syndrome, the one invariable clinical finding is small testes, associated with otherwise normal genitalia. The testes are less than half the normal size, measuring in length little more than 2 cm. Gynecomastia is present in less than half the cases. Most patients show evidence of lack of androgens, such as scant body and facial hair, poor recession of temporal hair, lack of libido and potency, and a small prostate. Patients tend to be taller than average with longer legs in relation to trunk lengths and wide arm span. These findings are apparent before puberty and are therefore not due to delayed epiphyseal fusion. The testicular defect is characterized by completely atrophic, hyalinized 'ghost' tubules devoid of elastic fibers alongside large masses of interstitial cells. In amongst the interstitial cells are occasional tubules lined solely by Sertoli cells, most of which are immature and undifferentiated. In rare cases, a single tubule may be found in which complete spermatogenesis is present. In the prepubertal testes, atrophic tubules are absent and spermatogonia may be found in a small proportion of tubules. Larger germ cells resembling oogonia at varying stages of calcification may occasionally be seen in prepubertal testes.

Patients with Klinefelter syndrome and more than two X chromosomes have greater physical and mental handicap associated with a number of malformations.

These include microcephaly, proximal radioulnar synostosis, undescended testes, congenital heart disease, cleft palate, and short incurved digit V. The facies is characteristic with prognathism, epicanthus, hypertelorism, myopia, strabismus, and mid-face hypoplasia. The maximum number of sex chromatin bodies per nucleus is always one fewer than the total number of X chromosomes, indicating that X-inactivation ensures that only one X chromosome is genetically active. However, abnormal dosage of X/Y homologous loci, which normally escape X-inactivation, is thought to be responsible for the level of clinical disability associated with additional X chromosomes. Male differentiation occurs irrespective of the number of X chromosomes and this attests to the dominant male-determining effect of the sex-determining region (SRY)-containing Y chromosome. The XXY condition has been observed in a number of other species including mouse, cat, horse, and sheep; in each case male differentiation is apparent.

Other variants of Klinefelter syndrome are known. SRY+ XX males, in whom the SRY locus has been transferred to the X by accidental recombination within the differential segments of the X and Y chromosomes, show little disability other than infertility (see Sex Reversal). Those with sex chromosome mosaicism, i.e., XY/XXY or XX/XXY, also tend to show less disability than XXY patients. XY/XXY patients are occasionally fertile and XX/XXY patients may rarely be found to be true hermaphrodites (see Hermaphrodite).

Intracytoplasmic sperm injection (ICSI) has been used increasingly to allow some Klinefelter patients to father children. In these cases, small numbers of viable sperm have been recovered by testicular or epididymal biopsy for the IVF procedure using ICSI. Many patients with Klinefelter syndrome benefit from routine therapy with small doses of testosterone.

See also: Fertilization; Hermaphrodite; Imprinting, Genomic; Infertility; Recombination, Models of; Turner Syndrome; X-Chromosome Inactivation

Knockout

L Silver

Copyright © 2001 Academic Press
doi: 10.1006/rwgn.2001.0733

A knockout is shorthand term used to describe a genetically manipulated organism that has had a specific gene eliminated or inactivated. A knockout allele is, thus, incapable of producing a gene product.

Knockout alleles are generated by an *in vitro* process of homologous recombination in embryonic stem cells.

See also: **Embryonic Stem Cells**

Kornberg, Arthur

T N K Raju

Copyright © 2001 Academic Press
doi: 10.1006/rwgn.2001.0734

The American scientist Arthur Kornberg (1918–) shared the 1959 Nobel Prize for Physiology or Medicine with the Spanish–American scientist Severo Ochoa (1905–1993). These scientists were honored "for their discoveries of the mechanisms of the biologic synthesis of ribonucleic and deoxyribonucleic acids." (Nobel Prize Foundation)

The son of Joseph and Lena (née Katz) Kornberg, Arthur was born in Brooklyn, New York. His parents had immigrated from Austria and his father operated sewing machines in sweatshops prior to owning a small hardware store. Arthur was a brilliant student, with a reputation as the "smart kid on the block." (Henerdson and Kornberg, 1991) His love of biology and biochemistry was sparked after he took a premedical course at the City College of New York. Enrolling himself for medical studies at Rochester University, New York, Kornberg earned his medical degree from there in 1941.

Following an internship and a brief period of service as medical officer in the US Coast Guards, Kornberg chose a career in biochemistry research rather than in medical practice. In 1943, he joined the National Institutes of Health in Bethesda, Maryland, where he was to conduct much of his prize-winning enzyme work. He also received brief, but valuable training under Severo Ochoa at New York University College of Medicine, in New York, in 1946, and under Gerty and Carl Cori at Washington University, in St. Louis, in 1947.

In 1955, Ochoa and Grunberg-Manago isolated a new enzyme from *Azobacter vinelandii* that was capable of synthesizing RNA in test tubes. They named the enzyme polynucleotide phosphorylase. Some years later it was shown that polynucleotides synthesized *in vitro* were also active as messengers in protein synthesis.

Working independently, Kornberg attempted to free enzymes from cells by using one of the latest physical methods – treating bacteria with sound waves. Subsequent steps in enzyme isolation were long and tedious, fraught with many technical difficulties. After isolating reasonably pure forms of the DNA polymerase enzyme from the bacterium *Escherichia coli*, Kornberg incubated them with radiolabeled thymine, one of the four bases of DNA. He then demonstrated that thymine had been incorporated into a chemical that had some of the properties of natural DNA.

However, to produce DNA artificially, Kornberg needed exquisitely pure forms of the enzyme. This required extensive experimentation that would take an additional 4 years. After succeeding in isolating the purest forms of polymerase enzyme, Kornberg showed that, in addition to the enzyme and the four base pairs of DNA as 'raw materials,' small quantities of 'primer' DNA were needed for artificial DNA synthesis.

Along with describing detailed enzymatic steps of DNA replication, Kornberg also presented the first experimental proof of how polymerase enzymes catalyzed reactions resulting in the production of new strands of DNA, which were virtually identical to the natural DNA.

Thus, nearly 100 years after the discovery of nucleic acids, DNA and RNA could be artificially synthesized. The findings of Ochoa and Kornberg were hailed as a milestone in the history of genetics. Hugo Theorell of the Royal Caroline Institute, the scientist who delivered the presentation address at the Nobel Prize ceremonies of 1959, prophetically predicted that just as the discovery of urea in the nineteenth century by Friedrich Wöhler, the discoveries of Ochoa and Kornberg's were the next major steps along the pathway of bridging the "first gap between the living and the dead." (Nobel Prize Foundation). As Theorell predicted, Ochoa and Kornberg's contributions were to play a central role in the technology of genetic engineering of the 1980s and in the Human Genome Project of the 1990s.

Kornberg once said with characteristic modesty, that he and Ochoa had simply opened up a tiny crack and tried driving a wedge – the hammer was the enzyme to understand the mystery of DNA molecule. When he was asked whether he and his colleagues had created life in a test tube, Kornberg replied that he might be able to answer the question "if you'd first care to define life."

Although Kornberg wrote extensively, he keenly appreciated the difficulties of good writing, which he referred to as "variety of mental torture" (Magner 1991). His autobiography, *For the Love of Enzymes: The Odyssey of a Biochemist*, was published in 1989.

References

Henerdson B and Kornberg A (1991) In: Magill, FN (ed.) *The Nobel Prizes Winners*, Vol. 2, *Physiology of Medicine* pp. 797–802. Pasadena, CA: Salem Press.

Kornberg A (1989) *For the Love of Enzymes: The Odyssey of a Biochemist.* Cambridge, MA: Harvard University Press.

Magner L (1990) Aurthur Kornberg, 1939 In: Fox DM, Meldrun M and Rezak I (eds) *Nobel Laureates in Medicine or Physiology: A Biographical Dictionary*, pp. 324–327. New York: Garland Publishing.

Nobel Prize Foundation. www.nobel.se/medicine/laureates/1959/press.html

See also: **DNA Polymerases; Genetic Recombination; Human Genome Project; Nucleotides and Nucleosides; Ochoa, Severo**

Kornberg Enzyme

See: **Kornberg, Arthur**

Kuru

M A Ferguson-Smith

Copyright © 2001 Academic Press
doi: 10.1006/rwgn.2001.0736

Kuru is a transmissible spongiform encephalopathy (TSE; see Transmissible Spongiform Encephalopathy), which reached epidemic proportions in the 1950s in Papua New Guinea among the Fore tribe. When it was first described in 1957, the disease was evident in about 1% of a population of more than 35 000 people. In some areas the disease was prevalent in as many as 5–10% of the population. Those affected first develop cerebellar symptoms with unsteadiness of gait, progressive trembling or shivering of the body (termed 'kuru' in the Fore language), and dysarthria. The ataxia becomes progressively worse and soon the patient is uable to walk or stand, muscle tremors and rigidity become pronounced, incontinence and dysphagia develop, and eventually the patient becomes mute and unresponsive. Death occurs within 1 year of the onset of the disease. Unlike other TSEs, severe dementia is not a feature of kuru.

Microscopic examination of the brain of affected patients revealed loss of neurons, particularly in the cerbellum, widespread astrocytosis, and spongiform change. Amyloid plaques were present in about 75% of cases. The cause of the condition was obscure until 1959, when the similarity of the neuropathology to scrapie was first noticed. This prompted attempts to transmit kuru to experimental animals. Intracerebral inoculation of brain tissue into chimpanzees led to a kuru-like disease within 1.5 years. Other animals also proved susceptible both by inoculation and by oral feeding, including Old World and New World monkeys and goats. Kuru does not transmit to sheep.

The early investigators of kuru noticed that the disease was common in women and children, but adult males were rarely affected. During the past 30 years, the condition has gradually disappeared except in a few elderly individuals. This correlates with the abandonment of ritual cannibalism in the early 1960s. Up to that time, it was the practice of local tribes to take part in consuming various tissues, including the brain of deceased relatives, partly as an act of respect and mourning. Women did the butchery and prepared tissues for consumption. This involved much bodily contamination with brain and body fluids, and it is likely that infection occurred through body sores in addition to oral ingestion. Men were not involved in handling the affected corpses and tended to eat the flesh rather than the brains, while women and children were much more exposed to the infection. Since the 1960s, the mortuary practices have been abandoned and this has been associated with a sharp decline in disease prevalence. At the time of writing, only a few elderly people develop the disease each year, and this suggests that, in these cases, the incubation period may be as long as 40 years. Children born to affected women in recent years, and since the cessation of cannibalism, have not developed the disease, suggesting that maternal transmission either *in utero* or via breast feeding does not occur to any extent.

It has been suggested that kuru might have originated from a sporadic case of Creutzfeldt–Jakob disease occurring early in the twentieth century, which spread to an increasing number of the population as a result of the practice of ritual cannibalism. The spread of bovine spongiform encephalopathy, via animal protein contained in commercial cattlefeed and thence to humans, has close similarity to the spread of kuru.

See also: **Transmissible Spongiform Encephalopathy**

L

lac Mutants

J Parker

Copyright © 2001 Academic Press
doi: 10.1006/rwgn.2001.0737

Lac mutants are organisms that contain mutations in some part of the *lac* operon or its controlling elements. Therefore, they contain some defect in the metabolism of the disaccharide lactose, or in the regulation of this metabolism, when compared with wild-type strains. Lac mutants are of historic interest because they helped to uncover the structure and regulation of the *lac* operon, the first operon discovered. They are also of interest because the techniques which were developed to screen or select these mutants are still used in the classroom and the laboratory.

Wild-type strains of the bacterium *Escherichia coli* are phenotypically Lac$^+$, meaning they have the ability to use lactose as a sole source of carbon. In order to be Lac$^+$, *E. coli* must be able to express a functional *lacZ* gene, which encodes β-galactosidase, and a functional *lacY* gene, which encodes the lactose permease. Mutants in which either of these genes have been inactivated are said to be Lac$^-$ and cannot utilize lactose. Joshua Lederberg and his associates were the first to isolate and map Lac$^-$ mutants of *E. coli*, beginning in the 1940s. Lac$^-$ mutants can be identified by their failure to grow when lactose is the sole carbon source or by the use of various types of indicator plates. Mutations in *lacZ* or *lacY* can be differentiated by a variety of techniques. For example, mutants which cannot produce the lactose permease also cannot grow on melibiose under certain conditions.

Mutations are also known in the regulatory genes or regions controlling the *lac* operon. Mutants with a mutation in the *lac* promoter will typically be Lac$^-$, that is, the promoter will no longer function or at least will show decreased expression. However, mutants which cannot make the lactose repressor, the product of the *lacI* gene, or which make a repressor that cannot bind the inducer, will remain Lac$^+$ but will constitutively express the products of the operon. Such mutants will grow on the sugar raffinose, which requires the lactose permease for entry into the cells but is not an inducer of the operon. Constitutive expression of β-galactosidase can also be monitored using the chromogenic compound X-gal (5-bromo-4-chloro-3-indolyl-β-D-galactosidase) which is also not an inducer of the operon.

However, *lacI* mutants are also known which lead to repressor binding to *lacO*, the lactose operation even in the presence of an inducer. These mutants will be phenotypically Lac$^-$, and the mutation will be dominant to the wild-type *lacI* allele. Similarly, most mutations in *lacO* should diminish or destroy the ability of this site to bind the repressor and lead to constitutive formation of the *lac* operon enzymes. However, some mutations in *lacO* lead to enhanced binding and the mutants are Lac$^-$. Note that because *lacO* is a noncoding regulatory region on the DNA, mutations in it will only have an effect on the operon of which they are a part; that is, they will only operate in *cis*. On the other hand, *lacI* mutations will function in *trans*. The ability to make partial diploid strains of *E. coli* was a very important tool in these Lac mutants.

The *lac* operon, like many others in *E. coli*, is also positively controlled by the level of cyclic AMP (cAMP) and the cAMP binding protein (catabolite activator protein, CAP), encoded by the *crp* gene. Mutations in the genes controlling the level of cAMP or the production of CAP will also be phenotypically Lac$^-$. However, such mutations will be very pleomorphic, and it would be unusual to refer to them as 'Lac mutants.'

Interestingly, amino acid residues can be added to the amino terminus of β-galactosidase without important effects on enzyme activity. Therefore *lacZ* is unusually insensitive to insertion mutations in this region if they maintain the correct reading frame. Because of this, many cloning vectors have been designed to contain a reporter which consists of a multiple cloning site, or polylinker, inserted into this region of the *lacZ* gene. Essentially all that is required is that the synthetic cloning site does not lead to a frameshift of termination of translation. DNA fragments which are subsequently inserted into such a multiple cloning site will typically introduce such mutations, and therefore clones which contain inserts can be readily identified by screening.

See also: Constitutive Expression; *lac* Operon; Lederberg, Joshua; Phenotype

lac Operon

J Parker

Copyright © 2001 Academic Press
doi: 10.1006/rwgn.2001.0738

The lactose or *lac* operon of *Escherichia coli* is a cluster of three structural genes encoding proteins involved in lactose metabolism and the sites on the DNA involved in regulation of the operon. The three genes are: (1) *lacZ*, which encodes the enzyme β-galactosidase (which splits lactose into glucose and galactose); (2) *lacY*, which encodes lactose permease; and (3) *lacA*, which encodes a lactose transacetylase. Functional β-galactosidase and lactose permease are required for the utilization of lactose by this bacterium. These proteins are present in the cell in very low amounts when the organism is grown on carbon sources other than lactose. However, the presence of lactose and related compounds leads to the induction of the synthesis of these proteins. Interest in understanding the induction of β-galactosidase by its inducer, lactose, led Jacques Monod and his associates to begin studying the regulation of lactose metabolism in the 1940s. These studies were aided by analogs of lactose that could also be synthesized. Of equal importance, genetic systems (conjugation and transduction) for *E. coli* were known which enabled genetic analysis of mutants with alterations in lactose metabolism.

Throughout the 1950s, Jacques Monod, François Jacob, and their colleagues performed physiological and genetic experiments on lactose metabolism in *E. coli* that led to important breakthroughs in our understanding of gene expression and regulation. It was found that some inducers were not substrates of β-galactosidase and some substrates were not inducers. Elegant genetic experiments involving *lac* mutants led in turn to the discovery of regulatory genes such as *lacI*, which encoded the *lac* repressor. These and other experiments led to the operon model of gene expression proposed in 1961. The power of this model was widely appreciated; Jacob and Monod won the Nobel Prize in 1966.

The genes in an operon are transcribed into a single, polycistronic messenger RNA (mRNA), in this case from the *lac* promoter *lacP*. The regulatory sites that are part of the operon also include the *lac* operator

lacO. When the lactose repressor binds to *lacO*, a region immediately upstream of the structural genes of the *lac* operon, it prevents transcription of the operon. This is an example of negative control. Inducers of the operon bind to the repressor and cause a conformational change that leads to the disassociation of the repressor from the operator. Transcription of the operon then begins. (Although the gene encoding the lactose repressor is not part of the *lac* operon, it is located next to it on the chromosome.)

Later it was discovered that there is another regulatory protein, which participates in positive control of the *lac* operon. This is the catabolite activator protein (CAP; also called the cAMP receptor protein, CRP), which, when bound to cAMP, itself binds to a region of the *lac* operon upstream of the promoter and allows RNA polymerase binding. The CAP protein is involved in regulation of many operons as part of a global control system, catabolite repression, which allows the efficient integration of the metabolism of different carbon sources.

The *E. coli lac* operon is of much more than historical importance. Not only has it proved extremely useful as a model for studies of gene regulation, it is also a powerful tool in genetic analysis. For example, the ease of assaying β-galactosidase, both *in vitro* using colorimetric assays and on plates using chromogenic substrates, has made *lacZ* an ideal reporter gene in a large variety of experimental situations. In addition, the regulatory system consisting of the *lac* repressor and *lac* operator is often incorporated into cloning vectors to provide an easily controlled regulatory system for cloned genes.

See also: Catabolite Repression; Cloning Vectors; Induction of Transcription; Jacob, François; *lac* Mutants; Monod, Jacques; Operators; Operon; Polycistronic mRNA; Promoters; Regulatory Genes

Lactose

J H Miller

Copyright © 2001 Academic Press
doi: 10.1006/rwgn.2001.0741

A disaccharide (two sugars joined by an O-glycosidic bond) commonly found in milk. Lactose is termed a β-galactoside because it consists of galactose joined to glucose via a β (1→4) glycosidic linkage. Lactose is cleaved by the enzyme β-galactosidase to yield galactose and glucose. The study of the regulation of β-galactosidase synthesis in bacteria by Jacques Monod and François Jacob led to the first breakthrough in understanding gene regulation, and resulted in the 'operon model' of gene regulation.

See also: **Beta (β)-Galactosidase**

Lagging Strand

The lagging strand of DNA elongates overall in the 3′–5′ direction, but is synthesized discontinuously in the form of short fragments (5′–3′) that are subsequently covalently linked.

See also: **Okazaki Fragment; Replication**

Lamarck, Jean Baptiste

G S Stent

Jean Baptiste, Chevalier de Lamarck (1744–1829), was the first person to develop a comprehensive theory of evolution. The essence of his theory, which he worked out at the end of the eighteenth century, was that the present-day diversity of living species arose via a gradual "transmutation" of ancestral species. Thus in conceiving the history of living forms in terms of "descent with modification," Lamarck's evolutionary theory was a precursor to that which Charles Darwin presented 50 years later in his *On the Origin of Species*.

Lamarck was born in the Picardy region of northeastern France in 1744. As a son of a family of impoverished aristocrats, he had only two alternative prospects for an honorable career: the Church or the Army, and Jean Baptiste tried both. After briefly studying for the priesthood with the Jesuits, he joined the Grenadiers and distinguished himself by his bravery in the battle at Bergen-op-Zoom in the Seven Years War. Suffering a head wound (not from hostile enemy fire but from friendly horseplay with his fellow Grenadiers), he was given a medical discharge and took up the study of medicine in Paris in 1766.

Lamarck did not become a physician, any more than he became a priest or professional soldier. Instead, he turned to the study of natural history, and in 1781, he was appointed to a junior curatorship in the King's Botanical Garden. This position gave him the opportunity to undertake field studies, and

in 1788, he published a definitive survey of the flora of France, presenting a dichotomous diagnostic method for the taxonomic classification of plants by scoring the presence or absence of alternative traits.

This novel procedure brought him to the attention of Georges Buffon, the foremost French naturalist of the time, who sponsored Lamarck's election to the French Academy of Science and his appointment to a professorship at the Museum of Natural History in Paris. Before long, Lamarck brought out his monumental *Dictionnaire de Botanique*, on which his scientific reputation would mainly rest during his lifetime.

When the Museum of Natural History was reorganized in 1793, in the aftermath of the political turmoil of the French Revolution, Lamarck was transferred to the chair of zoology and given the assignment of teaching the taxonomy of insects and worms. Being a botanist, he knew very little about animals, but he was a fast learner. Between 1815 and 1822, he published his great zoological treatise, *Histoire naturelle des animaux sans vertèbres*. This contained the first subdivision of the phyla of the animal kingdom into two grand categories, which he designated as "vertebrates" and "invertebrates," according to whether a vertebrate column was present or absent. Moreover, among the invertebrates (whose classification had flummoxed Linnaeus, the founder of modern taxonomy) Lamarck identified and named the phylum of annelids and the classes of arachnids and crustaceans of the arthropod phylum. His depth of knowledge of the natural history of both plant and animal kingdoms was highly unusual for its time, and led Lamarck to put forward another novel idea: that there exists a general science of living forms, for which he coined the (Greek-derived) compound neologism "biology."

In developing his theory of evolution (which is treated in these pages in a separate entry; see Lamarckism) Lamarck took into account the geological studies that indicated that the earth has existed for a very long time, during which its surface features underwent many very gradual changes. Moreover, he inferred from the character of fossils that animal life has been present for a large fraction of that long time, during which it too underwent gradual changes. Hence the species have to be transmutable rather than eternally fixed, as had been generally believed ever since Aristotle developed the species concept in the fourth century BC. As Lamarck pointed out, the seemingly empirical fact of the characterological permanency of the species is actually an illusion, attributable to the shortness of the human life span relative to the enormous length of the geological time scale.

Lamarck had moved evolutionary theory into the forefront of biological thinking, for which he received hardly any credit during his lifetime. This lack of appreciation was due in large measure to his having been overshadowed by his politically influential contemporary Georges Cuvier, the founder of comparative anatomy and leading authority on the classification of fossils. Despite his outstanding qualifications for the study of evolution, Cuvier was a creationist who believed in the literal truth of the story told in Genesis 1 of the Five Books of Moses. He firmly rejected Lamarck's theory and explained the origin of fossils in terms of a succession of catastrophes in the earth's history, each of which exterminated all extant forms of life and was followed by another round of *de novo* creation.

Lamarck was blinded by an infection for the last 17 years of his life, and fell into poverty, dying in 1829. Even posthumously, he never did receive the recognition he deserved as an important pioneer in the development of modern biology. Instead his name became the object of ridicule and the term 'Lamarckist' an invective because his evolutionary theory contained a fundamental flaw. Contrary to contemporary popular belief, Charles Darwin, who, it should be noted, did hold Lamarck in high regard, was no more able to provide a satisfactory explanation of the origin of novel hereditary traits than was Lamarck. Such an explanation had to await the rise of the science of genetics in the first part of the twentieth century and the development of neo-Darwinism.

See also: Darwin, Charles; Lamarckism

Lamarckism

G S Stent

Copyright © 2001 Academic Press
doi: 10.1006/rwgn.2001.0743

'Lamarckism' refers to the first comprehensive theory of evolution developed by the French natural historian Jean Baptiste Lamarck and set forth by him in his treatises *Recherches sur l'organization des corps vivants* (1802); and *Philosophie zoologique* (1809). Lamarck's theory was based on his lifelong direct observation of plants and animals, which provided him with a sense of the dynamic quality of life, as well as of the close interdependence of physical and vital processes in which life is grounded. As originally formulated, Lamarckism was part of an elaborate surmise about processes for whose operation Lamarck had no direct evidence.

Lamarckism asserted that all living things have arisen via a continuous process of a gradual modification throughout geologic history, as a vast sequence of life forms, ascending a staircase leading from the lowliest and simplest to the highest and most complex creatures. To account for this progressive movement Lamarck invoked what then seemed a reasonable hypothesis of the inheritance of acquired characteristics: that organisms develop new traits in response to needs created by their environment and pass them on to their offspring. The commonly cited example of Lamarckism is the evolution of the giraffe, whose ancestors were supposed to have acquired their long necks by stretching them to reach the upper leaves of trees and transmitted that gradually acquired neck length to their progeny. Lamarckism also provided for the permanent *loss* of *old* traits, in case a change in the environment eliminated the need for them. In *Philosophie zoologique* Lamarck summarized his theory in terms of two 'laws' governing the evolutionary ascent of life to higher stages. One stated that organs are improved with repeated use and weakened by disuse. The other stated that such environmentally determined acquisitions or losses of organs are preserved by transmission from parent to progeny.

Lamarckism was an important forerunner of the Darwinian theory of evolution, which, just as did Lamarckism, assigned a critical role to the environment in evolutionary processes. Contrary to a misconception held widely even among present-day biologists, Lamarckism is not in conflict with Darwin's theory of natural selection. According to Lamarckism, the offspring of those giraffes that did succeed in transmitting an acquired extension of their necks to the next generation could obtain more food than other members of their cohort. They would thus be more numerous, which, in turn, would result in an increase of the average neck length in successive generations. Thus Darwin's 'classical' Darwinism is an improvement over Lamarckism but not its refutation, since Darwin had no more clear idea than Lamarck had of the genetic basis of the hereditary variations that are at the root of the evolutionary process.

Lamarckism fell into disrepute only in the early years of the twentieth century, after the rediscovery of Mendel's laws of inheritance, the identification of genes as the atoms of heredity, and the recognition of gene mutation as the source of the novel hereditary features that are responsible for evolutionary change. These insights gave rise to the development of neo-Darwinism which accounts for evolution in terms of gene mutation, natural selection for traits, and the reproductive dynamics of conspecific populations.

By the middle of the twentieth century, the designation of someone as a 'Lamarckist' had become a term of abuse, partly because of its association with one of the few world-class monsters of twentieth century science: the Russian agronomist Trofim Lysenko, who dominated (not to say destroyed) genetics in the Soviet Union and its satellite popular democracies from the mid-1930s until the mid-1960s. Lysenko was not openly opposed to classical Darwinism, Karl Marx having been a great admirer of Darwin but he declared neo-Darwinism, with its reliance on Mendelian genetics and gene mutation, to be idealist–racist metaphysical speculations propagated by the Catholic Church and the Fascists to keep the proletariat intellectually enchained.

At first, in the 1930s, Lysenko denied that he was a Lamarckist and declared that "starting from Lamarckian positions, the work of remaking the nature of plants by 'education' cannot lead to positive results." Then, when he became director of the Institute of Genetics of the Soviet Academy of Sciences in 1940 and had Stalin's ear, Lysenko declared Mendelian genetics erroneous. By 1948, when he had ruthlessly silenced any Soviet geneticists who opposed him, he no longer concealed his adherence to Lamarckism, declaring that:

the well-known Lamarckian propositions, which recognize the active role of the conditions of the external environment in the living body and the inheritance of acquired characters, in contrast to the metaphysics of neo-Darwinism, are indeed scientific.

Lysenko was finally dismissed in 1965, after having gravely hampered scientific and agricultural progress in the Soviet Union for more than 25 years. Nevertheless, 'Lamarckist' remains a term of ridicule. This is a most regrettable affront to the memory of one of the great figures in the history of biology, to whom that discipline owes its very name.

See also: **Lamarck, Jean Baptiste; Lysenko, T.D./ Lysenkoism**

Lampbrush Chromosomes

H C Macgregor

Copyright © 2001 Academic Press
doi: 10.1006/rwgn.2001.0745

Lampbrush chromosomes (LBCs) are elongated diplotene bivalents in prophase of the first meiotic division in growing oocytes in the ovaries of most animals other than mammals and certain insects. Some LBCs reach lengths of a millimeter or more. The chromosomes go from a compact telophase form at the end of the last oogonial mitosis, become

lampbrushy, and then contract again to form normal first meiotic metaphase bivalents. They are characterized by widespread RNA transcription from hundreds of transcription units that are arranged at short intervals along the lengths of all the chromosomes.

LBCs were first seen in salamander oocytes by Flemming in 1882 and in oocytes of a dogfish by Ruckert in 1892. The name lampbrush originated from Ruckert, who likened the objects to a nineteenth-century lampbrush, equivalent to the modern test-tube brush. LBCs are delicate structures and they must be carefully dissected out of their nuclei in order to examine them in a life-like condition. The largest LBCs are to be found in oocytes of newts and salamanders, animals that have large genomes and correspondingly large LBCs.

The best oocytes for lampbrush studies are those that make up the bulk of the ovary of a healthy adult female at the time of year when the eggs are actively growing. They are about 1 mm in diameter and their nuclei are between 0.3 and 0.5 mm in diameter (**Figure 1**). The techniques for isolating and looking at LBCs from such oocytes are specialized but inexpensive and simple; details are available in the sources cited in the Further Reading section.

Since an LBC is a meiotic half bivalent, it must consist of two chromatids. The entire lampbrush bivalent will therefore have a total of four chromatids. The chromosome appears as a row of granules of deoxyribonucleoprotein (DNP), the chromomeres, connected by an exceedingly thin thread of the same material (**Figure 2**). Chromomeres are 0.25–2 μm in diameter and are spaced 1–2 μm center to center along the chromosome.

Each chromomere has two or a multiple of two loops associated with it. The loops have a thin axis of DNP surrounded by a loose matrix of ribonucleoprotein (RNP). The loops are variable in length, ranging from about 5 to 100 μm. Loops vary in appearance. Loops of the same appearance always occur at the same locus on the same chromosome within a species.

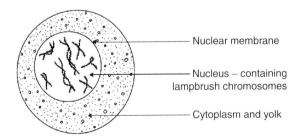

Figure 1 An oocyte (growing ovarian egg) showing the relative dimensions of the egg, its nucleus, and its lampbrush chromosome.

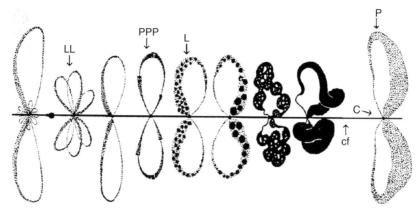

Figure 2 A region of a lampbrush chromosome showing the interchromomeric axial fiber (cf) connecting small compact chromomeres (c), chromomeres bearing pairs (L) or multiple pairs (LL) of loops, loops of different morphologies, polarization of thickness along individual loops, loops consisting of a single unit of polarization (P), and loops with several tandem units of polarization having the same or different directions of polarity (PPP).

Some particularly distinctive loops can be used for chromosome identification and the construction of LBC maps. Loops arising from the same chromomere have the same appearance and are usually, though not always, of the same length (**Figure 2**).

The general pattern of events during the lampbrush phase of oogenesis is one of extension followed by retraction of the lampbrush loops and there is a clear inverse relationship between loop length and chromomere size. The longer the loop, the smaller the chromomere, and vice versa.

Most lateral loops have an asymmetrical form. They are thin at one end of insertion into their chromomere and become progressively thicker towards the other end (**Figure 2**). If an LBC is stretched, breaks first happen transversely across the chromomeres so that the resulting gaps are spanned by the loops that are associated with the chromomeres (**Figure 3**). This demonstrates the structural continuity between the main axis of the chromosome – the interchromomeric fiber – and the axes of the loops.

Lampbrush loops are sites of active RNA synthesis and RNA is being transcribed simultaneously all along the length of the loop. In newts, there are more than 20 000 RNA-synthesizing loops per oocyte. Particular loops may be present or absent in homozygous or heterozygous combinations and the frequency of combinations within and between bivalents with respect to presence or absence of loops, signifies that these loops assort and recombine like pairs of Mendelian alleles. So there appears to be an element of genetic unity in a loop–chromomere complex.

By 1960, it was known that an LBC has two DNA duplexes running alongside one another in the interchromomeric fiber, compacted into chromomeres at intervals and extending laterally from a point within

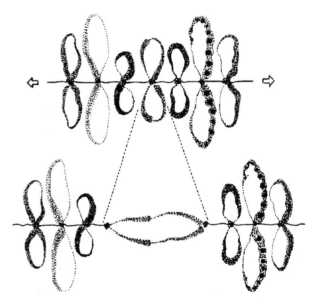

Figure 3 Breakage of a stretched lampbrush chromosome across a chromomere, such that loops associated with the chromomeres come to span the gap between the two halves of the chromomere.

each chromomere to form loops where RNA transcription takes place. Each duplex represents one chromatid (**Figure 4**). New technologies of the late 1970s confirmed this model and extended it.

A technique that removed most of the protein from chromosomes, leaving only the DNA and attached newly transcribed RNA, and then visualized what was left by electron microscopy, showed a lampbrush loop as a thin DNA axis with RNA polymerase molecules lined up and closely packed along its entire length. Each polymerase carried a strand of RNP. At one end of the DNA axis, the RNP strands are short.

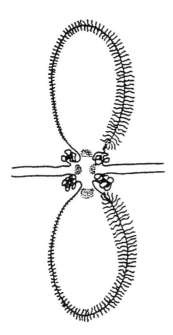

Figure 4 Accepted model of lampbrush chromosome organization showing the interchromomeric fiber consisting of two chromatids that separate from one another and become involved in RNA transcription in the regions of the loops.

At the other end, they are much longer and they show a smooth gradient in size from one end to the other. In essence, the entire region is polarized and asymmetric, in the same sense as a loop, as seen with the light microscope, is asymmetric. The DNA axis outside the region occupied by polymerases shows the structure that would be expected of nontranscribing chromatin. The lengths of the transcribed regions of the chromosome are about the same as the lengths of loops as seen and measured with light microscopy. Lampbrush loops are therefore polarized units of transcription. The polymerase moves on a stationary loop axis. A loop is formed by an initial 'spinning out' process, probably powered by the continuing attachment of more and more polymerases to a specific region of the chromomeric DNA. The loop remains and is transcribed as a permanent structure throughout the lampbrush phase. Towards the end of the lampbrush phase, transcriptive activity declines, polymerases detach from loop axes, and loops regress and disappear. The vast majority of the chromomeric DNA is never transcribed and a loop represents a short, specific part of the DNA in a loop–chromomere complex.

In situ nucleic acid hybridization is a means of locating specific gene sequences on chromosomes. Let us suppose that each loop represents 'a gene.' The RNA that makes up the loop matrix, the attached nascent transcripts, will all be or include transcripts of that 'gene.' In effect, the loop is a large object, consisting of hundreds of RNA copies of the gene, all clustered at one position on the chromosome set. Isolate and purify the DNA of that gene, and label it in some way, and it will be easy to make it single-stranded and bind it specifically to the complementary single-stranded RNA attached to the lampbrush loop. The technique is known as DNA/RNA transcript *in situ* hybridization (DR/ISH).

The end product of an experiment involving DR/ISH is a preparation showing one or more pairs of loops with label distributed along their lengths. It is not uncommon in DR/ISH experiments to find loops that are labeled over only part of their lengths. This is evidence that the DNA sequence of a loop axis can and does change from place to place along the length of the loop. Wherever there are partially labeled loops, it is usual to find the same partially labeled loops, with precisely the same pattern of labeling, in every oocyte over quite a wide range of size and stage. So loops are permanent structures that transcribe from the same stretch of DNA axis throughout the entire lampbrush phase. DR/ISH experiments prove that highly repeated short DNA sequences, commonly referred to as 'satellite' DNA, which could not possibly serve as a basis for transcription and translation into functional polypeptides, are abundantly transcribed on lampbrush loops along with more complex sequences that are definitely translated into functional proteins.

The current hypothesis for LBC function is as follows. At the thin base of each loop or the start of each transcription unit there is a promoter site for a functional gene sequence. RNA polymerase attaches to this site and moves along the DNA, transcribing the sense strand of the gene and generating messenger RNA molecules that remain attached to the polymerase (**Figure 5**). In the lampbrush environment there are no stop signals for transcription, so the polymerases continue to transcribe past the end of the functional gene and into whatever DNA sequences lie 'downstream' of the gene. This results in very long transcription units, very long transcripts, mixing of gene transcripts with nonsense transcripts in high molecular weight nuclear RNA, and lampbrush loops. This 'read-through' hypothesis predicts that the number of functional genes that are expressed to form translatable RNAs may be expected to equal the number of transcription units that are active in a lampbrush set. The hypothesis says, in effect, that the only unusual feature of an LBC, and the very reason for the lampbrush form, is that once transcription starts it cannot stop until the polymerase meets another promoter that is already initiated or some condensed chromomeric chromatin that is physically impenetrable and untranscribable.

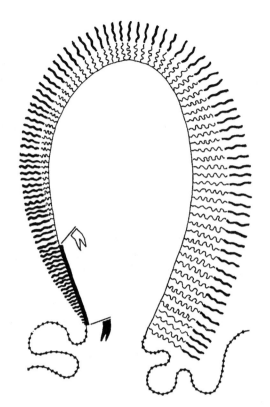

Figure 5 Transcription on a lampbrush chromosome loop where a gene (thick black line) is transcribed from its promoter (black flag) through to and past its normal stop signal (white flag) and into the normally nontranscribed DNA that lies downstream, thus generating very long transcription units with long transcripts that include RNA complementary to the sense strand of the gene (thick parts of the transcripts) and nonsense DNA that lies downstream of the gene (thin parts of the transcripts).

Evidence for the Read-Through Hypothesis for LBCs

LBCs dissected directly into a solution of the enzyme deoxyribonuclease-1 (DNase-1) fall to pieces and their loops break into thousands of fragments. This does not happen with ribonuclease or proteases. If breakage of the chromosome axis and the loops by DNase is watched and timed and the number of breaks plotted against time on a log scale, the slope of the plot for the chromosome axis is 4 and that for the loops is 2. This supports the model in which the axis consists of two chromatids – each a DNA double helix consisting of two nucleotide chains – and the loop is part of one chromatid – consisting of one double helix made up from two nucleotide chains.

A later experiment used restriction enzymes that cleaved DNA only at places along the molecule where there was a particular short nucleotide sequence. If a loop consisted entirely of identical tandemly repeated DNA sequences, all with a particular restriction enzyme recognition site, then the loop would be destroyed by that enzyme. If, on the other hand, the DNA sequences all lacked the enzyme recognition site, then the loop would be totally unaffected and would remain intact.

An experiment was set up using five enzymes and the LBCs from *N. viridescens*. The control enzyme was deoxyribonuclease-1. DNase-1 and three of the restriction enzymes destroyed everything. One enzyme, *Hae*III did likewise, except that it left one pair of loops completely intact. These *Hae*III resistant loops were big ones, 100 μm long, equivalent to at least 300 000 nucleotides. Their unique resistance to *Hae*III provided direct evidence that at least one pair of loops consisted of tandemly repeated short sequence DNA. At a later date, the effects of *Hae*III were tested again, with appropriate controls, on the *Hae*III resistant loops of *N. viridescens*. Breaks regularly occurred precisely at the thin beginnings of each loop, but the remainder of the loops remained intact, as would be predicted on the basis of the read-through hypothesis. The start of the transcription unit would be characterized by a long complex gene sequence that would almost inevitably include the *Hae*III recognition site. The remainder of the loop would consist entirely of repeat sequences that lacked the *Hae*III site.

Other Questions We Should Ask about Lampbrushes

Only a small fraction of the entire DNA of a loop–chromomere complex forms the transcription unit that makes the loop. What about the rest of the DNA? Is the DNA segment that makes the loop preferentially selected for transcription the same piece at the corresponding locus in every egg of every individual of a particular species? This question may be approached experimentally.

Why do loops have different morphologies that are heritable, locus-specific, and sometimes species-specific? The loop matrix is a site of processing, cleaving, and packaging of nuclear RNA, so most of the variation in gross structure may be expected to reflect different modes of binding and interaction involving quite a wide range of proteins and RNAs.

Do LBCs look the same in all animals? They do not. The relative lengths of LBCs at the time of their maximum development are the same as the relative lengths of the corresponding mitotic metaphase chromosomes from the same species. The overall lengths of LBC are broadly related to genome size. Birds, with their notably small genomes, have extremely small, but nonetheless very beautiful, LBC that present many extraordinary and hitherto unexplained features.

Some LBCs have long loops and others have very short ones. We have seen that the transcription units of LBCs are unusually long because they include interspersed repetitive elements of the genome. Structural genes in large genomes are more widely spaced than in small genomes as they are interspersed with noncoding DNA. One might therefore expect LBCs from large genomes to have longer loops (transcription units) than those of smaller genomes, and this is what has been observed.

Many of the very long loops that we see in LBC from animals with large genomes show multiple, tandemly arranged thin–thick segments (transcription units). The individual transcription units within one loop can have the same or opposite polarities and can be of the same or different lengths (**Figure 6**). This observation suggests that it is really the transcription unit that is the ultimate genetic unit in an LBC and not the loop/chromomere complex, as was once thought.

Why do LBCs exist at all? They are characteristic of eggs that develop quickly into complex multicellular organisms independently of the parent. A frog's egg is fertilized and develops into a complex tadpole within a few days. Much of the information and raw materials for this process are laid down during oogenesis through activity of LBCs and amplified ribosomal genes and the accumulation of yolk proteins imported from the liver. LBCs may therefore be regarded as an adaptive feature that has evolved to preprogramme the egg for rapid early development. The fact that they are not present in mammalian eggs could be regarded as an advanced feature that is consistent with the slow pace of mammalian development. A frog's egg, for example, will have completed gastrulation and the differentiation of its central nervous system and embryonic axis by the time a human embryo has only reached the 8-cell stage.

LBCs provide a uniquely powerful medium through which it has been possible to draw valid conclusions at the molecular level from observations and experiments carried out mainly with a light microscope. Their value extends into the fields of comparative molecular cytogenetics and systematics. Nowhere else is it possible to study genome structure, function, and diversity by actually looking at the genome itself with a light microscope. LBCs are technically challenging but not defeating. They are exceptionally beautiful to look at and fun to work with.

Further information on these remarkable structures can be found in the literature listed in the Further Reading section below and on the internet.

Further Reading

Callan G (1986) *Lambrush Chromosomes*. Berlin: Springer-Verlag.

Macgregor HC (1993) *An Introduction to Animal Cytogenetics*. London: Chapman & Hall.

Macgregor HC and Varley J (1988) *Working with Animal Chromosomes*. New York: John Wiley.

See also: Cytogenetics; Developmental Genetics

Late Genes

E Kutter

Copyright © 2001 Academic Press
doi: 10.1006/rwgn.2001.0746

During viral infection, late genes are those that are transcribed after the commencement of viral DNA synthesis. The bulk of these encode either components of the capsid, proteins aiding in morphogenesis or DNA packaging, or proteins that are to be carried with the DNA in the capsid.

See also: Virus

Leader Peptide

J Parker

Copyright © 2001 Academic Press
doi: 10.1006/rwgn.2001.0749

The term 'leader peptide' (or, less commonly, 'leader polypeptide') refers to a peptide encoded by a DNA

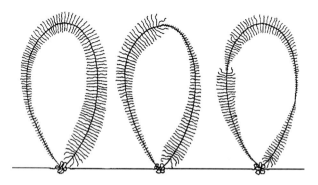

Figure 6 The various arrangements of transcription units that actually occur on lampbrush chromosomes. The loop on the left comprises a single transcription unit. In the middle loop there are two transcription units of the same size and polarity. The right-hand loop has four transcription units of different sizes and different directions of polarity.

sequence immediately upstream of the sequence encoding what eventually becomes a mature protein. However, this upstream peptide can be encoded in two quite different ways, and the term leader peptide is used to refer to both.

In most cases the leader peptide is produced as the amino terminus of a longer protein and is released from that protein by the action of a protease. That is, the sequence encoding the leader peptide is part of the open reading frame encoding the rest of the protein. Many of these leader peptides are also referred to as 'signal peptides' or 'signal sequences,' and these are involved in transport of the protein to or through cell membranes, transport to different membranous cellular compartments, or secretion of the protein from the cell. Signal peptides are removed from the mature protein during this process by a specific peptidase. Such signal peptides are composed typically of 16–30 amino acid residues. Signal peptides contain a hydrophobic core, which can span a membrane, a polar N-terminal region, and a hydrophilic C-terminal region. However, not all such leader peptides synthesized as part of a longer protein are signal sequences and, in some cases, e.g., the capsid proteins of certain viruses, their function remains unknown.

The other type of leader peptide is encoded by a short, but independent open reading frame immediately upstream of the beginning of certain polycistronic operons in some bacteria. Therefore, this leader peptide is produced independently of the following proteins. However, the peptide itself is apparently not functional. The efficiency of translation of the sequence encoding the leader peptide is coupled to the transcription of the downstream genes in a regulatory mechanism called 'translational attenuation.' Each of the short sequences encoding these peptides contains codons related to the function of the enzymes encoded by the polycistronic mRNA. For instance, the 16-residue leader peptide of the histidine operon of *Escherichia coli* contains seven consecutive histidine codons. If a ribosome can translate these codons, the transcription of the remainder of the message is terminated. However, if the ribosome stalls at one of the histidine codons, because of a low concentration of histidyl-tRNA, transcription of the rest of the operon proceeds. Note that both types of leader peptide are encoded at the 5′ end of the mRNA, the 'leader sequence.'

See also: Attenuation; Leader Sequence; Open Reading Frame

Leader Sequence

P S Lovett

Copyright © 2001 Academic Press
doi: 10.1006/rwgn.2001.0750

The mRNA region that precedes the coding sequence for a gene is called the leader sequence. Leader sequences can regulate downstream expression at the levels of transcription or translation in bacteria, and can modulate downstream translation in eukaryotes.

Transcription in Bacteria

Transcription attenuation comprises one level of regulation for most amino acid biosynthetic operons in enteric bacteria. Nucleotide sequences within the leader cause the formation of a domain of secondary structure which acts as a transcription termination signal for bacterial RNA polymerase. Transcription initiated in the upstream promoter terminates within the leader so as to prevent RNA polymerase from entering the structural genes of an operon. Transcription termination is relieved when the intracellular concentration of the end-product amino acid of the operon-specified enzymes falls below some minimal level. The level of the end-product amino acid is sensed by the translation of a short leader-encoded open reading frame (ORF) immediately upstream of the transcription termination signal; the open reading frame contains one or more codons for the operon end-product amino acid. Low intracellular levels of the end-product amino acid prevent high level charging of the cognate tRNA, resulting in ribosomal pausing at leader codons for the end-product amino acid. The paused ribosome interferes with the secondary structure of the transcription terminator causing the formation of a second configuration in the mRNA, the attenuator, which allows transcription to enter the downstream operon coding sequence.

Transcription antitermination involves the formation of a transcription termination structure in leader mRNA, which is either inhibited or facilitated by the interaction of a protein (or a tRNA molecule) with leader mRNA sequences. For example, the TRAP protein plus tryptophan binds to the leader sequence of the *Bacillus subtilis trp* operon causing the formation of a transcription terminator. In the absence of tryptophan, TRAP fails to bind to the leader sequence and an antiterminator structure forms allowing transcription to enter the operon. Other operons that follow this general pattern of regulation include the *bgl* operon of *Escherichia coli*, the *pur*, *pyr*, *hut*, *lic*, and *glp* operons and the *sac* regulon of *B. subtilis*, the *ami* operon of *Pseudomonas*, and the *nas* regulon of *Klebsiella*. Aminoacyl-tRNA synthetases in gram-positive bacteria are also regulated by antitermination. The uncharged tRNA interacts with the leader sequences to promote the formation of an antiterminator structure allowing transcription to enter the tRNA synthetase coding sequence.

Translation in Bacteria

Translation attenuation regulates several antibiotic inducible, antibiotic resistance genes (e.g., *cat*, *erm*). A domain of secondary structure in leader mRNA sequesters the ribosome binding site for the downstream resistance determinant, preventing translation initiation. Antibiotic-induced ribosome stalling in a short open reading frame within the leader causes destabilization of the secondary structure, which frees the ribosome binding site allowing translation of a coding sequence whose protein product can neutralize the antibiotic.

Translational repression is well exemplified by certain operons encoding bacterial ribosomal proteins. The translational repressor is a single ribosomal protein encoded by the operon; the nonregulatory function of this protein is to act as a structural component of the ribosome. In several examples, the binding target for the repressor protein in the operon leader sequence mimics the structure or sequence of the rRNA target for the same protein. Binding of the regulatory protein to leader mRNA is presumably of lower affinity than that for rRNA binding *in vivo*. Leader binding by the repressor interferes with translation of operon mRNA by occluding the ribosome binding site or by changing the secondary structure of leader.

Translation in Eukaryotes

Translation in eukaryotes is typically initiated by the scanning of a 40S ribosomal preinitiation complex. Scanning begins at the 5' capped end of the mRNA and halts at the first initiator codon, usually AUG, where translation begins. Translation initiation efficiency at any particular AUG is affected by the context of the leader sequence flanking the AUG codon; a preinitiation complex may ignore an AUG codon located in a region of poor context. Several features of the leader sequence can dramatically decrease translation of the main (downstream) coding sequence:

examples include a region of secondary structure proximal to the 5' cap site or a 5' proximal AUG codon lacking a following an open reading frame.

Regulation by Upstream Open-Reading Frames (uORFs)

In the leaders for many eukaryotic mRNAs, the first AUG initiates translation of an upstream open reading frame (uORF) which is typically short. Translation of the functional protein, therefore, requires translation of the uORF, followed by scanning of the ribosome to the next AUG and reinitiation of translation. Certain uORFs enhance downstream translation, probably because the uORF sequence facilitates reinitiation of translation at downstream AUG codons. uORFs which diminish downstream translation are believed to interfere with ribosome scanning beyond the uORF. Current evidence from studies of a cytomegalovirus uORF-encoded peptide indicate that the short peptide prevents ribosome release from the uORF termination codon. The stalled ribosome itself cannot continue scanning, and can block the movement of other ribosomes attempting scanning along the mRNA. The most extensively studied example of the effects of uORFs on downstream translation is seen in the regulation of the yeast gene GCN4.

Internal Ribosome Entry Site (IRES)

Certain eukaryotic mRNAs contain an internal ribosome entry site (IRES) prior to the coding sequence. An IRES presumably functions in an analogous manner to a bacterial ribosome binding site in allowing translation initiation by directly serving as a ribosome-binding target. The presence of an IRES preceding a coding sequence in a eukaryotic mRNA enables an mRNA that is not capped to be translated.

Further Reading

Henkin TM (2000) Transcription termination in bacteria. *Current Opinion in Microbiology* 3: 149–153.

Hinnebusch A (1994) Translational control of GCN4: an *in vivo* barometer of initiation-factor activity. *Trends in Biochemical Sciences* 19: 409–414.

Landick R, Turnbough CL and Yanofsky C *et al.* (1996) Transcription attenuation. In: Neidhardt FC *et al.* (eds) Escherichia coli and Salmonella, 2nd edn, pp. 1263–1286. Washington, DC: American Society for Microbiology Press.

Lovett PS and Rogers EJ (1996) Ribosome regulation by the nascent peptide. *Microbiological Reviews* 60: 366–385.

See also: Open Reading Frame; Transcription; Translation

Leading Strand

The leading strand of DNA is synthesized continuously in the 5′–3′ direction.

See also: Lagging Strand; Replication

Least Squares

W-H Li and K Makova

The least squares method is a well-established statistical method of parameter estimation. This method chooses predicted values e_i that minimize the sum of squared errors of prediction $\sum_i (d_i - e_i)^2$ for all sample points d_i (observed values). The least squares method has been utilized in molecular evolution to estimate the branch lengths in a phylogenetic (evolutionary) tree and to estimate the topology of a tree. The least squares estimates of the branch lengths b_i's are the estimates e_i's that minimize the following sum of squares: $\sum_{i,j} (d_{ij} - e_{ij})^2$, where d_{ij} is the observed evolutionary distance between taxa i and j, and e_{ij} is the sum of length estimates (e_i's) of the branches connecting taxa i and j. To choose the best topology according to the least squares criterion, the above sum is computed for each possible topology and the topology with the smallest sum is taken as the best tree.

See also: Phylogeny; Trees

Lederberg, Joshua

E Kutter

Joshua Lederberg (1925–) has made many major contributions to our understanding of the genetics of microorganisms. He was born in Montclair, New Jersey, and received the Nobel Prize just 33 years later (with George Beadle and Edward Tatum) for discovering the mechanisms of genetic recombination in bacteria. He has been a member of the National Academy of Sciences since 1957 and was a charter member of its Institute of Medicine.

Lederberg became intensely interested in studying biological mechanisms while still in High School and took advantage of a variety of opportunities in the New York area to work in laboratories from an early age. He studied at Columbia Medical School, including work on adaptation in mutants of *Neurospora*, and then did his PhD under Tatum at Yale, publishing "Gene recombination in *Escherichia coli*" in *Nature* in 1948. This work gave the first indication that bacteria can reproduce not only asexually, through binary fission, but also sexually, resulting in a complex shuffling of their genetic systems during the mating of bacteria. As he discusses in *Annual Review of Genetics* (1987; 21: 23–46), the choice of the K-12 strain was highly serendipitous; only about 1 in 20 *E. coli* strains would have given positive results in their experiments, and the key extrachromosomal elements bacteriophage lambda and the F (fertility) factor, important in recombination, were also isolated in that system.

Lederberg taught in the University of Wisconsin School of Agriculture from 1947 to 1959, making the key decision to join that strong center of research in microbiology and biochemistry rather than return to Columbia to complete his medical studies. He further helped lay the foundations of microbial genetics when he and student Norton Zinder discovered the phenomenon of phage transduction in *Salmonella*: they showed that certain bacteriophage strains could incorporate a piece of the bacterial genome and carry it to a different bacterium. There it could recombine into the new host's chromosome, thus providing a major new mechanism of lateral genetic exchange that has proven extremely important in understanding microbial ecology and evolution. These studies were soon extended to transduction of biochemical pathways in *E. coli* K-12, which was nonpathogenic and more extensively developed as a genetic system. With these discoveries, bacteria took their place along with *Drosophila* and *Neurospora* as key model organisms in understanding genetic principles.

In 1959, Lederberg moved to the new medical school at Stanford University, where he became the director of the Kennedy Laboratories of Molecular Medicine in 1962. He moved to Rockefeller University to become its President in 1978, continuing his research there as Sackler Foundation scholar and professor emeritus of molecular genetics and informatics after his retirement from the presidency in 1990. In addition to his work on the fundamental mechanisms of microbial genetics, he has been very interested in the expanding field of research in artificial intelligence and in the search for life on Mars.

Lederberg's interests extend well beyond basic science. He has played a number of important roles in the international health community, including spending years on the World Health Organization's Advisory Health Research Council and serving as chairman of the President's Cancer Panel and of the congressional Technology Assessment Advisory Council. He also chaired a UNESCO committee on improving global internet communications for science and helping third-world people get onto the internet so they can be more involved in the process. Family has played an important role in his life. His father, a Rabbi who emigrated from Israel shortly before his birth, had a strong impact. His French-born wife is a Clinical Professor of Psychiatry at Memorial Sloan Kettering Cancer Center, and he has two children, David and Annie. His life has exemplified the basic advice he gave to young people in a recent interview (www.almaz.com/nobel/medicine/lederberg-interview.htm):

Try hard to find out what you're good at, and what your passions are, and where the two converge, and build your life around that . . . and make deliberate choices.

See also: Bacterial Genetics; Conjugation, Bacterial; Phage (Bacteriophage); Transduction

Leguminosae

J J Doyle

Copyright © 2001 Academic Press
doi: 10.1006/rwgn.2001.1642

The legume or bean family (Leguminosae or Fabaceae), with over 650 genera and 18 000 species, is the third largest family of flowering plants (angiosperms), behind only orchids (Orchidaceae) and the composite or sunflower family (Asteraceae or Compositae). Morphologically and ecologically, it is a very diverse family, ranging from tiny alpine ephemerals to huge tropical rainforest canopy trees. As much as one-third of the family's species are concentrated in a handful of large genera, such as *Acacia*, *Astragalus*, and *Mimosa*, that have radiated abundantly in disturbed habitats.

The family is characterized by its distinctive (and eponymous) fruit, a two-valved pod whose halves separate to disperse the seeds; however, this form is modified into a wide variety of shapes and sizes, including indehiscent dry or fleshy forms. Symbioses with nitrogen-fixing soil bacteria (collectively called 'rhizobia'), which are housed in specialized organs called nodules, are common but not universal in the family, nor is nodulation limited to Leguminosae. The ability to nodulate is thought to be an important adaptation in the family, and is a major factor in the economic and ecological importance of legumes.

Phylogeny and Taxonomy

Relationships with Other Families
Molecular phylogenetic studies support the naturalness (monophyly, descent from a single common ancestor) of the Leguminosae (**Figure 1**). The legumes have their relationships with taxa of the broad 'rosid' alliance that includes a major portion of angiosperm diversity, among which are other families that participate in nitrogen-fixing symbioses. Within this large clade (the descendants of a single common ancestor), the relationships of the family are more controversial. Morphological and chemical data suggest affinities with families such as Connaraceae or Sapindaceae, but molecular results ally the legumes with families previously not suggested as close relatives: Polygalaceae (milk vetches), Surianaceae, and *Quillaja*, an anomalous member of the Rosaceae (rose family).

The Three Subfamilies
The family is typically divided into three subfamilies (Caesalpinioideae, Mimosoideae, Papilionoideae or Faboideae), though these are sometimes considered to be separate families (Caesalpiniaceae, Mimosaceae, Papilionaceae or Fabaceae). Two of the three subfamilies, Mimosoideae and Papilionoideae, are supported as natural groups, whereas Caesalpiniodeae is not (**Figure 1**).

Subfamily Caesalpinioideae comprises the earliest-diverging elements of the family, a group of separate evolutionary lineages, some more closely related to the other two subfamilies than to one another (**Figure 1**). The group is therefore very heterogeneous morphologically and ecologically, and is most easily characterized by the absence of the unique features that distinguish mimosoids and papilionoids. Its approximately 150 genera and 2500 species are mainly tropical in distribution and include a number of showy species that are planted as ornamentals.

Mimosoideae has fewer genera (around 65), but somewhat more species (around 3000) than Caesalpinioideae. Most of the genera are small, often with only a single species, and around two-thirds of mimosoids belong to a few speciose genera such as *Acacia* and *Mimosa*. Flowers in mimosoids typically are individually small but often form showy clusters; petals are inconspicuous but the stamens are colored and are

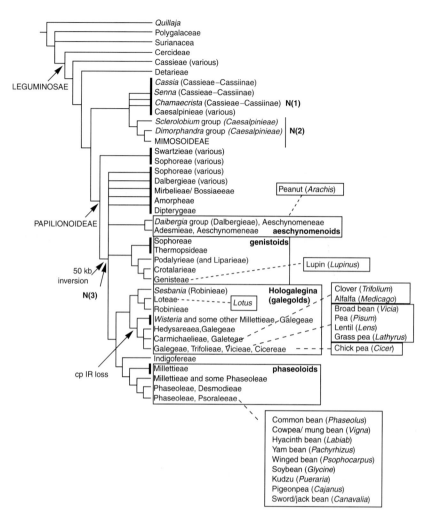

Figure 1 Phylogenetic relationships of legumes summarized from several phylogenetic studies, mostly using gene sequences from the chloroplast genome. Vertical heavy lines next to names indicate groups whose relationships are unresolved, or which themselves represent several lineages that are not closely related. The ancestor of all Leguminosae is indicated by an arrow. Two of the three subfamilies are also indicated: Mimosoideae as a terminal unit of the tree and Papilionoideae by an arrow pointing to its ancestor; all other legume taxa (genera or tribes) are Caesalpinioideae. Major lineages of papilionoid legumes referred to in the text are boxed, with the informal name of the group in bold type (e.g., 'phaseoloids') within the box. Dashed lines connect particular tribes to boxes containing economically or scientifically important representatives. 'N' indicates groups known to be capable of nodulation, followed by a number in parentheses that refers to one of three potentially independent origins of the syndrome. The vast majority of Papilionoideae derived from the ancestor indicated as 'N(3)' are known to nodulate, but some do not. Two chloroplast DNA structural mutations are indicated by arrows; the taxonomic distribution of the 50 kb inversion is not precisely known due to lack of sampling.

numerous in many species, providing the main floral display. Unlike other legumes, many mimosoids shed their pollen in polyads of 16 or 32 grains. Mimosoideae has some very close allies that are classified as Caesalpinioideae; this has been suspected for some time based on morphology, and the hypothesis has been supported by molecular data (**Figure 1**).

Papilionoideae is the subfamily most people visualize when legumes are mentioned. It is by far the largest and ecologically most diverse of the three subfamilies, with some 450 genera and 12 000 species. Its

members typically have bilaterally symmetrical flowers like those of pea (*Pisum*), with two wing petals, two keel petals, and a large standard petal. It is this butterfly-like ('papilionoid') floral morphology that gives its name to the subfamily. This morphological condition is derived, and some early-diverging members of the subfamily retain the radially symmetrical floral morphology of Caesalpinioideae and other rosid families. The large papilionoid radiation appears to have no particularly close allies among caesalpinioid taxa (**Figure 1**).

Relationships within Subfamilies: Economically and Scientifically Important Groups

Within each of the three subfamilies, genera are grouped in tribes, some of which are further subdivided formally into subtribes or informally into generic groups. A major focus of phylogenetic work has been to identify monophyletic groups of genera and to compare these with tribal boundaries, some of which have been suspected to be more taxonomically convenient than natural since the taxonomic foundation of the family was established in the late 1800s by George Bentham. Many of these suspicions have been confirmed, both for obviously unnatural amalgamations of genera with mostly ancestral morphologies, such as the papilionoid tribe Sophoreae, and for groups such as Phaseoleae, which has been considered to be among the most advanced papilionoid tribes. Molecular data, in particular, are revealing some unexpected relationships, and these findings are beginning to be reflected in the taxonomy of the family.

Genera containing species of economic or scientific importance are scattered unevenly throughout the three subfamilies and their constituent tribes. Apart from their use as ornamentals and tropical timber trees, Caesalpinioideae have only a few commonly known economic taxa, among them carob (*Ceratonia*). Relatively few genera are known from north temperate regions, exceptions being *Gleditsia* (honey locust) and *Cercis* (judas tree, redbud). The unfamiliarity and general inaccessibility of caesalpinioid genera is unfortunate from a scientific point of view, because, as noted, Caesalpinioideae represents the earliest diverging lineages of the family, and thus much of the genetic and evolutionary variation of Leguminosae. Simply put, it is impossible to make generalizations about any genetic phenomenon (genome organization, nodulation, floral development) for legumes without considering Caesalpinioideae. Although some caesalpinioid tribes appear to be natural, others clearly are not.

Mimosoideae includes some better-known genera, such as the large and ecologically important *Acacia*, one group of which are well known for housing and feeding ants that, in turn, protect the plants from predation. Species of *Mimosa* aptly named 'sensitive plants' are famous for the thigmotropic response of their leaves. *Neptunia* includes the only truly aquatic legumes. *Leucaena* is a fast-growing tree with promise for agroforestry. Other genera include *Prosopis* (mesquite) and *Parkia* (locust bean). Tribal boundaries in the subfamily are mostly uncertain.

Nearly all of the legumes familiar to inhabitants of the temperate northern hemisphere are members of Papilionoideae. For geneticists, these include both Mendel's pea (*Pisum sativum*) and the latest 'model organism' legumes, *Medicago truncatula* and *Lotus japonicus*, along with soybean (*Glycine max*) and lupin (*Lupinus* spp.). As with the legumes as a whole, a true appreciation of papilionoid diversity is not readily obtained from these model groups, all of which are relatively advanced in one sense or another. Many of the earliest diverging lineages of the family have features in common with caesalpinioids, and comprise a number of unrelated lineages whose relationships are not fully understood (**Figure 1**).

Molecular phylogenetic data suggest that there have been several major radiations in the subfamily, each of which includes some genera with scientifically or economically important species. Among these are an 'aeschynomenoid' group that includes members of tribes Dalbergieae, members of which provide rosewood, and Aeschynomeneae, among whose members are peanut (*Arachis hypogaea*). A 'genistoid' group includes Genisteae, with *Lupinus* and the familiar 'brooms' of the northern hemisphere, as well as southern hemisphere tribes such as the southern African Podalyrieae.

Two additional large groups apparently share a common ancestor. The 'Hologalegina' comprises two sister clades, one of which includes Robinieae, with genera such as *Robinia* (locust) and *Sesbania* (sesban, known for its stem nodulation), and Loteae (*Lotus*, including *L. japonicus*). The second galegoid lineage includes a group of mainly temperate, herbaceous tribes among which are: Vicieae, with *Pisum* (pea), *Lens* (lentil), *Lathyrus* (grass pea), and *Vicia* (vetch, broad bean); Trifolieae, with *Trifolium* (clover), *Medicago* (alfalfa and *M. truncatula*), and *Melilotus* (sweetclover); Cicereae (*Cicer*, chick pea); and Galegeae, among whose members is the huge genus *Astragalus*, with around 2000 species. Also part of this group are *Wisteria* and allied genera.

The final lineage, a 'phaseoloid' group, includes the largest (in number of genera) papilionoid tribe, Phaseoleae. Among the members of this tribe are *Glycine* (soybean), *Phaseolus* (common bean and other 'beans'), *Vigna* (cowpea, mung bean), *Cajanus* (pigeonpea), and *Canavalia* (sword or jack bean, the source of concanavalin). This lineage also includes many tropical, woody genera of the tribe Millettieae. Apparently sister to this entire clade is the small tribe Indigofereae, whose members include *Cyamopsis*, the source of guar gum, and *Indigofera*, the source of indigo dye. Neither Phaseoleae nor Millettieae are natural groups; tribes Desmodieae and Psoraleeae are nested among Phaseoleae genera and subtribes.

Fossil Record

Legumes have a well-developed macrofossil record in the Eocene (c. 50 million years ago) that includes flowers of each of the three subfamilies, suggesting that the major radiation of legumes occurred prior to that time. The ages of particular genera within each subfamily are more problematic, and thus it is difficult to say when, for example, pea diverged from soybean, or the various species of *Phaseolus* diverged from their common ancestor.

The divergence of the family from other rosid taxa is also difficult to determine with any precision. This is at least in part due to the fact that ancient legumes, like many modern Caesalpinioideae, were most probably fairly stereotypical rosids in much of their morphology. The rich Cretaceous floral fossil record from around 92 million years ago indicates that by that time most major lineages of flowering plants had diverged from one another. For example, fossil representatives of the lineage that includes *Arabidopsis* have been described from these Cretaceous deposits, indicating that legumes and *Arabidopsis* diverged at least this long ago. A possible ceiling for this divergence is around 110 million years ago, given the paucity of angiosperm fossils prior to this period, though this is in conflict with some estimates of angiosperm divergences based on molecular clock assumptions.

Evolution of Nodulation

Molecular phylogenies have identified a subgroup within the large rosid radiation that includes families that participate in nitrogen-fixing symbioses. However, within this 'nitrogen-fixing clade' the various nodulating families do not share a single common ancestor, suggesting that the ability to participate in these symbioses arose independently in different plant groups, but that an ancestor of the entire group may have evolved some key (but unknown) innovation that facilitates the formation of symbioses with diverse nitrogen-fixing microsymbionts. These are various 'rhizobia,' in legumes and *Parasponia* (Ulmaceae), or actinorhizal bacteria in other families. Some molecular similarities between mycorrhizal and nitrogen-fixing symbioses have been noted, and it may be that machinery of the pre-existing and more widespread mycorrhizal relationship was co-opted and modified in the evolution of nodulation. It now appears that genes encoding 'nodulins' (proteins that function in the nodule) are not, strictly speaking, novel or uniquely nodular, but have been recruited from other functions. Even the quintessential nodulin, leghemoglobin, whose presence in legumes was once considered so unexpected that it was thought to be a

case of horizontal gene transfer from the animal kingdom, is now known to be part of a plant gene family whose membership includes paralogous copies that are not associated with symbiosis.

The ability to nodulate is characteristic of most mimosoid and papilionoid legumes, but it is not universal in the family. Most Caesalpinioideae do not nodulate, nor do some early-diverging lineages of Papilionoideae. The phylogenetic distribution of nodulation in Leguminosae suggests that the ability to nodulate is not primitive in the family, and that nodulation may have arisen several times (**Figure 1**). There are also cases of loss of the ability to nodulate, which complicates the picture. Nodulation involves the production of a special organ, the nodule, and also what has been called a novel organelle, the symbiosome, consisting of nitrogen-fixing bacteroids enclosed in a primarily host-derived peribacteroid membrane. Independent losses of these structures seem more likely than does their independent origin, but the fact that nitrogen-fixing symbioses have almost certainly arisen multiple times elsewhere in the flowering plants is a mitigating consideration.

The details of nodulation vary even within Papilionoideae, where nodulation almost certainly had a single origin. The ancestral type of nodule appears to be an indeterminate, unbranched type that is also found in mimosoids and caesalpinioids. This 'caesalpinioid' type has been modified in several ways among papilionoids, including highly branched indeterminate types such as are common on Trifolieae, the large girdling indeterminate type of *Lupinus*, the clustered determinate types of aeschynomenoid taxa such as *Arachis*, and the single globular determinate 'desmodioid' nodules of Loteae and many phaseoloids. The desmodioid nodule appears to have originated independently in these two groups, and the determinate condition itself apparently arose yet another time in aeschynomenoids. Nitrogen is transported as amide compounds in many legumes, but as ureides in at least some phaseoloids.

Genomes

In some legumes the chloroplast genome departs from the common pattern of highly conserved gene content and order typical of photosynthetic angiosperms. Most notably, the large inverted repeat typical of land plant chloroplast chromosomes has been lost from all of the members of the major lineage of Hologalegina that contains Vicieae and allied temperate herbaceous tribes, plus *Wisteria* and allies. In some but not all species that have lost the inverted repeat there has been considerable subsequent rearrangement. Other major rearrangements include a 50 kb inversion found in most papilionoid species, and a 78 kb inversion found in *Phaseolus* and close allies (subtribe Phaseolinae). A number of chloroplast gene and intron losses have been reported from the family. Several related legume genera (e.g., *Medicago, Lens, Cicer*) depart from the typically maternal pattern of chloroplast inheritance found in angiosperms, and exhibit biparental or even predominantly paternal transmission.

What is known about mitochondrial genomes of legumes suggests that they are typical of angiosperms in being large relative to their counterparts in most animals, and exhibit a master/subgenomic structure due to recombination among direct repeats. Recent transfer of cytochrome oxidase subunit 2 (*cox2*) from the mitochondrial to the nuclear genome has occurred in Phaseoleae, with complex patterns of expression and, in some cases, subsequent loss from the mitochondrial genome.

Nuclear genomes of legumes vary greatly in size. The smallest legume genomes are only around twice the size of that of *Arabidopsis thaliana*, and are found in species of *Lablab, Scorpiurus, Trifolium,* and *Vigna*; genomes of the model legumes *Lotus japonicus* and *Medicago truncatula* are only slightly larger than these. At the other end of the spectrum, the genomes of some diploid (based on chromosome number) *Vicia* and *Lathyrus* species are nearly 100 times as large as that of *Arabidopsis*. Variation can be extreme within genera – diploid species of *Vicia* vary in their genome sizes from around 4 pg/2C (a haploid genome size of around 200 Mb) to over 50 pg/2C (2500 Mb). As is true for flowering plants in general, information on genome sizes is limited for legumes. A handful of papilionoid genera have been surveyed in detail, but otherwise there are few published values, with particularly sparse sampling in Caesalpinioideae and Mimosoideae. The sparse data from these subfamilies suggest that relatively small genome sizes are ancestral in the family as a whole. *Cercis* and *Bauhinia* of the caesalpinioid tribe Cercideae, which molecular data suggest is one of the earliest diverging legume groups, have genome sizes of 1.3 and 1.2 pg/2C, respectively.

The legumes as a whole are considered to have a base chromosome number of $x = 7$, but many groups of the family are thought to have experienced early polyploidization followed in many cases by aneuploid reduction. This is true, for example, of the entire subfamily Mimosoideae, and its closest allies in Caesalpinioideae, which as a group is thought to be tetraploid at $x = 14$. Similarly, the entire Detarieae–Amherstieae lineage of caesalpinioid legumes is considered to be based on $x = 12$. Within individual tribes there are relatively few genera that are wholly polyploid, among them *Glycine*, which is $2n = 40$, as compared with most Phaseoleae at $2n = 22$; however,

neopolyploidy is frequent, for example, *Medicago* includes both diploids (e.g., *M. truncatula*) and tetraploids (among them *M. sativa*). The same is true of *Glycine*, where neopolyploidy is superimposed on a fundamentally paleopolyploid base; other examples include *Lotus, Trifolium, Astragalus,* and *Lupinus*.

Linkage maps have been constructed for a handful of legumes, primarily cultivated papilionoid genera. Comparisons among published maps reveal synteny conservation among related groups such as the phaseoloid genera *Phaseolus, Vigna,* and *Glycine*, or the galegoids *Pisum, Lens,* and *Cicer*. Identifying conserved linkage blocks between such major groups or with more divergent taxa such as *Lupinus* has been more difficult, presumably in part because even in relatively close comparisons there are often many rearrangments. However, in light of increasing evidence of synteny conservation among angiosperms as a whole (e.g., between soybean and *Arabidopsis*) it seems likely that it will eventually be possible to trace linkage evolution across the entire family.

Further Reading

Crisp MD and Doyle JJ (eds) (1997) *Advances in Legume Systematics*, vol. 7, *Phylogeny*. London: Royal Botanic Gardens, Kew.

Herendeen PS and Bruneau A (eds) (2000) *Advances in Legume Systematics*, vol. 9. London: Royal Botanic Gardens, Kew.

Polhill RM (1994) *Classification of the Leguminosae*. In: Bisby FA, Buckingham J and Harborne JB (eds) *Phytochemical Dictionary of the Leguminosae*, vol. 1, *Plants and their Constituents*, pp. xxxv–lvii. London: Chapman & Hall.

Smartt J (1990) *Grain Legumes*. Cambridge: Cambridge University Press.

See also: Arabidopsis thaliana: The Premier Model Plant; Glycine max (Soybean); Nodulation Genes; Pisum sativum (Garden Pea)

Leiomyoma

See: Lipoma and Uterine Leiomyoma

Lejeune, Jérôme

P L Pearson

Copyright © 2001 Academic Press
doi: 10.1006/rwgn.2001.0753

Jérôme Lejeune (1926–94) is credited in 1959 with Gautier and Turpin as the first to identify trisomy of a small chromosome to be the cause of Down syndrome. The chromosome concerned was later designated chromosome 21 and Down syndrome was subsequently frequently referred to as trisomy 21 syndrome. The lead up to this discovery was the introduction of various improvements in the methods for studying mammalian chromosomes in the 1950s using cultured fibroblasts as a source of dividing cells and resulted in the first accurate count of the number of human chromosomes in 1956 by Tjio and Levan. The scene was set for wide-scale application of cytogenetic analysis in humans and in the same year as Lejeune's discovery of trisomy 21, Patricia Jacobs and colleagues independently confirmed trisomy 21 in Down syndrome and also demonstrated aneuploidy of the X chromosome in Klinefelter and Turner syndromes. These discoveries made an enormous impact in medical circles and overnight changed the perception of medical genetics as an obscure activity practised by a few hobbyists to a field with enormous potential for understanding the causes of mental retardation and other congenital abnormalities. In France, Lejeune was at the forefront of applying wide-scale cytogenetic analysis in the development of clinical genetics services. His group was one of the first to recognize the genetic importance of partial deletions of autosomes with their description in 1963 of a syndrome called the 'cri du chat' (cat cry) syndrome caused by deletion of part of the short arm of chromosome 5. Lejeune emphasized the relevance of recombination and segregation in balanced chromosome rearrangements in normal carriers to explain phenotypic defects in their progeny owing to cryptic duplications and deletions and coined the term 'aneusomie de récombination' to describe this. He coauthored many publications together with his long-term colleagues Rethoré and de Grouchy describing a wide range of phenotypes associated with various forms of partial trisomy or monosomy with perhaps trisomy 9p being one of the most notable.

At the time of the development of chromosome banding patterns at the beginning of the 1970s, Lejeune and Bernard Dutrilleaux were involved in developing a gallic form of chromosome banding which gave the reverse banding pattern to that developed by the rest of the world. This banding was termed reversed banding and appeared to be directly complementary to the G-banding technique used by many others. Besides using reversed banding for describing human chromosome abnormalities, Dutrilleaux and Lejeune went on to study the chromosome banding patterns in primates and to construct their karyotypic evolution.

The early 1970s saw the introduction of prenatal diagnosis on a wide scale in France. Lejeune was a devout Catholic and the concept of terminating genetically abnormal pregnancies was absolutely abhorrent to him. As an alternative, he advocated developing therapies for ameliorating mental retardation, particularly in Down syndrome, based on the surmise that neural development is compromised by a metabolic imbalance induced by the activity of genes present on the extra chromosome. In particular, Lejeune believed that there was disturbed monocarbon compound synthesis leading to an excess or deficiency of some amino acids in the plasma. Some of the metabolic features claimed by Lejeune to be characteristic of Down syndrome were the increased *in vitro* sensitivity to methotrexate and atropine. He advocated nutritional compensation for the amino acid deficiencies and folic acid medication for the increased methotrexate sensitivity. Although claims were made of an astonishing improvement in the mental capability of individual patients, these were anecdotal and in general the treatment strategies were not widely accepted by his medical colleagues. Gradually Lejeune became more and more isolated from the mainstream of human genetics activities in France and internationally. He devoted increasingly more time to running his clinic and cytogenetics laboratory according to pro-life principles. In the 1980s and early 1990s, Lejeune frequently appeared as a pro-life expert witness in court cases in North America.

Lejeune received many honors and awards in his lifetime. He was a member of the academies of sciences in the USA, Sweden, Italy, and Argentina, of the Royal Society of Medicine in London, the Academy of Medicine in France, and the Pontifical Academy of Science in Rome. In 1963 he received the Kennedy Prize for his discovery of the cause of Down syndrome and in 1969 the Memorial Allen Award medal from the American Human Genetics Society. Shortly before his death, he was appointed by the Pope to head the newly formed Pontifical Academy for Life.

His funeral, attended by 3000 people, was held in Notre Dame, Paris. The service was remarkable in that a Down syndrome patient spontaneously stood up and gave his personal thanks to Lejeune for giving him the courage and dignity not accorded him by French society. In 1996, family, friends and colleagues of Lejeune created the Jérôme Lejeune Foundation to carry on the work on mental retardation according to Lejeune's principles.

As in life, Lejeune was also a source of controversy in death. Four years following Lejeune's death, the Pope visited France and made a visit to pray at Lejeune's grave. The Holy Father received an unprecedented public rebuke from France's ruling Socialist Party who claimed that the Pope, merely by visiting his grave, was interfering in the legal right of the

French to abortion, a statute that had been in place since 1975.

See also: **Down Syndrome; Ethics and Genetics**

Lesch–Nyhan Syndrome

L De Gregorio and W L Nyhan

Copyright © 2001 Academic Press
doi: 10.1006/rwgn.2001.0754

The Lesch–Nyhan syndrome (LNS) is an X-linked severe disorder of purine metabolism, caused by an almost complete deficiency of the enzyme hypoxanthine-guanine phosphoribosyl transferase (HPRT). HPRT catalyzes the recycling reaction in which the free purine bases hypoxanthine and guanine are reutilized to form their respective nucleotides, inosinic and guanylic acids. This purine salvage mechanism provides an alternative and more economical pathway to *de novo* purine nucleotide synthesis. Uric acid is the end product of purine metabolism. In the absence of the salvage pathway, excessive amounts of uric acid are produced.

The classical Lesch–Nyhan disease is characterized by hyperuricemia, mental retardation, self-injurious behavior, choreoathetosis, and spasticity. However, there is wide phenotypic heterogeneity in the expression of HPRT deficiency. Three overlapping categories can be identified, in which the severity of clinical manifestations depends on the degree of residual enzyme activity:

1. **Classical Lesch–Nyhan syndrome (less than 1.5% of residual enzyme activity)** Male infants with Lesch–Nyhan disease appear normal at birth and usually develop normally for the first 6–8 months of their lives. Within the first few years of life, patients develop dystonia, choreoathetosis, spasticity, hyperreflexia, and extensor plantar reflexes. In established patients the overall motor defects are of such severity that they can neither stand nor sit unassisted. No patient with this disease has learned to walk. Most patients are cognitively impaired, but mental retardation is difficult to assess because of the behavioural disturbance and motor deficits. Many patients learn to speak, but atheoid dysarthria makes their speech difficult to understand. Self-injurious behavior is the hallmark of the disease and occurs in 100% of patients. The most characteristic feature is self-destructive biting of hands, fingers, lips, and cheeks. Hyperuricemia is present in almost all patients. The clinical consequences of the accumulation of large amounts of uric acid in body fluids are the classical manifestations of gout.

2. **Neurological variant (1.5–8% of residual enzyme activity).** The 'neurological' picture has been observed in a small but important group of patients and is characterized by a neurological examination that is identical to that of the classic Lesch–Nyhan patient (i.e., cerebral palsy or atheoid cerebral palsy). Patients are confined to wheelchairs and unable to walk. However, behavior is normal and intelligence is normal or nearly normal.

3. **Hyperuricemic variant (more than 8% of residual enzyme activity).** The phenotype of the patients with this partial variant enzyme consists of manifestations that can be directly related to the accumulation of uric acid in body fluids (acute attacks of gouty arthritis and tophi). Indeed, the central nervous system and behavior are normal.

The HPRT gene is located on the long arm of chromosome X (Xq26–q27). The gene has been cloned and its sequence determined: the entire locus spans more than 44 kb, the coding region consisting of 654 nucleotides in nine exons. The protein contains 218 amino acids. HPRT is expressed in all tissues, although at different levels, and the enzyme is particularly active in basal ganglia and testis. The incidence of LNS has been estimated to range from 1 in 100 000 to 1 in 380 000. Characterization of the molecular defect in the HPRT gene of a number of HPRT-deficient patients has revealed a heterogeneous pattern of mutations, with the same alteration rarely being found in unrelated pedigrees. About 63% of all the described molecular alterations represent point mutations, giving rise to either amino acid substitution in the protein sequence or stop codons, leading to truncated protein molecules. In some instances, the point mutation alters a splice site consensus sequence, activating an alternative, cryptic splice site, creating aberrant mRNA and protein products. It has not been possible to clearly correlate different types of mutations (genotype) with the various aspects of the clinical manifestations (phenotype). However, a rough guide predicts that mutations producing complete disruption of HPRT enzyme function (stop codons, deletions) are associated with classical LNS, while mutations allowing some residual HPRT enzyme activity (conservative amino acid substitutions) are associated with a less severe phenotype.

The excessive uric acid production in HPRT-deficient patients is effectively treated with daily administration of allopurinol. This is the unique and specific treatment available for all the patients diagnosed with HPRT deficiency, both classical Lesch–Nyhan and partial variants. Unfortunately, no medication has been found to be consistently effective in treating the neurological or behavioral manifestations of the disease in classical Lesch–Nyhan patients. The only successful approaches to the self-injurious behavior have been physical restraint and the removal of teeth, to prevent self-biting. Future approaches may include gene therapy: promising results have already been obtained *in vitro*.

See also: **Gene Therapy, Human; Genetic Counseling; Purine**

Lethal Locus

Copyright © 2001 Academic Press
doi: 10.1006/rwgn.2001.1890

A lethal locus is any gene in which a lethal mutation can be obtained.

See also: **Conditional Lethality; Lethal Mutation**

Lethal Mutation

M A Cleary

Copyright © 2001 Academic Press
doi: 10.1006/rwgn.2001.0755

Mutations result in permanent alterations or changes in DNA sequence. Such changes include point mutations, in which only single base pairs are affected, or chromosomal rearrangements, translocations, or deletions, in which larger regions of DNA are affected. When these alterations cripple a gene that is essential for an organism's survival and result in death, they are referred to as lethal mutations.

Leucine

E J Murgola

Copyright © 2001 Academic Press
doi: 10.1006/rwgn.2001.0757

Leucine is one of the 20 amino acids commonly found in all proteins. Its abbreviation is Leu and its

Figure 1 Leucine.

single-letter designation is L. As one of the essential amino acids in humans, it is not synthesized by the body and so must be provided in the individual's diet (**Figure 1**).

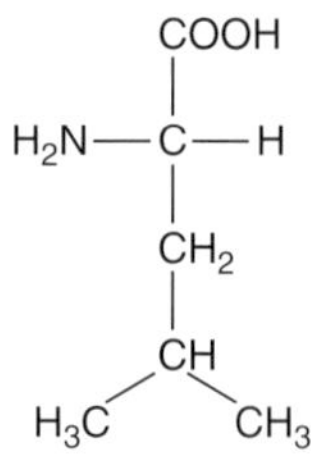

Leukemia

J D Rowley

Copyright © 2001 Academic Press
doi: 10.1006/rwgn.2001.0758

Leukemia is cancer of the blood which occurs in several forms. The disease can be chronic or acute; patients with the former live for a number of years whereas patients with the latter live for only a few weeks or months unless they receive appropriate treatment. In addition, leukemias are further subdivided by the type of cell that is involved. Common forms are chronic lymphatic leukemia or acute lymphoblastic leukemia (ALL) which affects lymphocytes of either the B- or T-cell lineage, and chronic myelogenous leukemia (CML) and acute myelogenous leukemia (AML), which affect bone marrow cells of the red cell, granulocytic, monocytic or megakaryocytic (platelet) lineages.

In this section, I will focus on acute leukemia, both ALL and AML. For reasons that are not understood at present, ALL occurs much more commonly in children and young adults whereas AML is more frequent in older adults. The genetic changes in ALL and AML are different and thus it is not surprising that the treatments differ as well. In general, children with ALL respond much better to present treatments and over 70% have very long survivals of more than 5 years. In contrast, adults with AML may respond initially to treatment but then relapse and die. In AML and, to a lesser extent, ALL, the length of survival is very closely associated with the types of genetic changes that are present in the leukemic cell. Present evidence indicates that these genetic changes occur *de novo* in an otherwise normal blood cell.

Table 1 Cytogenetic–immunophenotypic correlations in malignant B-lymphoid diseases

Phenotype	Chromosome abnormality	Involved genes
Acute lymphoblastic leukemia		
Pro-Pre-B	t(1;19)(q23;p13)	PBX1-TCF3 (E2A)
	t(12;21)(p13;q22)	TEL-AML1
B(SIg+)	t(8;14)(q24;q32)	MYC-IGH
	t(2;8)(p12;q24)	IGK-MYC
	t(8;22)(q24;q11)	MYC-IGL
B or B-myeloid	t(9;22)(q34;q11)	ABL-BCR
	t(4;11)(q21;q23)	AF4-MLL
	t(11;19)(q23;p13.3)	MLL-ENL
Other	50–60 chromosomes	
	t(5;14)(q31;q32)	IL3-IGH
	del(9p),t(9p)	?CDKN2(p16)
	t(9;12)(q34;p13)	TEL-ABL
	del(12p)	TEL;?p27^{KIP1}

Reproduced with permission from Rowley JD (1999) The role of chromosome translocations in leukemogenesis. *Seminars in Hematology* 36 (supp. 7): 59–72.

Table 2 Cytogenetic–immunophenotypic correlations in malignant T-lymphoid diseases

Phenotype	Chromosome abnormality	Involved genes
Acute lymphoblastic leukemia	t(1;14)(p34;q11)	LCK-TCRD
	t(1;14)(p32;q11)	TAL1-TCRD
	—	TAL1^{Del}
	t(7;9)(q35;q32)	TCRB-TAL2
	t(7;9)(q35;q34)	TCRB-TAN1
	t(7;7)(p15;q11)	TCRG-?
	t(7;14)(q35;q11)	TCRB-TCRD
	t(7;14)(p15;q11)	
	t(8;14)(q24;q11)	MYC-TCRA
	inv(14)(q11;q32)	TCRA-IGH
	t(14;14)(q11;q32)	TCRA-IGH
	t(10;14)(q24;q11)	HOX11-TCRA
	t(11;14)(p15;q11)	LMO1-TCRD
	t(11;14)(p13;q11)	LMO2-TCRD

Reprinted with permission from Rowley JD (1999).

That is, there is little evidence for predisposing genetic factors as may be found in breast cancer or colon cancer. All present evidence indicates that the transformation of a normal cell to a leukemic cell involves changes in a series of genes only some of which are presently known. Thus the challenge for the future is to identify all of the genetic changes that occur, the order in which they occur and the functional consequences of these changes.

Genetic Changes in Acute Leukemia

Chromosome Translocations

Most of our information about the genetic changes in all forms of human leukemia has come from an analysis of the chromosome pattern of the leukemic cells. The leukemic cells are obtained usually from a bone marrow sample or peripheral blood and the dividing cells which contain condensed chromosomes are

Table 3 Recurring structural rearrangements in malignant myeloid diseases

Disease	Chromosome abnormality	Involved genes
Chronic myeloid leukemia	t(9;22)(q34;q11)	ABL-BCR
CML blast phase	t(9;22), +8, +Ph, i(17q)	ABL-BCR
Chronic myelomonocytic leukemia	t(5;12)(q33;p13)	PDGFRB-TEL
Acute myeloid leukemia		
AML-M2	t(8;21)(q22;q22)	ETO-AML1
APL-M3, M3V	t(15;17)(q22;q12)	PML-RARA
atypical APL	t(11;17)(q23;q12)	PLZF-RARA-
AMMoL-M4Eo	inv(16)(p13q22) or	MYH11-CBFB
	t(16;16)(p13;q22)	
AMMoL-M4/AMoL-M5	t(6;11)(q27;q23)	AF6-MLL
	t(9;11)(p22;q23)	AF9-MLL
AMegL-M7	t(1;22)(p13;q13)	
AML	t(3;3)(q21;q26)	RPN1-EVI1
	or inv(3)(q21q26)	
	t(3;5)(q21;q31)	
	t(3;5)(q25;q34)	MLF1-NPM1
	t(6;9)(p23;q34)	DEK-CAN
	t(7;11)(p15;p15)	HOXA9-NUP98
	t(8;16)(p11;p13)	MOZ-CBP
	t(9;12)(q34;p13)	TEL-ABL
	t(12;22)(p13;q13)	TEL-NM1
	t(16;21)(p11;q22)	TLS(FUS)-ERG
	−7 or del(7q)	
	−5 or del(5q)	
	del(20q)	
	del(12p)	TEL, ?p27^{KIP1}
Therapy-related AML	−7 or del(7q) and/or	
	−5 or del(5q)	IRF1?
	t(11q23)	MLL
	t(3;21)(q26;q22)	EAP/MDS1/EVI1-AML1

Reprinted with permission from Rowley JD (1999).

processed according to standard techniques. Normal cells have 46 chromosomes, but leukemic cells can contain many abnormalities. Fortunately, a number of chromosome changes are recurring and many of these recurring changes are associated with certain subtypes of leukemia. Moreover, as will be discussed later, some chromosome changes provide physicians with very important information on the likely response of the leukemia cells to the treatment. In fact, certain chromosome changes only respond to certain types of treatment and thus analysis of the chromosome pattern (karyotype) of the leukemic cells helps the physician select the most effective treatment. The chromosome changes in leukemic cells involve both gains and losses of whole chromosomes or parts of chromosomes. In addition, chromosome translocations are important; in translocations, two chromosomes are broken and the broken ends are exchanged. Translocations are a very important mechanism of genetic change in leukemias, lymphomas, and a few solid tumors. Chromosome translocations have one of two consequences. In many of the malignant lymphoid tumors, the breaks occur in or near to the immunoglobulin gene in B cells or to the T cell receptor in T cells. The translocation joins these very highly active genes to a target gene that is then more actively expressed than in a normal cell. The protein produced by the target gene is a normal protein. In most of the myeloid leukemias, both acute and chronic, the two genes involved in the translocations are broken and two new genes may be formed as a result of a reciprocal exchange. In some situations, part of one gene is deleted so that there is only one new fusion gene; it is clearly the fusion that is critical for malignant transformation. These fusion genes and the resultant fusion protein are unique tumor-specific markers and they provide special targets for therapeutic intervention.

Chromosome Abnormalities in Acute Lymphoblastic Leukemia

All types of chromosome abnormalities are seen in ALL, often in combination. For the most part, the genes that are involved in gains or losses of chromosomes are unknown. Translocations or other rearrangements such as inversions that involve the immunoglobulin loci 14q32 (heavy chain), 2p12 (κ light chain), or 22q11 (λ light chain) or the T cell receptor loci, 14q11 (α/δ chain), 7q35 (β chain) or 7p13 (γ chain) are of the first type described above (section "Chromosome Translocations"). They alter the expression of the target genes but the target gene protein is normal. In fact, the first translocation identified in a B cell malignant disease was the 8;14 translocation in Burkitt lymphoma that subsequently was shown to involve the immunoglobulin gene at 14q32 and the *MYC* gene at 8q24. This translocation which is also seen in B cell ALL leads to the inappropriate expression of the *MYC* gene which is an important component of the pathway regulating cell growth. The immunoglobulin light chain genes are also involved in translocations with *MYC*. The other important chromosome changes are listed in **Tables I** and **2**.

Chromosome Abnormalities in Acute Myelogenous Leukemia

As with ALL, all forms of chromosome change are seen as recurring abnormalities in AML. The targets of these abnormalities are virtually unknown despite heroic efforts on the part of many investigators to identify the target genes. Identification of chromosome translocations in the 1970s showed that certain translocations were closely associated with particular subtypes of leukemia; in fact, the association is so important that the genetic changes are now used in morphologic classification of these leukemias. The first consistent chromosome translocation in any malignant cell was identified in 1972; it was the 8;21 translocation seen in AML. Since then several hundred different translocations have been identified and almost 100 of these have been cloned. The majority of translocations result in new fusion genes. The common recurring aberrations are listed in **Table 3**.

Clinical and Biological Importance of Chromosome Abnormalities in Leukemia

Studies of chromosome translocations will assume even greater importance in the future because the unique fusion genes and proteins that are identified in many of these rearrangements are tumor-specific markers for the malignant cells. With further understanding of the alterations in function of these genes and proteins, it should be possible to target cells with these fusion genes/proteins specifically and to spare the other normal cells in the patient. The major goal for the new millennium is to translate our increasingly sophisticated understanding of how the translocations interfere with the critical function of these genes to predict specific therapy that would likely be more effective and less toxic than current therapy. This requires that we identify the multiple genes that are involved in leukemogenesis.

See also: **Leukemia, Acute; Leukemia, Chronic; Translocation**

Leukemia, Acute

M J S Dyer

Copyright © 2001 Academic Press
doi: 10.1006/rwgn.2001.1538

The acute leukemias represent the malignant transformation of myeloid and lymphoid precursors within the bone marrow or thymus. All hematopoietic precursor cells can be transformed. The commonest leukemia in children from developed countries is B-cell precursor acute lymphoblastic leukemia (BCP-ALL), which express the neutral endopeptidase CD10; these are often referred to as 'common' ALL (cALL). BCP-ALL is the commonest malignancy of childhood. However, as with other malignancies, the incidence of other forms of acute leukemia and particularly acute myeloid leukemia (AML) increases with age. The etiologies of the acute leukemias remain unknown. Although many hypotheses have been advanced, particularly for childhood BCP-ALL, none have been proven, due in major part, to the rarity of the disease. Childhood leukemias may arise *in utero* (see below). Familial acute leukemia is rare, although when it occurs, it frequently exhibits genetic anticipation.

Diagnosis is based on a combination of morphology (particularly for the myeloid/monocytic leukemias), immunophenotype, and molecular cytogenetics. Most acute leukemias exhibit the immunophenotype of a single hematopoietic lineage, although some may coexpress molecules associated with two different lineages; these are known as biphenotypic acute leukemias. Cytogenetics, increasingly supplemented by fluorescent *in situ* hybridization (FISH) and molecular techniques such as reverse transcriptase polymerase chain reaction (RT-PCR), plays a

Table 1 some common chromosomal translocations and genes in acute leukemia

Diagnosis	Cytogenetic abnormality	Involved genes
B-cell precursor ALL	t(12;21)(p13;q21)	ETV6/AML1
	t(9;22)(q34;q11)	ABL/BCR
	t(1;19)(q23;p13)	PBX/E2A
T-cell precursor ALL	t(11;14)(p13–p15;q11)	LMO1/2-TCRD/A
	t(1;14)(p32;q11)	TAL1-TCRD/A
AML	Various 11q23 translocations	MLL fusions
	t(8;21)(q22;q22)	CBF α/ETO
	inv(16)(p13q22)	MYH11/CBF β
APL[a]	t(15;17)(q21;q22)	PML/RARα

[a]APL = acute promyelocytic leukemia.

prominent role. Detection of certain chromosomal translocation is of major prognostic significance and determines the intensity, duration, and type of therapy; patients with poor prognosis to conventional therapy may undergo allogeneic stem cell transplantation whilst in first remission. In other patients, it may be possible to reduce chemotherapy, without increasing the relapse rate.

Since the tumor can be readily accessed, and since it is possible to culture human hematopoietic cells *in vitro* and derive cell lines, cytogenetic analysis of the acute leukemias is advanced. Moreover, many clones are cytogenetically simple, containing only one chromosomal translocation, and lacking the cytogenetic complexity seen in solid tumors and lymphomas. Some of the common translocations are shown in **Table 1**. Much effort has been made to clone the recurrent chromosomal translocations and identify the involved genes, as these are intimately involved in the pathogenesis of the disease. This has been confirmed experimentally by the creation of 'knock-in' mice where the translocation is created in embryonic stem cells (Corral *et al.*, 1996). Most genes are involved with a single partner in a single disease. The *MLL* gene on 11q23 is remarkable for being involved with over 20 other genes in translocations, principally in AML.

In most instances, the consequences of translocation in acute leukemias are the generation of fusion genes derived from the coding regions of genes on the two chromosomes. These fusion transcripts are useful clone-specific markers, allowing the detection of disease with unprecedented sensitivity and redefining the criteria used for 'remission.' Translocations in T-cell precursor ALL in contrast involve the T-cell receptor (*TCR*) gene segments and result in deregulated expression of the incoming oncogene, through juxtaposition of transcriptional enhancers within the *TCR* loci. In both instances, the involved genes are transcription factor controlling development and differentiation. Further dissection of the transcriptional pathways involved may allow the rational introduction of new therapeutic strategies.

A fascinating observation is that many of the childhood leukemias may originate *in utero*. This conclusion was made initially on the basis of data from identical twins with concordant leukemia, the leukemic stem cell passing from one twin to another due to the shared placental blood supply. Such twins showed identical translocation breakpoints and antigen receptor gene rearrangements. These data have now been confirmed in other patients through the use of Guthrie blood spots, collected at birth for the screening for phenylketonuria. These spots contain sufficient DNA to allow retrospective analysis of the leukemic clone at birth by using long-range PCR methods to detect the t(12;21)(p13;q21) breakpoint. This clone may only present several years later, implying the necessity for other genetic/environmental events for its eventual appearance. However, at least some chromosomal translocations may occur in normal stem cells with normal capacity to differentiate without giving rise to overt leukemia.

Further Reading

Rowley JD (1998) The critical role of chromosome translocations in human leukemias. *Annual Review of Genetics* 32: 495–519.

Wiemels JL, Cazzaniga G, Daniotti M *et al.* (1999) Prenatal origin of acute lymphoblastic leukaemia in children. *Lancet* 354: 1499–1503.

Reference

Corral J, Lavenir I, Impey H *et al.* (1996) An *Mll-AF9* fusion gene made by homologous recombination causes acute leukemia in chimeric mice: a method to create fusion oncogenes. *Cell* 85: 853–861.

See also: Mouse Leukemia Viruses; Translocation

Leukemia, Chronic

M J S Dyer

Copyright © 2001 Academic Press
doi: 10.1006/rwgn.2001.1539

The chronic leukemias comprise a heterogeneous group of malignancies, representing the transformation of mature lymphocytes of B, T, and rarely NK lineages at specific points in their normal differentiation pathways. (Chronic leukemias of the myeloid lineage including chronic myeloid leukemia (CML) are discussed in the article *BCR/ABL* Oncogene.) Various subtypes may be recognized on the basis of cytology, immunophenotype, and molecular cytogenetic findings as summarized in **Table 1** (Catovsky, 1999). The causes of these diseases remain unknown, although progress has been made in the identification of key genes through the molecular cloning of chromosomal translocation breakpoints, principally involving the immunoglobulin (*IG*) or T-cell receptor loci.

B-cell chronic lymphocytic leukemia (CLL) is the commonest form of leukemia. It is a disease primarily of the elderly. CLL is a disease of CD5+ B cells, which may constitute a distinct B-cell lineage. A striking feature is the wide variation in biological behavior, some patients requiring no therapy for many years, others having rapidly progressive and chemotherapy-resistant disease. Correspondingly, there is no common genetic abnormality. Unlike other forms of both acute and chronic leukemia, about 5% of cases have a familial component and may exhibit genetic

anticipation. Abnormalities of chromosome 13q14 are common but, despite much work, the pathological consequences remain unclear. Patients with either deletions of chromosome 11q23 or mutations of *p53* have rapidly progressive/chemotherapy-resistant disease. The status of the immunoglobulin heavy chain variable (*IGHV*) region gene segments in CLL defines two biologically distinct groups of disease. B cells that have encountered antigen and passed through the germinal centre exhibit *IGHV* mutations within the DNA sequences that encode the antigen-binding loops of the antibody protein. Patients with unmutated *IGHV* segments have a worse prognosis than those with *IGHV* mutations.

Other forms of B-cell chronic leukemia are relatively uncommon. Splenic lymphoma with villous lymphocytes (SLVL) is generally a very indolent disease. Nevertheless, a subset exhibits translocations to the immunoglobulin loci involving either Cyclin D1 (*CCND1*) or *CDK6* genes involved in control of cell-cycle progression. B-cell prolymphocytic leukemia (B-PLL) is remarkable amongst hematological malignancies for having a high incidence of *p53* mutations. Moreover, the pattern of *p53* mutation is distinct from that seen in other diseases.

All T-cell malignancies are relatively rare. T-cell prolymphocytic leukemia (T-PLL) is clinically highly aggressive and is of interest, as the same disease is seen at increased incidence in patients with ataxia-telangectasia (AT). Sporadic T-PLL is characterized by enormous cytogenetic complexity. However, acquired *ATM* mutations and rearrangements are found in probably all cases of sporadic T-PLL. T-PLL

Table 1 Subtypes of chronic lymphoid leukemias

Disease	Immunophenotype	Recurrent cytogenetic changes	Molecular abnormalities
CLL	CD5+, CD22−, CD23+, FMC7− Surface Ig weak	Deletion/translocations of 13q14 Deletion of 11q23 Deletion of 17p13.1 Trisomy 12 (secondary) t(14;19)(q32.3;q13)	Unknown *ATM* mutation *p53* mutation Unknown *BCL3* (rare)
SLVL	CD5+/−, CD22++, CD23− FMC7+, sIg+/−	t(11;14)(q13;q32.3) t(2;7)(p12;q22)	*CCND1* overexpression *CDK6* overexpression
B-PLL	CD5+/−, CD22++, CD23− FMC7++, sIg++	deletion of 17p13.1	*p53* mutation
HCL*	sIg++, CD5−, CD22++, FMC7+ CD25+	No recurrent clonal abnormalities	
T-PLL	sCD3+, CD4+, CD25−	inv(14)(q11;q32.1) t(X;14)(q28;q11) deletion of 11q23	*TCL1* overexpression *MTCP1* overexpression *ATM* mutation
T-LGL	sCD3+, CD8+, CD25	No recurrent clonal abnormalities	
ATLL	sCD3+, CD4+, CD25+	No recurrent clonal abnormalities	
Sezary's	sCD3+, CD4+, CD25−	No recurrent clonal abnormalities	

*HCL = hairy cell leukemia.

also exhibits deregulated expression of two closely related genes of unknown function *TCL1* and *MTCP1*; the former locus on chromosome 14q32.1 contains a number of closely related genes. There are no consistent cytogenetic or molecular markers for the other T-cell leukemias including the leukemias of large granular lymphocytes (LGL), which may be of either T-cell or NK lineages, HTLV-1 associated adult T-cell lymphoma/leukemia (ATLL), or Sezary syndrome. Comparative genomic hybridization (CGH) has shown amplifications of chromosome 2p13 and 14q32.1 in ATLL.

Reference

Catovsky, D (1999) Chronic lymphoid leukaemias. In: Hoffbrand AV, Lewis SM and Tuddenham EGD (eds) *Postgraduate Haematology*, 4th edn, pp. 405–433. Oxford: Butterworth Heinemann.

See also: BCR/ABL Oncogene; Chromosome Aberrations; Immunoglobulin Gene Superfamily

Levan, Albert

M Hultén and K Fredga

Albert Levan (1905–98) is most famous for his discovery that the diploid chromosome number in humans is 46 and not 48 as had been the dogma since 1912. This discovery was made with Joe-Hin Tjio at the Institute of Genetics, University of Lund, Sweden. It resulted from the application to human cells of a methodology for chromosome preparation that Levan had pioneered in plants and animals.

Levan was born and grew up in the Swedish town of Gothenburg, where his father, who was Director of Post Services, passed to Albert an interest in classical languages and botany. After high school, Levan moved to the University of Lund where he graduated in botany in 1927. From 1926 to 1931, he held a post as Assistant at the Institute of Zoology. He was awarded a PhD in 1935. At the age of 30, he became Assistant Professor in Genetics, and subsequently, in 1947, also in Cytology. In 1961, he was awarded a personal chair in Cytology.

Levan published his first paper in 1929, on the chromosomes of onions. The large size and clear morphology of these species' chromosomes made them especially well suited for both descriptive and experimental studies.

In 1938, Levan published a very important paper, entitled "The effect of colchicine on root mitoses in *Allium*." This was the first in-depth study of the influence of colchicine on plant cell division, demonstrating its effect on the mitotic spindle and the concomitant condensation of metaphase chromosomes. This work, of course, paved the way to development of the methodology that would eventually lead to the correct identification of the chromosome number of humans. Since these first experiments, colchicine (or its synthetic derivative colcemid) has been a central component of the protocols used to obtain chromosome preparations from plants and animals.

Over a long period (1938–51), Levan studied the reactions of chromosomes to treatments by different chemicals. He devised the so-called 'Allium test' to evaluate the effects of both chemicals and ionizing radiation. This work merited for him an honorary doctorate from the Sorbonne University, Paris, in 1968. Levan also devoted himself with great success to practical plant breeding, and among other things, he produced the first tetraploid strains of sugar beet and red clover.

At the end of the 1940s, Levan became fascinated by the similarity between the chromosome aberrations caused by chemical agents and those described and illustrated in the literature of cancer genetics at that time. Levan showed that by applying the methods developed for plant chromosomes, he could produce first-class preparations of chromosomes from mouse ascites tumor cells. He realized that this advance opened up a completely new field for chromosome study, namely investigation of chromosome number and morphology during the transition of a normal cell to a cancer cell. His seminal work paved the way for another new, large field of applied research, namely the diagnosis and treatment of malignancies based on their underlying chromosome abnormalities.

The seminal paper by Tjio and Levan entitled "The chromosome number of man" was published in 1956 (Tjio and Levan, 1956). This publication had a dramatic input in genetics, becoming the starting-point not only for the new discipline clinical cytogenetics but also the rapid development of medical and human genetics. The paper also made a significant contribution in veterinary medicine and zoology. Tjio and Levan's study was performed on fetal lung fibroblasts cultured *in vitro* by Rune Grubb at the University's Medical Microbiology Department. These cells were induced to arrest at metaphase (the best stage for chromosome enumeration) by use of the mitotic spindle poison, colchicine. The chromosomes were fixed and stained with the dye acetic orcein, and squash preparations made. Remarkably, the millions of cytogenetic investigations performed to date each year are using basically the same methodology as pioneered by Tjio and Levan.

In 1953, Levan had set up the Cancer Chromosome Laboratory at the Institute of Genetics of the University of Lund. Initially, the laboratory had only a few scientists, but soon attracted researchers from all over the world. When Levan retired in 1976 at the age of 71, the laboratory had grown to 40 scientists, technicians and students. He and coworkers made many important contributions to our understanding of cancer cytogenetics, for example:

1. Chromosomal changes in tumor cell lines are not arbitrary but follow particular developmental patterns.
2. Measles virus and Rous sarcoma virus lead to an increased amount of chromosome breakage in human blood cells.
3. Patterns of chromosome rearrangements are related to tumor etiology. Thus, histologically identical tumors can show totally different patterns of chromosome aberrations dependent on whether they have been induced by virus or chemicals.
4. Burkitt's lymphoma is characterized by a particular chromosome aberration of chromosome 14; this was the second example of a specific chromosome abnormality in a human tumor cell.

Levan's interest and involvement in chromosome research continued after his retirement from the Directorship of the Cancer Chromosome Laboratory. In particular, he studied the phenomenon of 'double minutes' seen in some tumor types. These chromosomal elements result from the massive amplification of genes involved in malignancy.

The inspiring atmosphere that Albert Levan created in his laboratory made every member of staff do their best. He allowed much freedom of research. Although the key subject of the laboratory was cancer research, Levan accepted with much enthusiasm some of his younger colleagues devoting their time to the study of shrews, lemmings, hedgehogs, seals, and even whales. His curiosity for scientific matters never waned, and at the age of 85 he learnt to use the computer for writing and correspondence. As a researcher, Levan stands out as an intuitive and creative talent. He had an unusual attitude to work. Unlike many others in his laboratory, he adhered stringently to a 9-to-5 working day. This allowed him time for his many other interests, including playing the cello and writing music.

Reference

Tjio JH and Levan A (1956) The chromosome number of man. *Hereditas* **42**: 1–6.

See also: Human Chromosomes; Tjio, Joe-Hin

Lewis, Edward

K Handyside, E Keeling and S Brenner

Copyright © 2001 Academic Press
doi: 10.1006/rwgn.2001.1702

Edward Lewis (1918–), an American biologist, made substantial contributions to our understanding of the development of animal embryos through his studies of *Drosophila melanogaster* or fruit flies. His work won him the Nobel Prize for Physiology or Medicine in 1995 for "discoveries concerning the genes that control early embryonic development." He shared this award with Christiane Nusslein-Volhard of Germany and Eric F. Wieschaus of the United States, who were recognized for their independent studies.

Lewis gained a BA degree from the University of Minnesota in 1939 followed by a PhD in 1942 from the California Institute of Technology, where he spent his professional career. It was *Drosophila melanogaster*, a popular species for genetic experiments, on which Lewis based his studies. By use of crossbreeding experiments, Lewis demonstrated that the ordering of chromosomes that guide the development of the body segments generally matched the order of the corresponding body segments themselves, i.e., the first set of genes on the chromosome controlled the head and thorax, the middle set the abdomen, and the last set the posterior. This orderliness was termed the colinearity principle. He also discovered that genetic regulatory functions may overlap. For example, a fly with an extra set of wings has a defective gene not only in the abdominal region but also in the thoracic area which in a normal fly would act as a regulator of such mutations.

The results of his research helped to elucidate the mechanisms of biological development and shed light on the implications for congenital deformities in humans and other species.

See also: Colinearity; *Drosophila melanogaster*

Library

Copyright © 2001 Academic Press
doi: 10.1006/rwgn.2001.1891

A library (or gene or genomic library) is a set of cloned fragments that together represent the entire genome.

See also: Gene Library; Genomic Library

Ligation

doi: 10.1006/rwgn.2001.1893

Ligation is the formation of a phosphodiester bond to join two adjacent bases in DNA or RNA.

See also: **DNA Ligases**

Light Receptor Kinases

See: **Photomorphogenesis in Plants, Genetics of**

Light, Heavy Chains

See: **Immunoglobulin Gene Superfamily**

LIM Domain Genes

L W Jurata and G N Gill

doi: 10.1006/rwgn.2001.1586

LIM domains are composed of ~55 amino acids with the general sequence $CX_2 CX_{16-23}HX_2 CX_2CX_{16-23} CX_2C$ where C = cysteine, H = histidine, and X = any amino acid. LIM domains bind two atoms of Zn^{2+} with the most common tetrahedral coordination being S_3N and S_4. Modular LIM domains are found in both nuclear and cytoplasmic proteins where they function in molecular recognition to assemble multiprotein complexes. The name 'LIM' derives from the first three proteins found to contain two of these domains at their N-terminus and a homeodomain at their C-terminus (lin-11, Is11, and mec-3) (Freyd *et al.*, 1990). Nuclear LIM domains are found in homeodomain proteins (LIM-HD) and in small proteins containing little additional sequence (nuclear LIM-only, nLMO). Both LIM-HD and nLMO proteins are essential transcriptional regulators whose genetic disruption results in profound defects in hematopoiesis and development of the nervous system, endocrine system, and limbs. Cytoplasmic LIM proteins consist of variable numbers of LIM domains either alone (cLMOs) or in association with other functional modules, i.e., PDZ domains, kinase domains, α-actinin-binding sites, and GAP domains. Most, if not all, cytoplasmic LIM domain proteins are associated with the actin cytoskeleton and are essential to its structure and function. LIM domains are thus versatile protein molecular recognition modules that have essential functions in control of gene transcription and cytoskeletal architecture.

Structure of LIM Domains

Each LIM domain contains two Zn^{2+} fingers. The two Zn^{2+} atoms are bound independently in N- and C-terminal modules, which are packed together via a hydrophobic interface (Perez-Alvarado *et al.*, 1994). The structure consists of four antiparallel β sheets with the hydrophobic residues that constitute the core of the LIM domain being conserved among family members. The surfaces contain both basic and acidic residues. A short α helix is present in the C-terminus of the cytoplasmic LIM domains of CRP (cysteine-rich protein) and CRIP (cysteine-rich intestinal protein). The residues that coordinate the Zn^{2+} atoms are essential for LIM domain folding but it is not yet known how molecular targets are recognized by the overall structure.

LIM Domain Transcription Factors

LIM homeobox genes have been identified in *Caenorhabditis elegans*, *Drosophila*, and vertebrates and can be organized by homology into six subclasses; nuclear LIM-only genes have been isolated in flies and vertebrates, but not in worms (Hobert and Westphal, 2000). Through its ability to homodimerize, the nuclear LIM interactor (NLI) protein (the *Drosophila* ortholog is Chip) forms tetrameric complexes with nuclear LIM proteins (2NLI:2LIM) to coordinate their activity (Jurata *et al.*, 1998). Nearly all LIM-HD and nLMO proteins have unique patterns of expression throughout development and are required for the normal development of many tissue types, especially within the nervous and endocrine systems.

LIM Homeobox Subfamilies

Lhx1 subfamily
Members of the Lhx1 family, which includes *C. elegans* lin-11 and mec-3, *Drosophila* dlim1, and vertebrate Lhx1 and 5, are widely, but not ubiquitously, expressed throughout the nervous system. lin-11 and mec-3 are required for the specification of thermoregulatory interneurons and mechanosensory neurons, respectively, while Lhx5 is necessary for mouse hippocampal neuronal differentiation and migration. Early functions of Lhx1 in anterior patterning during gastrulation were revealed by gene deletion studies in the mouse, in which embryos developed without heads.

Lhx2 subfamily

Members of this group, C. elegans ttx-3, *Drosophila* apterous, vertebrate Lhx2 and Lhx9, are all expressed in subclasses of developing interneurons. ttx-3 is necessary for the development of a thermoregulatory neuron that functionally opposes the lin-11-expressing thermoregulatory neuron. In the fly, apterous is required for appropriate axon pathfinding of interneurons as well as in patterning and outgrowth of the wing. The function of the Lhx2 family is conserved from fly to vertebrates, as Lhx2 also plays a role in outgrowth of the chick limb. Additionally, in the mouse, Lhx2 is required for eye and forebrain development as well as erythropoiesis. While Lhx9 is highly expressed in the developing mouse brain and limbs, the major phenotype resulting from genetic disruption of this gene was failure of male gonad formation.

Lhx3 subfamily

C. elegans ceh-14, *Drosophila* dlim3, and vertebrate Lhx3 and Lhx4 comprise the third LIM-HD subfamily. ceh-14 was shown to specify a third type of thermoregulatory interneuron in worms, while dlim3, Lhx3, and Lhx4 were found to be expressed in and required for the normal axon trajectory of subclasses of motor neurons. These factors are also expressed in specific classes of interneurons. Lhx3 and Lhx4 are additionally expressed in the developing pituitary, where their coordinate functions are necessary for many aspects of pituitary formation. Mutations in human Lhx3 are associated with combined pituitary hormone deficiency disease.

Lhx6 subfamily

The members of this group are lim-4 in worm, arrowhead in fly, and Lhx6 and Lhx8 in vertebrates. lim-4 is necessary for specification of an olfactory neuron in worms, while in flies, arrowhead is expressed in neuroblasts and is involved in the development of abdominal and salivary imaginal cells. In the mouse, both Lhx6 and Lhx8 are expressed in the developing forebrain and branchial arches and loss of Lhx8 in the mouse resulted in cleft palate formation.

Islet subfamily

lim-7 in worms, fly dislet, and vertebrate Isl1 and Isl2 make up this LIM-HD subfamily. The fly and vertebrate genes are expressed in large classes of motor neurons, where they are required for appropriate axon pathfinding, neurotransmitter identity, and differentiation. In addition, Isl1 expression in the developing pancreas is involved in the formation of both exocrine and endocrine cells.

Lmx subfamily

C. elegans lim-6 and vertebrate Lmx1a and Lmx1b are members of the last group of LIM homeobox genes. lim-6 is expressed in subsets of neurons in the worm, and is necessary for differentiation of GABAergic motor neurons. In the vertebrate nervous system, Lmx1a is required for formation of the roof plate and dorsalization of the neural tube. Mutation of Lmx1b resulted in dorsal/ventral patterning defects within the limb as well as kidney defects, and was found to be a cause of the human genetic disease known as nail–patella syndrome.

The dLMO Family

Drosophila dLMO and vertebrate LMO1–4 are also expressed in the developing nervous system and limb. Genetic analysis of dLMO revealed that this factor functions in wing development to downregulate apterous activity by disrupting functional apterous/Chip complexes. In humans and mice, misexpression of LMO1 and LMO2 in T cells causes leukemia, while disruption of LMO2 function in mice resulted in failure of erythropoiesis.

Cytoplasmic LIM Domain Proteins

Most cytoplasmic LIM domain proteins are associated with and regulate the cytoskeleton (Dawid *et al.*, 1998). The cLMO proteins contain from one to more than five LIM domains. Adapter proteins contain one or more LIM domains in addition to protein-binding motifs such as PDZ domains and α-actinin-binding sequences. Both protein kinase and GTPase activating functions are found in cLIM domain-containing proteins.

LIM-kinase

There are two human LIM-kinases, each containing two N terminal LIM domains, a central PDZ domain, and a C-terminal Ser/Thr protein kinase domain. Hemizygous deletion of LIM-kinase is implicated in the neurological manifestations of Williams syndrome. LIM-kinases regulate the actin cytoskeleton by phosphorylating cofilin at Ser3. This phosphorylation blocks cofilin activity and thus decreases depolymerization of actin filaments thereby stabilizing them. LIM-kinase functions in a signal transduction pathway through which environmental signals are transmitted through the small GTPases of the Rho family via a protein kinase cascade to regulate actin cytoskeleton responses such as cell movement (Edwards *et al.*, 1999).

Adapter Proteins

Enigma family

The Enigma (ENG) family of adapters contains a single N-terminal PDZ and one to three C-terminal LIM domains. ENG, (LMP-1) cypher (Oracle), and the Enigma Homolog (ENH) contain closely related PDZ domains and three LIM domains. During development these proteins are preferentially expressed in cardiac and skeletal muscle. The PDZ domain of ENG binds to the skeletal muscle isoform of tropomyosin and the two are colocalized at the boundary between the Z line and I band. The PDZ domain of cypher binds to α-actinin2 and colocalizes with it at the Z line of cardiac myofibrils. A related family of proteins that contain a single LIM domain include Ril, CLP36, and α-actinin associated LIM protein (ALP). The PDZ domain of ALP binds to the spectrin-like motifs of α-actinin2 and colocalizes with it at the Z line of myofibers. The binding partners for the LIM domains of these proteins are incompletely defined but may include the protein kinase ret, the insulin receptor, and protein kinase C.

Focal adhesions

A number of LIM domain proteins are localized at focal adhesions, which are sites of integrin–extracellular matrix communication. The paxillin family of proteins (paxillin, leupaxin, Hic-5, and Pax B), contain four C-terminal LIM domains that target the proteins to focal adhesions. The N-terminus contains vinculin and focal adhesion kinase (FAK) binding sites. Paxillin is phosphorylated on tyrosine residues by FAK and thus binds both SH2 and SH3 domain proteins in macromolecular complexes present in focal adhesions. Zyxin is also located at focal adhesions and along actin filaments. This protein contains three C-terminal LIM domains and a proline-rich N-terminus that binds to α-actinin. Ajuba and LPP are related proteins that are localized to sites of cell–cell adhesion. The *C. elegans* unc-115 and human abLIM proteins contain three and four N-terminal LIM domains respectively and a C-terminal actin-binding domain related to the villin head piece and dermatin domains. unc-115 mediates axon guidance while abLIM is specifically expressed in the retina where it undergoes extensive phosphorylation. Both proteins are proposed to be molecular adapters that link the actin cytoskeleton to extracellular signals. *Drosophila* prickle, which contains three C-terminal LIM domains and an N-terminal PET domain, is necessary for the development of planar polarity in imaginal discs.

cLIM-Only Proteins (cLMO)

cLMO proteins consist of one to five LIM domains without other identifiable sequence motifs. The cysteine-rich protein (CRP) family members contain two LIM domains. CRP1-3 proteins bind to the LIM protein zyxin and colocalize with actin. Genetic deletion of muscle LIM protein (MLP(CRP3)) that is normally localized at Z lines results in disruption of cardiac myocyte cytoarchitecture and heart failure; skeletal muscle fibers are also abnormal. Other cLMO proteins are specifically expressed in smooth muscle (SmLIM) and in skeletal muscle. PINCH, which contains five LIM domains, is implicated in integrin-associated protein kinase signaling. The *C. elegans* ortholog of PINCH (unc-97) is necessary for structural integrity of the integrin containing muscle adherence junctions and contributes to the mechanosensory function of touch neurons.

Future Prospects

The catalog of nuclear LIM proteins is nearly complete. One high-affinity target, NLI, is the basis for combinatorial association of nuclear LIM proteins into a transcriptional 'code' underlying developmental choices. How these complexes operate in the context of other transcriptional regulators remains to be determined. The catalog of cytoplasmic LIM proteins is incomplete and other associated protein motifs are likely to accompany the LIM domains. Most, if not all, interact with and regulate the cytoskeleton. In contrast to nuclear LIM domains, a common high-affinity target for cytoplasmic LIM domains has not been identified and specific recognition sites remain to be determined. In both nuclear and cytoplasmic proteins LIM domains function as recognition modules for macromolecular assemblies.

Further Reading

Dawid IB, Breen JJ and Toyama R (1998) LIM domains: multiple roles as adapters and functional modifiers in protein interactions. *Trends in Genetics* 14: 156–162.

Edwards DC, Sanders LC, Bokoch GM and Gill GN (1999) Activation of LIM-kinase by Pak1 couples Rac/Cdc42 GTPase signalling to actin cytoskeletal dynamics. *Nature Cell Biology* 1: 253–259.

Freyd G, Kim SK and Horvitz HR (1990) Novel cysteine-rich motif and homeodomain in the product of *Caenorhabditis elegans* cell lineage gene lin-11. *Nature* 344: 876–879.

Hobert O and Westphal H (2000) Function of LIM-homeobox genes. *Trends in Genetics* 16: 75–83.

Jurata LW, Pfaff SL and Gill GN (1998) The nuclear LIM domain interactor NLI mediates homo- and heterodimerization of LIM domain transcription factors. *Journal of Biological Chemistry* 273: 3152–3157.

Perez-Alvarado GC, Miles C, Michelsen JW *et al.* (1994) Structure of the carboxy-terminal LIM domain from the cysteine rich protein CRP. *Nature Structural Biology* 1: 388–398.

See also: **Cell Lineage; Homeobox; Neurogenetics in *Caenorhabditis elegans*; Neurogenetics in *Drosophila***

Limb Development

R Johnson

Copyright © 2001 Academic Press
doi: 10.1006/rwgn.2001.0763

The tetrapod limb is a complex structure that exhibits considerable morphological diversity between species. For example, the forelimbs of bats and birds have been adapted for flying while the limbs of alligators retain features characteristic of the primitive tetrapod condition (Shubin *et al.*, 1997). Despite these variations, all tetrapod limbs exhibit a common organizational theme that reflects their conserved evolutionary origin (**Figure 1**). In this article, the cellular and molecular events that occur during embryogenesis to form this basic tetrapod limb structure are described.

The first morphological indication of limb development is a localized thickening of the lateral plate mesoderm in presumptive forelimb and hindlimb regions of the embryonic flank. This thickening is achieved through differential proliferation of the lateral plate mesoderm, with maintenance of high levels within limb bud forming regions and suppression of these high levels in interlimb regions. Available evidence suggests that members of the fibroblast growth factor (fgf) family are important mediators of limb induction. Implantation of beads soaked in recombinant FGF protein into the interlimb region results in the formation of an ectopic limb at the site of bead implantation (Cohn *et al.*, 1995). Several *fgf* family members, including *fgf-8* and *fgf-10*, are expressed in the intermediate mesoderm at the time of limb initiation, making them attractive candidates for mediating limb induction (Cohn *et al.*, 1995; Ohuchi *et al.*, 1997; Yonei-Tamura *et al.*, 1999). How these activities are deployed at specific axial levels is not known, but it is likely to involve the action of the clustered homeobox genes (Cohn *et al.*, 1997).

Once limb buds have formed, continued outgrowth depends on signaling from a specialized region of limb bud ectoderm, the apical ectodermal ridge (AER), to the underlying mesenchyme. The AER is a morphologically visible thickening of limb bud ectoderm occurring at the interface between dorsal and ventral ectoderm. The function of the AER has been determined by microsurgical manipulation of chick embryos (Saunders, 1948). Removal of the AER leads to limb truncations. The exact level of truncation depends on the time at which the AER is removed: removal at an early stage leads to proximal truncations while later removals lead to progressively more distal truncations (Summerbell, 1974b). Hence, the AER is

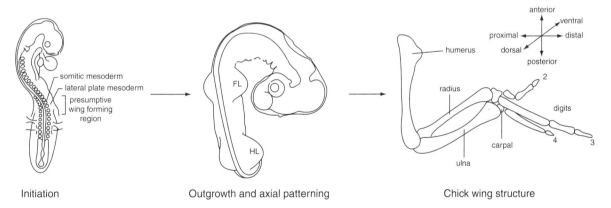

Initiation   Outgrowth and axial patterning   Chick wing structure

Figure 1 Key stages of vertebrate limb development as illustrated in the chick embryo. Limb initiation begins within the lateral plate mesoderm, which proliferates to form limb buds at the forelimb (FL) and hindlimb (HL) levels. At this stage outgrowth becomes dependent on the apical ectodermal ridge (AER) and axial pattern is specified through the action of multiple signaling centers (see text). After further development, a cartilage model of the adult limb is formed. Illustrated here is the skeletal pattern of a chick wing at 8 days of incubation. Note the conserved features of all tetrapod limbs: a single proximal long bone (humerus) followed by two long bones (raidus and ulna) and capped by carpals and digits. (Adapted from Johnson and Tabin, 1997.)

required for distal limb outgrowth. The AER is also sufficient for this process as grafting of additional AERs leads to the formation of supernumerary limbs (Johnson and Tabin, 1997).

As in limb initiation, *fgfs* figure prominently in AER formation and function. One of the *fgfs*, *fgf-10*, is expressed in the limb bud mesoderm prior to AER formation and mice that lack *fgf-10* fail to form the AER (Min *et al.*, 1998; Sekine *et al.*, 1999). Evidence that *fgfs* mediate the distal outgrowth activity of the AER first came from expression studies and explant experiments. The ectoderm of the AER expresses a number of *fgfs*, including *fgf-4* and *fgf-8* (Niswander and Martin, 1992; MacArthur *et al.*, 1995; Mahmood *et al.*, 1995; Vogel *et al.*, 1996). Culturing of mouse limb buds denuded of their AER in the presence of recombinant *fgfs* results in limited restoration of distal outgrowth (Niswander and Martin, 1993). These initial findings were extended by *in ovo* experiments using chick embryos (Niswander *et al.*, 1993). Removal of the AER followed by grafting a bead soaked in FGF protein to the distal limb bud results in near complete distal limb development. Hence *fgfs* are sufficient to replace the AER in directing limb outgrowth and expressed in the AER at appropriate times to mediate this function of the AER. Removal of fgf function in the AER through gene targeting methods has yet to reveal a requirement of *fgfs* in distal limb outgrowth most likely due to the fact that multiple *fgfs* are expressed in the AER (Moon *et al.*, 2000; Sun *et al.*, 2000).

Characteristic features of tetrapod limbs are asymmetries along the three cardinal limb axes (see **Figure 1**). Proximal–distal asymmetries are exemplified by the presence of a single long bone in most proximal regions followed by two long bones in more distal regions and a collection of small bones and digits in the distal-most regions.

Asymmetries along the anterior–posterior axis are most easily seen by comparing digit morphologies. For example, the anterior-most digit of the human hand is the thumb, which contains only two phalanges while other digits contain three phalanges. More pronounced differences can be seen in the three digits of the avian wing (**Figure 1**). Dorsal–ventral asymmetries can be found in both integument derivatives (nails and feathers for example) and internal tissues such as the arrangement of muscles and tendons. These latter asymmetries are essential for coordinated extension and flexion of the limb. The mechanisms that specify skeletal morphologies along the proximal–distal axis are not clear, although the action of several homeobox-containing genes are though to be essential for proper pattern formation along this axis (Capecchi, 1997; Rijli and Chambon, 1997; Capdevila

et al., 1999; Zakany and Duboule, 1999; Mercader *et al.*, 2000). In contrast, the mechanisms by which asymmetry along the anterior–posterior and dorsal–ventral axes are achieved is better understood and detailed below.

The anterior–posterior identity of limb tissues is controlled by a special group of mesenchymal cells located in the posterior limb called the zone of polarizing activity or ZPA (Saunders and Gasseling, 1968). The ZPA has the remarkable ability to induce the formation of mirror-symmetric digits when transplanted to the anterior margin of a host limb bud. This observation led to a model in which the ZPA produces a diffusible signal that gives cells their identity along the anterior–posterior limb axis (Tickle *et al.*, 1976). Recently, the molecule responsible for this activity has been identified as the product of the *sonic hedgehog* (*shh*) gene (Riddle *et al.*, 1993). *Shh* encodes a secreted factor that is expressed within the ZPA. Moreover ectopic expression of Shh in anterior limb bud tissues mimics the effects of ZPA transplantation, indicating that Shh can functionally substitute for the ZPA. Whether Shh acts as a diffusible morphogen in the limb is controversial, however recent studies suggest that its effects are long-range, consistent with the morphogen hypothesis (Yang *et al.*, 1997; Drossopoulou *et al.*, 2000; Wang *et al.*, 2000).

The dorsal–ventral polarity of the limb bud is achieved through a cascade of factors expressed within the limb bud ectoderm and mesenchyme. Rotation experiments indicated that positional identity along the dorsal–ventral limb axis is controlled by the ectoderm (MacCabe *et al.*, 1974). Inversion of the ectoderm, but not the mesenchyme, results in dorsal–ventral axis inversion. Hence, these studies suggest that the ectoderm sends a signal to the underlying mesenchyme to determine its identity along the dorsal–ventral axis. Gene targeting studies in the mouse and gain of function experiments in the chick indicate that three factors play critical roles in dorsal–ventral limb patterning. The secreted glycoprotein wnt-7a is expressed in the dorsal ectoderm and is both necessary and sufficient for dorsal pattern specification (Parr and McMahon, 1995; Riddle *et al.*, 1995; Vogel *et al.*, 1995). Localized expression of *wnt-7a* in the dorsal ectoderm is achieved by the action of *engrailed-1* (*en-1*) (Loomis *et al.*, 1996; Logan *et al.*, 1997). *En-1* is expressed in the ventral limb ectoderm where it directly or indirectly represses expression of *wnt-7a*. The function of *wnt-7a* is to induce the expression of a LIM-homeodomain class transcription factor, *lmx1b*, in dorsal limb mesenchyme (Cygan *et al.*, 1997; Loomis *et al.*, 1998). *Lmx1b*, in turn, is necessary and sufficient to modify a default ventral limb pattern to create a dorsal-specific arrangement of limb tissues

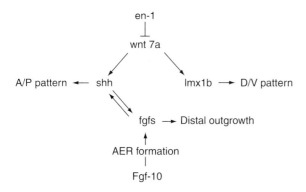

Figure 2 Important molecular regulatory interactions in vertebrate limb patterning. Apical ectodermic ridge (AER) formation, distal outgrowth, anterior–posterior (A/P), and dorsal–ventral (D/V) pattern are controlled by a network of transcription factors and signaling molecules. See text for details.

(Riddle *et al.*, 1995; Vogel *et al.*, 1995; Chen *et al.*, 1998). How *lmx1b* achieves this effect is not understood, but it likely modulates the expression of a number of genes within dorsal limb tissues.

Although it may appear that the three limb axes are specified through independent mechanisms, there are significant interactions among these pathways (**Figure 2**). Indeed, one might expect that to achieve the proper arrangement of limb tissues in three dimensions, the pathways that control patterning along each axis should be coupled to each other. A number of experiments indicate that this is the case. For example, transplantation of the ZPA at different times leads to duplication of tissues at different proximal–distal levels suggesting an integration of anterior–posterior and proximal–distal patterning mechanisms (Summerbell, 1974a). A second example is that both the outgrowth function of the AER and the polarizing activity of the ZPA are connected together through a reciprocal feedback loop whereby *fgf* expression in the AER and *shh* expression in the ZPA are codependent (Laufer *et al.*, 1994; Niswander *et al.*, 1994). Disruption of this loop is observed in the limbs of *shh* mutant mice (Chiang *et al.*, 1996), which exhibit distal truncations, resulting from the indirect modulation of AER function by *shh*. Finally, the dorsal–ventral and anterior–posterior pathways are linked through modulation of *shh* levels by *wnt-7a* (Parr and McMahon, 1995; Yang and Niswander, 1995). Removal of *wnt-7a* function leads to both dorsal–ventral patterning defects and loss of posterior-most digits. These examples provide only an indication of the degree to which axial limb patterning mechanisms are coupled and future experiments will provide additional complexities.

Once the axial pattern of the limb has been laid down, a series of additional events are necessary to achieve the final form of the limb. Prominent among these events is modulation of cell death, especially within interdigital regions (Zou and Niswander, 1996; Chen and Zhao, 1998). The spaces between digits are achieved through induction of cell death specifically within interdigital zones by a process that involves signaling by bone morphogenetic proteins (*bmps*). A second important event is endochondrial ossification (Karsenty, 1999). The initial skeletal structures of the limb are laid down as cartilagenous models that are replaced by bone through this process.

Endochondrial ossification is also important for embryonic and postnatal growth of the long bones of the limb and other tissues and is mediated through a complex signaling network involving bmps, Indian hedgehog, and parathyroid related peptide (Vortkamp *et al.*, 1996). Other important events during limb development that have received comparatively little attention are the formation of muscles, joints, tendons, and ligaments. Very little is known about the molecular basis of these processes, however, studies of mouse mutants are starting to reveal some key players that regulate these events (Storm *et al.*, 1994; Storm and Kingsley, 1996; Thomas *et al.*, 1996).

Many congenital malformations that affect limb development are known and it is becoming clear that most of these malformations can be interpreted as due to mutations in genes affecting pathways of limb initiation, axial patterning, or subsequent limb shaping events such as cell death, endochondrial ossification, and joint formation (Manouvrier-Hanu *et al.*, 1999). In cases where the genes responsible for limb defects have been identified, they can readily be integrated into known pathways. For example, Greig cephalopolysyndactyly (GCPS), in which affected individuals have a single extra pre-axial digit, results from mutations in the *GLI-3* gene (Vortkamp *et al.*, 1991). Studies of mice with *gli-3* mutations suggest that GCPS polydactyly is caused by ectopic anterior expression of Shh during early limb development (Vortkamp *et al.*, 1992; Hui and Joyner, 1998; Buscher *et al.*, 1997; Masuya *et al.*, 1997). Another example is nail–patella syndrome (nps), caused by mutations in the dorsal patterning gene *LMX1B* (Dreyer *et al.*, 1998; Vollrath *et al.*, 1998; Clough *et al.*, 1999). The limb phenotype of individuals with nps is small or absent patellae and misshapen or absent nails, each dorsal derivatives of the limb. These and other related studies highlight synergistic interactions between human genetics and developmental biology that has lead to an understanding of the etiology of many limb malformations. It is expected that in the future this synergy will continue, especially with the identification of novel regulators of limb patterning through the application of positional cloning methods to the many existing human and murine genetic limb malformations.

References

Buscher D, Bosse B, Heymer J and Ruther U (1997) Evidence for genetic control of Sonic hedgehog by Gli3 in mouse limb development. *Mechanisms of Development* 62: 175–182.

Capdevila J, Tsukui T, Rodriquez Esteban C, Zappavigna V and Izpisua Belmonte JC (1999) Control of vertebrate limb outgrowth by the proximal factor Meis2 and distal antagonism of BMPs by Gremlin. *Molecular Cell* 4: 839–849.

Capecchi MR (1997) Hox genes and mammalian development. *Cold Spring Harbor Symposium in Quantitative Biology* 62: 273–281.

Chen H, Lun Y, Ovchinnikov D et al. (1998) Limb and kidney defects in *Lmx1b* mutant mice suggest an involvement of LMX1B in human nail–patella syndrome. *Nature Genetics* 19: 51–55.

Chen Y and Zhao X (1998) Shaping limbs by apoptosis. *Journal of Experimental Zoology* 282: 691–702.

Chiang C, Litingtung Y, Lee E et al. (1996) Cyclopia and defective axial patterning in mice lacking Sonic hedgehog gene function. *Nature* 383: 407–413.

Clough MV, Hamlington JD and McIntosh I (1999) Restricted distribution of loss-of-function mutations within the LMX1B genes of nail–patella syndrome patients. *Human Mutation* 14: 459–465.

Cohn MJ, Izpisua-Belmonte JC, Abud H, Health JK and Tickle C (1995) Fibroblast growth factors induce additional limb development from the flank of chick embryos. *Cell* 80: 739–746.

Cohn MJ, Patel K, Krumlauf R et al. (1997) Hox9 genes and vertebrate limb specification. *Nature* 387: 97–101.

Cygan JA, Johnson RL and McMahon AP (1997) Novel regulatory interactions revealed by studies of murine limb pattern in *Wnt-7a* and *En-1* mutants. *Development* 124: 5021–5032.

Dreyer SD, Zhou G, Baldini A et al. (1998) Mutations in LMX1B cause abnormal skeletal patterning and renal dysplasia in nail–patella syndrome. *Nature Genetics* 19: 47–50.

Drossopoulou G, Lewis KE, Sanz-Ezquerro JJ et al. (2000) A model for anteroposterior patterning of the vertebrate limb based on sequential long- and short-range Shh signalling and Bmp signalling. *Development* 127: 1337–1348.

Hui CC and Joyner AL (1993) A mouse model of Greig cephalopolysyndactyly syndrome: the extra-toes J mutation contains an intragenic deletion of the *Gli3* gene. *Nature Genetics* 3: 241–246. [Erratum appears in *Nature Genetics* (1998) 19 (4): 404.]

Johnson RL and Tabin CJ (1997) Molecular models for vertebrate limb development. *Cell* 90: 979–990.

Karsenty G (1999) The genetic transformation of bone biology. *Genes and Development* 13: 3037–3051.

Laufer E, Nelson CE, Johnson RL, Morgan BA and Tabin C (1994) Sonic hedgehog and Fgf-4 act through a signaling cascade and feedback loop to integrate growth and patterning of the developing limb bud. *Cell* 79: 993–1003.

Logan C, Hornbruch A, Campbell I and Lumsden A (1997) The role of Engrailed in establishing the dorsoventral axis of the chick limb. *Development* 124: 2317–2324.

Loomis CA, Harris E, Michaud J et al. (1996) The mouse Engrailed-1 gene and ventral limb patterning. *Nature* 382: 360–363.

Loomis CA, Kimmel RA, Tong CX, Michaud J and Joyner AL (1998) Analysis of the genetic pathway leading to formation of ectopic apical ectodermal ridges in mouse Engrailed-1 mutant limbs. *Development* 125: 1137–1148.

MacArthur CA, Lawshe A, Xu J et al. (1995) FGF-8 isoforms activate receptor splice forms that are expressed in mesenchymal regions of mouse development. *Development* 121: 3603–3613.

MacCabe JA, Errick J and Saunders JWJ (1974) Ectodermal control of the dorsoventral axis in the leg bud of the chick embryo. *Developmental Biology* 39: 69–82.

Mahmood R, Bresnick J, Hornbruch A et al. (1995) A role for FGF-8 in the initiation and maintenance of vertebrate limb bud outgrowth. *Current Biology* 5: 797–806.

Manouvrier-Hanu S, Holder-Espinasse M and Lyonnet S (1999) Genetics of limb anomalies in humans. *Trends in Genetics* 15: 409–417.

Masuya H, Sagai T, Moriwaki K and Shiroishi T (1997) Multigenic control of the localization of the zone of polarizing activity in limb morphogenesis in the mouse. *Developmental Biology* 182: 42–51.

Mercader N, Leonardo E, Piedra ME et al. (2000) Opposing RA and FGF signals control proximodistal vertebrate limb development through regulation of Meis genes. *Development* 127: 3961–3970.

Min H, Danilenko DM, Scully SA et al. (1998) Fgf-10 is required for both limb and lung development and exhibits striking functional similarity to *Drosophila* branchless. *Genes and Development* 12: 3156–3161.

Moon AM, Boulet AM and Capecchi MR (2000) Normal limb development in conditional mutants of Fgf4. *Development* 127: 989–996.

Niswander L and Martin GR (1992) Fgf-4 expression during gastrulation, myogenesis, limb and tooth development in the mouse. *Development* 114: 755–768.

Niswander L and Martin GR (1993) FGF-4 and BMP-2 have opposite effects on limb growth. *Nature* 361: 68–71.

Niswander L, Tickle C, Vogel A, Booth I and Martin GR (1993) FGF-4 replaces the apical ectodermal ridge and directs outgrowth and patterning of the limb. *Cell* 75: 579–587.

Niswander L, Jeffrey S, Martin GR and Tickle C (1994) A positive feedback loop coordinates growth and patterning in the vertebrate limb. *Nature* 371: 609–612.

Ohuchi H, Nakagawa T, Yamamoto A et al. (1997) The mesenchymal factor, FGF10, initiates and maintains the outgrowth of the chick limb bud through interaction with FGF8, an apical ectodermal factor. *Development* 124: 2235–2244.

Parr BA and McMahon AP (1995) Dorsalizing signal *Wnt-7a* required for normal polarity of D-V and A-P axes of mouse limb. *Nature* 374: 350–353.

Riddle RD, Johnson RL, Laufer E and Tabin C (1993) Sonic hedgehog mediates the polarizing activity of the ZPA. *Cell* 75: 1401–1416.

Riddle RD, Ensini M, Nelson C *et al.* (1995) Induction of the LIM homeobox gene *Lmx1* by WNT7a establishes dorsoventral pattern in the vertebrate limb. *Cell* 83: 631–640.

Rijli FM and Chambon P (1997) Genetic interactions of *Hox* genes in limb development: learning from compound mutants. *Current Opinianism Genetics and Development* 7: 481–487.

Saunders JWJ (1948) The proximo-distal sequence of origin of the parts of the chick wing and the role of the ectoderm. *Journal of Experimental Zoology* 108: 363–403.

Saunders JWJ and Gasseling MT (1968) Ectoderm–mesenchymal interaction in the origins of wing symmetry. In: Fleischmajer R and Billingham RE (eds) *Epithelial–Mesenchymal Interactions*, pp. 78–97. Baltimore, MD: Williams & Wilkins.

Sekine K, Ohuchi H, Fujiwara M *et al.* (1999) Fgf10 is essential for limb and lung formation. *Nature Genetics* 21: 138–141.

Shubin N, Tabin C and Carroll S (1997) Fossils, genes and the evolution of animal limbs. *Nature* 388: 639–648.

Storm EE and Kingsley DM (1996) Joint patterning defects caused by single and double mutations in members of the bone morphogenetic protein (BMP) family. *Development* 122: 3969–3979.

Storm EE, Huynh TV, Copeland NG *et al.* (1994) Limb alterations in brachypodism mice due to mutations in a new member of the TGF beta-superfamily. *Nature* 368: 639–643.

Summerbell D (1974a) Interaction between the proximo-distal and antero-posterior co-ordinates of positional value during the specification of positional information in the early development of the chick limb-bud. *Journal of Embryology and Experimental Morphology* 32: 227–237.

Summerbell D (1974b) A quantitative analysis of the effect of excision of the AER from the chick limb-bud. *Journal of Embryology and Experimental Morphology* 32: 651–660.

Sun X, Lewandoski M, Meyers EN *et al.* (2000) Conditional inactivation of Fgf4 reveals complexity of signalling during limb bud development. *Nature Genetics* 25: 83–86.

Thomas JT, Lin K, Nandedkar M *et al.* (1996) A human chondrodysplasia due to a mutation in a *TGF-beta* superfamily member. *Nature Genetics* 12: 315–317.

Tickle C, Shellswell G, Crawley A and Wolpert L (1976) Positional signalling by mouse limb polarising region in the chick wing bud. *Nature* 259: 396–397.

Vogel A, Rodriguez C, Warnken W and Izpisua Belmonte JC (1995) Dorsal cell fate specified by chick *Lmx1* during vertebrate limb development. *Nature* 378: 716–720.

Vogel A, Rodriguez C and Izpisua-Belmonte JC (1996) Involvement of FGF-8 in initiation, outgrowth and patterning of the vertebrate limb. *Development* 122: 1737–1750.

Vollrath D, Jaramillo-Babb VL, Clough MV *et al.* (1998) Loss-of-function mutations in the LIM-homeodomain gene, *LMX1B*, in nail–patella syndrome. *Human and Molecular Genetics* 7: 1091–1098.

Vortkamp A, Gessler M and Grzeschik KH (1991) GLI3 zinc-finger gene interrupted by translocations in Greig syndrome families. *Nature* 352: 539–540.

Vortkamp A, Franz T, Gessler M and Grzeschik KH (1992) Deletion of GLI3 supports the homology of the human Greig cephalopolysyndactyly syndrome (GCPS) and the mouse mutant extra toes (Xt). *Mammalian Genome* 3: 461–463.

Vortkamp A, Lee K, Lanske B *et al.* (1996) Regulation of rate of cartilage differentiation by Indian hedgehog and PTH-related protein. *Science* 273: 613–622.

Wang B, Fallon JF and Beachy PA (2000) Hedgehog-regulated processing of Gli3 produces an anterior/posterior repressor gradient in the developing vertebrate limb. *Cell* 100: 423–434.

Yang Y and Niswander L (1995) Interaction between the signaling molecules WNT7a and SHH during vertebrate limb development: dorsal signals regulate anteroposterior patterning. *Cell* 80: 939–947.

Yang Y, Drossopoulou G, Chuang PT *et al.* (1997) Relationship between dose, distance and time in Sonic Hedgehog-mediated regulation of anteroposterior polarity in the chick limb. *Development* 124: 4393–4404.

Yonei-Tamura S, Endo T, Yajima H *et al.* (1999) FGF7 and FGF10 directly induce the apical ectodermal ridge in chick embryos. *Developmental Biology* 211: 133–143.

Zakany J and Duboule D (1999) *Hox* genes in digit development and evolution. *Cell and Tissue Research* 296: 19–25.

Zou H and Niswander L (1996) Requirement for BMP signaling in interdigital apoptosis and scale formation. *Science* 272: 738–741.

See also: **Developmental Genetics; Embryonic Development, Mouse; *Hox* Genes; LIM Domain Genes**

LINE

L Silver

Copyright © 2001 Academic Press
doi: 10.1006/rwgn.2001.0764

The general name coined for selfish genetic elements that disperse themselves through the genome by means of an RNA intermediate is retroposon. There are two classes of retroposons. The SINE family is made up of very small DNA elements that require other genetic information to facilitate their dispersion throughout the genome. The LINE family is derived from a full-fledged selfish DNA sequence with a self-encoded reverse transcriptase.

Selfish genetic elements of the LINE type have been around for a very long time. Homologous LINE elements have been found in a wide variety of organisms including protists and plants. Thus, LINE-related elements, or others of a similar nature, are likely to have been the source material that gave rise to retroviruses.

Full-length LINE elements have a length of 7 kb; however, the vast majority (>90%) have truncated sequences which vary in length down to 500 bp. But, of the many full-length LINE elements in any genome, only a few retain a completely functional reverse transcriptase gene which has not been inactivated by mutation. Thus, only a very small fraction of the LINE family members retain 'transposition competence,' and it is these that are responsible for dispersing new elements into the genome.

Dispersion to new positions in the germline genome presumably begins with the transcription of competent LINE elements in spermatogenic or oogenic cells. The reverse transcriptase coding region on the LINE transcript is translated into enzyme that preferentially associates with and utilizes the transcript that it came from as a template to produce LINE cDNA sequences. For reasons that are unclear, it seems that the reverse transcriptase usually stops before a full-length copy is finished. These incomplete cDNA molecules are, nevertheless, capable of forming a second strand and integrating into the genome as truncated LINE elements that are forever dormant.

The LINE family appears to evolve by repeated episodic amplifications from one or a few progenitor elements, followed by the slow degradation of most new integrants – by genetic drift – into random sequence. Thus, at any point in time, a large fraction of the cross-hybridizing LINE elements in any one genome will be more similar to each other than to LINE elements in other species. In a sense, episodic amplification followed by general degradation is another mechanism of concerted evolution.

See also: **Repetitive (DNA) Sequence; Retroposon; SINE**

Linkage

F W Stahl

In Mendel's crosses, diploids that were heterozygous at two loci produced the four possible kinds of haploid meiotic products in equal numbers – the hereditary factors (genes) assorted at random, appearing in the meiotic products without regard to the combinations (parental types) in which they entered the diploid. Deviations from random assortment occur when two loci are on the same chromosome. Such a deviation (linkage) is manifested as an excess of parental types over the new (recombinant) types. In a two-factor cross involving linked loci, the mutant alleles are said to enter the cross in coupling when the diploid is formed from the union of a wild-type gamete and a double mutant gamete. The mutant alleles are in repulsion if the diploid is formed by the union of gametes which are each mutant at one of the loci and wild-type at the other.

Detection of Linkage

When two loci on the same chromosome are far apart, they may fail to generate meiotic products whose frequencies differ significantly from the expectations of nonlinkage. Such a failure to demonstrate linkage of two loci may be overcome with the demonstration of their common linkage to a locus that lies between them on the chromosome.

Much of genetics involves determining the location in the genome of a newly identified gene (mapping the gene). The first step in such mapping is determining on which chromosome the locus is situated. Crosses of the new mutant by strains that carry mutant alleles at loci on each of the chromosomes may detect linkage to one of those loci. Since only a finite number of haploid meiotic products (or of meiotic tetrads) can be examined, the statistical test χ^2 is standardly employed to determine when deviations from random assortment should be taken seriously.

When tetrad data are available, an excess of parental ditype meiotic tetrads over nonparental ditype tetrads sensitively indicates linkage.

Linkage to a Centromere

Since homologous centromeres segregate in the first division of meiosis, relatively strong linkage of a locus to a centromere is indicated by an excess of first division over second division segregations; weaker linkage is implied as long as the frequency of second division segregation is less than 2/3, the frequency expected for random assortment of a locus from its centromere. In the presence of positive chiasma interference, second division segregation may exceed 2/3, indicating that most of the tetrads have a single exchange between that locus and its centromere.

Linkage Maps

Maps that reflect the degree of linkage between loci can be constructed from observed recombination frequencies (see Centimorgan (cM)). The distances on these linkage maps will accurately reflect physical distances on the chromosome only if exchange frequencies are constant along the chromosome.

Linkage in Prokaryotes

In bacteria with a single chromosome, loci that are close enough together on the chromosome to be transmitted together in phage-mediated transduction may be referred to as 'linked.' Similarly, when transformation is conducted with chromosomal DNA, markers are 'linked' if they are cotransformed as a result of sometimes being on the same fragment created by artifactual breakage of the chromosome. Unlinked markers are transduced or transformed into the same recipient cell at a frequency about equal to the product of the transduction or transformation frequencies of the individual markers.

In crosses with bacteriophages, as standardly conducted, recombination frequencies less than 50% cannot be taken as evidence of linkage because a fraction of the progeny phage particles has lacked the opportunity to assort its genes. Linkage is implied by a pair of loci that gives a significantly lower recombination frequency than do the loci with the largest observed values.

See also: **Centimorgan (cM); Genetic Recombination; Mapping Function; Tetrad Analysis**

Linkage Disequilibrium

N E Morton

Copyright © 2001 Academic Press
doi: 10.1006/rwgn.2001.0767

Dependence of gene frequencies at two or more loci is called allelic association, gametic disequilibrium, or linkage disequilibrium (LD). Whereas unlinked loci reach independence (Hardy–Weinberg equilibrium) in a single generation, linked loci with recombination rate $\theta < 0.5$ reduce initial LD in an infinite population to a proportion $e^{-t\theta}$ after t generations. The time required to go halfway to equilibrium is therefore $T = (\ln 2)/\theta$, or more than a million years if $\theta = 10^{-5}$ and there are 20 years per generation. A convenient but inaccurate rule of thumb is that $\theta = 0.01$ corresponds to about 1 megabase (Mb). By this approximation, $\theta = 10^{-5}$ corresponds to 1 kb. If θ is as small as 10^{-6}, the time since apes and hominids diverged is not long enough to go halfway to equilibrium. Therefore selection is not required to explain persistence of disequilibrium, which depends to a considerable extent on episodes of population contraction. There have been two major bottlenecks in human evolution. The first was when two chromosomes that are nonhomologous in apes fused to form the chromosome 2 inherited by our species. The second bottleneck was when we migrated out of Africa in the last 100 000 years. As a consequence, LD is least for sub-Saharan Africa. Lesser bottlenecks have occured in the history of particular populations.

LD may be measured in many ways. Some are confounded with significance tests, and therefore with sample size. All are to some degree confounded with allele frequencies. The most reliable and best validated is the association probability ρ_t, which is made up of two parts. Association that has diminished from an initial value ρ_0 in founders is $\rho_{rt} = \rho_0 e^{-(1/2N + \theta)t}$, where N is the effective size over t generations. Association that has built up by genetic drift since the founders is $\rho_{ct} = L\,(1 - e^{-(1/2N + \theta)t})$, and $\rho_t = \rho_{rt} + \rho_{ct}$. If N is constant, the equilibrium value as $t \to \infty$ is $L = 1/(1 + 2N\theta)$ if θ is small and $1/(1 + 2N)$ if $\theta = 0.5$. The latter is negligible in real populations. If $1/2N$ is small compared to θ, ρ_d follows the Malecot model for isolation by distance, equating $t\theta$ to εd, where d is distance between loci. On the genetic scale d measures recombination directly, with relatively larger sampling error over small distances. On the physical scale d is only indirectly related to recombination, but is more accurate if sequence-based. Choice should be based on goodness of fit to the best available maps. Analyzed as isolation by distance, LD provides a way to compare allelic association for chromosome regions in different populations, and therefore to detect variations in recombination, selective sweeps that reduced haplotype diversity, and effects of population history and structure. This information determines the optimal populations and density of markers for positional cloning of genes affecting normal physiology and disease. Localization is more precise by LD than by linkage. An alternative for multilocus haplotypes is cladistic analysis when its assumptions to reduce the number of independent variables are valid and the causal region has been made small by LD or other evidence.

See also: **Bottleneck Effect; Genetic Drift**

Linkage Group

M A Cleary

Copyright © 2001 Academic Press
doi: 10.1006/rwgn.2001.0768

Linkage among genes or genetic markers is determined by the frequency with which they are inherited together. Genes that are frequently inherited together tend to lie near each other on the same chromosome. The frequency with which genes or markers are inherited together is measured by the percentage of recombination that occurs between them. A linkage group is defined by all of the genes and markers for which linkage has been established. An entire chromosome is considered to be a linkage group.

See also: Crossing-Over; Independent Assortment; Linkage Map

Linkage Map

M F Seldin

Copyright © 2001 Academic Press
doi: 10.1006/rwgn.2001.0769

Linkage maps of the human and mouse genomes have provided the initial framework for genetic studies, including the positional cloning of disease genes and the scaffold for building physical maps and contiguous (contigs) stretches of cloned DNA. Although the usefulness of the human genetic maps to the completion of the Human Genome Project is nearing or at an end, the chromosomal positions of highly polymorphic markers that are necessary for many current studies still largely depend on these linkage maps. In other species, particularly the mouse, linkage maps have and continue to be the predominant tool for defining the chromosomal location of genes.

Linkage maps depend on the relationship between locations on a chromosome that are defined by crossovers that occur between homologous chromosomes during meiosis. The distance between two linked loci on a chromosome is defined by the recombination frequency, where 1% recombination is equal to 1 centimorgan (cM). This 'genetic distance' can provide an accurate relative positioning of genetic markers. However, it only roughly corresponds to actual physical distance (number of base pairs of DNA between loci), since different regions of the genome may have more frequent or less frequent crossover events during meiosis. In addition, genetic linkage maps must be viewed as reflecting a biologic process in which individual variation may be influenced by a large number of factors. For example, the recombination frequency between homologous chromosomes can be substantially different in oogenesis than in spermatogenesis. Although, in general, there is more frequent recombination in female meioses, different regions of the genome show different relationships and there are even chromosomal segments in which the recombination frequency is greater in male meioses. Other studies have indicated that recombination frequency is itself influenced by genetic factors; meiotic recombination frequency may differ in crosses between different strains of mice and to some extent in different human populations. These factors must be considered when utilizing information from mammalian linkage maps.

Construction of Linkage Maps

The actual linkage maps are derived by a process of linkage analysis or segregation analysis, in which the likelihood of a nonrandom relationship between various loci are measured and maps determined and/or verified by the application of sophisticated statistical algorithms. For human linkage maps, the use of the logarithm of the odds (LOD) score provides a measure of the strength of linkage relationships at the optimized recombination distance. The LOD score is the $\log_{10}$ of the likelihood ratio that two loci are linked and separated by a specific genetic distance, divided by the likelihood that the observed results would be obtained if the two loci were not linked. Although the concept is simple, the actual generation of the maps requires advanced algorithms that can determine LOD scores for a continuous range of possible recombination frequencies (termed θ) between markers in multiple complex pedigrees (e.g., Lathrop et al., 1985) (http://linkage.rockefeller.edu/bib/algorithms/). The entire human linkage map contains a sex-averaged map distance of about 3500 cM, with each of the 22 individual autosomal chromosomes and X chromosome varying considerably in genetic distance.

For the mouse, analysis of haplotypes in defined crosses has provided the most accurate relationship between markers. This analysis simply involves minimizing the total number of crossover events between linked loci. Here the observation that positive interference (the decreased frequency of crossover events occurring near other crossover events) is very strong in the mouse provides even more confidence in relative gene orders determined in a single cross within small (<10 cM) intervals. The mouse linkage map of each of 19 mouse autosomes and the X chromosome is approximately 1500 cM.

In general, the confidence in a map position can be estimated and described by a variety of algorithms. These include tests that determine: (1) the likelihood of alternative gene orders; (2) a LOD3 interval (indicating which positions are 1000-fold more likely than alternative positions); and (3) a Bayesian 95% interval to describe the limits of a particular genetic mapping. In the mouse a standard error formula, $[r(1-r)/n]1/2$ where r is the recombination frequency and n is the population number, is also commonly used.

Composite Linkage Maps

In both human and mouse, linkage maps have been compiled that contain thousands of markers. Linkage maps from most other mammalian species currently have limited numbers of precisely defined markers and will not be discussed further. Perhaps the most useful human and mouse linkage maps are those in which composite maps have been developed that contain information from a wide variety of studies, and include traits as well as genes and 'anonymous markers.' For humans, this information can be obtained from the internet in a variety of forms from multiple sites including: the Genome Data Base at http://gdbwww.gdb. org/gdb/gdbtop.html, the Genethon Human Genome Research Centre (http://www.genethon.fr/genethon_en.html/), the Cooperative Human Linkage Consortium (http://www.chlc.org/), and the Marshfield Center for Medical Research (http://www.marshmed. org/genetics/), as well as chromosome-specific web sites. In the mouse, this information can be obtained from the Mouse Genome Informatics (MGI) web site (http://www.informatics.jax.org/) including the mouse chromosome committee report composite maps (http://www.informatics.jax.org/bin/ccr/index).

The composite linkage maps include a wide array of markers defined by many different techniques. The most common method relies on examining length variation of polymerase chain reaction (PCR) amplified segments that contain microsatellite repetitive elements. These common repetitive elements are simple tandem sequence repeats (SSRs) of primarily di-, tri-, or tetranucleotides. Other assays that have been used for these linkage maps include detection of variable number of tandem repeat polymorphisms (VNTRs), restriction fragment length polymophisms (RFLPs), and other measurements of single nucleotide sequence variation including polymorphisms defined by random sequence oligonucleotide primers and many other methods.

For human maps, the use of the CEPH families (see http://www.cephb.fr/), a set of complex pedigrees that were distributed as part of a major effort in human chromosomal map building, provided a measure of integration between laboratories and quality control in the development of genome-wide linkage maps. For the mouse, several large mapping panels were developed, in which many markers were mapped relative to each other, and have been instrumental in the advent of reasonable, representative composite maps. However, whenever information from different sources and crosses or families is combined there is some uncertainty in the precise relationship between markers.

For human linkage maps, the vast majority of markers are microsatellite repetitive elements. Since these are highly polymorphic, informative meioses can be readily identified in a substantial percentage of the samples analyzed. Since relatively few genes have been characterized to date as having frequent polymorphisms, there is a scarcity of genes in present linkage maps derived from analysis of meiotic recombination. Attempts at integration of these linkage maps and the position of genes and expressed sequence tags (ESTs) will be discussed later.

In the mouse, several mapping panels of interspecific or intersubspecific backcross or intercross mice have been used to generate maps containing hundreds or thousands of markers (The data for many of these panels is available from MGI in the mapping data section (http://www.informatics.jax.org/crossdata.html)). In particular, crosses between laboratory strains of mice (predominantly *Mus musculus domesticus*) and the *Mus spretus* species have been used. These mouse species are estimated to have diverged over the course of 3 million years and are sufficiently different that polymorphisms can be detected in virtually all genes or cloned nonrepetitive genomic sequences by analysis of RFLPs. Thus, backcross progeny of these crosses can be typed using large numbers of markers in the same panel of potentially informative meioses. Resultant individual, interspecific cross-linkage maps provide the most accurate mammalian genetic maps and, more importantly, include genes. However, it must be stressed that the map positions in composite linkage maps in the mouse combine information from a wide variety of disparate crosses or other genetic techniques. These maps typically include data from recombinant inbred strain analyses, as well as backcross and intercross breeding schemes. In addition, since recombination frequency between disparate strains of mice may vary considerably, the relative position of many genes in composite maps cannot generally be regarded as definitive. Depending on the actual data as to how positions have been interpolated (including whether additional information, e.g., progeny testing data is available or utilized), there may be a large 95% confidence interval for which a marker is positioned relative to other markers, genes, or traits. A measure of the confidence of a particular chromosomal linkage map position for each entry is given in the Mouse Chromosome Committee reports (*Encyclopedia of the Mouse Genome VI*, 1997, or see http://www. informatics.jax.org/bin/ccr/index).

Utilizing Human Linkage Maps

At present, the prevailing technique utilized for linkage studies of human diseases is to map the trait with respect to highly polymorphic microsatellites. Most of these microsatellite markers are not specifically associated with coding sequences (i.e., they are not derived from relatively small clones containing known genes). However, for initial localization of a trait, the critical factors are how polymorphic a marker is, how easily it can be typed (i.e., for reasonably high throughput), and how well the marker has been previously mapped. Many of the internet sites discussed previously provide information on the heterozygosity scores of markers, as well as data indicating the two-point or multipoint linkage relationships between many of the markers. Thus, markers can be chosen that can enable genome-wide scans for susceptibility genes in both Mendelian and certain complex genetic diseases. In the future, it is possible that high throughput typing of single nucleotide polymorphism (Wang *et al.*, 1998, and see http://www.genome.wi.mit.edu/SNP/human/index.html) may also be used in either refining regional linkage studies or in genome-wide scans. However, the linkage or physical relationship between these polymorphisms and those currently used will be necessary for optimal utilization of linkage maps.

Integration of Human Linkage Maps with Genes and Physical Maps

As discussed above, the human linkage map is largely devoid of genes. It is, however, often very useful to know the relationship between genes and the anonymous polymorphic markers in the linkage map. A major effort utilizing radiation hybrid mapping has recently provided a good framework for integrating the position of genes with respect to the genetic linkage maps. Human radiation hybrids allow the development of a type of linkage map that is based on whether any specific segment of DNA from an irradiated human donor cell has been retained in cross-species somatic cell hybrids. In these maps, the distance between markers is measured in centirays (cR), where for each unit there is a 1% probability of X-ray-induced breakage for a specific dosage in rads. These maps include genes, as well as anonymous

sequences including microsatellites. The results of a consortium of many investigators allowed the relative ordering of several thousand genes, ESTs, and other sequence-tagged sites including the polymorphic microsatellites used in genetic linkage maps (Schuler *et al.*, 1996 and see http://www.ncbi.nlm.nih.gov/genemap/). Therefore, it is possible to determine a probable range of the Genethon or other marker positions with respect to ESTs. This can be extremely useful in searching for candidate ESTs for human diseases, if, for example, a critical interval containing the putative 'disease' gene has been defined using markers included in the Genethon map. If the markers are not in the Genethon map, then finding common markers between maps and interpolating may be necessary. The relationship among anonymous markers can also be examined in other radiation hybrid maps, including the large compilation of radiation hybrid data at the Stanford University and Whitehead Institutes genome sites (http://www-shgc.stanford.edu/RH/index.html, http://carbon.wi.mit.edu:8000/cgi-bin/contig/phys_map). It should be noted that for most radiation hybrid mapping results, the high confidence groupings, bins in which relative order is 1000-fold more likely than other orders, correspond to approximately 10 cM ranges of the meiotic recombination defined standard linkage maps discussed above.

Other ongoing efforts have incorporated many of the genetic linkage map markers in the development of contigs (e.g., http://www-genome.wi.mit.edu/, http://www.cephb.fr/bio/ceph-genethon-map.html, http://www.nhgri.nih.gov/DIR/GTB/CHR7/, http://gc.bcm.tmc.edu:8088/bio/yac_search.html). Thus, the polymorphic markers used in linkage can begin to be integrated within the physical map. As the human genome project proceeds through its current sequence-intensive step, establishing the precise position of these markers will become possible. However, linkage maps will, as discussed above, continue to be useful in the foreseeable future for efforts at positional cloning of genes corresponding to traits. The increasing availability of sequencing data will obviously provide another method for the integration of putative coding sequences and single nucleotide polymorphisms with the genetic linkage maps.

Utilizing Mouse Linkage Maps

As discussed above, linkage maps of the mouse genome contain a variety of markers. For microsatellites, the majority have been placed in a single-cross-defined linkage map in which there is strong confidence in the relative positions of most of the markers for even small genomic intervals (1–2 cM; http://carbon.wi.mit.edu:8000/cgi-bin/mouse/index).

Other linkage maps contain large numbers of genes and enough microsatellite markers to allow reasonable integration with other markers not included in these maps. The chromosomal positions of these markers and genes can be used to map specific traits in a manner similar to that employed in human genome screening. The cross-specific linkage maps and their derived composite maps (discussed above) have thus allowed a large number of traits to be placed in specific intervals and have facilitated positional cloning projects.

Using Linkage Maps for Defining Homology Relationships

Homology relationships can be very valuable in further utilizing linkage maps for linking genes to phenotypes. Many studies have indicated that mammalian genomes are mostly composed of chromosome segments that have been conserved over 100 million years of evolution. Review of human–mouse homology relationships suggests that there are over 200 such segments (DeBry and Seldin, 1996 and see http://www.ncbi.nlm.nih.gov/Homology/). For Mendelian traits, several examples can be cited in which information from either human or mouse studies has expedited the molecular definition of disease in the other species. Although it is less certain that these relationships will be as useful for complex genetic diseases, it can provide the first insight into whether or not animal models are likely to be a major adjunct to human studies.

Homology relationships can allow the use of what might be termed 'virtual maps,' in which all of the genes or ESTs located in a disparate species can be putatively placed in a linkage map of the species in question. This can suggest candidate genes for traits or markers that might be used to test for linkage in the other species. However, it is important to apply some critical evaluation of homology data to provide some assurance that orthologous (the same gene in both species) genes/ESTs are utilized, since related (paralogous) genes can result in incorrect interpretations. Finally, many of the borders of these homology relationships are not well defined and will require further resolution in one or the other species.

References

DeBry RW and Seldin MF (1996) Human/mouse homology relationships. *Genomics* 33: 337–351.

Encyclopedia of the Mouse Genome VI (1997) Mammalian Genome. 7 Spec. No.: S1–388.

Lathrop GM, Lalouel JM, Julier C and Ott J (1985) Multilocus linkage analysis in humans: detection of linkage and estimation of recombination. *American Journal of Human Genetics* 37(3): 482–498.

Schuler GD, Boguski MS, Hudson TJ et al. (1996) A gene map of the human genome. *Science* 274(5287): 547–558.

Wang DG, Fan JB, Siao CJ et al. (1998) Large-scale identification, mapping and genotyping of single-nucleotide polymorphisms in the human genome. *Science* 280(5366): 1077–1082.

See also: **DNA Cloning;** *Mus musculus*

Linker DNA

I Schildkraut

Copyright © 2001 Academic Press
doi: 10.1006/rwgn.2001.0770

Linker DNA is a short self-complementary palindromic DNA molecule which forms a blunt end duplex containing a recognition sequence for a restriction endonuclease. The linker DNA is generally blunt-end ligated between two blunt-ended DNA fragments to introduce a restriction site.

See also: **Restriction Endonuclease**

Lipoma and Uterine Leiomyoma

M M R Petit, W J M Van de Ven and E Jansen

Copyright © 2001 Academic Press
doi: 10.1006/rwgn.2001.1587

It is evident that genetics will have a major influence on everyday life in our modern society, and in areas such as predictive and therapeutic medicine and biotechnology the impact will be profound. Identification and characterization of the genes involved in genetic diseases have already made significant contributions to diagnosis and to both an understanding of therapy and suggestions for novel therapies (including gene therapy). It is well established that in benign as well as malignant tumors, recurrent genetic aberrations are regularly found by cytogenetic analysis, which are often translocations involving well-defined chromosome regions. Sometimes, such genetic lesions are characteristic of a particular tumor type suggesting the involvement of tumor type-specific genes, and occasionally the same chromosomal region is affected in a number of tumors and this possibly indicates a common genetic denominator in these diseases. Since such recurrent cytogenetic aberrations are often the sole chromosomal anomalies present, they are believed to represent critical molecular triggers of aberrant growth control in tumorigenesis.

Molecular evaluation of the chromosome breakpoint regions in two benign solid tumor types of mesenchymal origin, i.e., lipomas and uterine leiomyomas, has recently led to the identification of the first genes that are frequently targeted in these tumors by chromosomal defects. The architectural transcription factor genes, *HMGIC* and *HMGI(Y)*, appear to be important targets. Furthermore, preferential translocation partner genes of *HMGIC* have also been identified. In uterine leiomyoma, it is almost exclusively the *RAD51L1* gene on the long arm of chromosome 14. Structurally, this gene is listed as a member of the *recA/RAD51* recombination–repair gene family and its protein product displays protein kinase activity. The preferential translocation partner of *HMGIC* in lipoma is the *LPP* gene on chromosome 3, which encodes a LIM protein that enables communication between sites of cell adhesion and the cell nucleus. It should be noted that the *HMGIC* gene is also targeted by chromosomal aberrations in a variety of other benign solid tumors, including pleomorphic adenomas of the salivary glands, hamartomas of lung and breast, endometrial polyps, hemangiopericytomas, fibroadenomas of the breast, and chondromatous tumors. It clearly is a common genetic denominator in benign mesenchymal tumor formation. Their precise functions in tumor development remain to be established but these recently discovered genes form reliable starting points for further molecular genetic studies of these lesions. Molecular cytogenetic data of both uterine leiomyoma and lipoma are discussed in more detail in this article.

Uterine Leiomyoma

Pathology of Leiomyomas of the Uterus

Leiomyomas or myomas are benign tumors of smooth muscle cells and they are most frequently found in the genitourinary and gastrointestinal tracts and less frequently in the skin and in deep soft tissues (Enzinger and Weiss, 1995). Uterine leiomyomas (fibroids) represent the most common pathological growth in the female reproductive tract, occurring with a reported incidence of up to 77% of all women of reproductive age (Cramer and Patel, 1990). However, these mesenchymal tumors are rare in women below the age of 18 and, furthermore, they are more frequent in black than in white women. Affected women complain of fibroid-related symptoms, e.g., abnormal uterine bleeding, pelvic pain, or urinary dysfunction. Fibroids may also interfere with pregnancy, leading to premature delivery or even fetal wastage. Since the current long-term nonsurgical management of

leiomyomas (hormone replacement therapy) is associated with major side effects, more and more women are directly seeking some form of surgery to remove their fibroids. This has led to a situation in which in the United States alone, uterine leiomyomas are the leading indication for about 300 000 hysterectomies performed annually.

Cytogenetics of Uterine Leiomyoma

Besides a normal karyotype, which is being found in approximately 70% of the cases investigated, several cytogenetically abnormal subgroups (**Figure 1**) can be distinguished (Mitelman, 1998). Excluding the group with random changes, one of the largest cytogenetic subgroups (comprising approximately 25% of the cytogenetically abnormal tumors) is characterized by the involvement of 12q14–q15 and/or 14q23–q24, mainly as t(12;14)(q14–q15;q23–q24). Another subgroup, with a similar incidence, contains deletions involving the long arm of chromosome 7, with region q21–q22 being the commonly involved chromosomal segment. Another subset of uterine leiomyomas is characterized by numerical aberrations, mainly trisomy 12. This trisomy is found in approximately 10% of the cytogenetically abnormal cases. Furthermore, chromosome 6p21-pter has been found to be recurrently involved in roughly 5% of the cases studied. Finally, a small percentage (approximately 3.5%) of uterine leiomyomas shows t(1;2)(p36;p24). As will be discussed below, chromosome 12q13–q15 anomalies are frequently found in lipomas. In fact, as outlined above, they are encountered in a variety of other benign solid tumors as well. In general, these karyotypic changes are balanced and simple. The fact that these translocations are often the first or sole cytogenetically visible anomalies suggests that, pathogenetically, they are of critical importance in these tumors.

Genes Affected in Uterine Leiomyoma

Implication of the high mobility group protein genes HMGIC and HMGI(Y)

Using a classical positional cloning approach, the chromosome 12q14–q15 breakpoints in a number of uterine leiomyomas were mapped first within a 1.7 Mb DNA region on the long arm of chromosome 12. In subsequent FISH studies, it was conclusively demonstrated that many of the chromosome 12 breakpoints were clustering within a relatively small (175 kb) DNA segment, identifying it as a major target area. A single transcribed sequence was identified in this target area and it appeared to correspond to the human *HMGIC* gene (Schoenmakers *et al.*, 1995), which is a member of the high mobility group (HMG) protein gene family. The *HMGIC* gene (for

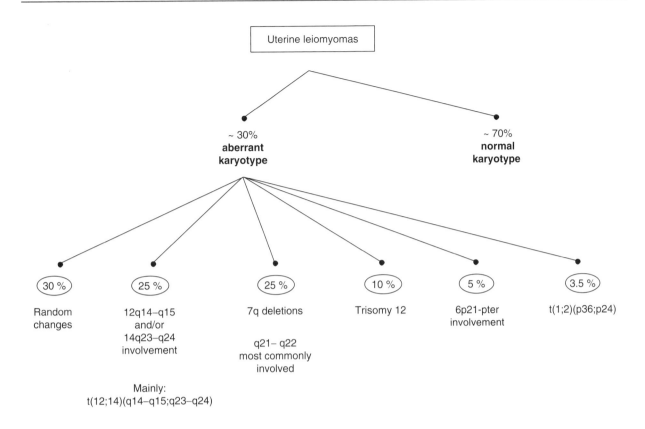

Figure 1 Cytogenetics of uterine leiomyomas. Schematic representation of the different cytogenetic subgroups of uterine leiomyomas.

review, see Jansen *et al.,* 1999) consists of five exons and spans about 175 kb (**Figure 2A**). The gene contains one large intron, i.e., intron 3, which spans about 140 kb. The HMGIC protein has three DNA-binding domains (a 9 basic amino acid DNA-binding motif, also referred to as the AT-hook) and an acidic C-terminal domain. The location of the breakpoints with respect to *HMGIC* is variable and has been found 5′ to the gene, in its 3′ nontranslated region, as well as in one of its introns. The intragenic breakpoints frequently occur in the large third intron. In such cases, the three DNA-binding domains in the N-terminal region of the protein become separated from the acidic, C-terminal domain. Furthermore, it is of interest to note that another member of the HMG protein gene family, i.e., the *HMGI(Y)* gene (**Figure 2B**), which maps at chromosome 6p21, is also implicated in uterine leiomyoma.

HMG proteins (Bustin *et al.,* 1990) are named after their fast electrophoretic migration at acidic pH, and were first discovered in the 1960s as contaminants in calf thymus histone H1 preparations. They are operationally defined as small (mol.wt < 30 kDa) and abundant, 2% TCA/2–5% perchloric acid-soluble, nonhistone proteins, extractable from chromatin with 0.35 M NaCl and having a high content of acidic and basic amino acid residues. Since this definition is based on physical and chemical rather than functional features, it may be clear that the HMG protein family is composed of an artificial group of proteins with possibly unrelated functions. Based on their primary structure, three subfamilies of HMG proteins can be distinguished, i.e., the HMG1/2, the HMG14/17, and the HMGI class, to which the proteins encoded by *HMGIC* and *HMGI(Y)* belong.

The HMGI subfamily consists of three members: HMGI, HMGY, and HMGIC (**Figure 2B**). HMGI and HMGY are isoforms resulting from differential processing of the same parental messenger RNA (mRNA). Except for a stretch of 11 contiguous amino acids, which are present in HMGI but not in HMGY, the two proteins, often referred to as HMGI(Y), are identical (**Figure 2B**). HMGI proteins (mol.wt around 10 kDa) have been shown to display a significant preference for the narrow minor groove of certain types of stretches of AT-rich, B-form DNA *in vitro*, and conserved (TATT)$_n$ motifs in the 3′ untranslated regions (UTR) of certain genes have been identified as preferential binding sites. Furthermore, HMGI proteins bind specifically to the AT-rich octamer sequence associated with a number of promoters and also to AT-rich regulatory elements of the ribosomal genes. However, it should

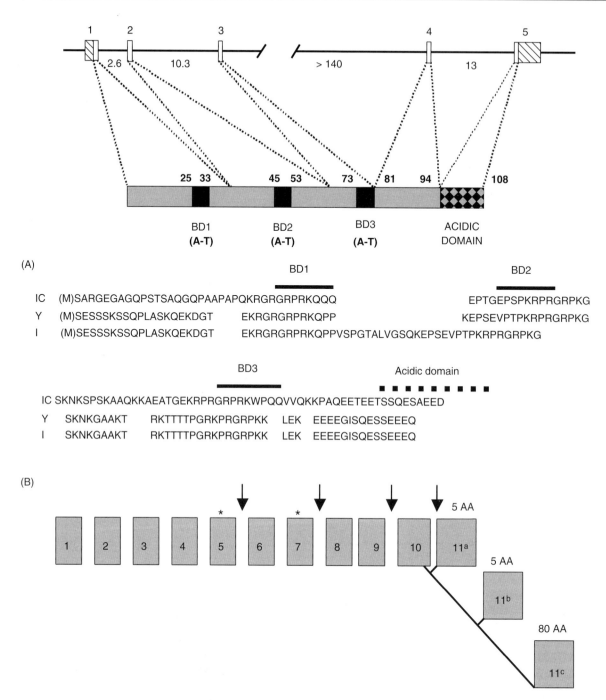

Figure 2 Structure of the human *HMGIC* gene and its protein product and the human *RAD51L1* gene products. (A) On the *HMGIC* gene map, the exons are depicted as boxes, with the 5'- and 3'-untranslated regions represented as shaded areas. The numbers below the map indicate the intron sizes in kilobase pairs. The dashed lines indicate which regions of the HMGIC protein are encoded by the individual exons and the amino acid numbering above the protein map marks the boundaries of the various DNA-binding and acidic domains. (B) HMGIC amino acid sequence aligned with HMGI and HMGY. (C) Schematic representation of the three alternative *RAD51L1* mRNA splice variants (exons are numbered). The relative position of two highly conserved nucleotide-binding Walker domains are marked by asterisks and the number of amino acids encoded by the three alternative terminal-coding exons are indicated. Arrows mark the positions of chromosome breakpoints found in the *RAD51L1* gene in various uterine leiomyomas.

be kept in mind that this preference for certain AT-rich stretches has been shown to be caused by recognition of substrate structure rather than nucleotide sequence.

As far as expression patterns of HMGI(Y) and HMGIC are concerned, there seems to be a link to cell proliferation. Expression of the *HMGIC* gene is tightly linked to growth, since it is mainly expressed during early development and in growing cells. Furthermore, it responds to serum induction as a delayed early response gene (Ayoubi *et al.*, 1999). Finally, homozygous disruption of the *Hmgic* gene in mice leads to the pygmy phenotype (Zhou *et al.*, 1995). The observation that the HMGI proteins are developmentally regulated and constitute abundant proteins might indicate that they could be involved in the regulation of many genes, some possibly involved in cell growth. The fact that HMGI(Y) is known to cause a more general regulatory effect on transcription through modification of chromatin structure by inducing DNA bends, thereby facilitating the assembly of transcriptionally active nucleoprotein complexes (Grosschedl *et al.*, 1994), has resulted in the definition of so-called 'architectural

transcription factors' (Wolffe, 1994; Lovell-Badge, 1995), of which HMGI(Y) is the founding member. Indeed, studies on the role of HMGI(Y) in the induction of *INF*β gene expression (Falvo *et al.*, 1995) point toward a more architectural role for HMGI(Y), and have resulted in the model that the HMGI proteins as a group, just like other architectural transcription factors, might function as 'facilitators' of gene expression (**Figure 3**). The intriguing question remains as to how particular genetic changes in such facilitators result in aberrant cell proliferation of a benign nature. Identification of the spectrum of their target genes is an important objective for future research of the HMGIC proteins.

Implication of RAD51L1, the Chromosome 14 Translocation Partner Gene of HMGIC

Using a positional cloning approach, the *RAD51L1* gene on human chromosome 14q23–q24 was recently identified as the almost unique translocation partner of *HMGIC* in uterine leiomyomas (Schoenmakers *et al.*, 1999). The *RAD51L1* gene (also known as *R51H2* and *hREC2*) is a member of the *recA/RAD51* recombination–repair gene family. The gene, which contains

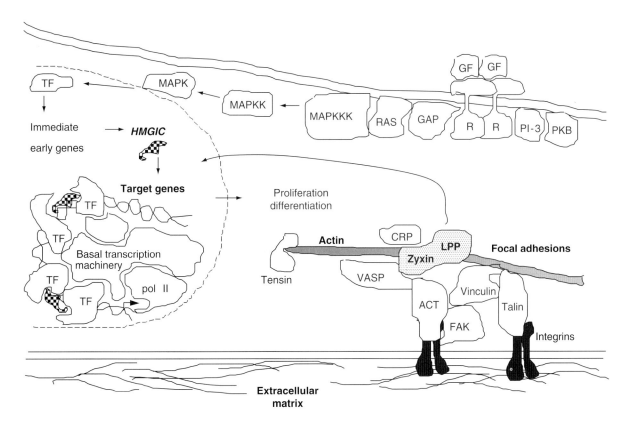

Figure 3 Signal transduction model for HMGIC and LPP. Schematic representation of the delayed, early response activation of the *HMGIC* gene via the growth factor (GF)–receptor (R)-mediated activation of MAP kinases (MAPK). The HMGIC protein will bind to regulatory regions of its respective target genes. The bottom part of this figure shows the localization of wild-type LPP in focal adhesions of which other structural components are also shown.

11 exons, expresses three distinct mRNA isoforms, which differ only in the sequences of their last exons (exons 11) (**Figure 2C**). Two isoforms are broadly expressed and their different last exon sequences encode only five amino acids. The third isoform displays a highly restricted expression pattern but is expressed in the uterus. Studies of uterine leiomyomas seem to indicate that the pathogenetically critical sequences reside in the last coding exon (encoding 80 amino acids including a putative membrane anchor) of this third *RAD51L1* isoform. It appears that allelic knockout of the third splice variant of *RAD51L1*, resulting in expression of truncated and C-terminally altered RAD51L1 proteins, is a tumor-specific feature of uterine leiomyomas with t(12;14)(q15;q23–q24) translocations. The precise physiological function(s) of the various isoforms of RAD51L1 in normal cells and the role(s) of their truncated variants in uterine leiomyoma remain to be elucidated. A highly related family member of RAD51L1, i.e., RAD51A, has been shown to promote ATP-dependent homologous pairing and strand transfer reactions *in vitro*, to play an essential role in mammalian cell viability, and to be linked etiologically to cancers, because of its interaction with p53, BCRA1, and BCRA2. However, until now, the typical recombinase activity of members of the *rec2A/RAD51* gene family could not be established for RAD51L1. On the other hand, it has been shown that overexpression of RAD51L1 in mammalian cells results in a delay in G1. Recently, it was reported that RAD51L1 exhibits protein kinase activity and is able to phosphorylate various substrates, including p53, cyclin E, and cdk2, but not a peptide substrate containing tyrosine residues only (Havre *et al.*, 2000).

Lipoma

Pathology of Lipomas

Lipomas are benign neoplasms of adipose tissue. Histologically, they belong to the group of lipomatous tumors that are classified as soft tissue tumors (Enzinger and Weiss, 1995). Several types of benign lipomatous tumors can be distinguished such as ordinary benign lipoma, angiolipoma, fibrolipoma, hibernoma, lipoblastoma, spindle cell/pleomorphic lipoma, and atypical lipomatous tumors. Lipomas are one of the most common soft tissue tumors and form part of the daily practice of many surgical pathologists. With rare exceptions they may occur at any age and at almost any anatomical location. However, in general, most of the lipomas become apparent between the fourth and sixth decade and most of these are found in the subcutaneous tissues of the upper back, neck, shoulder, and abdomen, followed in frequency by the proximal portions of the extremities. In a minority of cases, multiple lesions are observed, but mostly patients have one tumor. Ordinary lipomas (referred to as 'lipoma' throughout the rest of this article) are generally asymptomatic, and are mainly brought to the attention of a physician if they reach a large size or cause cosmetic problems or complications because of their anatomical site. As a consequence of this, the reported clinical incidence is probably much lower than the actual incidence. Microscopically, there is little difference between lipomas and surrounding fat tissue. Like fat tissue, lipomas are mainly composed of mature fat cells, but the cells vary slightly in size and shape and are somewhat larger. The tumors are usually thinly encapsulated and have a distinct lobular pattern. All tumors are well vascularized. Subcutaneous lipomas vary in size from a few millimeters to 5 cm or more. Occasionally, 'giant' cases are reported in the literature, measuring at least 20 cm (for review, Sanchez *et al.*, 1993). Deep-seated lipomas are very rare as compared to their cutaneous counterparts. These lipomas have been detected in numerous sites of the body. They are often detected at a later stage of development, and therefore tend to be larger than superficial lipomas.

Cytogenetics of Lipomas

In the past two decades, lipomas have been studied extensively by cytogenetic analysis (Sreekantaiah, 1998). These studies have demonstrated that more than 60% of solitary lipomas have an aberrant karyotype (Mitelman, 1998) (**Figure 4**). In two-thirds of these, chromosomal region 12q13–q15 is affected resulting from various types of chromosome aberrations, mainly translocations. In a quarter of these cases, chromosome 3 at bands q27–q28 was found as the translocation partner of chromosome region 12q13–q15. This means that the most consistent chromosomal aberration in lipomas is represented by t(3;12)(q27–q28;q13–q15), being present in about 10% of all solitary lipomas. Studies on the remaining cases indicated that most if not all chromosomes are able to act as translocation partner of 12q13–q15. The chromosome regions that are most frequently involved are 1p34–p32, 2p24–p21, 5q33, 21q21–q22, 2q35, 1p36, 11q13, and 13q12–q14. Finally, supernumerary ring chromosomes as well as complex karyotypes involving chromosome region 12q13–q15 have been reported.

Lipomas without involvement of chromosome region 12q13–q15, most often display chromosome 13q or chromosome region 6p23–p21 rearrangements (**Figure 4**). Abnormalities of 13q include deletions, with del(13)(q12q22) being the most frequently found, and translocations. Rearrangements of 6p23–p21 are usually due to translocations, inversions, or insertions. In addition, rearrangements of chromosome region 1p36 have been found as well as supernumerary ring chromosomes and complex karyotypes.

Apart from the fact that normal karyotypes are more common in patients younger than 30 years old, there appears to be no significant association between the cytogenetic pattern and patient sex, age, or tumor localization, size, or depth. Therefore, to date, the pathogenetic basis and clinicopathological relevance of the cytogenetic subtypes among lipomas remain unexplained (Willén *et al.*, 1998).

About 5–8% of all patients with lipomas have multiple tumors, varying in number from a few to several hundred lesions. These lipomas are indistinguishable from their solitary counterparts. They occur predominantly in the upper half of the body, usually in the back, shoulder, and upper arm. There is a definite hereditary trait in about one-third of patients with this condition. Cytogenetic analysis of these kinds of tumors revealed that most multiple lipomas (98%) have a normal karyotype.

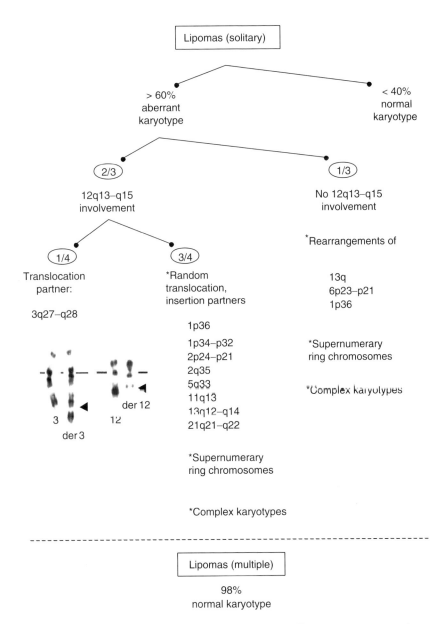

Figure 4 Cytogenetics of lipomas. Schematic representation of the different cytogenetic subgroups of lipomas. The inserted picture represents the partial karyotype from a lipoma showing a t(3;12)(q27–q28;q13–q15). Arrowheads indicate breakpoints.

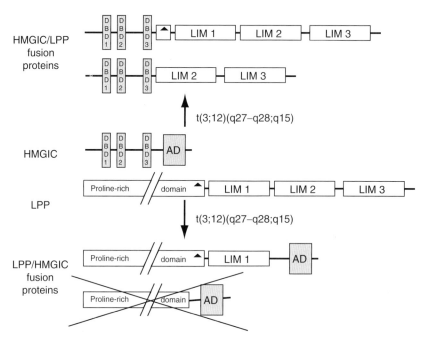

Figure 5 Schematic representation of wild-type HMGIC and LPP proteins and related fusion proteins predicted to be expressed in lipomas. The wild-type LPP protein is predicted to consist of a proline-rich N-terminal domain and three LIM domains in its C-terminal region. HMGIC consists of three N-terminal DNA-binding domains and an acidic C-terminal tail domain. Hybrid transcripts encoding the two variants of HMGIC/LPP fusion proteins (upper part) and the reciprocal LPP/HMGIC fusion protein (lower part) were detected in RT-PCR analysis of primary lipomas and lipoma cell lines. DBD, DNA-binding domain; AD, acidic domain; LIM, LIM domain.

Molecular Genetics of Lipomas with t(3;12)(q27–q28;q13–q15)

The most consistent chromosomal aberration in lipomas is represented by t(3;12)(q27–q28;q13–q15), found in about 10% of all solitary lipomas. It was established that the genes *HMGIC* at 12q15 and *LPP* (LIM containing lipoma preferred partner) at 3q27–q28 are affected by this preferential 3;12-translocation (Petit *et al.*, 1996). Furthermore, it was demonstrated that as a direct result of this, *HMGIC/LPP* fusion transcripts are expressed in these tumors (**Figure 5**). The HMGIC protein is described above see also (**Figure 2**).

The LPP protein belongs to a recently identified family of proteins, also comprising Zyxin and TRIP6 (Beckerle, 1997). They are all proline-rich in their N-terminal region while in their C-terminal region they have three LIM domains that are capable of mediating protein–protein interactions. In lipomas, two alternative *HMGIC/LPP* hybrid transcripts have been detected so far. They encode fusion proteins containing the three DNA-binding domains of HMGIC followed by: (1) part of the proline-rich domain and all three LIM domains of LPP; or more frequently (2) the two most C-terminal LIM domains (LIM 2–3) of LPP (**Figure 5**).

Recent findings suggest that LPP might play a dual role in the organization of the actin cytoskeleton and in gene regulation (Petit *et al.*, 2000). LPP is able to shuttle between the nuclear compartment and the sites of cell adhesion. At the sites of cell adhesion, more and more proteins are being identified that not only play a role in maintaining cell shape and motility but that, in addition to these structural functions, are also implicated in signaling events. In recent years, there has been increasing recognition that signaling events do not take place freely in the cytosol of the cell but, rather, occur in physically and functionally distinct signaling units. These signaling complexes may be organized around scaffold/adaptor proteins containing multiple protein–protein interaction motifs (Pawson and Scott, 1997). Because of this dual function, these proteins have to interact, via multiple binding motifs, with components of both the actin cytoskeleton and signaling pathways that regulate, for example, gene expression. Therefore, it is important to note that in contrast to wild-type LPP, the tumor-specific HMGIC/LPP fusion proteins are exclusively located in the nucleus and this may result in aberrant signaling of interacting proteins. In the case of the scaffold protein LPP, interacting proteins have been identified. One of these LPP partner proteins is also a scaffold protein, since it contains multiple protein–protein interaction domains. This protein is a member of the novel family of LAP proteins (Bilder *et al.*, 2000) and interacts via PDZ domains with the C-terminus of LPP. The detailed analysis of LPP and interacting proteins will reveal the exact nature of the signaling pathway in which LPP participates (**Figure 3**).

In summary, LPP participates in a novel signal transduction pathway between the sites of cell adhesion and the nucleus. The ectopic expression of tumor-specific HMGIC/LPP fusion proteins could deregulate this pathway, resulting in aberrant growth. Defining the physiological function of this signaling cascade in the regulating of growth and differentiation will provide insight into the molecular mechanism of benign solid tumor formation and may be instrumental in the development of potential therapeutic agents that interfer with tumor growth.

References

Ayoubi TAY, Jansen E, Meulemans SMP and Van de Ven WJM (1999) Regulation of HMGIC expression: an architectural transcription factor involved in growth control and development. *Oncogene* 18: 5076–5087.

Beckerle MC (1997) Zyxin: zinc fingers at sites of cell adhesion. *BioEssays* 19: 949–957.

Bilder D, Birnbaum D, Borg J-P et al. (2000) Collective nomenclature for LAP proteins. *Nature Cell Biology* 2: E114.

Bustin M, Lehn DA and Landsman D (1990) Structural features of the HMG chromosomal proteins and their genes. *Biochimica et Biophysica Acta* 1049: 231–243.

Cramer SF and Patel A (1990) The frequency of uterine leiomyomas. *American Journal of Clinical Pathology* 94: 435–438.

Enzinger FM and Weiss SW (1995) *Soft Tissue Tumors*. St Louis, MO: CV Mosby.

Falvo JV, Thanos D and Maniatis T (1995) Reversal of intrinsic DNA bends in the INFβ gene enhancer by transcription factors and the architectural protein HMGI(Y). *Cell* 83: 1101–1111.

Grosschedl R, Giese K and Pagel J (1994) HMG domain proteins: architectural elements in the assembly of nucleoprotein structures. *Trends in Genetics* 10: 94–100.

Havre PA, Rice M, Ramos R and Kmiec EB (2000) HsRec2/Rad51L1, a protein influencing cell cycle progression, has protein kinase activity. *Experimental Cell Research* 254: 33–44.

Jansen E, Petit MMR, Schoenmakers EFPM, Ayoubi TAY and Van de Ven WJM (1999) High mobility group protein HMGI-C: a molecular target in solid tumor formation. *Gene Therapy and Molecular Biology*. 3: 387–395.

Lovell-Badge R (1995) Living with bad architecture. *Nature* 376: 725–726.

Mitelman F (ed.) (1998) *Catalog of Chromosome Aberrations in Cancer '98*, version 1. New York: John Wiley.

Pawson T and Scott JD (1997) Signaling through scaffold, anchoring, and adaptor proteins. *Science* 278: 2075–2080.

Petit MMR, Mols R, Schoenmakers EFPM, Mandahl N and Van de Ven WJM (1996) LPP, the preferred fusion partner gene of HMGIC in lipomas, is a novel member of the LIM protein gene family. *Genomics* 36: 118–129.

Petit MMR, Fradelizi J, Goldsteyn R et al. (2000) LPP, a novel actin cytoskeleton protein related to the zyxin family harbours transcriptional activation capacity. *Molecular Biology of the Cell* 11: 117–129.

Sanchez MR, Golomb FM, Moy JA and Potozkin JR (1993) Giant lipoma: case report and review of the literature. *Journal of the American Academy of Dermatology* 28: 266–268.

Schoenmakers EFPM, Wanschura S and Mols R et al. (1995) Recurrent rearrangements in the high mobility group protein gene HMGI-C in a variety of benign mesenchymal tumors. *Nature Genetics* 10: 436–444.

Schoenmakers EFPM, Huysmans C and Van de Ven WJM (1999) Allelic knockout of novel splice variants of human recombination repair gene *RAD51B* in t(12;14) uterine leiomyomas. *Cancer Research* 59: 19–23.

Sreekantaiah C (1998) The cytogenetic and molecular characterization of benign and malignant soft tissue tumors. *Cytogenetics and Cell Genetics* 82: 13–29.

Willén H, Akerman M, Dal Cin P et al. (1998) Comparison of chromosomal patterns with clinical features in 165 lipomas: a report of the CHAMP study group. *Cancer Genetics and Cytogenetics* 102: 46–49.

Wolffe A (1994) Architectural transcription factors. *Science* 264: 1100–1101.

Zhou X, Benson KF, Ashar HR and Chada K (1995) Mutation responsible for the mouse pygmy phenotype in the developmentally regulated factor HMGIC. *Nature* 376: 771–774.

See also: **Adenoma; Chromosome Aberrations; DNA-Binding Proteins; Dwarfism, in Mice; Fusion Gene; Fusion Proteins; Gene Expression; Gene Rearrangement in Eukaryotic Organisms; Gene Rearrangements, Prokaryotic; Gene Regulation; Human Genetics; LIM Domain Genes; Oncogenes; Protein Interaction Domains; Transcription; Translocation**

Liposarcoma

See: **Myxoid Liposarcoma and *FUS/TLS-CHOP* Fusion Genes**

Little, Clarence

L Silver

Copyright © 2001 Academic Press
doi: 10.1006/rwgn.2001.0771

The major contribution of Clarence Little was the realization of the need for, and development of, inbred genetically homogeneous lines of mice. The first mating to produce an inbred line was begun by Little in

1909, and resulted in the DBA strain, so-called because it carries mutant alleles at three coat color loci – dilute (*d*), brown (*b*), and non-agouti (*a*). In 1918, Little accepted a position at the Cold Spring Harbor Laboratory, and with colleagues that followed – including Leonell Strong, L. and E. C. MacDowell – developed the most famous early inbred lines of mice including B6, B10, C3H, CBA, and BALB/c. Although an original rationale for their development was to demonstrate the genetic basis for various forms of cancer, these inbred lines have played a crucial role in all areas of mouse genetics by allowing independent researchers to perform experiments on the same genetic material, which in turn allows results obtained in Japan to be compared directly with those obtained halfway around the world in Italy. A second, and more important, contribution of Little to mouse genetics was the role that he played in founding the Jackson Laboratory in Bar Harbor, Maine, and acting as its first Director. The Jackson Laboratory has become a crucial center for the research, education, and the actual production of laboratory mice for other researchers around the world.

See also: BALB/c Mouse; Coat Color Mutations, Animals; Inbred Strain

LMO Family of LIM-Only Genes

T H Rabbitts

Copyright © 2001 Academic Press
doi: 10.1006/rwgn.2001.1589

The *LMO* family of genes (**Table 1**) was uncovered by the association of *LMO1* (previously called *RBTN1* or *TTG1*) with the chromosomal translocation t(11;14) (p15;q11) human T cell acute leukemia (T-ALL). Using *LMO1* probes, the two related genes *LMO2* and *LMO3* were isolated (previously called *RBTN2* or *TTG2* and *RBTN3*, respectively), of which *LMO2* is located at the junction of the chromosomal translocation t(11;14)(p13;q11) also in T-ALL. Subsequently a fourth member of the family was discovered, *LMO4*, but this gene, like *LMO3*, has no known association with chromosomal translocations. Although the *LMO* genes are evolutionary descendents, their exon structures vary; *LMO1* and *LMO4* have four coding exons, whilst *LMO2* and *LMO3* have three coding exons. Conservation between homologs in different species is extremely high, suggestive of defined and crucial roles for these genes. Each of the LIM-only genes encode a protein essentially consisting of two

Table 1   The *LMO* family of genes and chromosomal location

Gene	Chromosome		Human translocation
	Man	**Mouse**	
LMO1	11p15	7	t(11;14)(p15;q11)
LMO2	11p13	2	t(11;14)(p13;q11)
LMO3	12 p12-13	6	nd
LMO4	1p22.3	3	nd

The *LMO* gene family (LIM-Only genes and previously called *RBTN* and *TTG* genes) has three known members. *LMO1* (previously *RBTN1/TTG1*) was identified first and then *LMO2* (previously *RBTN2/TTG2*) and *LMO3* (previously *RBTN3*). Subsequently, a fourth member, *LMO4*, was identified. *LMO1* and *LMO2* are both located on the short arm of chromosome 11 and are both involved in independent chromosomal translocations in human T cell acute leukemia. As yet, *LMO3* nor *LMO4* have not been found in association with any chromosomal translocations.

zinc-binding LIM domains. Short stretches at the N-termini of *LMO1* and *LMO2* have transcriptional transactivation activity.

LMO Genes Encode Transcriptional Regulators in Development

The unique feature of the *LMO*-derived protein sequences is that they are small proteins comprising two tandem LIM domains. These zinc-containing finger-like structures have structural similarities to the DNA-binding GATA fingers but as yet no case of a direct, specific LIM–DNA interaction has been reported; rather the function of this domain appears to be restricted to protein–protein interaction. Gene targeting showed that the mouse *Lmo2* gene is necessary for yolk sac erythropoiesis in mouse embryogenesis. Further the use of embryonic stem (ES) cells with null mutations of both alleles of *Lmo2* in chimeric mice has shown that adult hematopoiesis, including lymphopoiesis and myelopoiesis, fails completely in the absence of Lmo2. In addition, *Lmo2* is required for the remodeling of existing blood capillary endothelium into mature blood vessels (the process of angiogenesis) but not in the *de novo* formation of capillaries (vasculogenesis).

The Role of the LIM Domain in Protein Interaction

The LIM domain acts as a protein interaction module. For instance, Lmo2 and Tal1/Scl proteins (the latter is a basic helix–loop–helix protein) could interact directly with each other mediated through the LIM domains.

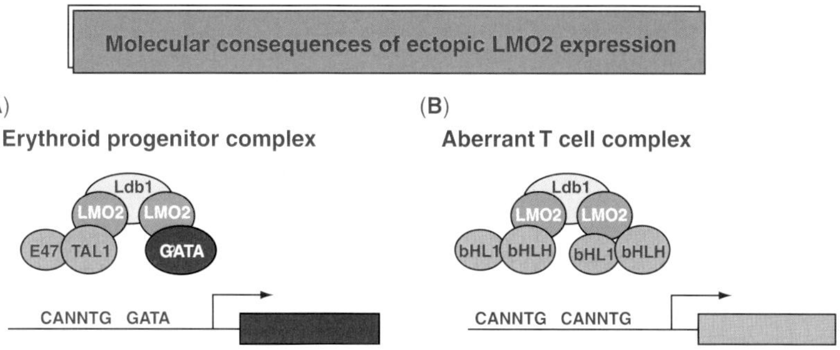

Figure 1 Lmo2 participates in DNA-binding complexes. (A) Erythroid Lmo2-containing complex. The Lmo2 protein interacts with Tall and with GATA1 in a complex comprising an Tall–E47 dimer, binding an E-box (CANNTG) and a GATA1 molecule, binding a GATA site, as part of an erythroid complex, which presumably regulates target genes. (B) T cell Lmo2-containing aberrant complex. An analogous DNA-binding complex comprises bHLH heterodimers linked by Lmo2 and Ldb1 proteins, binding to dual E-box sites.

The LIM domains of *LMO1* and *LMO2* can bind various proteins, such as GATA1, GATA2, and Ldb1/Nli1 protein. This array of interactions led to the observation that Lmo2 can be found in an oligomeric complex in erythroid cells which involves Tall, E47, Ldb1, and Gata-1 This complex is able to bind DNA through the GATA and bHLH components thereby recognizing a unique bipartite DNA sequence comprising an E-box separated by one helix turn from a GATA site, with Lmo2 and Ldb1 proteins seeming to bridge the bipartite DNA-binding complex (**Figure 1A**). Different Lmo2-containing complexes may exist in different hematopoietic cell types, which may differ in the types of protein factors expressed and may control distinct sets of target genes

Protein–protein interactions are crucial control points for normal cells and alterations in these are important components in tumorigenesis after chromosomal translocations have taken place. Gain-of-function transgenic mouse models of *LMO* gene expression induce clonal T cell leukaemia with a long latency, indicating that the transgenes are necessary but not sufficient to cause tumours. These mice show an accumulation of immature CD4⁻, CD8⁻, CD25⁺, CD44⁺ T cells in transgenic thymuses compared to nontransgenic littermates. Thus the role of Lmo2 in T-ALL is to cause an inhibition in T cell differentiation. T-ALL cells contain a Lmo2 complex which, like its analog in erythroid cells, binds to a bipartite DNA recognition site. Analysis of the components of this complex showed that E47–Tall bHLH heterodimeric elements were present as well as Lmo2 and the Ldb1 proteins (**Figure 1B**). A possible role for the E-box–E-box binding T cell complex is the regulation of specific sets of target genes, which, based on the difference in DNA-binding site, would differ from those putative genes controlled by the Lmo2–multimeric complex in hematopoietic cells.

Further Reading

Rabbitts TH (1994) Chromosomal translocations in human cancer. *Nature* 372: 143–149.

Rabbitts TH (1998) *LMO* T-cell translocation oncogenes typify genes activated by chromosomal translocations that alter transcription and developmental processes. *Genes and Development* 12: 2651–2657.

See also: Leukemia, Acute; Translocation

Locus

L Silver

doi: 10.1006/rwgn.2001.0772

A locus is any location on a chromosome, or any region of genomic DNA (of any length from a few base pairs to a megabase-size region containing a large gene family), that is considered to be a discrete genetic unit for the purpose of formal linkage analysis or molecular genetic studies.

See also: Alleles

LOD Score

doi: 10.1006/rwgn.2001.1898

The LOD score ('logarithm of the odds' score) is a statistical test for measuring the probability that there is linkage of loci. For non-X-linked genetic disorders in humans a LOD score of +3 (1000:1) is generally taken to indicate linkage (compared to the 50:1 probability that any random pair of loci will be unlinked).

See also: **Linkage**

Long-Period Interspersion

Long-period interspersion is a genomic pattern in which long stretches of moderately repetitive and nonrepetitive DNA alternate.

See also: **Genome Organization**

Long Terminal Repeats (LTRs)

Long terminal repeats (LTRs) are identical DNA sequences, several hundred nucleotides in length, found at the ends of transposons and retrovirus-derived DNA. LTRs contain inverted repeats and are thought to play an essential role in the integration of the transposon or provirus into the host DNA. In proviruses the upstream LTR acts as a promoter and enhancer and the downstream LTR as a polyadenylation site.

See also: **Provirus; Retroviruses; Transposable Elements**

Loss of Heterozygosity (LOH)

P Rabbitts

The development of tumors is associated with genetic damage confined to the cells of the tumor. This genetic damage can be visualized by examination of the tumor karyotype. Solid tumors, particularly those of epithelial origin, are characterized by highly aneuploid karyotypes with deletions as a common, frequently tumor-specific feature. If a patient's normal and tumor DNA are compared at a locus known to be heterozygous in that patient's normal DNA, it is possible to determine whether the tumor DNA has suffered genetic loss (deletion) encompassing that locus. If it has, only one of the two alleles will be detectable, and the locus will appear to be homozygous in the tumor and will show loss of heterozygosity (LOH).

Sources of Heterozygosity and their Detection

Within the mammalian genome, the majority of DNA is not involved in coding for proteins. Lack of selection pressure on this noncoding DNA allows inconsequential mutations to accrue. A locus at which the two parental alleles differ because of mutation is described as heterozygous/polymorphic. Single-nucleotide polymorphisms (SNPs) which form part of the recognition site for restriction enzymes were the first source of heterozygosity to be exploited for LOH analysis: first by Southern blotting, comparing normal and tumor DNA digested with the appropriate restriction enzyme, and then using PCR to amplify the region flanking the polymorphism followed by digestion of the PCR product with the restriction enzyme. However, the source of polymorphism most often used now exploits the observation that repetitive DNA occurs frequently in mammalian genomes. This DNA is often arranged in tandem repeat units, ranging in size from 8 to 50 bp, referred to as variable number of tandem repeats (VNTRs) or minisatellites. Of most value for LOH analysis are the repeat units ranging from 2 to 6 bp called 'microsatellites.' Human populations are highly polymorphous in the number of these repeats, such that the average rate of heterozygosity is more than 70%. Furthermore they are abundant and evenly distributed throughout the human genome, making them ideal genetic markers. They are detected by size fractionation after amplification by PCR using priming sites which flank the repeat region.

Recently there has been renewed interest in SNPs other than those involved in restriction enzyme sites. These are widely and evenly distributed throughout the human genome. Their information content is not as high as microsatellites, since they are biallelic, but the single base-change difference is much more amenable to high-throughput detection than the size differences of microsatellites, and they are likely to be the markers of choice for future genetic analyses, including LOH.

LOH and Location of Tumor Suppressor Genes

Tumor suppressor genes are recessive and require inactivation of both alleles for a phenotypic effect. Inactivation is frequently by mutation of one allele and loss, through chromosomal deletion, of the second. Chromosomal deletion is often first discovered by cytogenetic analysis of a few samples, usually of cell lines, and then confirmed by LOH analysis of paired tumor and normal DNA from a larger number of individual patients. This requires a group of polymorphic loci within and flanking the deleted region whose relative chromosomal positions are known. Many such loci have been identified and assigned a chromosomal location (D number in humans). By comparing the delineated stretch of LOH on the chromosome in individual patients in a large number of tumor/normal pairs, a common, minimally deleted region can be defined. This is sometimes small enough (less than 1 Mb) to allow the region to be investigated for genes which can be evaluated as tumor suppressor genes. This method of gene isolation, known as positional cloning, has been effective in the isolation or confirmation of a number of tumor suppressor genes.

Some tumors appear to have multiple but distinct regions of LOH on the same chromosome arm. It is uncertain whether all these regions of LOH indicate different tumor suppressor genes involved in the development of that tumor or whether some of the deletions occur as a consequence of the primary damage to the chromosome.

LOH Analysis and Clinical Research

Where tumor karyotyping is difficult, tumor DNA samples can be assessed for regions of allele loss by performing LOH analysis using evenly distributed markers for all chromosomes: 'allelotyping.' Different tumor types have regions of LOH in common, indicating a common defective gene in their etiology. This has been confirmed on isolation and mutation analysis of a gene within a deletion common to a variety of tumors. Despite this overlap, there are distinct patterns of LOH, sometimes associated with tumor progression, and thus loss of particular regions can have prognostic significance. The overall pattern of allele loss as determined by LOH analysis (together with any detected point mutations) can serve as a signature of an individual patient's tumor. The pattern of allele loss displayed by a tumor can be detected in material exfoliated from the tumor and sometimes in the patient's blood. This pattern, the signature, can be used as a means of following the course of disease during treatment and can indicate relapse before obvious clinical symptoms appear.

References

Mao L (2000) Microsatellite analysis. *Annals of the New York Academy of Sciences* 906: 55–62.

Human SNP Database: http://www- enome.wi.mit.edu/SNP/human/index.html

Wistuba II, Behvens C, Virmani AK *et al.* (2000) High resolution chromosome 3 allelotyping of human lung cancer and preneoplastic/preinvasive bronchial epithelium reveals multiple discontinuous sites of 3p allele loss and three regions of frequent breakpoints. *Cancer Research* 60: 1949–1960.

See also: **Chromosome Aberrations; Single Nucleotide Polymorphisms (SNPs); Tumor Suppressor Genes**

Lotus japonicus

J Stougaard

Copyright © 2001 Academic Press
doi: 10.1006/rwgn.2001.1665

Lotus japonicus is a model plant for the legumes. The Leguminosae (or Fabaceae) family is represented by approximately 18 000 species and is the third largest family of angiosperms. With around 700 genera divided into three subfamilies, Papilionoideae, Caesalpinioideae and Mimosoideae, the Leguminosae present a wealth of diversity. Several legumes, for example pea (*Pisum sativum*), soybean (*Glycine max*), peanut (*Arachis hypogaea*), and beans (*Phaseolus vulgaris*) are well-known and important crop plants. Others are cultivated as ornamentals, vegetables, pulses, or for production of protein, oil, and pharmaceuticals.

Lotus japonicus originates from East Asia and the species is distributed over the Japanese islands, the Korean peninsula, and east and central parts of China and has been reported from northern India, Pakistan, and Afghanistan. Two ecotypes 'Gifu' and 'Miyakojima' have been chosen for model studies. *Lotus japonicus* is a close relative of the tannin-containing tetraploid forage legume *L. corniculatus* (birdsfoot trefoil) cultivated for its antibloating properties. Phylogenetically, *L. japonicus* belongs to the tribe Loteae in Papilionoideae, the largest subfamily of the Leguminosae.

Many cultivated legumes like pea and soybean have complex genomes or are, for other reasons, not amenable to modern molecular genetic methods. Its favorable biological properties made *L. japonicus* the model plant of choice for classical and molecular genetic analysis of legumes. The qualities of *L. japonicus* are: a short seed-to-seed generation time, a small genome size of approximately 450 Mb, diploid genetics, six chromosome pairs, self-fertile flowers, ample seed production, small seeds, simple nonspiral seed pod, large flowers enabling manual crossing, described transformation procedures using *Agrobacterium tumefaciens* or *A. rhizogenes*, described *in vitro* tissue culture and regeneration procedures, effective nodulation and mycorrhization.

Most legumes develop root nodules in symbiosis with nitrogen-fixing soil bacteria belonging to the Rhizobiaceae, and nodulated legume plants can use atmospheric dinitrogen as their sole nitrogen source. The interaction between the bacterial microsymbionts and legumes is selective. Individual species of rhizobia have a characteristic host range allowing nodulation of a particular set of legume plants. *Mesorhizobium loti* and the broad host range *Rhizobium* sp. NGR234 induce nitrogen-fixing root nodules on *L. japonicus*. Roots of *L. japonicus* are also effectively colonized by symbiotic arbuscular mycorrhizal fungi, for example *Glomus intraradices* and *Gigaspora margarita*. These fungi invade the root tissue by intercellular and intracellular hyphal growth and form arbuscules in cortical cells where metabolic interchanges take place. Mycorrhizal hyphae increase the root surface and improves phosphor uptake.

Identification of single gene plant mutants impaired in both colonization by mycorrhizal fungi and rhizobial invasion demonstrates that the two interactions share common steps during the early infection processes. Extending this observation may open a broader approach to the understanding of plant–microbe interactions, where symbiotic studies not only contribute to realization of the potential of symbiosis, but also to our understanding of (for example) plant–pathogen interactions.

One of the interests of the plant science community is to use *L. japonicus* in the molecular genetic analysis of symbiosis. For this purpose, tools and resources for molecular analysis have been established. Insertion mutagenesis is possible with T-DNA or the maize transposon *Ac*, and EMS is effective for chemical mutagenesis. After mutant screening, more than 40 symbiotic loci have been identified. The phenotypes of these developmental plant mutants divide them roughly into three classes: non-nodulating mutants arrested in bacterial recognition or nodule initiation; nodule development mutants arrested at consecutive stages of the organogenic process; and autoregulatory mutants where the plant control of root nodule numbers is nonfunctional. Development of root nodules

can thus be divided into a series of genetically separable steps. For further studies the following genome resources are being developed: a general genetic map and bacterial artificial chromosomes (BAC) libraries for positional cloning of untagged mutants; recombinant inbred lines; and inventories of expressed sequence tags (ESTs) sampling the gene expression profiles from several tissues and growth conditions (www.Viazusa.or.jp/een/index.html).

Sequencing of the *L. japonicus* genome has been initiated. The sequences of the bacterial genes required for nodulation and nitrogen fixation located on the pSym plasmid of NGR234, and the complete genome of *Mesorhizobium loti*, are available, together with a wide selection of rhizobial mutants.

Like soybean, *L. japonicus* develops the determinate type of nodules. In contrast to for example pea nodules with a persistent meristem, the meristematic activity ceases early in determinate nodules developing on *L. japonicus*. After the initial phase with meristematic cell proliferation determinate nodules grows by expansion giving a typical spherical shape. All developmental stages from root hair curling to nodule senescence are consequently phased in time. Root nodule development is a rare example of induced and dispensable organ formation in plants. Nodulation mutants can be rescued on nitrogen containing nutrient solution and developmental control genes that would compromise plant development and completion of the life cycle in other organogenic processes could thus be identified from nodulation mutants. See www.mbio.aau.dk/nchp/table1.html for a list of literature on *L. japonicus*.

Further Reading

www.mbio.aau.dk/nchp/table1.html

See also: **Leguminosae; Nodulation Genes; Nodulins; Plant Development, Genetics of; Plant Embryogenesis, Genetics of;** *Rhizobium*; **Symbionts, Genetics of**

Lung Cancer, Chromosome Studies

P Rabbitts

Copyright © 2001 Academic Press
doi: 10.1006/rwgn.2001.1591

Lung tumors, like all common human epithelial tumors, have abnormal chromosomes, usually in both number and structure. Despite this polyploidy and aneuploidy, cytogenetic analysis has indentified a number of features which occur frequently in lung tumors, and further study of these abnormal regions has led to an understanding of molecular genetic changes underlying the development of lung cancer.

Cytogenetic Analysis

Tumor biopsies are a poor source of material for chromosome preparation, and most cytogenetic analysis has involved the use of cell lines established in tissue culture. Through the development of selective tissue culture media, hundreds of cell lines have been established, making lung tumors one of the most extensively studied types of tumor by karyotyping. Most work has used traditional G-banding, but more recently chromosome-specific paints have been used. Comparative genome hybridization in which tumor and normal DNA are competitively hybridized to normal chromosome spreads has been used to confirm and extend observations made by traditional cytogenetics. Molecular genetic analysis using DNA isolated from tumors has confirmed the existence of gene amplifications and chromosomal deletions and validated the cell lines as accurate representations of the tumors from which they were established.

Common Cytogenetic Abnormalities in Lung Cancer

Lung tumors are subdivided into histological subtypes which have a different clinical course and require different treatment. Nonetheless they are believed to a common histogenesis. Most cytogenetic abnormalities have been detected in all the different histological subtypes, although it is common for an abnormality to be seen in a higher proportion of small cell carcinomas than in non-small cell carcinomas. Common deletions are of 3p (associated genes are *FHIT* and others), 9p (associated gene, $p16^{INKA}$), 17p (associated gene, *TP53*), and 13q (associated gene, *RB*). Other regions have also been noted (e.g., 5q and 10q) but most studies now use loss of heterozygosity for revealing and defining deletions. Homogenously staining regions and double minutes are detectable in lung tumor karyotypes and are sometimes associated with amplification of members of the *MYC* gene family. Translocations have rarely been observed in lung tumors.

Further Reading

Girard L, Zochbauer-Muller S, Virmani AK, Gazdar AF and Minna JD (2000) Genome-wide allelotyping of lung cancer identifies new regions of allelic loss, differences between small cell lung cancer and non-small cell lung cancer, and loci clustering. *Cancer Research* 60: 4894–4906.

Wistuba II, Bryant D, Behrens C et al. (1999) Comparison of features of human lung cancer cell lines and their corresponding tumors. *Clinical Cancer Research* 5: 991–1000.

See also: Chromosome Aberrations; Tumor Suppressor Genes

Luria, Salvador
W C Summers

Copyright © 2001 Academic Press
doi: 10.1006/rwgn.2001.0773

Salvador Edward Luria (1912–91), an Italian-born American geneticist, was born 13 August 1912 in Turin. His research focused on the genetics of bacteria and bacteriophages, as well as the action of bacteriocines and bacterial membranes. Among many honors, Luria received the Nobel Prize for Physiology or Medicine in 1969, sharing it with Max Delbrück and Alfred Hershey.

Luria received his MD degree from the University of Turin in 1935. During his medical training, he became interested in physics and its applications to biology, leading him to do advanced work in radiology and physics in Rome, working with such teachers as Enrico Fermi and collaborating with Geo Riva, an Italian phage biologist. Leaving Italy because of Mussolini's "Racial Manifesto" in 1938, Luria moved to Paris where he collaborated with Elie Wollman and the well-know physicist Fernand Holweck at the Radium Institute on radiobiological experiments to determine the size of a bacteriophage. Again to avoid persecution, he left Paris and joined a group of radiobiologists under Frank Exner, a physicist at the College of Physicians and Surgeons of Columbia University from 1940 to 1942. He taught in the Biology Department of Indiana University (Bloomington) from 1943 to 1950, at the University of Illinois (Urbana) from 1950 to 1959, and then at the Massachusetts Institute of Technology until his death in 1991.

In 1941 Luria met Max Delbrück and they began a lifetime of collaboration and friendship. Luria secured a faculty position at Indiana University in 1943 and he and Delbrück along with Hershey, initiated the research school now known as the "American Phage Group." Much of Luria's early research was dominated by his orientation toward radiobiological target theories that fit well with Delbrück's attempts to make atomic physics relevant to genetics. One important line of work was a collaboration undertaken one summer at the Cold Spring Harbor Laboratory with Raymond Latarjet, a visiting French scientist. Latarjet was interested in the use of radiobiological target theory to follow the increase in intracellular infectious phage as a way to study phage multiplication (prior to the availability of radioisotopic tracers). Luria and Latarjet showed that this approach worked and for the first time obtained a detailed view of intracellular phage replication. This approach, later known as the Luria–Latarjet (or simply the L–L) experiment, was widely employed in the late 1940s and early 1950s.

An old problem in phage biology, that of the appearance of phage-resistant bacteria, interested Luria. He and Delbrück devised a way to test if the phage-resistant bacteria were produced spontaneously and subsequently grew out under selective conditions, or conversely, if the phage somehow induced the phage resistance to appear. Their approach was both sound and elegant, but indirect, relying as it did on probabilistic arguments similar to those they had often used in their radiobiological target theory work. This experimental approach, which came to be known as the Luria–Delbrück experiment, has been widely hailed at a landmark in the development of bacterial and molecular genetics.

While trying to better understand phage resistance and host-range mutations in bacteriophage, Luria and his collaborator Mary L. Human discovered that bacteriophages are subject to subtle "modification" by the last host in which they grew so that they might be "restricted" in their growth on hosts of different strains. In 1952 they described the phenomenon of host restriction–modification. The genetics of this phenomenon, as well as its biochemical explanations, were subsequently worked out by others. As is well known, this forms the basis for much current biotechnology.

In his later research, Luria turned to a phenomenon that was historically related to bacteriophage, namely that of bacteriocines. He investigated the physiology of these lethal molecules produced by some strains of bacteria that kill closely related strains, apparently to gain competitive advantages in natural environments. Luria and his collaborators focused mainly on the effects these proteins have on the functions of bacterial membranes, and they made substantial contributions to this field.

See also: **Bacteriophages; Delbrück, Max; Hershey, Alfred; Luria–Delbrück Experiment**

Luria–Delbrück Experiment

W C Summers

In 1943 Salvador Luria and Max Delbrück published "Mutations of bacteria from virus sensitivity to virus resistance" (Luria and Delbrück, 1943). In this paper they presented a novel experimental design aimed at answering two questions: Do mutations (to bacteriophage resistance) occur randomly in the absence of the selective agent, and if so, how can the mutation rate be estimated? The simplicity of its design and its wide applicability in microbial and cell genetics for the measurement of mutation rates has insured its eponymous status as a "classic experiment."

Since the early work on the existence of mutations in bacteria by Beijerinck, Neisser, and De Kruif, among others, it was unclear whether the conditions used to select or observe the mutations were actually inducing the altered state or simply allowing outgrowth of preexisting variants. Since mutations seemed to be rare events, it was difficult to observe the infrequent mutants in populations of bacteria prior to the application of some selection which inhibited the wild-type and permitted growth of the mutants. In the early 1930s this problem was taken up by I.M. Lewis who investigated a lactose-negative strain of *Eschenichia coli* designated *mutabile* (because it was noted to revert to lactose-utilization with some observable frequency). Lewis clearly formulated the problem and carried out careful plating experiments and concluded that the lactose-fermenting colonies that developed on lactose-containing medium came from the few variants that already existed in the culture which had been grown in glucose-containing medium.

When Luria and Delbrück investigated the process of bacteriophage multiplication, they observed the common phenomenon of phage-resistant variants. The origin of such phage resistance had been uncertain since its discovery almost as soon as phage had been discovered in 1917. Some experiments supported the notion that the phage resistance was acquired only after exposure to phage, and thus phage acted as a mutagen to change cell properties. Other experiments supported the idea that phage resistance occurs spontaneously even in the absence of exposure to bacteriophage. With subsequent deeper understanding, both of the genetics of bacteria and of the phenomenon of lysogenic immunity, it is now known that both mechanisms can occur. In these particular studies, their results clearly confirmed that the mutation to phage resistance had occurred spontaneously, prior to exposure to the phage.

What Luria and Delbrück realized was that because of the clonal, exponential growth of bacteria from a single cell (or at least a small homogeneous population), any mutation which appears at some stage in the exponential growth of the population is propagated exponentially as well, and thus a large population contains all the mutant progeny descended from each mutation event that occurred in the culture. If a mutation event occurred early in the history of a culture, a high fraction of the population would be mutant, whereas if a mutation event occurred late in the history of a culture, it would be represented by a very tiny proportion of the total population. Because of the rare occurrence of mutations, one would expect some populations to have a high fraction of mutants, some to have very few, and some to have in-between fractions, that is, in a series of replicate populations, the *variation* in the proportion of mutants would be great. Under the contrasting hypothesis, that is, if the selective condition imposed on the final population was causing the mutations, then, because the selection would be applied to nearly identical numbers of cells in the large, final populations, one would expect that the number of induced mutants would be about the same. So in this case, the expected variation would be very small. The difference in the two hypotheses, then, would appear in the size of the variations (fluctuation) in the proportion of mutants in multiple replicate populations grown up from pure wild-type parental organisms. Luria and Delbrück formalized the mathematical analysis of this process and observed that under the assumption that both the wild-type and mutant organisms grow exponentially at the same rate, one can calculate from the experimental parameters (number of generations, mutation frequencies) the actual mutation rate (as distinct from mutant frequency), that is, the number of mutation events per cell per generation.

The design of their experiment was extended to studies of both spontaneous mutation rates and induced mutations. The measurement of rates rather than frequencies of mutations greatly clarified this process and its genetic basis. The Luria–Delbrück 'Fluctuation Test,' as it is sometimes called, is indirect and statistical; because of the importance of the hypothesis of spontaneous mutation with subsequent selection (a basic principle of neo-Darwinism), additional research led to more direct confirmation of their findings. One such example was the replica-plating method of Lederberg and Lederberg for studying phage resistance.

Reference

Luria S and Delbrück M (1943) Mutations of bacteria from virus sensitivity to virus resistance. *Genetics* 28: 491–511.

See also: **Delbrück, Max; Lederberg, Joshua; Luria, Salvador**

Lutheran Blood Group

G Daniels

Copyright © 2001 Academic Press
doi: 10.1006/rwgn.2001.0775

The Lutheran blood group system is a complex system consisting of 18 red cell antigens, including four pairs of allelic antigens: Lu^a, Lu^b; Lu6, Lu9; Lu8, Lu14; Au^a, Au^b. The Lutheran glycoprotein, a member of the immunoglobulin superfamily of receptors and adhesion molecules, binds the extracellular matrix glycoprotein laminin.

See also: **Blood Group Systems; Immunoglobulin Gene Superfamily**

Lycopersicon esculentum (Tomato)

R Chetelat

Copyright © 2001 Academic Press
doi: 10.1006/rwgn.2001.1671

The cultivated tomato (*Lycopersicon esculentum* Mill.) and related wild species are members of the Solanaceae family, which includes potato, tobacco, and petunia, as well as the deadly nightshade. Though native to the Andean region of South America, tomato was first domesticated in Mesoamerica, to which it owes its common name, a derivation of the Nahuatl (Aztec) word 'tomatl.' Since its introduction to Europe in the early sixteenth century (**Figure 1**), tomato has assumed an increasingly important role in the diets of many cultures. Despite the relatively low nutrient content of its fresh fruit, tomato is a leading source of vitamins A and C and antioxidants such as lycopene, due in large part to its heavy consumption in either fresh and processed forms.

As an experimental organism for genetic studies, tomato presents many advantages. The cultigen and related wild species are true diploids, with a chromosome number of $2n = 2x = 24$. Eleven of the 12 chromosomes in its haploid nucleus are submetacentric,

*III. SOLANUM POMIFERUM, fructu rotundo, molli. *

Figure 1 Woodcut of tomato from P. A. Matthiolus (1554) *Commentarii in libris sex Pedacii Dioscoridis Anazarbei, de medica materia*. Venetiis. The 1544 edition of this herbal includes the first recorded mention of the tomato in Europe, which consists of a brief description of the plant and its culinary use in Italy at that time.

while chromosome 2 has an extremely short heterochromatic short arm consisting primarily of the nucleolus organizer region. Each chromosome is distinguishable from the others during pachytene by the pattern and length of chromatic and achromatic regions which are illustrated in corresponding cytological maps. The tomato genome is also well defined by genetic maps based on morphological and/or molecular markers; the high-density molecular marker map contains over 1000 restriction fragment length polymorphism (RFLP) markers comprising a total of 1276 map units (Tanksley *et al.*, 1992). In addition, the genetic maps have been integrated with cytological maps by the analysis of induced deletions.

The relatively low haploid DNA content of tomato, ~950 Mb, makes it well suited for molecular studies. Though larger than *Arabidopsis* or rice (about 145 and 425 Mb, respectively), the tomato genome is smaller than many other model plant species, such as maize or wheat (2500 and 16 000 Mb, respectively). The average ratio of physical to genetic distance is 750 kb/cM, a value low enough to enable the positional cloning of genes in most genomic regions.

Tomato is naturally self-pollinated, which simplifies the maintenance of stocks, yet hybridizations are easy to perform and yield large quantities of seed of controlled parentage. Tomato can be grown under a wide range of environmental conditions and propagated through seed or asexually via rooted cuttings. Its photoperiodic insensitivity and relatively short generation time permit the culture of three or more generations per year. The structure of the tomato plant, particularly its compound leaves and indeterminate sympodial growth habit, allows detection of an enormous array of hereditary variations; mutations that result in altered growth habit, leaf shape, texture and color, flower morphology, color, and function, and fruit size, shape, and color have been described. Tomato also provides a popular model for physiological and biochemical studies of fruit development, quality, and ripening.

Protoplasts are easily cultured, fused, and regenerated into whole plants. Transgenic plants are readily obtained by cultivation of cotyledon explants with *Agrobacterium tumefaciens*, followed by shoot regeneration. As a result of these and other advantages, the first transgenic food plant (GMO) to be marketed in the USA was a tomato (FlavrSavr®). For the analysis of gene function in tomato, there are several methods. The most widely applied are gene silencing, by transformation with antisense or cosuppression constructs, and complementation by transformation with sense constructs. Also, the maize transposable elements Ac and Ds, which are active in tomato and show the same preference for transposition to linked sites, can be used to produce insertional mutants. In contrast, insertional mutagenesis using the *Agrobacterium* T-DNA element is a relatively inefficient process in tomato, unlike *Arabidopsis*. Finally, the use of radiation-induced deletions is limited by their generally lethal affect during gametogenesis.

Research on tomato has depended to a large extent on genetic resources such as mutants, wild species populations, and other genetic stocks which are available to researchers through genebanks such as the C.M. Rick Tomato Genetics Resource Center (TGRC) at the University of California, Davis. The TGRC maintains over 1000 monogenic stocks, consisting of spontaneous or induced mutations at 600+ loci affecting most aspects of plant development and morphology. Over 1400 other genetic and cytogenetic stocks, including mutant combinations, translocations, trisomics, autotetraploids, Latin American cultivars, and derivatives of wild species such as alien additions, substitutions, and introgression lines, are maintained by the TGRC. Lastly, the collection also includes over 1100 wild species accessions, representing nine *Lycopersicon* and four *Solanum* species, of which all but two can be crossed to *L. esculentum*, albeit with varying degrees of difficulty.

These wild populations contain a vast amount of genetic diversity, in contrast to the cultigen which is severely depleted, and are important sources of enhanced disease resistance, yield, fruit quality, environmental stress tolerance, and other desiderata of interest to breeders. Resistances to over 42 diseases have been detected in the wild relatives, many of which have been bred into the cultivated tomato; cloning and sequencing of many of these resistance genes has contributed to our understanding of the molecular basis of plant–pathogen interactions. The wild *Lycopersicon* species are also tolerant of abiotic stresses encountered in their native habitats, which include extreme aridity (e.g., Atacama desert), flooding and high humidity (e.g., equatorial jungle), saline soils (e.g., coastal bluffs in Galapagos Islands), and freezing or chilling temperatures at high elevations in the Andes. Though bearing horticulturally unacceptable fruit, the wild species contain alleles that when bred into cultivated tomato confer desired characteristics such as increased soluble solids, fruit color, size, and yield.

Despite the complex genetic control of these fruit traits, the application of molecular marker maps has resolved quantitative trait loci (QTLs) for each of them. In the case of fruit size, a single QTL (*fw2.2*) accounts for a large portion of the difference between wild and cultivated forms; the recent cloning of this QTL (Frary *et al.*, 2000) has contributed to our understanding of the molecular basis of plant domestication, and has demonstrated that even genes for complex traits such as yield can be isolated through the use of molecular maps. Levels of diversity in *Lycopersicon* species vary greatly, due in large part to differences in mating systems, which include autogamy, facultative allogamy, and self-incompatibility of the gametophytic type; tomato is therefore a rich source of allelic variation for evolutionary and molecular studies of self-incompatibility, pollination biology, and many other reproductive characters.

Information on tomato germplasm and many types of genetic data are available through online databases. The TGRC database (http://tgrc.ucdavis.edu) provides search tools, gene descriptions, and photos of mutants and wild species from its collection. The GRIN database (http://www.ars-grin.gov) allows users to search the US Department of Agriculture's entire National Plant Germplasm System, which includes over 5000 accessions of tomatoes, mostly cultivated forms, maintained by the USDA at Geneva, New York. The SolGenes database (http://ars-genome.cornell.edu/solgenes/) interconnects genetic maps, gene sequences, probes, marker polymorphisms, QTLs, and other data on the tomato, potato, pepper, and eggplant genomes.

In conclusion, tomato has many favorable genetic and biological attributes, in addition to its status as a crop plant, which contribute to its usefulness as an experimental organism for genetic research. With excellent germplasm collections, databases, and molecular resources, tomato will likely remain an important tool for plant geneticists in the era of genomics.

References

Cornell University: SolGenes database. http://ars-genome.cornell.edu/solgenes/

Frary, Nesbitt CT, Grandillo S *et al.* (2000) fw2.2: A quantitative trait locus key to the evolution of tomato fruit size. *Science* 289: 85–88.

Tanksley SD, Ganal HW, Prince JP *et al.* (1992) High density molecular linkage mass of the tomato and potato genomes. *Genetics* 132: 1141–1160.

Tomato Genetics Resource Center: http://tgrc.ucdavis.edu

US Department of Agriculture: GRIN database. http://ars-grin.gov/

See also: *Solanum tuberosum* (Potato); Transfer of Genetic Information from *Agrobacterium tumefaciens* to Plants; Transgenes

Lyon Hypothesis

See: **X-Chromosome Inactivation**

Lysenko, T.D./Lysenkoism

W C Summers

Copyright © 2001 Academic Press
doi: 10.1006/rwgn.2001.0779

Trofim Denisovich Lysenko (1898–1976) (**Figure 1**) was prominent in the study of heredity in the Soviet Union, and a major political force in Soviet science under Joseph Stalin (from about 1934 to 1965). He believed in mechanisms of heredity that denied the primary importance of genes and mutations, and supported research predicated on his beliefs about the influence of environment on heredity. Because of his powerful political positions in the Soviet government,

Figure 1   T.D. Lysenko. (From Lysenko, 1954.)

he dominated the direction of Soviet genetic research for several crucial decades. His particular doctrine was termed 'agrobiology' and in the West came to be known as 'Lysenkoism.'

Lysenko

Lysenko was born in Karlovka, about 50 miles southwest of Kharkov, in Ukraine. His father was a small farmer. Lysenko graduated from the Kiev Agricultural Institute in 1925 and embarked upon a career of agronomical research, helped no doubt by the government policy (*vydvizhentsy*) of that time to bring young people of peasant and worker backgrounds into positions of leadership. He worked on practical breeding problems, especially the control of the growing periods of agricultural plants.

In 1925, immediately after his graduation, he went to work at the newly established experimental station at Kirovabad (in Azerbaijan) and he was entrusted with work on breeding legumes for fodder and silage. The need for such plants in that region did not correspond with the availability of reliable water from rains or irrigation, so he attempted to find ways to alter the growing seasons of legumes to produce fodder in the autumn and winter or early spring, when sufficient water was present. He sowed varieties of peas, vetch,

beans, and lentils in the fall and observed that some of the peas and vetch survived the winter and produced a crop early in the spring. From this research he concluded that "By changing the external conditions it is possible to change the behaviour of different plants of the same variety" (Lysenko, 1954, p. 18). In 1929 the term 'vernalization' was proposed for this plasticity of plant varieties. This work was extended to cereals and he claimed that spring-sown varieties could be transformed into winter-sown forms by the proper environmental manipulations. The results of this work was reported first at the All-Union Genetics Congress in Leningrad in January 1929.

Lysenko extended his experimental work to actual field studies by inducing his father, Denis N. Lysenko, to plant winter wheat in the spring. This crop was apparently successful and Lysenko reported that:

In the same summer (1929) the Soviet public learned from the press of the full and uniform earing of winter wheat sown in the spring under practical farming conditions in the Ukraine. (Lysenko, 1954, p. 23)

This well-publicized work apparently caught the attention of both agricultural policy planners as well as Marxist philosophers, because

the Soviet public came to the support of our explanation of the length of vegetative period in plants. By order of the People's Commissariat of Agriculture, a special laboratory, later a department, was established at the Ukrainian Institute of Selection and Genetics (Odessa) to study this problem. (Lysenko, 1954, p. 23)

Lysenko's theories of heredity drew on Darwinian pangenesis, Marxist ideology, and Lamarckianism. He wrote:

Whenever an organism finds the conditions (materials) in the external environment which are suitable for its heredity, its development takes the same course it took in the preceding generations. (Quoted in Dobzhansky, 1952, p. 4)

Heredity "is inherent not only in the chromosomes, but in every particle of the living body" (quoted in Huxley, 1949, p. 17).

By 1932, an agronomy journal, the *Bulletin of Vernalization*, began publication to report research in this field, and by 1935 Lysenko was its editor, a position he held until 1941. From the mid-1930s onward, Lysenko became increasingly involved in spreading his beliefs about agrobiology and vernalization in opposition to what he saw as the erroneous theories based on the work of Gregor Mendel, August Weissmann, and Thomas Hunt Morgan. His scientific work was intimately interwoven with political issues in the Soviet Union, and he was eventually relieved of most of his leadership roles by 1965. In 1966 he was relegated to directorship of the Lenin Hills Agricultural Experiment Station of the Academy of Sciences until his death in 1976.

Lysenkoism

Lysenko's beliefs and theories were so at odds with the rest of contemporary genetics, both inside (initially) and outside the Soviet Union, that his doctrines came to be known as Lysenkoism. He did not, however, claim sole credit for his position; Lysenko cited a rather obscure Russian horticulturist, plant breeder, and patriot, Ivan V. Michurin (1855–1935) as his inspiration, and intellectual forerunner. Thus, he usually presented his views as 'Michurinist' and he and his followers became known by that name. Michurin worked with fruit trees and developed a theory of 'mentoring.'

By grafting twigs of old varieties of fruit trees on the branches of a young variety, the latter acquires properties which it lacks, these properties being transmitted to it through the grafted twigs of the old varieties. (Lenin Academy, 1949, pp. 38–39)

Michurinist doctrine supposed that hereditary properties were transferred from graft to host and vice versa, clearly a belief inconsistent with chromosomal theories of genetics. Michurin was a protege of Lenin, and Lysenko canonized him as one of the founders of the new Soviet biology.

Soviet Genetics and Politics

Genetics in the Soviet Union developed along neo-Mendelian lines starting in the 1920s and H.J. Muller brought the first laboratory stocks of *Drosophila* to the USSR in 1922. In the 1930s Muller spent several years as Senior Geneticist in the Institute of Genetics of the USSR Academy of Sciences, but left in 1937 after becoming disillusioned by the political controls being exerted over genetics. For complex political and ideological reasons, Mendelian genetics came to be viewed as 'idealist' as opposed to 'realist,' a serious sin in the Marxist ideology of the time. The new Soviet emphasis on scientism and the belief that changes in the political environment would create the 'new Soviet man' led to the hope that similarly, in biology, changes in the environment of living organisms, including humans, could produce long-lasting, heritable changes (of course, all for the better) in the offspring. Thus, a version of Lamarckianism came to be

aligned with orthodox Marxist political philosophy. At the same time, internal political struggles in the Soviet governing bodies involved important issues such as agricultural planning and farm management. Lysenko, an ambitious person, allied himself with a skilled Marxist philosopher, Isaac I. Prezent, and together they attacked Mendelian genetics and its practitioners in the USSR in a book published in 1935. This attack marked the beginning of what later became known as "The Lysenko Affair." Lysenko skillfully employed the government press and entered politics in 1935 as a member of the Central Executive Committee of the Ukranian Communist Party. In 1936 he was appointed director of the Odessa Institute of Genetics and Breeding and that summer the presidium of the Lenin All-Union Academy of Agricultural Sciences (VASKhNIL) initiated public discussions on "issues in genetics." Although the supporters of Mendelian genetics dominated these discussions, just a few months later, under the Stalinist Great Terror, many senior geneticists were purged and Lysenko's supporters moved in to fill the voids in agricultural genetics. By 1938 Lysenko was president of VASKhNIL (having replaced Nikolai Vavilov, an internationally known geneticist), a member of the Supreme Soviet of the USSR, and a deputy head of the Soviet of the Union, the highest legislative body in the USSR.

Postwar central planning in agriculture in the USSR called for expansion of VASKhNIL which Lysenko opposed, and for a time between 1945 and 1947 there was a period of cooperation between the Soviets and the West, during which the geneticists recruited international opposition to the Lysenkoists. With the onset of the Cold War, however, science became part of the "patriotic campaign" and was exploited by Lysenko to clamp down on all foreign contacts.

In the summer of 1948 Lysenko staged his famous purge of Soviet genetics. Under the guise of open discussion of scientific views, he organized a meeting to debate "The situation in biological science." The meeting opened with Lysenko reading his carefully prepared paper outlining his theories of "Michurinist" biology as the basis of the "New Soviet Science." For about a week, many of the leading geneticists in the USSR debated and criticized Lysenko's position paper in the spirit of open scientific discussion. At the end of the meeting, Lysenko sprang his trap: In his concluding remarks, he said

The question is asked in one of the notes handed to me, What is the attitude of the Central Committee of the Party to my report? I answer: The Central Committee of the Party examined my report and approved it.

Thus, all the criticism of his secretly pre-approved position rendered the entire Mendelian genetics community as enemies of State policy, a serious, possibly fatal error at that time. At the next session, many of the previously critical geneticists fearfully recanted and realized that Lysenko had won the political battle for control of hereditary science in the USSR. Recent scholarship in newly available archives shows that Stalin, himself, worked with Lysenko on the draft of his talk to this meeting. The final draft has Stalin's handwritten editing and marginalia, a testimony to the importance attached to genetics in the Soviet Union at that time. It took almost two more decades before the cumulative failures of "Michurinist" biology, in the form of repeated crop failures and food shortages, led to the demise of Lysenkoism and the removal of Lysenko from his dictatorship of Soviet genetics by Nikita Khrushchev in 1965.

Further Reading

Krementsov N (1997) *Stalinist Science*. Princeton, NJ: Princeton University Press.
Medvedev ZA (1971) *The Rise and Fall of T.D. Lysenko*. Garden City, NY: Anchor.
Rossianov KO (1993) Editing Nature: Joseph Stalin and the "new" Soviet biology. *Isis* 84: 728–745.

References

Dobzhansky T (1952) Russian Genetics. In: Christman R (ed.) *Soviet Science*, pp. 1–7. Washington, DC: American Association for the Advancement of Science.
Huxley J (1949) *Heredity East and West: Lysenko and World Science*. New York: Henry Schuman.
Lenin Academy of Agricultural Sciences of the USSR (1949) *The Situation in Biological Science*. Moscow: Foreign Languages Publishing House.
Lysenko TD (1954) *Agrobiology*. Moscow: Foreign Languages Publishing House.

See *also*: Lamarckism; Muller, Hermann J

Lysine

E J Murgola

Copyright © 2001 Academic Press
doi: 10.1006/rwgn.2001.0776

Lysine is one of the 20 amino acids commonly found in proteins. Its abbreviation is Lys and its single-letter designation is K. As one of the essential amino acids in humans, it is not synthesized by the body and so must be provided in the individual's diet (**Figure 1**).

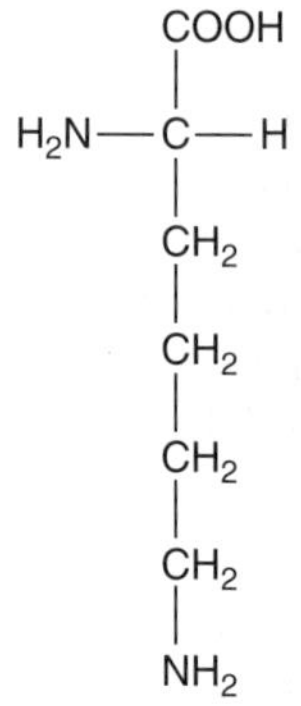

Figure 1 Lysine.

Lysis

E Kutter

Copyright © 2001 Academic Press
doi: 10.1006/rwgn.2001.0780

Lysis is the bursting of a bacterial cell by the breaking apart of its cell wall, leading to rupture of the cell membrane. An enzyme specialized in this function is called a lysozyme.

See *also*: Lysozyme

Lysogeny

B S Guttman

Copyright © 2001 Academic Press
doi: 10.1006/rwgn.2001.0781

Lysogeny is a condition in which a bacterial cell carries the genome of a virus in a relatively stable state. Investigators of bacteriophage growth during the 1920s and 1930s were often puzzled by a strange phenomenon: while some bacteria would produce phage shortly after infection, other bacteria yielded no phage and even appeared to be immune to infection by the phage. However, in a culture of such resistant bacteria, small amounts of phage appeared irregularly. These puzzling bacteria were termed lysogenic because it was supposed that some cells in a culture were capable of lysing and producing the observed phage. Max Delbrück refused to believe in the phenomenon and ascribed the appearance of phage to sloppy technique, even though some of the investigators – notably Eugene and Elisabeth Wollman – were known to be scrupulous workers.

In 1950, Andre Lwoff and Antoinette Gutmann demonstrated the reality of lysogeny through painstaking experiments with a strain of *Bacillus megaterium*. They followed individual cells in microdrops of broth by microscopic examination; each time a cell divided, the daughter cells were separated into their own drops by micromanipulation. Occasionally, a cell would disappear from a drop, leaving behind phage whose presence could be demonstrated by growth on susceptible bacteria. In later experiments, Lwoff demonstrated that when lysogenic cells are irradiated with UV light, they lyse uniformly and liberate phage, a phenomenon called phage induction. The hypothetical intracellular state of the phage in a lysogenized bacterium was called a prophage. Mapping experiments by Jacob and Elie Wollman (the son of the above Wollmans, who were killed by the Nazis) then demonstrated that the phage lambda prophage is located at a specific site, near the genes for galactose metabolism. Dale Kaiser provided strong evidence that the prophage genome is integrated into the bacterial DNA so that it is continuous with the bacterial DNA on either side. Thus, when bacterial DNA replicates during each round of reproduction, the prophage DNA is replicated as part of the whole genome. (The process of lambda integration is discussed in an article of its own.)

The lysogenic state is maintained by a control system intrinsic to the phage. Phage lambda, which has been most intensively studied, carries a single gene, *cI*, that encodes a repressor protein. In a stable lysogenic state, this protein binds to certain sites in the lambda genome and represses transcription of all other lambda genes. However, establishment of the lysogenic state is a complex process involving the products of several genes, binding to a series of regulatory sites. The heart of the molecular decision between the lytic and lysogenic states involves a competition between the repressor (cI) protein, which promotes lysogeny, and the Cro protein, which promotes lytic growth. The latter choice depends heavily on a complex process of antitermination (see Antitermination Factors). Furthermore, the decision involves proteins that measure the availability of energy, as signalled by the level of cyclic AMP (cAMP) (see Cyclic AMP (cAMP)). A cell with an adequate supply of glucose has a low level of cAMP, and a phage entering such a cell is likely to enter the lytic cycle; if the glucose level falls, the level of cAMP rises, and a phage entering such a cell is more likely to go lysogenic. In effect, the phage is determining whether the most prudent strategy for reproduction is a 'short-term tactic' of using the avilable energy for synthesis of a cellful of new phage or a 'long-term tactic' of producing more copies of its genome through bacterial growth.

See also: Antitermination Factors; Cyclic AMP (cAMP); Phage λ Integration and Excision

Lysozyme

E Kutter

Copyright © 2001 Academic Press
doi: 10.1006/rwgn.2001.0782

Bacterial cells are generally protected from lysis induced by such factors as osmotic shock by having a cell wall made of peptidoglycan, also called murein. The entire peptidoglycan sack around each bacterial cell is in fact one giant, covalently bonded bag-shaped molecule. Growth of the cell requires that links of this sack be opened up long enough to insert new links in between them; penicillin leads to the death of growing bacterial cells by interfering with the filling and resealing of these small gaps in the cell's armor. Lysozymes are a particular class of enzymes that are able to attack this murein structure and thus generally effect the destruction of the cell. In 1922, the Scottish physician Alexander Fleming showed that saliva, tears, and sweat all contined a substance that could destroy bacteria. What he was observing was in fact lysozyme – the first human secretion shown to have chemotherapeutic properties.

Peptidoglycans are composed of long polysaccharides that are alternating copolymers of N-acetyl glucosamine and N-acetylmuramic acid that are cross linked through unusual short peptides with structures such as (L-Ala)-(D-Glu)-(L-Lys)-(D-Ala).

In gram-negative bacteria, the peptidoglycan sack is generally only one layer thick and lies just inside an outer membrane. In gram-positive bacteria, it has no outer membrane cover but is many layers thick; this thick sack is able to take up and retain the Gram stain, giving these bacteria their name. In both groups of bacteria, lysozymes catalyze the hydrolysis of the glycosidic links between GlcNAc and MurNAc, dissolving the cell wall. Lysozymes are found throughout nature – in egg whites, in tears and sweat, and in mucus. A number of bacteriophages also encode lysozymes to help them get in and out of cells. Other phages make other endolysins – enzymes with peptidoglycan degrading activity. The others have somewhat different specificity but the same function as lysozyme, attacking the peptide crosslinker or the bond on the other side of MurNAc.

Four families of endolysins have been identified:

1. The true lysozymes (glycosidases) that have just been described, which include the products of the bacteriophage T4 *e* gene (for *e*ndolysin) and P22 gp19.
2. Transglycosylases, such as the phage lambda R protein and the product of the P2 phage K gene, which attack the same bond as lysozyme but conserve

the glycosidic bond energy by forming a cyclic 1,6-disaccharide product. They catalyze the intramolecular transfer of the O-muramyl residue to its own C6 hydroxyl group.

3. The amidases, such as bacteriophage T7 gp3.5, which degrade the peptide bond between MurNAc and the adjacent tetrapeptide crosslinker and endopeptidases, such as the *Listeria* monocytogenes A500 ply500, which degrade the peptide bond between two tetrapeptides, cutting between m-DAP Ala.

See also: **Lysis**

Lytic Phage

B S Guttman

Copyright © 2001 Academic Press
doi: 10.1006/rwgn.2001.0783

A virulent phage that cannot establish lysogeny and whose characteristic mode of multiplication is to produce rapidly a large number of new phage particles and lyse its host cell.

See also: **Lysogeny; Virulent Phage**